DDT

THE INSECTICIDE DICHLORODIPHENYLTRICHLOROETHANE
AND ITS SIGNIFICANCE

VOL. II

W0261629

DDT

THE INSECTICIDE DICHLORODIPHENYLTRICHLOROETHANE
AND ITS SIGNIFICANCE

DAS INSEKTIZID DICHLORDIPHENYLTRICHLORÄTHAN
UND SEINE BEDEUTUNG

EDITED BY PAUL MÜLLER, BASEL

VOL. II

CHEMISCHE REIHE · BAND 10
LEHRBÜCHER UND MONOGRAPHIEN AUS DEM GEBIETE
DER EXAKTEN WISSENSCHAFTEN

HUMAN AND VETERINARY MEDICINE

EDITED BY S. W. SIMMONS, ATLANTA, USA

AUTHORS

W. J. HAYES JR., SAVANNAH, USA; S. W. SIMMONS, ATLANTA, USA;
E. F. KNIPLING, WASHINGTON, USA

1959

Springer Basel AG

ISBN 978-3-0348-6796-2   ISBN 978-3-0348-6809-9 (eBook)
DOI 10.1007/978-3-0348-6809-9

All rights reserved
©Springer Basel AG 1959
Ursprünglich erschienen bei Birkhäuser Verlag Basel 1959.
Softcover reprint of the hardcover 1st edition 1959

PREFACE

In this second volume of 'DDT, The Insecticide Dichlorodiphenyltrichloro-ethane and Its Significance,' S. W. SIMMONS and his co-worker W. J. HAYES, Jr., and colleague E. F. KNIPLING have collected an exhaustive array of material on the pharmacology and toxicology of this compound and on its contribution to human and veterinary medicine through the control of disease-bearing insects. Although an increasing number of insects have become resistant to it, DDT is still the most commonly used insecticide and is being produced in large quantities (over 100,000,000 pounds per year in the United States).

Toxicologists, pharmacologists, physicians, veterinarians and entomologists will find this volume especially valuable, but it will also interest an even wider circle of workers in science and industry.

Dr. SIMMONS, as Chief of the Technology Branch of the Communicable Disease Center, U S Public Health Service, and as a member of the Expert Committee on Insecticides for the World Health Organization, is eminently qualified as an authority on the subject. With Dr. HAYES, also of the Communicable Disease Center, and Dr. KNIPLING, U S Department of Agriculture, he has presented a well-rounded, well-balanced account of the subject. The collected references alone should be of inestimable value to the readers of this volume.

The publisher and the editor both wish to express their congratulations and gratitude to the authors for their excellent contributions. They are confident that this work will find deserved appreciation.

Basel, December 1958   *The Editor*

TABLE OF CONTENTS

VI

PHARMACOLOGY AND TOXICOLOGY
OF DDT

BY

W. J. HAYES, JR.

1.

INTRODUCTION

It is customary in reviews of this sort to describe the basic pharmacology of a compound and to show how this understanding permits one to predict the clinical effects which the compound will have in different situations. However, although many of the pharmacological properties of DDT are known, its basic mode of action remains to be determined. For this reason it has appeared better to describe the symptomatology and toxicity which may be directly observed in experimental animals, and only then to consider the more detailed studies which have been made in the hope of discovering the mode of action. Finally, the practical problems involved in the toxicity of DDT to man and to useful animals will be discussed.

It must be recognized at the beginning that the purity of the DDT used in any given study has a bearing on the results which will be observed. The fact is, however, that in many studies the purity of the DDT sample used has not been recorded. This is not as serious as it might at first appear, although unpublished studies show that technical and pure DDT show slight differences in absorption and storage in the rat. A number of studies have been made on the toxicity of the highly refined p,p'-isomer and the results in these tests were not significantly different from those in which a relatively crude technical product was used (CAMERON and BURGESS [105][1]); WOODARD et al. [680]; DEICHMANN et al. [162]). This similarity of toxicity is undoubtedly due to the fact that the p,p'-isomer, which is the most active isomer from a pharmacological point of view, constitutes the greater part of the technical product. Although the different compounds in the technical mixture have certain qualitative as well as quantitative differences in toxicity, it remains true that they are essentially similar in action so that their effects are more additive than contradictory.

There is some indication that o,p-DDT has a smaller acute toxicity but a greater chronic toxicity than the p,p'-isomer. Thus DOMENJOZ [173] gives the acute oral LD_{50} for mice as 3,350 and 400 mg/kg, respectively. BROWNING et al. [91] found the acute toxicity of o,p-isomer relatively low. On the other hand, WOODARD et al. [680] reported that a dosage of 89 mg/kg/day of the o,p-isomer killed dogs in an average of 37 days while the p,p'-isomer given at the same rate required 55 days.

The formulation in which DDT is applied has proved to be much more important than the purity of the compound in determining its toxicity. This

---

[1]) Numbers in brackets refer to References, page 220.

W. J. Hayes, Jr.

Table 1

Units of Measure Used in this Chapter

Concept	Description of unit	Abbreviation
Dosage to animals	Milligrams of DDT per kilogram of body weight	mg/kg
Storage in tissue, residue in food, or concentration in water	Parts of DDT per million parts of tissue, food, or water by weight	ppm
Concentration in air	Milligrams of DDT per cubic meter of air	mg/m^3
Concentration in formulation	Parts of DDT per hundred parts of formulation (weight/volume)	per cent or %
Rate of application to surfaces	Milligrams of DDT per square meter of area	mg/m^2

Table 2

Conversion Table for the Units of Measure Frequently Used in Connection with the Toxicology of DDT. The Five Divisions of the Table Correspond to the Five Divisions in Table 1

1 ppm (by weight)	1	mg/kg
1 μg/100 g	0·01	ppm
1 μg/g	1	ppm
1 μg/100 mg	10	ppm
1 mg/100 g	10	ppm
1 grain/pound	142·9	ppm
1 mg/g	1,000	ppm
1 g/pound	2,204·6	ppm
1 % (concentration)	10,000	ppm
1 μg/l	1	mg/m^3
1 ppm (of DDT in air by volume)[1]	14·5	mg/m^3
1 g/1,000 cu. ft	35·3	mg/m^3
1 pound/gallon	119·8	g/l
1 pound/gallon	12·0	% w/v
1 pound/acre	10·4	mg/sq. ft
1 pound/acre	112·1	mg/m^2
1 mg/square foot	10·8	mg/m^2

[1] This is an expression for concentration frequently used in industrial hygiene. It is based on the assumption that the material in question exists as a gas or vapor and expresses the number of volumes of compound per million volumes of air. The value may be calculated by the formula:

$$ppm = \frac{\text{observed concentration (mg/l)} \times 24,450}{\text{molecular weight of compound}} .$$

The figure, 24,450 ml, is the gram molecular volume of a gas at a pressure of 760 mm of mercury and a temperature of 25°C.

difference depends largely, and perhaps entirely, on the absorbability of the compound when presented in different formulations. This, in turn, frequently depends on the amount and character of oil in the formulation. It appears possible that the degree of dispersion of the DDT may at times have a bearing on its toxicity to higher forms, especially fish and amphibia.

In order that the results of different authors may be compared, it has been necessary to adopt uniform units of measure throughout the chapter. Where possible, figures are presented in these units regardless of the method of presentation used in the original paper. The conventions used are shown in Table 1.

The terms 'milligram per kilogram' and 'parts per million' are, of course, equivalent (1 mg/kg = 1 ppm) and the two terms are used merely for convenience in distinguishing different kinds of measurements, as shown in the table.

For those more accustomed to other units of measurement, equivalents for some of the units encountered in the literature are given in Table 2.

2.

CLINICAL DESCRIPTION OF INTOXICATION IN ANIMALS

The toxicity of DDT was first studied by the Swiss workers. The description of intoxication in animals given by DOMENJOZ [172] remains one of the best. The first perceptible effect is abnormal susceptibility to fear, with violent reaction to normally subthreshold stimuli. There is definite motor unrest and increased frequency of spontaneous movements. As poisoning increases, hyper-irritability like that seen in strychnine poisoning develops, but convulsions do not appear at this time. A fine tremor, recognizable at first only as a terror reaction, is later present as an intention tremor in connection with voluntary movement, is then present intermittently without observable cause, and if finally present as a coarse tremor without interruption even for a period or days. Spontaneous movement is limited and food intake stops. In the lates stages, especially in some species, there are attacks of epileptiform tonic-clonic convulsions with opisthotonos.

All the signs are strengthened by external stimuli and become manifest at first through external stimuli. In all stages, the animals show normal position and labyrinth reflexes. The picture of poisoning in mammals recalls the disturbances of movement and tone such as are known in human pathology as the amyostatic syndrome.

Symptoms appear several hours after administration of the compound and death follows after 24 to 72 hours.

These findings have been very generally confirmed (NEAL et al. [424]; SMITH and STOHLMANN [532]; DRAIZE et al. [178], [179]; MOOSER [415]; WOODARD et al. [677]; WASICKY and UNTI [649]; LÄUGER et al. [348]; NEAL et al. [427]; ORR and MOTT [444]; ZEIN-EL-DINE [685]; PHILIPS and GILMAN [453]; LAUG and FITZHUGH [344]; McNAMARA et al. [392]; HAYMAKER et al. [268]; BING et al. [63]; EMMEL and KRÜPE [199]; DIAZ-JIMENEZ [171]; VELBINGER [632]; SPICER et al. [544]; DEICHMANN et al. [162]). The clinical picture has become so well known that many authors have spoken of 'typical signs of poisoning' without deeming it necessary to describe the signs.

NELSON et al. [432] showed that large daily doses, which were compatible with life, might produce tremors lasting for weeks or even intermittently for months. Tremors of variable duration were reported in rats, mice, guinea pigs, rabbits, dogs, and cows. The authors considered that appropriate dosage would produce tremors in most species of animals.

CAMERON and BURGESS [105] observed many of the same features of poisoning already mentioned and recorded certain others for the first time. In studies of rats, guinea pigs, and rabbits, CAMERON and BURGESS noticed that as animals became sick they became cold to the touch and showed ruffled fur. Some animals showed diarrhea. The authors found that muscular tremors were preceded by muscular weakness which occured first in the back and later in the hind legs. The front legs were relatively spared so that animals showing marked weakness of the hind quarters could still drag themselves about. Severely poisoned animals which survived long enough showed rapid loss of weight as a result of failure to eat, and some of them did not present nervous signs at the time of observation. Although the authors emphasized the liver pathology in certain animals and considered it to be an adequate explanation for death, they did not describe any clinical signs clearly indicative of liver dysfunction. Convulsions were observed but were rare. Death resulted from respiratory failure; it was observed that the heart continued to beat to the end and in some instances continued for a little while after respiration had stopped. Animals which did not die gradually recovered completely.

The selective weakening of the hind quarters of animals has been noted by others (ZEIN-EL-DINE [685]).

Contrary to the findings of CAMERON and BURGESS [105], several authors (HAYMAKER et al. [268]; PHILIPS and GILMAN [453]) have found that the tremor characteristic of DDT poisoning generally starts in the muscles of the face, including the eyelids and spreads caudally with variable severity until all the muscles are affected.

At appropriate dosage levels, tremors may appear and later may regress completely even though the daily dosage of DDT is continued (NELSON et al. [432]; DRAIZE et al. [179]; LAUG and FITZHUGH [344]; DEICHMANN et al. [162]; FITZHUGH and NELSON [208]). Less frequently, tremors may regress and then reappear following a single dose (DEICHMANN et al. [162]). Also, there is general agreement that the tremors, as well as the other nonfatal effects of DDT, are reversible even though severe or prolonged (WASICKY and UNTI [649]; CAMERON and BURGESS [105]; NEAL et al. [427]; HAYMAKER et al. [268]; PHILIPS and GILMAN [453]; BING et al. [63]; STOHLMAN and LILLIE [560]). This does not contradict the fact that violent convulsions are usually followed by death, especially with multiple doses of DDT.

Uncomplicated sensory changes are difficult to detect in animals but PLUVINAGE and HEATH [462] reported hyperesthesia in cats.

A number of authors (DOMENJOZ [172]; MOOSER [415]; DRAIZE et al. [178], [179]; WOODARD et al. [677]; ORR and MOTT [444]; HAYMAKER et al. [268]; BING et al. [63]) have mentioned anorexia or the inability to take food as being characteristic of severe DDT poisoning. Failure to take food may be properly ascribed to lack of appetite in animals receiving DDT by stomach tube or by any other route except voluntary ingestion. Food refusal requires special study when it involves a compound mixed intimately with the animal's regular diet. In this instance the refusal of food may represent only the refusal of a foul tasting or noxious material and not a systemic effect. Monkeys offered food containing 5,000 ppm of DDT usually eat far less than the normal amount and consequently lose weight severely. However, these animals will eagerly consume the same kind of food containing no DDT, even just after refusing the major portion of their daily ration of contaminated food (unpublished result).

Violent tremors may interfere mechanically with the ingestion of food even in the presence of normal appetite.

Anorexia found in severe poisoning is probably a nonspecific effect. On the contrary, appropriate repeated doses of DDT may cause rats to eat more food than comparable controls (LAUG and FITZHUGH [344]). This fact is discussed in detail below (p. 108).

Animals which have suffered severe weight loss as the result of DDT poisoning may become easy prey to secondary infection (DRAIZE et al. [179]). PLUVINAGE and HEATH [462] stated that the very first signs of intoxication in the cat, following repeated intermuscular injections, were congestion and hypersecretion of the eyes and nose, without fever. The authors were unable to determine whether this was an autonomic effect or was a result of secondary infection predisposed by intoxication.

Although there is a general similarity in the clinical behavior of all vertebrates poisoned by DDT, certain variations characteristic of different species were noted early. Cats show a greater extensor rigidity with opisthotonos than other laboratory animals and the duration of their illness tends to be longer (SMITH and STOHLMANN [532], [533]; PHILIPS and GILMAN [453]). PLUVINAGE and HEATH [462] detected stiffness in poisoned cats even before the onset of tremor and observed that the stiffness appeared first in the distal part of the extremities and later extended to the proximal part and to the trunk. In the final stages of severe poisoning, convulsions may become almost continuous (PHILIPS and GILMAN [453]). Poisoned cats show marked pilomotor activity (PHILIPS and GILMAN [453]).

Convulsions are also encountered in dogs, and were apparently the only sign observed in that species by CHOU et al. [125]. The presence of ataxia in the dog has been emphasized (McNAMARA et al. [392]; HAYMAKER et al. [268]; BING et al. [63]).

Clonic convulsions may occasionally occur in rats, but in that species tremors are so pronounced that convulsive episodes frequently are difficult to discern (PHILIPS and GILMAN [453]) although some have reported violent convulsions in rats (DEICHMANN et al. [162]).

Rats poisoned by DDT may show a reddish color about the eyes just as they do when ill from many other causes. The color has been ascribed to the over-secretion of a porphyrin by the Harderian glands (CAREY et al. [109]).

Frogs show hypersensitivity, tremors and convulsions followed by stupor and paralysis. In addition there is, during the early phase of poisoning, marked increase in cutaneous secretion, a lightening of color, and a tendency to cry out (TRIPOD [597]). (Although other animals may produce sounds during severe poisoning, this is not characteristic.) Poisoned chameleons usually turned from green to dark brown (CAREY *et al.* [109]).

Even fish show signs of poisoning similar to those observed in mammals. The fish is at first hyperirritable and hyperactive. Later it becomes incoordinated so that it cannot maintain a normal position in the water. Convulsive movements may occur from time to time. Weakness and prostration gradually ensue (ELLIS *et al.* [196]).

Signs of poisoning following intravenous administration of DDT are similar to signs following the administration of the poison by other routes. With the intravenous route, the latent period before the onset of illness is greatly reduced and poisoning progresses to a fulminating stage much more rapidly. Rats given an intravenous dosage of 50 mg/kg (LD_{94}) show a latent period of about 5 minutes and fully developed signs of poisoning in 30 minutes. Over 50% of the animals die within 3·5 hours and survivors are symptom-free within 18 to 24 hours. The time of onset and the duration of illness are not very different for other species (PHILIPS and GILMAN [453]). The findings of others are similar (STOHLMAN and LILLIE [560]; JUDAH [313]).

Death from DDT poisoning generally occurs after a relatively long period of muscular activity so that the animal is in a stage of exhaustion. Any spasticity which may have been present is replaced by a flaccid type of weakness or paralysis (SMITH and STOHLMAN [532]). It was pointed out early (CAMERON and BURGESS [105]) that death usually results from respiratory failure. However, some animals die suddenly soon after the onset of poisoning. In certain instances at least, these sudden deaths are caused by ventricular fibrillation (PHILIPS and GILMAN [453]; PHILIPS *et al.* [454], [455]).

DEICHMAN *et al.* [162] reported that a severely poisoned monkey was given artificial respiration manually when it stopped breathing for 30 to 45 seconds following each of 5 episodes of severe tonic convulsions. The monkey recovered. In the absence of a controlled experiment, it was not clear whether the monkey would have resumed respiration unaided. It does not seem possible to state whether the respiratory arrest in DDT poisoning is potentially reversible as it is in poisoning by certain organic phosphorus insecticides (BARNES [44]).

3.

DOSE-MORTALITY RELATIONSHIPS IN ANIMALS

Effect of a Single Dose of DDT

Insofar as possible, data on toxicity are presented in this review in the form of tables and graphs. Of course, it has not been possible to treat all the reports presented in the tables in the same way. Some authors have given complete protocols; some authors have given a final result in statistical form without indicating the method of statistical analysis or the degree of variability encountered; and some have discussed their findings in general terms without mention of complete protocol or statistical analysis. In those instances in which the author gave a complete account of his results, but did not analyze them himself, the LD_{50} values have been calculated by the reviewer according to the method of LITCHFIELD and WILCOXON [376].

Unfortunately, there is at present no satisfactory way to convert the results of all toxicity experiments into the same units and thus to compare them. For example, it is difficult to determine the milligram-per-kilogram dosage in experiments involving respiratory exposure in air, or in experiments involving aquatic animals which are completely immersed in a suspension or solution of a toxicant. Such a conversion would involve consideration of the volume of tidal air and the amount of inspired toxicant actually retained in the respiratory tract. In the case of aquatic animals, the total amount of toxicant removed from the medium by the animal may be determined by analysis. Such analyses and calculations have apparently not been attempted for DDT, and it is, therefore, necessary to present the results in terms of the original concentration of the compound in the air or water.

In fact, no matter what the route of administration, the dosage rate must be thought of as a measure of exposure and not necessarily as a measure of the amount or concentration which is physiologically active.

DDT shows its greatest observed toxicity when given by the intravenous route. This is undoubtedly because more of the material is made available to critical tissues in a shorter period of time. It does not follow that the total dose which eventually reaches critical tissues is larger when the compound is administered intravenously. The opposite may be true. The toxicity of an intravenous dose may represent an overwhelming of the system associated with relatively sudden absorption. This is true even though the greater part of each intravenous dose may be trapped in the lungs and reticulo-endothelial system within a few minutes after injection. However, in these depots, the DDT from the injected homogenized emulsion or colloidal suspension presents a large surface which presumably facilitates further mobilization.

Each route of administration has its own technical difficulties and, therefore, its own difficulties of interpretation. For example, in studies of dermal toxicity, many workers place animals in some kind of restrictive rack or binding so that they cannot lick themselves. The toxicant may even be introduced beneath a rubber sleeve in which the animal is encased. Other investigators prefer to apply toxicants to the shoulder area. The latter technique reduces the

Table 3

DDT Single-Dose Intravenous LD_{50} Values for Various Animals

Animal species	LD_{50} mg/kg	Formulation	Number of animals	Authority
Rat	47[1]	Homogenate	77	PHILIPS and GILMAN [453]
Rabbit	41[1] 30	Homogenate	21	PHILIPS and GILMAN [453], SMITH *et al.* [535]
Cat	32[1]	Homogenate	44	PHILIPS and GILMAN [453]
Dog	68[1]	Homogenate	16	PHILIPS and GILMAN [453]
Monkey	55	Homogenate	7	PHILIPS and GILMAN [453]

[1] Calculated by the reviewer.

chance that the poison will be ingested and, at the same time, insures against abnormalities associated with confinement, sweating, etc. In any event, the habit of licking is much more characteristic of some species, such as cats and cows, than of other species. In cattle sprayed with DDT for insect control, licking is the most important factor leading to loss of the poison from the hair. Skin secretions, growth and loss of hair, disturbance by body movements or wind or friction, flaking of the epidermis, absorption by the hair or skin, leaching by rain, and breakdown caused by solar radiation are minor factors in removing DDT from the hair (HACKMAN [254]; NORRIS [438]).

Before discussing the tables, it is interesting to observe that basic information on the toxicity of DDT became available very early. Practical experiments were reported by SCHMID [512] in 1943. Cows and sheep showed no injury when fed grass which had been sprayed with a 1 per cent solution. Lactating sheep and young lambs showed no ill effect when dipped in a 2 per cent emulsion. A 5 per cent suspension was shown to be harmless to the cornea of rabbits. By the end of 1944 the broad outlines of all that we know now about the acute and subacute toxicity of DDT were already published.

Table 4

DDT Single-Dose Oral LD$_{50}$ Values for Various Animals

Animal species	Approximate LD$_{50}$ (mg/kg)		Number of animals	Authority
	Water suspension or powder	Oil solution		
Mouse	1,600[3])	175[6])	5,5	Domenjoz [172]
		450[1])[10])	70	Woodard *et al.* [677]
		200–300[1])	12	Konst and Plummer [331]
		220[6])[10])	105	von Oettingen and Sharpless [641]
	300–450		—	Pulewka [473]
White rat		150[6])	57	Smith and Stohlman [532]
	500[3])	280[6])	5,5	Domenjoz [172]
		180[1])[10])	35	Woodard *et al.* [677]
		800[5])	—	Cameron and Burgess [105]
		185[7])[10])	—	Philips and Gilman [453]
	2,000–2,500[8])	100–150[1])	4,4	Konst and Plummer [331]
	1,000[3])		—	Hoffman and Lendle [285]
	1,400[4])	240[6])	—	Deichmann *et al.* [162]
		250	—	Lehman [360a]
Rattus rattus	622		—	Plague Research Lab. [459]
Tatera brantsii	1,180		—	Plague Research Lab. [459]
Tatera afra	about 1,000		—	Plague Research Lab. [459]
Mastomys coucha	> 1,000		—	Plague Research Lab. [459]
Peromyscus leucopus	1,500		16	Coburn and Treichler [131]
Guinea pig	2,000[3])		5	Domenjoz [172]
		560[1])[10])	80	Woodard *et al.* [677]
		400[5])	—	Cameron and Burgess [105]
		250	—	Serge [518]
		> 400[6])	—	Guareschi and Bini [248]

Animal species	Approximate LD$_{50}$ (mg/kg)		Number of animals	Authority
	Water suspension or powder	Oil solution		
Rabbit.	275[3])	300[6])	26	Smith and Stohlman [532]
			5	Domenjoz [172]
		> 400[1])	35	Woodard et al. [677]
		228–400	—	Lindquist et al. [369]
		300[5])	—	Cameron and Burgess [105]
		1,770[6])	—	Deichmann et al. [162]
Sylvilagus floridanus.	> 2,500		17	Coburn and Treichler [131]
Cat		200–300[6])	11	Smith and Stohlman [532]
		410[7])[10])	—	Philips and Gilman [453]
		100–300[8])	—	Velbinger [632]
Dog		> 300[2])	—	Velbinger [632]
Sheep	> 2,000		4	Orr and Mott [444]
Goat		> 1,500	2	Telford [580]
		> 1,000[9])	2	Spicer et al. [544]
Turkey	> 1,000	[11])	1	Kingscote and Jarvis [322]
Chicken		> 300[1])	30	Woodard et al. [369]
		300[1])	2	Konst and Plummer [331]
Colinus virginianus	300	60–85	84	Coburn and Treichler [131]
Anas platyrhynchos	> 2,000		—	Coburn and Treichler [131]
Sturnus vulgaris.	> 600		27	Coburn and Treichler [131]
Goldfish		63–200	—	Ellis et al. [196]
Smallmouth black bass . . .	< 50		6	Hoffmann and Surber [291]

[1]) Corn (maize) oil.
[2]) Fish oil.
[3]) Gum arabic.
[4]) Methylcellulose.
[5]) Mineral oil.
[6]) Olive oil.
[7]) Peanut oil.
[8]) Tragacanth.
[9]) Unspecified vegetable oil.
[10]) Calculated by the reviewer.
[11]) Birds receiving the same dosage (1,000 mg/kg) in peanut oil showed severe symptoms and some of them died.

Table 5

DDT Single-Dose Dermal LD_{50} Values for Various Animals

Animal species	Approximate LD_{50} (mg/kg)		Authority
	Water suspension	Oil solution	
Mouse	250,000–500,000[1]	250–500[2]	DOMENJOZ [172]
Rat	1,000,000[1]	250–500[2] 3,000[4]	DOMENJOZ [172] CAMERON and BURGESS [105]
Guinea pig	1,000,000–2,000,000[1] > 2,500	< 2,500[3] 1,000[4] > 1,000[5]	DOMENJOZ [172] WASICKY and UNTI [650] CAMERON and BURGESS [105] KONST and PLUMMER [331]
Rabbit	250,000–500,000[1]	300[4] > 2,820[6]	CAMERON and BURGESS [105] DOMENJOZ [172] DRAIZE et al. [179]

[1] Gum arabic.	[3] Cottonseed oil.	[5] Corn (maize) oil.
[2] Olive oil.	[4] Ether and kerosene, separately.	[6] Dimethylphthalate.

Table 6

DDT Single-Dose Subcutaneous LD_{50} Values for Various Animals

Animal species	LD_{50} (mg/kg)	Formulation	Authority
Mouse	1,000–1,500 300	Water suspension Corn oil	DOMENJOZ [172] KONST and PLUMMER [331]
Rat	> 2,000 1,500 200–300	Water suspension Mineral oil Corn oil	DOMENJOZ [172] CAMERON and BURGESS [105] KONST and PLUMMER [331]
Guinea pig	900 > 600	Mineral oil Mineral oil	CAMERON and BURGESS [105] KONST and PLUMMER [331]
Rabbit	250 > 3,200	Mineral oil Olive oil	CAMERON and BURGESS [105] DEICHMANN et al. [162]
Cat	< 650	Oil solution	BLANCO [76]
Frog	< 150	Olive oil solution	ELLIS et al. [196]

The toxicity of DDT administered as a single dose is given in Table 3 for the intravenous route, in Table 4 for the oral route, in Table 5 for the dermal route, in Table 6 for the subcutaneous route, and in Table 7 for the intraperitoneal route.

In spite of the divergence of method used in the different laboratories, comparable results reported in the tables are in rather good agreement. Furthermore, it is not safe to assume that those figures which appear to be out of line are entirely erroneous. It is true that the divergences apparently cannot be explained on the basis of the written record. However, experience indicates that divergent results reported by different laboratories frequently depend on

Table 7

DDT Single-Dose Intraperitoneal LD$_{50}$ Values for Various Animals

Animal species	LD$_{50}$ mg/kg	Formulation	Number of animals	Authority
Rat	80	Olive oil emulsion	48	CAREY *et al.* [109]
	200	—	—	HOFFMAN and LENDLE [285]
Guinea pig	150	Olive oil solution	—	SEGRE [518]
Rabbit	< 2,100	Corn oil solution	—	DEICHMANN *et al.* [162]
Chameleon (*Anolis caroliniensis*)	200	Olive oil emulsion	100	CAREY *et al.* [109]

differences in experimental materials and method which may not be recognized until years later. On the contrary, it cannot be said that all the data presented in the tables are of the same quality.

The tables are summarized in Figure 1. A tabular value expressed as a range has been plotted arbitrarily as the mid-point of that range. In the interest of graphic clarity, it has been necessary to plot the dosage values on a logarithmic scale.

Several conclusions appear justified from the figure and the tables.

(1) DDT, if properly dissolved, is absorbed through all portals.

(2) Formulation determines the greatest variation in the toxicity of DDT. This is especially true of the dermal route by which the compound in solution may be a thousand times more toxic than the undissolved material.

(3) DDT, as a dry powder, is absorbed through the skin to a negligible degree. It is frequently impossible to put enough DDT dust on the skin of animals to kill them so that an LD$_{50}$ value for this formulation has to be determined by extrapolation or cannot be calculated.

(4) The route of administration determines marked and consistent variation in the toxicity of DDT. Given intravenously, the compound is about ten times as toxic as when given in solution by other routes.

(5) Different species of mammals exhibit real but, with the apparent exception of the goat, small differences in susceptibility.

(6) Wide variation in individual susceptibility within the same species has been noted generally.

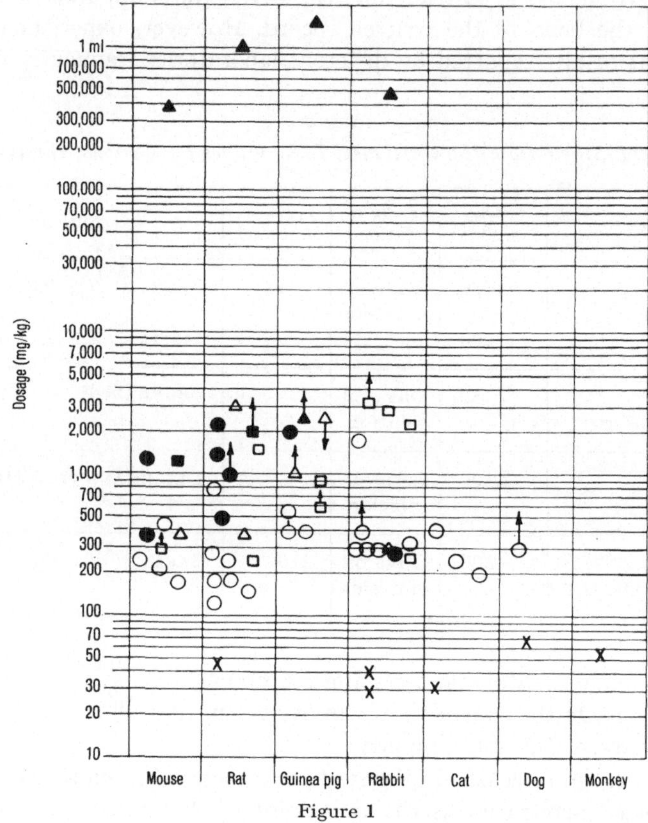

Figure 1

Approximate LD_{50} values for DDT administered by various routes to several species of animal. Tabular values and references are given in Tables 3–7, inclusive. An arrow on a point indicates that the true value is considered to be greater or less than the value plotted.

Formulation		Route
Solution	Undissolved	
○	●	Oral
△	▲	Dermal
□	■	Subcutaneous
X (homogenate)		Intravenous

So far as formulation is concerned, Figure 1 distinguishes only whether the toxicant is or is not dissolved. This matter of solution is the most obvious in determining the toxicity of formulations, and has been remarked on by many writers (WOODARD *et al.* [677]; WASICKY and UNTI [649], [650]; CAMERON and BURGESS [105]; PLUMMER [461]; ZEIN-EL-DINE [685]; VELBINGER [632]; LORY *et al.* [381]; HOFFMANN and SURBER [291]). Several related factors which may be important include: (1) Degree of dispersion of DDT, (2) absorbability of the solvent, and (3) toxicity of the solvent. The degree of dispersion of DDT undoubtedly plays a major role in the toxicity of the compound by the respiratory route and its toxicity to aquatic organisms. Apparently, the degree of dispersion has little effect on the oral toxicity of DDT to mammals. WASICKY and UNTI [648] found that a colloidal preparation had a rather low order of toxicity for rats.

In general, DDT appears more toxic as a solution in vegetable oil or animal fat than when given in some petroleum fraction. Petroleum may act as a laxative. The heavier fractions are never absorbed, and DDT dissolved in them has to be literally extracted from the solvent in order to show toxicity. Incidentally, in considering the effect of different solvents, it must be noted that some workers have used oils as a suspending medium, employing more of the toxicant than the volume of oil could dissolve. Obviously, under these circumstances, the effect of the oil must be limited.

Subcutaneous deposits of DDT are poorly absorbed and may become encapsulated (DOMENJOZ [172]). Subcutaneous deposits continue to give off the compound over an extended period.

There remain certain real differences between species which are apparently unrelated to absorption. For example, the goat is highly resistant to poisoning by the oral route. There is apparently nothing in the data which would allow one to guess more accurately about the susceptibility of man to fatal poisoning than to say that his susceptibility probably lies within the observed range of the susceptibility of other mammals.

Dermal Exposure

In addition to the information contained in Table 5 and Figure 1, and the foregoing discussion, other features of dermal exposure require comment.

DDT is peculiarly free of primary irritancy when applied as the pure compound or dissolved in bland solvents. This statement applies to the skin (WASICKY and UNTI [648]; DOMENJOZ [172]; CAMERON and BURGESS [105]), to the conjunctiva (HOFFMAN and LENDLE [285]), and to the cornea (HENNIG [271]) of all species studied. On the contrary, many solutions and some emulsions, with or without DDT, have proved irritating and sometimes fatal (DRAIZE *et al.* [179]; DOMENJOZ [172]; NEAL *et al.* [427]; CAMERON and BURGESS [105]; DEICHMANN *et al.* [162]). The susceptibility to different solvents varies from one species to another. DDT dust is definitely irritating to the conjunctiva and cornea, but not beyond the mechanical injury which would be expected from any inert, finely crystalline substance (DOMENJOZ [172]).

Similar principles regarding the irritancy of formulations have been found to apply to the subcutaneous route (NEAL *et al.* [427]; HOFFMAN and LENDLE [285]). It has been postulated that a part of the irritation caused by the intraperitoneal injection of DDT may be caused by the deposition of insoluble DDT on the peritoneal lining (NEAL *et al.* [427]).

DRAIZE *et al.* [179] found that solutions of DDT caused a mild but definite sensitization in the guinea pig. DUNN *et al.* [189] made a much more thorough study of the skin-sensitizing properties of DDT and concluded that the material was not sensitizing to the guinea pig. They used recrystallized material which they applied percutaneously and intracutaneously. Several solutions, as well as aqueous suspensions, were employed. The authors recognized that some of the compounds known to be contaminants of technical DDT are capable of producing sensitization, while the capacity of others had not been studied.

Information on the toxicity of DDT to aquatic animals exposed 'dermally' to the compound is considered below (Table 16, p. 94).

Respiratory Exposure

Determination of the toxicity of a single exposure to DDT by the respiratory route is complicated by the factor of time, as well as by the difficulty of determining the true dosage which is conditioned by particle size.

DOMENJOZ [172] reported no characteristic signs of poisoning in mice, rats, guinea pigs, or rabbits exposed for 12 hours to an atmosphere containing 1,700 mg of DDT per cubic meter of air; the DDT was dispersed as a mist of a 0·2 per cent emulsion.

In much more extensive studies, NEAL and his colleagues [424] found that the exposure of dogs, rats, and guinea pigs to aerosols with initial concentrations of 54,400, 12,440, and 6,220 mg of DDT per cubic meter of air for 45 minutes in a static atmosphere did not cause toxic signs. Mice tolerated only the lowest of these concentrations without injury. Even at the highest concentration, it was shown that the intoxication of the mice was not caused by respiratory exposure, but by DDT which adhered to their fur and which was subsequently licked off and perhaps was absorbed through the skin. Mice whose fur was protected during exposure did not develop signs of DDT poisoning.

Exposure of animals to the same concentration of Freon (1,200,000 mg/m³) as that used in the aerosol experiments failed to produce any injury even in mice. A number of different aerosol formulations were tested. One of them contained 10 per cent cyclohexanone. An initial concentration of 108,800 mg of cyclohexanone per cubic meter of air produced in the experimental chamber by the aerosol resulted in narcosis of both dogs and mice.

In a later study, NEAL [427] and his colleagues investigated aerosol formulations essentially similar to those in use today. Exposure for 45 minutes to initial concentrations of 26,400 to 32,900 mg of DDT per cubic meter was fatal to over 50% of immature mice, to about 15% of adult mice, and to about 15% of immature rats, but not to adult rats or adult guinea pigs. The residue of one of the aerosols containing all the constituents of the mixture except the propellent was absorbed by the skin of mice so that single doses of 0·1 ml, or repeated doses of 0·012 ml were fatal to approximately half of the mice tested. Dogs withstood repeated large doses of the same residue for many weeks; most of them finally died in severe depression not characteristic of uncomplicated DDT poisoning.

With regard to these results of NEAL and his colleagues, it should be observed that, considering different DDT aerosol formulations and different recommendations for their use, the initial concentration of DDT actually employed in the presence of human beings does not exceed 10 mg/m³ and is usually less.

Effect of Multiple Doses of DDT

In discussing the toxicity of a single dose of DDT, it was suggested that the simplicity of acute toxicity is more apparent than real. Even with a single dose, the rate of absorption may be different under different conditions, and it is difficult to evaluate the importance of the rate of absorption as compared with the total amount of toxicant absorbed. In considering the effect of multiple doses, the situation is further complicated by differences in the total duration of exposure and by differences in the interval between exposures.

In some experiments involving oral exposure, different authors have expressed the dosage in milligrams per kilogram of body weight, and others have expressed the dosage in terms of concentration of DDT in food. In order to compare both kinds of experiments directly, and in order to compare directly the results of oral and of dermal exposure, it has been necessary to convert all dosages to milligrams of DDT per kilogram of body weight per day. Figure 2 gives the approximate dosages resulting from maintaining animals on diets containing different concentrations of DDT. All of the points on the graph are derived from actual measurements of the food intake of rats and monkeys maintained on diets containing DDT at the indicated concentration. All of the lines on the graph for which no points are shown are based on direct observation of the food intake of the different species and the assumption that DDT at moderate dosages does not interfere with the amount of food which those species consume. This assumption is justified by the fact that data plotted for rats and monkeys indicate that the food intake of these animals was not affected by concentrations of DDT in the diet as high as 800 ppm. An exception appears to exist in the case of weanling rats. The curve for these young animals shows a slight reduction in slope indicating that their food intake was reduced by increasing concentrations of DDT.

The plotted data for male and female rats were taken from FITZHUGH and NELSON [208]. Data from a similar experiment in the reviewer's laboratory were in complete agreement with these plotted values.

It is interesting to note that female mice and rats regularly consumed more food in proportion to their weight than males of the same species. This means, of course, that when maintained on a diet containing the same concentrations of DDT females of these species receive a higher dosage than males. These statements apply to animals which have not been bred. It may be that the

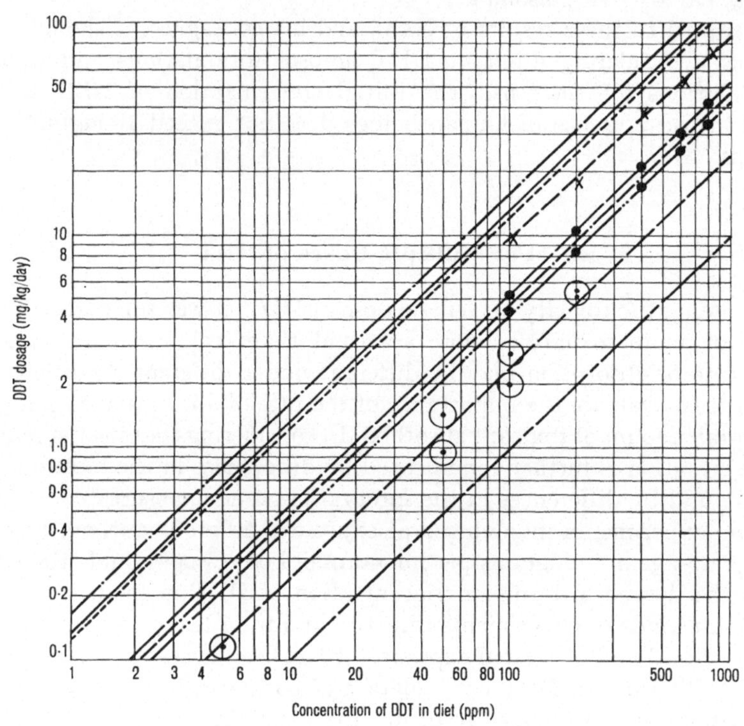

Legend in order from top to bottom:

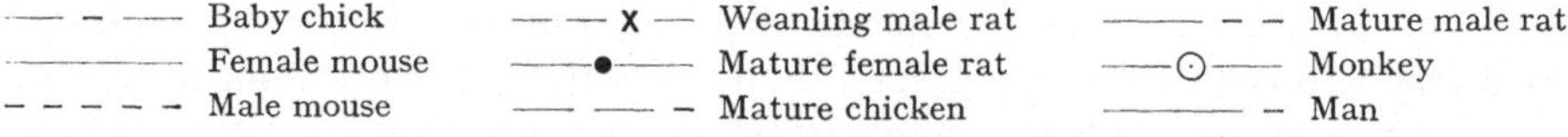

― ― ―― Baby chick	― ― X ― Weanling male rat	―― ― ― Mature male rat
――――― Female mouse	――●―― Mature female rat	――⊙―― Monkey
― ― ― ― ― Male mouse	―― ―― ― Mature chicken	――― ― Man

Figure 2

Dosage of DDT received by several species of animals when maintained on a diet containing a stated concentration of the compound. The position of each curve is determined by the relationship of food intake to body weight characteristic of the species.

difference is accentuated during pregnancy and lactation. It may be that females of other species also eat more in proportion to their weight than males, but information on the matter was not conveniently available.

Attention is called to the fact that young animals eat more in proportion to their body weight than do mature animals of the same species. This is illustrated in Figure 2, which includes curves for young and mature chickens and young and mature male rats. It may be seen that a one-month-old rat receives about the same dosage of DDT, when maintained on a diet of 100 ppm, as an adult rat would receive from a diet containing 200 ppm. The relation of sex and age in rats to dosage is shown in a different way in Figure 3, taken directly from FITZHUGH and NELSON [208].

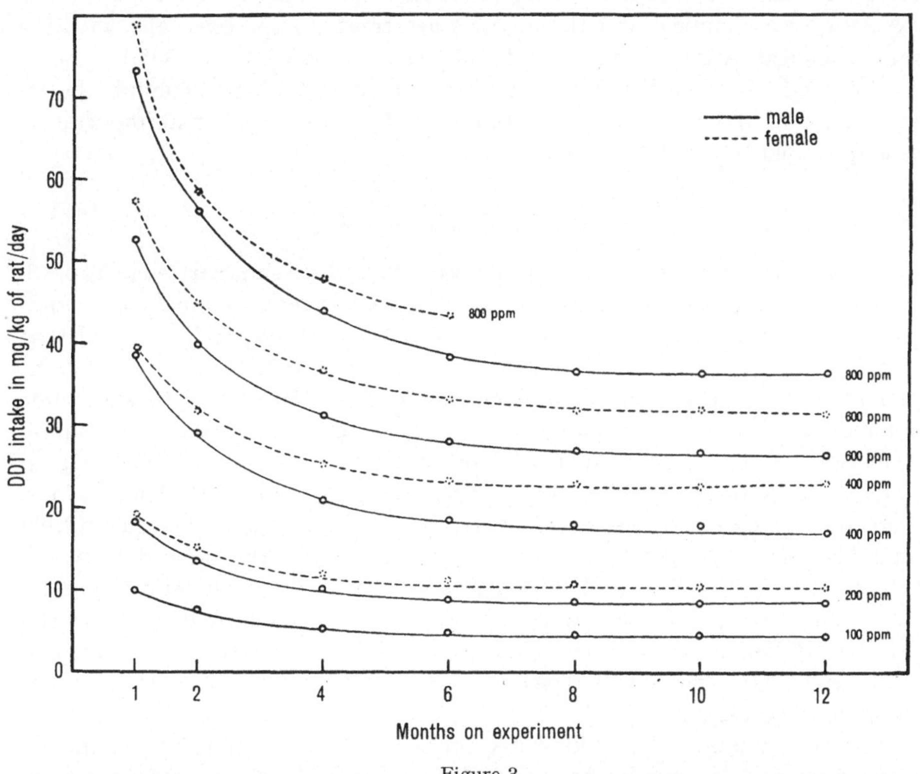

Figure 3

Calculated DDT intake in mg/kg of body weight in rats receiving various levels of DDT in the diet. The figure is taken from FITZHUGH and NELSON [208] and is reproduced by permission of the Journal of Pharmacology and Experimental Therapeutics.

It is worth noting that authors have rather generally misused the word 'chronic'. The assumption has frequently been made incorrectly that repeated exposure is *necessarily* the cause of chronic disease. Actually, this is true in regard to certain toxicants (e.g., lead) and apparently false in regard to others (e.g., parathion). Chronic disease resulting from exposure to chemicals may be associated with storage (e.g., lead), or it may be quite unrelated to storage (e.g., scar formation following caustics or cancer following 2-acetaminofluorene). The same material (e.g., lead) may cause acute or chronic illness and pathology according to the circumstances of administration. The mere fact of detectable storage does not necessarily constitute an injury. Furthermore, the mere fact of tissue change does not necessarily constitute chronic injury, for it must be shown that the tissue change is, in fact, an injury from the physiological standpoint and not an adaptive alteration. It is also important that conclusions reached with experimental animals should not be applied without consideration to man. For example, there is a poor correlation between the occurrence of cancer in man and in rodents in instances in which man receives occupational exposure to a chemical and the animals are treated experimentally with the same chemical. A few correlations do, of course, exist (SALTER [503]).

Although the qualifications just discussed must be kept in mind, the investigation of toxicity associated with repeated exposure is of great importance, as emphasized by DE EDS [158].

Oral Exposure

Data on the effect of multiple oral doses of DDT are summarized in Table 8.

The basic fact that the effects of DDT in experimental animals are cumulative was recognized very early. SMITH and STOHLMANN [532] reported that 80 g rats fed a diet containing 1,000 ppm of DDT died in from 18 to 80 days. Similar rats fed at a level of 500 ppm survived 3 months, though with some impairment of growth. The dosage of the former group would be approximately 70 mg/kg/day, or about 1,260 mg/kg in 18 days, and about 5,600 mg/kg in 80 days. These figures can be compared with an LD_{50} value of 150 mg/kg for a single dose given by the same authors for older and, therefore, less susceptible rats. The LD_{99} value for a single dose calculated from the same data would be only 700 mg/kg. This dosage relationship, which was not pointed out by SMITH and STOHLMANN, is typical of the results of later investigators. The relationship indicates that repeated doses of DDT are cumulative, but only to a limited degree, showing that some excretion or detoxification of the compound does occur.

Further examples of this kind may be found by examining the column in Table 8 headed 'Maximum tolerated dosage'. It will be seen that at intermediate or low dosage levels animals frequently tolerate a total dosage far in excess of the LD_{99} for a single dose by the same route. This general fact, although noted earlier (HAYMAKER *et al.* [268]) has not been properly emphasized.

The basic toxicity of the compound places some limitation on the uses which might otherwise be made of DDT. For example, it is entirely inadvisable to use DDT in treating cats for fleas. Cats lick their fur so extensively that the material cannot be used safely on them.

In a similar way, DDT is too toxic to be considered for the internal medication of man or animals (CHOU *et al.* [125]).

Far more interest and importance attaches to the lower limits of oral toxicity. The lowest level producing gross effects has been stated as 100 ppm (LEHMAN [355]), but the figure is probably conservatively low. FITZHUGH and NELSON [208] reported a decreased growth rate and increased mortality in

Table 8

Effect on Various Anima's of Repeated Oral Doses of DDT

| Dosage range mg/kg/day | Species | Details of dosage | | Total dosage (mg/kg) | | | Number of animals in test | Mortality % | Survivors intoxicated | Authority |
| | | Formulation (concentration of DDT/ vehicle) | Rate | Minimum | | Maximum tolerated | | | | |
				Fatal to one animal	Fatal to all animals					
0–2·5	Rat	Crystalline	50 ppm	—	—	175[1])	—	0	No	Domenjoz [172]
	Calf[2])	Spray residue	17·8 ppm	—	—	115[1])	4	0	No	Thomas et al. [585]
2·6–5·0	Rat (m)	10%/Oil	100 ppm	—	—	3,285[1])	12	58[3])	Yes	Fitzhugh and Nelson [208]
	Calf	Corn oil	104·6 ppm	—	—	493	2	0	No	Thomas et al. [585]
	Cat	Olive oil	5 mkd	?	—	130(?)	5	40[3])	Yes	Smith and Stohlman [533]
5·1–10·0	Mouse	Crystalline	50 ppm	—	—	448[1])	—	0	No	Domenjoz [172]
	Mouse[4])	Crystalline	> 125 ppm	—	> 36	—	—	100	—	Woodard et al. [677]
	Rat (m)	10%/Oil	200 ppm	—	—	6,300	24	50[3])	Yes	Fitzhugh and Nelson [208]
	Cat	Olive oil	10 mkd	—	—	260	5	0	Yes	Smith and Stohlman [533]
10·1–20·0	Rat (f)	10%/Oil	200 ppm	—	—	7,884	12	83	—	Fitzhugh and Nelson [208]
	Rat	Crystalline	250 ppm	—	—	4,380[1])	12	Cont.[3])	No	Woodard et al. [677]
	Rat[2])	2%/Alcohol	220 ppm	—	—	2,201[1])	20	0	No	Deichmann et al. [162]
	Rat (m)	10%/Oil	400 ppm	—	—	12,410[1])	24	92	Yes	Fitzhugh and Nelson [208]
	Turkey[2])	Oil	380 ppm	—	—	952[1])	6	0	No	Marsden and Bird [398]
20·1–40·0	Rat	Crystalline	500 ppm	—	—	6,760[1])	12	Cont.[3])	—	Woodard et al. [677]
	Rat	Crystalline	500 ppm	—	—	3,870[1])	10	0	Yes	Smith and Stohlman [533]
	Rat (f)	Crystalline	600 ppm	—	—	22,630[1])	12	92	Yes	Fitzhugh and Nelson [208]
	Rat (m)	Crystalline	600 ppm	—	—	18,980[1])	12	83	Yes	Fitzhugh and Nelson [208]
	Rat (f)	Crystalline	700 ppm	—	—	1,484[1])	16	13	—	Sarett and Jandorff [506]

Dosage range mg/kg/day	Species	Details of dosage		Total dosage (mg/kg)				Number of animals in test	Mortality %	Survivors intoxicated	Authority
		Formulation (concentration of DDT/vehicle)	Rate	Minimum		Maximum tolerated					
				Fatal to one animal	Fatal to all animals						
20·1–40·0 continued	Rat (m)	Crystalline	800 ppm	—	—	25,915[1])		12	83	Yes	FITZHUGH and NELSON [208]
	Rat (w)	10%/Pyrax	300 ppm	—	—	10,585[1])		7	Cont.[3])	No	HAAG et al. [252]
	Rat (w)	25%/Pyrax	300 ppm	—	—	10,585[1])		7	Cont.[3])	No	HAAG et al. [252]
	Rat (f)	10%/Oil	400 ppm	—	<16,060	—		12	100	—	FITZHUGH and NELSON [208]
	Rat (m)	10%/Oil	600 ppm	—	—	18,980[1])		12	83	Yes	FITZHUGH and NELSON [208]
	Rat (f)	10%/Oil	600 ppm	—	—	22,630		12	83	Yes	FITZHUGH and NELSON [208]
	Rat (m)	10%/Oil	800 ppm	—	—	25,915[1])		24	83	Yes	FITZHUGH and NELSON [208]
	Rat	1·25%/Oil	24 mkd	—	—	576[1])		15	0	No	DEICHMANN et al. [162]
	Guinea pig	2%/Corn oil	500 ppm	—	—	600		10	Cont.[3])	No	WOODARD et al. [677]
	Cat	—	36–43 mkd	—	—	250		—	—	—	WASICKY and UNTI [648]
	Turkey	Oil	750 ppm	—	—	—		6	17	Yes	MARSDEN and BIRD [398]
	Chicken[2])	2·5%/Peanut oil	620 ppm	—	—	—		10	Cont.[3])	No	RUBIN et al. [501]
40·1–80·0	Rat (f)	Crystalline	800 ppm	—	<15,695	—		12	100	—	FITZHUGH and NELSON [208]
	Rat	Crystalline	1,000 ppm	1,620	7,200	—		10	100	—	SMITH and STOHLMAN [533]
	Rat	Crystalline	1,000 ppm	—	—	—		12	> 50	Yes	WOODARD et al. [677]
	Rat (w)	10%/Pyrax	600 ppm	5,635	20,182	—		7	100	—	HAAG et al. [252]
	Rat (w)[5])	25%/Pyrax	600 ppm	—	—	19,162[1])		7	59	Yes	HAAG et al. [252]
	Rat	2%/Alcohol	630 ppm	—	—	8,008[1])		20	55	Yes	DEICHMANN et al. [162]
	Rat	2%/Alcohol	730 ppm	350	3,150	—		5	100	—	DEICHMANN et al. [162]
	Rat	2%/Alcohol	770 ppm	71	1,491	—		10	100	—	DEICHMANN et al. [162]
	Rat (f)	10%/Oil	800 ppm	—	—	—		12	100	—	FITZHUGH and NELSON [208]
	Rat	2%/Alcohol	1,080 ppm	1,120	—	14,560[1])		10	90	Yes	DEICHMANN et al. [162]
	Rat	5%/Oil	50 mkd	—	—	1,500[1])		10	0	No	CAMERON and BURGESS [105]
	Rat	2·5%/Oil	60 mkd	—	—	1,440[1])		10	0	Yes	DEICHMANN et al. [162]

| Dosage range mg/kg/day | Species | Details of dosage | | Total dosage (mg/kg) | | | Number of animals in test | Mortality % | Survivors intoxicated | Authority |
| | | Formulation (concentration of DDT/vehicle) | Rate | Minimum | | Maximum tolerated | | | | |
				Fatal to one animal	Fatal to all animals					
40·1–80·0 continued	Guinea pig	2%/Corn oil	1,000 ppm	—	—	1,200[1]	10	< 50	Yes	WOODARD et al. [677]
	Rabbit	Oil	50 mkd	750	1,250	—	6	100	—	SMITH and STOHLMAN [533]
	Rabbit	5%/Oil	50 mkd	400	—	1,000[1]	10	40	Yes	CAMERON and BURGESS [105]
	Cat[4]	Oil	50 mkd	300	—	—	1	100	—	SMITH and STOHLMAN [533]
	Dog	Crystalline	41–50 mkd	—	—	1,550[1]	2	0	No	WASICKY and UNTI [648]
	Dog	Oil	80 mkd	—	—	4,400[6]	—	100	—	WOODARD et al. [680]
	Turkey	Oil	1,500 ppm	—	—	—	6	84	Yes	MARSDEN and BIRD [398]
	Chicken	Crystalline	1,000 ppm	—	—	—	5	100	—	TELFORD and GUTHRIE [582]
	Chicken	2·5%/Peanut oil	1,250 ppm	—	—	—	10	90	Yes	RUBIN et al. [501]
80·1–160·0	Rat	—	2,000 ppm	776	970	—	4	100	—	PRILL et al. [470]
	Rat (w)	10%/Pyrax	900 ppm	< 1,260	—	—	7	86?	—	HAAG et al. [252]
	Rat (w)[4]	10%/Pyrax	1,200 ppm	—	< 1,461	—	7	100	—	HAAG et al. [252]
	Rat(w) [4][5]	25%/Pyrax	1,200 ppm	—	< 1,461	—	7	100	—	HAAG et al. [252]
	Rat	2%/Alcohol	1,400 ppm	100–700	3,500	—	5	100	—	DEICHMANN et al. [162]
	Rat	2%/Alcohol	1,410 ppm	3,375	—	11,553[1]	10	70	Yes	DEICHMANN et al. [162]
	Rat	5%/Oil	120 mkd	1,200	—	2,880[1]	10	60	Yes	DEICHMANN et al. [162]
	Dog[4]	Crystalline	100 mkd	300	1,400	—	3	100	—	CHOU et al. [125]
	Dog	Crystalline	100 mkd	—	—	4,200[1]	3	0	No	NEAL et al. [426]
	Dog	10%/Oil	100 mkd	—	—	—	—	0	Yes	BING et al. [63]
	Dog	—	100 mkd	300	900	—	2	100	—	McNAMERA et al. [392] ZEIN-EL-DINE [685]
	Sheep	Crystalline	100 mkd	—	—	2,700[1]	4	0	No	ORR and MOTT [444]
	Goat	Oil	100 mkd	—	—	4,000[1]	1	0	Yes	SPICER et al. [544]
	Cow	Crystalline	100[7] mkd	—	—	2,700[1]	4	0	Yes	ORR and MOTT [444]
	Horse	Crystalline	100[7] mkd	—	—	2,700[1]	2	0	No	ORR and MOTT [444]

| Dosage range mg/kg/day | Species | Details of dosage | | Total dosage (mg/kg) | | | | Number of animals in test | Mortality % | Survivors intoxicated | Authority |
| | | Formulation (concentration of DDT/vehicle) | Rate | Minimum | | Maximum tolerated | | | | | |
				Fatal to one animal	Fatal to all animals						
80·1–160·0 continued	Turkey	Oil	3,000 ppm	760	—	1,140[1])	6	67	Yes	Marsden and Bird [398]
	Chicken	Crystalline	500 ppm	—	—	—	10	100	—	Woodard et al. [677]
	Chicken[4])	2·5%/Peanut oil	2,500 ppm	—	—	—	10	100	—	Rubin et al. [501]
160·1–320·0	Dog	10%/Oil	150–350 mkd	1,350	—	12,000	8	50	Yes	Haymaker et al. [268]
	Dog	10%/Oil	250 mkd	—	—	—	—	0	Yes	{Bing et al. [63] / McNamera et al. [392]
	Goat	Crystalline	250–280 mkd	—	—	7,250	5	0	Yes	Telford and Guthrie [582]
	Chicken	Crystalline	1,000 ppm	483	1,610	—	10	100	—	Woodard et al. [677]
320·1–640·0	Dog	10%/Oil	300 mkd	—	—	—	—	0	Yes	{Bing et al. [63] / McNamera et al. [392]
	Chicken	—	400 mkd	—	—	2,000	1	0	No	Wasicky and Unti [648]
	Chicken	Crystalline	2,500 ppm	—	—	—	18	100	—	Rosenberg and Adler [499]
640·1–1,280·0	Rat	Suspension	650–710 mkd	3,900	—	4,260	6	17	No	Wasicky and Unti [648]
	Goat	Vegetat. oil	1,000 mkd	6,000	11,000	—	4	100	—	Spicer et al. [544]
	Chicken	Crystalline	5,000 ppm	—	—	—	18	100	—	Rosenberg and Adler [499]

[1]) Maximum tolerated was also maximum tested.
[2]) Lower dosages tested by the same authors were also harmless to this species.
[3]) Mortality same or less than that of control.
[4]) Higher dosages tested by the same authors were also fatal to this species.
[5]) Approximately same results for females.
[6]) Average required to kill all test animals.
[7]) Increased, in most instances, to 150 mg/kg/day during second week and to 200 mg/kg/day during third week.

(m) = Male. – (f) = Female. – (w) = Weanling. – mkd = Milligrams per kilogram of body weight per day. – ppm = Parts of DDT per million parts of diet.

female rats fed a diet containing 400 ppm or more of DDT. A concentration of 800 ppm was necessary to reduce the growth rate of male rats, but was not sufficient to cause a significant increase in their mortality. HAAG *et al.* [252] found that 900 ppm (but not 600 ppm) increased the mortality of male rats. Tremor and other signs of poisoning tend to appear earlier and to be more frequent and severe in female than in male rats.

Dermal Exposure

Data on the effect of repeated dermal applications of DDT are summarized in Table 9. This table emphasizes experiments in which evidence of absorption and toxicity were encountered. It is, therefore, deficient for showing the negligible absorption of powdered DDT from the skin. The essential failure of the dry material to be absorbed has great practical importance in regard to the use of louse powders or impregnated clothing for the control of epidemic typhus. In this connection, it has been very extensively studied (SMITH and STOHLMANN [532]; DRAIZE *et al.* [178], [179]; CAMERON and BURGESS [105]). Although some dry DDT is presumably absorbed from the skin, the amount is so small that it cannot be measured under practical conditions of dosage.

Solutions of DDT have to be used with caution. For example, the susceptibility of rats to mineral oil solutions of DDT used for the treatment of mange was reported by TAYLOR [577]. In a later communication (TAYLOR [578]), it was shown that a 5 per cent acetone solution of DDT which had been proposed for the treatment of mange in rabbits was fatal to both young and old rabbits when applied to the affected parts. Even a 0·5 per cent solution may be fatal to young rabbits, though apparently harmless to mature ones.

Respiratory Exposure

DOMENJOZ [172] reported that mice, rats, guinea pigs, and rabbits showed no adverse reaction when exposed for 1 hour a day for 30 days to 5 per cent DDT in talc dispersed in the air.

NEAL *et al.* [426] showed that exposure of dogs for 45 minutes a day for 6 days to an initial concentration of 12,220 mg of DDT per cubic meter of air (using an aerosol containing 1% of DDT, 6% of sesame oil, and 93% of Freon) caused no toxic effects. Mice exposed under the same conditions died. Mice given the same respiratory exposure, but protected from contamination of the fur, survived without sign of poisoning. In another test, 2 monkeys and 10 mice were exposed to initial concentrations of 183 mg/m^3 every 15 minutes for a period of 45 minutes, 5 days per week, for a total of 5 weeks; they suffered no toxic manifestations referable to DDT. In still another test, two monkeys were exposed to initial concentrations of 183 mg/m^3 every 15 minutes for 45 minutes

Table 9

Effect on Various Animals of Repeated Dermal Doses of DDT

Dosage range mg/kg/day	Species	Details of dosage		Total dosage (mg/kg)			Number of animals in test	Mortality %	Survivors intoxicated	Authority
		Formulation (concentration of DDT/vehicle)	Rate	Minimum		Maximum tolerated				
				Fatal to one animal	Fatal to all animals					
0–2·5	—	—	—	—	—	—	—	—	—	—
2·6–5·0	—	—	—	—	—	—	—	—	—	—
5·1–10·0	Mouse	Gum arabic	10 mkd	—	—	500	—	40	—	DOMENJOZ [172]
	Rat[2]	1%/Alcohol	10 mkd	—	—	140[1]	5	0	No	CAMERON and BURGESS [105]
	Rat	1%/Kerosene	10 mkd	—	—	140[1]	5	0	No	CAMERON and BURGESS [105]
	Guinea pig	5%/Pyro-phyllite	10 mkd	—	—	90[1]	3	0	No	CAMERON and BURGESS [105]
	Guinea pig	1%/Kerosene	10 mkd	110	—	140[1]	5	40	Yes[5]	CAMERON and BURGESS [105]
	Guinea pig	1%/Alcohol	10 mkd	—	—	140[1]	5	0	—	CAMERON and BURGESS [105]
	Guinea pig	1%/Acetone	10 mkd	70	—	140[1]	5	40	—	CAMERON and BURGESS [105]
	Guinea pig	1%/Ether	10 mkd	70	—	140[1]	5	40	—	CAMERON and BURGESS [105]
	Rabbit	1%/Kerosene	10 mkd	—	—	200[1]	5	0	No	CAMERON and BURGESS [105]
	Rabbit	1%/Alcohol	10 mkd	90	—	140[1]	5	40	—	CAMERON and BURGESS [105]
	Rabbit	1%/Acetone	10 mkd	70	—	140[1]	5	20	—	CAMERON and BURGESS [105]
	Rabbit	1%/Ether	10 mkd	130	—	140[1]	5	20	—	CAMERON and BURGESS [105]
10·1–20·0	—	—	—	—	—	—	—	—	—	—
20·1–40·0	Rat	H_2O Emulsion	25 mkd[3]	—	—	375[1]	5	0	No	CAMERON and BURGESS [105]
	Rat	5%/Pyro-phyllite	25 mkd·	—	—	225[1]	3	0	No	CAMERON and BURGESS [105]
	Guinea pig	H_2O Emulsion	25 mkd[3]	—	—	375[1]	5	0	No	CAMERON and BURGESS [105]

| Dosage range mg/kg/day | Species | Details of dosage | | Total dosage (mg/kg) | | | Number of animals in test | Mortality % | Survivors intoxicated | Authority |
| | | Formulation (concentration of DDT/vehicle) | Rate | Minimum | | Maximum tolerated | | | | |
				Fatal to one animal	Fatal to all animals					
20·1–40·0 continued	Rabbit	H$_2$O Emulsion	25 mkd[3])	—	—	375[1])	5	0	No	CAMERON and BURGESS [105]
	Rabbit	30%/DMP	37·5 mkd	—	—	2,437[1])	3	0	No	HAAG et al. [253]
40·1–80·0	Mouse	Gum arabic	50 mkd[4])	—	—	—	—	100	—	DOMENJOZ [172]
	Rat	1%/Kerosene	50 mkd	—	—	1,000[1])	5	20	Yes	CAMERON and BURGESS [105]
	Guinea pig	1%/Kerosene	50 mkd	—	900	—	5	100	—	CAMERON and BURGESS [105]
	Rabbit	Crystalline	50 mkd[3])	—	800	—	4	0	No	CAMERON and BURGESS [105]
	Rabbit	1%/Kerosene	50 mkd	—	800	—	5	100	—	CAMERON and BURGESS [105]
	Rabbit	30%/DMP	75 mkd	—	—	4,875[1])	3	0	Yes	HAAG et al. [252]
	Rabbit	30%/DMP	75 mkd	—	—	4,875[1])	3	33	Yes	HAAG et al. [253]
	Rabbit	Gum arabic	50 mkd	—	—	2,500[1])	—	40	Yes	DOMENJOZ [172]
	Cow	10%/H$_2$O	45 mkd	—	—	3,240[1])	1	0	No	TELFORD and GUTHRIE [583]
80·1–160·0	Rat	Gum arabic	100 mkd	—	—	5,000	—	0	Yes	DOMENJOZ [172]
	Rat	DMP	150 mkd	1,500	—	13,500[1])	3	33	Yes	DRAIZE et al. [179]
	Guinea pig	Gum arabic	100 mkd	—	—	5,000	—	0	No	DOMENJOZ [172]
	Guinea pig	DMP	150 mkd	4,050	—	13,500[1])	3	33	Yes	DRAIZE et al. [179]
	Rabbit	Gum arabic	100 mkd	—	—	5,000[1])	—	20	Yes	DOMENJOZ [172]
	Rabbit	5%/DMP	100 mkd	—	—	1,300[1])	3	0	Yes	SMITH and STOHLMAN [532]
	Rabbit	5%/DMP	100 mkd	—	—	1,500[1])	3	0	Yes	SMITH and STOHLMAN [532]
	Rabbit	10%/Kerosene	100 mkd	—	600	—	5	100	—	CAMERON and BURGESS [105]
	Rabbit	DMP	150 mkd	4,950	—	13,500[1])	3	66	Yes	DRAIZE et al. [179]
	Rabbit	30%/DMP	150 mkd	—	—	9,750[1])	3	67	Yes	HAAG et al. [252]
	Rabbit	30%/DMP	150 mkd	—	—	9,750[1])	3	0	Yes	HAAG et al. [252]
160·1–320·0	Rat	10%/Kerosene	200 mkd	—	2,800	—	5	100	—	CAMERON and BURGESS [105]
	Rat	DMP	300 mkd	14,700	—	27,000[1])	3	33	Yes	DRAIZE et al. [179]

| Dosage range mg/kg/day | Species | Details of dosage | | Total dosage (mg/kg) | | | Number of animals in test | Mortality % | Survivors intoxicated | Authority |
| | | Formulation (concentration of DDT/vehicle) | Rate | Minimum | | Maximum tolerated | | | | |
				Fatal to one animal	Fatal to all animals					
160·1–320·0 continued	Guinea pig	10%/Kerosene	200 mkd	—	2,400	—	5	100	—	CAMERON and BURGESS [105]
	Guinea pig	DMP	300 mkd	—	—	27,000[1]	3	0	No	DRAIZE et al. [179]
	Rabbit	10%/Oil	200 mkd	400	3,000	—	8	100	—	DEICHMANN et al. [162]
	Rabbit	DMP	300 mkd	3,000	—	27,000[1]	3	66	Yes	DRAIZE et al. [179]
	Rabbit	30%/DMP	300 mkd	< 4,500	—	—	3	100	—	HAAG et al. [252]
	Rabbit	30%/DMP	300 mkd	< 4,500	—	—	3	100	—	HAAG et al. [253]
	Goat	5%/H_2O	200 mkd	—	—	14,400	1	100[5]	—	TELFORD and GUTHRIE [583]
	Goat	5%/H_2O	300 mkd[6][7]	—	—	35,700[1]	1	0	No	TELFORD and GUTHRIE [583]
320·1–640·0	Rat	DMP	600 mkd[4]	3,000	18,000	—	3	100	—	DRAIZE et al. [179]
	Rat	Crystalline	350 mkd	—	—	16,800	—	0	Yes	TAYLOR and FRODSHAM [579]
	Guinea pig	DMP	600 mkd[4]	7,200	26,400	—	3	100	—	DRAIZE et al. [179]
	Rabbit	DMP	600 mkd[4]	3,000	3,600	—	3	100	—	DRAIZE et al. [179]
	Rabbit	30%/DMP	600 mkd	< 7,200	—	—	3	100	—	HAAG et al. [252]
	Goat	10%/H_2O Emulsion	485 mkd[7]	—	—	23,280[1]	1	0	—	TELFORD and GUTHRIE [583]
640·1–1,280·0	Dog	DMP	1,200 mkd[3]	—	—	108,000[1]	—	0	No	DRAIZE et al. [179]

[1] Maximum tolerated was also maximum tested.
[2] Similar groups of rats dosed with same dosage of DDT in acetone and ether solutions were not affected.
[3] Lower dosages tested by the same authors were also harmless to this species.
[4] Higher dosages tested by the same authors were also fatal to this species.
[5] Relationship to DDT not clear.
[6] Divided into 2 doses daily first 12 weeks and then 1 dose of 470 mg/kg/day last 5 weeks.
[7] Approximate dosage.

DMP = Dimethylphthalate. – mkd = Milligram per kilogram per day.

3 times daily, 5 times a week, for a total of 4 weeks; the monkeys showed no injury, but the same type of exposure was fatal to over half the mice studied.

The same authors investigated the respiratory toxicity of a dust containing 10% of DDT and 90% of Pyrax-ABB. They found that dogs exposed for 3 hours daily 5 days a week for a total of 4 weeks to a continuing concentration of 12,480 mg of DDT per cubic meter of air suffered no ill effects. Mice exposed in the same way to 13,900 mg/m³ showed toxic effects which, however, may have been caused by ingestion of some of the dust. The respiratory toxicity of pure DDT dust was also studied and compared to the oral toxicity of such dust. Ground DDT which had passed through a fine sieve was insufflated deeply into the nostrils of three dogs and fed by capsule to three others. In each instance, a dosage of 100 mg/kg/day, 6 days a week, was given for 7 weeks. An additional dog was insufflated at a rate of 100 and later 200 mg/kg/day during the same period. Only one animal, which was dosed at the lower rate, showed neuromuscular disturbances characteristic of DDT poisoning. Nineteen days before the onset of these characteristic disturbances, she had shown unmistakable signs of liver dysfunction and probably kidney complications. The authors concluded that the liver and kidney damage resulted from DDT and suggested, in effect, that the relationship between liver damage and neuromuscular signs was one of cause and effect.

Although this view appeared justified at the time, it would now seem open to question. At autopsy, the animal was found to have advanced, coarsely nodular cirrhosis with extensive scar tissue development. It is unlikely that the amount of scarring described by the authors would occur within the 50-day period of exposure, and especially within the 40-day period from the onset of detected liver disorder to autopsy. The greater susceptibility of animals with pre-existing chronic liver injury has been mentioned by JUDAH [314] and is probably the true explanation of the case just cited.

In a study of mists, using a 1 per cent solution of DDT in deobase, NEAL and his colleagues [426] found that rabbits exposed for 48 minutes daily, for a period of 4 weeks, developed no signs of DDT poisoning, but did suffer irritation of the exposed skin, of the eyes, and of the respiratory tract, caused by the solvent.

In further extensive experiments with aerosols, similar to those in use today, NEAL and his associates [427] found that single, daily, massive exposures of dogs to initial concentrations of 33,000 mg/m³ over a period of 8 weeks caused no harmful effect except moderate irritation and narcotic action resulting from the primary solvent. Monkeys exposed in the same general way to concentrations of 33,300 mg/m³ over a period of 22 weeks, or longer, failed to show definite nervous symptoms characteristic of DDT or laboratory signs of liver injury which it was thought might be the result of DDT. When the exposure of dogs was intensified by exposing them to the same concentration three times daily, they showed no signs of DDT intoxication but succumbed to pulmonary injury induced by the solvents.

It must be recalled that the initial concentration of DDT resulting from the recommended uses of aerosol does not exceed 10 mg/m³, and is usually less. The recommended concentration is, therefore, over a thousand times less than the concentrations studied by NEAL and his associates.

CAMERON and BURGESS [105] reported that exposure of rats, guinea pigs, and rabbits to mists containing an initial concentration of 15,000 mg/m³, for 2 hours daily for 11 days, produced clinical signs in all the animals and death in some of them. A dosage one-third as great did not produce signs of poisoning during six exposures.

Guinea pigs were exposed on alternate days to solutions of DDT atomized with a perfume sprayer. Two per cent solutions in alpha-chloronaphthalene, petroleum, petroleum ether (POTOSSI [466]), and methyl alcohol (POTOSSI [467]) were used. So much solvent was used that the experimental and control animals coughed and sneezed and showed severe irritation of the mucous membranes and the exposed skin. Although hyperexcitability and other nervous disorders were attributed to the DDT, the mortality of the experimental and control animals was similar in each instance.

HAAG *et al.* [252] used a constant-flow chamber to study the effect of respiratory exposure to dust and sprays containing DDT. Rats, dogs, and monkeys were exposed to 10 per cent dust administered at an equilibrium rate of 470 and 300 mg/m³ 2 hours a day 5 days a week for 5 weeks. The exposure was fatal to nearly all the rats and to some of the dogs. The dust without DDT was fatal to about half of the dogs and monkeys when administered at a high rate. In similar experiments, DDT in refined kerosene was given at the constant rate of 120, 190, and 300 mg/m³. The compound proved more toxic in solution than in the dust form, but dogs remained more resistant than rats. No special effort was made to prevent skin contamination. DDT was given in emulsions at the rate of 180 and 250 mg/m³; in this formulation it proved less toxic to rats than the kerosene solution; but the emulsion base without DDT was, for some unexplained reason, so toxic to dogs that no tests of the toxicity of DDT could be run on dogs. These results on respiratory toxicity were later confirmed by separate tests, HAAG *et al.* [253].

Intermuscular Exposure

PLUVINAGE and HEATH [462] reported that cats withstood two to nine intermuscular doses of DDT in oil, totaling 135 to 665 mg/kg, before the onset of symptoms. It was not possible to produce chronically poisoned animals; the cats either developed slight transient illness and recovered, or developed progressive illness and died.

4.

USE OF DDT FOR THE CONTROL
OF UNDESIRABLE VERTEBRATES

Since DDT is toxic to mammals generally, it has occurred to several workers that the compound might have a practical use as a rodenticide. It is possible that MACCHIAVELLO [386] was first to make the suggestion. The idea is especially attractive because of the extensive use of DDT for malaria, typhus, and plague control, and because murine typhus and plague depend on rodents as the normal host. If an attack could be made simultaneously on the host and on the vector using a single formulation, it would involve increased efficiency and a minimal cost.

With this idea in mind, a number of laboratory experiments have been conducted with mice (Fox [211]; SERGENT and SERGENT [519], [520]; JONES [312]), Norway rats (SERGENT and SERGENT [519], [520]), roof rats (WANSON and CAMPHYN [645]), and field rodents, including *Meriones shawii* (SERGENT and SERGENT [520]), and *Tatera angolae* (WANSON and CAMPHYN [645]). The animals have been forced to live in cages or go through passageways containing dust formulated with different concentrations of DDT. All of these tests have shown the various rodents to be susceptible to the poison, especially if a 50 per cent dust is used. As one would expect from more conventional tests, mice have proved more susceptible than Norway rats and, perhaps, more susceptible than the other species tested. However, failure to obtain complete mortality of mice under the highly artificial conditions of the tests has been reported (JONES [312]).

NELSON *et al.* [434] mixed DDT and a number of other compounds into bait and allowed rats to consume the baits voluntarily. DDT was considered to be the most toxic compound tested, but no evidence was given that it would be effective as a rodenticide under practical conditions.

In a more realistic test, Fox [211] found that house mice were partially though incompletely controlled by 50 per cent DDT dust scattered in their environment. Considerable mortality of mice has been reported following routine dusting for the control of murine typhus (U.S. Public Health Service [629]).

DENT *et al.* [168] released 113 *Rattus rattus* in a ratproof building with a floor space of 186 m², which was dusted with 4·77 kg of 10 per cent DDT dust. In spite of overcrowding and confinement to the area, the mortality reached only 36·3% in 11 weeks.

HILL and MORLAN [277] and DENT *et al.* [168] ascribed some rat deaths to 10 per cent DDT dust used in a large-scale, murine typhus control program.

                    W. J. Hayes, Jr.

They observed, however, that the roof rat, which was the predominant species present, avoided all but thin patches of the dust. It was estimated that the mortality under field conditions, in which rats can exercise some freedom in avoiding unfavorable situations, was less than controlled experiments would indicate (DENT et al. [168]).

Much better success in control of the same species (*R. rattus*) (WANSON and CAMPHYN [645]) and of an unidentified species (SERGENT and SERGENT [521]) has been claimed. WANSON and CAMPHYN [645] reported that DOUCET had observed effective rodent control as a result of routine DDT application for malaria control. The same authors also reported remarkable control of gerbiles through application of dust to their burrows.

Rodent control by means of DDT has been reviewed by POLLITZER [463]. He considered that the greatest opportunity for the use of DDT as a rodenticide lay in the application of high concentrations to the burrows of wild rodents which act as reservoirs of plague. It was considered practical to make applications in the immediate vicinity of settlements and thus prevent the transfer of plague to domestic rodents and to man. This is certainly a worthy objective, but the practicality of the method requires further confirmation.

In the reviewer's experience, moderately good mouse control may be obtained with DDT. Rat control is mediocre, even when 50 per cent dust is used. The safety of distributing this dust on a wide scale is doubtful and, since the advent of safe, effective rodenticides, the method appears contraindicated.

The complete failure of DDT to control two undesirable species of birds, the starling (*Sturnus vulgaris*) and the English sparrow (*Passer domesticus*), has been reported (JONES [312]).

5.

PATHOLOGY

The pathologic changes associated with exposure to DDT differ considerably depending on whether they result from a few massive doses or from many relatively small doses. In the first instance, agonal signs may be prominent or detectable pathology may be lacking entirely. In the second instance, at least following relatively high dosage rates, the changes are more constant in occurrence and anatomical location. For both types of poisoning, the vast majority of authors have failed to report changes sufficient to account for death. In fact, the absence of such serious lesions has been specifically pointed out (NEAL and VON OETTINGEN [425]). The few workers who considered that demonstrated anatomical lesions were sufficient to account for death have been in disagreement regarding the organ involved.

Pathology varies not only in relation to the number of doses but also in relation to the route of exposure and the formulation of the active ingredient. None of the lesions which may be present are diagnostic of DDT but, regardless of the route of administration, certain changes in the liver have been regarded as characteristic of the chlorinated hydrocarbon insecticides in general and of DDT in particular. Although it is generally agreed that the hepatic cells show the most striking histological changes caused by DDT, there is disagreement about whether these changes are characteristic, whether they occur at low dosage levels, and whether they are physiologically meaningful.

Effect of a Few Large Doses

Gross Findings

Gross findings are not outstanding. Those which occur with some regularity are as follows:
(1) Skin changes associated with dermal application of DDT and, especially, of solvents.
(2) Generalized congestion, foci of hemorrhage, or pulmonary edema associated with violent convulsions and the agonal state.
(3) Atrophy of muscles and viscera and evidence of loss of weight in animals that survive long enough to show the effects of inanition.
A few papers (CHOU et al. [125]; BLANCO [76]) describe marked hemorrhage of the gastrointestinal tract or other dramatic changes which have not been

generally described and which must have depended on factors peculiar to the experiments and not on DDT.

Neuromuscular System

Since the most prominent signs of DDT intoxication involve the neuromuscular system, pathologists have given considerable attention to that system.

Minor changes including edema, partial tigrolysis of neurones, pericellular or perinuclear vacuolation, loss of Nissl bodies, swelling of neurones, pycnosis, or destruction of occasional cells have been noted singly or in combination (LILLIE and SMITH [367]; CAMERON and BURGESS [105]; NEAL et al. [427]; DEICHMANN et al. [162]). These minor changes, when present at all, were generally confined to the brain stem and spinal cord (LILLIE and SMITH [367]) or to the spinal cord only (CAMERON [103]; CAMERON and BURGESS [105]).

An apparent increase in the frequency of a focal granulomatous encephalitis in rabbits histologically identical to an encephalitis caused by infection in that species has been reported (NELSON et al. [432]). The lesion showed no relation to the duration of exposure to DDT.

Several pathologists have remarked specifically on the absence of change or the presence of very minor changes in the central nervous system in spite of the pronounced neurological signs exhibited by animals (LILLIE and SMITH [367]; NELSON et al. [432]; CAMERON and BURGESS [105]; DIAZ-JIMENEZ [171]; GLOBUS [237]).

An entirely different result was reached by VIRGILI and MARCHIAFAVA [637] who described lesions of the central nervous system which they considered adequate to explain the grave neurological symptomatology commonly observed in animals poisoned with DDT, as well as the respiratory paralysis which they viewed as the usual cause of death. The studies involved guinea pigs and rabbits which were killed by a single massive oral dose (ranging up to 1,820 mg/kg) of DDT in olive oil. Cats accidently poisoned by the compound were also studied. The lesions observed included chromatolysis, vacuolization, pycnosis, and some cellular destruction; neurophagia and progressive alteration of the neuroglia; as well as inconstant meningeal and subependymal hyperemia and hemorrhages. These lesions differed from those observed by HAYMAKER et al. [268] chiefly in their distribution. (The studies of HAYMAKER and his colleagues are discussed in the section on pathology associated with repeated doses.) VIRGILI and MARCHIAFAVA reported that the most serious alterations were in the pontobulbar nuclei, while serious but inconstant alterations occurred in the cerebellum and slighter lesions appeared in the cerebrum. In the spinal cord only inconstant damage of the anterior horn cells was found. No alteration of the myelin, or nerve fibers, or of the sciatic nerve was reported. Interpretation of the experiment is made more difficult by the absence of controls.

Apparently the only morphologic study of the nerve endings and muscle fibers in relation to DDT toxicity is that of CAREY et al. [109]. The authors gave single intraperitoneal doses of 80 mg/kg to rats and 200 mg/kg to chameleons and reported a progressive discharge of auriphilic neurogenic substances from some motor end plates into the myoplasm of some voluntary muscle fibers and from the poles of the elongate nuclei in the dark muscle fibers. These changes were apparently considered to be the anatomical basis of the neuromotor dysfunction seen in the experimental animals.

Instances of acute damage to skeletal and myocardial muscle similar to the damage caused by repeated doses have been reported (NELSON *et al.* [432]; CAREY *et al.* [109]; DEICHMANN *et al.* [162]).

Liver

Liver changes are more prominent after a large number of smaller doses of DDT than after a few massive doses. After a single lethal dose the liver changes are slight and inconstant but of the same general type as those seen in severe poisoning following repeated doses (CAMERON and BURGESS [105]).

Irrespective of the number of doses of the compound, it is generally agreed, however, that the most striking tissue alterations caused by DDT are seen in the liver (LILLIE and SMITH [367]; NELSON *et al.* [432]; CAMERON [103]; CAMERON and BURGESS [105]; PLUMMER [461]; KONST and PLUMMER [331]; HAAG *et al.* [252]; CAMERON and CHENG [106]; GEREBTZOFF and PHILIPPOT [224]).

DIAZ-JIMENEZ [171] reported prominent changes not only in the liver but also in the kidney and lung. These changes resulted from the parenteral injection of solutions of DDT, but interpretation of the results is difficult because similar volumes of petroleum injected into control animals produced weight loss, death, and kidney pathology. Changes in the central nervous system in the absence of pathology of the liver or other organs have been reported rarely (PLUVINAGE and HEATH [462]).

On microscopic examination, fine-droplet fatty degeneration of the liver cells has frequently been encountered (LILLIE and SMITH [367]; NEAL *et al.* [427]; VIRGILI and MARCHIAFAVA [637]; DEICHMANN *et al.* [162]). This degeneration is confined to scattered cells and is small in amount when the number of doses leading to death is small. Albuminous degeneration has also been reported (DEICHMANN *et al.* [167]).

Hydropic degeneration, chiefly of centrolobular cells, has been encountered (LILLIE and SMITH [367]; NELSON *et al.* [432]).

Coagulative necrosis, chiefly of midzonal distribution has been reported (LILLIE and SMITH [367]). However, necrosis is generally found to be predominantly central (NELSON *et al.* [432]; CAMERON and BURGESS [105]; LILLIE *et al.* [368]; DEICHMANN *et al.* [162]). Proliferation of fibroblasts in the centrolobular area and even partial trabeculation of the parenchyma was observed in rats which survived several large doses of DDT (LILLIE and SMITH [367]; LILLIE *et al.* [368]). Other authors have pointed to a striking absence of fibrosis (CAMERON and BURGESS [105]).

Kidney

The presence of exudate or casts in the convoluted tubules of a few animals killed by DDT has been reported (LILLIE and SMITH [367]; NELSON *et al.* [432]), but casts are not present in many animals. When present, the casts may be hyaline or may contain coarse granules. A slight lymphocytic infiltration of the cortex has also been reported (LILLIE and SMITH [367]), as well as cloudy swelling and degenerative changes (NEAL *et al.* [427]; GUARESCHI and BINI [248]; VIRGILI and MARCHIAFAVA [637]; DEICHMANN *et al.* [162]), but the significance of the changes in the kidney is open to doubt. However, severe changes including necrosis have been reported (NELSON *et al.* [432]; DEICH-MANN *et al.* [162]), and it has even been claimed that changes in the kidney have exceeded the changes in the liver (ZEIN-EL-DINE [685]).

Other Organs

Congestion or edema of the lungs and occasional focal hemorrhages have frequently been reported (LILLIE and SMITH [367]; CAMERON and BURGESS [105]; VIRGILI and MARCHIAFAVA [637]; DEICHMANN *et al.* [162]). These changes are probably agonal in nature and some authors have specifically stated this to be true.

Changes in the spleen have consisted of pulp myelosis and hemosiderosis (LILLIE and SMITH [367]; NELSON *et al.* [432]).

Focal necrosis (NELSON *et al.* [432]) or hemorrhage (CAMERON [103]; CAMERON and BURGESS [105]; ZEIN-EL-DINE [685]) of the adrenal cortex has occasionally been seen.

Focal necrosis of the gall bladder of heavily dosed rabbits has been reported (NELSON *et al.* [432]).

Significant changes in the testis (NELSON *et al.* [432]; DIAZ-JIMENEZ [171]), bone marrow (NELSON *et al.* [432]), bone (NELSON *et al.* [432]) and pancreas (NELSON *et al.* [432]), have not been found.

Effect of Repeated Doses

Neuromuscular System

Some early authors (LILLIE and SMITH [367]; NEAL *et al.* [426], [427]; LILLIE *et al.* [368]) reported minor changes such as vacuolation around large neurones, vague peculiarities of staining, chromatolysis, or occasional karyolysis. These changes have been reported at one time or another in the spinal cord, the Gasserian ganglia, and the red nucleus (NEAL *et al.* [427]). Other early authors in studying a wide range of species and dosage schedules declared that, in spite of an effort to determine nerve cell changes in the brain and spinal cord of animals with tremors, no changes could be seen that were not present in controls similarly and concurrently fixed and stained (NELSON *et al.* [432]).

Differing from all earlier work, HAYMAKER *et al.* [268] reported that dogs which had received repeated, relatively large doses of DDT showed changes in the Purkinje cells and in the cells of the dentate and roof nuclei of the cerebellum. Dogs which received more than 10 doses were judged to show degeneration (in excess of artifacts seen in controls) of 7·4 to 49·8% of the cells of the cerebellar nuclei. The number of satellite glia was said to be increased, but no generalized gliosis was present. The same dogs showed only slight to moderate chromatolysis of some of the anterior horn cells of the spinal cord and no abnormality of the brain stem, cerebrum, eyes, or peripheral nerves.

PLUVINAGE and HEATH [462] administered repeated intramuscular doses of DDT to cats and reported diffuse degeneration of the ganglion cells throughout the brain, without any difference between the different areas. No significant pathology was encountered in the liver or other viscera of the cats.

Without doubt, the most important papers on the effect of DDT on the histopathology of the nervous system are those of GLOBUS [236], [237]. Monkeys were exposed orally, by inhalation, and by peritoneal injection. Dogs were given DDT in olive oil by stomach tube. Cats received the compound intravenously. Rats were fed various concentrations of DDT in the diet ranging as high as 800 ppm for 2 years. A wide variety of special histological techniques was used. Very minor changes were encountered in the brains of a few experimental animals, but similar changes occurred in controls. The author considered it especially significant that in no case was there an increase in the accumulation of free fat, nor an increase in glial elements. No pathological glial elements were encountered. It was concluded that all the brains and spinal cords of the animals studied failed to show pathological changes resulting from DDT.

NELSON *et al.* [432] have specifically called attention to the absence of gliosis or inflammatory cellular infiltrates in the brains of animals poisoned by DDT. Other authors would certainly have mentioned the important finding if gliosis had been present in their material.

Changes in the myelin of the cord and peripheral nerves are absent (LILLIE and SMITH [367]; HAYMAKER *et al.* [268]).

Occasional instances of focal hyaline degeneration, necrosis, interstitial hemorrhage, fibroblast proliferation and/or lymphocyte infiltration have been observed in the skeletal muscle (LILLIE and SMITH [367]; NELSON *et al.* [432]; CAMERON and BURGESS [105]). Changes in the myocardium which probably have the same origin have been reported (LILLIE and SMITH [367]; NELSON *et al.* [432]; CAMERON and BURGESS [105]; LILLIE *et al.* [368]; VIRGILI and MARCHIAFAVA [637]). The changes may include cloudy swelling, an increase in fat within the cells, slight focal necrosis, or subendocardial hemorrhages.

Liver

Significant gross findings have rarely been reported. Jaundice in one dog which had been exposed for 13 days has been reported by NELSON *et al.* [432], but it was not clear from the paper whether the jaundice was caused by the

DDT or was based on pre-existing disease. NEAL *et al.* [432] reported jaundice in a dog which was exposed for 49 days and which on autopsy showed a marked cirrhosis of the liver. In this instance there is strong evidence that liver disease preceded exposure to the compound. The presence of ascitic fluid and gross liver pathology quite unlike anything found by others in DDT poisoning was reported for one dog by ZEIN-EL-DINE [685].

The increased size and weight of the liver of rats maintained on a high dietary concentration of DDT is discussed below (p. 111).

Rats maintained on high dietary concentrations of DDT may also show a 'nutmeg' appearance or a slight yellowish or tan color of the liver (FITZHUGH and NELSON [208]).

Many of the histological changes which have been observed so far in the liver of rats subjected to small repeated doses of DDT were clearly set forth in the earliest paper dealing with the matter. LILLIE and SMITH [367] described hepatic alterations in seven rats sacrificed after 94 to 98 days on a diet containing 1,000 ppm of the compound. The liver cells showed fine-droplet fatty degeneration especially in the midzones but often extending into the periportal and central areas of the lobules. There was regularly an increase in the oxyphilic character of the cytoplasm in the centrolobular areas with a concurrent decrease in the normal, coarse, basophilic granulation within these cells. In some rats this process went on to the formation of hyaline oxyphil masses either contiguous with the remaining cytoplasm or suspended in clear vacuoles.

Similar lesions were reported in rabbits. In this species the nucleus of some hyaline liver cells was normal and in others there was karyolysis. Midzonal coagulation necrosis, sometimes associated with granulation tissue replacement or calcification, was described. Occasional hydropic changes were encountered. Essentially the same findings were reported in a second paper which appeared only a few weeks later in which NELSON *et al.* [432] reported on rats maintained for 2 months or more on dietary concentrations of 400 and 1,000 ppm. In addition, NELSON and his colleagues observed that some of the centrolobular cells showed a lytic type of necrosis while most of the surviving cells showed hypertrophy. Conversely, the peripheral cells were slightly atrophic. Occasional instances of slight bile duct proliferation were seen.

Other papers (NEAL *et al.* [426], [427]; KONST and PLUMMER [331]; HAYMAKER *et al.* [268]; LILLIE *et al.* [368]; FITZHUGH and NELSON [208]; STOHLMAN and LILLIE [560]; HAAG *et al.* [252]) reported findings essentially similar to those listed by LILLIE and SMITH and by NELSON and his colleagues. However, the pathology seen by HAAG *et al.* [252], who used recrystallized DDT, was very minor and confined to a small proportion of the experimental animals.

CAMERON and BURGESS [105] reported, in addition to fatty degeneration and cloudy swelling, focal necrosis and large areas of centrolobular necrosis in animals that died as the result of exposure to DDT. Although the animals were exposed repeatedly, the pathology was considered to be acute. The lesions were apparently more severe than those mentioned above and the authors considered them sufficient to account for death. The authors specifically noted that, if exposure were discontinued, the dead cells were removed by autolysis and phagocytic action and repair was complete without any fibrosis, although calcification was occasionally seen.

In other instances, slight fibroblastic trabeculation has been observed in an occasional liver (LILLIE and SMITH [367]). An apparent incipient cirrhosis in conjunction with necrosis has been observed in a rabbit that received repeated, intermittent doses of 100 mg/kg (LILLIE et al. [368]).

FITZHUGH and NELSON [208] described the occurrence of a minimal hepato-carcinogenic tendency in rats fed DDT for 2 years. The tumors were from 5 to 12 mm in diameter and paler than the surrounding tissue. Microscopically, they showed an almost complete loss of lobular architecture but were not sharply circumscribed from the rest of the liver tissue. Mitoses were not seen. Some of the cells showed typical DDT pathology. The authors felt that the tumors could be regarded as adenomas or as low grade hepatic cell carcinomas. Even if the carcinogenicity of DDT in rodents were established, it would be difficult to interpret the significance of the results, for SALTER [503] has shown that the epidemiology of neoplasms in man is parallel to the results of animal experimentation in only a few instances. Furthermore, neither benign nor malignant tumors were observed in strain C mice painted weekly for 52 weeks with 5% of DDT in kerosene. The treated mice showed acute and chronic inflammatory changes in the painted skin and minimal liver damage (BENNISON and MOSTOFI [54]).

DDT was reported to show some chemotherapeutic action against cancer, but not of a practical degree (HARTWELL et al. [266]).

Rats were maintained on a diet containing 1,000 ppm of DDT for 12 weeks and then placed on a similar diet which did not contain the compound. Those sacrificed after 1 and 2 weeks of 'recovery' showed typical pathology similar to that of rats sacrificed while they were receiving DDT. However, 4 to 6 weeks after DDT was discontinued only traces of injury could be detected and in 8 to 10 weeks the liver appeared normal (FITZHUGH and NELSON [208]).

The foregoing discussion is based on experiments in which animals were exposed repeatedly to relatively large amounts of DDT by different routes. For animals fed DDT as a part of their diet, the minimum concentration of the compound in the diet was 100 ppm which is equivalent to a dosage rate of about 4·3 mg/kg/day for male rats and about 5·4 mg/kg/day for female rats. Many of the dosage rates used were very much larger, ranging up to at least 50 mg/kg.

The following discussion is concerned with experiments in which at least some of the animals were given very small amounts of DDT by the oral route. These experiments, which were made as one approach to the study of residues on food, roused great interest when it was shown that characteristic histologic changes in the liver occur in male rats fed 5 ppm of DDT in their diet for 4 to 6 months (KUNZE et al. [334]; LAUG et al. [345]).

Equal interest followed the statement by other distinguished authors (CAMERON and CHENG [106]) that no significant changes whatever occur in the liver of rats maintained on even higher dosages. One must agree with CAMERON and CHENG that special importance must still be attached to histological evidence of tissue change resulting from DDT because the compound is extensively used in agriculture, appears as residues in human food, is stored in human fat, and has a mode of action which is not yet understood.

LAUG *et al.* [345], following a brief abstract (KUNZE *et al.* [334]), described experiments in which rats of both sexes were fed DDT at concentrations of 1, 5, 10, and 50 ppm in their total dry diet for 15 to 27 weeks. Their findings on the storage of DDT in the tissues are discussed below in connection with Figure 10. Histological changes in the rats were confined to the liver and were similar to but milder than changes in the liver after larger doses. No changes were seen in rats fed 1 ppm, but rats fed 5 ppm showed what was considered to represent the smallest detectable morphologic effect of DDT. Changes included cytoplasmic hypertrophy without nuclear enlargement, increased cytoplasmic oxyphilia with sometimes a semihyaline appearance, and margination of the basophilic cytoplasmic granules. The changes were most marked in the centrolobular area. The authors provided good photographs to illustrate the changes. Male rats consistently showed more hepatic cell changes than did females. The changes were regarded as characteristic. They were reported to occur in rabbits, mice, and guinea pigs, though not to the same extent as in rats. They were not observed at all, according to the authors, in chickens, dogs, cats, monkeys, or large domestic animals. The characteristic changes in the liver were not accompanied by nonspecific changes such as necrosis of the liver nor by changes in other organs.

In considering the work of CAMERON and CHENG [106] in which they denied the presence of histologic change in the liver of rats fed small amounts of DDT, it is of interest to review CAMERON's studies on high dosage rates. As mentioned above, CAMERON and BURGESS [105] reported changes in the liver of several species exposed to high dosages of DDT which were apparently more severe than the changes seen by other authors. In fact, CAMERON and BURGESS considered that the extensive hepatic cell necrosis which they saw was sufficient to account for death. However, although the descriptions by other authors mention occasional, limited necrosis, the description of CAMERON and BURGESS did not include the less severe changes which other investigations have regarded as characteristic of DDT. Furthermore, CAMERON and BURGESS denied the presence of any pathology whatever in the liver of rats dosed orally at the rate of 50 mg/kg/day for 30 doses, although rabbits showed necrosis after fewer doses at the same rate, and guinea pigs showed necrosis after an approximately equivalent dosage. It, therefore, might have been expected that their laboratory would find no pathology in rats when low dosage rates were studied.

At least one other group of workers (HAAG *et al.* [252]) in studies involving food concentrations of 100 ppm or higher specifically denied the presence of centrolobular hyperplasia, oxyphilia, and hyaline changes in the rat liver.

Likewise, GREENWOOD *et al.* [247], who fed rats at a great variety of concentrations of DDT ranging from less than 1 ppm up to 123 ppm, failed to detect any histological change in the liver of any of them.

CAMERON and CHENG [106] reported experiments in which rats of both sexes were given DDT in oil by stomach tube at the rates of 0·36, 3·6, and

36·0 mg/kg/day[1]). Rats at each dosage level were sacrificed for study after 33, 43, and 60 or 63 weeks of exposure to DDT. No changes described by others as characteristic of DDT poisoning were found in the livers of the experimental animals. Necrosis, considered by the authors to be characteristic of acute DDT poisoning, was found in only two rats which died during the experiment. It

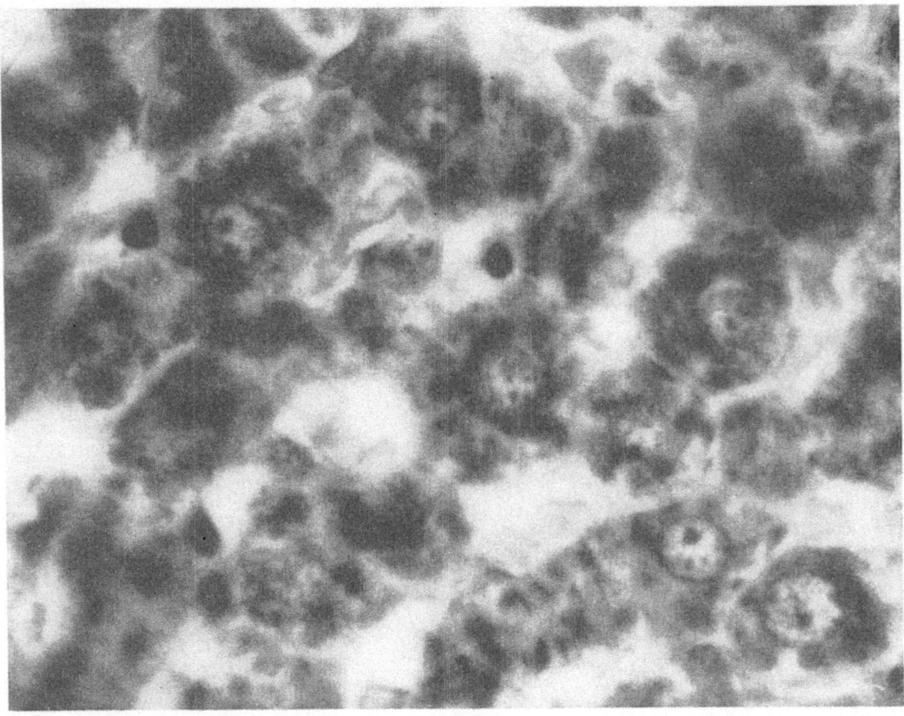

Figure 4

Normal male rat liver showing coarse basophilic granules scattered in the cytoplasm. The section was fixed with Zenker-formalin and stained with azure eosin. × 1,130.

was thought that the tolerance of these two animals had been acutely lowered by intercurrent infection. The authors concluded with the suggestion that the changes reported by others, and especially by LAUG *et al.* [345] were, in fact, fixation artifacts.

---

[1]) Based on calculations and an assumption on the food intake of rats, the authors stated that the three dosage rates correspond approximately to 3·5, 35, and 350 ppm of DDT in the total diet. The calculation seems to contain an error and the values based on the authors' assumptions should probably be 4·5, 45, and 450 ppm, respectively. Furthermore, the authors' assumption on the food intake of rats (consumption of 80 g/kg/day) reflects the behavior of very young rats. The dry food intake for adult rats is about 43 and 54 g/kg/day for males and females, respectively. Thus, the dosage rates used by CAMERON and CHENG [106] correspond to concentrations of about 7·5, 75, and 750 ppm of DDT in the total dry diet for adult rats–the concentrations being a little higher for males and a little lower for females.

The findings of DEICHMANN *et al.* [162] were apparently intermediate. Possible hypertrophy of the hepatic cells was reported for rats fed a diet containing 30 ppm of DDT, and hypertrophy, hyperplasia, oxyphilia, and focal coagulation necrosis were observed in rats fed 150 ppm. No changes were observed at a level of 10 ppm.

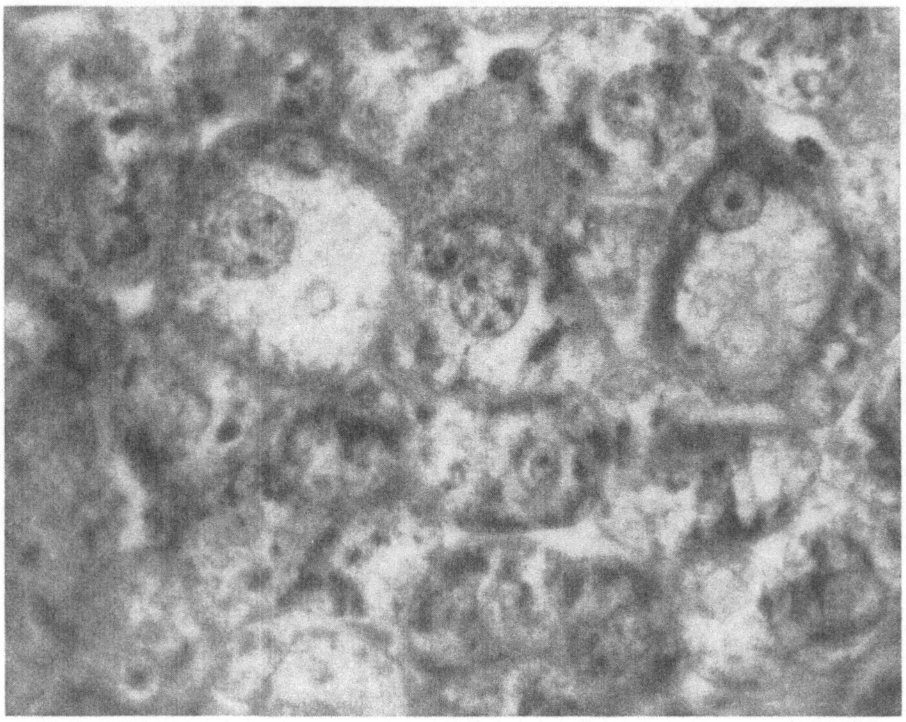

Figure 5

Liver from a male rat fed for 6 months on a diet containing 200 ppm of DDT. The cells show hypertrophy and a displacement of the basophilic granules to the cell margin. Some of the cells, notably the two largest ones, contain lipospheres but they are not prominent because no differential stain has been used. Fixation and staining as in Figure 4. ×1,130.

Work done by ORTEGA and others in the reviewer's laboratory has confirmed, in broad outline, the morphological observations of LAUG and his colleagues, and has given additional information on the character of one kind of cytoplasmic inclusion body found in the liver cells. The changes recognized were:
(1) Centrolobular cellular hypertrophy.
(2) Margination of the coarse, basophilic cytoplasmic granules. The cytoplasm in the central part of the cell is left free of large granules and therefore more optically transparent. Margination is initially more prominent in the centrolobular cells but, with higher levels, a midzonal pattern is most typical.

(3) Increase in fat as revealed by ordinary fat stains. When present, this change is more prominent in centrolobular cells.

(4) Cytoplasmic inclusion bodies which are complex in structure and have been referred to as lipospheres. Each liposphere is composed of an outer, more or less lobulated wall (which may be stained with dyes having an affinity

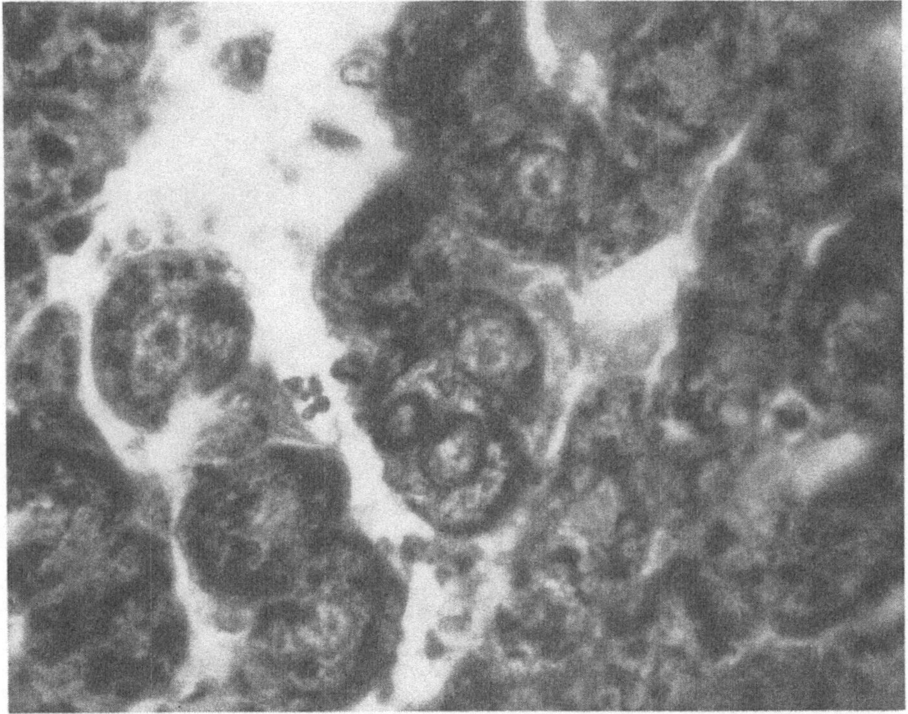

Figure 6

Liver from a male rat fed for 6 months on a diet containing 5 ppm of DDT. A few cells show very slight hypertrophy and margination, and one shows two lipospheres. The section was fixed with acid Zenkers and stained with BENSLEY's anilin blue-orange G-anilin acid fuchsin. × 1,130.

for phospholipids) and a foamy or reticular interior filled during life with fat (which can be stained with the common fat dyes). One or more lipospheres may be present in a single cell and they are most common in the midzonal area.

These changes, and especially the lipospheres, were observed in fresh, unfixed tissue using a phase contrast microscope. Their appearance in fixed tissue is shown in Figures 4 to 7.

It was also shown through the generous cooperation of Dr. ARTHUR A. NELSON of the U. S. Food and Drug Administration, and Drs. JOSEPH F. TREON

54 W. J. Hayes, Jr.

and FRANK P. CLEVELAND of the Kettering Laboratories, that lipospheres were
present in the livers of rats fed moderate or low dosages of DDT whose pathology had been previously investigated by the two laboratories and reported in
a number of papers. Dr. CLEVELAND indicated that the inclusion bodies had
been regarded as hyaline oxyphilic masses which, in fact, they appear to be
when stained with hematoxylin and eosin.

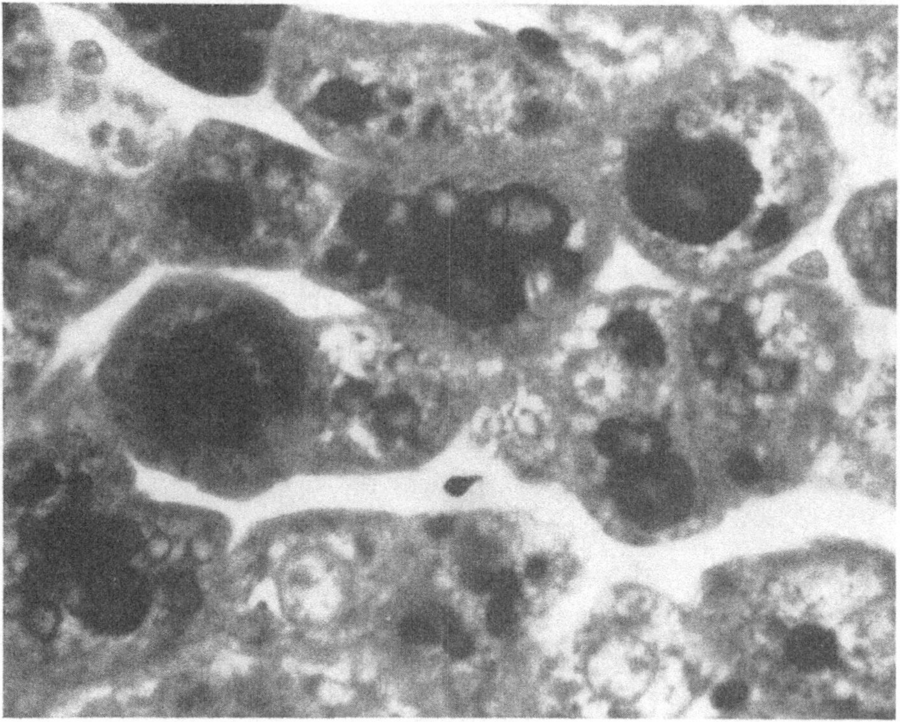

Figure 7

Liver from a male rat fed for 6 months on a diet containing 200 ppm of DDT. The cells show hypertrophy, margination, and a great many lipospheres. Fixation and staining as in Figure 6. ×1,130.

Therefore, not only the existence of the changes, but the details of their
character have been confirmed by different laboratories. The only remaining
questions would seem to involve the cause of variation in the intensity of the
lesions and the interpretation of their significance. Although ORTEGA and his
colleagues observed minimal changes in male rats fed a diet containing 5 ppm
of DDT, they have also observed a striking variation in the intensity of changes
among groups of rats which, at different times, were fed diets intended to be
identical and containing as much as 200 ppm of DDT. They did not observe a
total absence of changes in the liver of male rats which had been exposed to

50 ppm or more of DDT for 4 months or more. However, another laboratory which has frequently observed the changes has found them entirely absent in an occasional group of rats with significant exposure (TREON [595]). Several hypotheses (action of analogs in technical DDT, effect of feeding schedules, etc.) have been proposed to explain the observed variation in the intensity of the liver changes. The intensity of the changes does not correlate with the

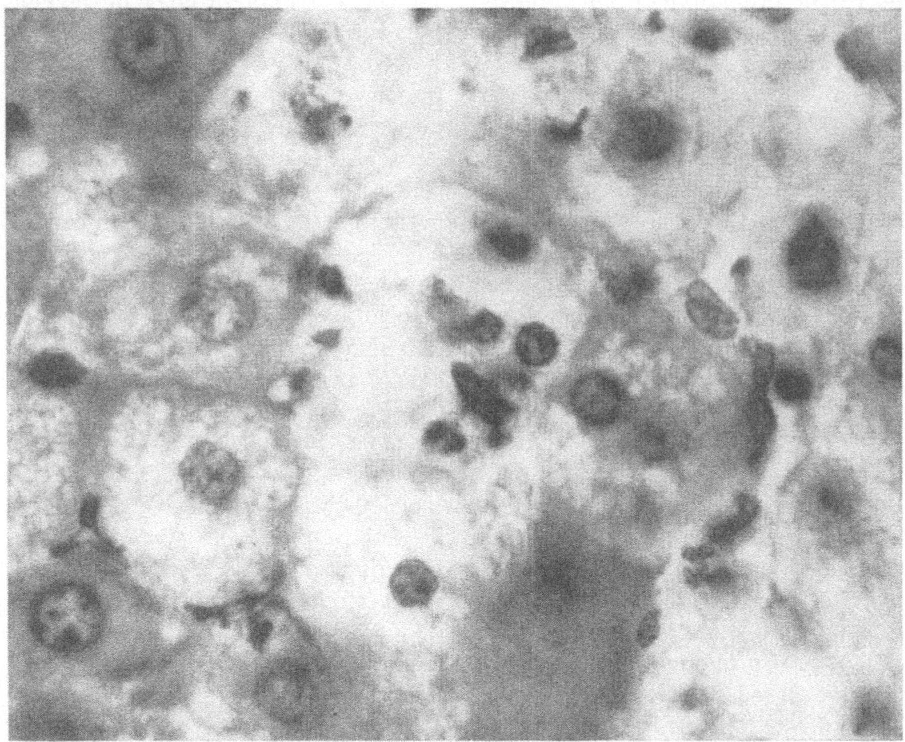

Figure 8

Liver from a male rat fed 1,000 ppm of DDT for 25 days and then 5,000 ppm for 5 days more. The cells show necrotic changes, including hyaline and hydropic degeneration of the cytoplasm and pycnosis of the nuclei. A few polymorphonuclear leucocytes have invaded the area of destruction. Fixation and staining as in Figure 1. × 1,130.

purity of the DDT used. Other possible explanations have apparently not been subjected to test. In conclusion, there is no reason to doubt the results of CAMERON and CHENG, although the suggestion that the changes seen by others represent artifacts is quite untenable. What is needed is further study to learn what factor or factors permit the occurrence of apparently contradictory results.

If the cause of variation were known, it might throw some light on the significance of the liver changes seen in rats fed small amounts of DDT. Although it has been customary to refer to the changes as injuries, the possibility that they are adaptive does not appear to be excluded. Certainly they cannot be regarded as early stages of the necrosis produced by large doses (see Figure 8). The liver alterations appear earlier and are more prominent in rats of the same sex as the dosage rate is increased up to a toxic level. This would be compatible with an adaptive change or an injury. In any event, any interpretation of the liver changes must take the following considerations into account. When male and female rats are maintained on the same concentration of DDT in the diet liver changes are always more prominent in the males, although the females receive a higher dosage of DDT because they consume more food in proportion to their body weight. On the contrary, toxicity and storage show a positive correlation with dosage. Females fed a diet containing 400 ppm or more of DDT show greater signs of poisoning than males fed on the same diet, and females on any diet containing DDT not only store more of the compound in their body fat than males fed the same diet, but store even more than can be accounted for by the difference in dosage.

The liver changes which are said to be characteristic are not necessarily 'chronic'. Under appropriate conditions they can be demonstrated in male rats which have been fed DDT for only 2 weeks. The minimum time required has not been established. Once formed, the changes are reversible.

Kidney

The same slight changes found in acute poisoning have also been encountered in animals exposed repeatedly (NEAL *et al.* [427]; DEICHMANN *et al.* [162]). In addition, fatty degeneration, necrosis, and calcification alone or in combination have been found in animals repeatedly exposed to relatively large doses of DDT (CAMERON and BURGESS [105]; LILLIE *et al.* [368]; STOHLMAN and LILLIE [560]). A slight brown pigmentation of the epithelium of the convoluted tubules was noted in rats fed DDT for 2 years (FITZHUGH and NELSON [208]). The glomerulus is not affected (CAMERON and BURGESS [105]). GEREBTZOFF and PHILIPPOT [224] reported plasmocytosis especially of the kidney but also of the spleen, the lymphoid structures, and the tissues generally in dogs which had received 25 to 30 doses at the rate of 100 mg/kg/day. In the kidney, the aggregations of plasmocytes and other leucocytes were visible grossly as white, irregular, radially-disposed nodules. No changes were observed in the circulating blood. Plasmocytosis was less marked in rats and absent in rabbits dosed at the same rate as the dogs.

Other authors have reported a complete absence of change in the kidney (LILLIE and SMITH [367]; NELSON *et al.* [432]; HAAG *et al.* [252]).

Other Organs

If DDT is introduced into the body in any way except by inhalation, it has no specific effect on the respiratory system. However, if DDT is inhaled either as a dust, mist, or aerosol, some degree of irritation of the respiratory tract may be expected. Congestion of the alveolar walls and of the blood vessels of the lamina propria of the trachea have been reported by NEAL et al. [426]. Animals actually killed by respiratory and dermal exposure to solvents may show, in addition to lesions of the respiratory tract and skin, congestion, hemorrhage, and other nonspecific injury of the viscera (POTOSSI [466], [467]).

Damage to the skin of experimental animals from the application of DDT solutions has been very severe in some instances. Such damage can be attributed to the solvent. NELSON et al. [432] have reported very mild changes in the skin of rabbits which received DDT solutions made up with solvents that produced no change when applied to control animals. The changes attributed to DDT included hyperkeratosis and thickening of the stratum spinosum. There was little or no cellular infiltration of the underlying corium. On higher dosage levels, a slight focal necrosis of the epidermis and a slight cellular infiltration of the subjacent corium was seen. Entirely similar mild injury has been attributed to the solvent alone (NEAL et al. [432]; DEICHMAN et al. [162]).

It is generally agreed that DDT does not cause severe injury to the skin. Ulceration, hemorrhage, and other severe damage may be caused by solvents (CAMERON and BURGESS [105]; ZEIN-EL-DINE [685]; POTOSSI [466]).

Minimal changes, sometimes including siderosis, have been reported in the spleen (LILLIE and SMITH [367]; DEICHMANN et al. [162]).

A slight mononuclear infiltration of the gall bladder has been observed (DRAIZE et al. [179]).

No change in the adrenal was observed by certain authors (LILLIE and SMITH [367]; HAAG et al. [252]; STOHLMAN and LILLIE [560]), while others (FITZHUGH and NELSON [208]) claimed that a slight generalized increase in the size of the organ occurred in rats fed for 2 years on DDT. LILLIE et al. [368] noted slight fatty changes in parenchymal cells in rats fed 1,000 ppm of DDT for a year.

In the thyroid some depletion of colloid may occur (NELSON et al. [432]; KONST and PLUMMER [331]; FITZHUGH and NELSON [208]), but the presence of any significant change has been specifically denied (CAMERON and BURGESS [105]; HAAG et al. [252]).

Hydropic degeneration and destruction of some of the 'waterclear' cells of the parathyroid have been reported (CAMERON and BURGESS [105]).

Changes not reported elsewhere have been observed in rats maintained on high dietary concentrations of DDT. The changes include: stromal fibrosis and cellular proliferation in the ovary and slight generalized increase in the number of interstitial cells in the testis (FITZHUGH and NELSON [208]). A much more dramatic change has been reported in the testes of cockerels given daily subcutaneous injections of DDT in chicken fat. The average weight of the testes of treated birds sacrficied when they were 80 to 90 days old was 1·05 g as compared to 5·63 g for normal controls of the same age. The intertubular tissue in treated birds was far in excess of that found in the controls, but the tubular development of the treated birds was very much retarded (BURLINGTON and LINDEMAN [94]).

6.

PHYSIOLOGY

Absorption

The toxicity of DDT when administered by different routes offers unimpeachable evidence for the absorption of the compound by those routes. The time of onset of typical signs of intoxication gives some idea of the relative rates of absorption. Thus, rats injected intravenously at the rate of 50 mg/kg showed convulsions in 20 minutes, while those receiving DDT orally at the rate of 500 mg/kg showed convulsions only after 2 hours. Both groups of rats died in approximately 2 to 3 hours after the first convulsion. Signs appeared in

Table 10

Absorption of DDT from the Gut of Rats, as Indicated by Recovery of the Compound from the Alimentary Tract 3 Hours After Administration by Stomach Tube. The Table is Taken from JUDAH *[313] and Was Reproduced by Permission of the British Journal of Pharmacology and Chemotherapy*

Rat No.	DDT administered mg	Found in gut mg
1	72·0	64·0
2	90·0	80·0
3	125·0	103·0
4	125·0	90·0
5	125·0	95·0
6	125·0	91·0

animals injected intraperitoneally at the rate of 1,000 mg/kg after about 6 hours (JUDAH [313]).

Absorption of DDT from the gut of rats is slow. Direct evidence is offered by Table 10, taken from JUDAH [313]. The table shows that 72 to 89% of the DDT administered could be recovered from the gut 3 hours after stomach tube administration. Absorption from the gut is irregular and, especially after large doses, very incomplete. This is suggested by the fact that as much as 50% or more of a single dose may be recovered in the feces (SMITH and STOHLMAN [533]).

In spite of progressive absorption, the concentration of DDT in the blood rapidly reaches a plateau following a single large dose. Two hours after an oral dosage of 550 mg/kg STIFF and CASTILLO [556] found a concentration of 68 ppm in the blood of rats; this value increased only to 98 ppm after 48 hours.

DDT is presumably absorbed in conjunction with lipoid materials. When administered by the dermal route the compound shows a tremendous increase in toxicity (which reflects absorption) when applied as an oily solution rather than as crystalline material in powder or suspension. The same trend has been shown in some experiments involving oral administration. Thus, WOODARD et al. [678] found that in relation to dosage, dogs stored more DDT in their fat when the compound was administered as a solution than when it was given as a solid. On the contrary, DEICHMANN et al. [162] found that, under the condition of their study, it required a slightly higher dosage of DDT dissolved in oil than of DDT mixed in the general diet to produce intoxication in rats. In any event, the absorption of DDT from the intestinal tract, as indicated by storage, is complex and involves unknown factors in addition to the degree of solution of the compound at the time of ingestion. See the discussion of papers by LARDY [341]; ELY et al. [197], [198]; HARRIS et al. [262], [263]; and GREEN-WOOD et al. [247] below (p. 159—162). It may well be that DDT must be dissolved in order to be absorbed from the intestinal tract. However, the average diet contains enough fat to dissolve very considerable amounts of DDT and at least partial solution may certainly occur after ingestion.

It appears that DDT dissolved in a nonabsorbable solvent is, itself, poorly absorbed either from the skin (LORY et al. [381]) or from the gastrointestinal tract (DEICHMANN et al. [162]).

No details are available regarding the actual mechanism of absorption, although the relatively high concentration of DDT in the mesenteric lymph nodes (SPICER et al. [544]) may suggest intestinal absorption by way of the lymphatics.

Distribution and Storage

The mere presence of stored material in the tissues does not necessarily represent an injury, and when injury is involved it may not correlate with the degree of storage. Thus, DDT is more toxic than TDE (DDD) either on the basis of a single dose or repeated doses. Even so, at comparable dosage levels, about the same amount (FINNEGAN et al. [204]) or even less (LEHMAN [358]) DDT is stored in the tissues than TDE.

In 1944 and later in 1945, SMITH and STOHLMAN reported experiments in which they determined DDT, or at least chlorides derived from it, by total organic chloride analyses. The authors presented evidence to show that 80 to 90% of the DDT present was measured by their method. Representative values from this work as well as values found by other authors are shown in Tables 11 and 12.

STIFF and CASTILLO [556] essentially repeated a part of the experiments of SMITH and STOHLMAN and, as shown in Table 11, obtained very nearly the same result by the method of total chlorides. When, however, the same tissues

Table 11

Accumulation of DDT in the Tissues of Different Species Following one or a few Doses. The Range of Values is Followed by the Mean in Parentheses

Species	Dosage mg/kg	No. doses	Route	Blood ppm	Liver ppm	Kidney ppm	Heart ppm	CNS ppm	Adrenal ppm	Fat ppm	Other ppm	Authority
Rat	50	1	IV	90–150 (116)	60–200 (128)	57–77 (63)	20–205 (76)	50–100 (55)	—	150–400 (250)	Muscle, 24–50 (41)	Judah [313]
	500	1	Oral	8–20 (15)	70–90 (80)	80–150 (106)	—	13–57 (36)	—	60–360 (214)	Muscle, 30–90 (60)	Judah [313]
	1000	1	IP	17–56 (34)	40–140 (87)	25–140 (46)	48	14–76 (53)	—	630–1100 (920)	Muscle, 20–100 (58)	Judah [313]
Rabbit	50	1	IV	—	120–210 (170)	54–64 (59)	75–120 (99)	45–50 (47)	170–250 (200)	—	—	Judah [313]
	200	1	Oral	5–50	10–30	20–50	—	10–30	—	100–500		Ofner and Calvery [441]
	350	1	Oral	1–2	0	2	—	0	—	—	—	Laug [343]
	400	1	Oral	8	12	13	—	—	—	—	Bile, 29	Laug [343]
	400–500	1	Oral	Tr-40 (13)	23–69 (51)	Tr-56 (28)	—	Tr-55 (27)	—	—	Bile, 79–333 (163)	Smith and Stohlman [533]
	550	1	Oral	107	63	39	800	160	—	—	—	Smith and Stohlman [532]
	550	1	Oral	68–98 (81)	150	172	—	—	—	—	Bile, 550	Stiff and Castillo [556]
Goat	500	3	Oral	—	351	215	2,670	155	2,877	11,210	Lung, 141 Spleen, 181 Nodes, 3068 Thymus, 1628	Spicer *et al.* [544]

Table 12

Accumulation of DDT in the Tissues of Different Species Following Repeated Doses. The Range of Values is Followed by the Mean in Parentheses

Species	Dosage mg/kg	No. doses	Route	Blood ppm	Liver ppm	Kidney ppm	Heart ppm	CNS ppm	Adrenal ppm	Fat ppm	Other ppm	Authority
Rat	1·5	735	Oral	—	1–3 (2)	2–5 (3)	0	0	—	68–90 (80)	Lung, 0–3 (2) Spleen, 0	DEICHMANN *et al.* [162]
	7·2	735	Oral	—	2–3 (3)	3–15 (10)	1–3 (2)	1–2 (1)	—	123–200 (163)	Lung, 2–5 (3) Spleen, 2–3 (2)	DEICHMANN *et al.* [162]
	9·5	730	Oral	—	3	14	—	—	—	158–313 (253)	—	LAUG and FITZHUGH [344]
	19	730	Oral	—	3	20	—	—	—	966–1091 (1028)	—	LAUG and FITZHUGH [344]
	38	730	Oral	—	23	64	—	—	—	4220	—	LAUG and FITZHUGH [344]
	48	15	Oral	—	60–170 (110)	30–70 (50)	—	20–60 (35)	260–470 (36)	—	—	LUDEWIG and CHANUTIN [382]
	96	11	Oral	—	140–365 (225)	50–210 (115)	—	40–200 (90)	440–950 (600)	—	—	LUDEWIG and CHANUTIN [382]
Rabbit	50	33–46	Oral	100	78–140 (109)	127–152 (139)	—	36–62 (49)	—	—	Bile, 115–267 (191)	SMITH and STOHLMAN [533]
	100	10–15	Oral	Tr-32 (16)	180–280 (230)	80–128 (104)	—	80–160 (120)	—	—	Bile, 210–530 (374)	SMITH and STOHLMAN [533]
Dog	25	25	Oral	—	0	7–14 (10)	3–8 (5)	1–4 (3)	62–83 (69)	200–910 (503)	Muscle, 12–18 (14) Mam. gl. 2–21 (12)	FINNEGAN *et al.* [204]
Cat	10	33	Oral	0	1330	450	—	83	—	—	Bile, 67	SMITH and STOHLMAN [533]
Turkey	8	56	Oral	—	—	266–753 (509)	—	—	—	1344–1606 (1475)	Muscle, 33–34 (34)	MARSDEN and BIRD [398]
	17	56	Oral	—	—	52–74 (63)	—	—	—	1640–1977 (1808)	Muscle, 28–41 (35)	
	34	56	Oral	—	—	80–84 (82)	—	—	—	4376–4500 (4438)	Muscle, 29–83 (56)	

and fluids were analyzed by the xanthydrol-KOH-pyridine method, no DDT was found. The authors interpreted this finding as an indication that DDT was not absorbed into the body but was changed by the alimentary tract into some unidentified derivative which was then absorbed, stored, and finally excreted. They considered that the compound might be di-(*p*-chlorophenyl)-acetic acid (DDA) demonstrated by others (WHITE and SWEENEY [663]) in the urine of poisoned rabbits. The failure of STIFF and CASTILLO to demonstrate DDT in tissues has been attributed (LAUG [343]) to the insensitivity of their method. However, a better explanation has been offered by JUDAH [313] who pointed out that for tissues the authors used a saponification with alcoholic alkali, a process shown by GUNTHER [249] to destroy DDT and form 2,2-bis-(*p*-chlorophenyl)-1,1-dichloroethylene (DDE), a product which gives no color by the xanthydrol-KOH-pyridine method.

It has been shown, moreover (LAUG [342]), that various analogs of DDT which might conceivably occur as a result of metabolic action show little insecticidal activity. The analogs tested included di-(*p*-chlorophenyl)-acetic acid [DDA], 2,2-bis-(*p*-chlorophenyl)-1,1-dichloroethylene [DDE], di-(*p*-chloro-phenyl)-ketone, 2,2-bis-(*o*,*p*-chlorophenyl)-1,1,1-trichloroethane [*o*,*p*, isomer of DDT], and 2,2-bis-(*p*-chlorophenyl)-1,1-dichloroethane [DDD]. LAUG [342] found good agreement between the DDT content of the perirenal fat of poisoned rats as determined chemically by the method of SCHECHTER and HALLER [510] and the DDT content of the same tissue estimated by bio-assay using flies. This indicated that the specificity of the Schechter-Haller method correlated well with insecticidal action. Although this might have been broadly true, it should be kept in mind that all of the known degradation products of DDT as well as many related compounds, whether insecticidal or not, will give Schechter-Haller colors.

In a second paper which followed immediately in the same journal, LAUG [343] reported the analyses of tissues from 10 rabbits which had been given one or more oral doses of DDT. The values which they found for liver, kidney, brain, and blood were very much lower than those found by the method of total organic chlorides (SMITH and STOHLMAN [532], [533]), although the rabbits had been given approximately the same or even higher doses of the drug. LAUG explained this result in terms of the nonspecificity of the total chloride method, especially for urine. However, SMITH and STOHLMAN analyzed the tissues and excreta of unpoisoned rabbits and also ran blanks on each sample analyzed. They established that a real increase in organic chlorides in various tissues occurs, and this has been confirmed by others. They clearly stated the possibility that DDT undergoes some degradation in the body. On the other hand, there is no evidence that under appropriate experimental conditions an increase in total organic chlorides in the tissues does not represent a summation of DDT and its metabolites. Two metabolites have been detected so far, and the existence of others is certainly not excluded. It is true that the method of total chlorides is nonspecific and therefore subject to error. It is partly for this reason and partly because of the greater sensitivity of the newer methods that the more recent investigations have been made with modifications of the Schechter-Haller method or with the use of radioactive DDT.

It was conceivable that the DDT found in body tissues following one or a few large doses did not represent storage but only a flooding of the organism. Evidence that storage does actually occur appeared early. It was shown that dogs maintained on daily dosages of 10, 50, and 80 mg/kg for 443 to 747 days stored up to 4,940 ppm of DDT in their fat (WOODARD et al. [678]). WOODARD and OFNER [676] reported that rats fed 1·65 mg/kg/day for 54 days or a total of 89·3 mg/kg stored the compound in their fat at a concentration of 300 ppm, indicating selective absorption of a part of the total dosage into the fat.

Data on the accumulation of DDT in the tissues of different species following one dose are shown in Table 11. Table 12 shows comparable data for repeated doses. The following conclusions appear justified:

(1) Under appropriate conditions, it has been possible to detect DDT in all tissues studied.

(2) There is considerable variation between individual animals subjected to supposedly identical conditions.

(3) After a single intravenous dose more DDT may be found in the tissues than after single oral or intraperitoneal doses at rates as much as 10 to 20 times larger.

(4) Irrespective of the route of administration or the number of doses, the amount stored in fat tends to be greater than the amount stored in the other organs of the same animal.

(5) At least in regard to adipose tissue, the degree of storage varies directly with the number of doses until a peak or plateau is reached. (This is illustrated for the rat in Figure 9.) This effect is so dramatic that repeated doses at a moderate rate (e.g. 20 mg/kg/day orally) result in greater total storage of DDT in the fat than a single dose at the highest rate which can be tolerated or even a single dose at a rate which is frequently fatal (e.g. 500 mg/kg orally).

(6) The time-dosage relationships which determine storage of DDT are much less well understood for other organs and tissues than for adipose tissue.

(7) When other factors are kept constant, the peak storage of DDT in each tissue varies directly with the daily dosage. (This is illustrated for the rat in Figure 10.)

(8) There is some evidence that the storage pattern is different in different species. Specifically, it has been claimed (SMITH and STOHLMAN [533]) that cats show a relatively higher concentration in the liver and kidneys and a lower concentration in the blood and bile than do rabbits. The concentration in the central nervous system of the two species was said to be similar. Some of these points are discussed below in greater detail.

There is some indication that the amount of DDT stored in different tissues is proportional to the lipid content (EMMEL and KRÜPE [199]). Independently LAUG and FITZHUGH [344], noting that the amount of DDT in the perirenal fat was roughly 50 to 100 times what they found in the viscera, investigated the possibility that all DDT in the viscera might be dissolved in the fat which the various organs contain. They found that the ether-soluble material of the liver, kidney, and spleen averaged about 2% of the fresh tissue weight. They observed that if the DDT actually present in these organs were restricted during life to the ether-soluble fraction, then its concentration in that medium would range from 250 to 1,000 ppm. They found that the perirenal tissue (95 per cent ether-soluble) contained 2,100 to 4,900 ppm in the same animals. This general correspondence was regarded as an indication that DDT may be dissolved in fat irrespective of the part of the body in which it is found, with the possible exception of muscle. The use of the factor, 1·1 (the value for the percentage of neutral fat in the liver given by BLOOR [77]), would result in an even closer correlation in this instance. Furthermore, the theory accounts for the observed fact that the DDT content of the adrenal of some animals is high, for BLOOR [78] has stated that in the guinea pig lipids constitute 15·8% of the adrenal and neutral fat constitutes 8·6%. BLOOR has stated that lipids constitute 6·5% of the brain in which essentially no neutral fat can be detected. Presumably, neutral fat is more effective in dissolving DDT than are other lipids. The theory might account for certain species differences mentioned above, for it has been shown (TURNER [601]) that the kidney of the cat contains more fat than the kidneys of many other species. However, it must be clearly understood that, in the present state of knowledge, the theory that DDT is stored in tissues by solution in the lipids of those tissues can be considered only as a guide. It will not explain the available data with any degree of precision. It may be that a more satisfactory correlation could be obtained if careful determinations of the various lipid fractions were made on the different tissues of the same animals in which DDT is measured. An organ of particular interest in this connection is the ovary. TAUBER and HUGHES [576] found almost as much DDT in the ovary of rats as other authors have reported in fat from rats on similar dosage levels.

There is little information on the variation in the concentration of DDT in samples of adipose tissue taken from different locations in the same animal. WOODARD et al. [678] found that the concentration of the compound in the subcutaneous fat and the intraperitoneal fat of a dog was the same. On the contrary, THOMAS et al. [585] found that the concentration of DDT in the perirenal fat of calves was higher than the concentration in the mesentery or in fat of other portions of the body. A similar result has been reported by others (CARTER [114]; CARTER et al. [117]). It is not clear in these instances whether the adipose tissues examined differed significantly in their neutral fat content.

If the intake in the diet remains constant, then the amount of the compound stored in the fat will gradually increase for a time, but will eventually reach a peak or plateau as illustrated in Figure 9. Evidence has been presented to show that rats on dietary concentrations of 200 ppm or less reach the plateau within about 50 days (WOODARD and OFNER [676]; WOODARD et al. [679]) or 133 to 161 days (KUNZE et al. [334]; LAUG et al. [345]). Certainly the greater part of the storage is accomplished within 50 days, especially in the male, which

was the only sex mentioned by WOODARD and his colleagues. Equilibrium
tends to be attained a little more slowly in the female or in animals fed at a
level of 400 ppm or higher (unpublished data). As mentioned below, there is
evidence that, at least in the male rat, the storage-plateau decreases after a
time so that, at the same dosage, the amount stored after 2 years of constant
exposure is smaller than the amount after 6 months of exposure.

The rate at which DDT accumulates in organs and tissues other than adipose
tissue of the rat under conditions of constant daily dosage has been little
studied. LUDEWIG and CHANUTIN [382] concluded that the concentration of

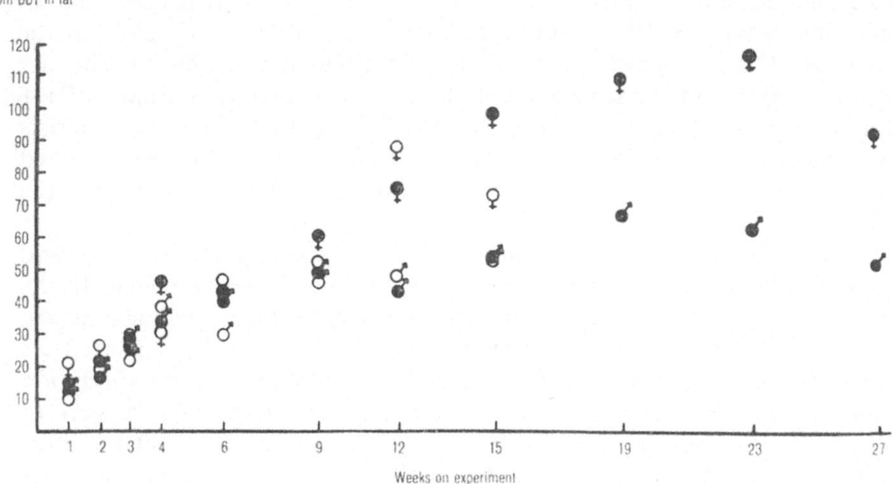

Figure 9

The increase in the storage of DDT in the fat of rats receiving 5 ppm of the compound in their diet.
The figure is taken from LAUG et al. [345] and is reproduced by permission of the Journal of
Pharmacology and Experimental Therapeutics.

♀ ♂ Rats were 15 weeks old when exposure to DDT began.
● ♂ Rats were 21-day-old weanlings when exposure to DDT began.

DDT in the liver, kidney, brain, and adrenal reached a maximum in a few
days and remained at a constant level in animals fed 1,000 ppm. However,
the concentration of DDT in the liver, kidney, and brain increased slightly
and progressively during an 11-day period of observation in rats fed 2,000 ppm.
The increase in concentration in the adrenal was more marked and showed no
indication of reaching a plateau during the 11-day period. Studies involving
smaller dosage rates and longer periods of exposure are needed to complete
the picture.

The rate of storage in other animals has been studied less than the rate of
storage in the rat. However, BRYSON et al. [93] showed that hens fed DDT at
a constant rate laid eggs showing an initial increase followed by a plateau in
the concentration of the compound. If one plots the concentration of DDT in

the eggs against time, it appears that a period of gradual rise in the concentration of DDT during the first month was followed by a more rapid rise during the next three months. The height of the plateau established at the end of the fourth month of DDT feeding depended on the concentration of DDT in the food. On higher dietary levels the concentration of DDT in the eggs tended to decrease somewhat after the fourth month. By far the greater proportion of DDT in an egg occurs in the yolk (DRAPER *et al.* [180]). It is of interest that the yolk is formed quickly. In chickens, the large diameter of the yolk increases from 6 mm to 35 mm in 6 days (ROMANOFF and ROMANOFF [498]).

The secretion of DDT in the milk of cows increases for a time after they are placed on a constant dietary level of DDT. See Figure 13. It is clear that an equilibrium is reached (R. F. SMITH *et al.* [539]; BIDDULPH *et al.* [59]) but the time at which it is reached appears to vary in different experiments. The fact that it requires about the same amount of time to reach an equilibrium of DDT output in eggs as in body fat suggests that the deposit of DDT in these materials may depend on the same mechanism. The fact that eggs and milk are eliminated by the organism which produces them emphasizes the dynamic quality of the DDT equilibrium.

Figure 10 summarizes the amount of DDT stored at equilibrium in several tissues by rats maintained on different dietary levels of the compound. In the figure, values for DDT storage in fat have been plotted for individual animals, but the values for storage in liver, kidney, and heart have been plotted for mean values given in Table 12. The curves for fat represent an interpretation by the reviewer. The curves for the liver, kidney, and heart merely serve to connect the observed points and therefore, in themselves, represent no interpretation. However, it is interesting to see that, broadly speaking, the slopes of the curves for these three organs are similar to the slope of the curves for fat. The curve for DDT storage in the liver shows a definite break, but it would be unwise without further investigation to suppose that this change of slope is meaningful and not merely a reflection of the paucity of data.

In addition to some of the points already enumerated, the following conclusions appear justified:
(1) Rats store DDT in the fat at all accurately measurable dietary levels.
(2) Storage in the fat and perhaps in other tissues is less complete, in relation to dosage, at·higher dietary levels. (This is indicated by the fact that the curves for fat and perhaps for other tissues have a slope of less than 45°.)
(3) The female rat consistently stores more DDT in the fat than the male when offered the same diet. (This is indicated by a comparison of the upper and middle curves representing, respectively, females and males exposed to DDT for 6 months or less.)
(4) The male rat tends to lose a part of the DDT it has stored in its fat at the peak level reached in about 6 months even though continued on the same dosage. (This is indicated by a comparison of the middle and lower curves for fat representing male rats exposed to DDT for 6 months or less, and for 2 years respectively.)

(5) There is a satisfactory agreement between different authors and, in fact, between different laboratories.

The fact that the female rat stores more DDT than the male when maintained on the same diet has been attributed to the fact that the female eats more food in proportion to body weight and therefore gets a higher dosage of

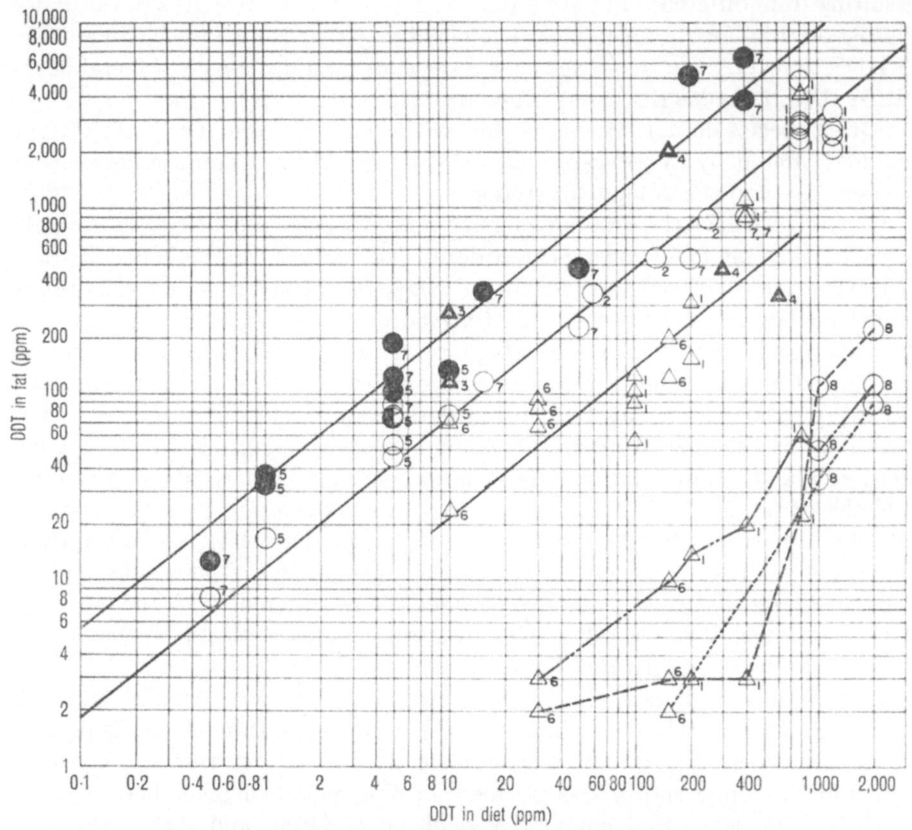

Figure 10

Storage of DDT in the tissue of rats maintained on diets containing different concentrations of the compound. The values are taken from the following papers: (1) LAUG and FITZHUGH [344]; (2) WOODARD et al. [679]; (3) FITZHUGH [207]; (4) FINNEGAN et al. [204]; (5) LAUG et al. [345]; (6) DEICHMANN et al. [162]; (7) Unpublished data; (8) LUDEWIG and CHANUTIN [382]. Adipose tissue is indicated by a solid line; heart, by a dotted line; liver, by a broken line; and kidney, by a line with alternate dashes and dots.

Legend for points

| | Sex | | Duration of exposure |
Male	Female	Unstated	
○	●	◓	6 Months or less
△	▲	◮	2 Years

the insecticide. However, the female eats only about 1·25 times as much food as the male but stores from 1·5 to 3·8 times as much DDT. This is illustrated in Figure 11. Also, Laug *et al.* [345] showed that rats stored DDT at the same rate and reached the same storage-equilibrium regardless of whether they began eating DDT as weanlings or as young adults. The difference in dosage, resulting from difference in food intake, is greater for the two ages of rat under discussion than for the two sexes. The possibility that the difference in storage of DDT shown by the two sexes of rats depends on a difference of metabolism rather than dosage is discussed below (p. 73).

Other species do not necessarily follow the same pattern. The limited data for dogs offered by Woodard *et al.* [678] do not show a definite correlation between sex and the amount of storage.

Figure 11 shows the amount of DDT stored by several species of animals in their body fat following repeated doses. The curves for the concentration of DDT in the fat of male and female rats have been converted on the assumption that male rats consume 43 g of dry food per kilogram of body weight per day, and female rats consume 54 g/kg/day. Observed values for other species are plotted on the same coordinates for comparison. In addition to findings already mentioned, it may be concluded that all species investigated so far store DDT in their fat at rates of the same order of magnitude when exposed repeatedly at the same dosage rate. Values given for DDT residues in the fat of lambs (Harris *et al.* [263]) and chickens (Bryson *et al.* [93]) experimentally fed DDT appeared consistent with this conclusion. This does not imply that the rate of storage is exactly the same for all species. In fact, it would appear from the graph that rats, monkeys, and dogs show different patterns (curves) of DDT storage. Possible differences in storage pattern would be defined if other species were investigated in the same detail as rats.

Considerably more information would be made available with little more work if those who study the storage of DDT in domestic animals would record the dosage which the animals receive instead of merely stating the concentration of the compound in a single food forming a part of their diet.

There is evidence that cows (Ely *et al.* [197], [198]) and sheep (Harris *et al.* [263]) deposit more DDT in milk and fat if the compound is fed as residues than if the compound is fed in a crystalline form (see Figures 14 and 15). The hypothesis has been offered that even in the alimentary tract of the animals, the crystalline material never becomes fully dissolved in the fat of the diet and is, therefore, poorly absorbed. Lardy [341] followed this line of reasoning to explain the fact which he observed that steers fed ordinary corn silage containing 6 to 9 ppm of DDT stored an average of 2·2 ppm in their fat, while similar steers fed husk and cob silage containing only 0·2 to 4 ppm stored an average of 17 ppm in their fat. However, Ely *et al.* [197], [198], in studying the secretion of DDT in the milk of cows, found that it made no consistent difference whether the crystalline DDT was dissolved in oil and fed in capsules or mixed in grain, or whether it was left undissolved and fed in capsules or mixed in grain.

The degree of solution of ingested DDT also fails to explain the fact observed by GREENWOOD *et al.* [247] that, in relation to dosage, rats stored about 6 times more DDT eaten as a residue in lamb muscle than DDT eaten as a residue in lamb fat or the butterfat of cows' milk. Most of the values observed by these authors for the storage of DDT in rat fat following feeding on lamb fat,

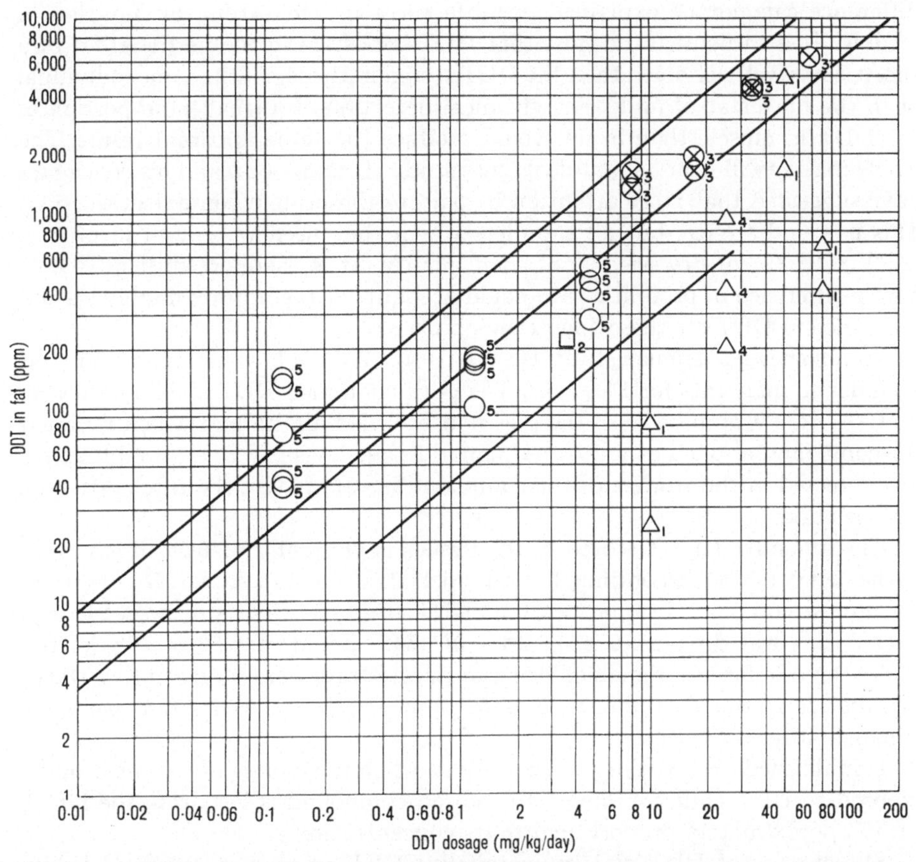

Figure 11

Storage of DDT in the adipose tissue of several species of animals given different daily dosages of the compound. The curves for rats developed in the preceding figure are shown for comparison. The values for other species, shown only by points, are taken from the following papers: (1) WOODARD *et al.* [678]; (2) WILSON *et al.* [667]; (3) MARSDEN and BIRD [398]; (4) FINNEGAN *et al.* [204]; (5) Unpublished data.

Legend for points

◯	Rhesus monkey
☐	Cow
△	Dog
⊗	Turkey

or on butterfat, would fall between the curves for male and female rats fed for 6 months or less if plotted in Figure 10. This is what one would expect because the values represented averages for male and female rats. However, most of the values (averages for males and females) for the storage of DDT in rat fat after feeding on lamb muscle would fall distinctly above the curve for female rats in Figure 10 and define a curve of about the same slope. The difference can not be explained by differences in age, strain, or duration of feeding. It is difficult to imagine that DDT deposited in the perirenal fat of a sheep is less well dissolved than the DDT stored in the leg of the same individual or in the butterfat of milk. The chemical detection of a smaller proportion of the DDT in muscle than in fat would produce the same apparent result. The whole matter will bear a great deal more study. GREENWOOD and his colleagues have suggested that more attention be paid to the protein content of the diet. This is sound advice, but all the diets which they used were high in protein (22·9 to 25·3% of dry weight) of good quality. It is therefore difficult to see how a difference in protein intake could account for the approximately sixfold difference in DDT storage which they observed.

In a somewhat similar study HARRIS et al. [265] fed swine fat, bacon, and shoulder to male rats for 14 weeks. The pork contained DDT residues varying from 0·1 to 3·9 ppm because the pigs had been fed alfalfa hay which retained the compound applied as a dust. At equal dosage levels of DDT, no difference was detected in the storage of the compound in rats fed on the three different cuts of pork.

The amount of DDT stored in the tissues is gradually reduced if exposure to the compound is discontinued or diminished. WOODARD et al. [678] observed the almost complete disappearance of DDT from the tissues of dogs 81 days after dosing was discontinued. LAUG et al. [345] found that 50 to 75% of the DDT stored in fat still remained one month after rats were returned to a DDT-free diet from diets containing 5 to 50 ppm of DDT. Twenty-five per cent remained after 3 months.

Unpublished observations indicate that, when the dosage of rats previously exposed to DDT is discontinued, the stored compound is lost from the fat at a rate approximately proportional to its concentration.

The storage of DDT and its metabolite DDE in man is discussed below (p. 145).

Metabolism, Detoxification, and Tolerance

DDA

The presence of DDA in tissue was mentioned by OFNER and CALVERY [441] who observed that the concentration ranges close to 10% of the concentration of DDT.

SPICER et al. [544] using the method of SCHECHTER and HALLER observed the presence of DDA in all the tissues of a goat which they examined. Relatively high concentrations were found in the mesenteric lymph nodes (84 ppm), liver (46 ppm), and kidney (71 ppm). The proportion of DDA as compared to total DDT-derived material was very much higher in the kidney (25%) than in any other organ. The proportion of DDA in the liver was 12% and that in the other organs 10% or lower being only 3% in the mesenteric lymph nodes. The authors concluded that the liver and kidney were probable sites of excretion for DDA and that some DDA might be formed in the gastrointestinal tract. They further concluded that it is unlikely that the degradation of DDT to DDA plays a role in the toxic reaction of the former since the relative and absolute concentration of DDA in the brain, heart, and adrenals was found to be low as is shown in Table 13. All three organs have been shown to enter into the toxic reaction of DDT.

The findings of SPICER and his colleagues were largely confirmed and were extended by JUDAH [313] who concluded that the only notable degradation

Table 13

Distribution of DDT and DDA in the Organs of a Goat After Three Oral Doses of 500 mg/kg of DDT Dissolved in Olive Oil, Four Days After the Last Dose. The Table is Modified After SPICER *et al.* [544] *and is Reproduced by Permission of the Publisher*

Organ	DDT ppm	DDA ppm	DDA % of total
Omentum fat	11,210	—	—
Lymph nodes, mesenteric .	3,068	84	3
Adrenals	2,877	31	1
Heart	2,670	15	< 1
Thymus	1,628	22	1
Testes	410	15	4
Liver	351	46	12
Kidney	215	71	25
Spleen	184	14	7
Spinal cord	174	12	6
Brain	155	17	10
Lungs	144	14	9

product of DDT is DDA and that it accumulates in the tissues to a small extent. An unsuccessful effort was made to demonstrate in the tissues of poisoned animals the presence of chlorobenzoic acid, dichlorobenzophenone, and dichlorodiphenyl-dichloroethylene (DDE). It was concluded that no evidence for a metabolite more toxic than DDT had yet been found, and that if such a material exists it must be active in very low concentration.

In acutely poisoned animals JUDAH failed to find DDA in all tissues examined but the compound was found in the liver, kidney, and skeletal muscle of rats. The highest concentration (2 to 110 ppm) observed was in the muscle of rats which had received DDT at the rate of 1000 mg/kg intraperitoneally. Incubation of liver, diaphragm, kidney, and brain with 100 μg of DDT for

2·5 hours resulted in the conversion of about 5% of the compound to DDA. No evidence for a specific site for the degradation was found. The disappearance of a small proportion of DDT incubated in a liver homogenate had been shown earlier (WOODARD et al. [679]) but without any demonstration of the degradation product.

That DDA does not represent the toxic factor was shown by the fact that rats tolerated intravenous doses of the sodium salt at the rate of 100 mg/kg. Eighteen hours after injection these rats showed higher concentrations of DDA in the tissues than were usually found in animals fatally poisoned by DDT (JUDAH [313]). The conversion of DDT to DDA thus constitutes a true detoxification in the mammal.

Other workers had earlier demonstrated the low toxicity of DDA. SMITH et al. [535] gave data on the acute oral toxicity of an olive oil solution to rats from which we may estimate an LD_{50} of about 1,900 mg/kg. They stated that the intravenous LD_{50} of the sodium salt to rabbits was 150 mg/kg. Three of 5 rabbits treated orally at the rate of 50 mg/kg/day survived 32 doses while other rabbits were killed by 25 or fewer doses of DDT given in the same way.

VON OETTINGEN and SHARPLESS [641] gave adequate figures from which one can calculate the acute oral LD_{50} of water solution of DDA to mice as 720 mg/kg which is in reasonable agreement with the value of 590 mg/kg given by DOMENJOZ [173], [174].

DDA produced rather less injury to the liver than DDT but, in higher dosage and especially when given intravenously, produced greater damage to the kidney (LILLIE et al. [368]).

SMITH et al. [535] found that during the first several days of oral dosing rabbits excreted DDA in the urine approximately 15 times as rapidly as DDT in relation to the daily dosage administered. Although the rate of DDA excretion increased somewhat, the rate of DDT excretion increased more rapidly so that the values differed by a factor of only 5 after the twentieth day of feeding.

Instead of causing multiple discharges and eventually spontaneous activity in the crustacean nerve like DDT, DDA causes a decrease in amplitude of muscle contraction (WELSH and GORDON [658]).

Unlike DDT, DDA does not cause hyperglycemia in rabbits, according to KAWANO [318].

JOHNSTON [311] has shown that DDA shows much less inhibitory action than DDT toward the succinoxidase and cytochrome oxidase systems of rat heart even at a concentration 50 times as great as that used for DDT.

DDE

The toxicity of DDE [p,p'-bis-(dichlorophenyl)-1,1-dichloroethylene], which has only recently been shown to be a metabolite, has not been studied as much as the toxicity of DDA. It is clear, however, that DDE is less toxic than DDT. SMITH et al. [535] found that a single dose of 1,000 mg/kg in olive oil killed only 46·6% of 15 rats. Rabbits given daily doses of DDE at the rate of 50 mg/kg/day died after 11 to 28 doses, whereas other rabbits given DDT in the same way died in 15 to 25 days so that, under the conditions of this test, no difference in toxicity was evident. Compared with DDT, DDE produced less clinical effect on the central nervous system, less injury to the liver, and slightly greater kidney damage. The pathology of the animals involved was described in detail by LILLIE et al. [368].

VON OETTINGEN and SHARPLESS [641] determined the oral toxicity to mice of DDE dissolved in olive oil. If their data are analyzed statistically an LD_{50} value of 700 mg/kg is obtained, as compared to 240 mg/kg for DDT under the same conditions. The authors observed that DDE produces less effect on the nervous system than does DDT. The LD_{50} value may be compared with 1,000 mg/kg for DDE given by DOMENJOZ [173], [174]. In a similar way, BROWNING et al. [91] found DDE less toxic than DDT to mice.

WOODARD et al. [680] found that different isomers of DDT fed at the rate of 80 mg/kg/day killed dogs in an average of 37 to 55 days, while dogs fed DDE at the same rate remained alive for more than 120 days.

OFNER et al. [442] reported that animals fed DDE excrete compounds (DDA and an unidentified material) indistinguishable by the methods available from those excreted after feeding DDT. The proportion of DDA and the unidentified metabolite was found to be the same after feeding DDT or DDE, but no information was given on the total rate of excretion characteristic of the two compounds.

SMITH et al. [534], [535] found that DDE was excreted in the urine even slower than DDT in relation to the dose administered. The authors considered that this militated against the view that DDE might be an intermediary in the degradation of DDT. Now that DDE has been shown to be a metabolite, it is difficult to understand how its excretion can be slower than the *final* or total excretion of the parent compound, unless DDT undergoes two kinds of degradation. The matter deserves re-study.

The fact that DDE, or a compound colorometrically and chromatographically indistinguishable from it, was stored in human fat was first shown by PEARCE et al. [449]. The method of analysis was described in greater detail by MATTSON et al. [400]. Unpublished work has shown that the average ratio of DDE to DDT is greater in human fat than in human food, suggesting that at least a part of the stored DDE in man is a true metabolite. The rat converts DDT to DDE, and especially the female stores a part of the DDE in its fatty tissue. Both sexes of rat store DDE when it is fed in the diet, but the male stores much less and may eliminate the store more rapidly than the female after dosage is discontinued.

It has been shown by JOHNSTON [311] that DDE inhibits the succinoxidase and cytochrome oxidase systems of rat heart, though not quite to the same degree as DDT.

General

The demonstration that the known derivatives of DDT are less toxic than the parent compound constitutes indirect proof that the compound itself is

    W. J. Hayes, Jr.

physiologically active. Direct proof may be found in the rapidity of its action following intravenous injection and the fact that in eviscerated and functionally hepatectomized cats the time of onset and subsequent course of DDT intoxication following intravenous injection is not altered (PHILIPS and GILMAN [453]). In the absence of the organs primarily concerned with the transformation of foreign substances, the possibility that such transformation may occur in the nervous system cannot be excluded. The possibility was, in fact, suggested by WITT [673] but without any supporting evidence.

A number of authors have observed that animals given relatively high dosages of DDT may promptly develop tremors and other signs of intoxication and later return to normal, even though dosage is continued in the same way. In the case of animals maintained on certain concentrations of DDT in the diet, it has been pointed out that the actual dosage of DDT (in milligrams per kilogram) decreases as young animals mature, because they eat less food in relation to body weight. Some workers have assumed that the recovery of animals maintained on a certain concentration of DDT in the diet is explained by the progressive decrease in actual dosage. However, in other experiments, recovery has occurred in the face of a truly constant dosage level (DEICHMANN et al. [162]). This finding demonstrates the presence of an acquired tolerance but throws no light on its mechanism.

It is entirely possible that other metabolites of DDT will be demonstrated. If so, some of them may prove to be analogs about which some toxicological information is already available. For this and other reasons, papers dealing with the toxicology of analogs not now known to be metabolites of DDT are listed in Table 14.

Materials previously observed in tissue, but unidentified, may have been DDA, DDE, or yet unidentified materials.

Excretion

The elimination of unabsorbed DDT provides a very real protection to a man or animal which has ingested a large amount of the compound. This kind of elimination does not depend entirely on the slow, normal movements of the alimentary tract. Under different circumstances, vomiting (MACKERRAS and WEST [388]; VELBINGER [631]; GARRETT [219]; JUDE and GIRARD [315]) or diarrhea (JUDE and GIRARD [315]) has been observed in human beings who ingested large doses of DDT. Vomiting has also been observed in dogs (NEAL et al. [426]; PHILIPS and GILMAN [453]) and in cats (SMITH and STOHLMAN [533]).

Urine

SMITH and STOHLMAN [532] observed the occurrence of increased organic chlorides in the bladder urine of a rabbit which had received orally 550 mg of

Table 14

Sources of Information on the Toxicology of Those Analogs of DDT Which Have not yet Been Isolated from the Tissues or Excreta of Animals Treated with DDT

| Analog | | Information | Authority |
Common name	Chemical name		
		Changes on aliphatic group only	
DDD or TDE	2,2-Bis-(*p*-chlorophenyl)-1,1-dichloroethane	Oral toxicity to mice	Domenjoz [173], [174]
		Acute oral toxicity to rats and chronic oral toxicity to rabbits—includes pathology	M. I. Smith *et al.* [534]
		Oral toxicity to mice	von Oettingen and Sharpless [641]
		Toxicity to fish and mosquito larvae	Ginsburg [230]
		Pathology in rabbits	Lillie *et al.* [368]
		Action on the arthropod nerve axon	Welsh and Gordon [658]
		Effect of sprays and dips on livestock	Bushland *et al.* [95]
		Chronic oral toxicity to rats, including pathology. Respiratory toxicity to dogs, rabbits, and rats, including pathology. Chronic dermal toxicity to dogs, rabbits, and rats, including pathology. Skin irritation and sensitization in man	Haag *et al.* [252]
		Acute oral and dermal and chronic oral and dermal toxicity with pathology in animals. Toxicity compared with other compounds	Lehman [354]
		Adrenal cortical atrophy and liver damage in dogs	Nelson and Woodard [431]
		Absorption, storage, and excretion in mammals	Woodard *et al.* [679]
		Chronic oral toxicity to dogs	Woodard *et al.* [680]
		Effect on 2 rabbits given chronic oral doses	De Meillon [166]
		Toxic effect on honeybees	Eckert [192]

Analog		Information	Authority
Common name	Chemical name		
DDD or TDE (continued)	2,2-Bis-(*p*-chlorophenyl)-1,1-dichloroethane	Storage and excretion in dogs and rats dosed orally	Finnegan *et al.* [204]
		Residues recovered from alfalfa	Ginsburg *et al.* [233]
		Residues on various crops	Hoskins [301]
		Major toxic action	Lehman [355]
		Precautions in use on crops for canning	Rohwer [497]
		Contamination of meat and milk from sprayed livestock	Bushland *et al.* [96]
		Excretion in milk of sprayed cows	Claborn *et al.* [128]
		Longevity of residues on corn	Ginsburg *et al.* [234]
		Suggested treatment of poisoning	Hough [303]
		Chronic toxicity to quail	Linduska and Springer [373]
		Acute and chronic toxicity to animals	Radeleff [475]
		Acute toxicity when sprayed on calves	Radeleff and Bushland [476]
		Toxicity to fish	Surber [565]
		Toxicity to fish when used as a mosquito larvicide	Tarzwell [572]
		Effect on animals, residues on crops	Webster [655]
		Precautions in use of poison	Bishopp [69]
		Storage in adipose tissue	Anonym [35]
		Effect on metabolism in rabbits	Kawano [318]
		Acute and chronic oral toxicity to rats, acute and chronic dermal toxicity to rabbits, and chronic oral toxicity to dogs	Lehman [360a, b, c]
		Production of insulin sensitivity in the dog	Nichols and Gardner [435]
		Influence on adrenal cortical function in adult rats	Brown [89]
		Storage in man, treatment of Cushing's syndrome	Sheehan *et al.* [522]

Analog		Information	Authority
Common name	Chemical name		
DDM	Di-(*p*-chlorophenyl)-methane	Acute oral toxicity to rats and chronic oral toxicity to rabbits—includes pathology Oral toxicity to mice Pathology in rabbits Effect of skin patch test on man Effect on metabolism in rabbits	M. I. Smith *et al.* [535] von Oettingen and Sharpless [641] Lillie *et al.* [368] Higgins and Kindel [276] Kawano [318]
	2,2-Bis-(*p*-chlorophenyl)-1-monochloroethane	Oral toxicity to mice	von Oettingen and Sharpless [641]
	Bis-(*p*-chlorophenyl)-sulfone	Oral toxicity in mice	Domenjoz [173]
	1,1-Bis-(4-chlorophenyl)-1-chloro-2,2,2-trichloro-ethane	Oral toxicity in mice Oral toxicity in mice	Domenjoz [173], [174] Browning *et al.* [91]
	1,1-Bis-(*p*-chlorophenyl)-2,3,3-trichloropropene	Oral toxicity in mice	Domenjoz [173]
	2,2-Bis-(*p*-chlorophenyl)-1,1,1-trifluoroethane	Lipoid affinity in rats	Kirkwood and Philips [324]
Tribromo DDT	2,2-Bis-(*p*-chlorophenyl)-1,1,1-ethane	Oral toxicity in mice Oral toxicity in mice Effect of skin patch tests on man	von Oettingen and Sharpless [641] Browning *et al.* [91] Higgins and Kindel [276]

Analog		Information	Authority
Common name	Chemical name		
		Changes on rings only	
m,p-DDT	2,2-Bis-(m,p'-chloro-phenyl)-1,1,1-trichloro-ethane	Oral toxicity to mice Oral toxicity to mice	DOMENJOZ [173], [174] VON OETTINGEN and SHARPLESS [641]
o,p-DDT	2,2-Bis-(o,p'-chloro-phenyl)-1,1,1-trichloro-ethane	Oral toxicity to mice Oral toxicity in mice Toxicity to fish and mosquito larvae Oral toxicity in mice Chronic oral toxicity to dogs	DOMENJOZ [173], [174], [175] VON OETTINGEN and SHARPLESS [641] GINSBURG [229] BROWNING *et al.* [91] WOODARD *et al.* [680]
Monochlor DDT	2-(p-Chlorophenyl)-2-phenyl-1,1,1-trichloro-ethane	Oral toxicity in mice Oral toxicity in mice Oral toxicity to mice	VON OETTINGEN and SHARPLESS [641] BROWNING *et al.* [91] DOMENJOZ [173], [174]
DHDT	2,2-Bis-(hydroxyphenyl)-1,1,1-trichloroethane	Oral toxicity in mice Estrogenic activity in ovariectomized rats	VON OETTINGEN and SHARPLESS [641] FISHER *et al.* [206]
Acetoxy DDT	2,2-Bis-(p-acetoxyphenyl)-1,1,1-trichloroethane	Oral toxicity in mice Effect of skin patch test on man	VON OETTINGEN and SHARPLESS [641] HIGGINS and KINDEL [276]
DMDT or Methoxychlor	2,2-Bis-(p-anisyl)-1,1,1-trichloroethane	Oral toxicity to mice Acute oral toxicity to rats and chronic oral tox-icity to rabbits, including pathology Oral toxicity in mice Toxicity to fish and mosquito larvae Toxicity to fish and mosquito larvae	DOMENJOZ [173], [174], [175] M. I. SMITH *et al.* [535] VON OETTINGEN and SHARPLESS [641] GINSBURG [229] GINSBURG [230]

Analog		Information	Authority
Common name	Chemical name		
DMDT or Methoxychlor (continued)	2,2-Bis-(*p*-anisyl)-1,1,1-trichloroethane	Pathology in rats	Lillie *et al.* [368]
		Action on arthropod nerve axon	Welsh and Gordon [658]
		Effect of sprays and dips upon livestock	Bushland *et al.* [95]
		Acute oral and dermal, and chronic oral and dermal toxicity with pathology in animals. Toxicity compared with other compounds	Lehman [354]
		Toxicity in sheep and cattle	Welch [656]
		Absorption, accumulation, fate, and excretion in mammals	Woodard *et al.* [679]
		Chronic oral toxicity to dogs	Woodard *et al.* [680]
		Toxicity to warm-blooded animals	Bishopp [67]
		Residues recovered from alfalfa	Ginsburg *et al.* [233]
		Effect of skin patch tests on man	Higgins and Kindel [276]
		Residues on various crops	Hoskins [301]
		Major toxic action	Lehman [355]
		Precautions in use on crops for canning	Rohwer [497]
		Contamination of meat and milk from sprayed livestock	Bushland *et al.* [96]
		Chronic oral toxicity to rats; chronic dermal toxicity to rabbits, dogs, and rats; respiratory toxicity to rabbits, dogs, and rats; includes pathology	Haag *et al.* [253]
		Suggested treatment of poisoning	Hough [303]
		Storage in the fat of the rat	Kunze *et al.* [335]
		Safety of use as a spray	Lehman [359]
		Oral toxicity in rats and dogs	Hodge *et al.* [282]
		Chronic toxicity to quail	Linduska and Springer [373]
		Acute and chronic toxicity to animals	Radeleff [475]
		Acute toxicity when sprayed on calves	Radeleff and Bushland [476]

Analog		Information	Authority
Common name	Chemical name		
DMDT or Methoxychlor (continued)	2,2-Bis-(*p*-anisyl)-1,1,1-trichloroethane	Toxicity to fish Effect on animals, residues on crops Effect of feeding dusted hay to dairy cows Precautions in use of poison Storage in adipose tissue *In vitro* effect on the succinoxidase system of rat heart Estrogenic activity in ovariectomized rats Acute and chronic oral toxicity to rats; acute and chronic dermal toxicity to rabbits and chronic oral toxicity to dogs Excretion in milk Chronic oral toxicity in rats and dogs Effect of feeding treated alfalfa hay to dairy cows	Surber [565] Webster [655] Biddulph *et al.* [60] Bishopp [69] Anonym [35] Johnston [311] Fisher *et al.* [206] Lehman [360a, b, c] Claborn and Wells [126] Hodge *et al.* [283] Biddulph *et al.* [61]
	2,2-Bis-(alphanophthyl)-1,1,1-trichloroethane	Oral toxicity in mice	Domenjoz [173], [174]
Propanoxy-carbinol DDT	1-(4-Chlorophenyl)-2,2,2-trichloroethylpropionate	Oral toxicity in mice	Browning *et al.* [91]
Carbinol DDT	1-(4-Chlorophenyl)-2,2,2-trichloroethanol	Oral toxicity in mice	Browning *et al.* [91]
DT or DPT	2,2-Bis-phenyl-1,1,1-trichloroethane	Oral toxicity to mice Toxicity to fish and mosquito larvae Acute oral toxicity to rats and chronic oral toxicity to rabbits, including pathology Oral toxicity to mice	Domenjoz [173], [175] Odum and Sumerford [440] M. I. Smith *et al.* [535] von Oettingen and Sharpless [641]

Analog		Information	Authority
Common name	Chemical name		
DT or DPT (continued)	2,2-Bis-phenyl-1,1,1-trichloroethane	Pathology in rabbits Action on the arthropod nerve axon Oral toxicity in mice Effect of skin patch test on man	LILLIE *et al.* [368] WELSH and GORDON [658] BROWNING *et al.* [91] HIGGINS and KINDEL [276]
	2,2-Bis-(3-chloro-4-methyl-phenyl)-1,1,1-trichloro-ethane	Oral toxicity in mice	DOMENJOZ [173]
	2,2-Bis-(2,4-dimethyl-phenyl)-1,1,1-trichloro-ethane	Oral toxicity in mice	DOMENJOZ [173]
	2,2-Bis-(3,4-dimethyl-phenyl)-1,1,1-trichloro-ethane	Oral toxicity in mice	DOMENJOZ [173], [174]
Ethoxy DDT	2,2-Bis-(*p*-phenethyl)-1,1,1-trichloroethane	Toxicity and pathology in rats Oral toxicity in mice Effect of skin patch test on man	PRILL *et al.* [470] VON OETTINGEN and SHARPLESS [641] HIGGINS and KINDEL [276]
Propoxy DDT	2,2-Bis-(*p*-propoxyphenyl)-1,1,1-trichloroethane	Oral toxicity in mice Effect of skin patch test on man	VON OETTINGEN and SHARPLESS [641] HIGGINS and KINDEL [276]
Butoxy DDT	2,2-Bis-(*p*-butoxyphenyl)-1,1,1-trichloroethane	Oral toxicity in mice Effect of skin patch test on man	VON OETTINGEN and SHARPLESS [641] HIGGINS and KINDEL [276]
Amyloxy DDT	2,2-Bis-(*p*-amyloxyphenyl)-1,1,1-trichloroethane	Oral toxicity in mice	VON OETTINGEN and SHARPLESS [641]

Analog		Information	Authority
Common name	Chemical name		
Methyl DDT	2,2-Bis-(*p*-tolyl)-1,1,1-trichloroethane	Oral toxicity in mice Oral toxicity in mice Oral toxicity in mice	DOMENJOZ [173], [174], [175] VON OETTINGEN and SHARPLESS [641] BROWNING *et al.* [91]
Tertiary Butyl DDT	2,2-Bis-(*p*-tert.-butyl-phenyl)-1,1,1-trichloro-ethane	Oral toxicity in mice	VON OETTINGEN and SHARPLESS [641]
DADT	*p,p′*-Diaminodiphenyl-trichloroethane	Acute oral toxicity in rats, mice, and rabbits	M. I. SMITH *et al.* [536]
	2,2-Bis-(*p*-nitrophenyl)-1,1,1-trichloroethane	Effect of skin patch test on man	HIGGINS and KINDEL [276]
DFDT	2,2-Bis-(*p*-fluorophenyl)-1,1,1-trichloroethane	Oral toxicity in mice Lipoid affinity in rats Toxicity to mice, rats, guinea pigs, and rabbits Toxicity to fish and mosquito larvae Action on the arthropod nerve axon Oral toxicity in mice Toxicity to rats and mice, includes pathology Human poisoning Acute and chronic oral toxicity in rats	DOMENJOZ [173], [174], [175] KIRKWOOD and PHILIPS [324] MÜHLENS [416] ODUM and SUMERFORD [440] WELSH and GORDON [658] BROWNING *et al.* [91] PIEKARSKI and HOLZ [457] KWOCZEK [336] LEHMAN [360a, b, c]
DIDT	2,2-Bis-(*p*-iodophenyl)-1,1,1-trichloroethane	Toxicity to fish and mosquito larvae Oral toxicity in mice Oral toxicity in mice Effect of skin patch test on man	ODUM and SUMERFORD [440] VON OETTINGEN and SHARPLESS [641] BROWNING *et al.* [91] HIGGINS and KINDEL [276]

Analog		Information	Authority
Common name	Chemical name		
DBrDT	2,2-Bis-(*p*-bromophenyl)-1,1,1-trichloroethane	Toxicity to fish and mosquito larvae Acute oral toxicity to rats and chronic oral toxicity to rabbits, includes pathology Oral toxicity in mice Pathology in rabbits Action on the arthropod nerve axon Oral toxicity in mice Fate of residues (radioactive) in food products	ODUM and SUMERFORD [440] M. I. SMITH *et al.* [535] VON OETTINGEN and SHARPLESS [641] LILLIE *et al.* [368] WELSH and GORDON [658] BROWNING *et al.* [91] WINTERINGHAM *et al.* [672]
Bromo-methyl DDT	1,1-Bis-(4-bromoethyl-phenyl)-2,2,2-trichloro-ethane	Oral toxicity in mice	BROWNING *et al.* [91]
Changes on rings and aliphatic groups			
o,p-DDA	1,1-Bis-(*o,p'*-chlorophenyl)-acetic acid	Oral toxicity in mice	DOMENJOZ [173], [174], [175]
DA	1,1-Bis-phenylacetic acid	Chronic toxicity and pathology in rabbits Pathology in rabbits	M. I. SMITH *et al.* [535] LILLIE *et al.* [368]
DE	1,1-Bis-phenylethane	Chronic toxicity and pathology in rabbits Pathology in rabbits Action on the arthropod nerve axon	M. I. SMITH *et al.* [535] LILLIE *et al.* [368] WELSH and GORDON [658]
Hydroxy DDT	2-(*p*-Chlorophenyl)-2-hydroxy-1,1,1-trichloro-ethane	Oral toxicity in mice Action on the arthropod nerve axon	DOMENJOZ [173] WELSH and GORDON [658]
	1-(4-Chlorophenyl)-2,2-dichloroethanone	Oral toxicity in mice	BROWNING *et al.* [91]

Analog		Information	Authority
Common name	Chemical name		
	2-(p-Chlorophenyl)-1,1,1,2-tetrachloroethane	Oral toxicity in mice	DOMENJOZ [173], [174]
Neotran	Bis-(p-chlorophenoxy)-1-methane	Toxicity to fish Toxicity to fish	LINDUSKA and SURBER [373] SURBER [565]
DHDE	1,2-Diphenylethane	Action of the arthropod nerve axon	WELSH and GORDON [658]
Methyl DDA	1,1-Bis-(p-hydroxyphenyl)-ethane 1,1-Bis-(p-tolyl)-acetic acid	Estrogenic activity in ovariectomized rats Oral toxicity in mice	FISHER *et al.* [206] DOMENJOZ [173], [174]
DDK	o,p'-Dichlorobenzophenone	Chronic toxicity and pathology in rabbits Pathology in rabbits Oral toxicity in mice	M. I. SMITH *et al.* [535] LILLIE *et al.* [368] BROWNING *et al.* [91]
DDS	p,p'-Diaminodiphenyl sulfone	Acute oral toxicity in rats, mice, and rabbits	M. I. SMITH *et al.* [536]
Fluoro-trifluoro DDT	1,1-Bis-(4-fluorophenyl)-2,2,2-trifluoroethane	Oral toxicity in mice	BROWNING *et al.* [91]
	1,1-Bis-(4-chlorophenyl)-1-fluoro-2,2-dichloro-2-fluoroethane	Oral toxicity in mice	BROWNING *et al.* [91]
	1,4-Dichlorobenzene	Oral toxicity in mice Oral toxicity in mice	DOMENJOZ [173], [174] BROWNING *et al.* [91]
	Hexachloroethane	Action on the arthropod nerve axon	WELSH and GORDON [658]

DDT per kilogram of body weight. Reported as DDT, the foreign substance amounted to 168 ppm. The authors clearly stated in their discussion that the analysis might represent some derivative of DDT and not the compound itself.

The same authors (SMITH and STOHLMAN [533]) later studied excretion following single and repeated doses of DDT. After a single oral dose the highest concentration of chlorides in the urine reported as DDT ranged from 100 to 652 ppm. Urinary elimination was detected as much as 12 days after ingestion of DDT although the greater part of the excretion was accomplished in 5 to 6 days and the peak reached in 2 to 3 days. The total amount eliminated in 5 experiments varied from 1·8 to 5·1% of the dose administered. There was some indication that diuresis favored the elimination of DDT or its metabolites. In the same experiments, the elimination of DDT in the feces varied from 5 to 50% of the dose administered. The authors assumed that this represented both unabsorbed DDT and DDT excreted in the bile and by the alimentary tract. A concentration of the compound in the bile ranging from 79 to 534 ppm was shown in other rabbits following single or repeated oral doses of DDT. However, no effort was made to show that DDT secreted in the bile is not reabsorbed or that true intestinal excretion exists.

In the presence of repeated oral doses, the daily urinary excretion of organic chlorides increased (SMITH and STOHLMAN [533]). The percentage of the intake which was excreted varied from 4·9 to 8·0 in two experiments. The DDT eliminated in the feces was determined in a rabbit which had received a dosage of 25 mg/kg/day. The proportion of each dose eliminated in the feces was slightly less than the amount eliminated in the urine. This was in marked contrast to the relation of fecal and urinary excretion found in rabbits which had received a single dose at the rate of 400 mg/kg.

Studies of cats given 5 and 10 mg/kg/day showed a concentration of organic chlorides in the urine usually ranging from 2 to 10 ppm in terms of DDT. In one instance the urinary concentration reached 37 ppm, and this was a bladder specimen obtained at autopsy. The urinary excretion in cats given 50 mg/kg/day was not significantly higher than the excretion in cats receiving smaller doses. Thus, the urinary excretion of DDT or its metabolites was low in cats as compared with rabbits. Dogs showed excretion similar to that of rabbits but of a more irregular pattern.

The chemical nature of the chief metabolite excreted in the urine was first elucidated by WHITE and SWEENEY [663]. Using an olive oil solution, rabbits were given 100 mg of DDT per kilogram of body weight, 6 days a week. Pure DDT melting at 107–108°C was used. It was observed that under these conditions the urine contained a considerable amount of organic chloride whereas normal rabbit urine did not. Using the organic chloride test to evaluate different methods of extraction, the authors were able to isolate a crystalline material containing 25·37% chlorine and melting at 166–166·5°C. The crystals were shown to be di-(*p*-chlorophenyl)-acetic acid (DDA). The product obtained from urine was identical to that synthesized from glyoxylic acid and chlorobenzene and with a compound obtained through the chemical degradation of DDT.

Identity of the three compounds and, therefore, their true chemical nature was established by determination of melting points, mixed melting points, elementary analysis, and X-ray powder diffraction patterns, as well as by demonstrating the similarity of the decarboxylation products of the three original materials. Only 80 to 85% of the total organic chloride of the rabbit urine was found soluble in alkali and in bicarbonate. For this and other reasons it was considered possible that DDA was not the only chlorinated organic compound present. Reason was given for believing that DDA was the principal metabolite present in the urine and that unchanged DDT, if present at all, was present in amounts less than 5%. The authors also suggested a possible mechanism for the degradation of DDT *in vivo*.

The absence of any significant quantity of DDT in the catheterized urine of dogs was established through bioassay by LÄUGER and his colleagues [349].

STOHLMAN [557], apparently unaware of the work of WHITE and SWEENEY, demonstrated independently the presence of an ether-soluble fraction and an alkaline-water soluble fraction in the residue of acid ether extracts of urine from rabbits fed DDT. From the ether soluble fraction STOHLMAN isolated crystals with a melting point of 106–107°C. A mixed melting point with DDT showed no depression and microcombustion analysis was consistent with DDT. Of the total organic chlorides, 35 to 60% was found in the ether soluble fraction and, according to the author, represented unchanged DDT. The alkali soluble fraction was considered to represent a different, unidentified compound. The proportions of the two fractions in the urine was believed to vary with the size of the dose of DDT administered and the length of time following administration.

Later in the same year STOHLMAN and SMITH [559] published a report on the alkaline-water soluble fraction of the residue of acid ether extracts of urine from rabbits fed DDT. They isolated DDA and established its identity by melting point, microcombustion analysis, and spectrophotometric analysis after nitration. Of the total organic chloride present, about 25%, considered to represent DDT, was in the ether soluble fraction while about 75%, considered to represent DDA, was in the alkali soluble fraction. The possibility of other chlorinated metabolites or even of completely dechlorinated metabolites was recognized.

OFNER et al. [442] confirmed the presence of DDA in the urine of dogs and rabbits, and reported, in addition, the presence of a smaller amount of a neutral material which, however, did not give a characteristic test for DDT.

OFNER and CALVERY [441] denied the presence of DDT in uncontaminated catheter-urine. They did find DDA and also two other metabolites which were present only to the extent of 25 to 30% of the total metabolic products of DDT. No detail of the nature of the two metabolites was given except that they were extractable in the same way as DDA and gave the same color reaction in the Schechter-Haller test. The urine concentration expressed as DDA was found to vary from 50 to 5 ppm during the first few days after a single dose at the rate of 200 mg/kg although small amounts appeared in the urine for 16 days.

LAUG [343], using a bioassay, was unable to demonstrate DDT in the urine of poisoned rabbits.

That DDT is excreted in man in the form of DDA was first shown by NEAL and his colleagues [428]. The authors used both the organic chloride method and the more specific colorimetric determination of SCHECHTER and HALLER. The two methods gave a satisfactory agreement on aliquots of 24-hour urine samples. The colorimetric method demonstrated the presence of DDA and the absence of DDT in the samples. The acidic character of the excreted material was proved by the fact that 75% could be extracted with alkali. The remaining 25%, which gave the same color reaction as DDA, was considered by the authors to represent some other unidentified metabolite. The final results were, however, expressed in terms of DDA. A volunteer, who had some previous experimental contact with DDT, ingested 770 mg of the pure, recrystallized compound in approximately 25 ml of olive oil on an empty stomach. The dosage rate was exactly 11 mg/kg. There followed a sharp rise in the excretion of DDA which reached its maximum during the second 24-hour period, decreased sharply on the third and fourth days, and thereafter decreased gradually. Although the authors did not make the calculation, their data appear to justify the assumption that they accounted by urinary excretion for approximately 2% of the total dose which was ingested.

The finding of DDA in the urine of man was confirmed 3 months later (M. I. SMITH [531]) in connection with a case of accidental poisoning. Analysis by the method of SCHECHTER and HALLER of a pooled specimen of 300 ml of urine voided under observation on the sixth and eighth days after ingestion showed the elimination of about 1 mg of a substance which gave an absorption spectrum similar to that of a mixture of 75% DDT and 25% DDA.

FINNEGAN et al. [204] confirmed that DDA is excreted in the urine of dogs and they failed to demonstrate DDT. They found the average urinary output of DDA in dogs receiving 25 mg of DDT per kilogram of body weight to be 0·13, 0·12, and 0·17 mg of DDA per kilogram of body weight at the end of the first, second, and fourth week of dosing. Thus, the urinary excretion amounted to approximately 0·5% of the ingested dose.

JUDAH [313] demonstrated DDA in the urine of rats. No unchanged DDT was found. WINTERINGHAM et al. [672] found that 94·3% of the excreted radioactive DBrDT behaved as the acetic acid derivative, DBrDA. When the radioactive material was fed to a man, 5% of the total amount was recovered from the urine excreted in the first 48 hours. Of the recovered material, 35% behaved as DBrDA, and the remainder behaved as DBrDT.

DDA has been reported in the urine of guinea pigs, also (WASICKY and UNTI [651]).

Feces

The passage of unabsorbed DDT in the feces of many species of animals following oral administration has been reported frequently (SMITH and STOHLMAN [533]; ORR and MOTT [444]; STOHLMAN [557]; LAUG [343]; FINNEGAN et al. [204]). True fecal excretion was first reported by WASICKY and UNTI [650], who reported recovery of the unaltered compound in the urine and feces of guinea pigs that had received DDT by intraperitoneal injection. The authors gave no indication of the method which they used to determine the presence of DDT, although it appears probable that they relied on the determination of total organic chlorides. More recent work indicates that identification of unchanged DDT in the urine may have been erroneous. Of much more importance than this detail, however, was the demonstration that DDT or some metabolite was excreted in the feces even when there was no possibility that oral ingestion would lead to the passage of unabsorbed DDT in the feces. The existence of true fecal excretion was confirmed by JUDAH [313], who detected both DDA and DDT in the feces of 6 rats which had received the compound intravenously at 25 mg/kg. More DDT than DDA was found in the feces, but in the urine only DDA was demonstrated. The total excretion in 5 days amounted to about 10% of the total dose. Rats injected intraperitoneally with 1000 mg/kg excreted only 2·5% of the dose in 5 days.

Although other authors have mentioned true fecal excretion as a logical possibility (SMITH and STOHLMAN [533]; NEAL and VON OETTINGEN [425]) they have generally failed to demonstrate its existence. When associated with oral intake, the presence of a metabolite of DDT in the feces could theoretically represent a decomposition within the intestine induced by enzymes of the host or of bacteria. However, the presence of a metabolite in the feces strongly suggests true excretion. WINTERINGHAM *et al.* [672] fed bread containing 11·2 ppm of radioactive DBrDT (bromine analog) to rats for 5 days. A higher concentration (23·1 ppm) was found in the feces than existed in the food. Twenty-three per cent of the material in the feces behaved as acetic acid derivative (DBrDA) while the remainder was unaltered.

Milk

Demonstration of the excretion of DDT in the milk was first published by WOODARD and his colleagues [678]. They reported finding 40 and 60 ppm, respectively, on two occasions in the milk of a dog which had received solid DDT at the rate of 80 mg/kg/day. Another dog was given 50 mg of the *o,p*-isomer per kilogram of body weight; 24 hours later a milk sample showed 50 ppm of that isomer. Four months later more extensive studies were published (TELFORD and GUTHRIE [581]). The authors showed that rats fed a diet containing 1000 ppm DDT produce milk which is toxic to their young. They also showed that milk from goats which had been fed on DDT was toxic to cats and to rats and that mother rats transferred this toxic principle to their own milk as evidenced by typical poisoning in suckling rats. The poison was still present in the milk of one goat 24 hours after the last dose. Goats were shown to be less susceptible than rats or cats, for a suckling goat showed no indication of poisoning when reared on milk similar to that which killed rats. The authors offered some evidence to show that DDT was concentrated in the fat globules of goat milk. When DDT was applied to the skin of a goat, an insufficient quantity was absorbed to render the milk toxic to rats.

In a later study TELFORD [580] showed that goat milk became sufficiently toxic to kill rats 29 to 31 hours after the administration to the goat of a single oral dose at the rate of 1,500 mg/kg. The goat produced toxic milk for approximately one week. The goat showed severe tremors but recovered, and about a month later gave birth to an apparently normal kid. In similar experiments the author also gave conclusive evidence that cream contained far more toxic material than skim milk. Using a bioassay with house flies, it was estimated that butter made from the milk of a goat fed DDT at the same rate of 1,500 mg/kg contained between 1,250 and 2,500 ppm.

SCHECHTER et al. [511] were apparently the first to demonstrate DDT in cow's milk. They stated that the concentration could rise to 25 ppm or higher, depending on intake. Butter made from milk containing 25 ppm showed a concentration of 532 ppm.

Later studies (SPICER et al. [544]) revealed marked but unexplained fluctuations in the concentration of DDT in the milk of a goat which received the compound dissolved in vegetable oil 5 days each week. Concentrations as high as 500 ppm and, on two occasions, as high as 1,000 ppm, were encountered. A concentration of 2,688 and 3,959 ppm were found in the cream; skim milk from the same samples showed traces of DDT within the limit of the blank.

Since these early laboratory studies the presence of DDT has been demonstrated repeatedly in the milk of cows. As tabulated below (p. 153), 10% or more of the total DDT ingested is commonly excreted in the milk, and amounts slightly in excess of 30% have been observed.

DDT has been demonstrated in the milk of lactating women (LAUG et al. [346]); see page 146.

Other Routes

Although it is not generally considered under the term 'excretion', passage of DDT into the fetus (FINNEGAN et al. [204]) or into the eggs of birds (RUBIN et al. [501]; CARTER [114]; BRYSON et al. [93]; DRAPER et al. [180]) constitutes an elimination of the compound from the parent organism.

Finally, DDT is excreted under certain conditions by the skin. WILSON et al. [667] used petroleum ether to wash a fatty secretion from the skin of a cow which had been maintained on an oral dosage of 24 g daily (about 53 mg/kg/day). The extract analyzed 115 ppm of DDT.

General Considerations

Although JUDAH [313] was able to recover 95·7% or more of the total material injected intravenously 3 hours earlier into 6 rats, the result is not typical. Most efforts to account for the fate of DDT in animals or man have been much less successful. This is especially true in connection with repeated small doses. After 50 to 150 days animals receiving a constant daily dosage do not store additional DDT and may even excrete a part of what they have already stored. In this state of equilibrium, the *total* daily elimination must equal or slightly exceed the total daily intake. So far it has been possible to account for only a small part of the material which must certainly be excreted in one form or another. This is a matter in urgent need of research.

Biological Sources of Variation

Effect on Isolated Cells

In determining the mode of action of a compound it is of great interest to know whether it is physiologically active when brought in contact with isolated cells. A distinction can be made between protoplasmic poisons and those which act by disturbing the integration of the organism. That DDT is not toxic to certain cells was first established by LEWIS and RICHARDS [365]. They found DDT inert when exposed to tissue cultures of heart, kidney, stomach, intestine, liver, and muscle from 7- to 9-day chick embryos, and of brain and spleen from a 1-day rat. The DDT was applied in different ways. It was deposited from an acetone solution on coverglasses on which hanging drop cultures were later grown; it was added in the form of crystals to roller tube cultures; and it was added in the form of an emulsion in which an olive oil solution of DDT, at a final concentration of 1%, formed the discontinuous phase. When the emulsion was used, a final concentration of approximately 600 ppm of DDT was achieved in the culture. It will be recalled that mosquito larvae die in water containing a concentration of 0·02 ppm or slightly less and certain fish die at about the same concentration under laboratory conditions. The authors observed fibroblasts, macrophages, endoderm, liver cells, kidney epithelium, nerve fibers, and muscle fibers. Under a variety of conditions, the cytology of the cells including the mytoses of fibroblasts was normal. The migration and extension of the various cells was unchanged. The authors stated that 'living fibroblasts as they moved about in the cultures sometimes touched or even migrated over DDT crystals without any appreciable injury to themselves during a period of several days'. Observations were carried out for periods as great as 21 days.

Wound healing is not affected by DDT. CAMERON and BURGESS [105] demonstrated this rigorously.

HOFFMAN and LENDLE [285] observed no harmful effect on red blood corpuscles from a concentration of 166,666 ppm. Under field conditions, LACKEY and STEINLE [338] failed to observe any effect on algae or protozoa from the use of DDT at the rate of 1·0 ppm, although fish and tadpoles, as well as many invertebrates, were killed in the same pools, and in other pools by one-fourth that concentration.

WASICKY and UNTI [650] observed no injury to the ciliate, *Opalina brasiliensis*, placed in a watery emulsion with concentrations up to 2,500 ppm.

HENNIG [271] found that the virus of vaccinia was inhibited by incubation with DDT.

LEGGIERI [353] could demonstrate no difference in the fermentation by yeast of flour containing 8 to 2,000 ppm.

If, then, the isolated cells of susceptible animals in tissue culture are immune to DDT, and if free-living animal and plant cells are frequently immune, it is of interest to learn whether any complex organisms are also immune. Quite obviously, the higher plants are not generally affected by DDT, a fact which cannot be explained entirely on the basis of their failure to absorb the compound, for some plants, at least, absorb and store high concentrations (REIBER and STAFFORD [483]). Fungi (NORRIS [437]; RICHARDS and CUTKOMP [486]) and algae (ODUM and SUMERFORD [440]), like other plants, are resistant.

The tubercle bacillus, *Mycobacterium tuberculosis*, was found susceptible to 1,1,1-trichloro-2,2-bis-(*p*-aminophenyl)-ethane, an analog of DDT (KIRKWOOD et al. [326]; KIRKWOOD and PHILLIPS [325]) but other workers (M. I. SMITH et al. [536]) failed to confirm this result.

Effect on Different Animal Phyla

In animals which are affected by DDT, the signs of illness caused by poisoning appear to be associated largely, or even exclusively, with the nervous system. It is of interest, therefore, to consider the effect of DDT on a wide range of animal phyla in which the nervous system exhibits different degrees of organization. It was suggested by BELKIN [51] that differences in the nervous system might account for the difference in susceptibility which he observed among certain protozoa. He found that *Paramecium aurelia* survived 2 hours at a concentration of 39 ppm, and only 45 minutes at a concentration of 62 ppm. A rotifer, *Hydatina senta*, was equally sensitive. However, *Amoeba proteus* and *A. dubia* immersed in similar suspensions survived 24 to 48 hours or longer, depending on the concentration. *Paramecium*, although unicellular, is considered to have a nervous system differentiated in its cytoplasm, while no such structure is recognized in *Amoeba*. In spite of this, one must accept the suggestion of BELKIN with reserve, for some other ciliates (e.g. *Opalina*) structurally similar to *Paramecium* are resistant to DDT. BELKIN found that *A. dubia* was more susceptible to DDT solution placed on the cell membrane than to DDT solution injected into the cytoplasm, but his statements do not permit one to judge the relative importance of the DDT and of the oil.

Probably the most complete study allowing such a comparison is that of RICHARDS and CUTKOMP [486]. These authors have presented their findings on some 40 genera of 12 animal phyla in great detail. DDT was used in the form of colloidal suspensions formed by adding acetone solutions of the compound to water. The authors' chief conclusion was that the presence of chitinous cuticle increased the apparent toxicity of DDT by facilitating its absorption, with the result that a higher concentration of the compound was reached inside animals with such a cuticle. This conclusion was supported not only by the comparative studies of animals but also by *in vitro* studies of DDT adsorption by chitin. LORD [380] confirmed the specific action of chitin in adsorbing DDT but failed to find any relation between adsorption and the insecticidal action of different analogs which proved to be as readily adsorbed as was DDT itself. RICHARDS and CUTKOMP further concluded that different organisms have an inherently different susceptibility to DDT over and above the factor of absorption. They found that injection of DDT into the snail, *Helisoma trivolva*, at a rate exceeding 1,000 mg/kg was without effect. The nematode, *Ascaris lumbricoides*, proved several times more resistant to injected DDT than the cockroach. Thus, the possession of a reasonably well developed nervous system does not always render the organism susceptible to injected DDT.

The findings of HOFFMAN and LENDLE [285] support the conclusion that the possession of a well-developed nervous system does not necessarily involve sensitivity to DDT, even when it is injected.

On the other hand, the susceptibility of certain arthropods and mammals is remarkably similar when DDT is injected within the body in such a way that the factor of primary absorption is reduced to a minimum. This is illustrated in Table 15. Furthermore, the susceptibility of certain arthropods and fish exposed in the same way is similar as shown in Table 16. There is some tendency for the figures for mammals and other vertebrates to be higher than those for insects and crustacea. Further, it has been claimed by HOFFMAN and LENDLE [285], whose results appear in Table 15, that a greater difference exists than the figures suggest. They found that frogs frequently developed severe signs of poisoning, but later recovered, while insects which showed the same apparent degree of effect always died. Thus, they speak of an 'all or nothing law' in this connection.

It is interesting to note that certain mammals may maintain an insecticidal concentration of DDT in their blood without themselves being affected. Thus, LINDQUIST *et al.* [369] showed that 2 species of bedbugs could be killed by feeding on rabbits which had, in turn, been fed on DDT. The rabbits were pretested to be sure that they were favorable hosts, then they were given doses by capsule or stomach tube in such a way that the skin was not contaminated. The dosage ranged from 228 to 537 mg of DDT per kilogram of body weight. The bugs were fed through a screen so that only the proboscis had any direct contact with the host. Some of the rabbits showed toxicity to bedbugs within 1·5 hours and all the animals tested 24 hours after dosing still showed activity. Mortality as high as 100% was obtained in bugs fed 3 to 5 hours after rabbits were dosed. Only the rabbit on the highest dosage (537 mg/kg) was killed by DDT although all the dosage levels were large. The same authors reported similar mortality of bedbugs fed on rabbits which had ingested pyrethrins at approximately the same dosages used with DDT.

MACCORMACK [387] ate 1·5 g of DDT, and found that lice experimentally fed on him 6 and 12 hours after the dose were killed.

EMMEL and KRÜPE [199] reported that, when enough DDT was administered to guinea pigs by mouth to kill 100% of bedbugs which received a single feeding on them, the cavies were either made ill or killed. It was implied that doses which produced no sign of poisoning in the guinea pigs were sufficient

Table 15

*Comparison of the Toxicity of DDT Injected Intra-Abdominally into Arthropods and Frogs
and Intravenously into Mammals*

Species	LD_{50} mg/kg	Vehicle	Reference
Periplaneta americana (roach)	5–8	Acetone solution	Tobias *et al.* [588]
	18	Peanut-oil emulsion	Tobias *et al.* [588]
	82	Peanut-oil solution	Tobias *et al.* [588]
	20	Oil emulsion	Dresden and Krijgsman [182]
Carausius morosus (fly)	60	Oil emulsion	Dresden and Krijgsman [182]
May beetle	20–30	Colloidal suspension from alcohol	Hoffman and Lendle [285]
Rana esculenta (frog)	12	Oil emulsion	Dresden and Krijgsman [182]
	20	Colloidal suspension from alcohol	Hoffman and Lendle [285]
Rana temporaria	20	Colloidal suspension from alcohol	Hoffman and Lendle [285]
Rat	47[1]	Homogenate	Philips and Gilman [453]
Rabbit	41[1]	Homogenate	Philips and Gilman [453]
	30	Homogenate	Smith *et al.* [535]
Cat	32[1]	Homogenate	Philips and Gilman [453]
Dog	68[1]	Homogenate	Philips and Gilman [453]
Monkey	55[1]	Homogenate	Philips and Gilman [453]

[1]) Calculated by the reviewer.

to kill a part of the bedbugs fed a single time on their blood, and all of the
bedbugs fed a number of times. The highest concentration of DDT in the blood
was reached in about 4 to 7 hours after a dose of 160 mg per animal.

Knipling *et al.* [329] found DDT ineffective for the control of lice (*Pediculus
humanus corporis*) or mosquitoes (*Aedes aegypti*) when the compound was fed to
rabbits. However, several indandione compounds showed marked activity against
lice and gamma benzene hexachloride was outstanding against mosquitoes. The
authors observed that there was little relationship between the value of the chemicals as contact insecticides and as internal therapeutic agents. They also noted
a high degree of specificity for different compounds against different insects.

Wilson [670] fed both benzene hexachloride and DDT to cattle. The
benzene hexachloride proved more effective as a systemic insecticide for the

W. J. Hayes, Jr.

Table 16

Comparative Susceptibility of some Aquatic Organisms to DDT as a Colloidal Suspension

Kind of organism	Concentration (ppm)			Authority
	Tolerated	Producing 50% mortality	Producing 100% mortality	
ANNELIDA Oligochaeta				
Lumbricus	100	—	—	HOFFMAN and LENDLE [285]
'Rain worms' . . .	100	—	—	RICHARDS and CUTKOMP [486]
MOLLUSCA Gastropoda				
Physa	0·1	—	.1	EIDE *et al.* [195]
Physa	10	—	100	RICHARDS and CUTKOMP [486]
ARTHROPODA Eucrustacea				
Daphnia	—	0·01	0·1	RICHARDS and CUTKOMP [486]
Gammarus	—	0·001	0·01	RICHARDS and CUTKOMP [486]
Insecta				
Aedes aegypti . . .	—	0·011	—	GINSBURG [229]
Aedes aegypti . . .	—	0·014	0·5	GINSBURG [230]
Aedes aegypti . . .	—	0·001	0·01	RICHARDS and CUTKOMP [486]
Anopheles albitarsis .	—	—	0·02	WASICKY and UNTI [648]
Anopheles strodei . .	—	—	0·02	WASICKY and UNTI [648]
Culex apicalis[1]) . .	—	—	0·0008[3])	ODUM and SUMERFORD [440]
Culex apicalis[2]) . .	—	—	0·024[3])	ODUM and SUMERFORD [440]
Culex quinque-faciatus	—	—	0·02	WASICKY and UNTI [648]
CHORDATA Osteichthyes				
Abramis lucidus . .	5	—	10	GÖTZ [243]
Carassius auratus .	—	0·1	0·20	GINSBURG [227]
Carassius auratus .			0·2	EIDE *et al.* [195]
Carassius auratus .			0·075	ODUM and SUMERFORD [440]
Carassius auratus .		0·054[3])		GINSBURG [229]
Carassius auratus .		0·145[4])		GINSBURG [229]
Carassius auratus .	0·05	0·125	0·25	GINSBURG [230]
Gambusia affinis . .	—	—	0·013	ODUM and SUMERFORD [440]
Lebistes reticulatus .	< 0·025	0·05–0·1	—	PAGAN and HAGEMAN [446]
Phallocerus caudi-maculatus (adult) . .	—	—	0·020	DE ALMEIDA [150]
Phallocerus caudi-maculatus (young) .	—	—	0·017	DE ALMEIDA [150]
Tilapia kafuensis . .	—	—	0·014	PIELOU [458]

[1]) Larvae.   [2]) Pupae.   [3]) p,p'-DDT.   [4]) Technical DDT.

control of tsetse flies and ticks. However, cattle tolerated oral doses of DDT at the rate of 250 and 500 mg/kg with no apparent ill effect. Of 36 tsetse flies which fed on the DDT-treated cattle, ten definitely engorged; a total of five died within 24 hours.

Of course, from a practical standpoint insects and mammals show a tremendous difference in their susceptibility to DDT. This is due primarily to the great difference in the ability of the two groups to absorb the compound, a fact first pointed out by LÄUGER *et al.* [347] and confirmed by all later research.

One must observe that the question of absolute toxicity can not be answered completely at this time. It is, of course, possible to determine the concentration of DDT in the tissues of larger animals but it is impossible at this time to determine its concentration on specific cell membranes or in specific cells or cell structures. This being true, one must interpret with caution the insusceptibility of tissue cultures and protozoa and, especially, the relative insusceptibility to injected DDT shown by certain metazoa with an organized nervous system.

Temperature

The effect of temperature on the toxicity of DDT for insects has been studied extensively and is discussed at length in the appropriate chapter. The finding by LINDQUIST *et al.* [370] and HOFFMANN and LINDQUIST [286] that in flies the toxicity of appropriate, small, topically applied doses of DDT is inversely proportional to temperature (negative temperature coefficient) has been abundantly confirmed by later research on flies, roaches, dipterous larvae, and one species of crustacean. In studies of dipterous larvae, FAN *et al.* [202] found that toxicity of DDT showed a negative temperature coefficient at lower concentrations of the toxicant and a positive temperature coefficient at higher concentrations. Apparently, this particular relationship has not been investigated further for DDT nor is it clear what relation this finding has to the positive temperature coefficient observed for certain other insecticides. However, the result with DDT was probably not associated with a greater absorption of small doses at low temperature as FAN and his associates supposed. Chemical and radiological measurements have indicated that the absorption of DDT is directly proportional to temperature even in insects in which the toxicity is indirectly proportional to temperature. Furthermore, the toxicity of injected DDT has given a negative temperature coefficient in experiments in which control insects survived properly.

It appears that at least the following factors may be inter-related in determining the toxicity of DDT to any particular species-and developmental stage-of insect: (1) absorption influenced by adsorption by chitin (negative temperature coefficient) and solution in lipids of the exoskeleton (positive temperature coefficient; (2) solubility in lipids within the body (positive temperature coefficient); (3) detoxification to DDE (positive temperature coefficient); (4) possible metabolism to other compounds.

In studies on the frog heart, which are described in the appropriate section below, HOFFMAN and LENDLE [285] observed that the preparation was affected

                                 W. J. Hayes, Jr.

by concentrations of 1 or 0·1 ppm in June at a room temperature of at least
22°C, while in December at a low room temperature a concentration of 300 ppm
was necessary to produce the same effect.

Studies of the susceptibility of mammals to DDT at different environmental
temperatures are probably not strictly comparable to those with arthropods
because mammals are homoiothermic. DEICHMANN *et al.* [162] found that rats
given a single LD_{50} dose and held at 5°C showed hyperexcitability and tremors
earlier than similar rats held at 20 and 33°C. The animals kept at 33°C and
shielded from sound and light developed, at most, only mild signs of central
nervous system stimulation. In spite of the differences in behavior, the mortality of the three groups was similar. The cold in this instance probably
represented a nonspecific environmental stimulation. It has been observed by
many workers that stimulation may increase the overt signs of intoxication in
animals, and EMMEL and KRÜPE [199] claimed that death was hastened by
stimulation. Effective stimuli include cold, sudden noise, the combination of
noise and vibration used to induce audiogenic seizures, disturbance of the fur
and vibrissae by air currents or direct contact, and olfactory irritation.

Nutrition

It has been shown that insects maintained on different foods show different
susceptibility to DDT (see, for example, MCGOVRAN and GERSDORFF [390]).
It is known that animals in good condition, especially those which are fat,
are more resistant to intoxication (SPICER *et al.* [544]). Like mammals, fish in
good condition are more resistant to poisoning (HOFFMANN and SURBER [291]).

SMITH and STOHLMAN [533] found only slightly greater mortality and liver
pathology in rats fed 500 ppm of DDT in a diet containing 8% protein than
in rats fed the same concentration of insecticide in a diet containing 28%
protein.

SAUBERLICH and BAUMANN [508] studied the susceptibility to recrystallized
DDT of mice and rats maintained on isocaloric diets in which the amount of
fat and protein was varied. The animals developed toxic symptoms and died
sooner when fed a diet containing 5% or more of fat than when fed a diet
containing only 0·5%. All fats tested, regardless of their degree of saturation,
increased the sensitivity of mice to DDT. A reduction in the level of protein
in the diet to 10% apparently decreased the resistance of mice to DDT whether
the percentage of fat in the diet was high or low. The authors attributed the
effect of fat primarily to the greater efficiency of absorption of the insecticide.
It is unfortunate that they did not make it clear whether the DDT was dissolved in the fat before the various diets were formulated. It is established
from other studies that this matter of solution may be important in determining toxicity even when the composition of the diet remains constant.

Findings similar to those of SAUBERLICH and BAUMANN have been encountered in far more extensive studies of nutritional factors affecting the
toxicity of the halogenated hydrocarbon solvents. This subject has been ably

reviewed by MILLER [410] who concluded, among other things, the following: (1) Without exception, a fatty liver is more easily injured than a nonfatty one. (2) High liver glycogen content *per se* apparently has no protective value. (3) The harmful effects of a high content of fat in the liver may be partially or completely counteracted by a good intake of protein or by supplement of methionine or cystine, other amino acids being entirely without specific protective action. For unknown reasons, choline does not appear to protect the liver, but it does reduce kidney damage. (4) Even when the fat content of the liver is essentially normal, animals in a good state of protein nutrition can withstand exposure to halogenated solvents fatal to animals in a poor state of protein nutrition. In this instance, too, the metabolic defect is related specifically to the sulfur-containing amino acids. (5) Any protective effect of carbohydrate probably depends on a protein sparing action.

The effect of DDT on the nitrogen utilization of calves is discussed under 'Protein Metabolism' (see p. 108).

Rats, which have stored large amounts of DDT and are then starved or for any other reason caused to mobilize their deposited fat, may show characteristic DDT tremors. FITZHUGH and NELSON [208] found that when 3 rats which had been fed diets containing 600 ppm or more of DDT were deprived of food completely they showed marked tremors. Those deprived of all food after being fed 200 and 400 ppm DDT showed increased irritability. Tremors did not develop in any group when they were fed one-fourth the normal requirement of food after being removed from a diet containing DDT. Poisoning in birds is also accentuated by starvation (see p. 216).

It should be recalled that the metabolic rate of small mammals is high. When starved, they mobilize their stored food more rapidly and, therefore, lose a higher percentage of their weight per day than do larger animals, including man.

Age

It is well established that young animals are more susceptible to DDT than mature animals of the same species, regardless of the route of administration (NEAL *et al.* [427]; TAYLOR [578]; VELBINGER [632]; SAUBERLICH and BAUMANN [508]). This fact, as it applies to the respiratory exposure of mice, is dramatically illustrated by Figure 12 taken from NEAL *et al.* [427]. Young fish (HOFFMANN and SURBER [291]; DARSIE [146]), like young mammals, are more susceptible than adults of the same species. To be sure, a few (DEICHMANN *et al.* [162]) have failed to find young animals more susceptible, probably because an insufficient number of animals were tested.

In this connection, it appears likely that the fetus is relatively highly susceptible, although there are, of course, many instances recorded in which animals receiving moderate doses of DDT have delivered normal young. WASICKY and UNTI [649] reported that pregnant guinea pigs maintained for a long time on 7·5 mg/kg/day aborted but showed no other signs of injury. TELFORD and GUTHRIE [583] described a goat which suffered a toxemia of

pregnancy after being sprayed with 150 cm³ of 5 per cent DDT eleven times
each week for 12 weeks. Two kids, delivered by Caesarean section, exhibited
tremors and convulsions suggesting DDT intoxication and died very quickly.
To test the relationship, a second goat was sprayed with 150 cm³ of 10 per cent
DDT for 8 weeks. At that time she gave birth to 2 kids which also showed
tremors and convulsions and died within 48 hours, although the mother was
unaffected. The authors pointed out that the possible toxic effects of the
solvents in the sprays had not been excluded.

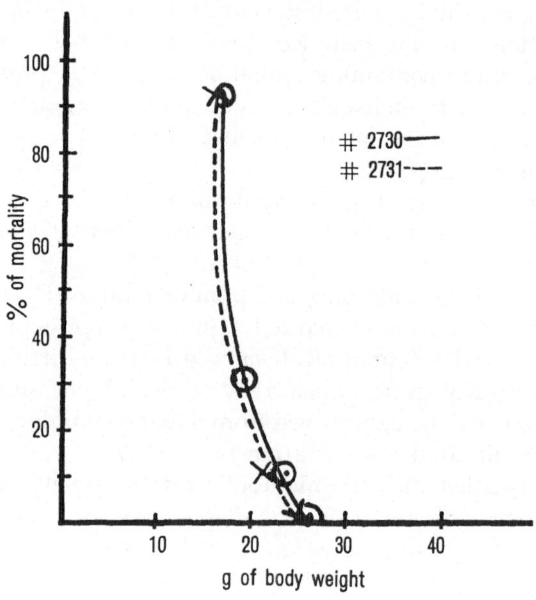

Figure 12

Mortality of mice in relation to their body weight following respiratory exposure to single high
concentrations of two aerosols. The figure is taken from NEAL *et al.* [427].

Less convincing was the abortion of a guinea pig which had received an
oral dose of olive oil solution at the rate of 125 mg/kg. Autopsy findings
suggested an infectious process (GUARESCHI and BINI [248]).

LAUG *et al.* [345] found that rats store DDT at the same rate regardless of
the age at which feeding is begun (see Figure 9).

Sex and Reproduction

Sex was found to exert little influence on the susceptibility of rats to a
single large dose of DDT dissolved in olive oil (DEICHMANN *et al.* [162]). On the
contrary, female rats are more susceptible to repeated doses than male rats are
(see p. 30, 35). It is not clear whether this difference exists in other species.

Male rats show much more frequent and extensive histological changes in the liver than female rats when both are exposed repeatedly at moderate dosage levels. The changes are peculiar to rodents.

All investigators who have studied the matter have found that female rats store more DDT in their fat than male rats fed at the same dosage rate (see Figure 10). The evidence for most species is not so clear. However, HARRIS et al. [265] found no difference in the DDT storage of male and female pigs. If a difference exists for monkeys or dogs, it is very small.

BURLINGTON and LINDEMAN [94] showed that DDT produced a striking inhibition of testicular growth and secondary sexual characters of cockerels. DDT was dissolved in chicken fat and administered subcutaneously daily at a rate which was increased from 15 mg/kg/day to 300 mg/kg/day. Testes of normal birds averaged 5 times as heavy as those of the experimental animals. The difference in weight was accounted for by a decrease in tubular development in the treated birds; intertubular tissue in the treated birds showed a relative and perhaps an absolute increase. The results were specific and not due to inanition, for the experimental animals grew at the same rate as the controls. The authors speculated that the DDT may have had an estrogen-like action, and called attention to the degree of similarity of the structural formulae of DDT and diethylstilbesterol.

FISHER et al. [206] demonstrated conclusively that DDT in total doses up to 45 mg has no estrogenic activity in ovariectomized rats, although an analog [2,2'-bis-(p-hydroxyphenyl)-1,1,1-trichloroethane] did show such activity in conformity with an hypothesis on the chemical nature of estrogens.

TAUBER and HUGHES [576] found that the cholesterol content of the ovaries of rats fed DDT was less than that of normal rats of the same age. Groups of rats were fed 0, 100, 300, and 600 ppm of DDT in their diet beginning when they were 30 days old. The rats were sacrificed when they were 90 days old. The ovaries of the controls showed 6,700 ppm of cholesterol, while the three experimental groups showed 3,900, 4,600, and 4,100, respectively. Using an organic chloride method, the authors found 364, 1,423, and 2,594 ppm of DDT in the ovaries of the different groups, the increasing values corresponding to increasing exposure. The authors speculated that the decrease of cholesterol might have some bearing on the formation of hormones, especially progesterone, in the ovary. No experimental evidence bearing on the point was presented.

FITZHUGH [207] reported briefly on the effect of DDT on reproduction. Rats fed diets containing 50, 100, and 600 ppm of DDT showed a progressive decline in the percentage of young successfully weaned, as compared with rats fed diets containing 0 or 10 ppm of DDT. However, the mortality of infant rats of mothers fed 50 and 100 ppm, although greater than the mortality of groups exposed to less DDT, was not greater than the mortality of infant rats in many laboratories. No effect on the number of rats born alive was evident in the first generation, but in the second generation rats fed 600 ppm produced very few living young, of which none survived the nursing period. Since 600 ppm is a toxic level for female rats, since young animals are especially susceptible, and since DDT is excreted in the milk at a high level, it is possible that the reported results do not indicate any specific effect of DDT on reproduction, but merely general systemic toxicity. This interpretation is consistent with the demonstration by DEICHMANN *et al.* [162] that the reproductive ability of surviving male and female rats was not affected by a single LD_{50} dose of DDT.

Effect on the Nervous System

The failure of DDT to have any marked effect on plants or on certain phyla of animals has thrown little light on the reason for its remarkable action on susceptible organisms. However, since the action, when present, is manifest largely or even exclusively through the nervous system, many studies have been made to discover the underlying cause of this relationship. Investigations in insects are reported in another chapter. In spite of the great differences between the two phyla, the effects of DDT on arthropod and chordate nervous systems have much in common: (1) Both central and peripheral effects have been observed. (2) No unifying theory is available which will explain the observed effects.

The signs of DDT poisoning in several species have been described in an earlier section. The most prominent signs are muscular tremor, incoordination, and, in severe poisoning, convulsions.

Morphological and Biophysical Studies

In one of the earliest communications on the subject, DOMENJOZ [172] observed that poisoning in animals resembles the disturbances of movement and tone such as are known in human pathology as the amyostatic symptom complex (this classification used by STRUMPELL includes paralysis agitans, progressive lenticular degeneration, and pseudosclerosis of WESTPHAL). Since the lesions responsible for these syndromes in man have been placed by neurologists in the central parts of the extrapyramidal system, the author implied that the effect of DDT might be in the same portion of the brain. As evidence that the effect of DDT is central, he offered the following: (1) Tremors can be stopped by curare, and, locally, by cutting the sciatic and femoral nerves. (2) Injection of 1 per cent procaine following spinal puncture at the level of the first lumbar disc stops the tremor in the hind extremities, while the peculiar hyperkinesis of the head and frontal extremities persists. (3) The local application of DDT formulations on the nerve, or intraneural injection of an oily solution is without any effect.

LÄUGER *et al.* [348], although agreeing that the primary effect of DDT was on the central nervous system, proposed that the point of attack lay in the sympathetic nervous system because they considered total mobilization of liver glycogen with a simultaneous hyperglycemia to be the earliest sign of DDT poisoning.

Because the signs are generally similar to those shown by decerebellate animals as described by various authors (e.g., FULTON and DOW [218]) it has been suggested (BING *et al.* [63]; HAYMAKER *et al.* [268]) that DDT selectively affects the cerebellum. In research attributed to PHILIP BARD (U.S. Chemical Warfare Service [603]) and to R. J. BING (HAYMAKER *et al.* [268]), the spinal cord of 3 cats was sectioned between the seventh thoracic and the second lumbar segments. Eight to 20 days later the animals received by stomach tube 300 mg/kg of DDT in peanut oil. Within 3 to 6 hours the cats showed tremors in the muscles innervated above the transection; no tremor appeared below the transection. Increased potentials were recorded from muscles above the cut but no potentials could be detected in the muscles below the cut. Three other cats were given DDT in the same way and, after tremor had developed, the brain stem was sectioned at the lower level of the midbrain; 10 to 20 minutes later the tremor recurred. It was concluded that DDT acts on the central nervous system below the midbrain and above the spinal cord and apparently does not act on the spinal cord, the myoneural junctions, or the muscles.

HAYMAKER and his coworkers [268] concluded, from the symptomatology of DDT poisoning and from the experiments attributed to BING which were just discussed, that the cerebellum is the part of the nervous system most affected by DDT. They supported this view by their own observations on histology which are reviewed in the section on pathology. They also believed that the findings of CRESCITELLI and GILMAN pointed to the predominant importance of the cerebellum; however, as will be shown below, this belief is not necessarily justified. It should also be observed that the findings of different pathologists regarding the effect of DDT on the nervous system have been very inconsistent.

DEICHMANN *et al.* [162] found that ablation of the cerebral hemispheres of rabbits, rats, and frogs, or of both the cerebrum and cerebellum of pigeons, failed to eliminate the hyperexcitability and tremor caused by DDT but destruction of the brain and spinal cord in frogs eliminated tremor. The recurrence of severe tremors in rats and rabbits was noted only in muscles innervated above the level of section in animals in which the spinal cord was cut usually just below the 5th cervical vertebra.

CRESCITELLI and GILMAN [143] observed alteration in the spontaneous electrical potentials recorded from the cerebral cortex and cerebellum of lightly anesthetized cats and monkeys following the intravenous administration of homogenized DDT emulsions. Unanesthetized animals in which movements were controlled by curare were also studied. Curare was injected intramuscularly at the rate of 4 mg/kg and additional doses of 1–2 mg/kg were given as required. In animals in which tremor and convulsion were suppressed by sodium pentobarbital (15–20 mg/kg) the cerebellar potential gradually increased in magnitude to a level 2 to 5 times normal while the pattern remained essentially unchanged except for a slight increase in frequency. The activity from the cerebral cortex increased slightly in magnitude and frequency but the chief effect, especially in cats, was to increase the regularity of the rhythm at a rate of 8–12 cycles per second. In nonanesthetized cats and monkeys fast waves appeared in the motor cortex and the pyramis vermis and portions of the lobulus simplex of the cerebellum. Within the limits of the recording instrument, the fast waves from the cortex and cerebellum appeared to be syncronous. The waves gradually increased in potential and frequency until characteristic periodic electrical seizures appeared in both the motor cortex and the cerebellum. The seizures were, in general, similar to those of *grand mal* epilepsy in man and to the tonic-clonic electrical manifestations induced in animals by convulsant drugs or electrical stimulation of the cortex. Within the cortex it was clear that the seizures began in the motor area and later spread to other areas. The seizures apparently arose in the brain independently of impulses arriving through the spinal cord, for the typical electrical manifestations were recorded in the cat after complete transection of the cord at the level of the atlanto-occipital membrane. The authors concluded that the attacks originated either in the cortex or in the cerebellum or in an unidentified group of neurones linked to both. No decision among these three possibilities was possible because of limitations of the method of recording and limitations of present knowledge of brain physiology. Although the study established that DDT causes a functional disorder arising in the brain, it did not, of course, rule out the occurrence of abnormality caused by DDT in other parts of the nervous system.

The paper just mentioned and many others dealing with the effect of drugs on the electrical activity of the brain have been reviewed by TOMAN and DAVIS [590].

PHILIPS and GILMAN [453], commenting on the experiments of CRESCITELLI and GILMAN, emphasized the predominant action on the motor cortex. They believed that the therapeutic effect of phenobarbital, a drug having a specific depressant action on the motor cortex, indicated the cortex as the point of action of DDT.

POLLOCK and WANG [464] fed undissolved DDT to cats at the rate of 300 to 500 mg/kg. On the following day most of the animals showed generalized fine tremors and marked ataxia but no *grand mal*-type convulsions. The animals were prepared for artificial respiration under divinyl ether anaesthesia, and electroencephalograms were taken after the anaesthesia had been discontinued and paralysis induced with 20 mg/kg of dihydro-β-erythroidine intravenously. In all the animals the amplitude of the cerebral cortical and cerebellar tracings was increased as compared with normal controls. Spikelike waves were seen but electrical fits were not observed. It was considered that the abnormal behavior of the cerebrum followed that of the cerebellum. The cats were made to breathe a mixture of 30% carbon dioxide and 70% oxygen for a few minutes. Some of them showed marked intensification of all electrical activity when the special CO_2–O_2 mixture was begun, but most of them showed a decrease in amplitude and increase in frequency first, and later showed the typical seizure pattern characterized by marked increase in amplitude when returned to normal air. In eight of the cats seizures were either observed in the cerebellar leads exclusively or appeared first in the cerebellum and then, after a short lag period, in the cerebral cortex. The authors failed to make clear what occurred in at least 24 other cats which were dosed for study. They considered synergism with carbon dioxide characteristic of compounds which act initially on the cerebellum. The hypothesis was offered that once the cerebellar convulsant has been introduced, the additional release from cortical inhibition by carbon dioxide permits seizures to start subcortically. The possible effect of increasing the oxygen tension was not discussed.

BROMILEY and BARD [88] found, in contradiction to earlier reports, that DDT can produce tremor in muscles innervated by the spinal cord and deprived of all neural connection with the brain. They also showed that DDT can produce a tonic-clonic convulsion in the decerebrate-decerebellate cat and in the decerebellate dog. All the studies involved acute DDT poisoning. The compound was given by stomach tube at the rate of 100 to 400 mg/kg in the form of a 10 per cent solution in peanut oil and intravenously at a rate of 60 mg/kg as an emulsion. The oral route was preferred because it was found that intravenous injection even of an emulsion containing no DDT produced neurological complications such as paralysis or tremor. In acutely decerebrate cats DDT produced tremor indistinguishable from that produced in the intact animal. The compound changed the reflex response from the response seen in unpoisoned decerebrate animals. The effect produced by DDT in the decerebrate cat was not essentially modified by removal of the cerebellum. In spinal animals, whether decerebrate or not, the tremor usually appeared first in the foreparts. In the presence of a slight but spontaneous tremor in the foreparts, the tremor might appear only reflexly in the hindparts in response to stimulation. With greater intoxication the tremor was spontaneous in all parts of spinal animals but was usually more marked above the transection. Sometimes, but not always, the tremor was coarser below the transection than above. Furthermore, DDT altered the reflex activity of the isolated lumbosacral cord, the exact effect differing somewhat in dogs and cats. The authors were careful to point out that their experiments have little bearing on the question of what part of the central nervous system is most susceptible to the action of DDT and chiefly responsible for the disturbances seen in the poisoned intact animal. The experiments do establish the functional effect of DDT outside the cerebrum and cerebellum.

An effect which was possibly similar was early observed by DOMENJOZ [172], but the phenomenon was interpreted differently. DOMENJOZ observed that if an appropriate amount of 1 per cent novocaine were injected intrathecally into poisoned rats and mice at the level of the first lumbar disc, the tremor in the hind legs stopped and then reappeared after 5 to 10 minutes in its original form. It is entirely possible that a segment of the cord remained blocked while reflex function returned in the caudal portion. (The cord in the rat extends to approximately the level of the fourth lumbar vertebra.)

104   W. J. Hayes, Jr.

TRIPOD [597], working with the frog, *Rana temporaria*, found that within the limits of the method which he used, DDT had an effect like strychnine on the sensory function of the nervous system and like phenol on the motor function. (The general similarity to phenol had been noted earlier by SMITH and STOHLMAN [532].) It is clear that in this reference to phenol the author had in mind a comparison of pharmacological action and not a suggestion that DDT acts by degradation to phenol. It was considered that, since curare removes the signs of DDT poisoning, an action of DDT directly on the muscle is excluded. Section of the femoral nerve removed signs in the affected leg, thus ruling out direct action of the compound (at the concentration existing in the tissue) on the motor nerve. Section of the dorsal roots removed signs of poisoning for a time but they returned after an hour or more. Tremors remained following section of the spinal cord below the bulb. The authors concluded that the point of attack was the spinal cord.

That transection of the cord or, in fact, decapitation of the frog was consistent with the reappearance of tremor after a period of shock was discovered independently by DRESDEN [181]. In the same paper, characterized by originality and clear presentation, the author offered convincing evidence that the frog, *Rana esculenta*, following subcutaneous injection of DDT at the rate of 60 mg/kg, shows facilitation of synaptic transmission in the spinal cord. The facilitation was followed by blocking which was at first dependent on activity and reversible and was later spontaneous and irreversible. This observation of synaptic facilitation followed by blocking was made the basis for an all-inclusive explanation of the symptomatology and cause of death in DDT poisoning. The arguments advanced in support of this interesting theory appeared a bit labored to the reviewer. The tendency to discount the objective findings of other workers simply because they failed to support the theory appeared particularly unjustified. The author's preoccupation with finding a single toxic effect of critical importance led him, in general, to avoid the use of higher concentrations of DDT. Under the limitations adopted, the following negative results were reported: Action potentials from the sciatic nerve indicated no stimulating or inhibiting influence on the proprioreceptors or tango-receptors of the frog. No increase or decrease in the threshold of conduction of the peripheral nerve or myoneural junction or in the reactivity of the muscle could be found. No effect of poisoning on the electroencephalcgram of the frog was observed; readings were taken from brains completely isolated from the body. Similarly, poisoning produced no change in the spontaneous electrical pattern of the spinal cord following transection of all the peripheral nerves. The results reported for the conduction of nerve are similar to those of DEICHMANN *et al.* [162] but are contradictory to those reported for insects and crustacea, and contradictory to the results of most workers for the rat. The results reported for reaction of the muscle and for the electroneurogram in the frog are contradictory to the results for the rat.

Following the demonstration of the veratrinic effect on the peripheral nerve of insects (ROEDER and WEIANT [494], [495]) and crustacea (WELSH and

GORDON [657], [658]) exerted by DDT, EYZAGUIRRE and LILIENTHAL [201] investigated the same phenomenon in the rat. The animals were injected intraperitoneally with DDT at the rate of 50 to 100 mg/kg using an homogenized emulsion. DDT produced in muscle a myotonic response similar both functionally and pharmacologically to the myotonia occurring spontaneously in man and goat. The response consisted of an increase in tension and duration of the twitch following a brief stimulus applied either through the sciatic nerve or directly into the curarized muscle. The electrical basis of the increased mechanical response was seen in the substitution of a train of spikes for the normal diphasic electroneurogram. A similar response followed mechanical stimulation by tapping. The response was seen in the isolated or intact nerve. Repeated stimulation at the rate of 12 per minute produced progressive waning of the myotonic response; this is the phenomenon of 'warm-up'. The effect of DDT on the nerve was augmented by potassium ions, and suppressed by quinine and by magnesium and calcium ions. The nerve showed an increased rate of recovery following stimulation. The authors commented on the difficulty of forming an hypothesis to explain the fact that a wide variety of compounds which produce repetitive responses in excitable tissue also generally induce the same altered state of excitability in other tissues of the same organism. They mentioned the attractive working hypothesis (WELSH and GORDON [657]; GORDON and WELSH [242]) which emphasizes the primary role of free and surface-bound calcium as a modulator of excitability in nerve. This matter of the role of calcium is discussed in the section on medication (p. 118).

Biochemical Studies

Biochemical studies of the nervous system in connection with DDT intoxication have not proved very illuminating, except for those related to the action of calcium. JANDORF *et al.* [309] found no change in the oxygen consumption, aerobic and anaerobic lactic acid production, or in the utilization of pyruvic acid by cerebral and by cerebellar homogenates. The brains were taken from rats which had received daily oral doses of DDT to which they responded by marked tremors lasting for only a part of each day. The rats were sacrificed at a time when the tremors were absent.

RIKER *et al.* [488] also failed to find any effect of repeated ingestion of DDT on the oxygen consumption of the rat brain.

JUDAH [313] likewise found no effect of DDT on the aerobic respiration of brain slices in the presence of substrate and no effect on anaerobic glycolysis. At a very low concentration of DDT ($2 \times 10^{-5} M$) and in the absence of substrate an increase in oxygen consumption up to 200% was recorded. No explanation could be offered; the increase was not caused by oxidation of the DDT. Since DDT produces a marked inhibition of anaerobic glycolysis of the liver and muscle, it was thought that the failure of this response by the brain might be explained by poor penetration of the compound into the brain slices. The hypothesis was tested by means of plain brain homogenates and homogenates fortified by ATP and coenzyme I, but the results obtained with brain slices were confirmed in each instance.

It has been shown that, during the late prostrate stage but not during the early hyperactive phase, the nervous systems and especially the connectives of the fly and cockroach and, to a smaller degree, the crayfish show an increase in free acetylcholine. The free ester is derived at the expense of the bound acetylcholine which often disappears entirely. DDT does not cause any increase in the actual synthesis of acetylcholine nor any inhibition of its action (TOBIAS *et al.* [589]). The same workers were unable to find any effect of DDT on the acetylcholine of the frog or rat brain, or on the rat submaxillary gland.

PRAJMOVSKY and WELSH [468] found that about 70% of the acetylcholine in normal rat spinal nerves was water-extractable and the ratio of free compound to total acetylcholine was constant. As in insects, acute DDT poisoning converted almost all the material to a water-extractable form, but there was no change in the total amount of acetylcholine.

No inhibitory action of DDT on brain or serum cholinesterase could be detected (LÄUGER *et al.* [350]). The same observation was made for the serum cholinesterase of the horse *in vitro* and of the guinea pig *in vivo* (TRUHAUT and VINCENT [599]; VINCENT and TRUHAUT [634]). This negative result was confirmed and extended to the specific cholinesterase of horse red cells and sheep brain *in vitro* (TRUHAUT and VINCENT [600]; VINCENT *et al.* [635]). A similar conclusion is justified on different grounds by the work of HOFFMAN and LENDLE [285]. It should be pointed out that the failure of DDT to inhibit cholinesterase is not inconsistent with the fact observed in some organisms that its action is augmented by acetylcholine or eserine and reduced by atropine.

That DDT is not transported by the peripheral nerves or their associated lymphatics was strongly suggested by the investigation of HOFFMAN and LENDLE [285].

Effect on the Liver and Metabolism

The finding that histological changes were more prominent in the liver of animals poisoned by DDT than in their other organs early led to study of the physiological effect of the compound on the liver and on carbohydrate metabolism.

Carbohydrate Metabolism

The earliest studies (LÄUGER *et al.* [348], [349]) indicated that, in the face of acute poisoning, the liver glycogen is quickly mobilized and later exhausted so that in some animals there is an early hyperglycemia followed by hypoglycemia. Increased but incomplete oxidation permits an increase in the basal metabolic rate but results in an increase of circulating lactic acid and a compensated acidosis. Although it has been shown experimentally (LÄUGER *et al.*

[349]) that some poisoned animals can be saved by glucose injections, it does not follow that disturbance of the carbohydrate metabolism represents more than a minor contributing factor in the cause of death. Similarly, various liver function tests fail to indicate sufficient abnormality to account for death or serious illness. LÄUGER *et al.* [348], [349] examined blood from a comatose poisoned animal and found the blood sugar level to be greatly depressed. To study this matter, they used an oil solution of DDT and poisoned dogs with 1,000 and 2,000 mg/kg. The blood sugar and blood lactic acid were determined at frequent intervals. In each instance the blood sugar increased from a normal concentration of about 100 mg% to something over 200 in 7 to 8 hours and then fell to a level below 50 mg% just before death. The concentration of lactic acid in the blood increased gradually during the entire poisoning period. The urine remained free of sugar and ketone bodies. In other poisoned animals it was found that although the pH of the blood was lowered, the amount of lowering was not great enough to explain the appearance of symptoms. The alkali reserve was reduced but not seriously so. The liver of many rats was found to be free of glycogen 2 hours after they received 1,000 mg/kg by mouth as an oil solution; all were glycogen-free in 6 hours. The authors felt that this loss of glycogen was not caused by motor activity, which was slight or absent at 2 hours, but was caused by sympathetic stimulation or the release of adrenalin.

KOSTER [332] reported that glucose given before or 1·5 hours after an intravenous LD_{33} dose of DDT reduced convulsions and mortality in cats. Tremors were reduced when the glucose was given before the DDT. Glucose increased the survival time but had no other effect when an LD_{95} dose of DDT was employed.

When DDT in corn oil was given orally to rabbits in well-tolerated doses of 300 mg/kg, the elevation in blood sugar and simultaneous drop in rectal temperature were approximately the same as the corresponding changes produced by the oil alone. However, a large dose of 600 mg/kg in a moderate amount of oil which produced severe symptoms and death within 18 hours caused a significant rise in the blood glucose and this rise occurred several hours earlier than a rise caused by the ingestion of oil (STOHLMAN [558]; STOHLMAN and LILLIE [560]). In general, when DDT was given by mouth, the maximum severity of the symptoms corresponded to the maximum elevation of blood sugar and maximum drop in rectal temperature. DDT injected intravenously produced hyperglycemia but, unlike the oral administration, resulted in a rise in body temperature. These authors (STOHLMAN and LILLIE [560]) were unable to confirm the antidotal effects of glucose in acute or chronic DDT poisoning either in terms of decreased mortality, increased survival time, or significantly decreased loss of body weight. Glucose did appear to reduce to some extent the severity of symptoms and the extent and frequency of histological lesions in the liver, kidneys, and spleen.

JUDAH [313] determined the blood glucose and blood lactic acid of acutely poisoned rats and rabbits. Some animals developed a marked hyperglycemia while others failed to do so despite pronounced convulsive seizures. The author felt that though hypoglycemia might contribute to a fatal end in any given animal, it did not explain death. Rats were found to be far more resistant than rabbits to derangement of the blood sugar level. In spite of the nearly normal blood sugar levels of some poisoned rats, it was found that the liver glycogen of all of them was exhausted. The effect is not direct, for in studies with tissue slices marked inhibition of anaerobic glycolysis in liver and muscle was found. The aerobic respiration of these tissues was little affected.

DEICHMANN *et al.* [162] found a marked hypoglycemia in all rats shortly before they died following oral doses of 1,000 to 3,000 mg/kg; early in poisoning, a mild hyperglycemia was present. No significant changes in blood sugar were found in rabbits which received a dosage apparently insufficient to cause convulsions or death.

No significant decrease in liver glycogen occurred in rats receiving 700 ppm of DDT in the diet for 36 to 40 days, although the regimen did decrease the growth rate of the animals and increase the weight and lipid content of the liver (SARETT and JANDORF [505], [506]).

In general, the changes in blood glucose and blood lactic are similar in acute DDT poisoning to changes seen following the administration of some other convulsant poisons (BÖMER [81]).

Protein Metabolism

BOHMAN *et al.* [79] presented figures which indicate that the rate of nitrogen storage in calves fed a diet containing 10·2% of protein was not altered by the feeding of DDT up to 75 ppm in the same diet. The authors confirmed earlier demonstrations that normal calves are able to utilize nitrogen added to a low protein diet in the form of urea. They showed that DDT interfered with the utilization of urea nitrogen presumably by interfering with the microorganisms of the rumen which have been shown to convert the urea to a usable form. There was no evidence that DDT interfered with the digestion or utilization of preformed dietary protein or with the digestion or organic matter, ether extract, cellulose, lignin, crude fiber, or other carbohydrates.

Metabolic Rate

Indication that the basal metabolic rate of poisoned animals is increased appeared early. LAUG and FITZHUGH [344] observed that rats receiving high concentrations of DDT in their diet grew more slowly than their litter mate controls although, unless severely poisoned, they ate more food. The authors did not confirm this indication of increased BMR in the intact animal by direct measurement, but did make tissue respiration studies. Liver slices taken from animals receiving 800 and 1,200 ppm of DDT in the diet showed an average decrease of 40% in their oxygen consumption as compared with controls.

It should be recalled that rats receiving DDT in the diet may eat considerably less than control animals even in the absence of overt poisoning (SARETT and JANDORF [506]) or the amount of food consumed may be essentially unchanged (FITZHUGH and NELSON [208]).

In more extensive experiments published later in the same year, RIKER and his colleagues [488] confirmed the findings of LAUG and FITZHUGH regarding the basal metabolic rate of the intact poisoned animal and clarified the question of the oxygen consumption of liver slices. The authors found that rats which had received DDT at the rate of 1,000 ppm in the diet consumed more food while losing weight. In taking the BMR, the rats were anesthetized so that tremors were abolished. The average basal metabolic rate rose sharply; it reached a peak 23% above the control on the third day and remained at about the same level through the eighth day, when the experiment was discontinued. On that day liver slices from the poisoned animals showed an oxygen consumption 22% above the control level. In a second experiment liver slices taken from rats poisoned in the same way but, on the average, for a longer period showed an average oxygen consumption of 11% above the control level. The authors attributed this difference to variations which they purposely introduced in their Warburg technique. However, if one analyzes their data for individual animals in terms of the relationship between the percent of change in oxygen consumption and the days of exposure to DDT, it appears possible that this time relationship may be important. This would be consistent with the authors' stated conclusion that the oxygen consumption of liver slices taken from portions of liver which appeared grossly altered in chronically poisoned animals was reduced as compared with controls. In another experiment, rats were sacrificed when symptoms became pronounced following the intravenous injection of a freshly homogenized 2 per cent emulsion at a rate of 50 mg of DDT per kilogram of body weight. The oxygen consumption of liver slices taken from these rats showed an insignificant difference from control values.

In the same issue of the same journal, JANDORF and his colleagues [309] reported similar experiments. They confirmed that liver slices taken from rats which received DDT in their regular diet for short periods showed an increased oxygen consumption. The average increase was 13% over controls in rats maintained for 1 to 10 days on a diet containing 2,000 ppm. There was no correlation between the values for oxygen consumption and the presence or absence of tremors. The oxygen consumption was increased even after one day of exposure to DDT. The same authors found no change in the oxygen consumption of liver slices taken from rats which had received 50 mg of DDT per kilogram of body weight in corn oil for 30 to 50 days. This again is consistent with earlier findings regarding the effect of repeated doses.

Enzymes

No effects of DDT on enzyme systems have been discovered which are sufficient to explain the mode of action of the compound.

LÄUGER *et al.* [350] found that DDT did not influence aminoxidases, diaminoxidases, or lipase systems but did inhibit phosphatidase (an enzyme associated with lecithin, cephalin, and sphingomyelin) and muscle phosphatases.

HOFFMAN and LENDLE [285] found in an exploratory experiment that DDT in fairly high doses does not limit the reduction of the nitro group in *o*-dinitrobenzene to the hydroxylamine group by frog muscle. The authors also found DDT has no specific inhibiting effect on catalase.

JUDAH [313] found no effect when DDT was tested against the following enzymes: aldolase, adenosine triphosphatase of brain and liver; choline oxidase, glutamic acid dehydrogenase, glycolytic cycle in brain homogenates; hexokinase, lactic dehydrogenase, malic dehydrogenase, pyruvic oxidase, succinoxidase, trios phosphate dehydrogenase, creatine phosphokinase, and aerobic phosphorylation. In a second paper, it was reported that DDT had no significant effect on the uptake of oxygen or inorganic phosphate by the kidney (JUDAH and WILLIAMS-ASHMAN [314]).

DDT and a variety of other convulsant agents were found to inhibit carbonic anhydrase (TORDA and WOLFF [591]), oxalacetic carboxylase, pyruvic carboxylase (TORDA and WOLFF [592]) and to increase the activity of cytochrome oxidase (TORDA and WOLFF [593]) and succinic dehydrogenase (TORDA and WOLFF [594]).

JOHNSTON [311] found that DDT inhibited rat heart succinoxidase at concentrations of from 10^{-4} to $10^{-5}\,M$; the degree of inhibition was about 70 to 90% at the higher concentration. This inhibition was also demonstrated for cytochrome oxidase but not for succinic dehydrogenase. The DDT was active when added to the enzyme in alcoholic solution so that a colloidal suspension of the insecticide was formed in the reaction vessel. Only slight inhibition occurred when DDT at 10 times the concentration was added as an oil emulsion in essentially the same manner as that used by JUDAH.

DEMUTH and LENDLE [167] found no effect on the muscle dehydrogenase of the frog.

Other studies of enzymes are discussed in the section on the nervous system (p. 105).

Liver Function Tests

The earliest reports of a liver-function test in animals poisoned by DDT was that of SMITH and STOHLMAN [532] who found an abnormal retention of injected rose bengal in rabbits receiving dermal applications of DDT in the form of a solution in dimethylphthalate. However, the same authors (SMITH and STOHLMAN [533]) found no abnormality by the same test in most cats showing a well established neurological syndrome of DDT poisoning and no abnormality in rats which had received DDT for 80 days. Some of the cats and rats showed fatty degeneration of the liver at autopsy.

NEAL *et al.* [426], [427] investigated liver function by measuring the bile pigments, urobilinogen, and urobilin of the urine and the bromsulphalein retention, cephalin-cholesterol flocculation, prothrombin time, and icterus index of the blood or serum. The subjects included human beings receiving massive exposure to aerosols, dogs receiving pure DDT by insufflation into the nostrils or orally in capsules, dogs exposed to DDT aerosols by the respiratory route or to aerosol residues by the dermal route, and monkeys exposed to air-borne aerosols. Of all the subjects, apparently only one dog showed any abnormality referable to the liver. While receiving DDT by insufflation into the nostrils at the rate of 100 mg/kg/day, the animal developed an increase in prothrombin time and icterus index. Both values returned to normal in spite of continued exposure to DDT. When the dog was killed, the liver showed grossly a marked cirrhosis.

Bing *et al.* [63] investigated hepatic function of 15 dogs following DDT poisoning by means of quantitative determination of serum bilirubin, the bromsulfalein test, and the formaldehyde-gel test. They considered the cephalin-flocculation test useless because of the high rate of positives in normal dogs. The formaldehyde-gel test which indicates alteration of the serum-globulin fraction of the blood revealed some degree of hepatic disfunction in 7 out of 10 animals and generally became positive in from 2 to 3 weeks after administration of the DDT was initiated. Four of 9 dogs showed a slight rise of serum bilirubin shortly before they died; the 5 animals which failed to show a rise were sacrificed and might otherwise have shown a terminal change. Significant retention of bromsulfalein dye did not occur in 10 dogs poisoned by DDT. Similarly, no change was found in the renal plasma flow as determined by the clearance of p-aminohippuric acid, or in glomerular filtration as determined by creatinine clearance, or in the maximal tubular excretory mass as determined by the p-aminohippuric acid TM even though the dogs were followed until persistent neurological symptoms had developed. The same studies have been mentioned very briefly by McNamara *et al.* [392].

Liver Size and Liver Lipids

Increased weight of the liver of poisoned animals has been observed (Laug and Fitzhugh [344]; Fitzhugh and Nelson [208]; Sarett and Jandorf [505], [506]). Hydration does not account for the change because there is an increase in the dry weight of the liver (Laug and Fitzhugh [344]; Sarett and Jandorf [506]). In an attempt to account for the increase in weight as well as histological changes, studies have been made of the liver constituents.

Sarett and Jandorf [506] found that chronic intoxication of rats with DDT led to an increase in the ether-soluble fraction of the liver. Studies were made in which the DDT was administered by stomach tube as a corn oil solution. In other studies, DDT without oil was added to the regular diet and in this instance an increase of 40% over the control value was observed in the weight of the liver. In the same livers, no increase in the percentage of water or glycogen was found. The increase in the phospholipid moiety was entirely proportional to the increase of the total ether-soluble fraction. Although the cholesterol showed an absolute increase proportional to the increase in liver size, the proportion of cholesterol in the total lipid fraction showed a slight but statistically significant reduction. Although the increase in total liver lipids was statistically significant it could hardly be said that the animals had fatty livers, the percentage based on wet weight being 5·7 and 5·5 in two groups, respectively, as compared with 4·5% in the controls. The addition of 0·2% choline to the diet in addition to DDT was without effect.

Miscellaneous Observations

Among various animals studied by JUDAH [313] were 4 rabbits and 2 rats which showed more severe symptoms and died more rapidly than expected. On autopsy, they were found to have extensive, long-standing liver damage. As an experimental procedure 12 rats were given subcutaneous injections of 2 ml carbon tetrachloride per kilogram and, 24 hours later, 50 mg of DDT per kilogram of body weight intravenously. These experimental rats survived longer and showed less severe symptoms than controls which received the same dosage of DDT. Autopsy showed that the experimental animals had suffered destruction of 30 to 90% of the liver tissue. Analysis showed that DDT had not been differentially trapped by the necrotic liver cells. The author remarked on the paradoxical effects of chronic and acute liver damage on DDT poisoning but was unable to offer an explanation.

Effect on Urinary and Blood Findings

NEAL et al. [426] found a small but consistent increase in the volume of urine excreted in 24 hours when dogs were dosed orally or by insufflation at the rate of 100 mg/kg/day. No other change in the urine and no change in kidney function was demonstrated. Apparently, other authors have not investigated the total urine output of animals exposed to DDT. The matter would bear further study, especially in relation to the finding of TORDA and WOLFF [591] that DDT inhibits carbonic anhydrase. The presence of carbonic anhydrase in the kidney was demonstrated by DAVENPORT and WILHELMI [148]. The physiology of this enzyme in the kidney and other organs has been reviewed by DAVENPORT [147].

JUDAH [313] found no change in the excretion of sulfur or glycuronic acid which he could attribute to DDT poisoning. No change in the excretion of amino acids was revealed by chromatography. Rats poisoned by intraperitoneal or intravenous injection of DDT showed no ketones, reducing substances, or protein in the urine. However, the total urinary nitrogen increased following intravenous injection of the poison at the rate of 25 mg/kg. The excretion of nitrogen increased rapidly for about 2 days and then fell. During the period of increase, the rats showed tremors and lost weight in spite of having free access to food. NEAL *et al.* [426] did not observe an increase in nitrogen excretion in dogs receiving DDT orally or by insufflation at the rate of 100 mg/kg/day.

In acute poisoning, the blood calcium may be normal or high (CAMERON and BURGESS [105]; JUDAH [313]). The blood potassium was almost trebled in rabbits receiving an acute intravenous lethal dose (JUDAH [313]) and this high potassium level may explain the cardiac irregularity observed by PHILIPS and GILMAN [453].

Serum proteins were normal 2·5 hours after symptoms appeared in lethally poisoned animals (JUDAH [313]).

NEAL *et al.* [426], [427] failed to demonstrate any definite change in the complete blood count of human beings, dogs, or monkeys exposed to DDT by a variety of routes.

DRAIZE *et al.* [179] reported that animals receiving dermal applications showed a moderate leucocytosis with a definite increase in the percentage of heterophiles. Serious anemia was not encountered except in moribund animals. DRAIZE *et al.* [178] reported a similar result for rabbits but found that rats, after unstated exposure, occasionally showed a fall in hemoglobin and a decrease in the absolute leucocyte count.

CAMERON and BURGESS [105] found that large dermal applications of DDT caused changes in the blood picture but that small repeated doses caused only slight changes. With a large dose, rabbits showed a decrease in hemoglobin but no change in the red cell count. Abnormal red cells were not found. Leucocytosis appeared usually on the second or third day after exposure. The increase, which concerned chiefly the neutrophilic cells, was considered an early sign of toxicity. Except for the increase in white cell production, no change was seen in the bone marrow.

Similar changes, including anemia, definite anisocytosis, inconstant poikilocytosis, and neutrophilic leucocytosis, were observed by TARSITANO [569] in guinea pigs poisoned by massive oral doses of DDT.

McNAMARA *et al.* [392] reported that repeated doses of DDT at the rate of 150 to 300 mg/kg produced a fall of hemoglobin in dogs without significant reduction in the red cell count.

NEAL *et al.* [428] reported on a man who voluntarily ingested 770 mg of DDT. No changes in the hemoglobin or in the blood cells were observed.

VELBINGER [631] found that oral doses of 500 to 1,500 mg in man did not cause either an immediate or a delayed leucopenia. Rather, there was in most instances a slight leucocytosis which was viewed as a defense mechanism. A decrease in both red cells and hemoglobin was observed. The hemoglobin dropped 15 to 18% in one day. Similar results were reported for cats (VELBINGER [632]).

DEICHMANN *et al.* [162] reported that changes in the hemoglobin or in the formed elements of the blood are not always present even in severe poisoning. Furthermore, NEAL and VON OETTINGEN [425] suggested that when a decrease in hemoglobin is present it may be the result of impaired nutrition and not the specific effect of DDT.

Effect on Respiration and Circulation

Death from DDT poisoning has most frequently been attributed to respiratory arrest. DEICHMANN *et al.* [162] recorded the respiratory movements of poisoned rabbits kymographically. A marked increase in the frequency and amplitude of respiration began about 5 to 40 hours after the dose was given and corresponded with the onset of hyperirritability. Later, with the occurrence of tremors, the depth of respiration frequently returned to a more normal level, but the rate remained high. In some animals respiration stopped suddenly after a deep inspiration during a fatal tonic convulsion. In other animals the rate and amplitude decreased progressively and finally ceased without any terminal spasm.

PHILIPS and GILMAN [453] reported that the majority of dogs killed by DDT died of ventricular fibrillation and that some fatally poisoned rabbits, cats, and monkeys died by the same mechanism. Detailed study of the phenomenon appeared in the same issue of the journal (PHILIPS *et al.* [455]). In dogs anaesthetized by sodium phenobarbital, fibrillation was produced in 2 of 3 which received DDT intravenously at the rate of 75 mg/kg and in 6 of 8 which received the insecticide at the rate of 100 mg/kg. The fibrillation was fatal to all of the animals which developed the arrhythmia; death followed the intravenous injection of epinephrine at the rate of 0·01 mg/kg in 5 dogs and at the rates of 0·015, 0·02, and 0·04 mg/kg in each of 3 dogs, respectively. Even larger dosages of epinephrine failed to cause any fibrillation or other unexpected abnormality in the electrocardiogram of control dogs which had previously been injected intravenously with an emulsion entirely similar to that used to carry the DDT except that it contained no insecticide. Additional studies were made to see whether the heart, sensitized by DDT, could be thrown into ventricular fibrillation by sympathetic discharges associated with a convulsion. For this purpose, dogs and monkeys were completely paralyzed by slow intravenous injection of curare (Intracostrin 4·0 mg/kg) and maintained by artificial respiration. Simultaneous electroencephalographic and electrocardiographic tracings were made. Of 6 dogs given 75 or 100 mg of DDT per kilogram of body weight, 3 died of ventricular fibrillation within 10 minutes after the injection of DDT and before a typical cortical electrical seizure appeared; 2 died of ventricular fibrillation induced by a seizure, and 1 dog failed to develop fibrillation even though it showed repeated seizures. One of 2 monkeys which received 75 mg of DDT per kilogram of body weight showed a marked cardiac arrhythmia during convulsive seizures, but no fibrillation, while the other monkey developed typical ventricular fibrillation during seizures. The monkey differed from the dogs studied in that the heart was able to recover from fibrillation and to resume a normal rhythm.

Besides fibrillation, premature systoles and changes in the T-wave, frequently involving inversion in all leads, were recorded in dogs in the absence of seizures.

It was concluded by the authors that DDT not only shares with other hydrocarbons and chlorinated hydrocarbons a tendency to sensitize the myocardium but also, through its action on the central nervous system, produces the stimulus necessary for the onset of ventricular fibrillation.

The effect on the heart is similar to that of the chlorinated hydrocarbon, chloroform, and some other anaesthetics (MEEK [401]).

Cardiac arrhythmias have also been observed in acutely poisoned rabbits (JUDAH [313]; DEICHMANN et al. [162]). Some authors, however (DEICHMANN et al. [162]), have attached no significance to the phenomenon, attributing it to mechanical stimulation incidental to experimental procedures.

An increase in the cardiac output associated with a decline in arteriovenous oxygen difference has been reported by McNAMARA et al. [392] for dogs which received daily oral doses at the rate of 150 to 300 mg/kg. The oxygen consumption of these animals showed no change, as compared to controls, although other workers have usually found an increase.

Vasomotor response to autonomic stimulation in rabbits and blood pressure and kidney blood volume in a dog were not affected until a few minutes before death resulting from single large doses of DDT (DEICHMANN et al. [162]).

HOFFMAN and LENDLE [285] studied the effect of DDT on the peripheral vessels and on the heart of the frog. Prolonged perfusion of a colloidal suspension of DDT at a concentration of 1 ppm in saline did not affect the sensitivity of the preparation to adrenalin but did cause periodic contractions of the vessels recognizable as a retardation in the rate of flow. At intermediate concentrations adrenalin sensitivity was interfered with and the effect was not reversible by washing out with RINGER's solution. With higher concentrations (100 ppm) spontaneous constrictions became so marked and constant that it became practically impossible to test the effect of adrenalin. The constrictions were not reversible by RINGER's solution.

The same authors found that DDT had a direct effect on the perfused frog heart. A seasonal difference associated with temperature was demonstrated. At low room temperature in December, a concentration of 300 ppm of DDT caused a decrease in the height of contraction with simultaneous rise in the basic point and, finally, the appearance of irregularities. At temperatures of at least 22°C in June, a slight decrease in stroke was suggested at a concentration of 1 to 0·1 ppm, and it was definite at a concentration of 10 ppm. The effect was entirely reversible by washing out with RINGER's solution, and the sensitivity of the preparation to acetylcholine remained normal throughout. The isolated frog rectus muscle placed in a saline bath with a concentration of 50 ppm of DDT showed no sign of contracture or fibrillary twitching and its sensitivity to potassium was unaffected.

These results were not confirmed for the turtle or rabbit heart (DEICHMANN et al. [162]), but it is doubtful whether the experimental technique used for these species was sufficiently refined.

                    W. J. Hayes, Jr.

Medication

SMITH and STOHLMAN [532] noted the possibility that narcotics in general may exhibit an antagonism to DDT. Rats survived on a diet containing 1,000 ppm of DDT for 90 days when they received cyclohexanone in the same diet at the rate of 2,000 ppm; rats were uniformly killed in a shorter period when they received DDT at the same rate but without cyclohexanone. Later, it was shown that cyclohexanone offers no protection when used as a solvent for single massive doses of DDT (DEICHMANN et al. [162]).

The possibility that narcotics may be antidotal was further investigated by SMITH and STOHLMAN [533] who showed that urethane and, to a lesser extent, sodium dilantin protected rats from poisoning. Sodium amytal gave slight benefit, sodium phenobarbital a doubtful benefit, and paraldehyde no protection at all. Experimental and control rats of both sexes were deprived of food for 18 hours and then given a uniform dose of DDT in oil by stomach tube (300 mg/kg in all but one test). The dose was sufficient to induce tremors within 3 to 5 hours. After signs of intoxication appeared, the drug to be tested was given in sufficient dosage to control the tremors and the treatment was repeated as often as necessary for as much as 1 to 3 days. Surviving animals were observed for 7 additional days. An exception arose from the fact that it proved impossible to control the tremors with paraldehyde even with doses approaching the toxic level. All drugs were given intraperitoneally except paraldehyde which was given by stomach tube. The mortality of rats treated with urethane was 12·5% and that of the controls was 80%. A total dosage of 1·2 to 2·5 gm/kg, spread over a period of 1 to 3 days, was found most satisfactory. Sodium dilantin gave a mortality of 46·7% as compared to 96·7% for the controls. The smallest effective dosage was 200 to 250 mg/kg, a value very close to the LD_{50} which is 300 mg/kg.

A slight reduction of mortality and a slight increase in the survival time of those which died were observed among rats poisoned by DDT and treated with large doses of luminal (sodium phenobarbital) (LÄUGER et al. [348], [349], [350]).

Drugs with sedative, depressant, or anticonvulsant properties were also investigated by PHILIPS and GILMAN [453]. Acutely poisoned rats, cats, dogs, and monkeys were used. The authors found phenobarbital by far the most outstanding remedy tested. In a dosage well below the anesthetic level it not only prevented death in many instances but also controlled tremor and convulsions. Signs of illness were more readily controlled in dogs and cats than in monkeys which required nearly a full anesthetic dosage before tremors completely disappeared. Magnesium sulfate did not reduce mortality although it did control tremors and convulsions briefly. Sodium bromide was entirely ineffective. Mortality was reduced with urethane but a full anesthetic dosage was required to control tremor and convulsion. Similarly, sodium barbital and sodium pentobarbital controlled symptoms only when given in full anesthetic doses and even then did not greatly reduce mortality. Dilantin when given to rats before they received DDT reduced the lethal action without showing a notable effect on the signs of poisoning; dilantin was not effective in cats.

Urethane has also been shown to oppose the effect of DDT in goldfish. The addition of 2,000 and 1,000 ppm of the compound to water containing a lethal concentration (0·2 ppm) of DDT reduced the mortality to 30 and 70%, respectively (MICKEY *et al.* [409]).

Intoxication by DDT and especially by certain other chlorinated hydrocarbon insecticides may be prolonged and require extended medication. For this purpose, the barbiturates would appear to be the drugs of choice among the sedatives and narcotics because of their effectiveness in higher animals and because of their wide margin of safety.

VAZ and his colleagues [630] were apparently the first to note the antidotal effect of calcium in DDT poisoning. Dogs were given DDT orally as a 10 per cent oily solution at a daily dosage of 100 mg/kg until signs of intoxication appeared. The same dosage could then be repeated to produce intense symptomatology from which the animals would recover spontaneously in 12 to 24 hours. For the actual tests a larger challenge dosage of DDT (150 to 200 mg/kg) was used. 30 ml of a 10 per cent solution of calcium gluconate was injected intravenously at each dose into dogs weighing 8 to 18 kg. Dogs which were injected with calcium gluconate daily for 4 days and challenged with a large dose of DDT on the fourth day developed no symptoms or only slight ones. Dogs receiving a single dose of calcium gluconate showed symptoms of short duration and survived following a dosage of DDT large enough to kill 2 control dogs. The authors made no measurement of blood calcium but assumed that their results demonstrated that DDT causes death by reducing blood calcium.

KOSTER [332] studied cats poisoned by the intravenous injection of a soya lecithin-corn oil emulsion of DDT. A comparison was made of several aspects of intoxication including number of convulsions, general severity (tremors, prostration, dyspnea), duration, and mortality. Calcium gluconate reduced mortality but not severity. However, sodium gluconate reduced mortality and, to a slight extent, convulsions, but not severity. Gluconic acid reduced mortality, but not convulsions or severity and it increased the survival time. Calcium chloride reduced convulsions, but not mortality or tremors. Molecular equivalent doses of the candidate antidotes were used. Gluconic acid and its 2 salts were effective against an LD_{95} dosage of DDT.

The effect on DDT-poisoned rats of intravenous injection of calcium gluconate at a rate of 40 mg/kg was investigated by JUDAH [313]. Three groups of 6 experimental animals were used along with an equal number of controls. The first group received calcium gluconate 1·5 hours after the intravenous injection of 50 mg of DDT per kilogram of body weight. No amelioration of symptoms was observed up to 1·5 hours after injection of the calcium gluconate but all the experimental animals survived whereas the controls died. In the second experiment the same dosage of calcium gluconate and of DDT were given but the calcium was given prophylactically a few minutes before the DDT. The medicated animals developed only mild symptoms and appeared almost normal 3 hours after the injection when the controls were moribund. In the third test, the same dosage of calcium gluconate failed to protect rats against 100 mg of DDT per kilogram of body weight given intravenously a few minutes later. Both the experimental and control animals died within 90 minutes. JUDAH found normal blood calcium values in most of the poisoned but unmedicated animals which he studied. One animal showed a high value. This finding was similar to that of CAMERON and BURGESS [105] who reported high blood calcium values in some animals. It has been suggested that increased blood calcium may be associated with acidosis caused by the accumulation of lactate. It is interesting that no significant variations in the calcium or potassium content of brain or muscle were found in the few observations which were made (JUDAH [313]).

Calcium has, then, an antidotal action against DDT in intact animals of several species. The suppression of the effect of DDT on the isolated nerve and muscle of the rat has been demonstrated (EYZAGUIRRE and LILIENTHAL [201]). The hypothesis has been advanced (WELSH and GORDON [657]; GORDON and WELSH [242]) that certain neurotoxins, including DDT, act by delaying the restoration of calcium ions to a surface complex, following breaking of the chelate linkage of calcium ions to surface polar groups by an initial exciting impulse. This action of the neurotoxin is conceived as depending largely on its physical rather than on its chemical properties. The hypothesis is helpful in explaining the fact that a wide variety of chemically unrelated compounds produce repetitive responses in excitable tissue and also the fact that many compounds that show a high toxicity for arthropods and mammals are fat soluble and chemically relatively inert. It has been pointed out that this hypothesis postulates a very localized action of calcium at the nerve-cell membrane; the hypothesis is not inconsistent with the finding that the blood calcium of poisoned animals may be unchanged or even increased.

Somewhat related is the hypothesis of LÄUGER *et al.* [350] that DDT has an affinity to cholesterol of the lipoid membrane of the nerve cell.

Having observed the effect of DDT on the metabolism of glucose and glycogen, LÄUGER and his colleagues [348], [349], [350] investigated the use of glucose as an antidote. All of 10 dogs given 2,000 mg of DDT per kilogram of body weight orally in the form of an oil solution died within 8 to 24 hours. Five of 10 dogs treated with 1 or more 20 ml doses of 20 per cent glucose survived the same dosage of DDT. The glucose was given intravenously in most instances. Adrenalin altered the character of the symptoms but did not influence the final outcome.

KOSTER [332] found that glucose given before or after an LD_{33} dosage reduced convulsions and mortality and, when given before the poison, reduced tremors, prostration, and dyspnea in cats. Glucose, unlike gluconic acid and its sodium and calcium salts, was ineffective against an LD_{95} dosage except to increase the time of survival. Insulin given intramuscularly 16 to 25 minutes before DDT increased the survival time and severity of poisoning but did not· affect mortality or convulsions. When given 53 to 130 minutes before DDT, insulin reduced convulsions in animals which died but increased convulsions, tremors, and other disorders in the survivors.

STOHLMAN and LILLIE [560] administered glucose orally, intravenously, and intraperitoneally to rabbits during a time when the animals were receiving repeated oral doses of DDT in oil. When the animals died, lesions in their organs were found to be less frequent and less extensive than in controls. The authors stated that the animals which received glucose did not survive, on the average, as long as DDT-dosed controls but they failed to connect this fact with the degree of histological change in the organs.

The failure of amino and sulfhydryl compounds to influence the action of DDT has been noted (VON OETTINGEN and SHARPLESS [641]). Likewise, the addition of 0·2% choline chloride to the diet of rats receiving repeated doses of DDT had no effect on the accumulation of lipids in their liver (SARETT and JANDORF [506]). Desoxycholic acid was not antidotal (LÄUGER *et al.* [350]).

A wide variety of other materials have been found to have no effect on the course of DDT poisoning as observed in aquaria studies with fish and frogs (HOFFMAN and LENDLE [285]). They include thiourea, cysthione, boviserine (serum protein), peptone, alanine, tyrosine, tryptophane, histidine, urotropin, phenylene-diamine, glycine, glucose, malonic acid, nicotinic acid, i-nicotinic acid, pyridine, nucleic acid, and sulfanilic acid.

7.

TOXICITY OF DDT TO MAN

Our direct knowledge of the toxicity of DDT to man is based on (1) intentional experimental exposure, (2) accidental poisoning, and (3) use experience.

Experimental Exposure

Oral

DOMENJOZ [172] stated that a man weighing 74 kg took 250 mg of pure DDT 3 times a day for 3 days without noting any effect on his well-being. It was found that, even after vigourous shaking, the compound imparted no recognizable taste or odor to drinking water.

MACCORMACK [387] reported eating 1,500 mg of DDT in butter without ill effect. Lice were killed when they were experimentally fed on his body 6 and 12 hours after the dose was taken. Lice fed 36 hours after the dose were not injured.

LAZAR [351] reported that an army officer, as the result of a bet, ate six pancakes in which DDT powder had been used in place of flour. The officer suffered no ill effects. This result is inconsistent with the results of tests done under conditions more conducive to scientific accuracy.

NEAL *et al.* [428] recorded an experiment in which a volunteer took 500 mg (7·1 mg/kg) of DDT in olive oil and 9 months later took 770 mg (11 mg/kg) in the same way. No subjective changes were noted. Control studies on the blood and urine and a detailed clinical examination including a teleoroentgenogram, an electrocardiogram, an electroencephalogram, one liver-function test, one blood sugar determination, and a variety of tests of nervous function were made during the 2 weeks before the last dose. The same studies were repeated after the DDT was ingested, but no changes were detected.

VELBINGER [631], [632] has reported on experiments in which he and two other healthy young men engaged. The pure substance was taken in milky suspension or in codliver-oil solution by mouth after a meal. The doses were 250, 500, 700, 1,000, and 1,500 mg. The studies extended over a 12-week period. A dose of 250 mg in milky suspension was almost without reaction except for a slight disturbance of sensitivity in the mouth. An oily solution of the same strength caused a far stronger, but variable hyperesthesia. A dose of 500 mg in oil given 4 weeks later produced no essentially different picture of poisoning. The symptoms were considerably greater after 750 mg. Disturbance of sensitivity involved the whole lower part of the face, and the gait became uncertain so that the volunteer walked with a reeling motion which he found especially unpleasant. The first reaction came in about 3 hours, The peak was reached in 6 hours after intake of the poison. At that time there was general malaise, the skin was moist and cold, the pupillary reactions and other reflexes were normal, and there was hypersensitivity to contact stimuli.

Twenty-four days later, 1,000 mg was administered in the same way. Surprisingly enough, there was no essential difference between the course of severity of the reaction between the last two doses (750 and 1,000 mg). One subject reacted after 2 hours, the other after 5 hours. It was specifically noted that joint pains, fatigue, fear, or difficulty in seeing or hearing were not present.

More severe reaction followed the ingestion of 1,500 mg of DDT. Two and a half hours after taking the compound, the volunteer noted prickling in the tip of the tongue, upper lip, and around the chin corresponding to the hyperalgesic zone described below. The area of paraesthesia gradually extended to the nostrils. Four to five hours after the dose, there was the first disturbance of equilibrium, and this was followed by dizziness, confusion, and tremor of the extremities. The different symptoms fluctuated somewhat in intensity. The peak was considered to come about 10 hours after the poison was ingested. At that time there was great malaise, headache, and fatigue as well as the symptoms already noted. The volunteer fell asleep but awakened about 11 hours after the dose and vomited actively. He experienced a sense of relief and thereafter slept for 12 hours. On the following day, he was in general good condition with only a slight remaining paraesthesia and disturbance of equilibrium.

Careful medical examination during the period of most severe symptoms showed the pupils equally dilated but normal in reaction to light and near vision. There was slight nystagmus. Sensitivity to touch and pain was exaggerated around the mouth but depressed in the remaining distribution of the cranial nerve V. The function of cranial nerves VII, VIII, and XII was normal. Movement of the arms was free, but there was extensive tremor. The finger-to-nose test was hesitating and uncertain but without tremor. The reflexes were not remarkable. There was difficulty in standing on one leg, and it was possible only for a few seconds with lively efforts to retain balance. Other difficulties of coordination were objectively noted. There was an hyperalgesia of the dorsum of the foot and the distal half of the lower leg.

On the following day, the only objective sign was a barely demonstrable increase in the sensitivity of the face and some incoordination in the finger-to-nose test.

Clinical tests were made in 18 to 24 hours and up to 70 days after the different doses of DDT. There was never any leukopenia. Conversely, in most cases, a leukocytosis was present; this was interpreted as a nonspecific defense mechanism. No constant deviation in the differential white count was observed. A decrease in hemoglobin was seen (15 and 18 per cent decrease following doses of 500 and 1,000 mg respectively). The pulse varied with the degree of poisoning. The blood pressure, temperature, and sedimentation rate remained normal.

Neither protein, sugar, urobilin, nor indican appeared in the urine, and urine output remained normal.

The author concluded that DDT as commonly used in vermin control is not dangerous to man, as long as he takes in no great amount.

The experiments of Domenjoz [172], Neal *et al.* [428], and Velbinger [631], [632] and the careful observations especially of Garrett [219] and Hsieh [306] make it possible to state the acute oral toxicity of DDT to man rather accurately. A single eating of 10 mg/kg produces illness in some but not all subjects even though no vomiting occurs. Smaller dosages generally produce no illness, although a dosage of 6 mg/kg produced perspiration, headache, and nausea in a man who was already sickly and who was hungry at the time of eating (Hsieh [306]). Those who have shown illness following ingestion of 10 mg/kg have not shown convulsions but convulsions have occurred frequently when the dosage level was 16 mg/kg or greater (Hsieh [306]). Rarely, a dosage as high as 20 mg/kg may be taken without apparent ill effect (MacCormack [387]). Dosages at least as high as 285 mg/kg have been taken without fatal result (Garrett [219]). However, large doses lead to prompt vomiting, so that the amount actually retained cannot be accurately determined.

According to Hoffman and Lendle [285], weak concentrations of aqueous colloidal DDT suspensions are tasteless, while a saturated alcoholic solution has a weak aromatic taste, or rather, odor. Some persons find that an alcoholic solution has a slight anaesthetic effect on the tongue.

Dermal

Domenjoz [172] was among the first to report tests on human beings. Gauze bandages encrusted with the pure substance were attached to the upper arm of fifteen persons for 10 days. In no instance was there any local reaction or systemic toxic effect. In a second experiment, eight women wore stockings impregnated with DDT at the rate of 0·4 to 0·5 g per pair. The stockings, which were retreated after each washing, were worn for 24 days. The women showed no evidence of local or resorptive effect.

Other investigators have found that DDT in different bland formulations causes no irritation of the skin (Wasicky and Unti [648]; Draize *et al.* [179]; Haag *et al.* [252]). In fact, Hoffman and Lendle [285] reported that the subcutaneous injection of colloidal suspensions in saline in concentrations up to 30 ppm of DDT caused no irritating effect.

Cameron and Burgess [105] reported on a group of 52 soldiers who wore clothing impregnated with 1% of DDT (dry weight basis) for 18 to 26 days. Some of the men wore the garments without change for the whole period and took exercise to insure that they became hot and sweaty so that the conditions for absorption were ideal. Some of the men were given a change of underwear once a week, but the clothing was all treated. A few of the men showed a transient dermatitis which cleared while they were still exposed to DDT and which was thought to come from other causes. There was no indication of systemic effect at any time. The blood and urine of 12 of the soldiers who were examined in this regard remained normal. Laboratory workers, some of whom had extensive exposure, showed no abnormality attributable to DDT.

ZEIN-EL-DINE [685] reported that DDT-impregnated clothing caused a slight, transient dermatitis. The method of impregnation was not stated.

One of the most provocative reports (WIGGLESWORTH [665]) concerned a laboratory worker about 30 years of age whose systemic disease was attributed to DDT poisoning. The subject voluntarily exposed himself to acetone solutions of the compound in order to determine whether it would cause any local irritation. He allowed small quantities of the solution to evaporate on the back of the hand and then wiped off the residual with absorbent cotton soaked in acetone. Later, he kneaded inert dust to form a slurry with a solution containing 25 g of DDT. Again the compound was removed by swabbing with acetone. A sense of heaviness and aching in all the limbs and weakness in the legs developed. Nervous tension was prominent. The account does not make it clear whether the onset was gradual over a period of 10 days immediately following exposure or was relatively sudden at some unstated time within 10 days of the exposure. In any event, the patient's condition improved during a short vacation but became much worse subsequently so that he took to his bed for 10 to 14 days. During this period he could not sleep. He suffered severe pain and acute mental anxiety. Muscular tremors over the whole body occurred about 6 days after he went to bed. The patient missed 10 weeks from work. Recovery was very slow and was not entirely complete in a year. No sign of local skin irritation ever appeared as a result of the exposure. It is reported that neither the patient nor his physician at first suspected a relationship between the systemic disease and the exposure to DDT. Later, the patient's condition 'immediately began to improve when he realized that his was almost certainly a case of poisoning by DDT'. The report makes no mention of any objective signs of illness or of any attempt at differential diagnosis.

DANGERFIELD [145] took exception to the conclusion of WIGGLESWORTH and suggested that an anxiety neurosis may have been present instead of DDT poisoning. DANGERFIELD reported that 6 volunteers underwent exposure somewhat more severe than that described by WIGGLESWORTH. The volunteers developed no toxic or irritant effect at all.

CASE [120] reported a very specialized experiment which has subsequently been submitted to the most generalized interpretation by other authors. Those who have associated illness with some exposure to DDT have recalled, as suited their purposes, the leukopenia, or the fall in blood pressure, or some other finding reported by CASE. Those who have failed to confirm his findings in completely different experiments have called attention to the highly artificial surroundings of the tests. Briefly, he was obliged to determine whether it would be entirely safe to cover interior surfaces with a paint containing 2% of DDT in the dry film in circumstances where the paint would inevitably be covered by a thin film of oil, where the temperature and humidity would be high, and where the military personnel exposed would be largely unclothed and their skins would be oily and sweaty. The circumstances under which an area so painted might be used did not permit regular bathing but required the highest efficiency of the personnel both for their own safety and for the success of their mission. The test situation was a steel chamber with a narrow steel seat around the inside and a small bench in the center. The chamber was maintained at 25–30°C and at a relative humidity of 88 to 94%. Two men were exposed for 48 hours after the chamber had been covered with a paint which contained no DDT. The chamber was then covered with the DDT paint and aired for 48 hours following which the men were again exposed for a 48 hour period. In each instance, the paint was covered with a film of oil. The men wore only shorts. By turns they sat with their backs against the painted walls or slept on the steel seat. Careful clinical observations, including neurological examinations, were made at intervals throughout the control and the experimental periods. Most of the changes noted after exposure to the DDT paint were subjective. Smarting of the eyes was caused by an unidentified substance, probably, according to the author, the chlorinated phenol used as a preservative in the paint. Both subjects suffered tiredness, heaviness, aching of the limbs, extreme irritability, a great distaste for work, and a feeling of mental incompetence. Slight changes in reflexes, hearing, and sensation, and a slight tremor were consistent with fatigue. Indoxyl sulfate appeared in the urine. There was a fall in systolic blood pressure, a decrease in the mean corpuscular hemoglobin, an increase in the siderocytes in the circulation, and a transient polymorphonuclear leukopenia. Certain other minor changes were claimed. Return to normal required 26 to 33 days. The author concluded that the paint should not be used under the conditions mentioned. He was careful to state that the experiments did not form a basis for any condemnation of the widespread use of DDT.

CHIN and T'ANT [123] applied small pads impregnated with different formulations of DDT to the inner surface of the forearm of 32 volunteers whose cutaneous sensation had previously been measured for a period of 5 weeks. Pads impregnated with all the elements of the formulation except DDT were applied to the corresponding position of the other arm as a control. Powdered DDT and 5 per cent solutions of DDT showed little effect. Ten and 20 per cent solutions in olive oil and in petrolatum showed no remarkable effect on sensation of pain, cold or heat, but reduced tactile sensation in most cases so that the minimal pressure which could arouse the tactile sensation was 1 to 2·5 g/cm² higher than in the control.

Respiratory

Like the work on dermal exposure, experiments on respiratory exposure were begun early. NEAL and his colleagues [426] reported detailed observations on two subjects exposed to aerosols. One 42- and one 54-year-old man were subjected to a calculated intermittent exposure of 35 μg of DDT per liter of air for 1 hour daily, for 6 days. During the hour, 10·4 g of aerosol were released into the exposure chamber of 14,750-liter capacity every 15 minutes. The aerosol consisted of 5% DDT, 10% cyclohexanone, and 85% Freon (difluoro-dichloro methane). In a second series of tests, the same amount of aerosol was released every 5 minutes during the hour of exposure on each of 5 succeeding days. With this more severe exposure, both subjects suffered moderate irritation of the nose, throat, and eyes starting 5 to 10 minutes after the beginning of each exposure and increasing moderately thereafter. The concentration was so great, however, that a white deposit of DDT formed on the nasal vibrissae of both men. One of the men was stripped to the waist, and the arms and shoulders of the other were bare. Except for the irritation already mentioned, neither of the subjects had any symptoms and there were no aftereffects. Laboratory tests and physical examination, including careful neurological evaluation done before and after the experiments, failed to reveal any significant changes.

FENNAH [203], in order to determine what consequences would result from frequent and indiscriminate use of DDT, daily inhaled 100 mg of pure DDT and drank water dusted at the rate of 3,240 mg/m². This exposure was continued for 13 months except for two interruptions totaling 1·5 months. At a later date the same individual received oral and dermal exposure. For a period of a month all food was sprayed after it had been prepared and exposed in dishes; DDT was applied as a 3 per cent emulsion and at a rate of 1,080 mg/m² over the surface of the food. In another experiment the arms and hands were repeatedly swabbed with DDT emulsion so as to leave a deposit of over 2,160 mg/m². No ill effects of any kind were observed.

Accidental Intoxication

Uncomplicated Poisoning

THOUNG [586] reported one of the earliest examples of what may have been uncomplicated DDT intoxication. Seventy-two men of a Frontier Force Constabulary suffered food poisoning on July 29, 1946. At least 27 of them began to be sick about one hour after the evening meal. All of them were sick on the following day when they were seen by the District Health Officer, Bhamo, Burma, who reported the matter to the author. The symptoms mentioned in the brief account were vomiting, diarrhea, a pulse rate of 40 to 50, giddiness on getting up, and dilatation of the pupils. Presumably all the men recovered, but the author failed to clarify the point. He also failed to mention any of the nervous signs and symptoms which later experience has shown to be characteristic. Two samples taken from a stock of rice used for the soldiers were analyzed for DDT; one showed 16% while the other showed none. No explanation for the divergence was given. No study to exclude the common causes of food poisoning was reported. It would appear impossible, therefore, to determine from the data given whether any of the men ate DDT.

MÜHLENS [416] reported that an interned soldier swallowed 3 to 4·5 g of DDT in the form of louse powder mixed with milk, sugar, and water. He became ill after an hour. He vomited, was restless, and suffered a headache. The heart was weak and slow but the patient recovered the next day.

MACKERRAS and WEST [388] reported an incident in which about 25 soldiers ate tarts in which, through error, DDT had been substituted for baking powder. After 1 to 2·5 hours, all of the men felt weak and giddy and 4 of them vomited. Two of the 25 were affected severely enough to be hospitalized. Both vomited 4 times but showed no diarrhea, perhaps because morphine was used in treatment. One staggered and collapsed as a result of muscular weakness and incoordination after a period of confusion but he felt well next day. His pulse rate during illness was 100. The second man was dizzy and felt palpitations of the heart and a slight numbness of his hands. Neither patient showed any tremor or convulsions and both were completely recovered in 48 hours.

The same authors reported the case of a man who got DDT powder in his eyes. He suffered pain requiring morphine and cocaine for 4 days and was unable to see for 2 weeks, but he recovered completely. In this instance, it is probably impossible to distinguish any specific toxicity which DDT may have exerted from the mechanical irritation of the dust. The authors observed that there was no reason to believe that DDT is more dangerous than many other insecticides in common use.

NAEVESTED [418] reported the illness of three young men who had eaten pancakes in which, by accident, louse powder had been substituted for meal. It was estimated that one man consumed 6 g and the others 5 g of DDT. They thought that the cakes tasted good but had noticed a gritty texture undoubtedly caused by the diluent in the insecticidal powder. Two to three hours after the meal, the men began to feel sick. They developed throbbing headaches, dizziness, incoordination, paraesthesias of the extremities, and an urge to defecate. All of the victims showed wide nonreacting pupils, reduced vision, dysarthria, facial paresis of a peripheral type, tremor, ataxic gait, reduced sensitivity to touch, reduced or absent reflexes, a positive Romberg sign, slightly low blood pressure, and persistently irregular heart action. Gastric lavage and other appropriate treatment was given. The symptoms receded in 2 to 3 days. Four to five days after the DDT was eaten, there was a slight jaundice with bile pigments in the urine; this lasted 3 to 4 days. When the men were examined 19 days after poisoning, there was nothing pathological except for irregular heart action in one case. Two other men who were thought to have eaten about 2 g of DDT did not become sick and showed only dilatation of the pupils.

GARRETT [219], [220] has reported vividly one of the most instructive accidents involving DDT. During World War II some Formosan military prisoners refused to carry out their assigned duties and they were punished by being deprived of their evening meal. About midnight they stole a box of powder which they thought was flour. Actually, the powder contained 10 per cent DDT with flour as the diluent. The prisoners used this dusting powder to make a dough which they cooked over a little gasoline burner. Mess kits were used for the baking and the men ate varying amounts up to as much as a whole pan filled with the half-baked dough. It was thought that 5 men ate as much as 20 g of DDT each. Twenty-eight men with varying degrees of poisoning were observed. It was estimated that the contaminated food was eaten 2 to 3 hours before the patients were first seen. It was further estimated that symptoms appeared in those most severely poisoned in 30 to 60 minutes after eating. When first seen 20 of the men were apprehensive and excited; the respiration was moderately rapid but the pulse was only 45 to 60 per minute. Severe vomiting had developed before the patients were seen and undoubtedly served as a protective mechanism. The vomiting made it impossible to estimate accurately the amount of DDT retained but the severity of poisoning was proportional to the quantity of dough which the men said they had ingested. Numbness and partial paralysis were most evident in the most distal portions of the extremities and the intensity was directly proportional to the amount of DDT ingested. Mild convulsions were observed in those most severely affected. Proprioception and vibratory sensation were diminished or lost in the fingers and toes but not in the more proximal joints. The knee jerk was hyperactive in 8 patients. Urinalysis and blood count were normal. Diarrhea was not present. There were no deaths. Treatment consisted of the use of emetics, phenobarbital, and other symptomatic medication. Within 48 hours only 8 men were suffering any ill effect. At the end of 2 weeks 3 patients still had weakness of both hands and feet. Five weeks after ingesting DDT these 3 had not recovered full use of their hands. At that time they were transferred by military order and thus were lost to science.

JUDE and GIRARD [315] described the poisoning of about 100 young women, all of whom had eaten in the same canteen at noon on July 16, 1946. The first symptoms appeared near 3.30 o'clock in the afternoon and consisted of gastro-intestinal difficulties including vomiting, abdominal pain, and later, diarrhea and nervous disturbances including a sensation of chilling and of prickling of the fingers. These symptoms were sufficiently severe in 37 of the ladies that they were hospitalized the same day and 4 others were hospitalized on the following day. About 50 had milder symptoms including headache, colic, and an ill-defined malaise which caused them to remain in their rooms. None of the patients were seriously ill. Of the 37 hospitalized on the 16th of July, 30 returned to their quarters the following day. Investigations showed that all of the patients had eaten vegetables or a tart or both, flour from the same source having been used to make the sauce for the vegetables and the dough for the pastry. The 'flour' had been taken from a container bearing the partially erased label: *'Insecticide Powder'*. More than 6 months earlier good flour had been placed in the partially empty container and had been used without incident. When the layer of insecticidal powder on the bottom was finally reached the poisoning occurred. Chemical analysis established the presence of DDT and showed that the concentration was 8·9%, corresponding to a mixture of 9 parts of 10 per cent insecticide and 1 part flour. The diluent in the insecticidal powder was aluminium silicate. It was not possible to estimate the amount of DDT which the young ladies had eaten but it was supposed, on the basis of their symptoms and the observations of others (NAEVSTAD [418]), that the dose was considerably less than 5 to 6 g.

STERNE [553] reported that a family ate cakes which had been prepared by mistake with DDT powder in place of flour. In another instance a nursing formula was made up with DDT powder in place of powdered milk and given to an infant. The only untoward effect in each instance was vomiting but there was no information on the dosage except that it was obviously large.

FRANCONE *et al.* [212] reported fourteen cases most of which were uncomplicated by any solvent. The dosage of DDT was not determined in any instance but those persons who ate DDT powder showed the same signs and symptoms which have been described in greater detail by other authors. The cases of illness described by FRANCONE and his colleagues associated with only trivial exposure may have represented unrelated, intercurrent infection. Some of them showed fever which is not characteristic of DDT poisoning.

HSIEH [306] has reported an accident involving an entire family of eleven members. The signs and symptoms were characteristic. The chief interest of the paper lies in the accuracy with which the different dosages were determined and this feature has been reviewed above.

Not all cases of uncomplicated DDT poisoning result in recovery. It must be emphasized that DDT can cause death. For example, a despondent 32-year-old woman said that she had taken 10 per cent DDT powder. She died in 13 hours after unstated treatment. Postmortem examination revealed congestion of the lungs and stomach. Analysis revealed a concentration of 180 ppm of DDT in the liver and 420 ppm in the stomach (Committee on Pesticides [132]). A summary table in the same paper lists at least 22 other cases new to the literature in which DDT (with or without solvents) was ingested. Three were fatal. Of the total, 15 were attempted suicide.

Poisoning Complicated by Other Agents

A great number of reports involve human poisoning by formulations containing other toxicants in addition to DDT. At least 3 sorts of cases have been described: (1) direct systemic poisoning, (2) systemic reactions based on allergy, and (3) dermatitis. Systemic intoxication has been ascribed to dermal absorption and inhalation as well as to ingestion. In addition, a certain number of cases ascribed to DDT poisoning undoubtedly represent unrelated organic or psychic disease.

In one instance a 32-year-old laborer who had been seen alive and well one morning was found dead lying face downward an hour later. Vomit found in two places nearby contained insecticide and necropsy done on the day of death revealed half a pint of fluid which included partly digested food and smelled strongly of insecticide. On the basis of analysis it was estimated that the victim had retained in his stomach about 180 ml of a concentrated DDT emulsion, a stock of which was found near the body. This accounted for a dosage of 500 mg DDT per kilogram of body weight in addition to the amount which was vomited and to an unknown amount which may have been aspirated or absorbed from the gastrointestinal tract. The retained emulsion included 72 ml of methylcyclohexanone as well as emulsifying agents. The only pathological findings except abrasions incurred in falling were pulmonary edema, dilatation of the stomach, and congestion of the stomach and the upper end of the intestine. The man's reason for drinking the formulation was unknown; he had suffered stomach trouble and insomnia 3 weeks earlier and had received a diagnosis of 'functional dyspepsia'. The authors (BIDEN-STEELE and STUCKEY [62]) after a review of certain literature, concluded that death must have been due primarily to the DDT. They were apparently unaware of the rapidity with which death frequently follows the aspiration of solvents into the lungs.

In a similar way HILL and ROBINSON [279] failed to distinguish the effect of DDT from that of kerosene used as a solvent. A 19-month-old infant drank about 30 ml of a 5 per cent solution, a dosage of about 150 mg of DDT per kilogram of body weight, and immediately began to cough and vomit violently. Within an hour and a half the child became comatose and had some sort of convulsion which was not witnessed by a medically trained person. The coma continued and the child died 4 hours after drinking the solution. Autopsy revealed vomit in the trachea, edema and hemorrhage of the lungs, and an odor of kerosene in the stomach; other findings were normal or insignificant. Pulmonary edema was considered the cause of death. The authors performed two experiments on baboons in each of which both animals received an equal dosage of kerosene, but only one received DDT dissolved in the kerosene. In the first experiment the animal which received DDT was slightly affected but recovered. In the second experiment the monkey which had received DDT at the rate of about 469 mg/kg died in 1 hour and on autopsy showed pathology similar to that of the child. The authors concluded that DDT had been the cause of the infant's death.

HILL [278], while admitting the toxicity of kerosene, argued that the observed death was caused by DDT because certain experimental animals withstood a larger volume of kerosene than the volume of a 5 per cent solution of DDT in kerosene necessary to kill similar animals. It would appear, however, from the clinical course and the pathology that the infant was moribund from kerosene poisoning before the more slowly acting DDT had had time to take effect.

The conclusion of HILL and ROBINSON was specifically criticised by BALABAN [40]. Attention was called to the fact that kerosene alone is frequently the cause of fatal accidents. In another specific criticism, PRATT-THOMAS and WARING [469] pointed out that the clinical symptoms and pathological findings reported by HILL and ROBINSON could in no way be distinguished from those resulting from the sudden aspiration of kerosene alone. The experiments on baboons were not designed so that a true distinction could be made between the toxicity of DDT and the toxicity of kerosene.

In spite of these criticisms, the views of HILL and ROBINSON were considered valid by N. J. SMITH [537], who described the fatal poisoning of a 58-year-old laborer who drank 120 ml of a 5 per cent solution of DDT which he followed by an unknown amount of beer. Besides DDT and kerosene, the formulation also contained xylene and Lethane (a thiocyanate). Vomiting began within an hour and continued intermittently until death. Part of the vomitus was bright red, and the patient passed two tarry stools. No urine was passed later than 2 hours after the poison was taken. In spite of these alarming symptoms, the patient did not seek medical attention until 6 days after drinking the formulation. At the time he entered a hospital where he had several tarry stools and repeated episodes of hematemesis. The liver was tender and enlarged. Signs or symptoms characteristic of DDT poisoning were not described. He died 30 hours after admission. Necropsy showed extensive necrosis of the liver, a toxic tubular degeneration of the kidneys, bronchopneumonia, and a variety of chronic conditions including syphilis and an old duodenal ulcer. While admitting the importance of the solvents, the author considered DDT to be the factor of prime importance.

It should be pointed out that in certain regions, kerosene is a common household material. It is used for fuel or for starting wood or coal fires. It is also frequently burned in lamps as a source of light. A number of reports (BOLOGNA and WOODY [80]; LUECK [383]) have indicated that, especially in children, kerosene intoxication is the most common single kind of accidental poisoning in those regions where kerosene is generally used. Dermal absorption of kerosene is not significant for systemic poisoning under ordinary conditions although dermal exposure is frequently the cause of a localized dermatitis.

Kerosene may be absorbed through the respiratory tract or through the gastrointestinal tract. The inhalation of the fumes in closed or poorly ventilated spaces may lead to fullness in the head, headache, blurred vision, dizziness, unsteady gait, and nausea. More massive exposure may cause collapse, nervous twitching, and coma apparently before the victim can become aware and seek fresh air (VON OETTINGEN [638], [640]; BROWNING [90]).

When taken orally, kerosene frequently leads to immediate and violent gagging and coughing and thus to aspiration of the oil. It has been known to produce pulmonary edema and death within a period as short as 2 hours (NUNN and MARTIN [439]). More commonly, in severe cases pneumonia develops and the patient dies more slowly or recovers after a prolonged illness. Survival has been reported following the ingestion of 1 l (SOLLMANN [541]) but death has followed a dose as small as 30 ml (NUNN and MARTIN [439]). This wide variability can be explained largely if not entirely by the decisive importance of aspiration. The aspiration of a very small quantity is highly dangerous, whereas large doses are relatively harmless provided no aspiration whatever takes place. Evidence supporting this interpretation may be drawn from clinical experience, from pathology, and from animal experiments (WARING [646]; NUNN and MARTIN [439]; LESSER *et al.* [364]; REED *et al.* [479]).

Clinically, respiratory dysfunction is prominent following aspiration although chest signs are likely to be few or absent even when X-ray reveals an extensive bronchopneumonia. In severe cases, liver and kidney damage may be manifest by hepatomegaly and by albumin, cells, and casts in the urine. Systemic action may also involve the central nervous system and produce signs and symptoms already mentioned in connection with respiratory exposure.

Autopsy frequently reveals irritation of the mucosa of the stomach and upper intestine and generalized visceral congestion. The characteristic pathology of kerosene poisoning, however, is a generalized bronchopneumonia with acute pulmonary edema and hemorrhage.

Animal experiments indicate that, when large doses are given, enough kerosene may be absorbed into the blood from the gastrointestinal tract to produce lung lesions even though aspiration is entirely prevented. The intravenous injection of kerosene causes the most severe damage, with complete loss of architecture of the lungs (DEICHMANN *et al.* [161]; RICHARDSON and PRATT-THOMAS [487]). In spite of these facts it remains true that experimental animals tolerate large doses of kerosene if aspiration is avoided but that small aspirated doses lead to severe illness or even death (RICHARDSON and PRATT-THOMAS [487]).

As has been seen, large doses of DDT alone frequently produce vomiting within an hour or so after ingestion. Kerosene, on the other hand, tends to cause coughing and vomiting immediately on ingestion. Probably a kerosene solution of DDT is somewhat more likely to be vomited and aspirated than is kerosene alone. For this and other reasons it is clear that, volume for volume, a DDT solution in kerosene is more dangerous than kerosene alone.

In an interesting case reported by STERLINGER [552], a 9·5-year-old boy drank what was estimated to be no more than a teaspoonful of a 5 per cent DDT solution in kerosene. He vomited immediately. Next morning he complained of pain in the stomach and coughed a little. It was not until the third day that he was hospitalized. He complained of headache and right upper quadrant pain. On examination, he showed extreme prostration, hyperpnea, and tenderness over the liver which was not enlarged. The temperature was 39°C, the pulse 110. The urine showed albumin, red cells, pus, and casts. The serum bilirubin was normal. The disease was characterized by exacerbations at night, but recovery was gradual and complete on the fifth day. The absence of neurological involvement and the striking visceral involvement especially of the lungs and kidneys suggests that the solvent was chiefly responsible for the illness.

Some authors have specifically recognized the predominant part played by solvents in the toxicity of dilute liquid formulations of DDT. Thus, REINGOLD and LASKY [484] ascribed the cause of death to kerosene poisoning in a case involving the ingestion of approximately 150 ml of a commercial solution containing 4% of DDT and 4% of Lethane. Ingestion of the formulation, presumably with suicidal intent, led to immediate and repeated vomiting and to severe epigastric pain. Within 2 hours the patient was comatose and flaccid; the respiration was slow and labored, the pulse slow and feeble, the pupils equally dilated. The patient did not respond to treatment and died within 3 hours after drinking the solution. Autopsy showed edema and hemorrhage of the lungs, especially of the lower lobes. The stomach was dilated and hemorrhagic and the upper small intestine was hyperemic and, like the stomach, contained some formulation which had not been removed by vomiting or by the gastric lavage which had been performed.

In a similar way, LURIE [385] reported a case of acute toluene poisoning in a laborer who was cleaning the inside of tanks coated with DDT-toluene emulsion.

It should not be supposed that the toxicity of the solvent always predominates. For example, the recurrent convulsions in a case reported by CUNNINGHAM and HILL [144], though unusual in DDT poisoning, were certainly not typical of solvent poisoning. A 2-year-old child drank an unknown quantity of fly spray of which 5% was DDT, but the nature of the other active ingredients or the solvent was unknown. About 1 hour after taking the material, the child became unconscious and had a generalized, sustained convulsion. Convulsions were present when the child was hospitalized 2 hours after taking the poison but the fits were controlled by barbiturates and other sedatives. Convulsions reoccurred on the fourth day and again on the 21st day but were stopped each time following renewal of treatment. On the twelfth day, it was noted that the patient was deaf. Hearing began to improve about the 24th day and was normal as were other neurological and psychic findings when the patient was seen about 2·5 months after the accident.

GIL and MIRON [226] reported a mass poisoning in which DDT formulated with thallium was mistaken for yeast and used in making bakery products. Twelve men died and 35 suffered severe poisoning which the author attributed principally to the thallium.

The most interesting single case in the literature is that reported by KLINGEMANN [328]. The interest lies in the fact that, after definite exposure, the patient showed early signs of intoxication very suggestive of those in uncomplicated DDT poisoning as described by VELBINGER, GARRETT, JUDE and GIRARD, and others, but the later course of the illness was entirely atypical. The patient gradually developed a severe polyneuropathy, and evidence of liver, kidney, and heart injury. Recovery was only partial 10 months after onset. An outline of the case is as follows: The patient was a hospital orderly and was frequently required to apply DDT to mattresses with a duster. On one occasion the duster was broken and he applied 10 per cent DDT powder to the three-part mattresses of 20 beds. He rubbed the powder into all the seams with his hands and then beat the mattresses to obtain an even and thorough distribution of the powder. The work required about 1·5 hours and used up about 4 kg of insecticide. During the time, the man was surrounded by a thick cloud of the dust. He had a disagreeable, slightly bitter taste in his mouth during the work, and after it was complete he washed out his mouth with hydrogen peroxide. Two or three hours later, paraesthesia appeared in the region of the lips and the tip of the tongue. After 2 or 3 more hours, general malaise and nausea set in. He vomited twice. During the evening there was general irritability, headache, and a slight transient tremor. To this point the history suggests DDT poisoning.

On the day following exposure, the patient complained of a feeling of heaviness in both legs and feet. During the first week the troubles increased; the gait became staggering, and the tempo of work decreased. On the second or third day the urine was dark. Pains appeared in the feet and increased daily. Foot drop developed, and the patient could no longer walk. A slight deafness which had been present for a long time became distinctly worse.

When hospitalized 10 days after onset the patient showed complete paralysis of the peroneus and tibialis bilaterally and evidenced associated neurological findings. The temperature was subnormal. The erythrocyte count was 3·75 million with a corresponding decrease in hemoglobin. The white cell count, differential, and sternal marrow smear were normal. The liver was somewhat enlarged and sensitive to pressure and several liver-function tests were abnormal to a moderate degree. The urine showed a trace of albumin and a few red and white cells. The blood pressure was 150/80. The electrocardiogram was normal. Psychically, the patient was depressed and showed increased irritability.

During the fifth week the condition and laboratory tests showed only minor changes. The patient showed paroxysms of weeping. The eyes showed a slight accomodation paralysis. Deafness continued. During the sixth and seventh week, the jaundice faded, although liver-function tests remained abnormal; the kidney picture improved; the blood pressure returned to normal; but the nerve deafness became worse. During the eighth week, the general condition became worse, the pulse rate increased to 90 to 100 per minute, the blood pressure increased to 170/95, and a little later there was electrocardiogram evidence of myocardial injury. By the eleventh week, the general condition was somewhat improved, and a very slow improvement continued through the tenth month. At that time, the polyneuropathy affecting the legs and the auditory portion of the eighth cranial nerve was still present though improved. The liver, kidneys, and heart were essentially normal, although the blood pressure was 160/100.

Because of the bilateral foot drop and some other features, this history suggests alcohol polyneuropathy which is now generally considered to involve a complex avitaminosis. For this reason, it is a pity that in an otherwise admirable account there is no information on the patient's drinking and eating habits.

CAMPBELL [108] has described seven cases involving peripheral or retrobulbar neuritis, or both, following one or a few gross skin contaminations with a proprietary insecticide. The exact formulation was unknown, but the preparation contained *ortho*- and *para*-dichlorbenzene, DDT, and pentachlorophenol. The author considered pentachlorophenol the most likely cause of the symptoms but could not exonerate DDT. Perhaps a synergistic action should also be considered.

A number of cases of systemic diseases presumably based on sensitization or idiosyncrasy to DDT have been reported. In one instance, a 32-year-old farmer, presumably in good health, spent about a week spraying his barn with DDT and lime. Then he noticed bleeding gums, sore throat, and red spots on his tongue and body. Examination confirmed the purpura and showed a completely aplastic bone marrow. The patient died as the result of a massive hemorrhage. The authors cautioned that although the disease was diagnosed as DDT poisoning, it may have been caused by the solvent (Committee on Pesticides [132]).

On two successive days, a 59-year-old farm laborer applied a dust containing 5% of DDT and 10% of benzene hexachloride to a total of 59 acres. On the evening of the second day of application, he felt unwell. He became progressively worse and when hospitalized 7 days later, he had a pronounced agranulocytopenia. In spite of appropriate treatment, the illness progressed to include severe anemia and thrombocytopenia. After 2·5 months the patient died, with a picture of septicemia. In the absence of any other demonstrable cause, the illness was attributed to one of the insecticides or to their combined action (FRIBERG and MARTENSSON [217]).

In other instances there has not been a depression of the entire bone marrow but an agranulocytosis or thrombocytopenia has been observed. Thus, WRIGHT *et al.* [682] reported a case of transient agranulocytosis which developed in a 22-year-old white male 10 days after he had used an aerosol bomb. The bomb contained DDT, pyrethrum extract, sesame oil, lubricating oil, and Freon. Contact with any of the common causes of agranulocytosis was denied. The patient reported to the clinic with a temperature of 102·6° F, a 4 mm ulceration on the tongue, and a hyperemic pharynx. The white count was 3,850 with 1% of neutrophiles, 2% of eosinophils, 75% of lymphocytes, and 22% of monocytes. He was treated with penicillin. Three days later the white cell count was essentially unchanged but the percentage of neutrophils had increased to 24. On the ninth day the total count was 15,600 and there were 73% of neutrophils reflecting the marked increase in myeloid cell series of the bone marrow demonstrated 4 days earlier. Recovery was uneventful.

A number of cases of purpura following exposure to DDT formulations have been reported. KARPINSKI [317] described five cases in children 1 to 3 years old. In each instance the child had been exposed to the residue remaining from household use of the insecticide, either as a spray or as an aerosol. Compared with the exposure of workers, the exposure of the children was small indeed. The evidence for a causal relationship was circumstantial, although a recent ingestion of drugs by the children was denied. Four of the cases showed a severe or moderate decrease in the blood platelets, but the platelet count in the remaining case was normal. Recovery was uneventful in each instance. Purpura in an adult was reported by CAMPBELL [108].

The matter of allergic reactions of the respiratory tract has been mentioned by WITTICH [674]. Quoting from HILDEBRAND, he has suggested that various dusts and sprays used in the orchards around Wenatchee, Washington, may aggravate or initiate asthma and other allergic disturbances. It is true that foreign matter generally is irritating to the respiratory tract. It would appear, however, that the author may have confused with asthma certain of the specific toxic effects of organic phosphorus insecticides. Part of the specific toxic effects which do occur in persons with extensive occupational exposure to organic phosphorus insecticides resemble asthma somewhat but do not have an allergic basis. Epidemiological and clinical laboratory investigations carried on for several years in Wenatchee under the reviewer's direction have failed to reveal cases of allergy which could be attributed to DDT.

A number of cases of dermatitis have been reported. NIEDELMAN [436] described the case of a 53-year-old housewife who sprayed a closet with a solution containing 5% of DDT in kerosene. On the same day as this first exposure, she experienced itching of the face. On the next morning there was distinct redness, swelling, and inflammation of the face, arms, and neck. Later the same day the eyes were closed by edema, and small vesicles appeared in the skin of the arms and neck. Itching was intense. The dermatitis was still present, though presumably improved, a week later. At that time the patient moved to another home for a week. The disease subsided but there was a temporary recurrence when the patient returned to her own home. She denied any dermatitis as the result of previous exposures to kerosene. On being patch tested, the patient was sensitive to DDT, to kerosene, and to a combination of the two. Although persons with dermatitis commonly react to a wide array of materials, it was concluded that DDT was the cause of the dermatitis.

In anonymous queries to the editor (Anonym [19]) a case of allergic rhinitis and a case said to resemble insulin shock were described. The consultant, who answered the queries, wisely discounted the possibility that the cases represented DDT poisoning.

Later in the same year, STRYKER and GODFROY [563] described 6 cases of dermatitis and mentioned a seventh which they attributed to DDT. In general, the patients presented macular lesions which showed some purpura and were accentuated at pressure points and in skin folds. The first lesions were said to appear on the ankles in 5 patients and on the hands and wrists in 1 patient. The face was not involved. Itching was the chief complaint in some of the cases. The clinical course was protracted. No evidence was presented that the disease was influenced in any way by treatment which consisted in most instances of the intravenous injection of large doses of ascorbic acid. The authors concluded that the dermatitis resulted from the absorption of the compound and its distribution by the blood and not directly from contact. One is forced to observe that of the 6 cases described in detail, 4 were exposed to a wide variety of chemicals in the course of chemical manufacture. After their dermatitis subsided, three patients returned to chemical manufacture in the same environment as previously without a recurrence of their disease. In each of the 6 cases there was, at best, only circumstantial evidence that DDT was involved in any way.

LEIDER [362] reported a case of nine months' duration involving eczematous eruption of the face, upper trunk, and extremities. The eruption appeared about a week after the patient sprayed a garden with DDT. A patch test with 5 per cent DDT in acetone produced an erythematous, papulo-vesiculai reaction.

A somewhat similar case which progressed to an exfoliative dermatitis and which presented positive patch tests was reported by HIGGINS and KINDEL [276]. A 59-year-old, obese, white machinist lived alone in a hotel room which was sprayed weekly with DDT. He developed a relatively mild dermatitis accompanied by paresthesia which responded rapidly to conservative therapy in a hospital. However, within 48 hours of his return to the hotel he got a generalized redness of the whole body with some lichenification, scaliness and petechial-like pinpoint macules, accompanied not only by burning and discomfort but also by a tight, pitting edema of the feet and legs and edema of the skin generally. Slow but steady improvement followed conservative measures. Patch tests were positive with nickel sulfate, DDT, and two DDT analogs but not with the solvents used or with a wide range of compounds including many other DDT analogs. It is interesting to note that the two analogs which were active were not degradation products of DDT; one was a nitro and the other a bromine compound. Further testing 2 months later indicated a loss of sensitivity to DDT. The author called attention to the rarity of reactions of sensitivity and toxicity involving DDT in spite of its extensive use.

HOLLANDER [294] reported an eczematoid dermatitis somewhat similar to that reported by LEIDER but much more severe. It was complicated by an erysipelas-like cellulitis. There had been only one slight exposure to DDT. Diagnosis was based on patch test. The author spoke of the very extensive use of DDT and called special attention to the small number of instances in which DDT is the true cause of dermatitis.

MARSHALL [399] reported a single case of dermatitis in which lesions were more abundant on the legs and lower part of the body and were purpuric in character. However, during an attack there was no anemia, the capillary fragility was normal, and other tests bearing on the existence of purpura were not reported. Of three attacks, the first two followed the use of sulfonamides, but this exposure was denied in connection with the third attack. At this time, the patient showed a strong patch test reaction to DDT and especially to DDT dissolved in kerosene. This reaction diminished with recovery. It should be pointed out, however, that in the presence of dermatitis, the reaction of the skin to nonspecific materials is generally increased.

SWINEFORD and RADFORD [568], in discussing their technique for the routine patch testing of patients with dermatitis, list DDT as one of a great many substances used. Some reactors were found, but the reactions were not as numerous or, on the average, as severe as those for soaps. There can be no doubt that DDT is an allergen, but it is an extremely weak one.

A number of cases of conjunctivitis and dermatitis previously unreported in the literature are included in the summary table given by the Committee on Pesticides [132]. Many of the new cases were reported from a state which pays compensation for agricultural as well as industrial injuries.

Disease Alleged to be Poisoning

It is the reviewer's opinion that DDT has not been established beyond the shadow of doubt as the cause of a single case of aplastic anemia, agranulocytosis, purpura, polyneuropathy, or dermatitis. On the contrary, these diseases are known to occur in a few susceptible individuals following sensitization to a wide variety of chemicals including some common drugs. It therefore appears reasonable to assume that DDT is capable of causing these conditions even though not every case attributed to that compound may be valid. Certainly one is forced to conclude (1) the incidence of these conditions following exposure to DDT is extremely low and (2) the incidence of these conditions irrespective of cause has shown no detectable increase since the introduction of DDT.

A different kind of logic must be applied to the contention that diseases never before proved related to chemical poisoning or sensitization are now caused by DDT. A few persons have carelessly ignored sound diagnostic and epidemiological principles in attributing heart disease, cancer and a variety of other illnesses to insecticides. NEAL and VON OETTINGEN [424] and NEAL [423] encountered this sort of thing while investigating alleged instances of DDT poisoning.

A few authors have abused epidemiology more directly, having attributed to DDT a marked increase of disease in situations in which the vast majority of physicians noticed no overall increase in sickness, and where official records, as well as special epidemiological study such as that of FOWLER [210], failed to show that any increase had occurred.

It is a general human weakness to seek for scapegoats. The neurotic patient is especially inclined to blame his disability on things outside his own body or personality. What better excuse could be found than a new poison!

A man under treatment for an unspecified heart disease died 5 days after he had sprayed his summer home with a 'concentrated solution of DDT'. Death was attributed to inhalation of the DDT vapor (Anonym [20]). No signs, symptoms, or autopsy findings were offered in support of the diagnosis.

HILL and DAMIANI [280] reported a case diagnosed clinically and pathologically as periarteritis nodosa which they attributed to DDT. The victim, a 47-year-old automobile electrician, worked in a room measuring about $10 \times 6 \times 3$ meters with 8 other men all of whom remained well. The room was sprayed with a 6 per cent DDT solution in kerosene a few hours before the patient reported to work on November 10, 1945. On the same day, he sprayed his part of the room some more, so that a total of 130 ml of the solution was used. On the very next morning, the man noticed an extensive pruritic eruption on the trunk and extremities and he complained of dyspnea and an unproductive cough. In an unstated period, less than 5 weeks, the patient lost 9·07 kg in weight and became so weak that he took to his bed. On December 12, 5 weeks after exposure, he was hospitalized, and on January 6, 1946, he died.

BABIONE [38] commented on the able clinical, laboratory, and pathological description of the case of periarteritis nodosa given by HILL and DAMIANI. However, he pointed out that sensitization had not been demonstrated and that only a very superficial resemblance existed between the tremor caused by DDT and that shown by the patient. The author concluded that, of all possible explanations of the case, the one assuming DDT to be the cause was the most imaginative and most likely to cause harm and confusion in the medical literature. The conclusion that the disease was related to DDT was also criticized by ANDREWS and SIMMONS [9] and by GIL and MIRON [226].

HERTEL [273] ascribed a variety of signs and symptoms to DDT poisoning in a 60-year-old man who was shown at autopsy to have tuberculosis. The author reported finding DDT crystals in the mesenteric lymph nodes. Since crystalline deposits of DDT have not been reported for man or found in animals on any of a great variety of dosage regimes, the finding would seem to require rigorous chemical proof. Furthermore, the tremor, pain, irritability, and anxiety commonly associated with chronic disease should not be confused with DDT poisoning.

A brief, nontechnical article (POMMERT [465]) mentions a man who was ill for 48 hours with severe nosebleed and kidney disturbance following exposure to DDT dust for 15 minutes.

In 1949, considerable alarm was caused by a series of articles (BISKIND [70], [71], [72]) which appeared in reputable scientific journals. The author claimed to have demonstrated that 'virus-X' in man and 'X-disease' in cattle (see p. 204) were, in fact, DDT poisoning. The human disease was described as consisting of all or some of the following: 'Acute gastroenteritis occurs, with nausea, vomiting, abdominal pain, and diarrhea usually associated with extreme tenesmus. Coryza, cough and persistent sore throat are common, often followed by a persistent or recurrent feeling of constriction or a lump in the throat; occasionally the sensation of constriction extends substernally and to the back and may be associated with severe pain in either arm. In some cases the hyoid bone becomes acutely painful to pressure for a few days. Pain in the joints, generalized muscle weakness and exhausting fatigue are usual; the latter are often so severe in the acute stage as to be described by some patients as "paralysis". Sometimes the initial attack is ushered in by vertigo and syncope. Intractable headache and giddiness are not uncommon. Occasionally herpes zoster appears. Paraesthesias of various kinds occur in most of the cases; areas of skin become exquisitely hypersensitive and after a few days this hyperesthesia disappears only to recur elsewhere, or irregular numbness, tingling sensations, pruritus or formication may occur. Erratic fibrillary twitching of voluntary muscles is common. Usually there is diminution of vibratory sense in the extremities' (BISKIND [72]).

It was claimed that over 200 cases had been seen in which illness immediately followed exposure to DDT. It was suggested that severe acute affections of this type had been confused with meningitis and poliomyelitis. A fourth paper (BISKIND and BIEBER [75]) emphasized the 'neuropsychiatric manifestations' of DDT poisoning including excitement, insomnia, hyperirritability, anxiety, confusion, inability to concentrate, forgetfulness, depression, extreme apprehensiveness, feeling of tension, and disorders of vision, taste, and hearing. The disease was believed to be persistent and recurrent. Some indication of the same derangements was indicated earlier (DEEDERER [157]). Additional papers (BISKIND [73], [74]) implied that DDT and compounds which the author considered chemically similar were the cause of alleged increases in the incidence of cardiovascular diseases, cancer, atypical pneumonia, retrolental fibroplasia, poliomyelitis, and hepatitis, as well as a number of specific diseases of animals.

It should be mentioned that 'virus-X' is a popular expression which was apparently coined and used for the first time during an epidemic in Los Angeles in December 1947 (Anonym [22], [24]). Virological studies begun during the outbreak showed that the disease was type A influenza (MEIKLEJOHN and BRUYN [402]). In later years, the term 'virus-X' was applied indiscriminantly to a variety of minor diseases (Anonym [24]). Several of the diseases which the authors have alleged to be increased have, in fact, not increased. Others have shown an increase which is adequately explained by known changes in the percentage of persons of different ages in the population. Finally, the few real increases in disease which have occurred are much better explained by other factors than by the use of DDT. For example, infectious hepatitis, a disease of proved infectious origin, increases when sanitary measures fail. It was a serious disease during the American Civil War as well as during World War II. The description given by BISKIND which would allow the diagnosis of a great number of diseases as DDT poisoning received only minor notice in the technical literature (JENKINS [310]) but got a much more extensive and sensational treatment in popular journals.

PLICHET [460] reviewed BISKIND'S statements in detail and proceeded on the assumption that they were valid. However, PLICHET observed that no increase in gastroenteritis occurred in France following the introduction of DDT and, in particular, that the striking syndrome described by BISKIND had not been seen in France. The author advised prudence in the use of the insecticide but cautioned that the fact that DDT saved Europe from epidemics at the end of the war should not be forgotten.

STONE and GLADSTONE [561] described a 4 year illness in a 24-year-old aerosol bomb maker. The authors consider that the illness was caused by DDT but admit the possibility that the other ingredients of the aerosol formulation may have played a major or contributing part in the disease. The man suffered weakness, poor appetite, and restless sleeping paroxysmally over a period of 4 years. For about 2 weeks before examination, weakness increased greatly and his speech became so indistinct that he was not understood by his companions. At the time of examination the patient complained of these difficulties and also of photophobia, blurring of vision, and a sense of floating. He was ataxic and showed tenderness along the large peripheral nerves. There was no clonus, pathological reflexes, sensory nerve changes, vestibulocerebellar signs, or spinal fluid abnormalities. The patient recovered completely following 14 days of absolute bed rest and vitamin therapy. After treatment the white count which had been slightly depressed was found to be normal. The patient returned to work, observed the recommended safety precautions which he had previously ignored, and remained well. Their account of chronic illness and rapid recovery is worthy of comment.

KEIZER [320] suggested that a single case of subcutaneous neonatal adiponecrosis may have been caused by excessive dermal exposure to DDT.

Use Experience

Dependable information on the toxicity of DDT has been obtained from the examination of persons to whom it was applied or, especially, the examination of workers who used it regularly.

In a mass delousing, several hundred persons, after a cleansing bath, were sprayed directly on the whole body with an emulsion of DDT. There was not a single instance of ill effect (DOMENJOZ [172]).

In one of the earliest reports ANGLEY [10] stated that about 56,780 l of 5 per cent DDT in kerosene (2,313 kg technical DDT) were used in eastern Italy between June and November, 1944. The solution was applied to tents and buildings using power paint sprayers and an air pressure of 25 lb. About 50 men did the work and of them 33 were available to the author at the time of the study. The men studied had worked an average of 6 hours a day, 6 days a week, and for periods varying from 1 week to 4·5 months. They often failed to wear respirators but usually did wear gloves, overalls, and hats and they usually bathed daily. History showed that none had suffered skin irritation. A few had felt slight vertigo, nausea, and loss of appetite but none had lost weight. No nervous or urinary symptoms had been present and evidence of tremor or abnormal reflexes was not found. The following laboratory tests or measurements were made: serum phosphorus, serum phosphatase, cephalin-cholesterol flocculation, urinalysis, hemoglobin, white count, and electrocardiogram. A few deviations from normal were found in these tests but the author presented evidence in each instance that the change was caused by intercurrent disease and was not associated with exposure to DDT.

GORDON [241] reported on 27 African spray operators who worked 5 days a week for 6 months using a 3·7 to 5·0 per cent solution of DDT in kerosene. The purity of the technical DDT was 52 to 61%. Little attention was paid to recommended safety precautions. Blood examinations, regular weighings, and skin inspections were made during the last 4 months of the exposure period. During the first month of exposure rashes were common but when the examinations were made only one rash was encountered. Patch tests done on the workers and on 30 control subjects showed that the irritation was primarily caused by kerosene but was aggravated by the technical DDT dissolved in it. DDT dissolved in dimethyl phthalate did not cause a rash. The blood counts revealed a slight increase in white cells. The body weight of the men remained normal.

Probably the most significant study of the practical problem of DDT toxicity under actual operating conditions is that of STAMMERS and WHIT-

FIELD[548], [549]. At Colombo, they studied 15 workers, 21 to 40 years of age, 10 of whom had been employed continuously for 9 months and 5 of whom had been employed for 7 months. One was a Tamil, the others Sinhalese. The work consisted of mixing and spraying a 5 per cent solution of DDT in grade 2 kerosene in ships and naval installations nearby. The technical material averaged 75 per cent p,p-isomer. Knapsack sprayers were used. On the average, each man was exposed to DDT 24 hours per week. The rest of the working time was taken up with travel and the assembling and repair of equipment. At the beginning, protective clothing was issued consisting of overalls, cap, tropical A.P.R. oiled skin gas cape, gloves, rubber boots, and gauze masks. The men soon discarded the capes, gloves, and masks because of the heat. For the most part they preferred to work with bare feet. The overalls were worn open at the neck and usually with the sleeves rolled up. They washed their hands for lunch and most of them bathed all over under a cold water tap at the end of the day. Most of the spraying was indoors so that the men worked in a confined atmosphere. The hands and arms and, to a lesser extent, the faces and feet were exposed to the solution dripping from the ceiling, splashing from walls, or leaking from faulty sprayers. At the end of each day a white frost of DDT crystals covered the exposed skin, and the overalls frequently were saturated by the solution. The total operation involved 7,490 man-hours of labor and the use of 35,650 l of insecticide formulation.

At the end of the exposure period, the men were examined with particular attention to the skin, mucous membranes, and nervous system. The urine was examined for albumin. Liver function was tested by the oral hippuric acid synthesis test and the hemoglobin measured by the method of SAHLI. Two months later additional tests (total erythrocyte count, total white cell count, differential white cell count, examination of the stools for parasites) were made of 9 of the same men who had sprayed for 11 months and of 3 who had worked for 9 months.

Some of the men suffered irritation, vesiculation, and patchy desquamation typical of kerosene dermatitis when they first went to work. This, however, cleared in a few days and did not recur except in the face of exceptionally severe contamination. The skin of all the workers was clear when they were examined except one who showed *Tinea cruris*. The mucous membranes appeared normal except that 2 men showed mild conjunctival congestion. The authors thought that this might have resulted from DDT and kerosene or from an addiction to hemp.

The chest was normal in each case. The systolic blood pressure ranged from 90 to 138 and averaged 111 mm of mercury. The diastolic pressure varied from 61 to 82 and averaged 70. A slightly enlarged liver in one man was the only abnormal abdominal finding. The authors pointed out that enlargement of the liver is common in the tropics. Significantly, there was no tremor or other abnormal neurological sign. In a period of 4 hours after a 6 g dose of sodium benzoate 14 of the subjects excreted more than 2·4 g of hippuric acid (calculated as benzoic acid) and averaged 3·7 g. The fifteenth subject excreted only 1·4 g and his result was similar on another test repeated a week later. Six weeks later he excreted well above the average amount. Fifteen control subjects gave an average excretion of 3·1 g. The same man who showed an initial abnormality of liver function also complained of frequent micturtion and showed pus and albumin in the urine and a total white blood cell count of 15,200. The authors recorded that this man was an opium addict who, at the time of the tests, was trying to substitute hemp (marijuana) in its place, a situation which may have had some bearing on his laboratory findings. The red cell counts and hemoglobin determinations were within the range of normal. With the exceptions noted, the total and differential white cell counts and urine examinations were normal. As expected, a variety of intestinal parasites was found. During the entire spray period, the work output of the group remained high and they showed a considerable *esprit de corps* consistent only with good health.

ANDERSON and KHORRAM [8] examined 32 men who had been working with DDT for 9 months. Some were engaged in spraying houses with a 3 to 4 per cent kerosene solution and were exposed by the dermal and respiratory routes. Some applied a 2·5 per cent oil formulation for larva control and were exposed dermally. The remaining workers crushed and dissolved DDT with rather primitive equipment to prepare the two formulations. All received dermal exposure. The examination included clinical tests and examination of the urine for organically bound chlorine. Thirty-six sanitary workers of the same economic class were examined as controls. The only abnormality which might have been due to DDT was tremor of the hands in nine exposed persons, but it was not possible to exclude bias. No organically bound chlorine was found in the urine. The authors concluded that DDT is safe to work with, but advocated the use of emulsions where practical to avoid the dermatitis associated with exposure to kerosene.

GIL [225] mentioned the absence of a single case of illness attributable to DDT used in the great insect control campaigns in such areas as Italy, Greece, Sardinia, Argentina, and Spain. However, GIL and MIRON [226] reported that some persons suffered irritability, anxiety, difficulty of concentration, fatigue, insomnia, and a sensation of heaviness of the body after exposure in the dusty atmosphere of a delousing station. The symptoms disappeared in a day or two, or could be prevented in the first place by the use of masks.

No functional or organic disturbances had been reported in factories where some of the employees were exposed to an atmosphere highly contaminated with powdered DDT. Certain illness had been reported among women exposed simultaneously to DDT and copper oxychloride, although neither compound alone gave any trouble (GIL [225]). Furthermore, no disease attributable to DDT was discovered in 5 spray workers who were studied in detail, especially in regard to liver function, blood changes, and the subjective and organic function of the nervous system (GIL and MIRON [226]).

MÜLLER *et al.* [417] reported that no resorptive injuries, local irritations, or allergic reactions had been observed among personnel of a Swiss factory engaged in the preparation of DDT, or its formulations, since production started in 1939.

Likewise, DEICHMANN *et al.* [162] reported that two groups of workers engaged in the manufacture of DDT for periods up to 13 months complained of no symptoms referable to the absorption of DDT and showed no signs of illness referable to the insecticide as indicated by physical examination, hematological data, and a battery of liver and kidney function tests.

DDT in an oily solvent has been recommended for the treatment of scabies (see, for example, DEGOS and GARNIER [159]). Quite aside from its effectiveness as a scabicide, the use of DDT in this way under medical supervision represents a critical test, and its failure to cause dermatitis or any systemic effect is noteworthy.

A case reported by CAMPBELL [107] resembles in some ways the illness which WIGGLESWORTH [665] claimed resulted from experimental exposure. A 32-year-old man had handled and, it is said, inhaled DDT powder for 6 years in the course of his employment. The exposure was presumably not greater than that of others in the same kind of work. The patient complained of extreme fatigue, migratory and inconstant pains in the limbs, and emotional instability. Physical examination revealed only incidental findings. Very extensive laboratory studies were done, all with normal results except that the urine contained albumin, red cells, and granular casts and the erythrocyte sedimentation rate was 24 mm in 1 hour. After hospitalization for 2 weeks, with undisclosed treatment, albuminuria and slight nervous tension were still present and the sedimentation rate had not yet returned to normal. The author reported no investigation of the cause or subsequent course of the nephritis. Although he recognized that the symptoms could not be ascribed with certainty to DDT he considered that to be the most likely explanation. The author was apparently unfamiliar with the more extensive experimental studies of human exposure or with the instances of accidental, uncomplicated DDT poisoning already published at that time. The author's interpretation of animal experiments involving extremely large doses of DDT is open to grave question.

DE LUCENA [165] stated that no toxic symptoms were observed among the inhabitants of communities treated for malaria control regardless of whether solution or emulsion had been used. However, workers who were applying the DDT solution usually developed after several days an irritation of the hands and lower arms. This difficulty apparently did not arise in the use of emulsion. The irritation sometimes required assignment to other tasks for a time and in several instances workmen had to be given sick leave for several days. Subjective symptoms including vertigo, headache, or nausea occurred but were less frequent than the dermatitis. Protection against the dermatitis was possible through the application of cocoa butter to the skin. It would appear that this dermatitis was caused by the solvent.

DE JONG and FIRTOS [163] and FIRTOS and DE JONG [205] reported a few cases of dermatitis bullosa in workmen who had used leaking sprayers for a long time. The localization was on the part exposed to the leaking spray, namely the flexural aspect of the forearm and the medial aspect of the upper arm on the right side. On the first day swelling and erythema developed, associated with burning and pain. On the second day vesicles and bullae developed. After a few days all symptoms subsided when the affected areas were treated with a soothing powder. The authors considered the lesions similar to those caused by petroleum and attributed them to kerosene which had been used as a solvent for the DDT.

GOMEZ [239] has reported a high incidence of myopia correctable by eye glasses in workers who have used waterwettable powder and occasionally kerosene solutions in connection with the malaria control program in Venezuela. It had not been possible to establish the cause of the trouble nor to rule out DDT as a cause. It is noteworthy that myopia has not been a common finding in workers using DDT elsewhere.

In the state of Sao Paulo, Brazil, clinical studies were reported on 160 persons including inhabitants of treated houses and workmen engaged in spraying DDT. Some of the subjects were under clinical observation for 5 years. Never so much as the slightest sign of intoxication was found in any of these persons and their urine was invariably free of DDT (WASICKY [647]; WASICKY and UNTI [651]).

WEBSTER [655], referring to agricultural use experienced in the State of Washington, noted that reports of injury to man appeared to be confined to those working at the spray tank and handling the insecticide in the form of a 50 per cent wettable powder. Bronchial irritation led to a persistent cough, usually lasting not more than 3 weeks. The part played by the auxiliary materials in the formulation was recognized.

CHIGNOLI and ILICETO [122] made a detailed study of 10 spray workers who had dermal and respiratory contact for 45 days with a petrol solution containing 3% of DDT and 2% of chlordane, and also with an emulsion containing 5% of DDT. The men showed an increase in the number of red cells (up to 7·4 million) while the hemoglobin remained essentially unchanged. The men at first showed a slight leucocytosis (up to 12,400), but later a leucopenia (lowest 4,050). The authors did not consider the formulations dangerous when properly used.

In the 4 year antimalarial campaign in Sardinia, whose extent is indicated by the cost of over 5 billion lire, there was no indication of injury to people. A few claims were made for alleged injury by larvicide to bees, fish, and livestock but most of them were voluntarily withdrawn following practical demonstrations to the claimant. It was concluded that only minor sums, if any, would be allowed by the courts for the very few cases brought to legal action. This stands in contrast to payments of over a million lire to cover property damage resulting from the drainage of land and other causes (LOGAN et al. [377]).

Although poisoning may not be involved at all, mechanical accidents associated with insecticide dispensers must be considered in evaluating the total health of those who use pesticides. Few accidents of this sort are reported, perhaps because mechanical accidents are common and often their occurrence is taken for granted. It has been suggested (BARNES [45]) that more farm workers are injured by the machines used to apply insecticides than are injured by the insecticide formulations.

A few curious mechanical accidents have been reported in connection with the use of DDT. In one instance the valve broke on an aerosol bomb so that the contents began to escape. A man put his hand over the opening in an attempt to stop the flow while he rushed the bomb out of a dwelling. When he dropped the bomb, his hand was frozen in the shape in which it had been held. He sustained injuries equivalent to a third degree burn (METZLER [408]).

Storage of DDT in Human Tissue

It is understandable that after the demonstration of the storage of DDT in the tissues of animals an attempt should be made to demonstrate storage in man. HOWELL [304] reported the storage of approximately 17 ppm of DDT in the fat of a man who had had extensive occupational exposure to the compound for over 4 years and who had eaten foods known to contain appreciable quantities of DDT.

PERRY and BODENLOS [451] studied 16 men assigned to control of malaria, insect pests, and rodents. These 16 men had mixed DDT-oil formulations and operated a fog generator, knapsack sprayers, and hand-operated DDT-dust dispensers for periods varying from 6 months to 5 years. None of the men showed any signs or symptoms attributable to DDT. No DDT was demonstrated in fat, blood, urine, or feces of the men.

LAUG et al. [346] analyzed samples of fat taken by biopsy from 75 persons without special occupational exposure and found that the average DDT content of all the samples was 5·3 ppm. Of the group, 20% showed no DDT and 9% showed 1·0 ppm or less. 28% fell in the range of 1·1 to 5·0 ppm, and another 28% fell in the range of 5·1 to 10·0 ppm so that 85% of the samples contained 10·0 ppm or less. The highest concentration observed was 34 ppm. No difference in concentration attributable to sex could be demonstrated.

LUIS [384] reports finding the following concentrations of DDT in the organs of a poisoned person: liver, 36 ppm; kidneys, 27 ppm; heart, 19 ppm. The nature of the accident or crime leading to the poisoning was not indicated.

The presence of DDT in human fat was confirmed by PEARCE et al. [449] and MATTSON et al. [400]. In the course of these detailed chemical studies the presence of a metabolite in addition to DDT was revealed. The metabolite was shown to be DDE or a compound with identical spectrophotometric and chromatographic properties. Most samples contained analyzable quantities of DDT and DDE. DDE constituted 39 to 86% of the total and most samples contained more DDE than DDT. Although it was pointed out that the samples used were not representative of any one population group, having been selected for chemical study, the greater number showed values for DDT plus DDE ranging from 2 to 9 ppm. The total values are, therefore, in general agreement with those of LAUG et al. [346], which failed to differentiate the metabolites.

MATTSON *et al.* [400] reported that fat from autopsy specimens collected prior to the advent of DDT showed no evidence of Schechter-Haller positive materials. This indicates that materials native to human fat do not give false positive results for DDT and its derivatives. This fact and the high degree of specificity of the Schechter-Haller method prove that the material stored in man is, in fact, DDT and its derivatives.

Of particular interest is a sample taken from a worker in a DDT formulation plant. Analysis showed 122 ppm of DDT and 127 ppm of DDE. This suggests that DDT is degraded to DDE in the human body, since it did not appear likely that the worker would be exposed to any significant amount of DDE (MATTSON *et al.* [400]). This worker volunteered for study. He had no complaints and careful examination in the hospital on two occasions failed to reveal any injury associated with DDT.

Excretion of DDT in Man

Excretion of derivatives of DDT in man was demonstrated earlier than storage of DDT. The experimental studies of NEAL and his colleagues [428] and their demonstration of the excretion of DDA have already been reported (p. 87).

SMITH [531] demonstrated DDT and DDA in the urine of a man who had accidentally ingested DDT. A 46-year-old farm hand chewed tobacco which had been contaminated by a small amount of 5 per cent DDT solution in kerosene. He was apparently unaware of the taste or odor of kerosene. About 2 hours later he became nauseated and apprehensive and had a feeling of stiffness and pain in the jaws and soreness of the throat. He took magnesium sulfate and, about 2 hours later, vomited. Next day he was well except for soreness of the throat which persisted for another day or two. The first sample of urine which was collected proved unsatisfactory, but a second uncontaminated combined sample obtained on the sixth and eighth day after exposure was satisfactory. Analysis of the acid ether extract by the method of SCHECHTER and HALLER indicated the excretion of about 5 mg of a mixture of about 75% DDT and 25% DDA per day on the sixth and eighth day after exposure. Urine from laboratory workers who had been working with the compound was entirely negative for DDT and DDA by the same test.

LAUG *et al.* [346] reported the analysis of 32 specimens of breast milk obtained from 32 different women. The average concentration was 0·13 ppm, and 66% of the samples fell in the range of 0·06 to 0·15 ppm. The highest value found was 0·77 ppm. This result might have been anticipated from the results of numerous experiments on animals. The excretion in human milk of chemical compounds including alcohol, caffeine or a derivative (from coffee or tea), and nicotine (from tobacco) has been reviewed by SAPEIKA [504].

8.

EXPOSURE OF WORKERS

There is little information on the degree of exposure to DDT which workers undergo, especially workers in manufacturing or mixing plants or in agriculture. This is unfortunate because it might be hoped that such information would make it possible to put to better use data already available from animal experiments either on DDT or, especially, on other, newer insecticides.

Some idea of the potentiality for exposure of workers may be gained from the production figures for a single year in several countries as shown in Table 17. The trend for the last several years is indicated by Table 18, which shows production by the United States for 1944 to 1952, inclusive.

Some information is available on the degree of exposure of workers engaged in applying residual deposits of DDT to houses for malaria control. According to STEPHENS [551], special study in 4 states indicated that workers spent from 39 to 43% of their time in actual spraying, and 6 to 19% of their time in filling cans, cleaning equipment, mixing, and other duties that involve potential exposure. The potential exposure time varied from 45 to 57% of the total time worked (in this instance the total time was 8 hours a day). The remainder of the time was spent in traveling between houses or in other duties which did not involve exposure to insecticides. In 9 other states the percentage of time devoted to different operations was also determined but not always with the same care as was used in the study in the 4 states just mentioned. In these 9 states, it was found that the percentage of time which involved potential exposure varied from 42 to 87%, and of the total time, 30 to 77% was spent in spraying.

The values given above are consistent with those given by BRIGHT [86].

BRIGHT [85] analyzed a somewhat different aspect of the same general spray program referred to above. In a study involving 1,055,503 house-spray applications, it was found that 544 g of DDT were used per application, and that 1·36 man-hours of labor were required for each application. This determined the expenditure of 2·51 man-hours of labor per kilogram of DDT used.

In previously unpublished research by Dr. FREDERICK FERGUSON and Mr. PETER SKALIY under the direction of the reviewer, tests were conducted to determine the approximate amount of insecticidal mists contacting the exposed areas of the skin of the individual workers using hand-spray equipment. These tests were performed on operators spraying under simulated working conditions indoors and outdoors, and on operators who were actively engaged in outdoor spraying as a part of a municipal fly control program. A 5 per cent

                    W. J. Hayes, Jr.

Table 17

Metric Tons of Technical DDT Produced and Exported by Different Countries in 1951 as Reported by the World Health Organization [681]

Country	Production	Exports
Australia	750	—
Belgium	90	—
Canada	140	—
Finland	100	0
France	2,000	500
Germany[1])	1,200	250
Italy	2,500	300
Japan	1,585	420
Netherlands	250	130
Portugal	120	0
Sweden	100	0
Switzerland	250	100
United Kingdom	2,000	160
United States	48,000	13,600
Total	59,085	15,460

[1]) Western Germany only.

Table 18

Metric Tons of Technical DDT Produced in the United States in Different Years. The Figures Are Based on U.S. Tariff Commission and Bureau of Census Stat stics

Year	Production	Year	Production
1944	4,366	1949	17,193
1945	15,079	1950	35,448
1946	20,707	1951	48,144
1947	22,499	1952	44,803
1948	9,181	1953	36,288[1])

[1]) Estimate for total year on production through August (SHEPARD [523]).

DDT water-xylene emulsion was used in all tests. Samples of spray mists contacting the exposed areas were collected by attaching a disc of filter paper to different parts of the individual's body. These patches were backed by a thin sheet of rubber and were placed on the back of the neck, the chest (over exposed skin with the collar unbuttoned), the forehead, the cheeks, the upper arm, and the back of each hand. All patches were exposed for a period of 1 hour of actual spraying time. Following exposure, a standard sample was cut out of the center of each patch and the insecticide extracted for quantitative analysis. The results are summarized in Table 19. Analyses revealed no statistical difference in the amount of contamination received in the simulated outdoor spraying and in the outdoor spraying which formed a part of an active campaign for fly control. For this reason the two groups of data on outdoor spraying have been combined in the table.

It is clear, from the record of ranges and standard errors, that the degree of contamination varied widely from one period of application to another. The variation between different tests on the same operator was as great as the variation between different operators working at the same time. It is clear, of course, that if a worker were careless he would suffer a greater average contamination than would a careful worker. There was a striking difference between the contamination encountered in indoor and outdoor spraying. Indoor spraying determined at least 5 times as much exposure as did outdoor spraying.

Table 19

Technical DDT Deposited on Different Unclothed Body Areas Under the Conditions of Indoor and Outdoor Spraying

Type of spraying	Part of body	DDT deposited		
		Range mg/m²/hour	Mean mg/m²/hour	Standard error mg/m²/hour
Inside	Face	32–562	257	82
	Hands	378–5252	2324	814
	Arms	226–7247	2665	1054
	Back of neck	43–637	248	85
	Chest	32–3467	1353	559
Outside	Face	0–313	116	27
	Hands	43–626	313	49
	Arms	119–745	392	65
	Back of neck	0–626	157	41
	Chest	0–518	206	42

This is explained in part by the fact that indoor spraying involves overhead application, and also by the fact that even if windows and doors are left open during indoor spraying, there is less average ventilation than in outdoor work. There is a tendency for droplets outdoors to be carried away by the slightest breeze, whereas droplets indoors undergo little lateral movement, and therefore remain longer in the environment.

It is evident from the table that the arms and hands are subject to greater contamination than are other parts of the body, such as the face and the neck, which are customarily unclothed. Using the average values given in the table, and accepted estimates for the area of different parts of the body (BERKOW [56]) the total amount of DDT which would impinge on the unclothed portion of the worker in the course of 1 hour of actual spraying was calculated. This value proved to be 543 mg/hour for inside spraying, and 84 mg/hour for outside spraying. It should be emphasized that these figures apply only to the almost imperceptible contamination resulting from fine spray droplets. It does not include the contamination which results from leakage around damaged shutoff valves, not to mention the contamination resulting from grosser accidents.

During the first portion of each exposure period, contamination is slight, but absorption of DDT from the oily skin undoubtedly continues between exposures until the compound is removed by washing. There appears to be no more accurate way to interpret the data than to assume that the total amount of DDT which reaches the skin during the working day is present during the entire period of potential exposure. Using 50% of an 8 hour day (STEPHENS [551]) as the period of potential exposure, one may calculate that the daily dermal dosage received by a worker doing inside spraying would be 31 mg/kg/day in the form of imperceptible droplets. Respiratory exposure and accidental contamination would be additive. Reference to Table 9 indicates that animals generally tolerate dosages higher than 30 mg/kg/day applied as an emulsion. However, inside spraying is sometimes done with solutions of DDT and, in the tropics, workers may wear little clothing. Repeated dosages less than 30 mg/kg/day may be fatal to some animals when applied as a solution and only slightly larger dosages are highly fatal to some species. Judging from animal experiments, the exposure of some workers would appear to approach the toxic level. Of course, the fact is that repeated studies have shown that the workers do not become poisoned. A similar apparent inconsistency has been observed in connection with exposure to parathion (BATCHELOR and WALKER [49]). The clinical observations are certainly more dependable than the extrapolation from animal experiments. It might be supposed that man is more resistant than laboratory animals, but the findings of VELBINGER make this unlikely. Perhaps an adequate reason for the apparent difference lies in the consistency with which animals are dosed once an experiment is begun. Another difference lies in the fact that workers encounter finely divided droplets from which DDT may crystalize out promptly, while laboratory animals receive dosages in an abundance of vehicle, so that the active ingredient can more readily penetrate the skin.

The estimate of the dermal dose of DDT in solution which would be dangerous for man given by LEHMAN [354], [355] is 9 g/day, or about 128 mg/kg/day. This estimate certainly appears conservative in view of the results with animal experiments. However, it appears likely that when all modes of exposure are taken into account, spray workers may equal or exceed this degree of dosage. This is especially true of workers in tropical countries who may have a large part of the legs and even the entire upper portion of the trunk directly exposed.

In addition to the studies already described, pieces of filter paper, backed by rubber sheeting, were placed on workers' arms beneath the single cotton shirt which clothed that part of the body. Eight of these samples were taken. In each instance the exposure was made during the entire working day. The average contamination was 5·4 mg/m²/day, beneath the shirt, as compared with a value of 1,285 mg/m²/day on an adjacent area outside the sleeve. Thus, ordinary clothing offers a remarkable protection against the imperceptible contamination involved in spraying.

9.

THE PROBLEM OF RESIDUES

Residues in Animal Products

DDT may find its way into foods of animal origin either through spray applications to the animal or to its environment, or through feeding the animal on some food which itself contains the compound. Although DDT has been added to harvested cereals in order to prevent infestation by insects, the addition of DDT to foods of animal origin has not been advocated.

Residues in Milk

The early recognition of the presence of DDT in the milk of experimental animals was discussed in the section on excretion, above (p. 88). Apparently, the earliest mention of DDT in milk which might conceivably become human food was that of HOWELL *et al.* [305]. The authors found that when dairy cattle were sprayed with 1·9 l of a 0·25 per cent DDT emulsion or suspension, once every two weeks, concentrations of less than 1 ppm appeared in the milk. When a 5 per cent suspension was applied daily at the same volume, then concentrations in the milk as high as 33·6 ppm were found. It was also noted that heavy spraying caused storage in the body because excretion in the milk continued for 19 weeks after the spraying was discontinued. When the cows became fresh again, no DDT appeared in the milk. Milk showing the highest average concentrations of DDT was fed to mice. No differences were observed in the test mice or their progeny, as compared with mice fed uncontaminated milk.

CARTER *et al.* [118] sprayed cattle with an average of 2·1 l of 0·5 and 0·25 per cent suspensions of DDT as frequently as was necessary for horn fly (*Siphona irritans*) control. The concentration of DDT in the milk was always less than 2 ppm and generally less than 1 ppm except for a short period after the cattle were sprayed. Concentrations in the milk after the application of the 0·25 per cent formulation averaged only slightly less than those after the application of the 0·5 per cent suspension. In a subsequent study of a single cow from a herd which was sprayed one time by a commercial applicator, the concentration of DDT in the milk was found to decrease from a maximum of 3·0 ppm two days after spraying to a minimum of 0·4 ppm a little over a month later (CARTER and MANN [115]).

152   W. J. Hayes, Jr.

SHILLINGER [525] reported that the application to cows of 150 g of a 5 per cent solution at intervals of 22 days resulted in a concentration of 2 to 3 ppm in the milk 11 to 12 days after the first application and a concentration of 5 to 6 ppm in 2·5 months.

CLABORN *et al.* [128] reported the experimental treatment of four dairies with DDT and of other dairies with methoxychlor. A 5 per cent DDT emulsion was applied one or more times to the barns for the control of house flies (*Musca domestica*) and stable flies (*Stomoxyx calcitrans*). In addition, the cattle were treated with 0·5 per cent emulsions for the control of the horn fly (*S. irritans*). In one of the dairies the cows received 1·9 l of emulsion at each application; in the other dairies only 0·9 l was applied. The average contamination of milk of all four herds was 0·21 ppm. No value was given for the concentration of DDT in the milk before any application was made. Although DDT was applied to the barns and to the cattle during the same period, the authors were able to show that a marked increase of DDT in the milk immediately followed barn spraying, but not in all instances. It was just after such spraying that the greatest contamination of milk (1·3 ppm) was produced. The data did not permit any distinction of the source–whether from application to the barn or from application to the cattle–when the final contamination of the milk was small. A single application of 1·9 l of emulsion resulted in more contamination than the application of half that volume but more of the small applications were required to achieve fly control so that the average contamination caused by the two procedures was not significantly different. Much of this information was made available before formal presentation through a news report on the oral presentation (Anonym [25]).

Apparently no papers were published on the contamination of milk as the result of barn spraying exclusively before the United States Department of Agriculture [609] recommended the substitution of other insecticides for the purpose. Some information was available to officials and, almost simultaneously with the recommendation, the matter was discussed at a scientific meeting (LAAKE [337]). The announcement by LAAKE that barn spraying led to contamination, and the recommendation by the Department of Agriculture aroused considerable comment and directly stimulated research.

The first paper to report this research was that of HARRIS *et al.* [261]. These authors found no DDT, or only very small amounts (0·05 ppm) in samples of milk taken from 4 herds before the experiment was begun. One herd showed no DDT after its barn was sprayed, two herds showed a transient and perhaps questionable contamination, while the fourth herd showed a consistent contamination of its milk (up to 0·05 ppm) for a week after the barn was sprayed. During the spraying the feed troughs and drinking fountains had been covered, and food either covered or removed. The barn from which contaminated samples of milk were consistently taken had relatively poor sanitation. The authors reasoned that the contamination must have come from careless handling of the milk rather than from inhalation, contamination of feed, the licking of sprayed surfaces by the cows, or from other causes. They concluded that when

Table 20

Relationship of Daily Oral Intake of DDT and Daily Secretion of the Compound in the Milk of Cows

Source of DDT	Residue concentration ppm	Portion of diet contaminated	No. cows	No. days fed	Average daily intake of DDT mg	Concentration of DDT in milk ppm	Average daily output of DDT mg	% Intake in milk During DDT feeding	% Intake in milk After DDT feeding	Authority
Capsule	—	—	1	150	24,000	44	—	—	—	Wilson *et al.* [667]
Pea vine silage	135	—	—	124	1,500	15	—	—	—	Wilson *et al.* [667]
Pea vine silage	< 1	—	—	141	—	0	—	—	—	Wilson *et al.* [667]
Peanut hay	184	—	—	90–105	—	10–36	—	—	—	Arant [36]
Pea vines	4·6	—	5	—	—	0	—	—	—	De Heus and Mann [160]
Pea vine silage	0–13	—	5	151	—	0–0·4	—	—	—	Lardy [341]
Corn silage	5–10	—	—	—	—	0·1–0·5	—	—	—	Lardy [341]
Alfalfa hay	7	Whole	6	79	126	2·3–3·0	24[1]	19·0[1])[2])	—	R. F. Smith *et al.* [539]
Pea vine silage	8–19	—	2	53	44–88	< 0·5	—	—	—	Carter *et al.* [119]
Alfalfa hay	74–134	Part	1	111	727	6·4	136	18·6	1·8	Shepherd *et al.* [524]
	74–134	Part	1	162	553	7·3	165	29·8	3·0	Shepherd *et al.* [524]
	74–134	Part	1	110	303	2·2	51	16·8	1·1	Shepherd *et al.* [524]
	4–30	Part	1	98	109	0·5	5	4·8	0·6	Shepherd *et al.* [524]

Source of DDT	Residue concentration ppm	Portion of diet contaminated	No. cows	No. days fed	Average daily intake of DDT mg	Concentration of DDT in milk ppm	Average daily output of DDT mg	% Intake in milk		Authority
								During DDT feeding	After DDT feeding	
Oil solution	—	— .	1	50	50	0·13	0·9	1·8	—	Ely *et al.* [197], [198]
	—	—	2	50–200	100	0·40	3·3	3·3	—	Ely *et al.* [197], [198]
	—	—	2	50	250	1·0	7·9	3·1	—	Ely *et al.* [197], [198]
	—	—	4	50–190	500	2·5	23·5	4·7	—	Ely *et al.* [197], [198]
	—	—	2	50–190	1,000	4·7	35·8	3·5	—	Ely *et al.* [197], [198]
	—	—	3	50–140	2,000	7·1	71·9	3·6	—	Ely *et al.* [197], [198]
Hay	8·7	—	1	100	109	0·5	5	4·8	—	Ely *et al.* [197]
	17·3	—	1	110	303	2·2	51	16·8	—	Ely *et al.* [197]
	23·8	—	1	120	237	3·2	33	13·9	—	Ely *et al.* [197]
	26·2	—	1	150	357	3·8	50	14·0	—	Ely *et al.* [197]
	33·1	—	1	160	553	7·3	165	29·8	—	Ely *et al.* [197]
	36·0	—	1	110	727	6·4	136	18·6	—	Ely *et al.* [197]
	38·9	—	1	130	405	4·9	60	14·8	—	Ely *et al.* [197]
	46·7	—	1	150	701	4·9	79	11·3	—	Ely *et al.* [197]
Alfalfa hay	0	Part	2	113	0	0·05	0	—	—	Biddulph *et al.* [59]
	0	Part	2	81	0	0	0	—	—	Biddulph *et al.* [59]
	9·0	Part	2	81	175[1]	0·7	6·6[1]	3·7[1]	—	Biddulph *et al.* [59]
	9·6	Part	2	113	137[1]	1·7	17·3[1]	12·6[1]	—	Biddulph *et al.* [59]
	12·1	Part	2	113	176[1]	3·3	34·6[1]	19·6[1]	—	Biddulph *et al.* [59]
	19·2	Part	2	81	375[1]	1·3	20·8[1]	5·5[1]	—	Biddulph *et al.* [59]
	30·0	Part	2	81	580[1]	2·2	3·6[1]	6·2[1]	—	Biddulph *et al.* [59]
	36·0	Part	2	113	585[1]	7·1	102[1]	17·5[1]	—	Biddulph *et al.* [59]

[1]) Calculated by the reviewer.
[2]) Certain cows on some days secreted up to 32·5%.

good dairy practices are followed, DDT may be applied with simple precautions to the barn without contamination of milk.

FREAR *et al.* [216] observed higher concentrations of DDT in milk than were reported by HARRIS and his colleagues. After applications to eight barns at the approximate rate of 2,160 mg/m^2, they found an average of 0·2 ppm on the day of spraying, 0·1 ppm on the first day after spraying, 0·3 ppm on the second day after spraying, and none on the seventh nor on the fourteenth day after spraying. They noted that the contamination on the day of spraying was caused by mechanical transfer, because the cattle were exposed for the first time immediately before the milking operation. It was not possible to distinguish the means of contamination of later samples. It was concluded from the study as a whole that the chance for serious contamination of commercial milk supplies from barn spraying with DDT is remote.

CLABORN *et al.* [127] briefly reported extensive experiments which showed: (1) DDT was actually secreted in the milk two days after barn spraying and, at that time, did not enter the milk through mechanical manipulation, (2) no contamination resulted through inhalation of the insecticide, and (3) much, if not all, of the contamination resulted from spray residues left on feed troughs and drinking cups. The maximum contamination of milk found was 1·4 ppm, and in the one instance where a test was made, DDT was still present in the milk one month after spraying. It was apparent that considerable care is necessary to prevent contamination.

Irrespective of the means of application, it appears that absorption following oral ingestion by the cow is the source of much of the DDT excreted in milk. Evidence to show that this is true in connection with barn spraying has just been mentioned. It has also been shown that, if cattle are prevented from licking themselves, DDT with which they have been directly sprayed does not appear in appreciable quantities in the milk (CLABORN *et al.* [128]).

A considerable number of studies have been made of the secretion of DDT in the milk of dairy cattle maintained on food known to contain the compound. The first was apparently that of ALLEN *et al.* [4], later presented in much greater detail by WILSON *et al.* [667]. This investigation involved a cow fed DDT by capsule, cattle fed artificially contaminated silage, and cattle fed silage made from pea vines which had previously been treated according to accepted agricultural practice. A portion of their data is summarized in Table 20. The authors were unable to demonstrate the presence of any DDT in the milk when recommended treatment of pea vines was used. They did observe a substantial, though apparently very inconstant, decrease in residues associated with the production of silage. This observation was later confirmed by LARDY [341]. The authors reported that the daily ingestion of something over 50 times as much DDT as the cattle took in eating pea vines treated in the recommended way resulted in no injury to cows nor to calves which they bore. These large intakes of DDT did result in concentrations of 15 ppm or greater in the milk. Laboratory rats maintained on milk containing 44 ppm of DDT, as well as an appropriate mineral supplement, showed a failure to gain weight normally.

The effect was corrected when uncontaminated milk was substituted in the diet. No gross pathology was observed at autopsy. No effect was found in rats maintained on milk containing 15 ppm of DDT. These results are probably consistent with those reported by others if the concentration of DDT in the total *dry* diet is taken into account.

ARANT [36] reported that cows fed hay containing 184 ppm of DDT gave milk containing from 10 to 36 ppm of the compound. A calf suckled on such milk stored 825 ppm in its fat. The author considered this an indication of the hazard which would be run by children if they should consume large quantities of milk heavily contaminated by DDT. It ought to be mentioned, however, that the animals studied by ARANT remained well, although the residues to which they were exposed were very much higher than those met in good agricultural practice.

WINGO and CHRISLER [671], using a bioassay, found that cows excreted DDT in their milk within 24 hours after a large dose, but varied widely in the amount secreted in the face of constant intake. No DDT could be detected in the milk two weeks after dosage was discontinued.

When cows were maintained on pea hay which showed on analysis 4·6 ppm of DDT, no contamination of the milk could be detected by DE HEUS and MAAN [160]. They used a modified Schechter-Haller method which was considered sensitive to 0·1 ppm.

The studies reported by LARDY [341] served to confirm earlier reports from the same laboratory as reported by ALLEN and WILSON and their colleagues. Unlike DE HEUS and MAAN, LARDY did detect concentrations of DDT up to 0·5 ppm in the milk of cows which had been maintained on pea and corn silage, which generally had a concentration of DDT well below the maximum of 19 ppm. The plants had been treated according to then acceptable agricultural practice.

CARTER [114] mentioned cooperative work at the University of Maryland in which cows fed pea vine silage which contained DDT at the rate of 2·7 to 5·4 ppm, failed to secrete appreciable amounts of the compound in the milk.

R. F. SMITH *et al.* [539], in an especially good paper, reported the study of 7 cows which were maintained for 79 days, during the winter, on a diet consisting entirely of alfalfa hay bearing a residue ranging from 1·9 to 13·9 ppm and averaging about 7 ppm. For 23 more days, the cows were fed a diet only about half of which consisted of the same hay. The animals had no other known source of DDT except the hay. It was estimated by the authors that during the first period an average of 16·3 kg of hay was consumed per cow and per day. This furnished approximately 126 mg of DDT per cow, or 0·25 mg of DDT per kilogram of body weight. Six cows produced milk throughout the experiment. Within about 10 days the concentration of DDT in the whole milk rose to a herd level of 2·3 to 3·0 ppm, which was maintained rather constant while the cows received only DDT treated hay. There was a reduction of about one third in the concentration of DDT in the milk when the cows were fed a mixed diet, only half of which was contaminated hay. Each cow showed the same pattern of secretion as did the herd but the range of variation was, of course, greater (1·7 to 4·8 ppm). All of the samples of milk taken 24 days after the contaminated hay was completely removed from the diet showed at least a trace of DDT, but not over 0·5 ppm.

Several observations by SMITH and his colleagues are of particular interest. The colostrum milk of a cow that had been maintained on DDT-contaminated hay for 79 days, and on this hay and other food for 9 days more, contained essentially the same concentration of DDT as did milk from the 6 cows which had given milk during the entire experiment. Similar observations have been made by others (LARDY [341]; SHEPHERD et al. [524]). The authors also found

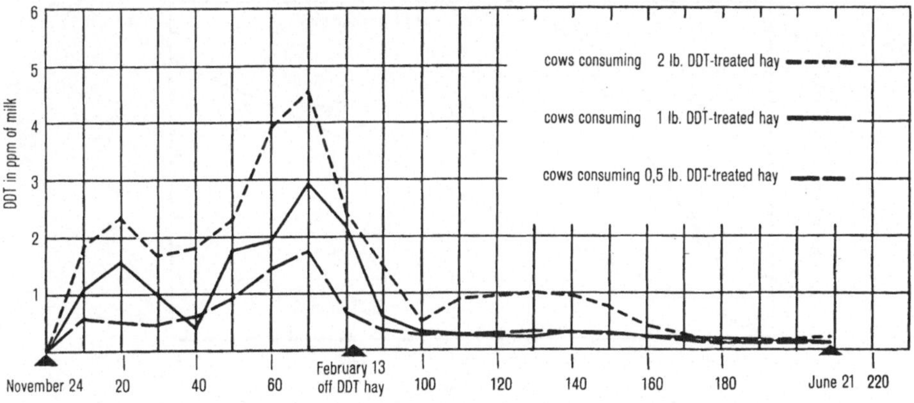

Figure 13

DDT in milk of cows fed alfalfa hay dusted with technical DDT. Feeding of the hay was begun on November 24, 1948, and discontinued on February 13, 1949, as indicated. The figure is taken from BIDDULPH et al. [59], and is reproduced by permission of the American Chemical Society.

that the cows secreted about one-fourth of their total daily intake of DDT in the milk (Table 20), and that a reduction in intake was associated with a reduction in secretion. From all these facts they concluded that absorbed DDT circulates in the blood long enough for a considerable fraction to be removed during the secretion of milkfat so that, under the practical conditions of the experiment, any exchange between circulating and stored DDT had only a minor effect on the amount in the milk.

SMITH and his associates found about ten times as much DDT in the milk of cows fed alfalfa hay as LARDY [341] reported for cows fed corn silage, even though the intake of DDT was approximately the same in each instance. SMITH and his colleagues speculated that the difference might be accounted for by the laxative character of the silage. They concluded that, whatever the explanation might be, the type of diet was very important in evaluating the hazard of DDT residues.

158 W. J. Hayes, Jr.

Shepherd *et al.* [524] reported on 3 cows fed a varied diet, including alfalfa hay bearing residues ranging from 74·2 to 134·5 ppm. This deposit resulted from the application of about four times as much DDT to the crop as was required for insect control. The milk contained up to 10·1 ppm of DDT, the average for each cow ranging from 2·2 to 7·3 ppm. A fourth cow was fed hay bearing residues ranging from 4·4 to 30·5 ppm resulting from an application to the crops considered practical for insect control. This cow excreted up to 0·9 ppm. When the feeding of large quantities of DDT was discontinued the compound was detected in the milk for 160 to 170 additional days but only for 30 to 40 additional days following the discontinuation of small quantities.

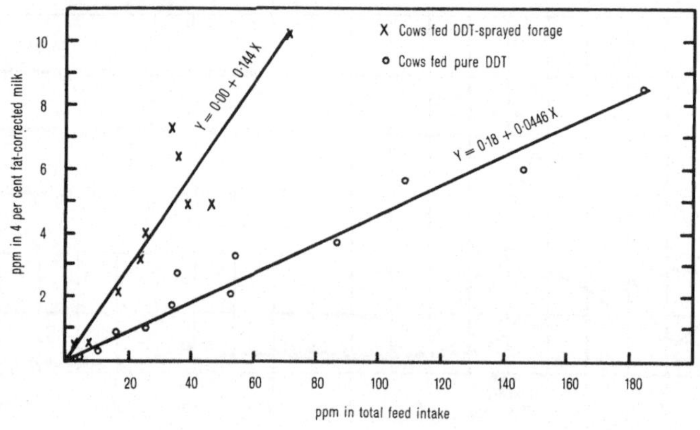

Figure 14

Relationship between DDT intake of cows and the concentration of the compound in their milk. The figure is taken with slight modification from Ely *et al.* [197].

Biddulph and his colleagues [59] reported well designed and well executed experiments which were replicated in succeeding years. Cows were fed a diet consisting, in part, of alfalfa hay which had been dusted during the growing season at the rates of 56, 112, 224, and 448 mg/m². The rates and the method of application were similar to those which might be used under practical conditions. Part of the results are shown in Figure 13 taken directly from the original publication. The results are essentially similar to those obtained during the first year of study, although the corresponding curves were a little lower than those for the second year. The curves differ slightly from those presented by Smith and his associates [539] in that the initial rise is less rapid. The concentration of DDT in the milk found by Biddulph *et al.* was slightly less for corresponding residues, but this is easily explained by the fact that the cows studied by Biddulph had supplementary feeding during the experiment, while those studied by Smith received contaminated hay as their only diet. During the second year, Biddulph *et al.* observed that the concentration of

DDT in milk showed an initial rise, followed by a decline, and then a second rise higher than the first. This peculiar finding would appear to be an experimental artifact. However, an effort was made to keep the intake constant, and the same thing was observed by R. F. SMITH *et al.* [539]. The phenomenon of a double peak may, therefore, be real but it cannot be explained at this time.

ELY *et al.* [197], [198] found that, when cows were fed 50 to 2,000 mg of DDT per day either in a crystalline or in a dissolved form, the DDT concentration of the milk was in direct proportion to the quantity fed. They presented a graph which indicated that the same was true when DDT was fed in the form of residues on hay. However, a greater proportion of the total DDT intake appeared in the milk when the compound was taken in the form of residues. Of course, the validity of this conclusion depends on the accuracy of the analysis of the residues. The presence of more DDT on the hay than could be detected chemically would tend to give the same apparent result regarding storage as would a greater facility of the cows to absorb residues. In this connection, it is interesting to note that there was no consistent difference in the excretion of DDT in the milk when the compound was fed as a soybean oil solution either in capsules or mixed with grain, or as crystalline DDT fed either in capsules or mixed with grain. Also, the addition of small amounts of detergents to the oil solution did not influence the excretion of DDT in the milk. Their results are shown in Table 20 and Figure 14.

Unfortunately, few reports have been made of the DDT content of food offered for general sale. Forty-seven samples of fluid milk taken in widely scattered states of the United States showed a maximum of 0·36 ppm of DDT, and 6 samples of evaporated milk showed up to 0·28 ppm. The average value was 0·08 and 0·09 ppm respectively. (This average was made by arbitrarily assuming that values reported as a 'trace' were 0·04 ppm, since the limit of accurate analysis and the lowest amount actually reported as a figure was 0·05 ppm.) In an earlier study, 120 samples of fluid milk had shown a range up to 0·32 ppm of DDT, and 37 samples of evaporated milk had shown up to 0·70 ppm (U.S. Food and Drug Administration [624]).

Residues in Butter

Early studies of DDT in relation to milk showed that the compound was concentrated in the cream (TELFORD [580]; SPICER *et al.* [544]). R. F. SMITH *et al.* [539] observed that the DDT content of a sample of butter which they analyzed was in close agreement with the figure calculated from the DDT content of the whole milk, and the percentage of butterfat, assuming all of the compound to be in the latter. They reported one sample of butter which contained 65 ppm of DDT.

The butterfat of milk often varies from 3·0 to 5·5%. In many places market milk is required to contain a minimum of 3·25% butterfat. Butter contains about 80% butterfat.

Five samples of butter offered for sale in June 1951 showed no more than a trace of DDT. In an earlier study concentrations as high as 2·0 ppm were found, while oleomargarine showed none (U.S. Food and Drug Administration [624]).

Residues in Meat

When studies were first made on meat which might form food for man, it was already known from laboratory experiments that DDT is stored preferentially in the fatty tissue. It is even possible to account for much if not all of the DDT stored in other tissues on the assumption that all the DDT in them is dissolved in the fat which they contain.

Many reports on the insecticide contamination of meat are confined to analyses of the fat. Furthermore, many of the studies involve the exposure of livestock to much larger amounts of DDT than they would ever encounter in agricultural practice. It should be remembered that edible meat as sold averages about 18% of fat (CLABORN *et al.* [129]). This figure may be used to obtain a reasonable estimate of the concentration of the compound in ready-to-cook meat.

Apparently, the earliest studies of the storage of DDT in livestock was that of ALLEN *et al.* [4], which was later reported in detail by WILSON *et al.* [667]. The authors found that the fat of a cow which had been fed DDT by capsule at the rate of 24 g per day (approximately 44 mg/kg) for five months contained 380 ppm of DDT about one month after the feeding period. A dairy cow which had received 3·61 mg of DDT per kilogram of body weight for 127 days showed 221 ppm in its fat. A calf, whose mother had been fed DDT contaminated silage for 102 days before the birth, was given milk from the same herd containing about 15 ppm for 33 days before slaughter; fat taken from the calf contained 305 ppm of DDT. Ewes which were fed silage containing an estimated 106 ppm of DDT showed 37 to 145 ppm of DDT in the fat and 1 ppm in the muscle when they were killed. A suckling lamb showed 43 ppm in the fat at the same time. Other lambs killed a month or more after feeding of the contaminated silage was discontinued showed 9 to 10 ppm in the fat. Control animals showed up to 12 ppm of DDT in their fat and up to 1 ppm in their muscles when analyzed in the same way.

ARANT [36] reported that steers fed hay containing 184 ppm of DDT for 90 to 105 days contained 60 to 160 ppm of DDT in the fat. A calf suckled by a cow fed on similar hay stored 825 ppm in its fat. A steer fed for 143 days on hay containing 48 ppm of DDT stored 80 to 88 ppm in its fat. Two steers fed unshucked maize with a residue of 15 ppm for 105 days stored 46 to 65 ppm of the compound in their adipose tissue, but pigs on the same diet for 102 days stored only 13 ppm.

LARDY [341] reported that steers fed corn (maize) silage containing about 10 ppm of DDT for 108 days stored an average of 12 ppm in their fat, while those fed silage made from only the husks and cobs of the corn stored an average of 17 ppm in the fat, although the silage contained only about 1 ppm when fed. The author believed that the high oil content of the germ retained in the cobs was responsible for the greater absorption of the small amount of DDT present in the husk-cob silage.

Groups of 6 pigs each were fed for 36 days on diets containing 5 and 3 ppm of DDT, respectively. In each instance, one-fourth of the diet was made up of beef (lean and fat ground together in the normal proportion). The DDT content resulted from the fact that the steers and calves from which the meat was taken had been maintained on hay containing 48 to 184 ppm of DDT (steers), or on milk from cows eating this hay (calves). The DDT content of the pigs is shown in Table 21. It is particularly noteworthy that 49 to 57% of the total dietary intake of DDT was recovered in the fat and muscle portions together (CARTER [114]; CARTER et al. [117]).

Table 21

DDT Content of Meat and Fat from Pigs Fed Beef Containing the Compound as Reported by CARTER [114] *and* CARTER *et al.* [117]

Material	DDT content (ppm)	
	From pigs fed 5 ppm of DDT	From pigs fed 3 ppm of DDT
Lean meat	2·0	1·7
External and intermuscular fat .	15·6	11·4
Leaf fat.	17·6	14·6

HARRIS *et al.* [262], [263] reported the results of feeding-DDT-contaminated alfalfa hay to lambs. The DDT had been applied to the alfalfa according to the agricultural practice common in the area. Other lambs received the same or greater amounts of the compound in capsules. Of particular interest is the fact, illustrated in Figure 15, that the lambs apparently stored a greater proportion of the DDT ingested in the form of residues than of the DDT ingested in capsules. However, the caution regarding chemical analysis must be considered here just as it must be considered in connection with the very similar figure, number 14. Even at the highest dosage levels studied, only very small concentrations of DDT were found in the muscle, liver, kidney, and brain of the lambs.

The exceptionally good experimental design of the study just mentioned was more completely described in another paper (HARRIS *et al.* [264]) in which further results were presented. No significant differences were noted in the hemoglobin, plasma inorganic phosphorus, or plasma calcium of lambs fed DDT as compared with lambs which did not receive the compound. No gross or histological changes were observed in the liver, gall bladder, kidney, cerebrum, cereelblum, medulla, hind and fore leg muscle, or thyroid of the DDT-treated animals.

Related to the papers just discussed was an experiment in which the fat and leg muscle from lambs and the butterfat from milk were fed to rats for 12 or more weeks (GREENWOOD *et al.* [247]). The observations regarding storage of DDT in the rats is described above (p.69). It may be mentioned here that

162   W. J. Hayes, Jr.

no pathology, decrease in growth rate, or other sign of injury was observed in
the rats whose diets contained up to 123 ppm of DDT.

Reporting on a related experiment, BIDDULPH *et al.* [59] stated that a cow which
had been fed alfalfa hay bearing residues of 9·6 ppm of DDT for 81 days stored
19·3 to 21·4 ppm in the fatty tissue. Residues of 19·2 ppm fed for 113 days resulted
in the storage of 89 to 90 ppm in fat. The same animals showed only traces of the

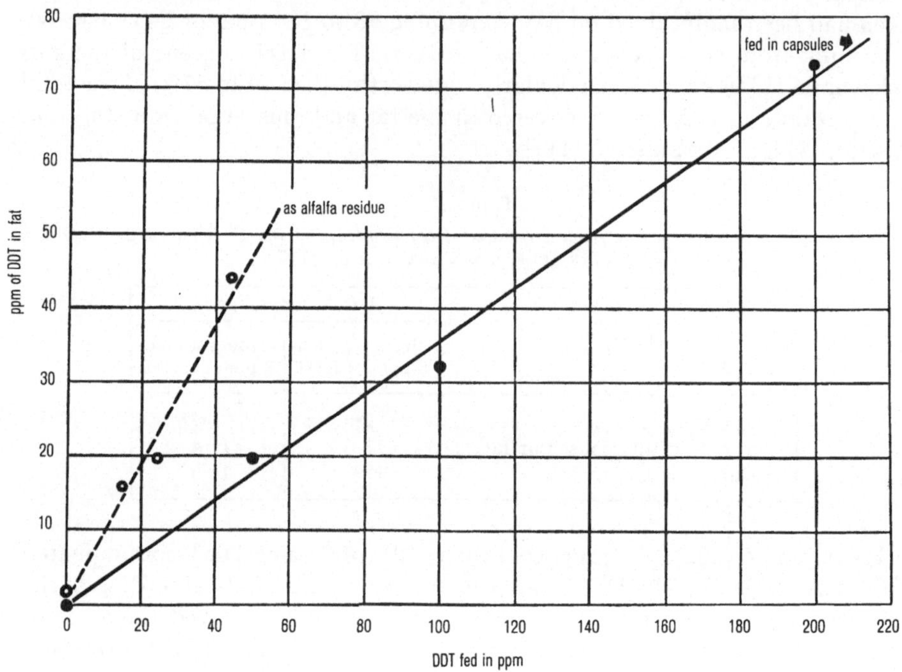

Figure 15

Relation between DDT ingested by sheep and the concentration of the compound in their
mesenteric and kidney fat. The figure is taken from HARRIS *et al.* [263] and is reproduced by
permission of the Archives of Biochemistry.

compound in the muscle, liver, or kidney. An animal fed the smaller residue
still showed 3 ppm of DDT in its fat 4 months after feeding was discontinued.

CLABORN *et al.* [129] in a paper devoted chiefly to other insecticides, showed
that sheep fed 10 ppm of DDT for 4 weeks stored 3·1 ppm in their fat, while
calves fed at the same rate stored 6·8 ppm.

A paper by MARSDEN and BIRD [398] is of very limited interest because the
authors studied food contaminated to an extremely high and unrealistic degree.
They reported that turkeys fed diets containing 190 and 1,500 ppm for 7 to
8 weeks stored 1,344 to 6,235 ppm as shown in Figure 11.

In a similar way, RUBIN *et al.* [501] reported that chickens fed 2,500 ppm
of DDT in their diet stored as much as 13,690 ppm in the fat, 122 ppm in the
muscle, 1,150 ppm in the kidney.

Of far greater interest were the studies of BRYSON *et al.* [93], in which chickens were fed a mash containing about the usual amount of alfalfa meal (15%) made from alfalfa which had been treated with DDT at rates commonly used in the area. Finished diets with final concentrations of 0, 2·3, 3·3, and 6·3 ppm were used in different tests. In addition, meals containing 50, 100, and 200 ppm of DDT were made by the direct addition of the compound. These were used to determine a possible margin of safety for feeding DDT-treated alfalfa. Hens maintained on the diets containing the different concentrations of DDT for one year were then killed. The leg and breast muscle and the fat were analyzed for DDT. The concentration in the fat was markedly higher than that in the diet. The values for fat are consistent with those for other species shown in Figure 11. Only small amounts of DDT were found in the leg or breast muscle of the hens.

Six samples of beef fat ready for market in June 1951, showed up to 2·4 ppm of DDT, but most samples showed only a trace. In the same study, lard analyzed up to 1·4 ppm. In an earlier study the highest concentration found in lard was 0·8 ppm (U. S. Food and Drug Administration [624]).

Residues in Eggs

RUBIN *et al* [501] reported that chickens fed a diet containing 310 ppm of DDT produced eggs which contained 180 ppm of the compound. Birds on higher dosages stored up to 360 ppm in their eggs.

More important, BRYSON *et al.* [93] showed that, when hens were maintained on a mash containing residues of DDT which might occur under practical conditions, the highest concentration of DDT in the eggs was small. The hens were maintained on mash containing 15% of alfalfa meal, which is about the usual proportion. The meal was made from alfalfa which had received a single application of 0, 112, 224, and 448 mg of DDT per square meter, respectively, during the growing season. Eggs produced by four groups of hens whose diets contained the different alfalfa meals showed a maximum concentration of 1·7, 2·1, 3·7, and 4·1 ppm, respectively. In a complete repetition of the experiment, similar groups of chickens produced eggs with 1·2, 3·6, 3·4, and 8·3 ppm, respectively, as maximal values. Other hens fed diets to which DDT had been added at the concentration of 50, 100, and 200 ppm produced eggs containing a maximum of 23·3, 18·7, and 65·6 ppm, respectively. Similar results were reported by DRAPER *et al.* [180].

Residues in Vegetable Products

The importance of the problem of residues resulting from the applications of insecticides to food plants has been generally admitted (National Research Council [420]; WILSON [666]; FLECK [209]; COX [141]). Some of the best discussions of the matter are those of DECKER [152] and HOSKINS [301].

Insecticide residues on plants generally arise from direct application of the poison to the growing plant some time before harvest. Residues from translocation out of the soil or from contamination of vegetable products after harvest must also be considered.

Direct application may be made by means of hand- or power-driven ground equipment or by aircraft. In any instance, DDT is frequently mixed with other insecticides or with fungicides so that the resulting residue is complex.

Some insecticides are actively absorbed by the leaves, stems, or roots of plants and translocated to all parts of the plant including those not exposed to the poison. DDT does not seem to be absorbed under ordinary circumstances and is not translocated. Applications to the soil do not form residues in plants and applications to leaves and stems do not form residues in roots, tubers, or other completely unexposed parts (CARTER [114]; HOSKINS [300], [301]; GINSBURG *et al.* [234]).

The translocation of DDT under highly artificial laboratory conditions has been reported. By using a wick to introduce a polyethylene glycol dispersion of DDT directly into the stem of bean plants at ground level for four weeks, the top leaves were rendered slightly toxic to a test insect so that 26 of a sample of 30 died in 12 days. The leaves were reported to contain 11·3 ppm of DDT but the method of analysis was not given (HUFFAKER [307]).

Contamination of foodstuffs after harvest may occur through accident or through intentional addition of the compound. Accidental contamination of human food with DDT is unlikely if appropriate care is used with this essentially nonvolatile chemical. However, the inadvertent contamination of animal food is often difficult to avoid if the barns where the food is stored are sprayed. As discussed below, DDT may be purposely added to grain to prevent its destruction by insects. The addition of DDT to whiskey of low alcohol content—to improve flavor—has been reported (KIVIRANTA [327]). The intentional addition of DDT to drinking water is discussed in a separate section below (p. 180).

When DDT is applied directly to living plants or to plant products, the amount of residue may be conditioned by one or more of the following factors:

(a) Character of the formulation.
(b) Number of applications.
(c) Weather between application and harvest.
(d) Time from application to harvest. (This involves two separate concepts; changes in the residue and growth of the plant.)
(e) Character of the plant.
(f) Method of harvest.

The influence of several of the enumerated factors is illustrated in Table 22. Considering the number and the variability of the factors involved, the results from different laboratories are in excellent agreement. However, the range of results for single experiments indicates that only rather broad generalizations may be drawn about the magnitude of residue to be expected from any particular procedure.

Table 22

Residues of DDT Remaining at Different Intervals on Fruit, Vegetables, Grains, and Forage Crops Following Various Methods of Application

Kind of food	Intended rate of each DDT dosage mg/m²	Formulation	No. app.	Last dose to harvest days	Residues (expressed in ppm) remaining at different numbers of days (0–27) after the last application of insecticide and residue remaining at harvest					Authority
					0	1–6	7–13	14–27	Harvest	
FRUITS										
Apples	—	0·06–0·18% Spray	4	—	—	—	—	—	1–4	HARMAN [259]
Apples	—	0·12% Spray	2	93	—	—	—	—	0·5–1	MANALO et al. [393]
Apples	—	0·12% Spray	3–7	—	—	—	—	1–3	—	MANALO et al. [393]
Apples	—	0·09–0·12% Spray	5	83	—	—	—	10–13	6–10	GRAHAM and CORY [244]
Apples	—	0·09–0·12% Spray	3	83	—	—	—	11–18	3–12	GRAHAM and CORY [244]
Apples	—	0·12% Spray	5	26	—	—	—	—	3–6[1]	HARMAN [260]
Apples	—	0·12% Spray	4	—	—	—	—	—	1–13	CARTER [114]
Apples	—	0·06–0·12% Spray	2	23	—	—	—	—	4	HOSKINS [301]
Apples	—	0·09% Spray	1	78	—	—	—	—	0·5	HOSKINS [301]
Apples	—	0·12% Spray	3	75	—	—	—	—	1	BARNES et al. [48]
Apples	—	0·06% Spray	4	50	—	—	—	—	0·3–2	BARNES et al. [48]
Apples	—	0·06% Spray	5–6	40	—	—	—	—	2–4	BARNES et al. [48]
Apples	—	0·12% Spray	6	60	—	—	—	—	5	BARNES et al. [48]
Apples	—	0·06–0·12% Spray	1–3	96–117	—	—	—	—	1–3	WESTLAKE and FAHEY [661]
Apples	—	—	6–7	34–47	—	—	—	—	4–12	WESTLAKE and FAHEY [661]
Apricots	—	0·09% Spray	1	64–81	19–23	—	—	—	<0·1–3	HOSKINS [301]
Apricots	—	0·09% Spray	2	81	31	—	—	—	<0·1	HOSKINS [301]
Apricots	224	5% Dust	1	52	2	—	—	—	0·3	HOSKINS [301]
Blueberries	—	10% Dust	2	—	—	—	—	—	—	MANALO et al. [393]
Currants	129	0·06% Spray	2[2]	37	—	—	—	—	2–3	ALLEN et al. [6]

Kind of food	Intended rate of each DDT dosage mg/m^2	Formulation	No. app.	Last dose to harvest days	Residues (expressed in ppm) remaining at different numbers of days (0–27) after the last application of insecticide and residue remaining at harvest					Authority
					0	1–6	7–13	14–27	Harvest	
Gooseberries . .	258	—	2[2])	39	—	—	—	—	6	Allen et al. [6]
Grapes	—	0·09% Spray	4	—	—	—	—	—	20[3])[4])	Frear and Cox [215]
Grapes	—	0·09% Spray	4	—	—	—	—	—	8	Frear and Cox [215]
Grapes	—	0·09–0·12% Spray	2–4	45	—	—	—	1–7	2–3	Manalo et al. [393]
Grapes	—	0·09% Spray and oil	1	—	8	—	—	3	—	Reiber and Stafford [483]
Grapes	—	5% Dust	1	> 30	—	—	—	—	0·7–2	Reiber and Stafford [483]
Grapes	—	0·09% Spray	1	—	—	—	—	—	0·9	Taschenberg [573]
Grapes	—	0·09% Spray	2	—	—	—	—	—	2–4	Taschenberg [573]
Grapes	—	0·09% Spray	3	—	—	—	—	—	4–10[3])[14])	Taschenberg and Avens [574]
Grapes	—	0·09% Spray	4	—	—	—	—	—	7–16[3])[14])	
Grapes	112	5% Dust	1	147	—	—	—	—	0	Hoskins [301]
Grapes	56–112	5% Dust	2	59	—	—	—	—	0·7	Hoskins [301]
Lemons[15]) . . .	—	0·24% Spray	1	—	—	14–24	—	—	2	Barnes et al. [48]
Lemons	—	0·24% Spray	1	—	—	16	—	6	2–3	Barnes et al. [48]
Lemons[16]) . . .	—	0·24% Spray	1	—	—	6–12	—	—	—	Carman et al. [111]
Olives	—	0·12% Spray	1	—	—	—	—	—	—	Reiber and Stafford [483]
Olive oil	—	0·09% Spray	1	74	—	—	—	—	116	Reiber and Stafford [483]
Olives	—	0·12% Spray	1	—	—	—	47	—	3	Hoskins [301]
Oranges[15]) . . .	—	0·24% Spray	1	141	—	9–17	—	—	2	Barnes et al. [48]
Oranges	—	0·24% Spray	2	25–88	—	—	—	—	6–8	Barnes et al. [48]

Kind of food	Intended rate of each DDT dosage mg/m²	Formulation	No. app.	Last dose to harvest days	Residues (expressed in ppm) remaining at different numbers of days (0−27) after the last application of insecticide and residue remaining at harvest					Authority
					0	1–6	7–13	14–27	Harvest	
Oranges[16]) . . .	—	0·09% Spray	1	60–210	—	—	—	—	0·6–5	CARMAN et al. [111]
Oranges	—	0·24% Spray	1	—	—	—	—	—	5–37[5])	CARMAN et al. [111]
Peaches	—	0·09–0·12% Spray	1–2	66	—	—	—	0·4–2	2–3	MANALO et al. [393]
Peaches	—	5–10% Dust	3	—	—	—	—	4	—	MANALO et al. [393]
Peaches	—	0·12% Spray	4	28	28	—	—	—	8	WHEELER and LAPLANTE [662]
Peaches	—	0·1% Spray	2	21	—	—	—	—	9[6])	HELSON [269]
Peaches	—	0·09% Spray	1	21	—	15	—	—	3	YETTER [683]
Peaches	—	0·12% Spray	2	—	—	—	—	—	6–23	CARTER [114]
Peaches	—	0·12% Spray	4	93	—	—	—	—	2	HAMILTON [258]
Peaches	—	0·06–0·12% Spray	2–3	36	—	—	—	—	7–8	HAMILTON [258]
Peaches	—	0·06% Spray	1	30	16	—	8	4	3	BARNES et al. [48]
Peaches	—	0·12% Spray	2–3	11–22	—	—	—	—	4–10	BRUNSON and KOBLITSKY [92]
Peaches	—	5% Dust	2	13	—	—	—	—	4	BRUNSON and KOBLITSKY [92]
Pears	561	50% Dust	2	—	7–9	—	—	—	0·4–11	BORDEN [82]
Pears	841–1121	50% Dust	2	—	8–13	—	—	—	0·5–2	BORDEN [82]
Pears	—	0·21% Spray	3	60	10	—	—	—	0·1	BORDEN [82]
Pears	—	0·12% Spray	4	40	—	—	—	10	4	HOSKINS [301]
Pears	—	0·12% Spray	1	22–34	—	11	—	—	3	HOSKINS [301]
Pears	—	0·12–0·24% Spray	1	110	—	—	—	—	0·4–0·9	BARNES et al. [48]
Pears	—	0·06% Spray	2–3	40–85	—	—	—	—	0·9–2	BARNES et al. [48]
Pears	—	0·06% Spray	4	14	—	—	—	—	3	BARNES el al. [48]
Prunes.	—	0·12% Spray	1–2	79	—	—	—	—	0·7–1	HOSKINS [301]

Kind of food	Intended rate of each DDT dosage mg/m²	Formulation	No. app.	Last dose to harvest days	Residues (expressed in ppm) remaining at different numbers of days (0–27) after the last application of insecticide and residue remaining at harvest					Authority
					0	1–6	7–13	14–27	Harvest	
VEGETABLES										
Roots and tubers	—	—	—	—	—	—	—	—	—	—
Leaf and stem vegetables										
Asparagus . . .	—	5% Dust	1	3	—	—	—	—	6[7]	ROBINSON [491]
Broccoli	—	5% Dust	1	1–3	—	—	—	—	5–10	ROBINSON [491]
Cabbage[17] . . .	—	5% Dust	1	—	—	0·4	—	21[8]	—	MANALO *et al.* [393]
Cabbage[17] . . .	—	3% Dust	2	36	—	—	—	—	4	REIBER and STAFFORD [483]
Cabbage[18] . . .	—	5% Dust	1	—	—	0·9	—	—	—	MANALO *et al.* [393]
Cabbage[18] . . .	—	3% Dust	2	36	—	—	—	—	1	REIBER and STAFFORD [483]
Cabbage	22	1·5% Dust	6	31	—	—	—	—	0·2–16	CLARK [130]
Cabbage	45	0·1% Spray	6	31	—	—	—	—	0·5–2	CLARK [130]
Cauliflower . . .	—	5% Dust	1	—	—	—	—	0·6–0·8	—	MANALO *et al.* [393]
Celery[17]	—	5% Dust	1	—	—	13	—	2	—	MANALO *et al.* [393]
Celery[17]	118	3% Dust	3	13	—	—	—	—	1–3	ALLEN and BERCK [5]
Celery[17]	118–280	3–5% Dust	7	4	—	—	—	—	55–33[9]	ALLEN and BERCK [5]
Celery[18]	—	5% Dust	1	—	—	0·5	—	1	—	MANALO *et al.* [393]
Celery[18]	118–280	3–5% Dust	7	4	—	—	—	—	2–26[10]	ALLEN and BERCK [5]
Lettuce	—	3% Dust	3–5	2–17	—	—	—	—	4–211[11]	ASHDOWN and WATKINS [37]
Lettuce	—	3% Dust	1–2	24–31	—	—	—	—	0·2[11]	ASHDOWN and WATKINS [37]
Lettuce	208–280	5% Dust	3–5	16–24	7–62	—	—	—	0·0–0·5[11]	SLOAN *et al.* [530]
Lettuce	56	W.W. Spray	4	14–25	9–68	—	2–6	0·5–3	0·0–0·4[11]	SLOAN *et al.* [530]

Kind of food	Intended rate of each DDT dosage mg/m²	Formulation	No. app.	Last dose to harvest days	Residues (expressed in ppm) remaining at different numbers of days (0–27) after the last application of insecticide and residue remaining at harvest					Authority
					0	1–6	7–13	14–27	Harvest	
Lettuce	56	W.W. Spray	5	11–17	25–572	—	4–10	10	0·1–0·4[11]	Sloan et al. [530]
Lettuce . . .	56	W.W. Spray	6	4–6	31–100	—	—	—	2–5[11]	Sloan et al. [530]
Lettuce	56	0·1–1·0% Emulsion	4	18–23	9–82	43–51	5–13	0·1–3	0·0–0·8[11]	Sloan et al. [530]
Lettuce	56	0·1–1·0% Emulsion	5	20–28	32–192	—	3–9	0·7–2	0·0–0·5[11]	Sloan et al. [530]
Turnips (tops) .	—	5% Dust	1	—	—	15–17	—	5–6	—	Manalo et al. [393]
Flower, fruit, and seed vegetables										
Beans, green . .	—	1% Dust	2	—	—	—	0·3	—	—	Manalo et al. [393]
Beans, green . .	—	3% Dust	2	—	—	—	2	—	—	Manalo et al. [393]
Beans, green . .	—	5% Dust	2	—	—	—	3	—	—	Manalo et al. [393]
Beans, green . .	—	0·12% Spray	2	3	—	11	—	—	—	Reiber and Stafford [483]
Beans, green . .	—	0·12% Spray	2	15	—	—	—	3	—	Reiber and Stafford [483]
Peas, green . . .	224	5% Dust	1	—	—	—	—	—	0	Lardy [341] Wilson et al. [668]
Peas, green . . .	56–112	Dust	1	—	—	—	—	—	0	Carter [114]
Peas, green . . .	—	5% Dust	1	7	—	—	—	—	Trace	Robinson [491]
Tomatoes . . .	—	3% Dust	1	7	—	—	—	—	2	Robinson [491]
Tomatoes . . .	168	0·18% Spray	1	—	—	1	0·3	—	—	Hoskins [301]
Tomatoes . . .	168	50% Dust	1	—	—	0·8–1	0·2	—	—	Hoskins [301]
Grain products										
Corn, fresh . . .	112	Spray and dust	4	—	—	—	—	—	0·1	Decker et al. [154]
Corn, fresh . . .	561	3% Dust	5	12	—	—	—	—	0	Ginsburg et al. [234]
Corn, dry . . .	—	3% Dust	4	8–19	—	—	—	—	0[12]	Ginsburg et al. [234]
Corn, dry . . .	224	5% Dust	3	59	—	—	—	—	0	Ginsburg and Filmer [232]
Corn, dry . . .	224	Emulsion	1	54	—	—	—	—	0	Ginsburg and Filmer [232]

Kind of food	Intended rate of each DDT dosage mg/m²	Formulation	No. app.	Last dose to harvest days	Residues (expressed in ppm) remaining at different numbers of days (0–27) after the last application of insecticide and residue remaining at harvest					Authority
					0	1–6	7–13	14–27	Harvest	
FORAGE CROPS[13])										
Alfalfa	112–224	Dusts	1	—	—	—	—	—	2–48	CARTER [114]
Alfalfa	101	3% Dust	1–3	—	117	—	68	41	—	EDEN and ARANT [193]
Alfalfa	51	3% Dust	1–3	—	38	—	35	22	—	EDEN and ARANT [193]
Alfalfa	28	0·6% Spray	—	—	24	—	0–8	0–12	—	R. F. SMITH et al. [538]
Alfalfa	—	3% Dust	1	—	18–35	—	—	0–3	—	GINSBURG et al. [234]
Alfalfa	56	Dust	1	—	—	—	—	—	9[12])	LIEBERMAN et al. [366]
Alfalfa	112	Dust	1	—	—	—	—	—	9–19[12])	LIEBERMAN et al. [366]
Alfalfa	224	Dust	1	—	—	—	—	—	12–30	LIEBERMAN et al. [366]
Alfalfa	448	Dust	1	—	—	—	—	—	36–42	LIEBERMAN et al. [366]
Corn (maize) . .	140	5% Dust	4	—	71	40	—	8–22	—	DECKER et al. [154]
Corn (maize) . .	785	Spray	4	—	—	—	—	—	190	QUESTEL and COUNIN [474]
Corn (maize) . .	112	Spray	4	—	—	—	—	—	18	QUESTEL and COUNIN [474]
Corn (maize) . .	—	3% Dust	4	—	—	—	18	0·4	—	GINSBURG et al. [233]
Corn (maize) . .	392	10% Dust	3	21	—	—	—	—	32	HOSKINS [301]
Corn (maize) . .	—	3% Dust	5	12	—	—	—	—	13	GINSBURG et al. [234]
Corn (maize) . .	224	5% Dust	3	59	—	—	—	—	7–24	GINSBURG and FILMER [232]
Corn (maize) . .	224	Oil emulsion	1	54	—	—	—	—	21–51	GINSBURG and FILMER [232]
Pea vines . . .	211	2·5% Dust	1	—	5	—	—	—	2	
Pea vines . . .	196	5% Dust	1	—	4	—	3	—	2	WILSON et al. [668]
Pea vines . . .	224	5% Dust	1	—	6	—	3	4	—	LARDY [341] (part)
Pea vines . . .	280	5% Dust	1	—	9–13	—	—	—	—	
Pea vines . . .	34–56	Aerosols	1	—	—	—	—	—	15–50	CARTER [114]
Pea vines . . .	56–112	Dusts	1	—	—	—	—	—	2–10	CARTER [114]

Kind of food	Intended rate of each DDT dosage mg/m²	Formulation	No. app.	Last dose to harvest days	Residues (expressed in ppm) remaining at different numbers of days (0–27) after the last application of insecticide and residue remaining at harvest					Authority
					0	1–6	7–13	14–27	Harvest	
Pea vines . . .	—	5% Dust	1	—	—	—	—	—	0·8–22	Robinson [491]
Pea vines . . .	112	Spray	1	31	—	—	—	3	0·4	Ginsburg et al. [233]
Peanut hay . .	81	2% Dust	1	9	—	—	—	—	9	Vinson and Arant [636]
Peanut hay . .	64	2% Dust	1	14	—	—	—	—	3	Vinson and Arant [636]
Peanut hay . .	64	2% Dust	1	25	—	—	—	—	2	Vinson and Arant [636]
Peanut hay . .	73	2% Dust	1	35	—	—	—	—	0	Vinson and Arant [636]

1) Based on apples harvested in the usual way; when special care was taken not to remove the residue, the values ranged from 6·1 to 9·1 ppm.
2) Postbloom sprays.
3) Juice from these grapes contained no DDT.
4) Dried pomace from these grapes contained 164 ppm of DDT.
5) Pulp of these same citrus fruits contained no DDT.
6) Similar peaches showed no residue after canning.
7) The asparagus showed 1·7 ppm after processing.
8) Taken from a part of the plant generally discarded for human food but sometimes fed to animals.
9) When commercially washed and ready for market, this celery showed 1·1–4·5 ppm.
10) When commercially washed and ready for market, this celery showed 0·6–4·4 ppm.
11) Values given for harvest are for trimmed heads. In most instances, it was shown by analysis that the outer leaves, usually left in the field, had somewhat higher residues.
12) Dry weight basis.
13) Food for cattle and other animals.
14) Jam from these grapes contained no DDT.
15) Whole fruit.
16) Peel only.
17) Outer portion.
18) Inner portion.

Character of the Formulation

A number of values in Table 22 illustrate increased residues resulting from the use of formulations of increasing concentration.

The physical character of the application helps to determine the amount of residues. Under practical conditions dusts usually give less residue than liquid formulations as illustrated in Table 23.

Table 23

Residues of DDT 20 Days After the Last of Four Applications on Sweet Corn (Maize) Using Three Formulations at the Same Rate of Approximately 112 mg/m². The Figure is Taken from DECKER [152] *and is Reproduced by Permission of the Journal of Economic Entomology*

Formulation used	DDT residues (ppm)		
	Leaves	Husks	Kernels
Oil solution	117·3	2·1	0·08
Water emulsion	33·2	1·2	0·04
5% Dust	14·9	0·3	0·07

The importance of oils in increasing residue at harvest has been observed (FREAR and COX [215]; STEINER *et al.* [550]; DECKER *et al.* [154]; BORDEN [82]; GRAHAM and CORY [241]; ROBINSON [491]; WALKER [642]; TASCHENBERG [573]; TASCHENBERG *et al.* [575]; GINSBURG and FILMER [232]) and this, too, is illustrated by Table 23. However, at least some investigators have found that

Table 24

Recovery of DDT from Alfalfa Hay After Treatment at 10-Day Intervals of Wet Foliage with 3 per cent DDT Dust at the Rate of 3,363 mg/m². The Calculated Residue of 543 ppm per Application is Based on the Rate of Application and the Final Yield of 750 kg of Dry Alfalfa per Acre. The Table is Taken from EDEN and ARANT [193] *and Reproduced by Permission of the Journal of Economic Entomology*

No. of applications	Calculated total application ppm	DDT recovered after stated period following last application of dust									
		0 days		10 days		20 days		30 days		40 days	
		ppm	%	ppm	%	ppm	%	ppm	%	ppm	%
3	1629	162	9·9	70	4·3	40	2·5	36	2·2	32	2·0
2	1086	147	13·5	70	6·4	65	6·0	60	5·5	50	4·6
1	543	132	24·3	60	11·0	50	9·2	41	7·5	32	5·9
Average		147	15·9	67	7·3	52	5·9	46	5·1	38	4·1

formulations containing oil are not more effective for insect control than those without oil (GRAHAM and CORY [244]). Other adhesive agents besides oil are, of course, used in the formulation of DDT. Dew or other moisture on the leaves serves as an adhesive for dust and results in a larger residue immediately after application (EDEN and ARANT [193]).

It must be recalled that, irrespective of the nature of the formulation, only a small portion of the DDT applied actually reaches the plants for which it is intended. Part of the material is lost by drift and part sifts down between the leaves and falls to the ground. Thus, when WILSON *et al.* [668] applied DDT dust at the rate of about 175 ppm in terms of the weight of the crop at harvest they recovered only 13 ppm (7·5%) from the plants immediately after the application. The high recovery (up to 24·3%) of DDT dust reported in Table 24 was probably made possible only because the insecticide was put on by a special experimental method which would not be practical for agricultural use.

It is interesting to observe that in the case of insecticides more volatile than DDT a smaller proportion can be recovered from treated vegetation (DECKER *et al.* [155]).

Number of Applications

The amount of residue remaining at harvest does not bear any direct quantitative relationship to the number of applications (MANALO *et al.* [393]). As a matter of fact, the proportion of the total applied DDT which can be recovered immediately after the last application decreases as the number of applications increases. Of course the actual amount of residue present just after the last application increases as the number of applications increases. These relationships are illustrated in Table 24. If the number of applications is increased and the period from application to harvest is decreased then it may occur that the final residue is directly proportional to the number of treatments as found by TASCHENBERG and AVENS [574].

Weather

The removal of residues by rain has been frequently observed (GINSBURG *et al.* [233]; DECKER *et al.* [155]; SLOAN *et al.* [530]). Some authors (GUNTHER *et al.* [250]; ALLEN *et al.* [6]) have failed to observe any effect of rain on residues. The difference appears to be largely one of time. Rain, shortly after the application of DDT, dislodges a significant proportion of the compound but rain late in the season has little effect on the more firmly attached residues which persist at that time.

Time

The amount of residue does not decrease proportionately with the length of time elapsing between the last application and harvest (MANALO *et al.* [393]; BRUNSON and KOBLITSKY [92]). The initial loss is much more rapid and later loss is slower (DECKER *et al.* [155]; SLOAN *et al.* [530]). It has been suggested that loss of DDT soon after application results largely from mechanical effects whereas chemical alteration in the residue plays a relatively more important role in later loss (EDEN and ARANT [193]).

There is some evidence that DDT undergoes partial decomposition when exposed on plant surfaces under field conditions so that the residue at harvest consists partly of DDT and partly of one or more DDT derivatives (CARMEN et al. [111]).

The elapse of time is accompanied by growth of the plant which plays an important part in reducing the concentration of residue. If pods or fruits are small or absent at the time DDT is applied, then the residue remaining on them at harvest will be small even if the residue on other parts of the plant is great. Forage crops may gradually lose lower leaves while the plants grow higher and increase in weight. The leaves which are shed may carry a high proportion of the total residue while the new growth has none (WILSON et al. [668]).

The various factors influenced by the amount of time after the last application may combine to equalize residues present after a month or more irrespective of the concentration of the formulation used or the number of applications. Such a relationship was observed by EDEN and ARANT [193] and is illustrated by Table 24 which summarizes a part of their findings. The relationship may also be inferred from data of various authors listed in Table 22.

Character of the Plant

The texture of a plant surface influences the residue which it will retain. Waxy surfaces are difficult to wet. Hairy leaves such as those of soy beans retain a relatively high proportion of the spray or dust which reaches them.

It is interesting that DDT is preferentially stored in plant oils just as in animal fats. In the plant it is, of course, necessary that the oil be present at the surface or that some specialized mechanism be present which allows absorption to occur. Most oils stored within seeds or in other protected places show no deposits of DDT. On the contrary olive oil may show high concentrations of the compound (STAFFORD and HINKLEY [547]). REIBER and STAFFORD [483] reported a concentration of 116·0 ppm in oil from olives which showed a surface residue of 7·9 ppm. Others (HOSKINS [300], [301]) have observed the high uptake of DDT by olives. DDT is absorbed into the peel of citrus fruit (BARNES *et al.* [48]; CARMEN *et al.* [111]) and presumably the oily character of the peel is responsible. The avocado, which is an oily fruit, also absorbs an excessive amount of DDT (HOSKINS [300]).

Roots and tubers accumulate no DDT either from the soil or from the stems or leaves.

Other things being equal, plants which offer an extensive surface will receive more DDT than those with less surface. If only a part of the plant is used for food then the surface-volume relationships of that part are important.

Compared to leaves, fruits have a relatively small surface in respect to their volume. As a result of this and other factors, the residues on fruits are frequently much smaller than the residues on the leaves from the same plant (MANALO et al. [393]; DECKER et al. [154], [155]). GRAHAM and CORY [244] found leaf residues of approximately 900 to 4,500 ppm; the maximum recorded residue for fruit from the same orchard was 18·7 and the average fruit residue was much smaller. Thus residues on fruit present a relatively small difficulty. Likewise, most flower-fruit-and-seed vegetables offer little complication because they too have a small surface in relation to their weight and frequently show significant growth after the last application of DDT and before harvest.

Leaf-and-stem vegetables, because of their great surface in relation to their weight, are likely to produce a residue problem. In structure, these leaf-and-stem vegetables resemble the forage crops which serve as an important source of animal food. With good agricultural practices, DDT may be used on cabbage and lettuce and other vegetables which form heads because most of the compound lodges on the outer leaves which are discarded before the product is marketed.

The application of DDT to forage crops constitutes a hazard not only because of the high residues which frequently result but because animals fed on the crops store the compound in their tissues and excrete it in their milk. The situation is further complicated by the fact that several crops such as green peas customarily serve for both human and animal food. Even though the shelled peas which are eaten by people contain no DDT, the presence of the insecticide in the vines may limit use of the vines for silage.

DDT is used extensively on corn (maize) and to a much smaller extent on other growing grains. Usually no residue reaches the grain as the result of applications to the growing plant. Such small residues as do sometimes occur may be the result of contamination in handling.

Method of Harvest

The ordinary handling of crops associated with harvesting may remove appreciable quantities of DDT. Thus HARMAN [260] found that when apples were gathered with great care to preserve the residue they showed a contamination of 6·1 to 9·1 ppm while similar apples harvested in the usual way showed only 3·0 to 6·4 ppm.

Post-Harvest Application of Insecticide

The direct addition of DDT to threshed grain and other foods has been advocated in order to prevent insect damage during storage. Because of the inherent danger, certain countries with an abundant food supply and with adequate equipment for the storage of grain do not practice the addition of DDT to grain. On the contrary, under emergency conditions, this addition has been an alternative to the loss of precious food and consequent starvation.

The amount of DDT necessary to control grain insects is relatively large. LE PAGE and GIANNOTTI [363] found that a final concentration of 60 ppm gave protection to wheat and beans. Potatoes were protected by dusting with 3 per cent powder. The authors showed that the majority of DDT on these products, as revealed by bioassay, was removed by industrial washing procedures. Experiments with several species of animals convinced them that the amount of DDT involved in treatment of the foodstuffs was not injurious. GAY [221] found a final concentration of 10 ppm was insufficient to completely eliminate an infestation of *Rhizopertha dominica*. PARKIN [448] stated that for practical purposes a concentration of 30 to 50 ppm must be used to protect grain from infestation. Grain contaminated to this degree retained about 12 ppm after being cleaned by the methods now in use in Great Britain.

Earlier, ADAM and ZUST [2] claimed that only traces of DDT remained in the flour, the greater part being left on the bran and other byproducts. They used an original contamination of 12,000 ppm and recorded less than 300 ppm in the flour. Because of the magnitude of the figures they cannot be applied directly to a practical situation.

WINTERINGHAM *et al.* [672] used a radioactive bromine analog (DBr*DT) to study the problem. Whole grain, with a residue of 40·2 ppm, was milled to form flour which contained 14·6 ppm. The bran contained 127·4 ppm while wheat feed (another byproduct) contained 134·7 ppm. Preliminary cleaning resulted in products containing more than 450 ppm. The authors presented evidence that a part of the compound, or a derivative, combined with the flour and probably with the protein to form a product which could be dissociated by alkali digestion.

If grain is ground by cruder methods, or merely crushed, a high percentage of the DDT in the whole grain will remain in the dark flour. It is for this reason, among others, that the addition of DDT to grain is prohibited in Italy (ALLESSANDRINI [7]).

It may be possible to treat the surface of bags containing grain or the outer layer of piles of loose grain without the production of high final residues (PARKIN [448]). However, the contamination of grain under similar conditions is greater than might be expected. HOSKINS [301] reported on the contamination of grain stored in elevators which had been sprayed when empty. Barley taken from the center of a 2,540 m³ elevator contained 1·5 ppm of DDT, and wheat taken from the center of a 1,480 m³ elevator contained 2·4 ppm after 3 and 4 months, respectively. The result was attributed to volatility of the insecticide, other sources of contamination presumably having been ruled out.

In a similar way, BUTTERFIELD *et al.* [98] found that relatively large amounts of DDT impregnated into cotton and jute bags were transferred in a few months from the bags to the food stored in them. The rate of transfer was greater when the food was finely ground or had a high fat content. Experiments showed that the transfer could be accounted for, at least in part, by the diffusion of DDT vapor.

In the United States, the addition of DDT to grain is recommended only if the grain is to be used for seed. It is advised that grain intended for feed should be fumigated if control of insects becomes necessary. Infestation may be kept at a minimum or avoided entirely by proper cleaning of the elevators, boxcars, barges, or ships, followed by spraying of the walls with DDT (U.S. Department of Agriculture [612], [613]).

The addition of DDT to stored grain in England and in certain South American countries has been reported (PACKARD [445]) and the practice has been discussed in Germany (MARQUARDT [397]).

The use of DDT residual treatment in food handling establishments need not contaminate food if proper precautions are taken (ROWE [500]). In such situations, DDT is undoubtedly preferable to sodium fluoride and other dangerous poisons which have been required in the past (HOLMES and SALATHE [296]).

Reduction of Residues by Cleaning or Storage

MANALO et al. [394] considered the effect of polishing, peeling, dry brushing, and washing. Dry polishing removed one fourth to one third of the DDT present. Peeling polished apples removed all the residue and peeling unpolished fruit removed practically all the DDT. Brushing removed about a third of the residue. Washing with tap water removed about a third of the residue. The use of different detergents and different temperatures did not make the procedure any more efficient and a 1 per cent aqueous solution of hydrochloric acid made no improvement. Weak solutions of solvents including alcohol and acetone removed only about one tenth of the residue. Except for peeling, no method was found which would reduce high residues (10 to 42 ppm) below the informal legal tolerance of 7 ppm. Even with the same washing procedure, the results were erratic.

HARMAN [260] showed that trees which had had five cover sprays during the season produced apples with an average residue of 7·2 ppm. It was necessary to pick the apples with some care to preserve the residue and when apples from the same trees were put through the usual harvest procedure and taken for analysis from the sorting table they showed a residue of only 3·0 ppm.

BORDEN et al. [84] examined a number of common washing materials and recommended a commercial product containing a water soluble sodium alkyl aryl benzene sulfonate plus a small amount of an oil soluble oleate soap. This product removed from about one-third to one-half of DDT present in residues from nonoily sprays. Attempts to remove DDT applied with oil were not successful.

WALKER [642] showed that the recommended application of DDT to apples and pears resulted in excessive residues on only 7·6% of apples and on no pears. Additional spray applications or improper techniques resulted in an increased proportion of high residues. Furthermore, a variety of commercial fruit cleaning agents, sodium silicate, sodium hydroxide, and several washing temperatures were tried. It was found that fruit, which had been sprayed with oil late in the season (an operation reflecting the earlier practice with acid lead arsenate and cryolite sprays), had an unusually high residue at harvest and it was not possible to reduce these higher residues to an acceptable level by any method of washing. Up to about one third of the DDT applied in the accepted way could be removed by washing. In a second related paper (WALKER [643]), the author pointed out that no correlation could be established between the percentage of DDT residue removed and the amount originally present. The average removal by hydrochloric acid treatment was 21·3%, by sodium silicate treatment was 29·5%, and by the combined treatment 35·6%.

GUNTHER et al. [251] reported far greater success in decontaminating apples and pears than did earlier authors. They did not discuss the reason for the difference and it is open to question whether a reason is apparent. It is true that they used a washing machine of advanced design and that they worked with relatively small initial residues (0·9 to 7·2 ppm). Chemical removal was attempted with alkaline and halogen-carrier media, such as sodium silicate, trisodium phosphate, ferric chloride, sodium carbonate and bicarbonate, and alkaline soaps. Mechanical removal was sought by using emulsifying agents, detergents, pressure sprays, and scrubbing brushes. Solvents including kerosene, mineral oil, xylene, and polymethylated naphthalenes were used. With apples and pears sodium silicate or a combination with a proprietary mixture containing lauryl sulfate and salts of substituted aromatic sulfonic acids proved best in most instances, although trisodium phosphate was also effective.

Routine removal of 50% of the residue was achieved and the removal sometimes approached 90 per cent effectiveness. The usual packing house treatment for citrus, which includes a thorough scrubbing with alkaline soap, proved sufficient to remove all but traces of residual DDT from oranges and lemons.

HALLER and CARTER [255], in a study which lasted 4 years, found that some of the washing methods which were tried for apples were less effective than had been reported by some investigators. They concluded that none of the available washing treatments was sufficiently effective to be of practical value. The authors pointed out that it is possible to remove a high percentage of a residue before the DDT has had time to become dissolved in the natural wax of the fruit peel—that is, a short time after the DDT is applied. However, removal of residues of this kind has no practical importance.

The removal of DDT from stored grain has already been considered (p. 176).

Alfalfa hay does not lose an appreciable amount of DDT during storage. In one instance, hay which was dried but which had not yet been removed from the field showed a residue of 5 to 8 ppm. The hay was stored in bales and sampled at intervals. Counting from the day on which the mature dry hay was sampled, the following residues were found: 9 days, 10 ppm; 79 days, 7 ppm; 94 days, 2–3 ppm; 105 days, 13–14 ppm; 133 days, 6–8 ppm; 147 days, 8 ppm; and 168 days, 10 ppm (R. F. SMITH et al. [538]). The same authors found that, in general, there was some reduction in DDT on alfalfa straw following threshing for seed.

In contrast to hay which is stored dry, silage may lose DDT during storage. WILSON *et al.* [667], [668] found that silage which contained 500 ppm when it was stored in June contained only 108 ppm in December. No DDT was found in silage made from pea vines treated at the rate of 4,484 mg/m² with 5 per cent dust, and presumably contaminated at the time they were placed in the silo. In another instance, fresh silage showed 5 to 7 ppm in September, 5 ppm in October, and 4 ppm in November.

From the same laboratory, LARDY [341] confirmed the results with pea vine silage and showed that the concentration of DDT in corn (maize) silage gradually decreases with storage. Corn which showed 19, 9, and 4 ppm 40 days after ensiling in different silos showed only 3, 6, and 0·2 ppm, respectively, after 100 additional days of storage.

In the absence of other evidence, it would appear that large residues of DDT are considerably reduced through some unidentified process associated with silage fermentation. The change in small residues, of the magnitude which might be expected to occur in actual practice, is apparently variable. The entire subject deserves further study.

The special problem of removing DDT from public water supplies has been considered in another section (p. 183).

Effect of Heating on DDT Residues in Food

Although it probably has little to do with heating, it must be kept in mind that other processes associated with canning may leave on fruit only a small fraction of the DDT which it carried when fresh. Thus HELSON [269] found that, when peaches bearing a residue of 9 ppm were peeled and canned, they did not show any trace of the compound in the syrup or fruit. Even when the contaminated and uncontaminated portions are intimately mixed during processing, separation can sometimes be achieved. For example, uncontaminated juice may be obtained from grapes bearing residues (TASCHENBERG and AVENS [574]). This is true even if the residues are much greater than those resulting from good agricultural practices (FREAR and COX [215]). Uncontaminated jam may be made from contaminated grape pomace (TASCHENBERG and AVENS [574]).

In 1947 TRESSLER studied the effect of canning on DDT added to food for the purpose of study. He developed a method for rapidly extracting the compound from foods and, using an adaptation of the Schechter-Haller method, was able to recover as high as 89% of DDT added to strained peaches but only 61% in the case of apple sauce. The partial failure to recover the insecticide from the foods was regarded as a defect of the extraction method and not a specific effect of the food on the compound. This loss in unprocessed food was treated as a blank in the calculations. When DDT was held at 212°F for 4 hours in sealed Pyrex test tubes either in water or in a phthalate buffer, little destruction of the insecticide occurred; when sealed tin cans were used instead of Pyrex tubes, about 20% of the added DDT was destroyed. When DDT was added to foods which were then placed in cans and processed according to the usual commercial method, the loss in excess of the blank was about 27% in strained peaches, 20% in apple sauce and 60% in tomatoes. Further study showed that, when pure p,p'-isomer was canned with tomato juice, degradation products were formed which produced atypical colors on analysis.

The analysis of beef taken from an animal which had been fed DDT residues for 3 months and then put on uncontaminated pasture for slightly over a month before slaughter has been reported (CARTER *et al.* [116]; CARTER [114]). During the period of exposure the animal received daily 4·5 kg of clover hay bearing a residue ranging from 84 to 184 ppm. Typical cuts of meat with the normal amount of fat were analyzed raw and after cooking. Both organic chloride determinations and Schechter-Haller colorometric determinations were made. Five methods of cooking; roasting, broiling, pressure cooking, braising, and frying; were used. The highest residue in a cooked portion was 33 ppm in fat drippings; the lowest residue was 7 ppm. Most of the determinations indicated that cooking caused a loss of DDT ranging from 11 to 46%. However, the meat which was broiled showed an apparent increase of 11 and 16%, respectively, in DDT content when analyzed by the two chemical methods. No investigation of this increase was made and the authors concluded that the DDT in beef was not materially decomposed or lost during cooking.

The effect of baking on the DDT content of bread and macaroni has been studied by LEGGIERI [353]. Flour containing 8 ppm was used to make the bread. About half of the DDT in bread was destroyed by baking. The crust, which reached 200°C in baking, was found to contain less DDT than the moister bread in the center of the loaf which reached a temperature of only 100°C in baking. Macaroni made from hard wheat containing 7 ppm of DDT lost almost none of the compound during boiling. Macaroni made from soft wheat containing 8 ppm of DDT broke during boiling and it lost a greater portion of the compound than did other macaroni.

Pasteurization of milk containing small amounts of DDT did not decompose the compound (CARTER *et al.* [118]; MANN *et al.* [395]).

Residues in Water

It was early observed that the use of DDT as an indoor residual spray for the control of *Anopheles* also served to control *Aedes aegypti* when that species was present. The same method of residual spraying was also at times practical when *A. aegypti* constituted the major problem (see, for example, DE CAIRES [151]). However, another method involving the addition of DDT to water has been developed and is widely used in those areas of Central and South America where yellow fever offers a constant potential threat to the population. The compound may be added to a reservoir from which a town draws its supply, or it may be added directly to small cisterns or jars in which water is held in individual homes. In either instance, the practice of adding DDT to drinking water creates a very specialized residue problem which deserves consideration.

Apparently, WASICKY and UNTI [649] were the first to mention the possibility of treating potable water directly. They stated that mosquito larvae of several species died in 4 to 6 hours when placed in water containing 0·04 ppm of DDT colloidally dispersed. The water remained free of larvae for some months, although the oviposition was not disturbed. Water treated in this way showed no change in taste or smell. The authors studied the toxicity of DDT by several routes and in several species of animals. They concluded that the concentrations needed for mosquito control were harmless and stated specifically that water containing 0·04 ppm was suitable for drinking.

NEGHME et al. [430], after determining that the organoleptic character of the water was not changed, added to a 4,000 m³ reservoir tank in Iquique, Chile, 400 l of an aqueous suspension containing 2% of wettable DDT. They thus produced a concentration of 1 ppm of DDT in the reservoir water which was led by pipes to fountains from which housewives filled large earthen jars for home use. A second reservoir, from which water for the city was drawn on the following day, was also treated, thus insuring that the earthen jars in which the mosquitoes passed their larval life would contain DDT for 24 hours. The authors reported that the results were spectacular. In less than a day the larval foci of *Aedes*, as well as foci of *Culex*, practically disappeared. The success of the procedure was made possible by the fact that Iquique, a city of 48,000 population, is in a very arid region. Its water is piped from an oasis and held in the city in reservoir tanks pending further distribution. Mosquito breeding was limited to water supplied by man.

NEGHME [429] discussed the expanded use of the technique in different areas of Chile where it was appropriate. He believed that, under the arid conditions mentioned above, the elimination of *A. aegypti* could be achieved by 3 applications of DDT. He discussed the possibility of reducing the concentration of insecticide in the reservoir water to 0·2 or 0·1 ppm, and extending each application to about 4 days. He discounted the chance that this concentration of DDT would do any harm to the health of the population, especially if the applications were sufficiently spaced (one, two, or even six months apart, and over a period of one to two years).

RODRIGUEZ [493] reported studies in El Salvador similar to those of NEGHME in Chile. On the basis of the studies it was determined to use the technique in the dry season in suitable communities. One month was taken arbitrarily as the interval between treatments with the idea that experience might later reveal a better timing.

The technique used by NEGHME in Chile has been considered from the standpoint of hazard. After reviewing the literature of the toxicology of DDT, GOMEZ [238] concluded that there was no basis for thinking that the addition of 1 ppm of DDT to central water supplies once a month or twice a year presented any hazard whatever to man.

The method for *Aedes* control by which DDT is added directly to home cisterns or earthen jars has apparently found wider use than the addition of DDT to central supplies. This undoubtedly results from the fact that most of the places where control is required have many breeding places supplied by rainfall or by other sources unrelated to the central water system. Actually, in some places no central water supply exists.

Sasse [507] conducted laboratory tests with DDT against *A. aegypti*, using various water containers common in homes. He concluded that a dosage of 1 ml of a 2 per cent DDT-alcohol solution per 6 l of water capacity should be applied every 4 to 5 weeks. This recommendation was made, even though the tests showed that a smaller dosage was effective for 6 weeks. The larger dosage was considered advisable to provide a margin of safety in mosquito control. It was the author's opinion that, even at the higher dosage, the amount of DDT was still so very small that it did not represent a danger to the health of the people. The dosage recommended by Sasse is 3·3 ppm in the water to which it is originally applied, if one assumes that it is equally distributed, and that it remains suspended. The assumption is entirely unjustified, however, for Sasse found that vessels remained lethal to larvae exposed in them for 42 hours 11 weeks after the original treatment. This was true, although the water was emptied from each container and replaced with a fresh supply once a week during the entire test. It cannot be assumed that the larvicidal action of the water after 11 weeks resulted from DDT remaining after direct dilution of the original sample. (If 1 ml remained in a 6 l vessel each time it was emptied, then the final concentration at 11 weeks would be $5·5 \times 10^{-13}$ ppm, a concentration not lethal to larvae.) One is forced, then, to assume that a portion of the DDT was adsorbed on the walls of the vessel and that the death of mosquito larvae, especially during the latter weeks of the test, was a result of their contact with DDT on the walls and not in the water. This view is supported by the fact that vessels treated with DDT when dry, and then stored for 13 or more weeks, were not larvicidal when first tested but became larvicidal when the dust they had accumulated was washed off. It follows that no reasonable guess can be made regarding the concentration of DDT in water which people drink from household containers except that it must be smaller than the calculated dosage.

Berti [57], [58] has reported a practice in Venezuela similar to that used in Peru. In those parts of Venezuela where malaria was prevalent and residual sprays were applied to all houses, it was found that *A. aegypti* disappeared. In other regions where malaria was not prevalent and residual spraying was unnecessary it was found more suitable to add DDT directly to household water containers. At first, this was done using a 2 per cent alcoholic solution added dropwise to reach a final concentration of 5 ppm in drinking water. A larger dosage in a kerosene solution was used for waste water. Later, it was found cheaper, more practical, and longer lasting to use, for both purposes, a suspension of 50 per cent water wettable DDT containing 7·5% of the active material. The final concentration of DDT remained 5 ppm. Each cistern was treated, on the average, about 3 times a year, the actual number of treatments being determined by inspections for larvae.

SA ANTUNES [502] reported the treatment of 31,794 cisterns in Brazil, beginning in 1948. He stated that, in spite of the tremendous number of applications made in the various regions of Brazil, the people who utilize this water containing DDT have not shown the slightest symptoms of intoxication.

No record has been found of a careful study of the final concentration of DDT in water stored in cisterns following the intentional addition of the compound, nor of the concentration in natural water which might reasonably be expected to be used for human consumption. DDT may be sprayed on pools or streams, or may be added to streams at a predetermined rate by various mechanical devices in order to kill the larvae of various vectors or pests. LOGAN *et al.* [377] described an original method by which porous masses of seaweed encased in cement were filled with a solution of DDT in fuel oil and placed in isolated pools, wells, and water holes in order to maintain larvicidal treatment of the water for a minimum of two weeks. Irrespective of the method of application, traces of the DDT used against aquatic larvae may potentially find their way into an unfiltered water supply for human use. There is no evidence that the amount of such DDT is of the slightest public health significance. An excellent study has been made of the effect of storage, sedimentation, and various kinds of filtrations on the DDT content of natural water intended for general distribution in a city water system. CAROLLO [113] found that, on storage, clear water gradually lost its ability to kill *A. aegypti* larvae. The water was transferred to clean vessels after different periods of storage so that the toxicity of the water and not the toxicity of the vessel was tested. The following times were required to inactivate different original concentrations of DDT:

DDT concentration ppm	Storage time hours
0·0015	48
0·003	96
0·01	120
0·04	288
0·1	312+

The inactivation was considered not due to sedimentation nor to the effect of sunlight, but adsorption on the walls of the storage vessel was not ruled out. Turbidity in water had little effect on its ability to kill larvae, but settling removed some of the DDT suspended in turbid water. Any of the conventional water treatment processes involving coagulation, sedimentation, and filtration would remove from 80 to 98% of DDT from a water supply containing from 0·1 to 10·0 ppm. Complete removal (to less than 0·001 ppm) of DDT from a water supply was possible if intimate contact with activated carbon was provided for 15 minutes after coagulation and sedimentation, but before filtration. The last traces of DDT could also be removed from water by storage for 4 or 5 days, or by filtration through anion and cation zeolites. Determination of

DDT was made by bioassay, by the xanthydrol-KOH-pyridine method, and by nitration followed by the addition of alcoholic KOH.

In the same paper, CAROLLO reported experiments with mice in which the animals showed no ill effect from drinking only water containing 10 ppm of DDT for 75 days or, in another test, water containing 50 ppm of DDT for 58 days.

No authentic reports of intoxication caused by the addition of DDT to water have been noted. Even so, it must be remembered that the compound is toxic and its addition to drinking water constitutes a sort of medical treatment which ought to be under constant professional supervision. Although the report is unconfirmed, it has been stated that this has not always been true. FRANK [213] reported the informal recommendation of final concentrations as high as 24 ppm, the crude and careless measure of materials, and an increase in illness which he thought might be caused by these errors. Whether the report was well founded or not, it is certain that the use of an excess of DDT offers no improvement in *Aedes* control, but does involve unnecessary economic waste and hazard to health.

The toxicity of water bearing DDT to fish and other useful aquatic life has been treated in the section on wildlife (p. 205) and also in Table 16.

Residues in the Total Human Diet

The study of residues on individual food products is of great importance in helping to arrive at effective and safe agricultural practices. The other types of residue studies which have been reviewed have their own interest. However, of much more importance for public health is a knowledge of the amount of residue actually consumed by the general population. This knowledge cannot be obtained with sufficient accuracy by calculation or extrapolation. More specifically, it is not valid to suppose, as ROBINSON [492] apparently did, that items of the diet which are not commonly sprayed are dependably free of DDT.

A study of the residues in complete meals was made by WALKER *et al.* [644]. The individual items of typical meals were analyzed using an adaptation of the method described by MATTSON *et al.* [400] for DDT and DDE in human fat. Eighteen complete table-d'hôte restaurant meals were selected with due care to avoid regional items or extremes of luxury or cheapness. Through the cooperation of the managers of the restaurants, ordinary portions of the separate items of the meals were served into chemically clean beakers. Seven meals were obtained in a similar manner from a correctional institution. At the laboratory, each portion was weighed after inedible materials such as bone were removed. After the individual portions were analyzed, the results were combined into the original 25 meals, including 8 breakfasts, 9 noon meals, and 8 evening meals. Judging from the weight of the individual portions and values from standard dietetic tables, the three typical meals for a single day had a wet weight of 2,634 g and a dry weight of about 588 g. Study of a few meals indicated a daily caloric value of about 2,974 cal. No single meal failed to contain a trace of DDT, but the amount present was small in each instance. Many single items of food contained no detectable insecticide; the highest concentration of DDT found in any single item (butter) was 4·0 ppm, and in any single whole meal was 0·2 ppm based on wet weight. The average concentration of DDT in all the meals was 0·07 ppm based on wet weight. In individual meals, the DDE content ranged from 18 to 61% of the total DDT plus DDE and averaged 34%. Thus only a part of the residue involved the more toxic compound, DDT. The results indicate that, on the average, a man would eat about 184 μg of DDT per day. This is equivalent to a dosage of 0·0026 mg/kg/day for a man of average size, or a concentration of DDT in the total dry diet of about 0·31 ppm. By comparison, a rat would have to eat a diet containing only about 0·05 ppm (dry weight basis) to obtain the same dosage because the rat eats more food than man in proportion to body weight.

It would appear that the findings of WALKER and his colleagues are consistent with the degree of storage of DDT which is known to occur in the fat of the general population. This statement regarding consistency is based, by necessity, on studies of DDT storage made in rats, monkeys, and other animals (see Figure 11). It is clear that the amount of DDT ingested and stored by man under ordinary conditions is much less than the amount which may be ingested and stored by experimental animals without observed injury to growth, survival, or other functions. The dosage taken by man is about 1/100 of the dosage which produces minimal liver changes of questionable significance in rats but not in nonrodents.

10.

PROTECTION FROM EXCESSIVE EXPOSURE

Residue Tolerances

Obviously, under ordinary conditions, the residues of agricultural chemicals which will occur on food will be determined by agricultural practices and food processing methods. If pesticides are properly used on the farm and in the food processing factory, excessive residues should be avoided. Directions for use appear on the label of each container of DDT or other pesticide. Those laws which regulate labeling (and which have been discussed below under that heading) form the first line of defense against excessive residues and against other misuses as well. In fact, several nations have found it necessary to regulate labeling, but unnecessary to establish tolerances.

It must be recognized, further, that the chief usefulness of labeling laws and tolerance regulations is education. They should serve to guide but not to limit commerce. Different nations have different legal mechanisms for limiting the contamination of food by potentially injurious substances. The most simple arrangement is that in which cases are brought to court under some kind of 'poisons law'. This is the method in use in England, where it appears to operate satisfactorily. One must hasten to add, however, that geographical, technologic, political, and social differences between nations may require different legal mechanisms to accomplish the same objective.

Without doubt, the most complex system of law for regulating the presence of chemicals in food is that in the United States. The number of insecticides, fungicides, herbicides, preservatives, conditioners, and other chemicals which find their way directly or indirectly into food is increasing. For this reason, it is often urged that the legal controls of these materials should be increased. Although the newer chemicals do not differ in any respect, except possibly in potency, from the older ones, there are reasons (including a certain organic quality of law) to review with reference to DDT, the philosophy, history, and proposed expansion of the American regulations. Those who suppose that these matters are the narrow interest of one nation should consider, for example, the attitude of the Spanish investigator, GIL [226].

Residue tolerances are established and administered in the United States under the Federal Food, Drug, and Cosmetic Act (U.S. Laws [626]). Many of the individual States have similar laws for the regulation of commerce within their borders. The philosophy of the law, especially as it applies to DDT and other pesticides, has been expounded by DUNBAR [185]. Briefly, the relevant portion of the law states that any poisonous or deleterious substance added to food shall be deemed unsafe, regardless of the amount added, with the following exception: When it can be shown that a pesticide chemical is required in the production of a food or cannot be avoided in good manufacturing practice, the Secretary is directed to establish tolerances. As first written, the law required that public hearings be held before tolerances were fixed. In its present form, the law requires no public hearings unless an applicant wishes to have an unfavorable decision reviewed, but this and other forms of review are provided. Under ordinary circumstances, tolerances are now fixed within a short time by simple administrative action. The law recognizes the peculiar economic, agricultural, and public health problems which are important in the regulation of pesticide chemicals. The determination of questions of agricultural usefulness and probable residue levels involved in the establishment of tolerances is made a function of the Department of Agriculture; while the determination of questions of a public health nature remains a function of the Department of Health, Education, and Welfare.

In the United States, DDT became available for rather extensive experimental work in agriculture early in 1945, and it was commercially available in limited quantities early in the autumn of the same year (U.S. Department of Agriculture [610], [611]). Before 1945, the compound was all used by the military services for various medical and public health uses. In 1946, and especially by 1947, the material was extensively employed in agriculture. A record of the production of DDT in the United States is shown in Table 18.

Before DDT became available for commercial use in the United States, the problem of residues was considered by the responsible officials. In January 1945, DUNBAR, as Commissioner of the Food and Drug Administration, addressed a letter to those interested in carrying on experimental work involving the spraying of relatively large plots. It was realized that a considerable quantity of foodstuff from such plots would enter interstate commerce and, therefore, the matter of a temporary tolerance was discussed in the letter. It was stated that there was general agreement among pharmacologists that DDT was not more toxic than fluorine or lead. At that time a formal tolerance of 7 ppm existed for fluorine, and an informal tolerance of 7 ppm existed for lead. The commissioner stated that under the circumstances no regulatory action would be taken during the coming year against commodities containing 7 ppm or less of DDT. An article appearing about the same time emphasized the need of tolerances, and suggested that the tolerance for DDT could not exceed 10 ppm (CALVERY [101]; Anonym [12]).

The general policy mentioned in the last paragraph received further publicity in administration releases (U.S. Food and Drug Administration [621], [622]). After a time it was tacitly assumed by those outside the administration that a tolerance level of 7 ppm represented the established policy of the Food and Drug Administration (MANALO *et al.* [393]; DECKER [152]; HARMAN [259], [260]; STEINER *et al.* [550]; CLARK [130]; PARKER and ESHBAUGH [447]; YETTER [683]; WALKER [642]; ROBINSON [491]; ASHDOWN and WATKINS [37]; QUESTEL and CONNIN [474]; EDEN and ARANT [193]; ARANT [36]; WESTLAKE and FAHEY [661]; BARNES *et al.* [48]; BRUNSON and KOBLITSKY [92]).

At about the same time a provisional tolerance of 10 ppm was in use in Australia (CLARK [130]) and Canada (HELSON [269]).

WASICKY [647] recommended a tolerance of 4 ppm for milk, and a tolerance of 8 ppm for other foods.

Authors, especially those connected with the U.S. Food and Drug Administration, continued to emphasize the potential problem involved in the repeated ingestion of residues of insecticides in foods. The results of laboratory experiments were repeatedly summarized and interpreted as indicating a possible hazard to man, although it was generally admitted that danger from a single ingestion of food sprayed with DDT would occur only as the result of gross carelessness (COX [140]; Anonym [23]; FITZHUGH [207]; LEHMAN [356], [358]; CRAWFORD [142]; THIEMANN [584]).

LEHMAN [357], in a paper read before the meeting of the American Chemical Society in March 1949, proposed tentative residue tolerance for DDT, as well as a number of other new pesticides. Although the proposed tolerances were not official, they were backed in principle by the Commissioner of the Food and Drug Administration (DUNBAR [184], [185]). The level suggested for DDT was 1·0 ppm, if all items of the diet were contaminated, but 5 ppm for single items if it could be shown that most other foods contained only traces of the compound. It was further suggested that no tolerance should be established for DDT in milk. This was virtually the same as saying that milk should not contain even a chemically detectable trace of the compound.

This point of view regarding milk had already been expressed at a meeting of the National Committee of Food Sanitarians in Chicago on January 28, 1949 (Anonym [25]). The views of the Food and Drug Administration were stated again on February 10, 1949, at the Thirteenth Annual Purdue Pest Control Operators' Conference (Anonym [26]). The entire subject received a great deal of informal and formal publicity (DUNBAR [184], [185], [186]; U.S. Food and Drug Administration [623]).

On March 24, 1949, the U.S. Department of Agriculture [609] declared that DDT should not be used on dairy cows or in dairy barns. It was suggested that methoxychlor might be substituted for this use. Earlier cautions were repeated against the use of forage treated with DDT, or other chlorinated hydrocarbon insecticides, as food for dairy animals or livestock being finished for slaughter. No change was made in the recommendation for the use of DDT on livestock other than dairy cattle. DDT was still recommended for use in and around farm buildings other than dairy barns. Residual sprays of methoxychlor and space sprays of pyrethrum were recommended for barns and other buildings where milk is processed. The importance of sanitation in fly control was emphasized. This statement received further publicity in trade journals (Anonym [28], [29]). Although these changes were not universally accepted at first (Anonym [29]), they rapidly became a part of all official recommendations and of general practice. All labeling was promptly brought into conformity. At tremendous expense to the industry, labels already in existence had to be changed before DDT products previously intended for use on dairy cattle or dairy barns could be shipped in interstate commerce. The notice to manufacturers, registrants, and distributors of insecticides containing DDT was dated April 7, 1949 (REED [480]).

At about the same time that the changes in recommendations and regulations were being carried out, there began an alarming series of releases in a certain segment of the popular press (DEUTSCH [170a–k]). The first article which appeared on March 30, 1949, under a bold headline featuring a skull and crossbones, began with the following statement:

'DDT, the great bug-killer, may turn out to be one of the most devasting biological weapons ever loosed by a people upon themselves.'

The author drew heavily on the publications of BISKIND, mentioned above (p. 138). Although the articles had been highly critical of government agencies, the journal modestly assumed credit for the change in labeling practice (Anonym [30]; DEUTSCH [170l]).

Without special study, one can only speculate on the motives behind this inflammatory type of journalism. Were people not accustomed to sensationalism from a certain segment of the press, such a series of articles might originate a crusade or some other demonstration of mass hysteria. As it was, alarm was sufficient that it was necessary for reassurance to be given. A number of government agencies issued a statement informing the public regarding the wholesomeness of the milk supply, and stating that the new order by the Department of Agriculture was intended merely as a precaution against even the smallest amount of DDT in this important staple food (U.S. Federal Security Agency [617]). The joint statement received wide circulation in French (Anonym [34]), as well as in English (Anonym [32], [33]). DUNBAR [185] indicated that apprehension was unfounded, as shown by spot checks of market milk throughout the United States. This reassurance was repeated in the annual report for 1950 (U.S. Food and Drug Administration).

In spite of these assurances, a few articles, more or less calculated to destroy public confidence, continued to appear (CARLETON [110]; Anonym [31]; DELANEY [164]; MERKIN [405], [406]; SCOTT [515]).

Bringing an honest, clear presentation of any scientific subject to the public through popular journals which so frequently demand the absolute and the sensational is a problem which has been ably discussed by HARWOOD [267]. Although his article was partly inspired by what HARWOOD justly called 'the DDT scandal', the problem is a general one. It is a moral problem which the reviewer feels may be of greater importance to the general welfare than are technical problems which can usually be solved in an orderly fashion by a few trained people.

The situation is far from hopeless. Articles written for the laity can give valuable information and even deal with controversial topics without sacrifice of interest or accuracy. A striking example is the able and fascinating account of DDT by RICE [485].

Legal hearings on which official tolerances for DDT on fresh fruits and vegetables would be based were begun January 17, 1950 (U.S. Federal Security Agency [618], [619]). The portion of the hearings related to DDT has been summarized in detail with many direct quotations from the testimony (PICKE-RING *et al.* [456]). Most of the exhibits presented in evidence were, of course, published papers which are also the subject of this review. In the brief, the authors concluded that DDT is required in the production of all fresh fruits and vegetables, and that a single over-all tolerance of 7 ppm for DDT on fruits and vegetables should be established.

At the time of writing, no legal tolerance for DDT has been established for the Federal Government of the United States.

The first suggestion for extension of the Federal Food, Drug, and Cosmetic Act to include the pre-testing of insecticides and fungicides to make sure that they will not produce harmful residues has been credited to DUNN [188], although that author stated in the reference cited that the matter was already under consideration. The suggestion received the active support of the Food and Drug Administration (DUNBAR [187]). About a year later, a select com-mittee was authorized to study food additives, pesticides, and fertilizers, and to make recommendations for legislation (U.S. Congress [604]). The committee, headed by The Honorable JAMES J. DELANEY, produced some 2,681 pages of testimony (U.S. Congress [605]). A number of reputable investigators, and many persons with a lay interest, were heard. The committee submitted a majority report and a minority report (U.S. Congress [606]). The extensive amendment of the Federal Food, Drug, and Cosmetic Act, which became law on July 22, 1954, instituted a more workable mechanism for the establishment of tolerances than had existed previously. The origin and nature of the changes have been reported (U.S. Congress [607]).

Quite aside from the details of any existing or proposed legislation, the scientific basis of legal control in this field is of great concern, both to specialists and to the public. The broad aspects of the problem have been stated in masterly fashion by SEEVERS [517] in the chairman's address presented before the Section on Experimental Medicine and Therapeutics of the American Medical Association in June 1953. He has reminded us that it is impossible to guarantee safety, that animal experiments are merely a prelude to human studies, and that chemical compounds should be chosen not only in terms of their toxicity under the conditions of actual use, but also in terms of their usefulness. In determining the toxicity of compounds to man, there is absolutely no substitute for use experience and studies made directly with man. Both science and the public are losers when legal controls are extended beyond the limits justified by a truly scientific appraisal. Other points and many details worthy of serious thought may be found in Dr. SEEVER's article, which should be read in its entirety.

Labeling and Other Regulation of Use

Labeling

The term 'labeling' as used in connection with economic poisons is defined as including all written, printed, or graphic matter on, or attached to, or meant to apply to, a product offered for sale, or any container or wrapper for such a poison, or any device used in connection with such a poison. The labeling is provided by the manufacturer, formulator, sales organization, or any of their agents.

Besides providing legitimate advertising, labeling may give the following information:

(1) List of constituents and their concentration in the formulation.
(2) Directions for mixing or diluting.
(3) Directions for actual use.
(4) Prohibition of certain hazardous uses.
(5) Warning regarding the nature of any danger.
(6) Suggested treatment.
(7) Name of manufacturer.

In some countries, the power to regulate labeling is apparently inherent in the authority of the Ministry of Agriculture. In other nations and states, the power is established by a specific law. Some features of the labeling regulations of 10 countries are outlined in Table 25. The presence or absence of laws involving residues of economic poisons in foods is noted in the same table. It is usually understood that, where specific legislation does not exist, the contamination of food by any toxic substance may be prosecuted under general health legislation.

The government agricultural advisory services frequently have no direct connection with the authorities who regulate labeling. It should be pointed out, however, that these advisory services make a tremendous contribution to the safe and effective use of pesticides. They make repeated contacts with farm organizations, custom applicators, and individual farmers. This gives an unparalleled opportunity for safety education where it will do the most good. In this connection, we should not overlook the efforts of the more farsighted manufacturers to educate not only their own workers, but also the pest control operators, individual farmers, and others who use their products.

The classification of poisons for the purpose of labeling is a technicality which deserves some attention. The so-called 'International Scheme' has been proposed by a committee of the International Congress of Phytopharmacy (or Crop Protection). According to this method, poisons are divided into three categories: Toxic, hazardous, and 'other'. Labeling for the toxic compounds is printed in red; the word 'POISON' is clearly written, and the skull and cross-bones insignia is used. Labeling for hazardous chemicals is printed in green, and appropriate cautions are given. Labels for other compounds are printed in any color other than red or green. The method is used in Belgium and, with some modification, in France.

Table 25

Legal Control of Insecticides in Different Countries. This Table is Based Largely on Information Provided by Dr. BARNES [43], [45]

Country	Pre-sale registration required	Approval of labeling required	Responsible agency	Classification of poisons	Estimated part of pesticides used applied by contract sprayers	Education as an extension of labeling	Specific regulations of residues
Belgium	Yes	Yes	Institute of Phytopharmacy (a Government Body)	International Scheme	80%	Yes	None
Canada	Yes Annual renewal	Yes	Plant Products Division, Department of Agriculture	Similar to U.S.A.	Very little	Extensive cooperation between Government and growers associations in devising treatment programs	None except milk and bread must contain no insecticides
Denmark	Yes[1]	Informal[3]	Chemical Control Division of Department of Agriculture	Rules of the Ministry of Interior	80%	Close cooperation between Government and powerful farmers cooperatives	None

Country	Pre-sale registration required	Approval of labeling required	Responsible agency	Classification of poisons	Estimated part of pesticides used applied by contract sprayers	Education as an extension of labeling	Specific regulations of residues
France	Yes	Yes	Ministry of Agriculture	Rules of an interdepartmental committee	Extensive	Well developed agricultural advisory service	None[1]
Germany	No[1]	No	—	Certain old compounds, e.g., cyanides, on poison list	Not widely used	Active advisory service	None
Greece	No[2]	No[2]	—	—	None	Advisory service takes up problems after they arise	None
Italy	Yes	—	Ministry of Agriculture	None	Not developed	Field service active especially in cooperation with growers' cooperatives	None

Müller II/13

Country	Pre-sale registration required	Approval of labeling required	Responsible agency	Classification of poisons	Estimated part of pesticides used applied by contract sprayers	Education as an extension of labeling	Specific regulations of residues
Switzerland	No federal law. Regulations in some cantons	—	—	Unofficial poison commission	Not developed	—	None
United Kingdom	No	Informal[3][4]	Special board under the Home Office Ministry of the Interior	Rules of board	Fairly widely used	Well developed advisory service	Unofficial tolerances recognized. Any food can be condemned if it contains a 'toxic' chemical
U.S.A.[5]	Yes	Yes	Department of Agriculture	Rules (see text)	Varies widely between different areas	Well developed at the state and county level	Yes

[1]) New legislation is being considered.
[2]) Special regulations for parathion.
[3]) Certain procedures have official approval and suitable preparation can carry an official approval mark.
[4]) Certain general regulation of labeling and sale.
[5]) 33 of the 48 states have pesticide laws similar to the federal regulations.

Some other methods of classification for labeling which, at present, involve a much greater tonnage of pesticides, are not very different in spirit from the 'International Scheme'. For example, the method used in the United States and Canada recognizes four classes of materials:

(1) Those which are highly toxic.
(2) Those which may be fatal if swallowed.
(3) Those which may be harmful if swallowed.
(4) Other formulations.

This classification applies to formulations rather than compounds, and takes into account the concentration of the active ingredients. The highly toxic class includes formulations of which:

(1) The oral LD_{50} is 50 mg/kg or less (mice, rats, or rabbits).
(2) The inhalation LD_{50} result from an exposure to a concentration of 200 ppm for 1 hour or less (mice, rats, or rabbits).
(3) The dermal LD_{50} is 200 mg/kg (rabbits only, using continuous exposure of the bare skin for 24 hours).

The upper limit for the second class is an oral LD_{50} of 250 mg/kg, and the upper limit of the third class is an oral LD_{50} of 2,500 mg/kg; in each instance the respiratory and dermal toxicity is taken into account. The fourth class, with essentially no toxicity, contains few useful pesticides. Classes (3) and (4) in this system are approximately equivalent to class (3) under the international system.

Many of the more complex features of labeling regulations are illustrated by the law in the United States known as 'The Federal Insecticide, Fungicide, and Rodenticide Act' (U.S. Laws [627]). This Act has been implemented by necessary regulations (U.S. Department of Agriculture [614]) and interpretations (U.S. Department of Agriculture [615]). The operation and philosophy of the labeling law have been explained by PERRY [450] and by REED [482]. Very similar laws exist in most of the states of the United States (Chemical Specialties Manufacturers', Inc. [121]).

The manufacturers and packagers of pesticides may have to conform not only to labeling laws, but also to laws regarding the transportation of dangerous materials (U.S. Laws [628]), and laws regarding caustic poisons (U.S. Laws [625]; U.S. Federal Security Agency [620]). The person who uses some of these materials, including DDT, in agriculture must be concerned with laws regulating chemical residues (U.S. Laws [626]; U.S. Federal Security Agency [616]).

A useful discussion of labeling and its related problems has been given by KAY [319]. The details concern Canada, but the paper contains much of general interest.

It is, of course, of mutual interest to the manufacturer, the user, and the public, that all laws regulating the labeling, sale, or use of economic poisons be as simple and unitary as possible. Uniformity in the laws of all states or provinces of a nation greatly facilitates the flow of trade while increasing the efficiency of necessary regulations. International trade is facilitated by regulations which permit products for export to be marked according to the laws of the country to which they are to be sent.

A useful guide for the preparation of warning labels for hazardous chemicals has been prepared by the Manufacturing Chemists' Association, Inc. [396].

The factors leading to the persistence of residues at harvest have been taken into account by agriculturists in making their recommendations. These specific and detailed recommendations have become a part of labeling so that growers are kept advised of proper methods. When these suggestions are followed, acceptably low residues will generally result at harvest. WALKER [642] collected 127 samples of apples from 49 different private growers, some of whom had followed the officially advised procedures for the area (Washington State College [652]), and some of whom had not. Table 26, which is simplified from one given by WALKER, shows that when the recommended program was followed, 92·4% of the samples were below the tentative tolerance, whereas only 33·3% of samples were satisfactory when the recommendations were ignored.

Direct Regulation of Use

A number of laws and regulations regarding the use of economic poisons form a direct extension of labeling regulations. For example, Great Britain [246] in the Agriculture (Poisonous Substances) Act has set up legal responsibilities of employers and their employees who use specified economic poisons. The purpose of the act is to guard the health of workers by providing for the use of protective clothing, proper equipment, appropriate working hours, and many other details of safety. The act was drawn up to meet existing problems as revealed by a study (Great Britain [245]) of agricultural requirements and practices, and of accidents which had occurred. It specified the dinitrophenols and their salts, organophosphorus compounds, formulations of these materials, and such other materials as the regulatory authority should designate. DDT is not specified in the act. However, the act is of interest in this discussion because it is among the earliest legislation attempting to extend to agricultural workers direct protection similar to that enjoyed by industrial employees who work with hazardous chemicals.

In a similar way, California [100] has found it necessary to regulate the use of pesticides by issuing permits to individuals or companies for specific applications at particular places and at particular times. The basic legislation is contained in the Agricultural Code (Division 5, Chapter 7, Article 4, Paragraph 1080) and rulings appear under the Administrative Code (Title 3, Group 2, Articles 20 and 21). The regulations have only a limited bearing on the health of agricultural workers, but are more directly concerned with preventing damage to persons, livestock, or crops which might be exposed unintentionally if appropriate precautions were not taken. The California regulations, like the British regulations, make no mention of DDT, which, under practical conditions of agricultural use, has not proved sufficiently hazardous to require direct control.

Some states and municipal governments have regulated the use of residual insecticide sprays in food-handling establishments (HOLMES and SALATHE [295]). A few highly specialized uses of DDT have necessarily required special warnings. The dispersal of DDT by continuous thermal vaporization has had extensive commercial promotion. The method involves contamination of the air and, in certain situations, involves continuous respiratory exposure and continuous exposure of food.

Table 26

Frequency with which Residues of Different Magnitude Occurred in Unwashed Apples Following Recommended and Nonrecommended Spray Programs

DDT residue ppm	Number of samples showing stated residue	
	Recommended spray programs	Nonrecommended spray programs
0– 0·0	2	—
1– 1·9	10	—
2– 2·9	29	—
3– 3·9	19	2
4– 4·9	17	1
5– 5·9	16	2
6– 6·9	5	2
7– 7·9	3	6
8– 8·9	3	5
9– 9·9	—	2
10–10·9	1	—
11–11·9	1	—
18–18·9	0	1
Total	106	21

STAMMERS and WHITFIELD [549] showed that DDT put out by vaporizers soon condensed to the liquid phase. Near the vaporizer the droplets had a median diameter of 0·5 μ and at various points in the room the droplets had median diameters up to 5·0 μ. They reported that continuous exposure of animals for 28 days and intermittent exposure of human beings for as much as 3 months was harmless.

SPEAR and SWEETMAN [542] reported that the compound remained in the liquid phase at room temperature for 4 to 5 days before crystallizing into very fine needle-like crystals. Under the conditions of their experiment, the distribution of the insecticide was rather uniform in the air, and it was finally deposited almost entirely on horizontal surfaces. Others have shown that if vaporizers are installed near the ceiling, a large amount of DDT will be deposited in delicate crystals above the device. Part of the deposit may be dislodged, and thus contaminate the area below. Furthermore, some electrical heating units have been made without thermostats or other proper controls. For these reasons, the Interdepartmental Committee on Pest Control recommended minimal standards for the construction and installation of vaporizers (HALLER and SIMMONS [256], [257]) and later recommended that they not be used in living quarters or in places where food is prepared or served (SIMMONS and ALPERT [528]).

The method of protecting workers in manufacturing plants is chosen directly by the operators and frequently is not published. PRINCI [471] has given a very general description of precautions which are required. Directions for industrial hygiene are occasionally contained in specifications (U.S. Defense Department [608]) or contracts.

General Considerations

Many of the papers which have offered an evaluation of the hazard of DDT to man or suggested treatment of poisoning have been summaries covering one or more topics including physiochemical properties, method of analysis, animal studies, experiments with man, accidents, use experience, etc. These summaries have been widely circulated and have done a great deal of good in pointing out real dangers, allaying unfounded fears, and calling attention to specific aspects of the subject on which research was especially needed. Part of the summaries have been written by investigators actively engaged in the study of the toxicology and pharmacology of DDT, and some have contained reports of research not published elsewhere (VON OETTINGEN [639]; DOMENJOZ [172]; WASICKY and UNTI [649], [650]; CAMERON [101]; NEAL [421]; NEAL and VON OETTINGEN [424], [425]; PHILIPS [452]; CALVERY and NEAL [102]; DRAIZE and WOODARD [177]; STAMMERS and WHITFIELD [549]; CAMERON [104]; GLASSMAN and BUCHAN [235]; TRUHAUT [598]; MÜLLER et al. [417]; VELBINGER [633]; GIL [225]; GIL and MIRON [226]; HEYROTH [275]). Other valuable reviews on toxicology have been written by editors or investigators who were personally involved in general public health, or other fields with an interest in DDT (Anonym [13], [14], [15], [16]; BISHOPP [66]; MILLER [411]; WEST and CAMPBELL [659], [660]; ABRAMS [1]; HOLT [298]; SCHMIDT [513]; DECKER [153]; PRINCI [472]).

Without exception, these writers have pointed out the inherent toxicity of the compound, but have concluded that when used according to directions DDT presents little or no hazard.

Other papers less directly concerned with toxicology have expressed the same general attitude toward DDT (Anonym [13]; BISHOPP [65]; LEARY et al. [352]; Anonym [18]).

A number of papers have given directions for the proper use of DDT, or have warned against one or more specific misuses, but have implied or stated that appropriate uses are safe (BUXTON [99]; Anonym [22]; ROHWER [497]; BISHOPP [67], [68]; FREAR [214]; BISHOPP [69]; BUSVINE [97]).

Professional groups have taken a lively interest in the newer pesticides, including DDT, either through editoral notices, news items, or even the formation of special committees. Notable among the groups are the British Medical Society (Anonym [11], [14], [15], [17], [18]), the American Medical Association (Council on Pharmacy and Chemistry [138], [139]; CONLEY [133]; CONLEY and WILSON [134]; Anonym [16], [22], [23], [32], [35]), the American Public Health Association (Anonym [33]), the National Research Council [420], (BEARD [55]), and the World Health Organization (BARNES [45]).

A number of investigators have outlined in greater or less detail the tests which they considered desirable or even necessary for evaluating the toxicity of pesticides (LEHMAN *et al.* [361]; DE COURT *et al.* [156]; NALE [419]; BARNES and DENZ [47]). A useful discussion of the importance of diet in determining the toxicity of repeated doses of chemicals has been given by WILSON and DE EDS [669] and by BARNES and DENZ [47]. As pointed out by BARNES [46], acute tests are more satisfactory than tests for chronic toxicity. No matter what combination of tests may be decided on, it is axiomatic that the final evaluation must rest on human studies (BARNES [46]; SEEVERS [517]). Carefully controlled investigations on volunteers are valuable, but use experience is necessary to obtain the definitive answer.

Experience with DDT as well as the other newer pesticides shows that the type of illness which characterizes poisoning in man frequently can be predicted from animal experiments. Minor exceptions arise from the greater frequency of allergy in men and from the fact that subjective symptoms such as headache and apprehension can only be guessed at from animal studies. Some poisons, for example beryllium, produce a distinctly different kind of poisoning in man and in animals. The possibility must be kept in mind that man may show at least some difference in symptomatology from animals. Aplastic anemia and similar disorders of the hematopoietic system, as well as dermatitis, are possibilities especially worthy of consideration. However, evidence should be gathered very carefully before a causal relationship is considered as established between a chemical and a human disease that differs significantly from the disease produced in animals by the same chemical. Finally, it can be firmly predicted that the degree of injury, if any, will correspond to the degree of exposure. Biological variation must, of course, be taken into account; but when large samples are considered, poisoning can be expected to appear most quickly, most frequently, most diversely, and most severely in those persons most extensively exposed.

The severity of disease in man which will follow a given exposure cannot be predicted as confidently from animal experiments as can the kind of disease. This results not so much from differences in the susceptibility of man as compared with animals, as from difficulty in estimating the degree of human exposure under practical conditions. Thus, actual use experience is of paramount importance in determining the existence of any real hazard.

11.

HAZARDS TO DOMESTIC ANIMALS

Domestic Insects

No papers have been found on the effect of DDT on silkworms under practical conditions. A number of studies of the effect of the compound on honeybees have been made. According to HOLST [297], the earliest preliminary work (WIESMANN [664]) indicated that 0·05 per cent DDT was not a stomach poison for bees but did kill them if sprayed directly on them. On the contrary, HOLST found DDT more effective against bees as a stomach poison than as a contact poison. An early indication of the low toxicity of DDT to bees came from the great slowness with which colonies were eliminated when purposely sprayed or dusted (WOLFENBERGER [675]; KULASH [333]). High concentrations of the compound (20 per cent dust or 5 to 10 per cent spray) applied directly into the entrance of the hive have been recommended for the elimination of unwanted colonies (DAVIS [149]). However, according to McGREGOR and VORHIES [391] the application of DDT in pyrophyllite and in kerosene around hive entrances did not damage the colonies within. The treatment of large acerages of cotton with DDT under practical conditions was not hazardous to beekeeping.

More detailed study showed that at room temperature the oral LD_{50} of DDT dispersed as a colloidal suspension in 20 per cent sugar syrup was 4·6 μg/bee (about 33 mg/kg) (ECKERT [190]). A similar result was obtained by WAY and SYNGE [654] using a colloidal suspension in which the particles were about 0·5 μ in diameter, but an LD_{50} of 20 mg/bee was obtained using a suspension in which DDT existed in the form of flat plates $60 \times 15\ \mu$.

ECKERT found that bees were killed by queen-cage candy containing small amounts of DDT but tolerated pollen paste containing much larger quantities. No explanation could be found for the inconsistency. Fifty per cent of caged bees withstood the direct application of 1 per cent DDT dust; stronger formulations were highly toxic under the same conditions. Although some bees and larvae were killed, colonies were not decimated even when the inside of the hive was dusted with 2 per cent DDT.

LINSLEY [375] reported on work done in connection with the practical control of lygus bugs on alfalfa-seed crops. In the different areas investigated, it was found that at least 12 to 19 species of solitary wild bees were active in pollinating the alfalfa. These wild bees were more important as pollinators than honeybees. When DDT was applied as 3, 4, 5, and 10 per cent dusts, formulated with a variety of inert ingredients, the bee population in the treated fields dropped markedly on the day following dusting and then gradually built up over a period of 3 or 4 days. This was true of both wild bees and honeybees. After dusting, the population frequently reached a peak higher than the level reached before dusting and usually higher than the peak reached in corresponding undusted check areas. In general the initial reduction of population was not accompanied by recognizable abnormality in the behavior of the bees remaining in the field or by the presence of many dead or dying bees. It was considered likely but not established that the initial reduction of field population was due to repellent action of the formulations used.

McGregor and Vorhies [391] also noted a reduction of bees for 1 or 2 days after treatment but failed to find a significant amount of repellency. In a similar way Eckert [192] considered that DDT did not have enough repellency to prevent the loss of bees but no evidence was offered. R. F. Smith *et al.* [540] using marked bees, found a decline in bee population in a dusted field immediately after treatment. They gave evidence that the decline was not caused to any significant degree by mortality and interpreted the data as indicating that the decline was largely caused by repellent action of the DDT dusts. No direct evidence of repellency was offered. An alternative explanation for the temporary decline in the field population was a temporary intoxication of the bees which the authors observed but failed to emphasize.

Eckert [190] emphasized the potential hazards of DDT to bees, but suggested that in extensive trials no significant damage to colonies had resulted. By contrast, noticeable reduction or even elimination of colonies was attributed to cryolite, dinitrocresol dust, and especially calcium arsenate. The devastating effects of arsenical dusts have also been observed by others (Eckert [191]; Eide [194]; McGregor and Vorhies [391]). In careful population studies using marked bees R. F. Smith *et al.* [540] determined that the application of DDT dust in practical amounts to blooming alfalfa produced a small but definite injury to the production and activity of honeybees. However, the detrimental effect was more than offset by the increased nectar flow following control of lygus bugs on the plants.

In other experiments bees were marked only after they came to control flowers or to flowers sprayed or dusted with commercial DDT formulations. No repellency or increase in mortality was caused by the DDT (Way and Synge [654]). As one might expect, the small dosage rates used for anopholene larviciding cause no adverse effect on bees (Scudder and Tarzwell [516]).

Domestic Birds

Kingscote and Jarvis [322] found that 6-months-old turkeys showed no symptoms when given as much as 10 g of technical DDT, but were poisoned by 10 g dissolved in oil. 90% of young birds were killed by the solution in oil given at the same rate. The authors concluded that dusts and water dispersable formulations were safe and that oil formulations, although inherently more toxic, could be used with reasonable care.

MARSDEN and BIRD [398] reported that turkeys were unaffected by feeding on mash containing 190 and 380 ppm of DDT for 8 weeks. At a level of 750 ppm the compound caused a depression of growth and some mortality. At a level of 1,500 ppm, the birds developed tremors after about 10 days and few of them survived more than about a month.

RUBIN et al. [501] stated that toxicity was not apparent when chickens were fed a diet containing 310 and 620 ppm of DDT except that egg production was reduced. Hatchability was not affected by 310 ppm but was reduced by 620 ppm. Higher concentrations (1,250 and 2,500 ppm) produced weight loss, molting, tremors, incoordination, and death.

No signs of injury were detected in chickens fed 200 ppm or less of DDT (BRYSON et al. [93]; DRAPER et al. [180]).

MELIS [403] stated that a concentration of 100 ppm of DDT was recommended by entomologists for the control of insects in stored grains. When grain containing this and a higher concentration of DDT was fed to chickens which were also exposed to the compound sprinkled in their environment for a long while, the mortality of the experimental birds was greater than that of the controls. In a second paper [404] the author found no injury in rabbits fed a diet containing 200 ppm of DDT for 15 months.

SCHWARTE and BIESTER [514] in a chapter on 'Poisons and Toxins' have summarized the hazard of a wide variety of chemicals to poultry. They concluded that under proper conditions of use DDT offers no danger to poultry, but RADELEFF et al. [478] have warned against spraying chickens with DDT or dipping them in it.

Domestic Mammals

There have been very few reports of DDT poisoning in either farm animals or pets and even in some of these instances the diagnosis is highly questionable. Cats, to be sure, are susceptible, but this fact is generally known, so that the compound is seldom used on cats. The greater part of the relevant literature is concerned with the purposeful exposure of different species to amounts of DDT far greater than would be used in practical insect control.

An alleged case of DDT poisoning in a goat has been described (HOF-FERBER [284]).

KINGSCOTE et al. [323] reported tests in which cattle were sprayed excessively and their food and water were grossly contaminated by the same formulations. No harm to the animals was observed and the authors concluded that the much smaller exposure involved in recommended use must be safe.

HITCHCOCK and MACKERRAS [281] estimated that a well grown cow might retain about 3·8 l of fluid on the skin and hair and thus, following dipping in a 0·5 per cent formulation, be exposed to 20 g of the compound or about 50 mg/kg. To test the possible ill effects of such use, two animals were given an aqueous suspension of DDT by stomach tube. One received seven 25 g doses in 14 weeks and one received three 50 g doses in 6 weeks. Another animal received three 50 g doses of DDT dissolved in oil. All of the animals remained well. Eighteen young Hereford steers were inuncted along the back with 22·5 g of DDT in peanut oil each week. After 51 weeks no ill effect was detectable.

BUSHLAND *et al.* [95] applied 1·5 per cent emulsions and suspensions of wettable powders as sprays and dips to cattle, sheep, goats, hogs, and horses. Two series of tests were made, the treatments being applied 8 times at 4 day intervals. No injury was noted following the use of DDT although some other compounds studied at the same time did cause injury.

WELCH [656] studied the oral dosage of both sheep and cattle.

A number of studies have been made of residues on forage crops and of the occurrence of DDT in the fat and milk of farm animals which consumed the forage. Although the residues have frequently been excessive, no injury has occurred to the animals themselves (WILSON *et al.* [667]; ARANT [36]; R. F. SMITH *et al.* [539]; HARRIS *et al.* [265]) except that WINGO and CRISLER [671] observed transient nervousness in cows fed at the rate of 12·2 mg/kg/day or at higher rates. Transient tremor was reported in a calf receiving only 2·2 mg/kg/day (BOHMAN *et al.* [79]).

Excellent summary articles on the toxicity of DDT to domestic animals have been prepared by RADELEFF [475] and by RADELEFF *et al.* [477], [478]. DDT in the form of emulsions or suspensions can be used safely in concentrations as high as 8% on cattle, horses, pigs, sheep, and goats. The compound should not be used on cats, for they may be susceptible to the small amounts necessary for insect control. The storage of DDT in fat and its secretion in milk preclude the use of the compound in dairy barns, or as a spray on dairy cows or on animals being finished for slaughter or on crops intended for forage.

ROBERTS and MOULE [490] reported that sheep showerdipped to saturation with 2·0 per cent p,p'-DDT and 0·1 per cent lindane showed a definite though not severe ataxia for several days. It was not determined which insecticide caused the effect. However the same and larger concentrations of lindane commonly kill calves (RADELEFF and BUSHLAND [476]).

ROGERS [496] reported the death of 500 of 2,200 lambs after they had been sheared and then dipped in a 0·1 per cent DDT formulation of which the other components were not identified. The cause of death was definitely established as infection with a species of the genus *Clostridium*. There was epidemiological evidence and some clinical evidence that the dipping formulation had irritated wounds caused by shearing and that these irritated wounds served as portals of entry for the bacteria. Similar wounds in other sheep which were not dipped did not become infected although the animals were exposed to the same environment. The irritant effects on sheep of certain solvents has been pointed out (WATERHOUSE and SCOTT [653]).

Cases of alleged DDT poisoning have been reported without proper consideration of the relationship between exposure and disease. In some instances no evidence has been presented that the animals which became sick ever had any contact whatever with DDT. In other instances no evidences have been given that the animals which became sick had sufficient exposure to produce illness in experimental animals or had greater exposure than animals in the same herd which remained well. Some of the illnesses attributed to poisoning by DDT undoubtedly were caused by chlorinated naphthalenes while others involved uncomplicated infectious disease.

BISKIND [70] attributed X disease of cattle to DDT poisoning. The idea was apparently accepted by KOGER [330]. The term 'X disease' has been used to refer to four distinct entities in veterinary medicine (hyperkeratosis, virus diarrhea, leptospirosis, and encephalitis), but was presumably used by BISKIND to refer to hyperkeratosis. This disease certainly occurred in the United States at least as early as May 1941 (OLAFSON [443]) or perhaps as early as 1939 (SIMPSON [529]). In any event, it antidates the use of DDT in that country. Oddly enough, hyperkeratosis has been demonstrated to be a form of intoxication. It is caused by chlorinated naphthalenes which the animals encounter in a number of products, especially lubricating oils. The disease has been produced experimentally in cattle (BELL [52], [53]; SIKES and BRIDGES [526]; SIKES et al. [527]). Other species appear to be resistant. The present status of hyperkeratosis has been the subject of an excellent review by LINK [374]. The word 'hyperkeratosis' was used by NELSON et al. [432] to describe skin changes in rabbits which had received applications of DDT at the rate of 150 mg/kg/day or greater. The term was used in a descriptive sense and was not intended to indicate a specific disease entity comparable to the syndrome in cattle.

BOHMAN et al. [79] reported that some but not all calves fed DDT at levels up to 75 ppm in the diet developed marked wrinkling of the skin on the side of the neck and jaw and the back of the withers. The wrinkling was accompanied by thickening of the skin, scaling, and thinning of the hair in the same area. Some of the affected calves showed small proliferations and lack of pliability of the skin and a discharge from the eyes and nostrils. The weight gain of experimental and control animals was similar. The skin of calves receiving no DDT was said to be normal but a photograph of one of the controls leaves one in doubt on this important matter. The subject would bear reinvestigation especially in view of the more recent work on hyperkeratosis in cattle mentioned above.

Most of the cases reported by BRITO-BABAPULLE [87] are clearly spurious, but a calf which he mentioned apparently suffered acute poisoning after being treated with 16 ounces of a saturated solution in kerosene. Recovery of the calf was attributed to medication with riboflavin, but many animals poisoned experimentally by DDT to about the same degree recover spontaneously and rapidly.

12.

HAZARDS TO WILDLIFE

A great number of publications have appeared regarding the hazard to wildlife resulting from the use of DDT. Many of the papers involve the use of dosage levels higher than those likely to be used under field conditions. Also many of the papers are based on rather casual observations and not on careful population studies. On the contrary, several quantitative studies have been made of wildlife under practical conditions of malaria or forest-insect control (STEWART *et al.* [554]; HOFFMANN *et al.* [292]; KENDEIGH [321]; TARZWELL [570]; HOFFMANN and SURBER [289]; HOFFMANN *et al.* [293]).

The details of these excellent papers are beyond the scope of this review but the papers are highly recommended to readers with a specialized interest in the field. Other readers may be more interested in papers summarizing one or more aspects of the effect of DDT on wildlife (STORER [562]; COTTAM and HIGGINS [135], [136]; NELSON and SURBER [433]; TARZWELL [571]; HOFFMANN and MERKEL [288]; KALMBACH and LINDUSKA [316]; LINDUSKA and SURBER [372]; LINDUSKA [371]; HOFFMANN and LINDUSKA [287]; SURBER [565]).

Wildlife may become exposed to DDT chiefly through three kinds of uses:

(1) Applications directly to water.
(2) Applications to uncultivated lands.
(3) Applications to agricultural lands.

These three kinds of application differ in their justification and their locale and they frequently differ in their extent and technique. For this reason, they will be treated separately. Simply for convenience, the different species of wild animal life that have been studied in regard to DDT will be considered in approximate taxonomic order.

Applications to water are usually made for the control of malaria or pest mosquitoes, although applications have also been made for mosquitoes as vectors of virus encephalitis, filariasis, dengue, and yellow fever and for other purposes. Applications of DDT to water may be done by hand equipment, by power equipment mounted on trucks or boats, or by aircraft.

Applications to uncultivated lands are usually made for the control of a single dominant species of destructive or disease bearing arthropod such as the spruce budworm (*Choristoneura fumiferana*) or ticks infesting prairies. This circumstance and the ecology of each arthropod may account at one time, for the treatment of hundreds of square kilometers of forest and, at other times, for the treatment of isolated areas known to be the only vicinities infested. In any event, forest treatment may affect any land animals in the treated area and also any water animals in streams which pass through the area. Finally, rain falling on the treated land conceivably may carry DDT into the streams and in this way affect the water forms. A small amount of uncultivated land is treated by hand, especially along the edges of inhabited areas. By far the greatest amount of uncultivated land is treated by aircraft.

Applications of DDT to agricultural lands are frequently made for the control of several species of destructive insects even when only a single crop is involved. Croplands may harbor birds which are considered desirable. Mammals and reptiles in the same situations are less frequently considered desirable and efforts may even be made to control them. Little attention has been given to the direct effect of agricultural insecticide applications on aquatic life. More attention has resulted from washing of insecticides into streams during heavy rains. Applications to agricultural lands differ in an important aspect from the two other sorts of application mentioned: that is, a number of different insecticides are frequently used on the same plot either as a mixture or at different times during the season. Agricultural lands are treated by hand, by power sprayers, and by aircraft.

Whatever the reason for applying an insecticide or the method by which it is applied, the effect on wildlife may be complex and extensive. Some species may be directly affected by the dosage of insecticide which they contact or which they get in their food. Other species which are not directly affected may suffer indirectly because their food supply is reduced or because their enemies are no longer subject to the same degree of natural control. However, this reasoning can be carried too far. Nature has great stability. Thorough study may show that some species are increased while others are decreased. Thus TARZWELL [570] in a careful population study with adequate controls, found that the total number of surface forms increased in treated ponds. The increase occurred in spite of a considerable decrease in aquatic insects, especially Chironomidae and depended on the abundance of nematodes, oligochaetes, and copepods. The increase in the 3 latter groups was thought to reflect a reduction of their predators or competitors. The significance of the population shift from the standpoint of fish production was not clear in the absence of any information on the real value of any of the forms as fish food. In any event, the author considered that the total volume of food was not greatly changed.

The practical use of DDT may even be compatible with an increase of clearly desirable forms which are susceptible to the poison. Thus KENDEIGH [321] indicated that a bird population which had gradually increased in response to an outbreak of spruce budworm (*C. fumiferana*) remained at the abnormally high level in spite of the use of DDT.

Differences between formulations account for considerable variation in the response of wildlife to DDT. It has been demonstrated that differences in solvents have some bearing on the mammalian toxicity of DDT. Fish and, to a smaller degree, amphibia and reptiles, are influenced not only by the type of formulation but also by the degree of its dispersion. There is no clearcut evidence that mammals are significantly influenced by the degree of dispersion of DDT at a constant dosage rate.

In general, solutions have been found more destructive to wildlife than dusts or suspensions of wettable powders (METCALF *et al.* [407]; SURBER [564]; TARZWELL [570]; HOFFMANN and SURBER [289]; HOFFMANN and MERKEL [288]; SURBER and FRIDDLE [566]; SURBER and HOFFMANN [567]). EIDE *et al.* [195], SURBER [564], and TARZWELL [571], [572] found solutions more toxic than dusts but found stable emulsions even more toxic than solutions to fish, amphibia, larger crustacea, and several orders of aquatic insects. This is consistent with the laboratory findings of GINSBURG [227] who found dust least toxic, emulsion more toxic, and colloidal suspension most toxic to goldfish. Colloidal suspensions are seldom if ever used in situations in which wildlife would be affected.

SURBER and FRIDDLE [566] found that, although an oil solution killed more fish than a suspension applied at the same rate, the effect of the oil solution did not extend so far downstream beyond the area of application. He attributed the difference to the immiscibility of the oil film with water and the observation that the oil film easily drifted ashore. On the contrary, in the presence of rapids or wave action, oil sprays may be emulsified to a considerable degree. This accounts in part for the higher mortality of insects in rapids as compared to the mortality of insects in quiet pools (HOFFMANN and MERKEL [288]; SAVAGE [509]).

In some instances, DDT solutions appear to be toxic not so much because the DDT in them is in solution but because of the direct effect of the oil which they contain. Thus METCALF *et al.* [407] found that a 5 per cent solution of DDT in kerosene applied as a spray at a rate of $11\cdot2$ mg/m^2 was quite destructive to aquatic insects living in close contact with the water surface although a 20 per cent solution applied as a thermal aerosol at a rate of $13\cdot5$ mg/m^2 was significantly less destructive to aquatic organisms. Furthermore, the peak mortality may be reached much sooner with an oil solution than with water wettable powder suspensions. TARZWELL [570] found that the application of $9\cdot4$ l of fuel oil per hectare killed more surface insects than the application of DDT dust at the rate of $11\cdot2$ mg/m^2. The importance of this stems from the fact that $11\cdot2$ mg/m^2 of DDT is sufficient for anopheline larva control whereas in the past 140 to 374 l of oil per hectare have been used for routine larviciding.

Velsicol NR–70 (a tetramethyl naphthalene) was found to be more toxic to surface organisms than fuel oil and fuel oil more toxic than kerosene. Ethyl alcohol and acetone were found to have little toxicity (TARZWELL [570], [572]).

Only a portion of the DDT applied can be recovered from the treated surface. For example, HESS and KEENER [274] found that thermal aerosols applied at the rate of approximately $11\cdot2$ mg/m^2 resulted in a recovery rate of only $1\cdot4$ mg/m^2 at the center of the swath and, of course, the recovery of even less material at increasing distances from the center. Similar findings have been reported by others (ERICKSON [200]; SCUDDER and TARZWELL [516]).

A higher percentage of material applied as a spray can be recovered at ground level. For example, SCUDDER and TARZWELL [516] reported the recovery of 13 to 88% of the DDT discharged from a plane as a spray. When spray was applied at the rate of 11·2 mg/m² the average recovery was 5·2 ± 2·4 mg/m² as compared with a recovery of 0·9 ± 0·2 following the application of a thermal aerosol. In a distinct but related study, TARZWELL [572] reported the recovery of a seasonal average of 56 to 76% of DDT applied by aircraft as a spray and only 10 to 12% of DDT applied as a thermal aerosol. Slightly smaller deposits from the application of spray have been reported by others; for example, 39% with a range of 0 to 110% (HOFFMANN and SURBER [289]) and 27% with a range of 0 to 186% (SURBER and FRIDDLE [566]).

In a forest with a close canopy only 2·5% of the DDT applied as an oil spray was recovered although 30% was recovered in open areas (HOFFMANN and MERKEL [288]). In another study the recovery under a forest canopy varied from less than 1 to 10% and averaged 2·5%. When there was an additional cover of weeds, grass, or shrubs, the maximum recovery was 3% and the average recovery was 0·4% (STEWART et al. [554]; STICKEL [555]). In a more open forest, the recovery of DDT averaged 4·6% (GOODRUM et al. [240]).

DDT may fail to reach the intended surface because of a heavy cover of low vegetation irrespective of the presence of a forest canopy. Protection of animal life by vegetation has been observed frequently (HESS and KEENER [274]; SPRINGER and WEBSTER [545], [546]; SAVAGE [509]).

Observations on the recovery of DDT made in connection with residue studies are recorded above (p. 173).

Effect of Application of DDT Directly to Water

As shown in Table 16 some fish may suffer 100 per cent mortality when exposed to a concentration of 0·02 ppm of DDT in laboratory aquaria. The laboratory methods involve no loss of insecticide during application and the formulations used are frequently the most toxic available. Even so, it is of some interest to mention the conditions of field application which are at least theoretically comparable to the laboratory conditions. It is frequently recommended that DDT be applied at the rate of 11·2 mg/m² (0·1 lb/acre) for larviciding in malaria control. If DDT were applied at this rate without loss and became evenly dispersed in water 0·56 m deep, the concentration would be 0·02 ppm. Obviously many ponds in which anopheline larviciding needs to be carried on have an average depth of less than half a meter. Besides the factors of formulation and of loss during application which have already been discussed, there are at least two other factors which allow susceptable organisms to withstand higher calculated concentrations of DDT under field conditions than they are able to tolerate under laboratory or other highly artificial conditions. The organism may actively escape from a contaminated environment or the DDT may be adsorbed and thus made unavailable.

A distinct avoidance reaction to DDT or some other component in the formulation was observed in the small fish, *Gambusia* (LACKEY and STEINLE [338]). Fish have also been observed to migrate from streams to adjoining lakes, not so much to avoid DDT as in search of food after the insects of the streams had been severely reduced (HOFFMANN and SURBER [290]).

LACKEY and STEINLE [338] seem to have also been the first to observe the apparent detoxification of DDT in natural ponds. A similar finding was made by PIELOU [458]. TARZWELL [572] found that under similar treatment, fish mortality occurred first in barren sand-bottom ponds and was smallest in clay- or silt-bottom pools which had considerable vegetation and organic matter on the bottom and which received muddy water during rains.

The adsorption of DDT on organic matter has been studied by UPHOLT [602].

Although the detoxification of DDT in natural ponds and streams is quite real, it must be re-emphasized that a part of the difference between laboratory and field results arises from drifting and loss of insecticide during airplane distribution as described above.

Organisms Serving as Fish Food

The most important conclusion which has been drawn repeatedly from careful observations under the conditions of malaria control is that anopheline mosquitoes may be controlled by DDT without undue injury to those organisms considered important as fish food (METCALF *et al.* [407]; HESS and KEENER [274]; TARZWELL [570], [571], [572]).

Plankton organisms in smaller ponds were so little affected by DDT, as used for malaria control, that any changes which may have been caused by the insecticide were overshadowed by seasonal variations in the populations. No drastic killing of any group occurred and the total number of organisms supported by a given volume of water remained essentially constant during the summer months both in ponds to which DDT was applied and in control ponds (BISHOP [64]).

Besides mosquito larvae, the high degree of susceptibility of surface Diptera (Chironomidae), Hemiptera, Coleoptera, Ephemeroptera, and amphipod crustacea has been observed (EIDE *et al.* [195]; METCALF *et al.* [407]). A complete season of routine treatments almost entirely eliminated surface Hemiptera but other forms were not significantly affected so that the population of fish food organisms as a whole was not injured by thermal aerosols at an application rate of 11·2 mg/m² (HESS and KEENER [274]).

In a study of the effect of routine anopheline mosquito larviciding by airplane, SCUDDER and TARZWELL [516] found no significant over-all reduction in the insect population of areas adjoining those routinely treated except for a reduction of mosquitoes, deer flies, and sand flies and a questionable reduction of caddis flies (Trichoptera). Although a considerable number of dead May flies, damsel flies and certain beetles were found floating on water surfaces, there was no evidence of a real effect on the population level of these aquatic insects. Population levels of strictly terrestrial insects were not affected. Particular studies of aphid colonies, aphid parasites and predators, and of bee colonies failed to indicate any damage. Population studies made with light traps in treated and untreated areas revealed no decrease of any group with the possible exception of the Trichoptera.

It has frequently proved necessary to use higher application rates for the control of pest mosquitoes generally (HENDERSON [270]), and of saltmarsh mosquitoes in particular, than are required for the control of anopheline mosquitoes. This fact and the chemical and ecological differences between brackish and fresh water demand a separate consideration of the use of DDT under saltmarsh conditions.

Extensive mortality of the commercially important blue crab (*Callinectes sapidus*) was reported by SPRINGER and WEBSTER [545], [546]. Application at the rate of 22·4 mg/m², the lowest dosage tested, produced a mortality of 60% in one week. Greater dosage rates produced more rapid and more complete kills. Other species of crabs and other crustacea suffered severely as did Hemiptera, Homoptera, and Diptera. Snails and muscles were unaffected.

Fiddler crabs (*Uca crenulata*) on exposed beaches suffer a severe mortality (GOODRUM *et al.* [240]). On the contrary, crabs (*C. sapidus*) in water about 2 m deep remained uninjured when DDT was applied to the surface at the rate of 336·3 mg/m² (TILLER and CORY [587]).

Most of the concern over fish-food organisms has been that their depletion would lead to the gradual and relative starvation of fish. The possibility has also been considered that fish would be injured by ingesting DDT on insects killed by the compound. It is not astonishing that fish were killed by eating a number of insects which had been directly sprayed with DDT (SURBER [564]; HOFFMANN and SURBER [291]). Of more interest is the fact that fish were uninjured when maintained on a diet of *Aedes aegypti* larvae that had been killed by DDT under conditions simulating those in nature (GINSBURG [231]). During larviciding operations the intoxication of fish seems to have been direct and not secondary to the ingestion of poisoned insects.

Incidentally, an interesting use for DDT may have been found in oyster culture. In some regions, shells from which oysters have been removed are put back in the water so that larval oysters may attach to them and thus have a suitable base on which to grow. It was found that the application to the shells of a spray containing 5% of DDT in kerosene at rates up to 168 mg/m² had no adverse effect on the subsequent attachment of oyster larvae. Furthermore, it was found that this application of DDT interfered greatly with the attachment of barnacles which compete with oysters for food and space. It was considered that, in areas where barnacles are a serious problem in oyster growing, DDT could be used to reduce barnacles and increase the harvest of oysters (LOOSANOFF [379]).

Fish

HESS and KEENER [274] found that fish were not injured following a season in which 16 weekly applications were made with thermal aerosol at the rate of 11·2 mg/m². The analysis, which was based on adequate controls, involved the total number of fish, their weight and condition, the species composition,

and the abundance of very young fish. An average population per acre of
12,549 fish weighing 318 kg was encountered. Similar results were reported by
TARZWELL [572] following 3 years of study. The following procedures were
found essentially harmless to fish and effective for malaria control and were
therefore recommended: hand application of oil solutions at the rate of
5·6 mg/m² of DDT and 3·8 l of solution, or the hand application of dusts at
the rate of 11·2 mg of DDT per square meter, or the application by aircraft
of sprays or thermal aerosols at the rate of 11·2 mg/m². Other procedures involving different formulations or higher doses or both were destructive to fish.
The paper contains a number of detailed population studies of great interest
to specialists in wildlife management. Of general interest is the fact that
different species of fish showed a considerable range of susceptibility to DDT
under similar circumstances.

As might be expected, excessive application of DDT leads directly to the
mortality of fish (DEONIER et al. [169]; EIDE et al. [195]; GINSBURG [227],
[228]; PIELOU [458]; SPRINGER and WEBSTER [545], [546]).

Of particular interest is the report of HERALD [272] regarding the loss of
young milkfish in commercial ponds following routine malaria control operations. The death of the fish caused a considerable financial loss for their owner.
The DDT was applied by aircraft as a 5 per cent solution in fuel oil. The
exact dosage was unknown but was thought to be greater than 22·4 mg/m².
The same author reported another situation in which fish killed by DDT in
the course of malaria control operations were used for human food without
any ill effects being reported or observed.

Under some conditions, the nature of which has not been clearly defined,
fish may withstand high application rates without apparent harm. Thus TILLER
and CORY [587] reported no injury to top minnows or frogs in a fresh-water
pond sprayed at the rate of 336·3 mg/m² with an oil spray and again at the
same rate with an emulsion.

The same authors made a similar finding in regard to fish in water a meter
or more in depth and subject to tidal flow. The dosage rate and formulations
were the same. The protection was not caused by brackish water as shown by
the completely different results observed by SPRINGER and WEBSTER [545],
[546]. Heavy losses of fish occurred when a saltmarsh was sprayed at low tide
at a rate of 179·4 mg/m² and a moderate mortality when the spraying was done
at the rate of 89·7 mg/m². Some mortality of small fish in shallow pools occurred
when the marsh was treated at the rate of 28·0 mg/m².

Birds and Mammals

Routine Anopheline larviciding does not affect birds or mammals living
within the treated area. ERICKSON [200] reported extensive studies of birds
and mammals exposed to routine larviciding in areas totaling 330 ha. Weekly
airplane applications were made over a 17 week period using a 20 per cent

solution of DDT in a highly methylated naphthalene at the rate of 11·2 mg/m². In part of the areas all the applications were made in the form of a spray; this allowed an actual recovery of DDT at the rate of 5·1 ± 0·8 mg/m² per treatment, or about 87·4 mg/m² per season. In other areas, all the applications were made at the same rate in the form of a thermal aerosol; this permitted a smaller recovery of DDT; 0·9 ± 0·2 mg/m² per application or about 14·1 mg/m² per season. The recovery studies were made on the low islands and dykes which formed the habitat of the birds and mammals studied. No dead birds or mammals were found in any of the areas. Careful population studies revealed no significant effect of DDT on some 32 species of birds and 4 species of mammals in treated areas as compared with the same species in ecologically comparable, untreated areas. Random fluctuations and fluctuations associated with season were similar in the populations in both treated and untreated areas.

The spraying of a saltmarsh resulted in the death of many crabs and other crustacea and a decrease in the insect population. Gulls (*Larus atricilla*), herons, and perhaps rails (*Rallus longirostris*) entered the area to feed on the dead arthropods. Swallows (*Iridoprocne bicolor*) temporarily left the area but returned as the number of flying insects increased. The death of a young rail was ascribed to the application of DDT at the rate of 28 mg/m². Other rails successfully reared young in areas treated at the rate of 112 mg/m². Song birds were unaffected and reared young successfully (SPRINGER and WEBSTER [545], [546]).

Effect of the Application of DDT to Nonagricultural Lands

Applications of DDT to nonagricultural lands are made principally for the control of forest insects. Oil solutions of DDT at the rate of 112 mg/m² are adequate for the control of many foliage infesting insects. Dusts and especially granular formulations have been applied at rates as high as 2,800 mg/m² for the control of soil-infesting grubs such as the white fringed beetle (*Graphognathus leucoloma*).

There has been a great deal of study of the effect on wildlife of those formulations which adhere to foliage. There has been little study of the effect of granular formulations on terrestrial organisms.

Aquatic Invertebrates

The application of 112 mg or more of DDT per square meter, which is required for the control of forest insects, constitutes a real threat to many organisms in streams passing through forested areas. This might be predicted from the studies of wildlife made in connection with malaria control.

The application to ponds of DDT at the rate of 112 mg/m² results in the extermination of practically all the surface insects and those breathing at the surface and a severe reduction of free-swimming and crawling species. The bottom forms are not so severely affected.

The application of DDT to streams at the rate of 112 mg/m² may kill as much as 90% of the insect population and locally exterminate a third of the species present (HOFFMANN *et al.* [292]). Crayfish, like insects, are highly susceptible (COUCH [137]).

Experimental evidence indicates that streams carry lethal amounts of DDT only for short distances (HOFFMANN *et al.* [292]). On the basis of this and other facts, it has been concluded (HOFFMANN *et al.* [292]) that, when streams and their watersheds are sprayed by aircraft, most if not practically all the damage to the stream's fauna is done by insecticide deposited directly on the stream's surface and not by insecticide washed into the stream. However, BANDT [42] has reported that fish occasionally are killed when DDT dust used for forest trees is washed into streams. It has been suggested that, where fish are of some importance, pilots making forest treatments should avoid direct applications of insecticides to streams (LANGFORD [339], [340]). However, the practicality of doing this with large aircraft is doubtful (HOFFMANN and SURBER [290]).

Although ponds tend to hold an insecticidal dosage of DDT longer than streams, the ponds undergo repopulation quicker than the streams after the effect of DDT has worn off (HOFFMANN *et al.* [292]). The difference is ascribed to the greater proportion of pond-inhabiting species with a short life cycle or a long flight range or both. Insect eggs and pupae are both relatively resistant to DDT (SAVAGE [509]).

A further factor in the rate of repopulation may be latitude–involving both environmental temperature and length of life cycle. Streams at approximately 40° north latitude have shown good or even complete recovery of insects one year after a single application of DDT (HOFFMANN and MERKEL [288]). On the contrary, streams at roughly 50° north latitude showed in one year a poor recovery of insects, ranging from 5 to 50% (SAVAGE [509]).

Of particular interest is the study of HOFFMANN and SURBER [290] in a 21,044 ha area treated under practical conditions at the rate of 112 mg/m² for control of the gypsy moth (*Porthetria dispar*). Although 70 to 80% of the stream insects were killed within 3 days and although fish were killed in one stream over a period of a month, the loss of fish was small in comparison to the total population.

Terrestrial Invertebrates

Earthworms (*Lumbricus*) which feed on fallen leaves are susceptible to DDT under practical conditions of use. In the late summer, part of a dense stand of young elm trees was sprayed to runoff with a 0·25 per cent emulsion using ground equipment. During the following spring, earthworms were only about one third as abundant in the sprayed area as in a control area. They had been equally abundant in the two plots in previous years. Although some worms remained in the treated area, their activity was so reduced that at one time during the spring it was possible to identify the sprayed plot accurately merely by inspection because there the worms had not yet eaten the fallen leaves while worms in the surrounding area had already cleared the ground (BAKER [39]).

Insects are by far the most important land invertebrates and, as a group, they are highly susceptible to DDT. Their survival following the application of DDT to forests depends largely on their opportunity for actual exposure as determined by ecology and life history. Species living in burrows or in leaf mold or beneath bark or in other protected situations may escape unharmed. Individuals which are in the egg stage at the time of application may not emerge until the deposit of DDT has been reduced below an effective level. A few insects appear relatively immune as a result of their structure. Thus very hairy forms, such as bees, fared better than the nonhairy members of the same order and moths were also relatively resistant (HOFFMANN and MERKEL [288]; HOFFMANN *et al.* [293]).

The application of DDT in an oil spray at the rate of 112 mg/m² or higher causes an almost immediate and pronounced effect on many species of insects inhabiting different kinds of forests. Mortality of exposed forms, as determined by carefully controlled population studies, frequently reaches 90% and some species appear to be eliminated temporarily in the treated area. The residual effect of DDT applied as an oil solution at the rate of 112 mg/m² lasted about a week; the effect of a 224 mg/m² dosage lasted about 2 weeks; the effect of a 560 mg/m² dosage was severe after 6 weeks and there was little recovery until 3 months after application (HOFFMANN and MERKEL [288]; HOFFMANN *et al.* [293]).

The results of some less completely described studies appear consistent with the above discussion (COUCH [137]).

High dosage levels, for example, 560 mg/m², may so destroy natural enemies of aphids and mites that these forms are able to multiply to an astonishing degree (HOFFMANN and MERKEL [288]; HOFFMANN *et al.* [293]). Obviously, the economic damage done by aphids and mites must be carefully considered in evaluating the effect of DDT on forest trees and in determining the most beneficial dosage rate. The need for reasonable restrictions on the use of residual insecticides has been discussed by BALACHOWSKY [41].

Fish

In a number of experiments (SURBER [564]; HOFFMANN *et al.* [292]; HOFFMANN and SURBER [289]; SURBER and FRIDDLE [566]; LANGFORD [340]) DDT has been applied to streams in forested areas by aircraft flying directly above the stream. Although the applications were made at rates used for the control of forest insects, it has been suggested that the dosage reaching the water was greater than would have occurred in routine forest treatments (HOFFMANN and MERKEL [288]). In any event, some mortality of fish has regularly been observed in these tests but it has not been excessive. The same is true of studies with experimental ponds (SURBER [564]) and of studies of ponds under conditions of practical use (GOODRUM et al. [240]).

A number of studies (SURBER [564]; HOFFMANN and SURBER [289]; SURBER and FRIDDLE [566]; SURBER and HOFFMANN [567]; ADAMS et al. [3]) have emphasized the wide differences in susceptibility of different species. For example, bluegills (*Helioperca incisor*) suffered a mortality of 79 to 97% while the mortality of yellow perch (*Perca flavescens*) confined in the same live-boxes did not exceed 10% (SURBER [564]). In general, the mortality of fish, in contrast to fish food organisms, has not been severe (HOFFMANN and SURBER [289]; SURBER and FRIDDLE [566]).

Death of fish from DDT poisoning appears to be the result of direct contact with the compound. Trout showed only a small mortality following heavy feedings on insects paralyzed by DDT applied to the stream by aircraft (SURBER [564]; LINDUSKA and SURBER [372]). Trout were killed when they fed on terrestrial insects which had fallen into a surface film of DDT in oil. Minnows in the same stream were not killed although they fed on aquatic insects affected by DDT and carried downstream by the current (LANGFORD [340]).

Under practical conditions of forest treatment even with dosages of 280 mg/m², the effect of DDT on fish populations may be slight although some individuals are killed (LINDUSKA and SURBER [372]; ADAMS et al. [3]).

Amphibians and Reptiles

The application of DDT at a rate suitable for the control of forest insects has little effect on frogs, salamanders, snakes or turtles (HOFFMANN and SURBER [289]; GOODRUM et al. [240]; HOFFMANN and SURBER [290]; SPEIRS [543]). The death of some snakes and other reptiles (HOFFMANN and SURBER [290]; GEORGE and STICKEL [223]) and of frogs (SPEIRS [543]; GOODRUM et al. [240]) has been reported. The high degree of inherent susceptibility of frogs, toads and snakes is evident from the studies of LOGIER [378].

Birds

In a well controlled experiment passerine-bird nests containing eggs or young were sprayed by hand at the rate of 560 mg/m² with an oil solution of DDT. The application had no detrimental effect on the hatching of the eggs or the development of the young, nor did it cause the adult birds to abandon the nests (MITCHELL [412]). Similar results were obtained by HOPE [299]. These negative results support the frequent assumption that birds which die in treated areas are killed by eating insects contaminated by DDT rather than by direct contact with the compound.

The effect of feeding DDT-treated insects to nestling birds has been investigated by GEORGE and MITCHELL [222]. When insects sprayed at the rate of 112 mg/m² were fed to young birds at levels of 25 and 50% of the weight of the birds, there was no indication of DDT toxicity. During the study the birds remained in their nests and received additional food and other necessary care from their parents. When an attempt was made to maintain nestlings in the laboratory on a diet consisting entirely of insects sprayed at the rate of 112 mg/m², and the birds were then subjected to partial starvation, many of them died. The percentage of mortality in the checks was 60% and in the experimental birds was 85%. Some of the nestlings fed DDT-treated insects in the laboratory showed signs of poisoning which were described in detail. Similar results of less extensive tests were reported by HOPE [299].

MITCHELL et al. [413] studied the effect on nesting birds of DDT applied by aircraft at the rate of 336 mg/m². A total of 293 nest boxes were placed in the abandoned fields and along the hedgerows where the study was made. Most of the information obtained concerned birds nesting in these boxes. Spraying caused only a slight decrease in the total bird population, no avoidance of the area by nesting birds, no increase in the number of cases of desertion, and no significant differences in feeding ranges even though insects for food were reduced immediately after spraying to 5·9% in the first year and to 15·9% in the second year. The application caused considerable mortality to wren nestlings of the first brood one year but not the next year. The mortality of second-brood wrens was not increased and the number of nests and eggs in the sprayed area was greater the second year than the first. The average weight of first-brood wrens in the sprayed area was significantly lower than the average weight in the control area both years. The relative importance of starvation and of DDT poisoning in causing this effect could not be determined. Chemical analysis of nestlings thought to have died from poisoning revealed levels of 2·6 to 77·0 ppm while analysis of birds thought to have died of other causes showed levels from 14·3 to 38 ppm. Nestlings sacrificed when signs of DDT poisoning were present contained 3·6 ppm; however, an apparently normal bird sacrificed for analysis showed 9·0 ppm. Controls showed up to 16·2 ppm of DDT.

It is certain that adult birds feed eagerly on affected insects and may enter a treated area for that purpose (STEWART et al. [554]; HOPE [299]; BENTON [55]).

Under practical conditions, the application of DDT to forests by aircraft at the rate of 112 mg/m² (HOTCHKISS and POUGH [302]; KENDEIGH [321]; LINDUSKA and SURBER [372]) or 224 mg/m² (STEWART et al. [554]) produced little or no effect on the bird population, the hatchability of eggs, the mortality of young, or the abandonment of nests. This was true although in one instance the application of 112 mg/m² produced a small, apparently negligible, immediate mortality. Several birds with signs of poisoning were captured and they later died. The DDT was applied for the control of the spruce budworm (*Choristoneura fumiferana*) and produced a mortality of 50 to 60% among the larvae. There was evidence that the gradual increase of this forest pest over a period of years had permitted an increase in the bird population and that the birds remained more abundant than normal during the season in which DDT was applied (KENDEIGH [321]).

STEWART *et al.* [554] found evidence that one species, the redstart (*Sedophaga ruticilla*), was affected by a dosage of 224 mg/m², although there was no change in the bird population as a whole.

Even the use of 336 mg/m² (GOODRUM *et al.* [240]) and 448 mg/m² (SPEIRS [543]) was reported to cause no apparent effect on bird populations under careful ecological study.

On the contrary, the use of more than 448 mg/m² frequently leads to poisoning so that birds showing clear-cut signs of intoxication may be captured. These birds almost invariably die (HOTCHKISS and POUGH [302]; LINDUSKA and SURBER [372]; ROBBINS and STEWART [489]; GEORGE and STICKEL [223]; SPEIRS [543]; ADAMS *et al.* [3]). In one study heavy application of DDT for the purpose of gypsy moth control led to an almost complete elimination of the bird population which 2 weeks after spraying had returned to only about 16% of normal (HOTCHKISS and POUGH [302]). In another study, the application of DDT at 560 mg/m² for experimental purposes reduced the bird population to about 35% of normal. It was concluded that the decrease was caused by death of the birds rather than by their migration from the area (ROBBINS and STEWART [489]). In a third study, DDT was applied as a dust to an area of mixed tall grass, prairie, and woodland at the rate of 493 mg/m² for the control of the Lone Star tick (*Amblyomma americanum*). Used as a dust, DDT reduced the bird population only to 40 or 50% of normal, although the populations of certain ground- and bush-feeding birds were reduced to less than 10% of normal (GEORGE and STICKEL [223]). In a fourth study a marked decrease in the bird population was observed after four experimental applications of DDT totaling 1,121 mg/m² (SPEIRS [543]). In a fifth study, the application of 560 mg/m² was followed by an increase in the bird population but the application of 841 mg/m² was followed by a decline of about 70%.

Under certain circumstances, the exact nature of which has not been defined, applications of DDT to forests at rates as high as 840 to 1,121 mg/m² have produced little effect on bird populations which were carefully studied (LINDUSKA and SURBER [372]; ADAMS *et al.* [3]).

Extremely heavy applications of DDT varying from 680,388 to 1,360,777 mg per individual tree for the control of Dutch elm disease produced effects which could be predicted from the results of earlier work. BENTON [55] reported the death of a number of birds, some of which showed tremors and other characteristic signs before death. The author gave no conclusive evidence for any effect on the bird population as a whole. Some effect on the survival of nestlings was claimed but, because of the small sample, is open to question. In any event, it is clear that the application of insecticide to a single species of tree causes less dramatic effects on wildlife than application of the same formulation at the same rate over an extensive area.

Mammals

The application of DDT to forest as an oil solution at the rate of 224 mg/m^2 produced no effect on the population of deer mice (*Peromyscus leucopus*) (STICKEL [555]). Even the use of a much larger amount of DDT for tick control on a prairie caused no injury to deer, racoons, armadillos, striped skunks, or rabbits (GEORGE and STICKEL [223]). Chiefly on the basis of laboratory tests, MACKIE [389] concluded that concentrations of DDT of the order of 673 mg/m^2 had no direct harmful effect on mammals. Under practical conditions, the application of 336 to 1,121 mg/m^2 produced no harmful effects (LINDUSKA and SURBER [372]; GOODRUM *et al.* [240]; SPEIRS [543]; ADAMS *et al.* [3]) on population levels although there was equivocal evidence that applications of 560 to 841 mg/m^2 caused at least a transient effect in some individuals (LINDUSKA and SURBER [372]; ADAMS *et al.* [3]).

Although the applications involved market and warehouse buildings it is interesting to record that bats flying through a heavy spray mist were forced to land on the floor where they remained for 15 or 20 minutes and then flew off uninjured. The effect was apparently due to wetting and irritation caused by the emulsion, for the same effect was later produced by a similar emulsion containing no DDT (FENNAH [203]).

The oral LD$_{50}$ for several wild mammals as well as for common laboratory mammals is given in Table 4.

Effect of the Application of DDT to Agricultural Lands

There is very little information on what effect, if any, agricultural uses of DDT have on wildlife. In the first place, the wild animals on farms which are generally regarded as desirable are not primarily inhabitants of highly cultivated land although they may live in orchards or meadows and they may enter cultivated fields to obtain a part of their food. This would lead one to expect that no general effects on populations would occur although a few individuals, especially birds, might be injured by feeding on insects affected by DDT.

There have been a number of instances in which fish in small streams or drainage ditches have been killed by insecticides washed from treated fields by heavy rains (YOUNG and NICHOLSON [684]). The effect has not been observed in the larger streams into which the temporarily poisoned water flowed. Apparently fish kills resulting from the agricultural use of DDT alone have not been reported. The kills which have been reported have involved two or more insecticides and frequently toxaphene. Fish are more susceptible to aldrin and much more susceptible to toxaphene than they are to DDT (LINDUSKA and SURBER [372]; TARZWELL [572]; DOUDOROFF *et al.* [176]).

LINDUSKA and SURBER [372] reporting work done by ODUM and NORRIS, stated that the application of DDT at the rate of 224 to 729 mg/m^2 for control of the pecan weevil produced no change in the bird population not observed in a comparable, untreated area. No dead or poisoned birds were seen. An apple orchard which had received 3,363 to 5,044 mg/m² of DDT at the time of study contained more birds than a neglected orchard used as a check, but the authors gave no evidence that this condition was caused by DDT or was not characteristic of the two situations.

On the contrary, MOHR *et al.* [414] concluded that the rate of application of DDT used in the apple and pear orchards of eastern Washington could either cause the death of ring-necked pheasants or contribute to their death. Under practical conditions, the rate was 5,000 to 6,100 mg/m²/year. Following spraying, birds were observed showing signs considered typical of DDT poisoning. The average period between application of DDT in combination with parathion, and the death of several species of birds was 7 days. Chemical analyses of the livers of pheasants which became sick in the orchards showed 15 to 326 ppm of DDT. Pheasants intentionally fed DDT mixed with their food at the rate of 320 to 1,000 ppm for 39 days showed little mortality but stored 20 to 94 ppm of DDT in their livers. In those instances in which DDT is used in agriculture in combination with parathion or other insecticides, there remains much to be learned about the importance and interaction of the different materials in regard to their effect on wildlife.

In a very thorough study, JACKSON [308] found that the application of DDT and parathion to vegetable crops had no harmful effect on wood mice (*Peromyscus leucopus*) living in the edges of forests which bordered the fields.

It has been shown that DDT accumulates in the soil under sprayed fruit trees reaching concentrations as high as 176 mg of DDT per kilogram of surface soil after 5 years of spraying. In orchards which were sampled for three or more years the concentration increased in each succeeding year. In uncultivated orchards, about 90% of the insecticide was found in the surface debris. In cultivated soil the depth to which DDT penetrated seemed to be determined by the depth of cultivation (CHISHOLM *et al.* [124]). It has also been shown that DDT persists in the soil for long periods (CARNE [112]).

REFERENCES

[1] ABRAMS, H. K., *Public Health Aspects of Agricultural Chemicals*, Calif. Health *6* (13), 97–102 (1949).

[2] ADAM, F., and ZUST, A., *Die Bestimmung des 'Geigy 33' in Mahlprodukten und das Verhalten des Insektizids beim Mahlprozess*, Gebiete Lebensmittelunters. Hygi. *38* (6), 371–375 (1947).

[3] ADAMS, L., HANAVAN, M. G., HOSLEY, N. W., and JOHNSTON, D. W., *The Effects on Fish, Birds, and Mammals of DDT Used in the Control of Forest Insects in Idaho and Wyoming*, J. Wildlife Mgmt. *13* (3), 245–254 (1949).

[4] ALLEN, N. N., LARDY, H. A., and WILSON, H. F., *The Effect of Ingestion of DDT upon Dairy Cows*, J. Dairy Sci. *29* (8), 530–531 (1946).

[5] ALLEN, W. R., and BERCK, B., *DDT Residues on Celery Resulting from Dust Treatment for Control of the Tarnished Plant Bug*, Sci. Agric. *30*, 375–379 (1950).

[6] ALLEN, W. R., RICHARDSON, H. P., BERCK, B., and ROBINSON, A. G., *DDT Residues on Currants and Gooseberries*, Sci. Agric. *30*, 380–383 (1950).

[7] ALLESANDRINI, MARIA, *DDT Used for the Control of Grain Insects*, Conference at Savannah, Ga. (1952).

[8] ANDERSON, A., and KHORRAM, M. A., *Exposure to DDT*, Brit. med. J. *1*, 1132–1134 (1948).

[9] ANDREWS, J. M., and SIMMONS, S. W., *Developments in the Use of the Newer Organic Insecticides of Public Health Importance*, Amer. J. publ. Hlth. *38* (5), 613–631 (1948).

[10] ANGLEY, J. C., *Toxicity of DDT in Kerosene*, Air Surg. Bull. *2*, 77–78 (1945).

[11] ANONYM, *A New Synthetic Insecticide*, Brit. med. J. *2*, 217 (1944).

[12] ANONYM, *Tolerance Figure Suggested for DDT*, A. I. F. News *3* (3), 3 (1945).

[13] ANONYM, *Toxicity of DDT*, Lancet (6339) *1945*, 248.

[14] ANONYM, *Toxicology of DDT*, Brit. med. J. *1*, 338 (1945).

[15] ANONYM, *Further Results with DDT*, Brit. med. J. *2*, 260 (1945).

[16] ANONYM, *Toxicity of DDT*, J. Amer. med. Ass. *129* (11), 741 (1945).

[17] ANONYM, *DDT in Common Use* (Medical notes in parliament), Brit. med. J. *1*, 338 (1946).

[18] ANONYM, *DDT in General Use*, Brit. med. J. *2*, 203–204 (1946).

[19] ANONYM, *Queries and Minor Notes. DDT Poisoning*, J. Amer. med. Ass. *130* (1), 56 (1946).

[20] ANONYM, *Human Death Attributed to DDT*, J. Amer. med. Ass. *132* (3), 160 (1946).

[21] ANONYM, *Danger of DDT in Kitchens*, J. Amer. med. Ass. *133* (1), 73 (1947).

[22] ANONYM, *'Virus X' and Influenza A*, J. Amer. med. Ass. *136* (13), 884 (1948).

[23] ANONYM, *Pesticides: Chemical Contaminants of Foods*, J. Amer. med. Ass. *137* (18), 1604–1605 (1948).

[24] ANONYM, *Queries and Minor Notes. Virus-X Infection*, J. Amer. med. Ass. *138* (9), 720 (1948).

[25] ANONYM, *No DDT for Dairies?*, Agric. Chemic. *4* (2), 53 (1949).

[26] ANONYM, *Stephens Says 'No Tolerance for DDT in Food' in Purdue Conference*, Pests *17* (3), 34 (1949).

[27] ANONYM, *Use of DDT on Lactating Animals*, Nat. Agr. Chem. Ass. News *7*, 3, 5 (1949).

[28] ANONYM, *No DDT in Dairies; U. S. D. A.*, Agric. Chemic. *4* (4), 55 (1949).

[29] ANONYM, *DDT Discredited for Dairies*, Chem. Engng. News *27* (15), 1058 (1949).

[30] ANONYM, *U. S. Acts to Keep DDT out of Milk Supply* (Newspaper clipping), N.Y. Post Home News *1949* (April 22).

[31] ANONYM, *DDT, too much Spraying, too Little Government Action*, Consumer Repts. *14* (6), 275–277 (1949).

[32] ANONYM, *Statement on the Toxicity of DDT*, J. Amer. med. Ass. *139* (18), 1285 (1949).

[33] ANONYM, *Possible Hazards from the Use of DDT*, Amer. J. publ. Hlth. *39* (7), 925–927 (1949).

[34] ANONYM, *Toxicité du dichlor-diphényl-trichloréthane*, Méd. trop. *10* (1), 135–136 (1950).

[35] ANONYM, *Insecticide Storage in Adipose Tissue*, J. Amer. med. Ass. *145* (10), 735–736 (1951).

[36] ARANT, F. S., *Status of Velvetbean Caterpillar Control in Alabama*, J. econ. Ent. *41* (1), 26–30 (1948).

[37] ASHDOWN, D., and WATKINS, T. C., *Control of the Lettuce Yellows Disease in New York*, J. econ. Ent. *41* (2), 252–258 (1948).

[38] BABIONE, R. W., *DDT 'Poisoning'*, New Engl. J. Med. *236*, 643–644 (1947).

[39] BAKER, W. L., *DDT and Earthworm Populations*, J. econ. Ent. *39* (3), 404–405 (1946).

[40] BALABAN, J. B., *DDT Poisoning*, Brit. med. J. *1*, 147 (1946).

[41] BALACHOWSKY, M., *Vœux concernant l'utilisation des insecticides non spécifiques à longue durée d'action dans les zones non cultivées et les «stations-refuges»*, C. R. Acad. Agr. France *39*, 318–321 (1953).

[42] BANDT, H. J., *The Toxic Action of Gesarol, Which is Used for the Protection of Forest Trees, on Fish*, Beitr. Wasser-Chem. *1946*, 42–43.

[43] BARNES, J. M., Personal communication (1952).

[44] BARNES, J. M., *The Reactions of Rabbits to Poisoning by p-Nitrophenyldiethylphosphate (E–600)*, Brit. J. Pharmacol. *8* (2), 208 (1953).

[45] BARNES, J. M., *Toxic Hazards of Certain Pesticides to Man*, Bull. World Hlth. Org. *8* (4), 419–490, 535–589 (1953) (Reprinted: WHO Monograph Series No. 16).

[46] BARNES, J. M., *Crop Protection as a Health Hazard*, Chem. & Ind. *1953*, 625–627.

[47] BARNES, J. M., and DENZ, F. A., *Experimental Methods Used in Determining Chronic Toxicity. A Critical Review*, Pharmacol. Rev. *6* (2), 191–242 (1954).

[48] BARNES, M. M., CARMAN, G. E., EWART, W. H., and GUNTHER, F. A., *Fruit Surface Residues of DDT and Parathion at Harvest*, Adv. Chem. Ser. *1*, 112–116 (1950).

[49] BATCHELOR, G. S., and WALKER, K. C., *Health Hazards Involved in the Use of Parathion in Fruit Orchards of North Central Washington*, A. M. A. Arch. Ind. Hyg. occup. Med. *10* (6), 522–529 (1954).

[50] BEARD, R. L., Symposium, *Toxicity of DDT*, J. econ. Ent. *39* (3), 425 (1946).

[51] BELKIN, MORRIS, *The Effect of DDT (2,2-Bis-[p-chlorophenyl]-1,1,1-trichloroethane) on Protozoa*, Fed. Proc. *4*, 112 (1945).

[52] BELL, W. B., *Further Studies on the Production of Bovine Hyperkeratosis by the Administration of a Lubricant*, Virginia J. Sci. *3*, 169 (1952).

[53] BELL, W. B., *The Relative Toxicity of Chlorinated Naphthalenes in Experimentally Produced Bovine Hyperkeratosis (X Disease)*, Vet. Med. *48* (4), 135–140, 146 (1953).

[54] BENNISON, B. E., and MOSTOFI, F. K., *Observations on Inbred Mice Exposed to DDT*, J. nat. Cancer Inst. *10* (4), 989–992 (1950).

[55] BENTON, A. H., *Effects on Wildlife of DDT Used for Control of Dutch Elm Disease*, J. Wildlife Mgmt. *15* (1), 20–27 (1951).

[56] BERKOW, S. G., *The Value of Surface-Area Proportions in Prognosis of Cutaneous Burns and Scalds*, Amer. J. Surg. *2*, 315–317 (1931).

[57] BERTI, A. L., *Routine Consumption of DDT in Drinking Water*, Direct communication (1949).

[58] BERTI, A. L., *Practice of Adding DDT to Drinking Water in Venezuela*, Personal communication (1950).

[59] BIDDULPH, C., BATEMAN, G. Q., BRYSON, M. J., HARRIS, J. R., GREENWOOD, D. A., BINNS, W., MINER, M. L., HARRIS, L. E., and MADSEN, L. L., *DDT in Milk and*

Tissues of Dairy Cows Fed DDT-Dusted Alfalfa Hay, Adv. Chem. Ser. *1*, 237–243 (1950).

[60] BIDDULPH, C., GREENWOOD, D. A., HARRIS, L. E., DRAPER, C. I., BATEMAN, G. Q., MADSEN, L. L., BINNS, W., MINER, M. L., SORENSON, C. L., and LIEBERMAN, F. V., *Toxicity of DDT and Methoxychlor*, Fm. Home Sci. *12* (3), 45, 55, 56, 65 (1951).

[61] BIDDULPH, C., BATEMAN, G. Q., HARRIS, J. R., MANGELSON, F. L., LIEBERMAN, F. V., BINNS, W., and GREENWOOD, D. A., *Effect of Feeding Methoxychlor-Treated Alfalfa Hay to Dairy Cows*, J. Dairy Sci. *35*, 445–448 (1952).

[62] BIDEN-STEELE, K., and STUCKEY, R. E., *Poisoning by DDT Emulsion. Report of a Fatal Case*, Lancet *1946*, 235–236.

[63] BING, R. J., McNAMARA, B., and HOPKINS, F. H., *Studies on the Pharmacology of DDT (2,2-Bis-parachlorophenyl-1,1,1-trichloroethane). The Chronic Toxicity of DDT in the Dog*, Johns Hopk. Hosp. Bull. *78* (5), 308–315 (1946).

[64] BISHOP, E. L., *Effects of DDT Mosquito Larviciding on Wildlife. III. Effects on the Plankton Population of Routine Larviciding with DDT*, U. S. publ. Hlth. Repts. *62* (35), 1263–1268 (1947).

[65] BISHOPP, F. C., *The Medical and Public Health Importance of the Insecticide DDT*, Bull. N.Y. Acad. Med. *21*, 561–580 (1945).

[66] BISHOPP, F. C., *Present Position of DDT in the Control of Insects of Medical Importance*, Amer. J. publ. Hlth. *36* (6), 593–606 (1946).

[67] BISHOPP, F. C., *Precautions in the Use of Insecticides for Mosquito Control*, Mosquito News *9* (4), 137–142 (1949).

[68] BISHOPP, F. C., *Latest on Safety of DDT and Substitutes*, Fmrs' Fed. News *29* (10), 9, 12 (1949).

[69] BISHOPP, F. C., *Using Insecticides Safely*, Pest Control *19* (1), 9–12, 14 (1951).

[70] BISKIND, M. S., *DDT Poisoning and X Disease in Cattle*, J. Amer. vet. med. Ass. *114* (862), 20 (1949).

[71] BISKIND, M. S., *DDT Poisoning a Serious Public Health Hazard*, Amer. J. dig. Dis. *16* (2), 73 (1949).

[72] BISKIND, M. S., *DDT Poisoning and the Elusive 'Virus X': A New Cause for Gastro-Enteritis*, Amer. J. dig. Dis. *16* (3), 79–84 (1949).

[73] BISKIND, M. S., *The New Insecticides and the Public Health*, Harefuah *44* (1), 9–13 (1952).

[74] BISKIND, M. S., *Public Health Aspects of the New Insecticides*, Amer. J. dig. Dis. *20* (11), 331–341 (1953).

[75] BISKIND, M. S., and BIEBER, I., *DDT Poisoning—a New Syndrome with Neuro-Psychiatric Manifestations*, Amer. J. Psychother. *3* (2), 261–270 (1949).

[76] BLANCO, J. L., *Estudio farmacológico del D.D.T. en animales de experimentación*, Farmacoter. act. *3*, 595–603 (1946).

[77] BLOOR, W. R., *Distribution of Unsaturated Fatty Acids in Tissues. III. Vital Organs of Beef*, J. biol. Chem. *80*, 443–454 (1928).

[78] BLOOR, W. R., *Biochemistry of the Fatty Acids and Their Compounds, the Lipids* (Reinhold Publ. Corp., New York 1943), p. 204, 221.

[79] BOHMAN, V. R., CHI, I. A., HARRIS, L. E., BINNS, W., and MADSEN, L. L., *The Effect of DDT upon the Digestion and Utilization of Certain Nutrients by Dairy Calves*, J. Dairy Sci. *35* (1), 6–12 (1952).

[80] BOLOGNA, N. A., and WOODY, N. C., *Kerosene Poisoning*, N.O. Med. Surg. J. *101* (6), 256–260 (1948).

[81] BÖMER, M., *Über die Wirkung einiger Krampfgifte auf Blutzucker und Blutmilchsäure*, Arch. exp. Path. Pharmak. *149*, 247–256 (1930).

[82] BORDEN, A. D., *DDT Dust Deposits on Pears*, J. econ. Ent. *40* (6), 926–927 (1947).

[83] BORDEN, A. D., *Control of Codling Moth on Pears with a DDT Spray*, J. econ. Ent. *41* (1), 118–119 (1948).

[84] BORDEN, A. D., HOSKINS, W. M., and FULLMER, O. H., *The Removal of DDT from Pears*, The Blue Anchor *24* (2), 19 (1947).

[85] BRIGHT, J. H., *Residual Spray Program Memorandum No. 23*, Personal communication (1949).

[86] BRIGHT, J. H., *Exposure Time of Spray Labor to DDT Spray*, Personal communication (1949).

[87] BRITO-BABAPULLE, L. A. P., *An Antidote for DDT Poisoning – a Record of 35 Cases Seen*, Brit. vet. J. *107*, 106–122 (1951).

[88] BROMILEY, R. B., and BARD, P., *Tremor and Changes in Reflex Status Produced by DDT in Decerebrate, Decerebrate-Decerebellate, and Spinal Animals*, Johns Hopk. Hosp. Bull. *84* (5), 414–429 (1949).

[89] BROWN, J. H. U., *Influence of the Drug DDD on Adrenal Cortical Function in Adult Rats*, Proc. Soc. exp. Biol., N.Y. *83* (1), 59–62 (1953).

[90] BROWNING, E., *Toxicity of Industrial Organic Solvents* (Chem. Publ. Co., New York 1953), 411 p.

[91] BROWNING, H. C., FRASER, F. C., SHAPIRO, S. K., GLICKMAN, I., and DÛBRULE, M., *The Biological Activity of DDT and Related Compounds*, Canad. J. Res. [D] *26*, 282–300 (1948).

[92] BRUNSON, M. H., and KOBLITSKY, L., *A Study of DDT Deposits on Peach Foliage and Fruit Treated for Control of the Oriental Fruit Moth*, U. S. Dept. Agric., Bur. Ent. Plant Quarantine, E–809 (1950).

[93] BRYSON, M. J., DRAPER, C. I., HARRIS, J. R., BIDDULPH, C., GREENWOOD, D. A., HARRIS, L. E., BINNS, W., MINER, M. L., and MADSEN, L. L., *DDT in Eggs and Tissues of Chickens Fed Varying Levels of DDT*, Adv. Chem. Ser. *1*, 232–236 (1950).

[94] BURLINGTON, H., and LINDEMAN, V. F., *Effect of DDT on Testes and Secondary Sex Characters of White Leghorn Cockerels*, Proc. Soc. exp. Biol., N.Y. *74* (1), 48–51 (1950).

[95] BUSHLAND, R. C., WELLS, R. W., and RADELEFF, R. D., *Effect on Livestock of Sprays and Dips Containing New Chlorinated Insecticides*, J. econ. Ent. *41* (4), 642–645 (1948).

[96] BUSHLAND, R. C., CLABORN, H. V., BECKMAN, H. F., RADELEFF, R. D., and WELLS, R. W., *Contamination of Meat and Milk by Chlorinated Hydrocarbon Insecticides Used for Livestock Pest Control*, J. econ. Ent. *43* (5), 649–652 (1950).

[97] BUSVINE, J. R., *Current Therapeutics. XXXVIII. The Use and Abuse of Insecticides*, Practitioner *166*, 179–188 (1951).

[98] BUTTERFIELD, D. E., PARKIN, E. A., and GALE, M. M., *The Transfer of DDT to Foodstuffs from Impregnated Sacking*, J. Soc. chem. Ind. Vict. *68*, 310–313 (1949).

[99] BUXTON, P. A., *Use of the New Insecticide DDT in Relation to the Problems of Tropical Medicine*, Trans. roy. Soc. trop. Med. Hyg. *38* (5), 367–400 (1945).

[100] CALIFORNIA, *Agricultural Code of California Relating to Use and Application of Injurious Materials*, Agricultural Code, Ch. 7, Art. 4, paragraph 1080, as amended (1953).

[101] CALVERY, H. O., *DDT is Poisonous. Tolerance Level Deemed Necessary when it is Used on Human Food*, Food Packer *1945* (26), 61–62.

[102] CALVERY, H. O., and NEAL, P. A., *DDT and Other Insecticides and Repellents Developed for the Armed Forces. X. The Toxicity of DDT, Solvents, Emulsifiers, Activators, and Other Adjuvants Mentioned in this Report*, U. S. Dept. Agric., Misc. Publ. No. 606, 68–71 (1946).

[103] CAMERON, G. R., *Risks to Man and Animals from the Use of 2,2-Bis-(p-chlorophenyl)-1,1,1-trichloroethane (D.D.T.): with a note on the toxicology of benzene hexachloride (666, gamma)*, Brit. med. Bull. *3*, 233–235 (1945).

[104] CAMERON, G. R., *Toxicity of D.D.T.*, Pharm. J. *105*, 68–69 (1947).

[105] CAMERON, G. R., and BURGESS, F., *The Toxicity of 2,2-Bis-(p-chlorophenyl)-1,1,1-trichloroethane (D.D.T.)*, Brit. med. J. *1*, 865–871 (1945).

[106] CAMERON, G. R., and CHENG, M. B., *Failure of Oral D.D.T. to Induce Toxic Changes in Rats*, Brit. med. J. *2*, 819 (1951).

[107] CAMPBELL, A. M. G., *DDT Poisoning in Man. A Suspected Case*, Lancet *257*, 1178 (1949).

[108] CAMPBELL, A. M. G., *Neurological Complications Associated with Insecticides and Fungicides*, Brit. med. J. *2*, 415–417 (1952).

[109] CAREY, E. J., DOWNER, E. M., TOOMEY, F. B., and HAUSHALTER, E., *Morphologic Effects of DDT on Nerve Endings, Neurosomes, and Fiber Types in Voluntary Muscles*, Proc. Soc. exp. Biol., N.Y. *62*, 76–83 (1946).

[110] CARLETON, R. M., *DDT Stored in Your Fat?*, Bett. Homes & Gard. *1949* (May), 139.

[111] CARMAN, G. E., EWART, W. H., BARNES, M. M., and GUNTHER, F. A., *Absorption of DDT and Parathion by Fruits*, Adv. Chem. Ser. *1*, 128–136 (1950).

[112] CARNE, P. B., *Persistence of DDT in the Soil*, Nature, Lond. *162*, 743 (1948).

[113] CAROLLO, J. A., *The Removal of DDT from Water Supplies*, J. Amer. Wat. Wks. Ass. *37* (7), 1310–1317 (1945).

[114] CARTER, R. H., *DDT Residues in Agricultural Products*, Ind. Engr. Chem. *40* (4), 716–717 (1948).

[115] CARTER, R. H., and MANN, H. D., *The DDT Content of Milk from a Cow Sprayed with DDT*, J. econ. Ent. *42* (4), 708 (1949).

[116] CARTER, R. H., HUBANKS, P. E., and MANN, H. D., *Effect of Cooking on the DDT Content of Beef*, Science *107* (2779), 347 (1948).

[117] CARTER, R. H., HUBANKS, P. E., MANN, H. D., ZELLER, J. H., and HANKINS, O. G., *The Storage of DDT in the Tissues of Pigs Fed Beef Containing this Compound*, J. Anim. Sci. *7* (4), 509–510 (1948).

[118] CARTER, R. H., WELLS, R. W., RADELEFF, R. D., SMITH, C. L., HUBANKS, P. E., and MANN, H. D., *The Chlorinated Hydrocarbon Content of Milk from Cattle Sprayed for Control of Horn Flies*, J. econ. Ent. *42* (1), 116–118 (1949).

[119] CARTER, R. H., HUBANKS, P. E., MANN, H. D., SMITH, F. F., PIQUETTE, P. G., SHAW, S. C., and DITMAN, L. P., *DDT Content of Milk from Cows Fed Pea Vine Silage Containing DDT Residues*, J. econ. Ent. *42* (1), 119–122 (1949).

[120] CASE, R. A. M., *Toxic Effects of 2,2-Bis-(p-chlorphenyl)-1,1,1-trichlorethane (D.D.T.) in Man*, Brit. med. J. *2*, 842 (1945).

[121] CHEMICAL SPECIALTIES MANUFACTURERS', INC., *Compilation of Federal and State Economic Poison Laws* (C. S.M., Inc., New York 1947).

[122] CHIGNOLI, V., and ILICETO, N., *Controllo sanitario di operai addetti allo spargimento di insettici di sintesi*, Riv. Igiene Sanit. pubbl. *9* (1–2), 46–54 (1953).

[123] CHIN, Y., and T'ANT, C., *The Effect of DDT on Cutaneous Sensation in Man*, Science *103* (2682), 654 (1946).

[124] CHISHOLM, R. D., KOBLITSKY, L., FAHEY, J. E., and WESTLAKE, W. E., *DDT Residues in Soil*, J. econ. Ent. *43* (6), 941–942 (1950).

[125] CHOU, C., KWANG, D. R., and CHU, C. H. U., *On the Toxicity of DDT (Dichlorodiphenyl trichloroethane) in Mammals*, Proc. Chin. physiol. Soc. Chengta Branch *2*, 167–171 (1945).

[126] CLABORN, H. V., and WELLS, R. W., *Methoxychlor, DDT, CS–708 and Dieldrin – Their Rates of Excretion in Milk*, Agric. Chemic. *7* (10), 28–29, 115 (1952).

[127] CLABORN, H. V., BECKMAN, H. F., and WELLS, R. W., *Contamination of Milk from DDT Sprays Applied to Dairy Barns*, J. econ. Ent. *43* (5), 723 (1950).

[128] CLABORN, H. V., BECKMAN, H. F., and WELLS, R. W., *Excretion of DDT and TDE in Milk from Cows Treated with These Insecticides*, J. econ. Ent. *43* (6), 850–852 (1950).

[129] CLABORN, H. V., BOWERS, J. W., WELLS, R. W., RADELEFF, R. D., and NICKERSON, W. J., *Meat Contamination from Pesticides*, Agric. Chemic. *8* (8), 37–39, 119, 121 (1953).

[130] CLARK, P. J., *D.D.T. Residues on Cabbages*, N.Z. J. Sci. Tech. [A] *29* (1), 1–4 (1947).

[131] COBURN, D. R., and TREICHLER, R., *Experiments on Toxicity of DDT to Wildlife*, J. Wildlife Mgmt. *10* (3), 208–216 (1946).

[132] COMMITTEE ON PESTICIDES, *Pharmacologic and Toxicologic Aspects of DDT (Chlorophenothane U.S.P.)*, J. Amer. med. Ass. *145*, 728–733 (1951).

[133] CONLEY, B. E., *Why the American Medical Association is Interested in the Pesticide Problem*, Pests *17* (4), 18–22 (1949).

[134] CONLEY, B. E., and WILSON, J. R., *The Physician's Role in the Pesticide Problem*, Adv. Chem. Ser. *1*, 61–64 (1950).

[135] COTTAM, C., and HIGGINS, E., *DDT: Its Effect on Fish and Wildlife*, J. econ. Ent. *39* (1), 44–52 (1946).

[136] COTTAM, C., and HIGGINS, E., *DDT: Its Effect on Fish and Wildlife*, U.S. Fish & Wildlife Service, Circ. No. 11, 1–14 (1946).

[137] COUCH, L. K., *Effects of DDT on Wildlife in a Mississippi River Bottom Woodland*, Trans. N. Amer. Wildl. Conf. *11*, 323–329 (1946).

[138] COUNCIL ON PHARMACY AND CHEMISTRY, *Toxicity of Pesticides*, J. Amer. med. Ass. *139* (6), 378 (1949).

[139] COUNCIL ON PHARMACY AND CHEMISTRY, *Committee on Pesticides*, J. Amer. med. Ass. *142* (13), 989–990 (1950).

[140] COX, A. J., *Drug Aspects of DDT*, Calif. Citrogr. *30* (1), 6, 22–23, 26–27 (1944).

[141] COX, J. A., *Control of Cherry Fruit Flies*, J. econ. Ent. *40* (4), 588–590 (1947).

[142] CRAWFORD, C. W., *Insecticides and the Food Law*, J. econ. Ent. *42* (3), 564–566 (1949).

[143] CRESCITELLI, F., and GILMAN, A., *Electrical Manifestations of the Cerebellum and Cerebral Cortex Following DDT Administration in Cats and Monkeys*, Amer. J. Physiol. *147* (1), 127–137 (1946).

[144] CUNNINGHAM, R. E., and HILL, F. S., *Convulsions and Deafness Following Ingestion of DDT*, Pediatrics *9*, 745–747 (1952).

[145] DANGERFIELD, W. G., *Toxicity of DDT to Man*, Brit. med. J. *1*, 27 (1946).

[146] DARSIE, R. F., jr., *Effect of DDT and Lindane Sprays on Goldfish*, Proc. N. J. Mosq. Ext. Ass. *39*, 175–179 (1952).

[147] DAVENPORT, H. W., *Carbonic Anhydrase in Tissues Other than Blood*, Physiol. Rev. *26* (4), 560–573 (1946).

[148] DAVENPORT, H. W., and WILHELMI, A. E., *Renal Carbonic Anhydrase*, Proc. Soc. exp. Biol., N.Y. *48*, 53–56 (1941).

[149] DAVIS, J. J., *DDT to Control Household and Stored Grain Insects*, J. econ. Ent. *39* (1), 59–61 (1946).

[150] DE ALMEIDA, P. R., *Ensaios de Laboratório sôbre a toxidez do 'DDT' aos peixes guarús (Phallocerus caudimaculatus hensil)*, Arch. Inst. biol., S. Paulo *18*, 31–37 (1948).

[151] DE CAIRES, P. F., *Aedes aegypti Control in the Absence of a Piped Potable Water Supply*, Amer. J. trop. Med. *27* (6), 733–743 (1947).

[152] DECKER, G. C., *Agricultural Applications of DDT, with Special Reference to the Importance of Residues*, J. econ. Ent. *39* (5), 557–562 (1946).

[153] DECKER, G. C., *Insecticide Toxicity Problems*, Mod. Sanit. *5* (11), 22–25, 65–66 (1953).

[154] DECKER, G. C., APPLE, J. W., WRIGHT, J. M., and PETTY, H. B., *European Corn Borer Control on Canning Corn*, J. econ. Ent. *40* (3), 395–400 (1947).

[155] DECKER, G. C., WEINMAN, C. J., and BANN, J. M., *A Preliminary Report on the Rate of Insecticide Residue Loss from Treated Plants*, J. econ. Ent. *43* (6), 919–927 (1950).

[156] DE COURT, P., DUPOUX, R., and PELLOUX, A., *Sur la toxicité des parasiticides et des molluscocides pour les vertébrés et les coefficients de sécurité nécessaires à leur emploi*, Bull. Soc. Pat. exot. *42*, 33–38 (1949).

[157] DEEDERER, C., *DDT Toxicity*, Med. Record, N.Y. *161* (4), 216–220 (1948).

[158] DE EDS, F., *The Significance of Chronic Toxicity Studies*, Food Tech., Champaign *4*, 40–42 (1950).

[159] DEGOS, R., and GARNIER, G., *Traitement de la gale par le diphényl dichlorotrichloro-

226   W. J. Hayes, Jr.

éthane (*D.D.T.*) *en dissolution dans un solvant organique*, Ann. Derm. Syph., Paris 5, 322 (1945).

[160] DE HEUS, J. G., and MAAN, W. J., *Bepaling van de melkgift, vetproductie en DDT-gehalte in de melk van koeien, gevoerd met stro van met DDT bespoten erwtenplanten*, Mededeling uit het Inst. voor Moderne Veevoeding 'De Schothorst' te Hoogland bij Amersfoort, 8 p. (1948).

[161] DEICHMANN, W. B., KITZMILLER, K. V., WITHERUP, S., and JOHANSMANN, R., *Kerosene Intoxication*, Ann. internal Med. 21, 803–823 (1944).

[162] DEICHMANN, W. B., WITHERUP, S., and KITZMILLER, K. V., *The Toxicity of DDT. I. Experimental Observations* (The Kettering Lab. Univ. Cincinnati. Dept. Prev. Med. & Indus. Health, Coll. Med. 1950), 71 p. plus 33 tables and 15 figures.

[163] DE JONG, J. C. M., and FIRTOS, S. C., *Ervaringen met DDT*, Ned. Tijdschr. Geneesk. 92 (43), 3404–3408 (1948).

[164] DELANEY, J. J., *Peril on Your Food Shelf*, Amer. Mag. *1951* (July), 19, 112–114.

[165] DE LUCENA, D. T., *Profilaxia da Malaria com DDT insecticida no Vale do Baixo São Francisco. Resultados do 1-ano de experiencia*, Rev. bras. Malariol. 2 (2), 186–208 (1950).

[166] DE MEILLON, B., *Experiments with DDD and Chlordane as Systemic Insecticides*, Insect. Abstr. News Sum. No. 10 (1949).

[167] DEMUTH, G., and LENDLE, L., *Prüfung insektizider Stoffe auf Dehydrase-Systeme*, Arch. Pharm., Berl. 284 (5/6), 298–305 (1951).

[168] DENT, J. E., MORLAN, H. B., and HILL, E. L., *Effects of DDT Dusting on Domestic Rats Under Colony and Field Conditions*, U. S. publ. Hlth. Repts. 64 (21), 666–671 (1949).

[169] DEONIER, C. C., BURRELL, R. W., MAPLE, J. D., and COCHRAN, J. H., *DDT as an Anopheline Larvicide: Preliminary Field Studies*, J. econ. Ent. 38 (2), 244–249 (1945).

[170] DEUTSCH, A., *DDT and You!*, N.Y. Post Home News *1949* (clippings as follows): a) March 30; b) March 31; c) April 1; d) April 3; e) April 4; f) April 5; g) April 6; h) April 7; i) April 8; j) April 10; k) April 11; l) April 25.

[171] DIAZ-JIMINEZ, C., *Contribucion al estudio de la toxicidad del DDT (dichloro-difenil-tricloro-metil-metano)*, Med. colon. 9 (1), 51–74 (1947).

[172] DOMENJOZ, R., *Experimentelle Erfahrungen mit einem neuen Insektizid (Neocid Geigy), ein Beitrag zur Theorie der Kontaktgiftwirkung*, Schweiz. Med. Wschr. 74 (36), 952–958 (1944).

[173] DOMENJOZ, R., *Zur biologischen Wirkung einiger DDT-Derivate*, Arch. int. Pharmacodyn. 73 (1–2), 128–146 (1946).

[174] DOMENJOZ, R., *Action biologique de quelques dérivés du DDT*, Helv. chim. acta 29 (5), 1317–1322 (1946).

[175] DOMENJOZ, R., *Kontaktgifte*, Arch. exp. Path. Pharmak. 208 (2/3), 144–167 (1948).

[176] DOUDOROFF, P., KATZ, M., and TARZWELL, C. M., *Toxicity of Some Organic Insecticides to Fish*, Sewage industr. Wastes 25 (7), 840–844 (1953).

[177] DRAIZE, J. H., and WOODARD, G., *Le DDT – Ses propriétés – Ses applications*, Actualités chirurgicales No. 12, 5–47 (1946).

[178] DRAIZE, J. H., WOODARD, G., FITZHUGH, O. G., NELSON, A. A., SMITH, R. B., jr., and CALVERY, H. O., *Summary of Toxicological Studies of the Insecticide DDT 2,2-Bis-(p-chlorophenyl)-1,1,1-trichloroethane*, Chem. Engng. News 22, 1503 (1944).

[179] DRAIZE, J. H., NELSON, A. A., and CALVERY, H. O., *The Percutaneous Absorption of DDT (2,2-Bis-[p-Chlorophenyl]-1,1,1-trichloroethane) in Laboratory Animals*, J. Pharmacol. 82 (2), 159–166 (1944).

[180] DRAPER, C. I., HARRIS, M. R., GREENWOOD, D. A., BIDDULPH, C., HARRIS, L. E., MANGELSON, F., BINNS, W., and MINER, M. L., *The Transfer of DDT from the Feed to Eggs and Body Tissues of White Leghorn Hens*, Poult. Sci. 31, 388–393 (1952).

[181] DRESDEN, D., *Physiological Investigations into the Action of DDT. 2,2-Bis-(p-chlorophenyl)-1,1,1-trichloroethane* (G. W. Van der Wiel & Co., Arnhem 1949), 114 p.

[182] DRESDEN, D., and KIRJGSMAN, B. J., *Experiments on the Physiological Action of Contact Insecticides*, Bull. ent. Res. *38* (4), 575–578 (1948).

[183] DUNBAR, P. B., Mimeographed letter dated January 26, 1945.

[184] DUNBAR, P. B., *The Chief of F.D.A. Discusses Insecticide Chemicals*, Agric. Chemic. *4* (6), 42–44 (1949).

[185] DUNBAR, P. B., *The Food and Drug Administration Looks at Insecticides*, Food, Drug, Cosmetic Law Quart. *1949* (4), 233–239.

[186] DUNBAR, P. B., *Report from the Food and Drug Administration*, Quart. Bull. Ass. Food & Drug Officials U. S. *13* (3), 82–93 (1949).

[187] DUNBAR, P. B., *Amendment of Federal Food, Drug, and Cosmetic Act re Chemicals Used in Manufactured Foods, or as Pesticides on Growing Foods*, Food, Drug, Cosmetic Law Quart. *1949* (4), 296.

[188] DUNN, C. W., *The Federal Food, Drug, and Cosmetic Act and the Food Industry*, Food, Drug, Cosmetic Law Quart. *3* (2), 166–177 (1948).

[189] DUNN, J. E., DUNN, R. C., and SMITH, B. S., *Skin-Sensitizing Properties of DDT for the Guinea Pig*, U. S. publ. Hlth. Repts. *61* (45), 1614–1620 (1946).

[190] ECKERT, J. E., *The Effect of DDT on Honeybees*, J. econ. Ent. *38* (3), 369–374 (1945).

[191] ECKERT, J. E., *Effect of Certain Insecticides on Beekeeping*, Calif. agr. exp. Sta. Circ. *365*, 22–25 (1946).

[192] ECKERT, J. E., *Determining Toxicity of Agricultural Chemicals to Honeybees*, J. econ. Ent. *42* (2), 261–265 (1949).

[193] EDEN, W. G., and ARANT, F. S., *DDT Residues on Alfalfa*, J. econ. Ent. *41* (3), 383–387 (1948).

[194] EIDE, P. M., *Experiments with Insecticides on Honeybees*, J. econ. Ent. *40* (1), 49–54 (1947).

[195] EIDE, P. M., DEONIER, C. C., and BURRELL, R. W., *Toxicity of DDT to Certain Forms of Aquatic Life*, J. econ. Ent. *38* (4), 492 (1945).

[196] ELLIS, M. M., WESTFALL, B. A., and ELLIS, M. D., *Toxicity of Dichloro-diphenyl-trichlorethane (DDT) to Goldfish and Frogs*, Science *100* (2604), 477 (1944).

[197] ELY, R. E., MOORE, L. A., CARTER, R. H., and POOS, F. W., *The DDT, Toxaphene, and Chlordane Content of Milk as Affected by Feeding Alfalfa Sprayed with these Insecticides*, U. S. Dept. Agric. BDIM-Inf-85 (December 1949).

[198] ELY, R. E., MOORE, L. A., CARTER, R. H., MANN, H. D., and POOS, F. W., *The Effect of Dosage Level and Method of Administration of DDT on the Concentration of DDT in Milk*, J. Dairy Sci. *35*, 266–271 (1952).

[199] EMMEL, L., and KRÜPE, M., *Beiträge zur Kenntnis der Wirkungsweise des 4,4'-Dichlordiphenyl-trichlormethyl-methan beim Warmblüter*, Z. Naturf. *1*, 691–695 (1946).

[200] ERICKSON, A. B., *Effects of DDT Mosquito Larviciding on Wildlife. II. Effects of Routine Airplane Larviciding on Bird and Mammal Populations*, U. S. publ. Hlth. Repts. *62* (35), 1254–1262 (1947).

[201] EYZAGUIRRE, C., and LILIENTHAL, J. L., jr., *Veratrinic Effects of Pentamethylene-tetrazol (Metrazol) and 2,2-Bis-(p-chlorophenyl)-1,1,1-trichloroethane (DDT) on Mammalian Neuromuscular Function*, Proc. Soc. exp. Biol., N.Y. *70* (1), 272–275 (1949).

[202] FAN, H. Y., CHENG, T. H., and RICHARDS, A. G., *The Temperature Coefficients of DDT Action in Insects*, Physiol. Zool. *21* (1), 48–59 (1948).

[203] FENNAH, R. G., *Preliminary Tests with DDT Against Insect Pests of Food-Crops in the Lesser Antilles*, Trop. Agriculture *22* (12), 222–226 (1945).

[204] FINNEGAN, J. K., HAAG, H. B., and LARSON, P. S., *Tissue Distribution and Elimination of DDD and DDT Following Oral Administration to Dogs and Rats*, Proc. Soc. exp. Biol., N.Y. *72* (2), 357–360 (1949).

[205] FIRTOS, S. C., and DE JONG, C. M., *Gunstige uitkomsten van DDT-toepassing*, Ned. Tijdschr. Geneesk. *94* (16), 1105–1110 (1950).

[206] FISHER, A. L., KEASLING, H. H., and SCHUELLER, F. W., *Estrogenic Action of Some DDT Analogues*, Proc. Soc. exp. Biol., N.Y. *81* (2), 439–441 (1952).

[207] FITZHUGH, O. G., *Use of DDT Insecticides on Food Products*, Ind. Engng. Chem. *40* (4), 704–705 (1948).

[208] FITZHUGH, O. G., and NELSON, A. A., *The Chronic Oral Toxicity of DDT (2,2-Bis-[p-chlorophenyl]-1,1,1-trichloroethane)*, J. Pharmacol. *89* (1), 18–30 (1947).

[209] FLECK, E. E., *Residual Action of Organic Insecticides*, Ind. Engng. Chem. *40* (4), 706–708 (1948).

[210] FOWLER, R. E. L., *Insecticide Toxicology—Manifestations of Cotton-Field Insecticides in the Mississippi Delta*, J. Agric. Food Chem. *1* (6), 469–473 (1953).

[211] FOX, I., *DDT to Control House Mice in Buildings*, Bol. Col. Quim. P. Rico *5* (1), 27–29 (1948).

[212] FRANCONE, M. P., MARLANI, F. H., and DEMARE, C., *Clínica de la intoxicación por D.D.T.*, Rev. Asoc. méd. argent. *66*, 56–59 (1952).

[213] FRANK, J. E., Personal communication (1952).

[214] FREAR, D. E. H., *Newer Pesticides; Formulations, Hazards, Precautions, Compatibility*, Pennsylvania State College School of Agriculture, Agr. Expt. Sta. Prog. Rept. No. 29, 14 p. (1950).

[215] FREAR, D. E. H., and COX, J. A., *DDT Spray Residues on Grapes and in Grape Juice*, Food Pack. *27*, 64 (1946).

[216] FREAR, D. E. H., BRUCE, W. N., and RAGSDALE, A. C., *DDT in Milk Following Barn Spraying*, J. econ. Ent. *43* (5), 656–657 (1950).

[217] FRIBERG, L., and MARTENSSON, J., *Case of Panmyelophthisis After Exposure to Chlorophenothane and Benzene Hexachloride*, Arch. industr. Hyg. *8* (2), 166–169 (1953).

[218] FULTON, J. F., and DOW, R. S., *The Cerebellum: A Summary of Functional Localization*, Yale J. Biol. Med. *10*, 89–119 (1937).

[219] GARRETT, R. M., *Toxicity of DDT for Man*, J. med. Ass. Ala. *17* (2), 74–76 (1947).

[220] GARRETT, R. M., Personal communication (1950).

[221] GAY, F. J., *Studies on the Control of Wheat Insects by Dust: The Use of DDT- and 666-Impregnated Dusts for the Control of Grain Pests*, Commonwealth of Australia Council for Scientific and Industrial Research, Bull. No. 225, 33–38 (1947).

[222] GEORGE, J. L., and MITCHELL, R. T., *The Effects of Feeding DDT-Treated Insects to Nestling Birds*, J. econ. Ent. *40* (6), 782–789 (1947).

[223] GEORGE, J. L., and STICKEL, W. H., *Wildlife Effects of DDT Dust Used for Tick Control on a Texas Prairie*, Amer. Midl. Nat. *42* (1), 228–237 (1949).

[224] GEREBTZOFF, M. A., and PHILIPPOT, E., *La réaction plasmocytaire dans l'intoxication des mammifères par le DDT et par le diphénylméthane*, Experientia 8 (10), 395–396 (1952).

[225] GIL, G. P., *Un problema de suma actualidad y importancia. La toxicidad del dicloro-difenil-trichlorometil-metano (DDT) para el hombre*, Med. colon. *14* (5), 427–433 (1949).

[226] GIL, G. P., and MIRON, B. F., *Investigacioness acerca de la toxicidad del DDT para el hombre*, Med. colon. *14* (5), 459–470 (1949).

[227] GINSBURG, J. M., *Toxicity of DDT to Fish*, J. econ. Ent. *38* (2), 274–276 (1945).

[228] GINSBURG, J. M., *Toxicity of DDT to Subsurface-Feeding Species of Mosquitoes*, J. econ. Ent. *38* (4), 494–495 (1945).

[229] GINSBURG, J. M., *Comparative Toxicity of DDT Isomers and Related Compounds to Mosquito Larvae and Fish*, Science *105* (2722), 233 (1947).

[230] GINSBURG, J. M., *Tests with New Toxicants, in Comparison with DDT, on Mosquito Larvae and Fish*, Proc. N. J. Mosq. Ext. Ass. *34*, 132–135 (1947).

[231] GINSBURG, J. M., *Results from Feeding Mosquito Larvae, Killed by DDT, to Goldfish*, J. econ. Ent. *40* (2), 275–276 (1947).

[232] GINSBURG, J. M., and FILMER, R. S., *DDT Recovered on Corn Plants from Different Treatments Against Corn Earworm*, J. econ. Ent. *44* (4), 620 (1951).

[233] GINSBURG, J. M., FILMER, R. S., REED, J. P., and PATERSON, A. R., *Recovery of Parathion, DDT and Certain Analogs of Dichlorodiphenyl-dichloroethane from Treated Crops*, J. econ. Ent. *42* (4), 602–611 (1949).

[234] GINSBURG, J. M., FILMER, R. S., and REED, J. P., *Longevity of Parathion, DDT and Dichlorodiphenyl-Dichloroethane Residues on Field and Vegetable Crops*, J. econ. Ent. *43* (1), 90–94 (1950).

[235] GLASSMAN, J. M., and BUCHAN, R. F., *2,2-Bis-(p-chlorophenyl)-1,1,1-trichloroethane (DDT)*, Occup. Med. *5*, 536–560 (1948).

[236] GLOBUS, J. H., *DDT Poisoning*, Trans. Amer. neurol. Ass. *73*, 202–208 (1948).

[237] GLOBUS, J. H., *DDT (2,2-Bis-[p-chlorophenyl]-1,1,1-trichlorethane) Poisoning*, J. Neuropath. *7* (4), 418–431 (1948).

[238] GOMEZ, H. A., *Inocuidad de la adición al agua potable de 1 p.p.m. de DDT para la erradicación del Aedes aegypti*, Bol. Ofic. sanit. pan-amer. *30* (3), 330–337 (1951).

[239] GOMEZ, JUAN, *DDT*, Conference at Savannah, April 20 (1950).

[240] GOODRUM, PHIL, BALDWIN, W. P., and ALDRICH, J. W., *Effect of DDT on Animal Life of Bull's Island, South Carolina*, J. Wildlife Mgmt. *13* (1), 1–10 (1949).

[241] GORDON, I., *The Occupational Hazard of DDT Spraying*, Brit. J. industr. Med. *3*, 245–249 (1946).

[242] GORDON, H. T., and WELSH, J. H., *The Role of Ions in Axon Surface Reactions to Toxic Organic Compounds*, J. cell. comp. Physiol. *31* (3), 395–420 (1948).

[243] GÖTZ, BRUNO, *Die Wirkung von Gesarol und anderen DDT-Präparaten auf Fische*, Anz. Schädlingsk. *21* (3), 39–40 (1948).

[244] GRAHAM, C., and CORY, E. N., *Codling Moth and European Red Mite Control, and Seasonal Analysis of Spray Deposits*, J. econ. Ent. *40* (5), 752–754 (1947).

[245] GREAT BRITAIN. MINISTRY OF AGRICULTURE AND FISHERIES, *Toxic Chemicals in Agriculture*, Report of the Working Party, 15 p. (1951).

[246] GREAT BRITAIN. MINISTRY OF AGRICULTURE AND FISHERIES, *Agriculture (Poisonous Substances) Act*, p. 1–9 (1952).

[247] GREENWOOD, D. A., HARRIS, L. E., BIDDULPH, C., BATEMAN, G. Q., BINNS, W., MINER, M. L., HARRIS, J. R., MANGELSON, F., and MADSEN, L. L., *Feeding Rats Tissues from Lambs and Butterfat from Cows that Consumed DDT-Dusted Alfalfa Hay*, Proc. Soc. exp. Biol., N.Y. *83* (3), 458–460 (1953).

[248] GUARESCHI, G., and BINI, G., *Ricerche sperimentali sulla tossicità del D.D.T. con particolare riguardo all'emuntario renale*, Ateneo parmense *19*, 72–81 (1948).

[249] GUNTHER, F. A., *Quantitative Estimation of DDT and of DDT Spray or Dust Deposits*, Ind. Eng. Chem. (Anal. ed.) *17*, 149–150 (1945).

[250] GUNTHER, F. A., LINDGREN, D. L., ELLIOT, M. I., and LaDUE, J. P., *Persistence of Certain DDT Deposits under Field Conditions*, J. econ. Ent. *39* (5), 624–627 (1946).

[251] GUNTHER, F. A., BARNES, M. M., and CARMAN, G. E., *Removal of DDT and Parathion Residues from Apples, Pears, Lemons, and Oranges*, Adv. Chem. Ser. *1*, 137–142 (1950).

[252] HAAG, H. B., FINNEGAN, J. K., LARSON, P. S., DREYFUSS, M. L., MAIN, R. J., and RIESE, W., *Comparative Chronic Toxicity for Warm-Blooded Animals of 2,2-Bis-(p-chlorophenyl)-1,1,1-trichloroethane (DDT) and 2,2-Bis-(p-chlorophenyl)-1,1-dichloroethane (DDD)*, Industr. Med. *17* (12), 477–484 (1948).

[253] HAAG, H. B., FINNEGAN, J. K., LARSON, P. S., RIESE, W., and DREYFUSS, M. L., *Comparative Chronic Toxicity for Warm-Blooded Animals of 2,2-Bis-(p-chlorophenyl)-1,1,1-trichloroethane (DDT) and 2,2-Bis-(p-methoxyphenyl)-1,1,1-trichloroethane (DMDT, methoxychlor)*, Arch. int. Pharmacodyn. *83* (4), 491–504 (1950).

[254] HACKMAN, R. H., *The Persistence of DDT on Cattle*, J. Coun. sci. industr. Res. Aust. *20* (1), 56–65 (1947).

[255] HALLER, M. H., and CARTER, R. H., *DDT Spray Residues on Apples and the Effect of Removal Treatments*, Proc. Amer. Soc. hort. Sci. *56*, 116–121 (1950).

[256] HALLER, H. L., and SIMMONS, S. W., *A Statement on the Health Hazards of Thermal Generators as Used for the Control of Flying Insects*, J. econ. Ent. *44* (6), 1027 (1951).

[257] HALLER, H. L., and SIMMONS, S. W., *A Revised Statement on the Health Hazards*

of Insecticide Vaporizers as Used for the Control of Flying Insects, J. econ. Ent. *46* (1), 181 (1953).

[258] HAMILTON, D. W., *Oriental Fruit Moth Control with DDT,* J. econ. Ent. *42* (1), 25–28 (1949).

[259] HARMAN, S. W., *DDT for Codling Moth Control in Western New York in 1945,* J. econ. Ent. *39* (2), 208–210 (1946).

[260] HARMAN, S. W., *An Analysis of the DDT Spray Program for Controlling Codling Moth,* J. econ. Ent. *40* (2), 256–258 (1947).

[261] HARRIS, H. J., HANSENS, E. J., and ALEXANDER, C. C., *Determination of DDT in Milk Produced in Barns Sprayed with DDT Insecticides,* Agric. Chemic. *5* (1), 51–52 (1950).

[262] HARRIS, J. R., BIDDULPH, C., GREENWOOD, D. A., BRYSON, M. J., HARRIS, L. E., MINER, M. L., BINNS, W., and MADSEN, L. L., *The Concentration of DDT in the Blood and Tissues of Sheep Fed Varying Levels of DDT,* Fm. Home Sci. *9* (3), 13–14 (1948).

[263] HARRIS, J. R., BIDDULPH, C., GREENWOOD, D. A., HARRIS, L. E., BRYSON, M. J., BINNS, W., MINER, M. L., and MADSEN, L. L., *Concentration of DDT in the Blood and Tissues of Sheep Fed Varying Levels of DDT,* Arch. Biochem. *21* (2), 370–376 (1949).

[264] HARRIS, L. E., MYINT, T., BIDDULPH, C., GREENWOOD, D. A., BINNS, W., MINER, M. L., and MADSEN, L. L., *Effect of Feeding DDT-Dusted Alfalfa Hay to Fattening Lambs,* J. Anim. Sci. *10* (3), 581–591 (1951).

[265] HARRIS, L. E., HARRIS, J. R., MANGELSON, F. L., GREENWOOD, D. A., BIDDULPH, C., BINNS, W., and MINER, M. L., *Effect of Feeding DDT-Treated Hay to Swine and of Feeding the Swine Tissues to Rats,* J. Nutr. *51* (4), 491–505 (1953).

[266] HARTWELL, J. L., SHEAR, M. J., JOHNSON, J. M., and KORNBERG, S. R. L., *Present Status of a Joint Institutional Research Program on Chemotherapy of Cancer. VI. Selection and Synthesis of Organic Compounds,* Cancer Res. *6* (9), 488–490 (1946).

[267] HARWOOD, P. D., *Pseudoscience and the DDT Scandal,* Science *114* (2970), 583–584 (1951).

[268] HAYMAKER, W., GINZLER, A. M., and FERGUSON, R. L., *The Toxic Effects of Prolonged Ingestion of DDT on Dogs with Special Reference to Lesions in the Brain,* Amer. J. med. Sci. *212,* 423–431 (1946).

[269] HELSON, G. A. H., *Investigations on the Control of Oriental Peach Moth, Cydia molesta* Busck., *in the Goulburn Valley, Victoria,* J. Coun. sci. industr. Res. Aust. *20* (1), 17–24 (1947).

[270] HENDERSON, J. M., *Mosquito Control Practice with DDT,* Engng. News Rec. *140* (26), 1034–1037 (1948).

[271] HENNIG, E., *Über den Einfluss des insektiziden Lauseto auf das Vaccina-Virus,* Dtsch. Gesundheitsw. *2* (1), 1–2 (1947).

[272] HERALD, E. S., *Notes on the Effects of Aircraft-Distributed DDT-Oil Spray upon Certain Philippine Fishes,* J. Wildlife Mgmt. *13* (3), 316–318 (1949).

[273] HERTEL, H., *Chronische DDT-Intoxikation,* Dtsch. Arch. klin. Med. *199* (3), 256–274 (1952).

[274] HESS, A. D., and KEENER, G. G., jr., *Effect of Airplane-Distributed DDT Thermal Aerosols on Fish and Fish Food Organisms,* J. Wildlife Mgmt. *11* (1), 1–10 (1947).

[275] HEYROTH, F. F., *The Toxicity of DDT.* II. *Survey of the Literature,* The Kettering Laboratory, Univ. Cincinnati, Dept. of Prev. Med. and Ind. Health, College of Med. 72–233 (1950).

[276] HIGGINS, E. L., and KINDEL, D. J., *Exfoliative Dermatitis from Contact with DDT,* J. invest. Derm. *12* (4), 207–209 (1949).

[277] HILL, E. L., and MORLAN, H. B., *Evaluation of County-Wide DDT-Dusting Operations in Murine Typhus Control,* U. S. publ. Hlth. Repts. *63* (51), 1635–1653 (1948).

[278] HILL, K. R., *DDT Poisoning,* Brit. med. J. *1,* 255 (1946).

[279] HILL, K. R., and ROBINSON, G., *A Fatal Case of DDT Poisoning in a Child: With an Account of Two Accidental Deaths in Dogs*, Brit. med. J. *2*, 845–846 (1945).

[280] HILL, W. R., and DAMIANI, C. R., *Death Following Exposure to DDT*, New Engl. J. Med. *235*, 897–899 (1946).

[281] HITCHCOCK, L. F., and MACKERRAS, I. M., *The Use of DDT in Dips to Control Cattle Tick*, J. Coun. sci. industr. Res. Aust. *20* (1), 43–55 (1947).

[282] HODGE, H. C., MAYNARD, E. A., THOMAS, J. F., BLANCHET, H. J., jr., WILT, W. G., jr., and MASON, K. E., *Short-Term Oral Toxicity Tests of Methoxychlor (2,2-Di-[p-methoxyphenyl]-1,1,1-trichlorethane) in Rats and Dogs*, J. Pharmacol. *99* (1), 140–148 (1950).

[283] HODGE, H. C., MAYNARD, E. A., and BLANCHET, H. J., jr., *Chronic Oral Toxicity Tests of Methoxychlor (2,2-Di-[p-methoxyphenyl]-1,1,1-trichlorethane) in Rats and Dogs*, J. Pharmacol. *104* (1), 60–66 (1952).

[284] HOFFERBER, O., *DDT-Vergiftung bei einer Ziege*, Mh. Vet.-Med. *5*, 15 (1950).

[285] HOFFMAN, I., and LENDLE, L., *Zur Wirkungsweise neuer insektizider Stoffe*, Arch. exp. Path. Pharmak. *205*, 223–242 (1948).

[286] HOFFMAN, R. A., and LINDQUIST, A. W., *Effect of Temperature on Knockdown and Mortality of House Flies Exposed to Residues of Several Chlorinated Hydrocarbon Insecticides*, J. econ. Ent. *42* (6), 891–893 (1949).

[287] HOFFMANN, C. H., and LINDUSKA, J. P., *Some Considerations of the Biological Effects of DDT*, Sci. Mon., N.Y. *69*, 104–114 (1949).

[288] HOFFMANN, C. H., and MERKEL, E. P., *Fluctuations in Insect Populations Associated with Aerial Applications of DDT to Forests*, J. econ. Ent. *41* (3), 464–473 (1948).

[289] HOFFMANN, C. H., and SURBER, E. W., *Effects of an Aerial Application of Wettable DDT on Fish and Fish-Food Organisms in Back Creek, West Virginia*, Trans. Amer. Fish. Soc. *75* (1945); 48–58 (1948).

[290] HOFFMANN, C. H., and SURBER, E. W., *Effects of an Aerial Application of DDT on Fish and Fish-Food Organisms in Two Pennsylvania Watersheds*, Progr. Fish Cult. *11* (4), 203–211 (1949).

[291] HOFFMANN, C. H., and SURBER, E. W., *Effects of Feeding DDT-Sprayed Insects to Fresh-Water Fish*, U.S. Dept. Interior Special Sci. Rept.: Fisheries *1949* (3), 1–9.

[292] HOFFMANN, C. H., TOWNES, H. K., SAILER, R. I., and SWIFT, H. H., *Field Studies on the Effect of DDT on Aquatic Insects*, U.S. Dept. Agric., Bur. Ent. Plant Quarantine, E–702, 1–26 (1946).

[293] HOFFMANN, C. H., TOWNES, H. K., SWIFT, H. H., and SAILER, R. I., *Field Studies on the Effects of Airplane Applications of DDT on Forest Invertebrates*, Ecol. Monogr. *19* (1), 1–46 (1949).

[294] HOLLANDER, L., *Dermatitis Caused by DDT*, Arch. Derm. Syph., N.Y. *62*, 66–68 (1950).

[295] HOLMES, E. L., and SALATHE, L. J., *State and Municipal Health Department Requirements for Use of Common Residual Insecticide Sprays*, Adv. Chem. Ser. *1*, 25–27 (1950).

[296] HOLMES, E. L., and SALATHE, L. J., *Use of Residual Spray Materials in a Typical Food Industry*, Adv. Chem. Ser. *1*, 28–30 (1950).

[297] HOLST, E. C., *DDT as a Stomach and Contact Poison for Honeybees*, J. econ. Ent. *37* (1), 159 (1944).

[298] HOLT, J. P., *DDT Poisoning in Man*, Med. Bull. Standard Oil, N.J. *9* (1), 1–3 (1949).

[299] HOPE, C. E., *Effect of DDT on Birds and the Relation of Birds to the Spruce Budworm, 1944*, Canada, Dept. Lands and Forests, Div. Research Biol. Bull. No. 2, 57–62 (1949).

[300] HOSKINS, W. M., *How Pest Control Programs Affect Residues in Foods*, Agric. Chemic. *4* (10), 22–27 (1949).

[301] HOSKINS, W. M., *Deposit and Residue of Recent Insecticides Resulting from Various Control Practices in California* J. econ. Ent. *42* (6), 966–973 (1949).

[302] HOTCHKISS, N., and POUGH, R. H., *Effects on Forest Birds of DDT Used for Gypsy Moth Control in Pennsylvania*, J. Wildlife Mgmt. *10* (3), 202–207 (1946).

[303] HOUGH, J. W., *Important Organic Insecticides, Physiological Effects and Suggested Treatment*, Industr. Hyg. (News Lett.) *10* (4), 10–16 (1950).

[304] HOWELL, D. E., *A Case of DDT Storage in Human Fat*, Proc. Okla. Acad. Sci. *29*, 31–32 (1948).

[305] HOWELL, D. E., CAVE, H. W., HELLER, V. G., and GROSS, W. G., *The Amount of DDT Found in the Milk of Cows Following Spraying*, J. Dairy Sci. *30*, 717–721 (1947).

[306] HSIEH, H. C., *D.D.T. Intoxication in a Family of Southern Taiwan*, A. M. A. Arch. Ind. Hyg. occup. Med. *10* (4), 344–346 (1954).

[307] HUFFAKER, C. B., *A Technique for Translocation of DDT in Plants*, J. econ. Ent. *41* (4), 650–651 (1948).

[308] JACKSON, W. B., *Populations of the Wood Mouse (Peromyscus leucopus) Subjected to the Applications of DDT and Parathion*, Ecol. Monogr. *22* (4), 259–281 (1952).

[309] JANDORF, B. J., SARETT, H. P., and BODANSKY, O., *Effect of Oral Administration of DDT on the Metabolism of Glucose and Pyruvic Acid in Rat Tissues*, J. Pharmacol. *88* (4), 333–337 (1946).

[310] JENKINS, G. L., *DDT Poisoning and 'Virus-X' Infection the Same*, Ind. Pharm. *31*, 133 (1949).

[311] JOHNSTON, C. D., *The in vitro Effect of DDT and Related Compounds on the Succinoxidase System of Rat Heart*, Arch. Biochem. Biophys. *31* (3), 375–382 (1951).

[312] JONES, J. C., *Some Experiments with DDT*, Pest Control *17* (8), 20–22 (1949).

[313] JUDAH, J. D., *Studies on the Metabolism and Mode of Action of DDT*, Brit. J. Pharmacol. *4*, 120–131 (1949).

[314] JUDAH, J. D., and WILLIAMS-ASHMAN, H. G., *Some Inhibitors of Aerobic Phosphorylation*, Biochem. J. *44* (4), xl–xli (1949).

[315] JUDE, A., and GIRARD, P., *Toxicité du DDT – intoxication collective par ingestion accidentelle*, Ann. Méd. lég. Criminol., Police Sci. Med. Soc. Toxicol. *29* (4), 209–213 (1949).

[316] KALMBACH, E. R., and LINDUSKA, J. P., *Controls Beyond Control*, Trans. N. Amer. Wildl. Conf. *13*, 112–128 (1948).

[317] KARPINSKI, F. E., jr., *Purpura Following Exposure to DDT*, J. Pediat. *37*, 373–379 (1950).

[318] KAWANO, T., *Effects of DDT and Its Related Compounds on Metabolism*, Folia pharm. jap. *47*, 6–8 (1951).

[319] KAY, K., *Health Aspects of the New Organic Insecticides*, Agric. Inst. Rev. *1951* (January).

[320] KEIZER, D. P. R., *DDT-Intoxicatie bij een Zuigeling*, Ned. Tijdschr. Geneesk. *97* (2), 882–884 (1953).

[321] KENDEIGH, S. C., *Bird Population Studies in the Coniferous Forest Biome During a Spruce Budworm Outbreak*, Canada, Dept. Lands and Forests, Div. Research Biol. Bull. No. 1, 100 p. (1947).

[322] KINGSCOTE, A. A., and JARVIS, C. H., *Report upon Experiments Conducted to Establish the Tolerance of Turkeys to DDT*, Canad. J. comp. Med. *10* (8), 211–218 (1946).

[323] KINGSCOTE, A. A., HENDERSON, J. A., and McINTOSH, R. A., *Observations upon the Effect of Spraying Cattle with Excessive Quantities of DDT and Black Disinfectant*, Canad. J. comp. Med. *10* (11), 322–324 (1946).

[324] KIRKWOOD, S., and PHILLIPS, P. H., *The Relationship Between the Lipoid Affinity and the Insecticidal Action of 1,1-Bis-(p-fluorophenyl)-2,2,2-trichloroethane and Related Substances*, J. Pharmacol. *87*, 375–381 (1946).

[325] KIRKWOOD, S., and PHILLIPS, P. H., *The Antitubercular Action of 1,1,1-Trichloro-2,2-bis-(p-aminophenyl)-ethane*, J. Amer. chem. Soc. *69*, 934–936 (1947).

[326] KIRKWOOD, S., PHILLIPS, P. H., and McCOY, E., *The Antitubercular Action of 1,1,1-Trichloro-2,2-bis-(p-aminophenyl)-ethane*, J. Amer. chem. Soc. *68*, 2405 (1946).

[327] KIVIRANTA, U. K., *Liquor Adulterated with Lye as a Cause of Dysphagia and Stricture of the Esophagus*, New Engl. J. Med. *243* (2), 220–222 (1950).

[328] KLINGEMANN, H., *Die DDT-Vergiftung*, Ärztl. Wschr. *4* (29/30), 465–469 (1949).

[329] KNIPLING, E. F., BUSHLAND, R. C., BABERS, F. H., CULPEPPER, G. H., and RAUN, E. S., *Evaluation of Selected Insecticides and Drugs as Chemotherapeutic Agents Against External Blood-Sucking Parasites*, J. Parasit. *34* (1), 55–70 (1948).

[330] KOGER, L. M., *DDT Poisoning in a Cow*, J. Amer. vet. med. Ass. *115*, 22–23 (1949).

[331] KONST, H., and PLUMMER, P. J. G., *Studies on the Toxicity of DDT for Domestic and Laboratory Animals*, Canad. J. comp. Med. *10* (5), 128–136 (1946).

[332] KOSTER, R., *Differentiation of Gluconate, Glucose, Calcium, and Insulin Effects on DDT Poisoning in Cats*, Fed. Proc. *6* (1), 346 (1947).

[333] KULASH, W. M., *DDT and 'Control' of Honeybees*, J. econ. Ent. *38* (5), 609–610 (1945).

[334] KUNZE, F., NELSON, A. A., FITZHUGH, O. G., and LAUG, E. P., *Storage of DDT in the Fat of the Rat*, Fed. Proc. *8*, 311 (1949).

[335] KUNZE, F., LAUG, E. P., and PRICKETT, C. S., *Storage of Methoxychlor in the Fat of the Rat*, Fed. Proc. *9*, 293 (1950).

[336] KWOCZEK, J., *Über die Toxizität der DDT- und Hexachlorzyklohexanpräparate*, Med. Mschr. *4*, 25–28 (1950).

[337] LAAKE, E. W., *Chlorinated Hydrocarbons and Their Toxicity to Domestic Animals*, Paper presented at North Central States Branch of Amer. Ass. econ. Ent., Milwaukee, Wisconsin (March 24–25, 1949).

[338] LACKEY, J. B., and STEINLE, M. L., *Effects of DDT upon Some Aquatic Organisms Other than Insect Larvae*, U. S. Suppl. publ. Hlth. Repts. No. 186, 80–89 (1945).

[339] LANGFORD, R. R., *The Effect of DDT on the Natural Fauna of the Forest Exclusive of Insect Pests*, Canada, Dept. Lands and Forests, Div. Research Biol. Bull. No. 2, 13–18 (1949).

[340] LANGFORD, R. R., *The Effect of DDT on Freshwater Fishes, 1944–45*, Canada, Dept. Lands and Forests, Div. Research Biol. Bull. No. 2, 19–38 (1949).

[341] LARDY, H. A., *Experiments with Peas and Sweet Corn Treated with DDT Insecticides*, Ind. Eng. Chem. *40* (4), 710–711 (1948).

[342] LAUG, E. P., *A Biological Assay Method for Determining 2,2-Bis-(p-chlorophenyl)-1,1,1-trichloroethane (DDT)*, J. Pharmacol. *86* (4), 324–331 (1946).

[343] LAUG, E. P., *2,2-Bis-(p-chlorophenyl)-1,1,1-trichloroethane (DDT) in the Tissues, Body Fluids and Excreta of the Rabbit Following Oral Administration*, J. Pharmacol. *86* (4), 332–335 (1946).

[344] LAUG, E. P., and FITZHUGH, O. G., *2,2-Bis-(p-chlorophenyl)-1,1,1-trichloroethane (DDT) in the Tissues of the Rat Following Oral Ingestion for Periods of Six Months to Two Years*, J. Pharmacol. *87* (1), 18–23 (1946).

[345] LAUG, E. P., NELSON, A. A., FITZHUGH, O. G., and KUNZE, F. M., *Liver-Cell Alteration and DDT Storage in the Fat of the Rat Induced by Dietary Levels of 1 to 50 ppm DDT*, J. Pharmacol. *98* (3), 268–273 (1950).

[346] LAUG, E. P., KUNZE, F. M., and PRICKETT, C. S., *Occurrence of DDT in Human Fat and Milk*, A. M. A. Arch. Ind. Hyg. occup. Med. *3*, 245–246 (1951).

[347] LÄUGER, P., MARTIN, H., and MÜLLER, P., *Über Konstitution und toxische Wirkung von natürlichen und synthetischen insektentötenden Stoffen*, Helv. chim. acta *27*, 892–928 (1944).

[348] LÄUGER, P., PULVER, R., and MONTIGEL, C., *Über die Wirkungsweise von 4,4'-Dichlordiphenyl-trichlormethylmethan (DDT-Geigy) im Warmblüter-Organismus*, Experientia *1* (4), 120–121 (1945).

[349] LÄUGER, P., PULVER, R., and MONTIGEL, C., *Über die Wirkungsweise von 4,4'-Dichlordiphenyl-trichlormethylmethan (DDT-Geigy) im Warmblüter-Organismus*, Helv. physiol. acta *3*, 405–415 (1945).

[350] LÄUGER, P., PULVER, R., MONTIGEL, C., WIESMANN, R., and WILD, H., *Mechanism of Intoxication of DDT Insecticides in Warm-Blooded Animals* (Geigy Co., New York 1946), 24 p.

[351] LAZAR, T., *DDT Pancakes*, Brit. med. J. *1*, 932 (1946).

[352] LEARY, J. C., FISHBEIN, W. I., and SALTER, L. C., *DDT and the Insect Problem* (McGraw-Hill Book Co., Inc., New York 1946), 175 p.

[353] LEGGIERI, G., *Il D.D.T. nel pana e nella pasta ottenuti con sfarinati di grano disinfestato. Azione dell'insetticida sul «Saccharomyces cerevisiae»*, Ric. sci. *20* (4), 478–481 (1950).

[354] LEHMAN, A. J., *The Toxicology of the Newer Agricultural Chemicals*, Quart. Bull. Ass. Food & Drug Officials U. S. *12* (3), 82–89 (1948).

[355] LEHMAN, A. J., *Toxic Actions of Insecticides*, Bull. N.Y. Acad. Med. *25* (6), 382–387 (1949). (Reprinted: Soap & San. Chem. Blue Book 1949, 173–175.)

[356] LEHMAN, A. J., *The Pharmacology of DDT*, Advanc. Food Res. *2*, 201–217 (1949).

[357] LEHMAN, A. J., *Pharmacological Considerations of Insecticides*, Quart. Bull. Ass. Food & Drug Officials U. S. *13* (2), 65–70 (1949).

[358] LEHMAN, A. J., *The Toxicity of the Newer Agricultural Chemicals*, Proc. Insect. Conf. *42*, Nat. Canners Ass., Atlantic City, N. J., January 17, 1949.

[359] LEHMAN, A. J., *Some Toxicological Reasons why Certain Chemicals May or May not be Permitted as Food Additives*, Quart. Bull. Ass. Food & Drug Officials U. S. *14* (3), 82–98 (1950).

[360] LEHMAN, A. J., *Chemicals in Foods: A Report to the Association of Food and Drug Officials on Current Developments. II. Pesticides Section I: Introduction. Section II: Dermal Toxicity. Section III: Subacute and Chronic Toxicity*, Quart. Bull. Ass. Food & Drug Officials U.S.: a) *15* (4), 122–125 (1951); b) *16* (1), 3–9 (January 1952); c) *16* (2), 47–53 (April 1952).

[361] LEHMAN, A. J., LAUG, E. P., WOODARD, G., DRAIZE, J. H., FITZHUGH, O. G., and NELSON, A. A., *Procedures for the Appraisal of the Toxicity of Chemicals in Foods*, Food, Drug, Cosmetic Law Quart. *1949*, 412–434.

[362] LEIDER, M., *Allergic Eczematous Contact-Type Dermatitis Caused by DDT*, J. invest. Derm. *8*, 125–126 (1947).

[363] LE PAGE, H. S., and GIANNOTTI, O., *Experiências com o DDT*, Biológico *10*, 353–366 (1944).

[364] LESSER, L. I., WEEMS, H. S., and McKEY, J. D., *Pulmonary Manifestations Following Ingestion of Kerosene*, J. Pediat. *23*, 352–364 (1943).

[365] LEWIS, W. H., and RICHARDS, A. G., jr., *Non-Toxicity of DDT on Cells in Cultures*, Science *102* (2648), 330–331 (1945).

[366] LIEBERMAN, F. V., SNOW, S. J., and SORENSON, C. J., *Control of Alfalfa Weevil in Hay Alfalfa with DDT Dust*, J. econ. Ent. *43* (3), 374–376 (1950).

[367] LILLIE, R. D., and SMITH, M. I., *Pathology of Experimental Poisoning in Cats, Rabbits, and Rats with 2,2-Bis-(para-chlorphenyl)-1,1,1-trichlorethane*, U. S. publ. Hlth. Repts. *59* (30), 979–984 (1944).

[368] LILLIE, R. D., SMITH, M. I., and STOHLMAN, E. F., *Pathologic Action of DDT and Certain of Its Analogs and Derivatives*, Arch. Path. *43*, 127–142 (1947).

[369] LINDQUIST, A. W., KNIPLING, E. F., JONES, H. A., and MADDEN, A. H., *Mortality of Bedbugs on Rabbits Given Oral Doses of DDT and Pyrethrum*, J. econ. Ent. *37* (1), 128 (1944).

[370] LINDQUIST, A. W., WILSON, H. G., SCHROEDER, H. O., and MADDEN, A. H., *Effect of Temperatures on Knockdown and Kill of Houseflies Exposed to DDT*, J. econ. Ent. *38* (2), 261–264 (1945).

[371] LINDUSKA, J. P., *DDT and the Balance of Nature*, Proc. & Papers, Intern. Tech. Conf. Protection Nature, No. VIII, 362–371 (1949).

[372] LINDUSKA, J. P., and SURBER, E. W., *Effects of DDT and Other Insecticides on Fish and Wildlife. Summary of Investigations During 1947*, U. S. Fish & Wildlife Service, Circ. No. 15, 19 p. (1948).

[373] LINDUSKA, J. P., and SPRINGER, P. F., *Chronic Toxicity of Some New Insecticides to Bob-White Quail*, U. S. Fish & Wildlife Service, Special Sci. Rept. No. 9, 11 p. (1950).

[374] LINK, R. P., *Bovine Hyperkeratosis (X Disease)*, J. Amer. vet. med. Ass. *123* (920), 427–430 (1953).

[375] LINSLEY, E. G., *Effect of DDT on Insect Pollinators of Alfalfa*, Calif. agr. exp. Sta. Circ. *365*, 18–22 (1946).

[376] LITCHFIELD, J. T., jr., and WILCOXON, F., *A Simplified Method of Evaluating Dose-Effect Experiments*, J. Pharmacol. *96* (2), 99–113 (1949).

[377] LOGAN, J. A., AITHEN, T. H. G., CASINI, G. V., KNIPE, F. W., MAIER, J., and PATTERSON, A. P., *The Sardinian Project: An Experiment in the Eradication of an Indigenous Malarious Vector*, Amer. J. Hyg. Monograph. Ser. No. 20 (Johns Hopkins Press, Baltimore 1953), 415 p.

[378] LOGIER, E. B. S., *Effect of DDT on Amphibians and Reptiles, 1944*, Canada, Dept. Lands and Forests, Div. Research Biol. Bull. No. 2, 49–56 (1949).

[379] LOOSANOFF, V. L., *Effects of DDT upon Setting Growth and Survival of Oysters*, Fish. Gaz. *64* (May), 94, 96 (1947).

[380] LORD, K. A., *The Sorption of DDT and Its Analogues by Chitin*, J. Biochem., Tokyo *43*, 72–78 (1948).

[381] LORY, R., CHARTIER, P., and SERREAU, J. P., *Sur quelques conditions de toxicité du DDT*, Ann. pharm. franç. *7*, 101–104 (1949).

[382] LUDEWIG, S., and CHANUTIN, A., *Distribution of 2,2-(p-Chlorophenyl)-1,1,1-trichlorethane (DDT) in Tissues of Rats After its Ingestion*, Proc. Soc. exp. Biol., N.Y. *62*, 20–21 (1946).

[383] LUECK, W. W., *Poisoning in Children, Newer Methods of Treatment*, Lancet *69* (5), 155–159 (1949).

[384] LUIS, P., *A Case of Detection of DDT in Viscera*, Rev. Asoc. bioquim. argent. *17*, 334–338 (1952).

[385] LURIE, J. B., *Acute Toluene Poisoning*, S. Afr. med. J. *23*, 233–236 (1949).

[386] MACCHIAVELLO, A., *Plague Control with DDT and '1080': Results Achieved in a Plague Epidemic at Tumbes, Peru, 1945*, Amer. J. publ. Hlth. *36* (8), 842–854 (1946).

[387] MacCORMACK, J. D., *Infestation and DDT*, Irish J. med. Sci. *6* (238), 627–634 (1945).

[388] MACKERRAS, I. M., and WEST, R. F. K., *'DDT' Poisoning in Man*, Med. J. Aust. *1*, 400–401 (1946).

[389] MACKIE, R. E., *Effect of DDT on Mammals, 1944*, Canada, Dept. Lands and Forests, Div. Research Biol. Bull. No. 2, 63–70 (1949).

[390] McGOVRAN, E. R., and GERSDORFF, W. A., *The Effect of Fly Food on Resistance to Insecticides Containing DDT or Pyrethrum*, Soap, N.Y. *21* (12), 165 (1945).

[391] McGREGOR, S. E., and VORHIES, C. T., *Beekeeping Near Cotton Fields Dusted with DDT*, Univ. Arizona agr. exp. Sta. Bull. *207*, 1–19 (1947).

[392] McNAMARA, B. P., BING, R. J., and HOPKINS, F., *Studies of Chronic Toxicity of DDT in the Dog*, Fed. Proc. *5*, 67–68 (1946).

[393] MANALO, G. D., HUTSON, R., and BENNE, E. J., *DDT Residues on Fruits and Vegetables*, Mich. agr. exp. Sta. Quart. Bull. *28* (4), 1–9 (1946).

[394] MANALO, G. D., HUTSON, R., and BENNE, E. J., *Removal of DDT-Spray Residues from Apples*, Mich. agr. exp. Sta. Quart. Bull. *29* (1), 1–8 (1946).

[395] MANN, H. D., CARTER, R. H., and ELY, R. E., *The DDT Content of Milk Products*, J. Milk Tech. *13* (6), 340–341 (1950).

[396] MANUFACTURING CHEMISTS' ASSOC., INC., *Warning Labels: A Guide for the Preparation of Warning Labels for Hazardous Chemicals*, Manual L–1, 3d rev., 98 p. (1953).

[397] MARQUARDT, P., *Besteht nach Lindane-Anwendung zur Bekämpfung von Getreideschädlingen eine Vergiftungsgefahr für den Brotkonsumenten?*, Wschr. Mühle *89* (29), 1 (1952).

[398] MARSDEN, S. J., and BIRD, H. R., *Effects of DDT on Growing Turkeys*, Poult. Sci. *26* (1), 3–6 (1947).

[399] MARSHALL, J., *Contact Dermatitis Due to DDT*, S. Afr. med. J. *24*, 300–301 (1950).

[400] MATTSON, A. M., SPILLANE, J. T., BAKER, C., and PEARCE, G. W., *Determination of DDT and Related Substances in Human Fat*, Analyt. Chem. *25*, 1065–1070 (1953).

[401] MEEK, W. J., *Some Cardiac Effects of the Inhalant Anesthetics and the Sympatho-mimetic Amines*, Harvey Lect. *36*, 188–226 (1941).

[402] MEIKLEJOHN, G., and BRUYN, H. B., *Influenza in California During 1947 and 1948*, Amer. J. publ. Hlth. *39* (1), 44–49 (1949).

[403] MELIS, R., *Contributo alla conoscenza dell'azione che il DDT esercita sugli animali da cortile e su alcuni artropodi loro parassiti. I. Azione del DDT sulla gallina di razza bianca livornese*, Gac. Mal. infett. parassit. *3* (5/6), 280–287 (1951).

[404] MELIS, R., *Contributo alla conoscenze dell'azione che il DDT esercita sugli animali da cortile e su alcuni artropodi loro parassiti. II. Azione del DDT sul coniglio e sull'acaro suo parassita (Sarcoptes scabiei var. cunicoli L.)*, Gac. Mal. infett. parassit. *3* (5/6), 288–294 (1951).

[405] MERKIN, S., *Warning: Poison in Milk!*, Nat. Police Gaz. *159* (3), 8–9 (1954).

[406] MERKIN, S., *DDT Is Poisoning Your Food!*, Nat. Police Gaz. *159* (5), 20–21, 29 (1954).

[407] METCALF, R. L., HESS, A. D., SMITH, G. E., JEFFERY, G. M., and LUDWIG, G. L., *Observations on the Use of DDT for the Control of Anopheles quadrimaculatus*, U. S. publ. Hlth. Repts. *60* (27), 753–774 (1945).

[408] METZLER, D. F., *Injury Resulting from Use of Aerosol Bomb*, U. S. publ. Hlth. Repts. *61* (15), 546 (1946).

[409] MICKEY, G. H., SUMERFORD, W. T., and JONES, J. P., *Urethane as an Antidote for DDT*, Anat Rec. *99* (4), 617–618 (1947).

[410] MILLER, L. L., *Nutritional Factors Affecting the Toxicity of Halogenated Hydro-carbons*, Occup. Med. *5*, 194–209 (1948).

[411] MILLER, S. E., *DDT in Medical Practice*, Curr. Med. Digest *1946* (July), 27–33.

[412] MITCHELL, R. T., *Effects of DDT Spray on Eggs and Nestlings of Birds*, J. Wildlife Mgmt. *10* (3), 192–194 (1946).

[413] MITCHELL, R. T., BLAGBROUGH, H. P., and VAN ETTEN, R. C., *The Effect of DDT upon the Survival and Growth of Nesting Songbirds*, J. Wildlife Mgmt. *17* (1), 45–54 (1953).

[414] MOHR, R. W., TELFORD, H. S., PETERSON, E. H., and WALKER, K. C., *Toxicity of Orchard Insecticides to Game Birds in Eastern Washington*, Wash. agr. exp. Sta. Circ. *170*, 22 p. (1951).

[415] MOOSER, H., *Die Bedeutung des Neocid Geigy für die Verhütung und Bekämpfung der durch Insekten übertragenen Krankheiten*, Schweiz. med. Wschr. *74* (36), 947–952 (1944).

[416] MÜHLENS, K., *Über die Bedeutung der Dichlor-diphenyl-trichlormethylmethan-Prä-parate als Arthropodengift in der Seuchenbekämpfung, unter Berücksichtigung eigener Erfahrungen*, Dtsch. med. Wschr. *71* (20), 164–169 (1946).

[417] MÜLLER, P., DOMENJOZ, R., WIESMANN, R., and BUXTORF, A., *I. Dichlordiphenyl-trichloräthan als Insektizid und seine Bedeutung für die Human- und Veterinär-hygiene*, Ergebn. Hyg. Bakt. Immunitätsforsch. exper. Therap. *26*, 5–138 (1949).

[418] NAEVESTED, R., *Forgiftning med DDT-Pulver og et par andre forgiftningstilfelle*, Tidsskr. norske Lægeforen. *67*, 261–263 (1947).

[419] NALE, T. W., *Current Method of Determining Safety in the Application of New Chemicals*, Ind. Med. Surg. *20* (11), 501–506 (1951).

[420] NATIONAL RESEARCH COUNCIL, BIOLOGY SUBCOMMITTEE, *Conference on Toxicity of DDT to Man and Other Mammals and the Hazards Involved in the Agricultural Use of DDT* (May 24, 1946), 29 p.

[421] NEAL, P. A., *DDT Toxicity—a Report on the Toxicity to Warm-Blooded Animals of Aerosols, Mists, and Dusting Powders Containing DDT*, Soap, N.Y. *21* (1), 99–101, 111 (1945).

[422] NEAL, P. A., *The Effects of Exposure to DDT in Man*, Conference on toxicity of DDT, National Research Council (May 24, 1946), p. 11–15.

[423] NEAL, P. A., Personal communication (1951).

[424] NEAL, P. A., and VON OETTINGEN, W. F., *The Toxicity and Potential Dangers of DDT to Humans and Warm-Blooded Animals*, Med. Ann. D.C. *15* (1), 15–19 (1946).

[425] NEAL, P. A., and VON OETTINGEN, W. F., *Toxicity of DDT... A Report on Experimental Studies*, Soap, N.Y. *22* (7), 135, 137, 139, 141, 143 (1946).

[426] NEAL, P. A., VON OETTINGEN, W. F., SMITH, W. W., MALMO, R. B., DUNN, R. C., MORAN, H. E., SWEENEY, T. R., ARMSTRONG, D. W., and WHITE, W. C., *Toxicity and Potential Dangers of Aerosols, Mists, and Dusting Powders Containing DDT*, U.S. Suppl. publ. Hlth. Repts. S. *177*, 1–32 (1944).

[427] NEAL, P. A., VON OETTINGEN, W. F., DUNN, R. C., and SHARPLESS, N. E., *Toxicity and Potential Dangers of Aerosols and Residues from Such Aerosols Containing Three Percent DDT*, U.S. Suppl. publ. Hlth. Repts. *183*, 1–32 (1945).

[428] NEAL, P. A., SWEENEY, T. R., SPICER, S. S., and VON OETTINGEN, W. F., *The Excretion of DDT (2,2-Bis-[p-chlorophenyl]-1,1,1-trichloroethane) in Man, Together with Clinical Observations*, U.S. publ. Hlth. Repts. *61* (12), 403–409 (1946).

[429] NEGHME, A., *Control del Aedes aegypti en Chile*, Bol. Ofic. sanit. pan-amer. *29* (4), 389–396 (1950).

[430] NEGHME, A., MONTERO, R., VILLALON, J., GUTIERREZ, J., ALBI, H., and REYES, A., *Método de lucha anti Aedes aplicada en Iquique*, Bol. parasit. chil. *4* (2), 4–6 (1949).

[431] NELSON, A. A., and WOODARD, G., *Adrenal Cortical Atrophy and Liver Damage Produced in Dogs by Feeding 2,2-Bis-(parachlorophenyl)-1,1-dichloroethane (DDD)*, Fed. Proc. [P] *7*, 277 (1948).

[432] NELSON, A. A., DRAIZE, J. H., WOODARD, G., FITZHUGH, O. G., SMITH, R. B., jr., and CALVERY, H. O., *Histopathological Changes Following Administration of DDT to Several Species of Animals*, U.S. publ. Hlth. Repts. *59* (31), 1009–1020 (1944).

[433] NELSON, A. L., and SURBER, E. W., *DDT Investigations by the Fish and Wildlife Service in 1946*, U.S. Fish & Wildlife Service, Special Sci. Rept. *41*, 1–8 (1947).

[434] NELSON, J. W., EDWARDS, L. D., CHRISTIAN, J. E., and JENKINS, G. L., *The Study, Testing and Synthesis of Certain Organic Compounds, for Rodenticidal Activity*, J. Amer. pharm. Ass., Sci. Ed. *1947*, 349–352.

[435] NICHOLS, J., and GARDNER, L. I., *Production of Insulin Sensitivity with the Adreno-Corticolytic Drug DDD (1,1-Dichloro-2,2-bis-[p-chlorophenyl]-ethane)*, J. Lab. clin. Med. *37*, 229 (1951).

[436] NIEDELMAN, M. L., *Contact Dermatitis Due to DDT*, Occup. Med. *1*, 391–395 (1946).

[437] NORRIS, D. O., *The Evaluation of D.D.T. as a Fungicide*, J. Coun. sci. industr. Res., Aust. *17* (4), 289–290 (1944).

[438] NORRIS, K. R., *The Use of DDT for the Control of the Buffalo Fly (Siphona exigua [de Meijere])*, J. Coun. sci. industr. Res., Aust. *20* (1), 25–42 (1947).

[439] NUNN, J. A., and MARTIN, F. M., *Gasoline and Kerosene Poisoning in Children*, J. Amer. med. Ass. *103* (7), 472–474 (1934).

[440] ODUM, E. P., and SUMERFORD, W. T., *Comparative Toxicity of DDT and Four Analogs to Goldfish, Gambusia, and Culex Larvae*, Science *104* (2708), 480–482 (1946).

[441] OFNER, R. R., and CALVERY, H. O., *Determination of DDT (2,2-Bis-[p-chlorophenyl]-1,1,1-trichloroethane) and Its Metabolite in Biological Materials by Use of the Schechter-Haller Method*, J. Pharmacol. *85* (4), 363–370 (1945).

[442] OFNER, R. R., WOODARD, G., and CALVERY, H. O., *Studies on the Metabolism of 2,2-Bis-(p-chlorophenyl)-1,1,1-trichloroethane (DDT)*, Fed. Proc. *4*, 132–133 (1945).

[443] OLAFSON, P., *Hyperkeratosis (X Disease) of Cattle*, Cornell Vet. *37* (4), 279–291 (1947).

[444] ORR, L. W., and MOTT, L. O., *The Effects of DDT Administered Orally to Cows, Horses, and Sheep*, J. econ. Ent. *38* (4), 428–432 (1945).

[445] PACKARD, C. M., *Treatment of Wheat with DDT: Not Recommended*, Letter dated July 11, 1950, to Prof. A. MISSIROLI.

[446] PAGAN, C., and HAGEMAN, R. H., *Determination of DDT by Bioassay*, Science *112* (2904), 222–223 (1950).

[447] PARKER, R. L., and ESHBAUGH, E. L., *Codling Moth and Mite Control in Kansas with New Insecticides*, J. econ. Ent. *40* (6), 861–864 (1947).

[448] PARKIN, E. A., *The Contamination of Stored Wheat with DDT*, Insect. Abstr. News Sum. No. 25, 87 (1952).

[449] PEARCE, G. W., MATTSON, A. M., and HAYES, W. J., jr., *Examination of Human Fat for the Presence of DDT*, Science *116* (3010), 254–256 (1952).

[450] PERRY, D. P., *Explanation of the Principal Features of the Federal Insecticide, Fungicide, and Rodenticide Act*, Quart. Bull. Ass. Food & Drug Officials U. S. *12* (2), 64–68 (1948).

[451] PERRY, W. J., and BODENLOS, L. J., *2,2-Bis-(p-chlorophenyl)-1,1,1-trichloroethane (DDT) Determinations in Tissues, Body Fluids and Excreta of Human Subjects*, Mosquito News *10* (1), 1–3 (1950).

[452] PHILIPS, F. S., *Insecticides and Rodenticides*, Fed. Proc. *5* (2), 292–298 (1946).

[453] PHILIPS, F. S., and GILMAN, A., *Studies on the Pharmacology of DDT (2,2-Bis-[parachlorphenyl]-1,1,1-trichloroethane). I. The Acute Toxicity of DDT Following Intravenous Injection in Mammals with Observations on the Treatment of Acute DDT Poisoning*, J. Pharmacol. *86* (3), 213–221 (1946).

[454] PHILIPS, F. S., GILMAN, A., and CRESCITELLI, F. N., *The Sensitization of the Myocardium to Sympathetic Stimulation During Acute DDT Intoxication in Animals*, Fed. Proc. *5*, 80 (1946).

[455] PHILIPS, F. S., GILMAN, A., and CRESCITELLI, F. N., *Studies on the Pharmacology of DDT (2,2-Bis-[parachlorphenyl]-1,1,1-trichloroethane). II. The Sensitization of the Myocardium to Sympathetic Stimulation During Acute DDT Intoxication*, J. Pharmacol. *86* (3), 222–228 (1946).

[456] PICKERING, J. H., RATHBUN, H. T., BUNNELL, M. J., and ALEXANDER, C. C., *DDT Insecticides: Tolerances for Poisonous or Deleterious Residues on or in Fresh Fruits and Fresh Vegetables*, FDA Docket No. FDC–57 (Geigy Company, Inc., Washington, no date).

[457] PIEKARSKI, G., and HOLZ, K., *Über die Wirkung des kontakt-insektiziden Handelspräparates Gix auf Ratten und Mäuse*, Arch. exp. Path. Pharmak. *210*, 71–90 (1950).

[458] PIELOU, D. P., *Lethal Effects of D.D.T. on Young Fish*, Nature *158*, 378 (1946).

[459] PLAGUE RESEARCH LABORATORY, DEPT. OF HEALTH, JOHANNESBURG, *Preliminary Report on Investigations on the Dosage-Mortality Relations of Several Poisons to Some Rodents*, Special Rept. No. 3/49, 1–12 (1950).

[460] PLICHET, A., *Le DDT serait-il responsable de certaines gastro-entérites?*, Pr. méd. *57* (76), 1121–1122 (1949).

[461] PLUMMER, P. J. G., *Toxicity of DDT for Laboratory and Domestic Animals*, Pests *14* (2), 28 (1946).

[462] PLUVINAGE, R. J., and HEATH, J. W., *Neural Effects of DDT Poisoning in Cats*, Proc. Soc. exp. Biol., N.Y. *63* (1), 212–214 (1946).

[463] POLLITZER, R., *A Note on the Possibility of Using DDT as a Rodenticide*, World Health Organization mimeographed release (September 25, 1950).

[464] POLLOCK, G. H., and WANG, R. I. H., *Synergistic Actions of Carbon Dioxide with DDT in the Central Nervous System*, Science *117* (3048), 596–597 (1953).

[465] POMMERT, R., *DDT–Caution*, Gladiolus Mag. *11* (3), 15 (1947).

[466] POTOSSI, O., *Ricerche igienico-sanitarie su disinfestanti a base di D.D.T. in rapporto alla maggiore o minore tossicità del solvente*, Farmaco *1* (6), 405–419 (1946).

[467] POTOSSI, O., *Quadri anatomo-istopatologici nell'intossicazione sperimentale da D.D.T.*, Boll. Soc. ital. Biol. sper. *23*, 1084–1087 (1947).

[468] PRAJMOVSKY, M., and WELSH, J. H., *Total and Free Acetylcholine in Rat Peripheral Nerves*, J. Neurophysiol. *11*, 1–8 (1948).

[469] PRATT-THOMAS, H. R., and WARING, J. T., *DDT Poisoning*, J. Amer. med. Ass. *131* (16), 1384 (1946).

[470] PRILL, E. A., HARTZELL, A., and ARTHUR, J. M., *Insecticidal Activity of Some Alkoxy Analogs of DDT*, Science *101* (2627), 464–465 (1945).

[471] Princi, F., *The Chlorinated Hydrocarbon Insecticides, Hazards and Precautions*, Mod. Sanit. *5* (12), 49–52 (1953).

[472] Princi, F., *DDT Toxicity Up-To-Date*, Soap Chem. Specialties *30* (6), 167, 169, 171 (1954).

[473] Pulewka, P., *Versuche über die Wirkung von Dichlordiphenyl-Trichloräthan-Präparaten auf Insekten und Warmblüter mit Rücksicht auf ihre Anwendungsform*, Turk. Z. Hyg. exp. Biol. *7* (1), 5–38 (1947).

[474] Questel, D. D., and Connin, R. V., *Investigations of Sprays for Control of the European Corn Borer, Toledo, Ohio, 1945–1946*, U. S. Dept. Agric., Bur. Ent. Plant Quarantine E–743, 10 p. (1948).

[475] Radeleff, R. D., *Insecticide Toxicology for Veterinarians*, Proc. Amer. Vet. Med. Ass., Miami Beach, *87*, 68–72 (1950).

[476] Radeleff, R. D., and Bushland, R. C., *Acute Toxicity of Chlorinated Insecticides Applied to Livestock*, J. econ. Ent. *43* (3), 358–364 (1950).

[477] Radeleff, R. D., Claborn, H. V., Wells, R. W., and Nickerson, W. J., *Effects on Beef Cattle of Prolonged Treatment with a DDT Spray*, Vet. Med. *47* (3), 94–96 (1952).

[478] Radeleff, R. D., Bushland, R. C., and Claborn, H. V., *Toxicity to Livestock*, U. S. Dept. Agric. Yrbk., Insects, *1952*, 276–283.

[479] Reed, E. S., Leikin, S., and Kerman, H. D., *Kerosene Intoxication*, Amer. J. Dis. Child. *79* (4), 623–632 (1950).

[480] Reed, W. G., *Notice to Manufacturers, Registrants and Distributors of Insecticides Containing DDT*, U. S. Dept. Agric. Prod. Marketing Adm. Release (April 7, 1949).

[481] Reed, W. G., *Registration Procedure Under the New Federal Insecticide Law*, Soap Sanit. Chemicals Blue Book *1949*, 183–186.

[482] Reed, W. G., *Labeling Requirements for Insecticides and Other Economic Poisons Under Federal Law*, Adv. Chem. Ser. *1*, 17–20 (1950).

[483] Reiber, H. G., and Stafford, E. M., *DDT Deposits and Residues on Olives, Grapes, Cabbage, Green Beans, and Peas*, Calif. agr. exp. Sta. Circ. No. 365, 104 (1946).

[484] Reingold, I. M., and Lasky, I. I., *Acute Fatal Poisoning Following Ingestion of a Solution of DDT*, Ann. internal Med. *26* (6), 945–947 (1947).

[485] Rice, R., *DDT*, New Yorker *30* (22), 31–41, 43–51 (1954).

[486] Richards, A. G., and Cutkomp, L. K., *Correlation Between the Possession of a Chitinous Cuticle and Sensitivity to DDT*, Biol. Bull., Woods Hole *90* (2), 97–108 (1946).

[487] Richardson, J. A., and Pratt-Thomas, H. R., *Toxic Effects of Varying Doses of Kerosene Administered by Different Routes*, Amer. J. med. Sci. *221*, 531–536 (1951).

[488] Riker, W. F., jr., Huebner, V. R., Raska, S. B., and Cattell, McK., *Studies on DDT (2,2-Bis-[parachlorophenyl]-1,1,1-trichlorethane), Effects on Oxidative Metabolism*, J. Pharmacol. *88* (4), 327–332 (1946).

[489] Robbins, C. S., and Stewart, R. E., *Effects of DDT on Bird Population of Scrub Forest*, J. Wildlife Mgmt. *13* (1), 11–16 (1949).

[490] Roberts, F. H. S., and Moule, G. R., *A Preliminary Report on the Value of DDT and BHC for the Control of Body-Strike in Sheep*, Aust. vet. J. *27* (2), 35–39 (1951).

[491] Robinson, R. H., *Harvest Analyses of DDT Residues*, Food Pack. *29* (11), 50, 52–53 (1948).

[492] Robinson, R. H., *Spray Residues on Food Crops and Their Relation to Total Food Consumption*, Adv. Chem. Ser. *1*, 49–51 (1950).

[493] Rodriguez, J. A., *Ensayo de control del Aedes aegypti mediante la aplicación de DDT en los tanques de captación de aguas*, Bol. Ofic. sanit. pan-amer. *29* (11), 1150–1151 (1950).

[494] Roeder, K. D., and Weiant, E. A., *The Site of Action of DDT in the Cockroach*, Science *103* (2671), 304–306 (1946).

[495] Roeder, K. D., and Weiant, E. A., *The Effect of DDT on Sensory and Motor Structures in the Cockroach Leg*, J. cell. comp. Physiol. *32* (2), 175–186 (1948).

[496] Rogers, E. S., *Losses Associated with Dipping in DDT*, Year Book Inst. Insp. Stk. N. S. W. *1948*, 59–60.

[497] Rohwer, S. A., *Use and Misuse of DDT, BHC, DDD (DDE), Chlordan, Toxaphene, Methoxychlor, and Parathion for Canning Crops*, Proc. Insecticide Conf., Natl. Canners Ass., Atlantic City, N. J. *42* (January 17, 1949).

[498] Romanoff, A. L., and Romanoff, A. J. S., *The Avian Egg* (John Wiley & Sons, Inc., New York 1948), p. 202–203.

[499] Rosenberg, M. M., and Adler, H. E., *Comparative Toxicity of DDT and Chlordan to Young Chicks*, Amer. J. vet. Res. *11* (38), 142–144 (1950).

[500] Rowe, J. A., *Pest Control in Food Plants*, Quart. Bull. Ass. Food & Drug Officials U. S. *12*, 3–13 (1948).

[501] Rubin, M., Bird, H. R., Green, N., and Carter, R. H., *Toxicity of DDT to Laying Hens*, Poult. Sci. *26* (4), 410–413 (1947).

[502] Sa Antunes, W. S., *Campaign for Eradication of Aedes aegypti*, Personal communication (December 4, 1951).

[503] Salter, W. T., *Chemistry of Carcinogens*, Occup. Med. *5* (5), 441–465 (1948).

[504] Sapeika, N., *Excretion of Drugs in Human Milk–a Review*, J. Obstet. Gynaec. Brit. Emp. *54*, 426–431 (1947).

[505] Sarett, H. P., and Jandorf, B. J., *Effects of Chronic Intoxication of Rats with DDT on Lipids and Other Constituents of Liver*, Fed. Proc. *5*, 151–152 (1946).

[506] Sarett, H. P., and Jandorf, B. J., *Effects of Chronic DDT Intoxication in Rats on Lipids and Other Constituents of Liver*, J. Pharmacol. *91* (4), 340–344 (1947).

[507] Sasse, B. E., *DDT in Alcohol as a Larvicide for Aedes aegypti Control*, Mosquito News *10* (3), 144–149 (1950).

[508] Sauberlich, H. E., and Baumann, C. A., *Effect of Dietary Variations Upon the Toxicity of DDT to Rats or Mice*, Proc. Soc. exp. Biol., N.Y. *66*, 642–645 (1947).

[509] Savage, J., *Aquatic Invertebrates: Mortality Due to DDT and Subsequent Re-Establishment (1944–45)*, Canada, Dept. Lands and Forests, Div. Research Biol. Bull. No. 2, 39–48 (1949).

[510] Schechter, M. S., and Haller, H. L., *Colorimetric Tests for DDT and Related Compounds*, J. Amer. chem. Soc. *66*, 2129–2130 (1944).

[511] Schechter, M. S., Haller, H. L., and Pogorelskin, M. A., *Colorimetric Determination of DDT in Milk*, Agric. Chemic. *1* (6), 27, 46 (1946).

[512] Schmid, G., *Versuche über Läusebekämpfung beim Pferd*, Arch. Tierheilk. *85*, 206–209 (1943).

[513] Schmidt, H. W., *Zur Frage der toxischen Wirkung von DDT auf den Menschen*, Zbl. allg. Path. path. Anat. *88* (10/11), 409–411 (1952).

[514] Schwarte, L. H., and Biester, H. E., *Disease of Poultry*, 2nd ed. (Iowa State College Press, Ames 1948), p. 1009–1020.

[515] Scott, A. E., *Health Foods Doorway to Death*, Sir. *11* (5), 44–45, 62–64 (1954).

[516] Scudder, H. I., and Tarzwell, C. M., *Effects of DDT Mosquito Larviciding on Wildlife. IV. The Effects on Terrestrial Insect Population of Routine Larviciding by Airplane*, U. S. publ. Hlth. Repts. *65* (3), 71–87 (1950).

[517] Seevers, M. H., *Perspective Versus Caprice in Evaluating Toxicity of Chemicals in Man*, J. Amer. med. Ass. *153* (15), 1329–1333 (1953).

[518] Segre, G., *Sulla tossicità della DDT nell'uomo*, Minerva med., Roma *1* (8), 224–226 (1947).

[519] Sergent, E., and Sergent, E., *Toxicité du DDT pour les souris et les rats*, C. R. Acad. Clerm.-Ferrand *227*, 416–419 (1948).

[520] Sergent, E., and Sergent, E., *Le DDT peut servir à la destruction de rongeurs nuisibles (souris, rats d'égout, mérions)*, C. R. Acad. Agric. *34*, 954–960 (1948).

[521] Sergent, E., and Sergent, E., *Nouveau procédé de dératisation basé sur l'emploi du DDT*, Arch. Inst. Pasteur Algér. *27*, 18–30 (1949).

[522] Sheehan, H. L., Summers, V. K., and Nichols, J., *DDD Treatment in Cushing's Syndrome*, Lancet *1953* (264), 312–314.

[523] Shepard, H. H., *Production Data, DDT*, Personal communication dated November 17, 1953.

[524] SHEPHERD, J. B., MOORE, L. A., CARTER, R. H., and POOS, F. W., *The Effect of Feeding Alfalfa Hay Containing DDT Residue on the DDT Content of Cow's Milk*, J. Dairy Sci. *32* (6), 549–555 (1949).

[525] SHILLINGER, Y. I., *Passage of DDT into the Milk of Cows Treated with DDT to Combat the Gadfly*, Vop. Pitan. *12* (2), 68–73 (1953).

[526] SIKES, D., and BRIDGES, M. E., *Experimental Production of Hyperkeratosis ('X Disease') of Cattle with a Chlorinated Naphthalene*, Science *116* (3019), 506–507 (1952).

[527] SIKES, D., WISE, J. C., and BRIDGES, M. E., *The Experimental Production of 'X Disease' (Hyperkeratosis) in Cattle with Chlorinated Naphthalenes and Petroleum Products*, J. Amer. vet. med. Ass. *121* (908), 337–344 (1952).

[528] SIMMONS, S. W., and ALPERT, M., *A Statement on the Health Hazards of Insecticide Vaporizers Used as Fumigators for the Control of Insects*, J. econ. Ent. *46* (4), 739 (1953).

[529] SIMPSON, C. F., *Hyperkeratosis or X Disease in Florida*, Vet. Med. *44*, 51–52 (1949).

[530] SLOAN, M. J., RAWLINS, W. A., and NORTON, L. B., *Residue Studies on DDT and Parathion Applied to Lettuce for Control of Six-Spotted Leafhopper*, J. econ. Ent. *44* (5), 691–701 (1951).

[531] SMITH, M. I., *Accidental Ingestion of DDT, with a Note on Its Metabolism in Man*, J. Amer. med. Ass. *131*, 519–520 (1946).

[532] SMITH, M. I., and STOHLMAN, E. F., *The Pharmacologic Action of 2,2-Bis-(p-chlorophenyl)-1,1,1-trichlorethane and Its Estimation in the Tissues and Body Fluids*, U. S. publ. Hlth. Repts. *59* (30), 984–993 (1944).

[533] SMITH, M. I., and STOHLMAN, E. F., *Further Studies on the Pharmacologic Action of 2,2-Bis-(p-chlorophenyl)-1,1,1-trichlorethane (DDT)*, U. S. publ. Hlth. Repts. *60* (11), 289–301 (1945).

[534] SMITH, M. I., BAUER, H., STOHLMAN, E. F., and LILLIE, R. D., *The Pharmacologic Action and Metabolism of a Series of Compounds Chemically Related to DDT*, Fed. Proc. *5*, 203–204 (1946).

[535] SMITH, M. I., BAUER, H., STOHLMAN, E. F., and LILLIE, R. D., *The Pharmacologic Action of Certain Analogues and Derivatives of DDT*, J. Pharmacol. *88* (4), 359–365 (1946).

[536] SMITH, M. I., JUNGE, J. M., and McCLOSKY, W. T., *The Pharmacologic and Chemotherapeutic Action of p,p'-Diaminodiphenyltrichloroethane*, J. Amer. pharm. Ass., Sci. Ed. *37* (11), 461–464 (1948).

[537] SMITH, N. J., *Death Following Accidental Ingestion of DDT. Experimental Studies*, J. Amer. med. Ass. *136* (7), 469–471 (1948).

[538] SMITH, R. F., FULLMER, O. H., and MESSENGER, P. S., *DDT Residues on Alfalfa Hay and Seed Chaff*, J. econ. Ent. *41* (5), 755–758 (1948).

[539] SMITH, R. F., HOSKINS, W. M., and FULLMER, O. H., *Secretion of DDT in Milk of Dairy Cows Fed Low-Residue Alfalfa Hay*, J. econ. Ent. *41* (5), 759–763 (1948).

[540] SMITH, R. F., MacSWAIN, J. W., LINSLEY, E. G., and PLATT, F. R., *The Effect of DDT Dusting on Honeybees*, J. econ. Ent. *41* (6), 960–971 (1948).

[541] SOLLMANN, T., *A Manual of Pharmacology and Its Application to Therapeutics and Toxicology*, 7th ed. (W. B. Saunders Co., Philadelphia 1948), p. 105.

[542] SPEAR, P. J., and SWEETMAN, H. L., *Continuous Vaporization of Insecticides with Special Reference to DDT*, J. econ. Ent. *45* (5), 869–873 (1952).

[543] SPEIRS, J. M., *The Relation of DDT Spraying to the Vertebrate Life of the Forest, 1946*, Canada, Dept. Lands and Forests, Div. Research Biol. Bull. No. 2, p. 141 (1949).

[544] SPICER, S. S., SWEENEY, T. R., VON OETTINGEN, W. F., LILLIE, R. D., and NEAL, P. A., *Toxicological O_servations on Goats Fed Large Doses of DDT*, Vet. Med. *42* (8), 289–293 (1947).

[545] SPRINGER, P. F., and WEBSTER, J. R., *Effects of DDT on Saltmarsh Wildlife: 1949*, U. S. Fish & Wildlife Service, Special Sci. Rept. No. 10, 25 p. (1949).

[546] SPRINGER, P. F., and WEBSTER, J. R., *Biological Effects of DDT Applications on Tidal Salt Marshes*, Mosquito News *11* (2), 67–74 (1951).

[547] STAFFORD, E. M., and HINKLEY, H. S., *DDT and Related Compounds for Control of Black Scale on Olives*, Calif. agr. exp. Sta. Circ. *365*, 91–93 (1946).

[548] STAMMERS, F. M. G., and WHITFIELD, F. G. S., *Toxicity of DDT to Man*, Nature *157*, 658 (1946).

[549] STAMMERS, F. M. G., and WHITFIELD, F. G. S., *Toxicity of DDT to Man and Animals*, Bull. ent. Res. *38* (1), 1–73 (1947).

[550] STEINER, L. F., SUMMERLAND, S. A., and FAHEY, J. E., *Experiments in 1945 with DDT for Control of the Codling Moth*, Proc. Ohio hort. Soc. *79*, 105–130 (1946).

[551] STEPHENS, P. A., *Exposure Time Spray Labor to DDT Spray*, Personal communication (November 17, 1949).

[552] STERLINGER, R., *Intoxication with DDT*, Harefuah *35* (8), 59 (1948).

[553] STERNE, J., *Etat actuel du traitement du paludisme*, Gaz. méd. France *58* (15), 927–931 (1951).

[554] STEWART, R. E., COPE, J. B., ROBBINS, C. S., and BRAINERD, J. W., *Effects of DDT on Birds at the Patuxent Research Refuge*, J. Wildlife Mgmt. *10* (3), 195–201 (1946).

[555] STICKEL, L. F., *Field Studies of a Peromyscus Population in an Area Treated with DDT*, J. Wildlife Mgmt. *10* (3), 216–218 (1946).

[556] STIFF, H. A., jr., and CASTILLO, J. C., *The Determination of 2,2-Bis-(p-chlorophenyl)-1,1,1-trichloroethane (DDT) in Organs and Body Fluids After Oral Administration*, J. biol. Chem. *159* (2), 545–548 (1945).

[557] STOHLMAN, E. F., *Preliminary Report on the Identification of 2,2-Bis-(p-chlorophenyl)-1,1,1-trichlorethane (DDT) in the Excreta of Poisoned Rabbits*, U. S. publ. Hlth. Repts. *60* (13), 350–353 (1945).

[558] STOHLMAN, E. F., *The Effects of Corn Oil and Olive Oil on the Blood Sugar and Rectal Temperature of Rabbits*, J. Pharmacol. *93*, 346–350 (1948).

[559] STOHLMAN, E. F., and SMITH, M. I., *The Isolation of Di-(p-chlorophenyl)-Acetic Acid (DDA) from the Urine of Rabbits Poisoned with 2,2-Bis-(p-chlorophenyl)-1,1,1-trichlorethane (DDT)*, J. Pharmacol. *84* (4), 375–379 (1945).

[560] STOHLMAN, E. F., and LILLIE, R. D., *The Effect of DDT on the Blood Sugar and of Glucose Administration on the Acute and Chronic Poisoning of DDT in Rabbits*, J. Pharmacol. *93* (3), 351–361 (1948).

[561] STONE, T. T., and GLADSTONE, L., *DDT (Dichlorodiphenyl-trichloroethane)*, J. Amer. med. Ass. *145*, 1342 (1951).

[562] STORER, T. I., *DDT and Wildlife*, J. Wildlife Mgmt. *10*, 181–183 (1946).

[563] STRYKER, G. V., and GODFROY, B., *Dermatitis Resulting from Exposure to DDT*, J. Mo. med. Ass. *43*, 384–386 (1946).

[564] SURBER, E. W., *Effects of DDT on Fish*, J. Wildlife Mgmt. *10* (3), 183–191 (1946).

[565] SURBER, E. W., *Toxicities of Some Chemical Substances to Fish*, Presented at the 6th Annual Pollution Abatement Conf., Manufacturing Chemists Ass., New York City (April 17, 1950).

[566] SURBER, E. W., and FRIDDLE, D. D., *Relative Toxicity of Suspension and Oil Formulations of DDT to Native Fishes in Back Creek, West Virginia*, Trans. Amer. Fish. Soc. *76* (1946), 315–321 (1949).

[567] SURBER, E. W., and HOFFMANN, C. H., *Effects of Various Concentrations of DDT on Several Species of Fish of Different Sizes*, U. S. Fish & Wildlife Service, Special Sci. Rept. No. 4, 19 p. (1949).

[568] SWINEFORD, O., jr., and RADFORD, P. J., *Contact Dermatitis. Results of 312 Patch Tests, with Observations on Technique*, Sth. med. J., Nashville *41* (8), 667–672 (1948).

[569] TARSITANO, F., *Ricerche sull'avvelenamento da D.D.T.*, Folia med., Napoli *31* (7), 297–306 (1948).

[570] TARZWELL, C. M., *Effects of DDT Mosquito Larviciding on Wildlife. I. The Effects on Surface Organisms of the Routine Hand Application of DDT Larvicides for Mosquito Control*, U. S. publ. Hlth. Repts. *62* (15), 525–554 (1947).

[571] TARZWELL, C. M., *Effects of Routine DDT Mosquito Larviciding on Wildlife*, J. nat. Malar. Soc. *7* (3), 199–206 (1948).

[572] TARZWELL, C. M., *Effects of DDT Mosquito Larviciding on Wildlife. V. Effects on Fishes of the Routine Manual and Airplane Application of DDT and Other Mosquito Larvicides*, U. S. publ. Hlth. Repts. *65* (8), 231–255 (1950).

[573] TASCHENBERG, E. F., *Evaluation of Spray Programs for the Control of the Grape Berry Moth, Polychrosis viteana* Clemens, N.Y. State agr. exp. Sta. Tech. Bull. No. 283, 3–70 (1948).

[574] TASCHENBERG, E. F., and AVENS, A. W., *DDT Residue Studies of Fresh Grapes, Juice, and Jam*, N.Y. State agr. exp. Sta. Tech. Bull. No. 286, 3–18 (1949).

[575] TASCHENBERG, E. F., AVENS, A. W., and COX, J. A., *Supplements in DDT Sprays for Control of Grape Berry Moth*, J. econ. Ent. *43* (2), 152–157 (1950).

[576] TAUBER, O. E., and HUGHES, A. B., *Effect of DDT Ingestion on Total Cholesterol Content of Ovaries of White Rat*, Proc. Soc. exp. Biol., N.Y. *75* (2), 420–422 (1950).

[577] TAYLOR, E. L., *Benzenehexachloride. A Promising New Acaricide*, Vet. Rec. *57* (18), 210–211 (1945).

[578] TAYLOR, E. L., *Danger of Innunction with DDT*, Lancet *1945*, 2, 320.

[579] TAYLOR, H., and FRODSHAM, J., *Assay of Toxic Effect of 'Gammexane' on Man and Animals*, Nature *158* (4016), 568 (1946).

[580] TELFORD, H. S., *DDT Toxicity*, Soap, N.Y. *21* (12), 161, 163, 167, 169 (1945).

[581] TELFORD, H. S., and GUTHRIE, J. E., *Transmission of the Toxicity of DDT Through the Milk of White Rats and Goats*, Science *102* (2660), 647 (1945).

[582] TELFORD, H. S., and GUTHRIE, J. E., *Effect of Oral Dosages of DDT on Certain Vertebrates*, J. econ. Ent. *39* (3), 413 (1946).

[583] TELFORD, H. S., and GUTHRIE, J. E., *Toxicity of DDT Sprays to Livestock*, Soap, N.Y. *22* (9), 124 (1946).

[584] THIEMANN, H. A., *New Insecticides and Rodenticides and Their Health Aspects*, Amer. Ind. Hyg. Ass. Quart. *10* (1), 10–15 (1949).

[585] THOMAS, J. W., HUBANKS, P. E., CARTER, R. H., and MOORE, L. A., *Feeding DDT and Alfalfa Sprayed with DDT to Calves*, J. Dairy Sci. *34* (3), 203–208 (1951).

[586] THOUNG, C. U., *Poisonous Effects of DDT on Humans*, Indian med. Gaz. *81* (10), 432 (1946).

[587] TILLER, R. E., and CORY, E. N., *Effects of DDT on Some Tidewater Aquatic Animals*, J. econ. Ent. *40* (3), 431 (1947).

[588] TOBIAS, J. M., KOLLROS, J. J., and SAVIT, J., *Relation of the Absorbability to the Comparative Toxicity of DDT for Insects and Mammals*, J. Pharmacol. *86*, 287–293 (1946).

[589] TOBIAS, J. M., KOLLROS, J. J., and SAVIT, J., *Acetylcholine and Related Substances in the Cockroach, Fly and Crayfish and the Effect of DDT*, J. cell. comp. Physiol. *28* (2), 159–182 (1946).

[590] TOMAN, J. E. P., and DAVIS, J. P., *The Effects of Drugs Upon the Electrical Activity of the Brain*, J. Pharmacol. *97* (4), 425–492 (1949).

[591] TORDA, C., and WOLFF, H. G., *Effect of Convulsant and Anticonvulsant Agents on the Activity of Carbonic Anhydrase*, J. Pharmacol. *95* (4), 444–447 (1949).

[592] TORDA, C., and WOLFF, H. G., *Effect on Convulsant and Anticonvulsant Agents on the Activity of Oxalacetic and Pyruvic Carboxylase*, J. Pharmacol. *98* (4), 358–365 (1950).

[593] TORDA, C., and WOLFF, H. G., *Effect of Convulsant and Anticonvulsant Agents on the Activity of Cytochrome Oxidase*, Proc. Soc. exp. Biol., N.Y. *74*, 744–746 (1950).

[594] TORDA, C., and WOLFF, H. G., *Effect of Convulsant and Anticonvulsant Agents on the Activity of Succinic Dehydrogenase*, Fed. Proc. *10* (1), Part 1 (1951).

[595] TREON, J. F., Personal communication (1953).

[596] TRESSLER, C. J., jr., *Recovery of DDT from Canned Foods and Its Stability During Processing*, J. Ass. off. agric. Chem. *30* (1), 140–144 (1947).

[597] TRIPOD, J., *Le point d'attaque du DDT (4,4'-dichlor-diphényl-trichloroéthane) chez la grenouille*, Arch. int. Pharmacodyn. *74* (3–4), 343–363 (1947).

[598] TRUHAUT, R., *Les dérivés organiques halogénés doués d'activité insecticide* (Sedes 99, Boulevard Saint-Michel, Paris Ve, 1948).

[599] TRUHAUT, R., and VINCENT, D., *Sur le mécanisme d'action toxique du D.D.T. D.D.T. et cholinestérase du sérum*, Ann. pharm. franç. *5* (3), 159–163 (1947).

[600] TRUHAUT, R., and VINCENT, D., *Contribution à l'étude du mécanisme de l'action toxique de l'insecticide D.D.T. D.D.T. et systèmes cholinestérasiques chez les animaux supérieurs et les insectes*, Bull. Soc. Chim. biol., Paris *30* (9–10), 694–699 (1948).

[601] TURNER, K., *The Fatty Acids of the Cat's Kidney*, Part I, Biochem. J. *25*, 49–56 (1931).

[602] UPHOLT, W. M., *The Inactivation of DDT Used in Anopheline Mosquito Larvicides*, U.S. publ. Hlth. Repts. *62* (9), 302–309 (1947).

[603] U.S. CHEMICAL WARFARE SERVICE (INFORMATION SERVICE), *Localization of the Site of Action of DDT*, C.W.S. Monthly Progress Report on Insect and Rodent Control No. 2, 69–70 (1945).

[604] U.S. CONGRESS, *Select Committee to Investigate the Use of Chemicals, Compounds, etc., in the Production of Food Products*, House Resolution 323, Congr. Record *69* (121), 9060–9063 (1950).

[605] U.S. CONGRESS. HOUSE, *Hearings Before the House Select Committee to Investigate the Use of Chemicals in Foods and Cosmetics*, 81st–82nd Sessions, 5 pts. (1950–1952).

[606] U.S. CONGRESS. HOUSE, *Investigation of the Use of Chemicals in Food and Cosmetics*, 82nd Congress, 2nd Session, Rept. 2356, pats. 1 and 2 (1952).

[607] U.S. CONGRESS. HOUSE, *Amending the Federal Food, Drug, and Cosmetic Act with Respect to Residues of Pesticide Chemicals in or on Raw Agricultural Commodities*, 83rd Congress, 2nd Session, Rept. 1385 (1954).

[608] U.S. DEFENSE DEPARTMENT, Joint Army-Navy specification JAN–I–180, *Insecticide, Powder, Louse*, Washington, Govt. (January 31, 1945).

[609] U.S. DEPARTMENT OF AGRICULTURE, AGRICULTURAL RESEARCH ADMINISTRATION, *U.S.D.A. Entomologists Recommend Substitute Insecticide for DDT to Control Insects on Dairy Cattle and in Dairy Barns*, News release (March 24, 1949).

[610] U.S. DEPARTMENT OF AGRICULTURE, BUREAU OF ENTOMOLOGY AND PLANT QUARANTINE, Release (February 13, 1945).

[611] U.S. DEPARTMENT OF AGRICULTURE, BUREAU OF ENTOMOLOGY AND PLANT QUARANTINE, *Outlook for Supplies of DDT Insecticides*, Release (August 6, 1945).

[612] U.S. DEPARTMENT OF AGRICULTURE, BUREAU OF ENTOMOLOGY AND PLANT QUARANTINE, *Insect Control in the Country Wheat Elevator*, U.S. Dept. Agr., Bur. Ent. Plant Quarantine, EC–24 (1952).

[613] U.S. DEPARTMENT OF AGRICULTURE, BUREAU OF ENTOMOLOGY AND PLANT QUARANTINE, *Insects in Farm-Stored Wheat. How to Control Them*, U.S. Dept. Agr. Leaflet No. 345 (1953).

[614] U.S. DEPARTMENT OF AGRICULTURE, PRODUCTION AND MARKETING ADMINISTRATION, *Regulations for the Enforcement of the Federal Insecticide, Fungicide, and Rodenticide Act*, Service and Regulatory Announcement No. 166 (1948).

[615] U.S. DEPARTMENT OF AGRICULTURE, PRODUCTION AND MARKETING ADMINISTRATION, *Interpretations of the Regulations for the Enforcement of the Federal Insecticide, Fungicide, and Rodenticide Act* (Title 7, Ch. I, Pt. 162 of the Code of Federal Regulations), Service and Regulatory Announcement No. 167, Revised (1950).

[616] U.S. FEDERAL SECURITY AGENCY, *Federal Food, Drug, and Cosmetic Act and General Regulations for Its Enforcement*, Service and Regulatory Announcements, Food and Cosmetic No. 1, Rev. No. 3, 57 p. (1949).

[617] U.S. FEDERAL SECURITY AGENCY, *FSA 551*, Newspaper (April 2, 1949).

[618] U.S. FEDERAL SECURITY AGENCY, *Tolerances for Poisonous or Deleterious Residues on or in Fresh Fruits and Vegetables*, Federal Register *14* (180), 5724 (1949).

[619] U.S. FEDERAL SECURITY AGENCY, *Tolerances for Poisonous or Deleterious Residues on or in Fresh Fruits and Vegetables*, Federal Register *14* (250), 7810 (1949).

[620] U.S. FEDERAL SECURITY AGENCY, *The Federal Caustic Poison Act, Regulations for Its Enforcement, and Antidotes for Caustic and Corrosive Substances*, Service and Regulatory Announcements, Caustic Poison No. 1, 2nd rev., 20 p. (1952).

[621] U. S. Food and Drug Administration, *Statement on DDT*, Release (September 5, 1945).

[622] U. S. Food and Drug Administration, *F.D. & C. Act Trade Correspondence*, Release (November 5, 1945).

[623] U. S. Food and Drug Administration, *Annual Report of the Federal Security Agency...*, *1949*, pp. 6, 11, 44 (1950).

[624] U. S. Food and Drug Administration, *DDT in Fluid Milk and Certain Other Staple Foods*, Mimeo. Rept. (August 16, 1951).

[625] U. S. Laws, Statutes, etc., *An Act to Safeguard the Distribution and Sale of Certain Dangerous Caustic or Corrosive Acids, Alkalies, and Other Substances in Interstate and Foreign Commerce*, approved March 4, 1927, 44 Stat. 1406–1410 c. 489, Publ. No. 783, 69th Congress.

[626] U. S. Laws, Statutes, etc., *Federal Food, Drug, and Cosmetic Act*, approved June 25, 1938, 52 Stat. 1040–1050 c. 657, Publ. No. 717, 75th Congress.

[627] U. S. Laws, Statutes, etc., *Federal Insecticide, Fungicide, and Rodenticide Act*, approved June 25, 1947, 61 Stat. 163–173 c. 125, Publ. No. 104, 80th Congress.

[628] U. S. Laws, Statutes, etc., *Explosives and Combustibles*, Chapt. 39, Title 18, U. S. Code, approved June 25, 1948, 62 Stat. 738–739: 18 U. S. C. 831–835; Public Law 772, 'Sec. 831–835', 80th Congress.

[629] U. S. Public Health Service, Communicable Disease Center, *Rat-Borne Disease Prevention and Control*, Federal Security Agency, Atlanta, Georgia, p. 263 (1949).

[630] Vaz, Z., Pereira, R. S., and Malheiro, D. M., *Calcium in Prevention and Treatment of Experimental DDT Poisoning*, Science *101* (2626), 434–436 (1945).

[631] Velbinger, H. H., *Zur Frage der 'DDT'-Toxizität für Menschen*, Dtsch. Gesundheitsw. *2* (11), 355–358 (1947).

[632] Velbinger, H. H., *Beitrag zur Toxikologie des 'DDT'-Wirkstoffes Dichlor-diphenyl-trichlormethylmethan*, Pharmazie *2* (6), 268–274 (1947).

[633] Velbinger, H. H., *Über die unterschiedliche Wirkung der neuzeitlichen 'Insektizide' DDT, Gammexan und E 605*, Pharmazie *4* (4), 165–176 (1949).

[634] Vincent, D., and Truhaut, R., *Contribution à l'étude du mécanisme de l'action physiologique de l'insecticide D.D.T. D.D.T. et cholinesterase du serum*, C. R. Soc. biol. *141*, 65–66 (1947).

[635] Vincent, D., Truhaut, R., and Abadie, A., *Contribution à l'étude du mécanisme de l'action toxique de l'insecticide D.D.T. D.D.T. et systèmes cholinestérasiques chez les animaux supérieurs et les insectes*, C. R. Soc. biol. *142*, 1500–1502 (1948).

[636] Vinson, E. B., and Arant, F. S., *Parathion, Toxaphene and DDT Residues on Peanut Hay*, J. econ. Ent. *43* (6), 942–943 (1950).

[637] Virgili, R., and Marchiafava, G., *Quadri anatomo-patologici nell'intossicazione acuta sperimentale da DDT con particolare riferimento al sistema nervoso*, Riv. Malariol. *28* (2), 107–124 (1949).

[638] von Oettingen, W. F., *Toxicity and Potential Dangers of Aliphatic and Aromatic Hydrocarbons*, U. S. publ. Hlth. Bull. *255*, 55–57 (1940).

[639] von Oettingen, W. F., *Experimental Hepatic Injury Due to Chemicals*, Trans. Conf. Liver Injury 2, 64–67 (1944).

[640] von Oettingen, W. F., *Poisoning, a Guide to Clinical Diagnosis and Treatment* (Paul B. Hoeber, Inc., New York 1952), 524 p.

[641] von Oettingen, W. F., and Sharpless, N. E., *The Toxicity and Toxic Manifestations of 2,2-Bis-(p-chlorophenyl)-1,1,1-trichloroethane (DDT) as Influenced by Chemical Changes in the Molecule. A Contribution to the Relation Between Chemical Constitution and Toxicological Action*, J. Pharmacol. *88* (4), 400–413 (1946).

[642] Walker, K. C., *DDT Residue and Its Removal from Apples and Pears*, Proc. Amer. Soc. hort. Sci. *51*, 85–89 (1948).

[643] Walker, K. C., *Problems Relating to the Removal of DDT Spray Residue from Apples*, J. agric. Res. *78* (10), 383–387 (1949).

[644] Walker, K. C., Goette, M. B., and Batchelor, G. S., *Dichlorodiphenyltrichloro-*

ethane and Dichlorodiphenyldichloroethylene Content of Prepared Meals, J. Agric. Food Chem. *2* (20), 1034–1037 (1954).

[645] WANSON, M., and CAMPHYN, R., *Propriétés raticides du D.D.T.*, Ann. Soc. belge Méd. trop. *29* (4), 549–556 (1949).

[646] WARING, J. I., *Pneumonia in Kerosene Poisoning*, Amer. J. med. Sci. *185* (3), 325–330 (1933).

[647] WASICKY, R., *Modernos insecticidas–toxicidade para o homem e metodos de dosagem*, Ann. Fac. Farm. Odontol., S. Paulo *7*, 263–296 (1948–1949).

[648] WASICKY, R., and UNTI, O., *Dicloro-difenil-tricloro-etano (DDT) no combate às larvas de Culicídeos*, Arch. Hig., S. Paulo *9* (21), 87–102 (1944).

[649] WASICKY, R., and UNTI, O., *O DDT na Profilaxia de malária e no combate a alguns insetos de Interêsse Sanitário*, Rev. XXV Janeiro *1944* (Nov.-Dez.), 12 p.

[650] WASICKY, R., and UNTI, O., *Dicloro-difenil-tricloro-etano (DDT). Ulteriores pesquisas sôbre as suas propriedades e aplicações*, Arch. Hig., S. Paulo *10* (23), 49–64 (1945).

[651] WASICKY, R., and UNTI, O., *Cinco anos de observações sôbre a toxidez do DDT*, Ann. Fac. Farm. Odontol., S. Paulo *7*, 49–56 (1951).

[652] WASHINGTON STATE COLLEGE, *Control of Codling Moth, Orchard Mite, Aphid and Scale in Eastern Washington*, Wash. Ext. Bull. *1947*, 279.

[653] WATERHOUSE, D. F., and SCOTT, M. T., *Insectary Tests with Insecticides to Protect Sheep Against Body Strike*, Aust. J. agric. Res. *1* (4), 440–455 (1950).

[654] WAY, M. J., and SYNGE, A. D., *The Effects of D.D.T. and of Benzene Hexachloride on Bees*, Ann. appl. Biol. *35* (1), 94–109 (1948).

[655] WEBSTER, R. L., *New Insecticides: Their Use, Limitations and Hazard to Human Health*, Wash. agr. exp. Sta. Circ. No. 64 (3d rev.), 51 p. (1950).

[656] WELCH, H., *Test of the Toxicity to Sheep and Cattle of Certain of the Newer Insecticides*, J. econ. Ent. *41* (1), 36–39 (1948).

[657] WELSH, J. H., and GORDON, H. T., *The Mode of Action of DDT*, Fed. Proc. *5* (1), 1 p. (1946).

[658] WELSH, J. H., and GORDON, H. T., *The Mode of Action of Certain Insecticides on the Arthropod Nerve Axon*, J. cell. comp. Physiol. *30* (2), 147–172 (1947).

[659] WEST, T. F., and CAMPBELL, G. A., *DDT, the Synthetic Insecticide* (Chapman & Hall, Ltd., London 1946), 301 p.

[660] WEST, T. F., and CAMPBELL, G. A., *DDT and Newer Persistent Insecticides* (Chem. Publ. Co., New York 1952), 632 p.

[661] WESTLAKE, W. E., and FAHEY, J. E., *DDT and Parathion Spray Residues on Apples*, Adv. Chem. Ser. *1*, 117–122 (1950).

[662] WHEELER, E. H., and LAPLANTE, A. A., jr., *DDT and Ryanex to Control Oriental Fruit Moth: Their Effect Upon Parasite Populations*, J. econ. Ent. *39* (2), 211–215 (1946).

[663] WHITE, W. C., and SWEENEY, T. R., *The Metabolism of 2,2-Bis-(p-chlorophenyl)-1,1,1-trichloroethane (DDT). I. A Metabolite from Rabbit Urine Di-(p-chlorophenyl)-Acetic Acid; Its Isolation, Identification, and Synthesis*, U. S. publ. Hlth. Repts. *60* (3), 66–71 (1945).

[664] WIESMANN, R., *Über neue, wirksame Arsenersatzstoffe im Obstbau und ihre Bedeutung für die Bienenzucht*, Schweiz. Bienenztg. *65* (5), 227–229 (1942).

[665] WIGGLESWORTH, V. B., *A Case of D.D.T. Poisoning in Man*, Brit. med. J. I, 517 (1945).

[666] WILSON, H. F., *Will DDT Residues be Dangerous?*, Food Pack. *27* (4), 63–64 (1946).

[667] WILSON, H. F., ALLEN, N. N., BOHSTEDT, G., BETHEIL, J., and LARDY, H. A., *Feeding Experiments with DDT-Treated Pea Vine Silage with Special Reference to Dairy Cows, Sheep, and Laboratory Animals*, J. econ. Ent. *39* (6), 801–806 (1946).

[668] WILSON, H. F., SRIVASTAVA, A. S., HULL, W. B., BETHEIL, J., and LARDY, H. A., *DDT Residues on Pea Vines and Canned Peas from Fields Treated with DDT Dusts*, J. econ. Ent. *39* (6), 806–809 (1946).

[669] WILSON, R. H., and DE EDS, F., *Importance of Diet in Studies of Chronic Toxicity*, Arch. industr. Hyg. *1*, 73–80 (1950).

[670] WILSON, S. G., *The Feeding of 'Gammexane' and DDT to Bovines*, Bull. ent. Res. *39* (3), 423–434 (1949).

[671] WINGO, C. W., and CRISLER, O. S., *Effect of DDT on Dairy Cattle and Milk*, J. econ. Ent. *41* (1), 105–106 (1948).

[672] WINTERINGHAM, F. P. W., HARRISON, A., JONES, C. R., McGIRR, J. L., and TEMPLETON, W. H., *The Fate of Labeled Insecticide Residues in Food Products. I. Studies with a Radioactive Bromine Analogue of DDT*, J. Sci. Food Agr. *1* (7), 214–219 (1950).

[673] WITT, P. N., *Ein Test zur Prüfung der Wirksamkeit insektizider Substanzen und ein Beitrag zum Mechanismus der Wirkung von DDT und HCC*, Z. Naturf. [B] *2*, 361–366 (1947).

[674] WITTICH, F. W., *Respiratory Tract Allergic Effects from Chemical Air Pollution*, Arch. industr. Hyg. *2* (3), 329–334 (1950).

[675] WOLFENBARGER, D. O., *DDT for 'Out of Place' Honeybee Colonies*, J. econ. Ent. *37* (6), 849–850 (1944).

[676] WOODARD, G., and OFNER, R., *Accumulation of DDT in the Fat of Rats in Relation to Dietary Level and Length of Feeding*, Fed. Proc. *5*, 215 (1946).

[677] WOODARD, G., NELSON, A. A., and CALVERY, H. O., *Acute and Subacute Toxicity of DDT (2,2-Bis-[p-chlorophenyl]-1,1,1-trichloroethane) to Laboratory Animals*, J. Pharmacol. *82* (2), 152–158 (1944).

[678] WOODARD, G., OFNER, R. R., and MONTGOMERY, C. M., *Accumulation of DDT in the Body Fat and Its Appearance in the Milk of Dogs*, Science *102*, 177–178 (1945).

[679] WOODARD, G., DAVIDOW, B., and LEHMAN, A. J., *Metabolism of Chlorinated Hydrocarbon Insecticides*, Industr. Engng. Chem. *40* (4), 711–712 (1948).

[680] WOODARD, G., DAVIDOW, B., and NELSON, A. A., *Effects Observed in Dogs Following the Prolonged Feeding of DDT and Its Analogues*, Fed. Proc. *7*, 266–267 (1948).

[681] WORLD HEALTH ORGANIZATION, *Critical Shortage of Insecticides for Public Health Purposes. Report of the Working Party on Insecticides DDT and BHC Under Council Resolution 377 (XIII)*, United Nations Economic and Social Council, fourteenth session, item 8 (March 10, 1952).

[682] WRIGHT, C. S., DOAN, C. A., and HAYNIE, H. C., *Agranulocytosis Occurring After Exposure to a DDT Pyrethrum Aerosol Bomb*, Amer. J. Med. *1* (5), 562–568 (1946).

[683] YETTER, W. P., jr., *New Insecticides for Control of the Oriental Fruit Moth*, J. econ. Ent. *40* (2), 274–275 (1947).

[684] YOUNG, L. A., and NICHOLSON, H. P., *Stream Pollution Resulting from the Use of Organic Insecticides*, Progr. Fish Cult. *13* (4), 193–198 (1951).

[685] ZEIN-EL-DINE, K., *The Insecticide DDT*, J. Egypt. med. Ass. *29* (1/2), 38–54 (1946).

VII

THE USE OF DDT INSECTICIDES
IN HUMAN MEDICINE

BY

SAMUEL W. SIMMONS

1.

INTRODUCTION

The insecticide DDT (2,2,bis-[*p*-Chlorophenyl]-1,1,1-trichloroethane) ranks in the top bracket of important developments that have occurred in the control of communicable diseases. This can be appreciated by the fact that approximately 30 diseases, some of them the most important in the world, can be directly controlled, either partially or completely, by the use of DDT. These diseases are as follows:

Malaria	Dysentery and diarrhea	Cholera
Filariasis	Shigellosis	Chagas' disease
Dengue	Amebiasis	Scrub typhus (tsutsugamushi)
Urban yellow fever	Leishmaniasis	Scabies
Virus encephalitis	Bartonellosis	Rickettsialpox
Louse-borne typhus	Onchocerciasis	Tick-borne diseases
Louse-borne relapsing fever	Sandfly fever	Relapsing fever
Trench fever	Trypanosomiasis	Rocky Mountain
Plague	Yaws	spotted fever
Murine typhus	Infectious conjunctivitis	Tularemia

The efficacy of DDT in controlling these diseases varies widely. It is theoretically and, in some instances, actually, possible to eliminate from large areas some of these diseases, such as malaria, yellow fever, dengue, filariasis, louse-borne typhus, and louse-borne relapsing fever by the use of DDT. This is because these diseases are transmitted only by arthropod vectors, which are susceptible to DDT, and which, with few exceptions, are environmentally vulnerable to attack. Other diseases, such as dysentery, diarrhea, and yaws, can be controlled to a certain extent by DDT; these cannot be completely controlled because their dissemination is only partially dependent upon arthropod vectors. The practical difficulty of controlling the vectors of trypanosomiasis, tularemia, Rocky Mountain spotted fever, tick-borne relapsing fever, and other tick-borne infections, is sufficiently difficult to limit the value of insecticides for this purpose. Other diseases listed fall into various intermediate categories.

Until the advent of resistance to DDT among certain arthropod disease vectors, this insecticide was a virtual panacea for the control of many important tropical diseases. Irrespective of the resistance problem, however, DDT and its offspring, the members of the chlorinated hydrocarbon family, are still the principal weapons in the global fight against malaria, typhus, plague, and a host of other arthropod-borne diseases. As an example, India had under way (1954) one of the largest malaria control programs which had ever been attempted. During 1953, over 14 million lb of 75 per cent DDT wettable powder was shipped to India, at a cost of $ 4,750,000. This was sufficient to protect 125 million people against malaria. During 1953 alone, the homes of some 70 million Indian people were treated (PRICE [455][1])).

---

[1]) Numbers in brackets refer to References, page 475.

The total value of DDT to mankind is inestimable, and is comprised of health, economic, and social benefits. Health benefits are both direct and indirect and fall into three principal categories: (1) direct control of vector-borne diseases such as malaria, typhus, etc., and of insect pests, by the use of DDT for destroying the insects concerned; (2) use of DDT in agriculture for crop pest control, resulting in an increased food supply, often where malnutrition is the principal health problem; and (3) an increase in resistance to non-vector-borne diseases through better health as a result of freedom from malaria and other vector-borne diseases and malnutrition. As Sir MALCOLM WATSON has stated, when malaria was cleared out of the Malay States, dysentery wards were closed (SOPER [538]). If data were available, it is predicted that they would show a decrease in tuberculosis and some other diseases in areas where malaria and malnutrition, particularly, have been largely conquered by the use of DDT.

Raising the health standards of the people has resulted in better agricultural and industrial production. Much land has been reclaimed, new factories have been built, and more goods have been made available for sale as well as home consumption. There has been a significant decrease in absentee workers in countries where malaria has been controlled, and this has enabled higher earnings, with an increase in economic status for both individuals and the community.

The sociological changes brought about by DDT are only beginning to be apparent. In some countries, Madagascar for example, the population has doubled since 1947, although it had been practically stationary for years previously. A DDT malaria campaign was initiated in Madagascar in 1949, and is largely credited with the population increase. This is no isolated phenomenon, and the full sociological signifiance of this trend will be much greater than is at present anticipated.

It was estimated in 1953 that no less than 5 million lives had been saved and no less than 100 million illnesses prevented through the use of DDT for controlling malaria, typhus, dysentery, and other arthropod-borne diseases, since it became available about 1942 (KNIPLING [312]). These figures have now increased, the latter one especially. In a complete treatise on the value of DDT to the public health, it would be necessary to consider the entire group of chlorinated hydrocarbon insecticides in order to give full credit to this parent compound. The extent of the information on all chlorinated hydrocarbon insecticides is such, however, that it would not be feasible to include in a work of this nature. It should not be forgotten, nevertheless, that insecticides such as chlordane, dieldrin, aldrin, and toxaphene were developed as a result of the development of DDT, and no doubt other effective materials will be forthcoming.

Except for the antibiotics, it is doubtful that any material has been found which protects more people against more diseases over a larger area than does DDT. Most of the peoples of the globe have received some measure of benefit from this compound, either directly by protection from infectious diseases and pestiferous insects, or indirectly by better nutrition, cleaner food, and increased disease resistance. Irrespective of future developments, the discovery of DDT will always remain an historic event in the fields of public health and agriculture.

2.

MOSQUITO-BORNE DISEASES

Mosquitoes are probably responsible for the dissemination of more diseases than any other group of arthropods. No less than a dozen important diseases including malaria, yellow fever, encephalitis, filariasis, and dengue fever are transmitted by mosquitoes. With the advent of DDT, major emphasis was logically placed on tests designed to determine the efficacy of the new compound in mosquito control. These efforts are reflected in thousands of publications on the subject, of which only a selected portion could possibly be reviewed in the present work.

The first test of DDT on mosquitoes was apparently conducted by VON EMMEL [615]. Working in a Swiss laboratory with ROSE, VON EMMEL exposed *Anopheles maculipennis* mosquitoes to DDT and described in some detail the locus of action, physiological response, and subsequent death of the exposed specimens. MANDEKOS [357] treated walls of stables with 0·05 g of DDT per square meter, which resulted in freedom from *Anopheles* for one to five weeks. Increasing the dosage of DDT did not increase the period of protection from *Anopheles*, but it is not stated how much the dose was increased. ROSE [478] further demonstrated the residual qualities of DDT when he found that a one per cent spray applied in a room rendered it mosquito-free for a period of two weeks. When a two per cent spray was used, the period of protection was increased to three or four weeks.

GAHAN *et al.* [205] initiated tests in April 1943 on the effect of DDT residues on mosquitoes using *A. quadrimaculatus* and *Aedes aegypti*. This work was done at the Orlando, Florida laboratory of the Bureau of Entomology and Plant Quarantine, of the United States Department of Agriculture, where DDT was first tested in the United States. Wooden boxes and canvas cages were sprayed with various formulations of DDT in both solutions and water emulsions at the rate of 50 mg of DDT per square foot. At the end of 16 weeks, the treatments were still completely effective. Increasing the doses to 400 mg slightly decreased the knockdown period.

In August 1943, GAHAN *et al.* [205] extended their tests to the field. Ten buildings were treated in the vicinity of Tallahassee, Florida, where the anopheline population ranged from 50 to 300 specimens per building prior to treatment. Both kerosene solutions and water emulsions of DDT were applied to give a deposit which ranged from 68 to 225 mg of DDT per square foot. In two of the four buildings where examinations were made in the afternoons, live mosquitoes were not seen for a 70-day period subsequent to spraying. A few live specimens were encountered at various intervals in other buildings but the numbers were very small compared with untreated buildings.

Following these small-scale field tests, GAHAN and LINDQUIST [203] conducted a larger scale test to determine the operational effectiveness and duration of DDT residual sprays when used against *A. quadrimaculatus*. This work was undertaken at Stuttgart, Arkansas, in the summer of 1944. DDT spray was applied to the inside walls and ceilings of practically all buildings in two 9-square-mile areas. One area was treated at the rate of 56 mg of DDT per square foot and the other at the rate of 208 mg. Results were determined by comparing the number of *A. quadrimaculatus* mosquitoes found in the treated buildings with the number in buildings in a comparable untreated area. The heavy treatments reduced the number of mosquitoes in the buildings by over 99%, and the light treatments by 91%. Larvae counts made in rice fields revealed a reduction of 63 and 57% respectively, in the heavily and lightly treated areas. These results definitely indicated the feasibility of utilizing DDT residual sprays for the control of malaria.

In April 1944, work was initiated at The Henry R. Carter Memorial Laboratory of the United States Public Health Service in Savannah, Georgia, to determine, among other things, operational techniques and practical limitations for the large-scale use of DDT for malaria control in the southern United States.

A critical evaluation of the net mortality of *A. quadrimaculatus* mosquitoes was made by liberating specimens in occupied and unoccupied rooms treated with 200 mg of DDT per square foot. Released mosquitoes could escape from the room into window traps where they where collected and held for mortality determination (SIMMONS and staff [521]). Complete kills were obtained up to 24 weeks after treatment and good kills were obtained for a period of more than 11 months. Mortality was found to be less in occupied dwellings due mainly to the large untreated surfaces presented by furniture and other household effects (TARZWELL and STIERLI [574]). For anopheline control, a dosage of 200 mg of DDT applied as a xylene-water emulsion to all surfaces practical within a building was recommended (SIMMONS [518]).

All of the three principal types of DDT formulations (emulsions, solutions, and wettable powders) are employed in mosquito control. The formulation of choice will depend upon availability of materials and types of surfaces to be treated. Large-scale residual treatment of premises in the United States has been mainly with 5 per cent DDT-xylene-water emulsions prepared from a 25 per cent DDT-xylene concentrate containing an emulsifier. A 5 per cent concentration of DDT in kerosene has been extensively used throughout the world and, in general, little difference in effectiveness has been noted with such solutions and water emulsion sprays utilizing various aromatic hydrocarbon solvents. Oil sprays, however, are more uncomfortable to use, may cause skin burning, and present a greater toxic and fire hazard than the emulsions. DDT wettable powders give best results on porous surfaces such as adobe walls. In fact, emulsions and solutions are absorbed so rapidly by most adobe surfaces that they have relatively little residual qualities. Wettable powders produce a more unsightly residue, however, and are not usually tolerated in better homes. The wettable powders also appear to be more susceptible to being removed from outside surfaces by rain than are emulsions or solutions. In arid climates, however, wettable powders are satisfactory for outside residual use. Emulsions and solutions may be used with most knapsack type compressed air sprayers but for the effective application of wettable powders, sprayers have to be equipped with special strainers to prevent clogging of the nozzle.

Species of mosquitoes undoubtedly vary in their susceptibility to intoxication by DDT, but apparently all species will succumb to appropriate exposure unless they have acquired resistance. The effectiveness of DDT for the control of mosquito vectors of disease also varies considerably due to differences in habits of individual species. Endophilic species, particularly those which are also anthropophilic, are ordinarily easier to control since they frequent human habitations which can be readily treated with residual sprays. Therefore, the relative effectiveness of DDT in the control of different mosquitoes depends to a great extent upon the use of techniques which are devised to take advantage of the characteristic behavior of the species against which control is directed. For example, knowledge of the resting habits of mosquitoes permits application of residual spray where a large portion of the mosquito population will be affected. Work with *A. quadrimaculatus* in the United States showed that the degree of control was largely dependent upon the completeness of coverage by the residual spray. Spot treatment in dwellings was not satisfactory, since this species shows no decided preference for resting surfaces within a suitable shelter. Effective reduction of *A. quadrimaculatus* depended upon thorough treatment, including areas behind furniture and along baseboards, as well as open walls and ceilings (McCauley *et al.* [368]).

A great deal of work has been done on dosage of residual sprays for mosquito control, particularly for control of malaria vectors. In most instances, an effective dosage is 200 mg of technical grade DDT per square foot of surface area, roughly equivalent to 2 g/m². The proper dosage, however, depends on many factors including the nature of the surface treated, residual duration desired, meteorological conditions, and the economics of malaria control. In countries where material is the principal cost involved, and labor is abundant and cheap, application of small dosages at short intervals may be advisable. If funds are available to purchase enough DDT to control adequately 10 million cases of malaria but there are 40 million cases in the country, a hypothetical alternative is to apply two 200 mg treatments per year (400 mg) with essentially 100 per cent control of 10 million cases or to apply two 50 mg treatments for the control of 40 million cases with a theoretical 80 per cent suppression of transmission. Naturally, the difference in labor costs and in duration of effectiveness does not make the choice this simple. In areas where labor is the principal cost, the tendency is toward the use of larger dosages of

DDT applied over a longer time interval. Oftentimes, one treatment per season of 2 g/m² will suffice (SIMMONS [515]).

It has been calculated that, regardless of individual dosage and treatment interval, most treatments represent about 0·33 g of DDT per square meter for each month of effective action desired (MACDONALD and DAVIDSON [347]).

The method of estimating results of residual treatments has been based mainly on the presence or absence of mosquitoes from treated buildings. Obviously, this is not an accurate evaluation of mosquito mortality, but it has proved to be a rather practical procedure. Even if mosquitoes are not killed but are driven from treated residences, there is a considerable degree of protection from biting and consequent transmission of disease. Investigations have shown that all mosquitoes which enter treated buildings do not succumb. A small percentage is able to obtain a blood meal, escape, and survive. The number that escape, among other things, depends upon the age of the insecticide and the resting habits of the vector species involved. Work by the U. S. Public Health Service on *A. quadrimaculatus* showed that about 15 % of specimens which enter treated buildings are able to escape and live a sufficient time for further feeding (QUARTERMAN [459]).

DDT is of course toxic to immature as well as to mature mosquitoes and its use as a larvicide has been very effective. Early experiments by MANDEKOS [357] portrayed the potential value of the larvicidal qualities of this chemical. In the United States, early work by the U. S. Department of Agriculture proved the effectiveness of DDT as an anopheline larvicide (WISECUP *et al.* [635]; EIDE *et al.* [167]).

Subsequent experiments by the U. S. Public Health Service showed that DDT oil mists were very effective and economical as anopheline larvicides (FERGUSON *et al.* [181]). The treatment employs 0·05 lb of DDT per acre on a routine basis where fish life is of importance. The recommended formula consists of 0·625 per cent DDT and 0·5 per cent emulsifier in No. 2 fuel oil applied at the rate of 1 gal of solution per acre at intervals of about ten days. The material is equally effective against all larval instars. The small amount of material distributed results in less fatigue to the applicator and is usually more economical than where larger quantities of larvicides are used. DDT can also be used quite successfully as a residual larvicide in areas where fish and other aquatic life are of no importance. When used in this manner DDT is applied at the rate of 3 lb/acre (2 gal of finished oil spray) but it should never be employed where aquatic life must be preserved.

The relative value of DDT as a larvicide and as an adulticide depends upon the specific field conditions. Where the human population is dense and the breeding area of mosquitoes is relatively small, larviciding may be the method of choice. Oftentimes, the treatment of breeding areas adjacent to municipalities is more practical than trying to treat all the residences. Conversely, when the human population is relatively sparse, such as in rural areas and in small villages, and where the breeding places of mosquitoes are extensive, the most economical and effective method of controlling mosquitoes, particularly malaria vectors, is residual treatment of residences. In some instances combined use of larvicides and adulticides is indicated, particularly where an epidemic threatens or is in progress. Ordinarily, however, it is best not to use DDT for both purposes since there is a tendency for the exposed mosquito population to build up resistance more rapidly if control is directed against both larvae and adults.

Malaria

Europe

Italy

'Malaria' and 'Italy' have been companion words for centuries. Italy has been preeminent in its malaria morbidity and mortality in the Western world since the dawn of history. It was in Italy that the disease obtained its name 'malaria', from the words 'mala' (bad), and 'aria' (air). Together these became 'malaria', and the bad air of Rome was associated with the fever for centuries prior to the identification of the disease. VARRO (116–27 B.C.) in his *Rerum Rusticarum* (Lib. I) suggested that the marshes were conducive to the growth of small animals that the eye could not see, but which entered into the nostrils and caused fevers (RUSSELL [485]).

The prevalence of malaria in Italy made it a great natural laboratory for the study of the disease, and many now famous malariologists studied and worked there. The sequence of events leading to the elucidation of the etiology, transmission, epidemiology, and control of malaria carry with it the names of many Italian malariologists. GIOVANNI LANCISI in 1717 advocated the drainage of marshes for malaria control. Following LAVERAN's recognition of the malaria plasmodia in 1880, ETTORE MARCHIAFAVA and ANGELO CELLI in 1883–4 described unpigmented parasites within red corpuscles and found that they multiplied by fission. CAMILLO GOLGI in 1885–6 demonstrated the relation between the cyclical development of the parasite and the periodical succession of febrile paroxysms, and differentiated between the quartan and benign tertian plasmodia, on a morphologic basis. GIOVANNI BATTISTA GRASSI in 1898 allowed mosquitoes from a malarious area to feed on a subject who subsequently contracted the disease. BASTIANELLI, BIGNAMI, AND GRASSI demonstrated the development of a malaria parasite in the gut wall of *A. claviger* in 1898. The first cinchona bark in Europe was brought to Rome by Father ALONSO MESSIAS VENEGAS in 1632, and TORTI observed its effects on malaria. Many other natives contributed to the elucidation of the disease. Among them were CELLI, who observed phagocytosis of plasmodia by white blood cells, and TOMASELLI, who demonstrated the relationship between haemoglobinuria and the taking of quinine (RUSSELL [485]). Probably the first society for the study of malaria was founded in Italy in 1898 (ATTI SOCIETA PER GLI STUDI

DELLA MALARIA). In more recent years, MISSIROLI has carried forward the work of his eminent predecessors, and was responsible for the initiation of the five year plan which essentially converted Italy to a nonmalarious country.

One of the enigmas of malaria in Italy prior to about 1930 was the paucity of the disease in some areas having dense populations of *A. maculipennis*, the supposed vector; and the prevalence of the disease in other areas supporting a relatively sparse population of this mosquito.

SWELLENGREBEL in 1924 noted differences in the hibernating characteristics of *A. maculipennis*, and VAN PHIEL observed that the non-hibernating forms had shorter wings (RUSSELL [485]). MISSIROLI and HACKETT in 1930 determined the presence of animal blood in *A. maculipennis* collected in houses. In 1931, MARTINI ET AL. proved that *A. maculipennis* was a complex of species, or varieties, varying in their vector capabilities. There are at least seven varieties of *A. maculipennis*, three of which, *A. m. labranchiae*, *A. m. atroparvus*, and *A. m. sacharovi (elutus)*, readily bite man and thus are capable malaria vectors (HACKETT [242]; MISSIROLI [383]).

Malaria in Italy — in fact, in Europe, South America, the United States, and some other countries — has been on the decline since about the beginning of the twentieth century. In Italy, there were 177,946 reported cases with 9,908 deaths in 1902 and only 92,301 cases with 488 deaths in 1940. The morbidity rate per 1,000,000 inhabitants thus dropped from 5,442 to 2,149 during the period (PAMPANA [424]). However, due to an influx of people, the breakdown of drainage systems, and the general outdoor activities of the population during World War II, the disease showed a sharp increase. In 1940 there were only 978 cases reported in the lower Ferrara district, but in 1945, 5,997 cases occurred and this increased to 6,888 cases in 1946 (SANI [498]). In 1944, an epidemic of malaria occurred in Agro Romano. In only two years between 1933 and 1943 were there over 2,000 cases of malaria reported from this area. In 1944, there were 9,063 cases, of which 7,986 were primary infections. In 1945, there were 11,025 with only 2,965 primary infections (JERACE [296]).

By the combined measures of drainage, screening, larviciding, and the use of quinine, malaria in Italy had been reduced to 55,000 cases by 1939, but by 1945 there were 411,600 cases in the country (RUSSELL [486]).

Following cessation of hostilities, the decline of malaria to near its prewar incidence normally would have been expected, but the practical elimination of the disease from its high wartime peak could hardly be anticipated by anyone, by the use of any methods previously known.

Although the war produced an increase in the disease, it also resulted indirectly in a means for its elimination. The Allied Control Commission, through the Rockefeller Foundation Health Commission, in 1944 initiated DDT residual spraying of human habitations in the Volturno area north of Naples (SOPER *et al.* [541]). The walls and ceilings of all buildings in specified areas of the Bonifica di Castel Volturno were treated during the week of May

17 with 5 per cent DDT in kerosene, at the rate of about 60 mg of DDT per square foot. In other areas, 10 per cent DDT in pyrophyllite was dusted on the walls and ceilings at an estimated dosage of 20 mg/ft². The buildings which were first dusted on May 17 were redusted during the week of July 9 and the sprayed buildings were retreated between August 7 and 27 with 80 mg of DDT per square foot. Weekly captures of adult *Anopheles* were made prior and subsequent to treatment in designated houses and stables in both treated and untreated areas. Spleen and parasite surveys were made of all school children 6–14 years of age prior to and after the treatments, and blood smears were taken for parasites. Captures of mosquitoes prior to treatment in the dusted areas were ten times those in the sprayed areas so that the effectiveness of the spraying was not immediately apparent. By September, however, the *Anopheles* density in the sprayed sections had dropped to less than one specimen per premises. The ineffectiveness of the DDT dust was apparent, however, when compared with untreated areas. *Anopheles* involved were *A. labranchiae*, *A. sacharovi*, *A. melanoon*, and *A. messeae*. During the survey period from June to October, the spleen rates varied from 43% to 39%, and the parasite rates dropped from 21% to 8%.

In 1945, the zone was enlarged to approximately 95 square miles, and treatment consisted of the application of DDT spray at the rate of approximately 85 mg of DDT per square foot. Table 1 shows the results obtained in vector control.

Table 1

Comparative Densities of Mosquitoes per Square Yard of Surface Examined, by Months, in Treated and Untreated Sections of Castel Volturno, Italy

Area	April	May	June	July	August	September
Treated	0·001	0·003	0·05	0·13	0·04	0·06
Untreated	9·87	41·84[1])	167·04[1])	118·50[1])	185·54[1])	185·54[1])

[1]) Based on estimates of over 1,000 mosquitoes.

The spleen rates in the areas treated in 1944 in Castel Volturno showed no significant drop in 1945, but the parasite rate dropped significantly from 21% to 1·4%.

The splenic indices in the Sessa Bonifica, an untreated area, increased from 50% to 63% between May and August 1945, and the parasitic index from 18% to 41%. *Plasmodium falciparum* accounted for 61% of the infections (AIT-KEN [7]).

In the summer of 1944, 5 per cent DDT in kerosene was applied to troop buildings in Ostia Libo and to farm buildings in Isola Sacra in the Tiber delta. The area treated was of sufficient size so that little infiltration was possible, particularly since buildings on the east and south of the area were treated by military control units. The houses were treated with 200 mg of DDT per square foot. A total of 5,869 buildings were treated, covering 286 km² containing 35,293 persons. The results obtained were described by MISSIROLI [381] as follows: 'Not one case of primary malaria has been verified in the Delta of the Tiber, and Ostia has achieved a healthiness the like of which has not been seen there for some two thousand years, that is, since the invasion of Italy by malaria.'

Searches for *A. labranchiae* adults and larvae revealed a greatly reduced density. No *Anopheles* were found in buildings a year after treatment. It was concluded that the program effected a greater reduction in *Anopheles* than was needed to prevent transmission of malaria, and where malaria control only is considered, a more limited program might be planned.

Following this striking success with DDT, MISSIROLI conceived in 1945 a campaign for the eradication of malaria from Italy by DDT residual spraying.

On June 5, 1945, MISSIROLI initiated the first experiment of his five-year plan to eradicate malaria from Italy by DDT residual spraying. Treatment was started in the southeastern area of the Pontine marshes in the province of Latina, a notorious malaria region (MISSIROLI [379], [378]). The country was divided into four zones: (1) *A. labranchiae atroparvus* was dominant, which under favorable conditions could maintain a slight degree of endemicity. Spraying was carried out only in those areas where malaria occurred; (2) *A. sacharovi* was the dominant vector and *A. messeae* was present. Spraying was confined to the areas infested with *A. sacharovi;* (3) *A. labranchiae labranchiae* was the vector and infested the plains of Latium and Tuscany. All dwellings of the area were treated with DDT; and (4) *A. l. labranchiae* was again the vector and the area comprised the south of Italy and the islands, where all dwellings were treated with DDT.

In 1946, DDT spraying started on March 5 and ended in the middle of May. A 5 per cent solution of DDT in kerosene was initially used, at the rate of about 2 g/m² of surface treated.

By 1947, there was a practical disappearance of *A. l. labranchiae* and *A. sacharovi*. No *sacharovi* larvae were found, and larvae of *A. l. labranchiae* were greatly diminished or absent. In the province of Latina, the parasitic index, which was taken every March, dropped from 10·32% in 1946 to 0·49% in 1947, and to 0·24% in 1948.

Figure 1 demonstrates the reduction of *Anopheles* in Latina from 1945 to 1949, while Figure 2 shows the corresponding decrease in malaria morbidity (MOSNA and ALESSANDRINI [398]). Figure 3 portrays the decline, or practical elimination, of the disease from Italy over a five-year period.

The magnitude of the program is illustrated from 1948 data when 4,000,000 persons were protected over an area of 191,155,000 m², involving 2,850,000 premises. A total of 335 t of DDT were used at the average treatment rate of 1·50 g/m² of surface treated (PATRISSI [430]).

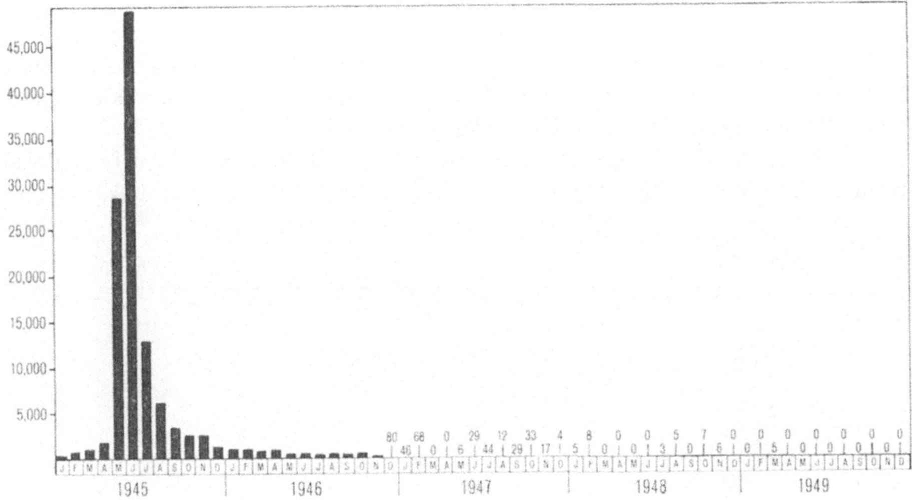

Figure 1

Control of anophelines in the Province of Latina, 1945–1949
(Numbers per month from capturing stations).

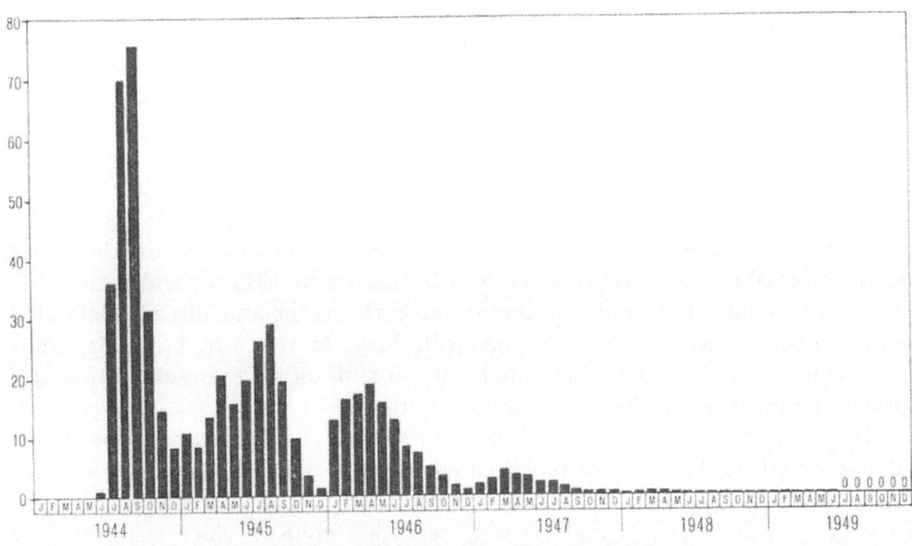

Figure 2

Number of malaria cases per 1,000 inhabitants reported monthly from 1944 to 1949, inclusive,
in the Province of Latina.

The cost of the campaign in Latina, an intensely treated area, was in 1949 only 3 lire/m² of area treated, and 187 lire per inhabitant protected (MOSNA and ALESSANDRINI [398]). Over the entire area of Italy, taking into acount higher cost during the first part of the campaign, the cost per person was about 350 lire (about $ 0.55) (PAMPANA [424]).

In the epidemic of Agro Romano, previously cited, a reduction in malaria from 11,025 cases in 1945 to 435 in 1947, to 135 in 1948, to 16 in 1949, to zero

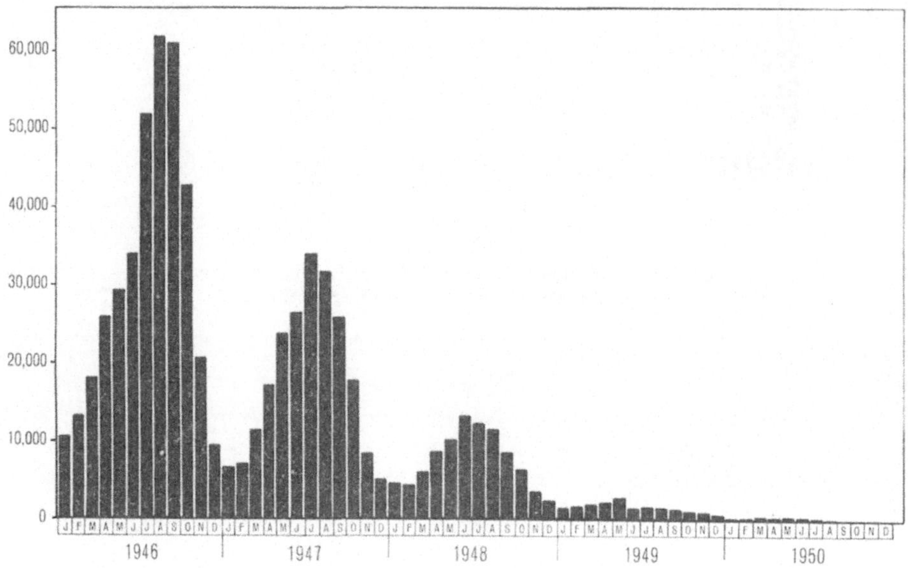

Figure 3
Reported malaria morbidity in Italy, 1946–1950
(Number of cases)

in 1950, was effected; and *Anopheles* collections dropped to nil in the latter year. Some DDT was used in 1944, and beginning in 1947 a regular spraying program was initiated, and a reduction in both vector and disease coincided with the progressive use of this chemical. Prior to the war, however, adult anopheline catches persisted at a high rate in spite of all the reclamation and larviciding measures in force (SABATINI [489]).

The epidemic in Lower Ferrara District showed a similar decline with the advent of DDT. There were 6,888 cases in 1946. DDT house spraying was initiated in June of that year and was repeated in the two following years. In 1948, only 265 cases of malaria were reported, none of which were primary infections (SANI [498]).

Various other wartime outbreaks in endemic areas were controlled with DDT, following the war. BUONOMINI and MACCARRONE [77] and BUONOMINI *et al.* [76] described the control of an epidemic on the Island of Elba from 1946–8. DDT residual spraying was initiated in 1947 using both a kerosene solution and a xylene-water emulsion at the rate of 2 and 3 g of DDT per square meter. Paris green and oil larviciding was also employed. In 1946, there were 201 primary infections and 323 relapses, while in 1948 there were no primary infections and only 22 relapses. It was believed that discontinuance of control measures would reestablish the former epidemiological conditions (BUONOMINI and MACCARRONE [77] and BUONOMINI *et al.* [76]).

RAFFAELE and COLLUZI [462] used DDT in the Province of Frosione (Lazio) beginning in 1946, and as a result, interrupted transmission to the extent that infections of both *Plasmodium vivax* and *P. falciparum* were rapidly coming to an end.

Epidemiologic data since 1887 reveals a gradual decline in the malaria incidence of Italy, but irrespective of all efforts toward its elimination, the disease persisted in important proportions until the advent of DDT. Many of the most capable present-day malariologists were engaged in the DDT malaria control program in Italy, and it is abundantly clear from their writings that this chemical was responsible for the practical elimination of the disease. In the short period of six years, large fertile areas were converted from sub-marginal status to healthy, prosperous farmlands. When one views the Pontine Marshes today with its more than 100,000 healthy people, it is difficult to visualize that the area had been an uninhabitable morass of malaria since, or before, the dawn of Christendom.

Not a single death from malaria has occurred in Italy since 1948, and with only 392 cases in 1951, this country can look forward to the complete defeat of its ancient enemy.

Recent developments do, however, suggest a note of caution in this optimistic outlook. House flies, *Musca domestica*, were found by SACCA [492] to have acquired resistance to DDT, and MOSNA (397) detected resistance of *Culex pipiens autogenicus*. Although malaria vectors in Italy have not at present shown this trend, malariologists must keep a constant vigil for such a manifestation in order to institute corrective measures that would preclude the recurrence of potent endemic foci of the disease.

Sardinia[1]) (*The Sardinian Eradication Project*)

Because of the success achieved in the control of malaria through the use of DDT as a residual spray during the latter part of World War II, a number of nation-wide projects were initiated on the conclusion of hostilities in such widely scattered parts of the world as Greece, Venezuela, and the mainland of Italy. While residual spraying had shown great promise, it was not certain

---

[1]) The account herein presented was prepared by Dr. JOHN A. LOGAN who was in charge of the Sardinian program during the major portion of its existence.

at that time whether it would prove to be an economic or a practical means of control on a national basis.

As DDT had also proved to be an effective larvicide, it was believed that a combination of DDT residual spraying plus DDT larviciding would greatly increase the effectiveness of the species eradication technique which had been used so successfully against the invading malaria vector *A. gambiae* in Brazil (SOPER and WILSON [543]) and Egypt (SHOUSHA [508]). An experimental project was, therefore, initiated in Sardinia against the indigenous vector *A. labranchiae* (LOGAN [336]). As residual spraying was being planned for the remainder of Italy, an adequate basis of comparison for the effectiveness of the eradication technique was readily available.

The entire island was used for the experiment because many of the problems and difficulties inherent to eradication could only be discovered by full-scale operations. Funds and equipment were available on a generous scale and this made it possible to employ any method or equipment which might reasonably be expected to assist in eradication. Thus drainage operations were employed, as well as DDT residual sprays and larvicides.

Sardinia proved to be a particularly difficult area in which to test the species eradication technique, because of its topography, its size, and its primitive state of development. While several anopheline species were known to exist, only one of these, *A. labranchiae*, was known to be a malaria vector and MISSIROLI [382] believed Sardinia to be the epicenter of its range. While Italy has been notorious for malaria in the Western world since the time of the Greek and Roman empires, Sardinia, at least in modern times, has been considered as the most malarious part of Italy (HACKETT [242]). The disease is believed to have existed in the island since the 12th century, B.C., but the first recorded evidence appears to be 236 B.C., during the Roman Conquest (LOGAN [338]). The island is also the least cultivated part of the country and has the lowest population density.

Sardinia is the second largest island in the Mediteranean (following Sicily) with an area of 23,896 km^2 and a population of 1,250,000. It has a semitropical climate, with relatively hot summers and with moderate rainfall and occasional snow storms during the winter. It is a rough and primitive land of hills and mountains with scattered villages and large sections not served by either road or railway. A considerable portion of the coastal area is flat and fringed with swamps and salt- or brackish-water lagoons, and the river systems have shallow, tortuous beds choked with vegetation. Mountain streams also contain heavy growths of vegetation, and upland and mountain swamps are common. Small-scale irrigation consisting of stream blockage and flooding is widely practiced. While a number of planned reclamation projects have been developed, many of these have been neglected and abandoned (LOGAN [338]).

On October 1, 1945, the International Health Division of the Rockefeller Foundation assumed technical direction of the proposed eradication project, and on April 12, 1946, a special agency of the Italian High Commission for

Hygiene and Public Health, known as ERLAAS[1]), was established to carry it out. Finance was arranged through counterpart funds made available by the United Nations Relief and Rehabilitation Administration (UNRA) and later by the Economic Cooperation Administration (ECA).

Based on the experience of Brazil and Egypt, the island was divided into sectors, each comprising an area of approximately 4·5 km². These areas served as the basis for both larviciding and scouting operations. Due to a lack of trained personnel, training schools were established for instructing residual spray operators, larviciders, scouts, and auxiliary personnel. The operations were carried out as follows:

1946	Entomological survey.
1946–47	Residual spraying campaign against overwintering *A. labranchiae* for training purposes and to suppress malaria transmission.
1947	Larviciding in test area.
1947–48	Winter residual spraying of all man-built structures in Sardinia with 2·0 g of DDT per square meter of wall and ceiling surface.
1948	Larviciding of all fresh-water surfaces on the island on a weekly-cycle basis using a DDT-oil-Triton larvicide.
1948–49	Mop-up residual spraying.
1949	Mop-up larviciding.
1950	Check on eradication and additional larviciding.

In residual treatments, a 5 per cent DDT spray was applied at the rate of 2·0 g/m² of surface treated, and larviciding was with 2·5 per cent DDT in fuel oil, containing an emulsifier, and applied at the rate of 0·1 lb/acre.

Due to the extent of undeveloped and uninhabited area, the problem of logistics proved particularly difficult, and it was found necessary to build large numbers of field camps to serve these areas. Supplies and men were transported by a fleet of over 250 motor vehicles, aided by animal transport. Fog generators and DDT smoke candles were used to supplement the residual spraying operations; and airplanes, two helicopters, boats, rafts, and specially designed larvicide 'bubblers' were used in the larviciding program.

Because of the necessity of preparing water surfaces for efficient larviciding, considerable clearing and drainage work was necessary. Most of this was done by hand, but dynamite ditching and tractor- and animal-drawn ditchers were also used. This part of the program required a large labor force and the number of ERLAAS workers at one time (August 1948) amounted to more than 33,500. The cost of the project amounted to more than six billion lire, extending over a period of four and a half years.

The program eliminated malaria as a public health problem from Sardinia. The number of cases, including primary infections, reinfections, and relapses fell from a high of 78,173 in 1944 to 44 in 1950 and 9 in 1951. No new cases were verified in 1950, and of the 9 cases reported in 1951, only one is considered

---

[1]) Ente Regionale per la lotta Anti Anofelica in Sardegna.

as a possible primary infection. The number of cases (unverified) reported by provinces for the years 1944–51 is indicated in Table 2 (Logan [336]).

The reduction in spleen and parasite indices, although following the same downward trend as that of malaria itself, was relatively slower, which is probably an indication of the previous degree of endemicity of the disease in the island. A summary of the results of surveys in 66 villages (Group I Villages), which represent a geographical cross section of the island, and in 13 of the more malarious of these villages (Group II Villages), made during the period

Table 2

Cases of Malaria Reported in Sardinia, By Province and by Year, from 1944–1951
(*From Monthly Reports of Provincial Medical Officers*)

Year	Island total	
	New cases[1])	Total cases
1944	11,177	78,173
1945	8,519	74,641
1946	10,149	75,447
1947	2,968	39,303
1948	341	15,121
1949	6	1,314
1950	4	44
1951	3	9

[1]) The term 'new cases' includes all primary cases and reinfections reported. In 1950 and 1951 these cases were later investigated by the ERLAAS medical staff for confirmation.

1946 to 1950, is indicated in Tables 3 and 4 (Logan [337]). While specific data for 1951 are not available, the downward trend has continued.

Beginning in the fall of 1948, blood examinations were made of all infants between two months and two years of age in the test villages. Five infants out of the 871 examined during the first survey were found to be parasite-positive; thereafter no positives were found in infants.

Unfortunately, the success achieved in the reduction of malaria was not accompanied by the eradication of *A. labranchiae*. After apparently clearing most of the island in 1948, following the island-wide program, the species continued to appear in 1949, 1950 and 1951, although always in extremely low numbers. During the entomological survey in 1946, *A. labranchiae* was by far the dominant species; 63% of all inspections were positive for anophelines and over 90% of these were for *A. labranchiae*. By 1950 it took more than 330 scout-days to find a collection, or, in other words, it would have taken a trained scout over a year to locate a single *labranchiae* infestation. There was also a

Table 3

Summary of Results of Spleen Surveys in Sardinian Villages, 1947–1950

Survey period	Number examined	Spleen class (HACKETT)						Total number enlarged	Spleen rate (% ± S.E.[1])	Average spleen size	Average enlarged spleen size
		0	1	2	3	4	5				
Group I: 66 villages, total population 190,000											
Winter 1947–1948	12,915	10,000	1,745	809	292	67	2	2,915	22·6 ± 0·37	0·35	1·55
Winter 1948–1949	15,625	13,047	2,118	369	83	8	0	2,578	16·5 ± 0·30	0·20	1·22
Winter 1949–1950	15,509	13,466	1,741	264	34	3	1	2,043	13·2 ± 0·27	0·15	1·17
Group II: 13 of the most malarious of the group I villages, total population 35,000											
Winter 1947–1948	2,376	1,564	409	242	125	35	1	812	34·2 ± 0·97	0·60	1·74
Spring 1948 . . .	1,999	1,312	351	188	143	5	0	687	34·4 ± 1·06	0·59	1·71
Autumn 1948 . .	2,400	1,641	557	152	46	4	0	759	31·6 ± 0·95	0·42	1·34
Winter 1948–1949	2,924	2,184	564	126	46	4	0	740	25·3 ± 0·80	0·33	1·31
Spring 1949 . .	3,041	2,296	496	191	53	5	0	745	24·5 ± 0·78	0·35	1·42
Autumn 1949 . .	2,294	1,682	461	112	36	3	0	612	26·7 ± 0·92	0·35	1·32
Winter 1949–1950	3,092	2,487	490	102	11	2	0	605	19·6 ± 0·71	0·24	1·21
Spring 1950 . . .	2,863	2,354	399	75	30	5	0	509	17·8 ± 0·71	0·23	1·29
Autumn 1950 . .	2,282	1,954	272	45	8	2	1	328	14·4 ± 0·73	0·17	1·22

[1] Standard error (S.E.)

Table 4

Summary of Results of Parasite Surveys in Sardinian Villages, 1947–1950

Survey	Number examined	Number positive						Total number positive	Parasite rate (Per cent $\pm$ S. E.[2])
		Plasmodium vivax	Plasmodium falciparum	Plasmodium malariae	Mixed				
					vf[1]	vfm[1]	vm[1]		
Group I: 66 villages, total population 190,000									
Winter 1947–1948 . .	12,665	232	147	3	0	0	0	382	3·02 $\pm$ 0·15
Winter 1948–1949 . .	15,634	76	12	19	3	1	1	112	0·72 $\pm$ 0·068
Winter 1949–1950 . .	15,517	32	6	8	4	0	0	50	0·32 $\pm$ 0·045
Group II: 13 of the most Malarious of the group I villages, total population 35,000									
Winter 1947–1948 . .	2,290	51	46	1	0	0	0	98	4·28 $\pm$ 0·42
Spring 1948	1,999	19	11	0	0	0	0	30	1·50 $\pm$ 0·27
Autumn 1948	2,400	10	3	0	0	0	0	13	0·54 $\pm$ 0·15
Winter 1948–1949[3] .	2,924	34	6	5	2	0	1	48	1·64 $\pm$ 0·23
Spring 1949	3,041	19	4	3	0	0	0	26	0·85 $\pm$ 0·17
Autumn 1949	2,294	9	0	2	0	0	0	11	0·48 $\pm$ 0·14
Winter 1949–1950 . .	3,092	10	1	3	1	0	0	15	0·49 $\pm$ 0·13
Spring 1950	2,863	14	7	3	0	0	0	24	0·84 $\pm$ 0·17
Autumn 1950	2,282	6	7	2	0	0	0	15	0·66 $\pm$ 0·17

[1] vf *Vivax-falciparum;* vfm *Vivax-falciparum-malariae;* vm *Vivax-malariae.*
[2] Standard Error (S. E.).
[3] Beginning with this survey, a more intensive technique of slide examination was adopted.

great decrease in the size of the collections. In 1946, collections of immature stages ran from 3 to 12 per dip. In 1950, of the 420 *labranchiae* collections made as a result of over 2,200,000 larval inspections, a total of 1,379 specimens were collected and many of the collections consisted of a single specimen only. Twenty-eight *labranchiae* adults were found in twelve collections following 178,279 inspections.

A. labranchiae, however, continued to exist even though in extremely small numbers, and it was evident eradication had not been achieved. Often sectors were found positive after they had apparently been free of *A. labranchiae* for as much as two years. Although the extent of the initial destruction was comparable to that achieved in Brazil and Egypt, the population curve did not approach zero, as in these former campaigns, but appeared to flatten out as eradication appeared imminent.

In carrying out the project, over 30,000 hectares of land were reclaimed by drainage operations. The planning of drainage, larviciding, and residual spraying work made it necessary to consider the island as a geographic whole and this led to a consideration of its over-all problems and potentialities. From this point of view, the extent of the uncultivated and unused land, the sparseness of the population, and the lack of knowledge of the island's social and economic resources were evident. The program made it possible for the first time to live and work anywhere on the island and opportunities for development existed which had never before been possible. With the pressure of excess population existing on the Italian mainland, it seemed logical that a study be made to examine the possibility of Sardinian development over a long-range period, with particular reference to its ability to absorb people from the continent. This social economic survey is now under way, emphasizing the fact that ERLAAS has meant more to Sardinia than just malaria control; it has also served as a rehabilitation project. A bronze plaque has been placed in Cagliari to commemorate the liberation of Sardinia from malaria.

The DDT residual spray program controlled flies and other domestic insects as well as mosquitoes, and lowered the infant mortality rates and the incidence of intestinal infections (BARNARD [36]).

It is believed that the most economic method of control is a continuation of eradication pressure on a maintenance basis over a period of years, which would result in the possibility of eventual eradication, and such a program is now being carried out by the Regional Government.

From the evidence furnished by this program, the eradication of indigenous anophelines, such as *A. labranchiae*, is a difficult and expensive proposition. In Sardinia, from the standpoint of malaria control, the residual spraying method would have been both cheaper and easier to operate. The use of eradication may, nevertheless, offer some advantages under certain special conditions, particularly if it is incorporated in a general conservation and development program, where part of the cost of eradication could be charged to rehabilitation. Such a program should not, however, be rigidly confined by time, finance, or methods.

Corsica

Malaria has long been a deterrent to the economic and social development of Corsica. Until recently the disease drove many of the inhabitants to the mountains during the summer, leaving undeveloped the potentially productive lowlands and producing overcrowding in the higher areas. A reduction in the population of the Island from 322,854 in 1936 to 276,873 in 1946 has been attributed in large measure to malaria (CAVAILLON *et al.* [93]).

As a result of diverse antimalaria measures, progress in the control of the disease had been made prior to World War II. As a result of this conflict, however, the disease increased and extended westward into areas which had long been free of it.

A DDT antimalaria campaign was inaugurated in March 1948. A 5 per cent DDT-kerosene residual spray was generally employed, although an aqueous suspension made from 50 per cent DDT water-wettable powder was sometimes used. Larviciding was carried out with 5 per cent DDT in fuel oil containing an emulsifier. In 1950, anti-adult operations extended from mid-March to mid-July and antilarval measures from mid-June to the end of October (TRINQUIER and JAUJOU [585] and JAUJOU *et al.* [291]).

A. maculipennis labranchiae and *A. m. sacharovi* are the chief vectors and they were drastically reduced during the first season of operations. There was likewise a marked decline in the spleen and parasite indices in the treated areas. Also a decline of the disease in the nontreated areas was observed, but to a much lesser degree. There were 1,443 positive bloods found in the Department of Health Laboratories in 1947 but only 675 in 1948, the first year of operations (JAUJOU *et al.* [290]). *Plasmodium vivax* is the most frequently encountered parasite in Corsica, although *P. falciparum* and *P. malariae* have been present.

After the second year of the program, CAVAILLON *et al.* [93] stated that there had been a reduction in malaria in the proportion of 20 to 1 as a result of anti-anopheline measures. Furthermore, it was estimated that an economy of 100 million francs had been realized in 1949 as compared to 1947.

JAUJOU *et al.* [291] reported the results of three years of the DDT program, involving the use of both residual sprays and larvicides. Some drug prophylaxis was also employed.

The spleen index was 36·3% in 1947, 3·5 in 1948, 2·0 in 1949, and only 1·0% in 1950. The parasite index showed a similar downward trend from 23·4% in 1947 to 1·5 in 1948 to 0·5% in 1950. The number of positive bloods encountered in the Department Laboratories decreased from 1,443 in 1947 to 43 in 1950, and of the latter only one was a primary infection.

The Corsica Anti-Malaria Program may continue until the disease is eradicated, but it has already liberated the population from the tentacles of its worst enemy, malaria. In addition, the program in its first year practically eliminated flies and other household insects, and reduced infant mortality in the treated areas. The acquisition by flies of resistance to DDT resulted, however, in poorer control in 1949–50 and, as in other countries, decreased the popularity of the DDT program. As a result, chlordane was used for fly control in 1950, but there is indication that this also will soon be ineffective.

Spain

MORALES and JUAREZ [392] and MORALES [391] have reported on a residual spray campaign in the malarious zone of the Tiétar and Jerte River Valleys. The campaign was initiated in 1949 against *A. maculipennis* using both benzene hexachloride and DDT. The dosages used were BHC at 0.5 g/m^2 and DDT at 1 g/m^2. In the case of BHC an emulsion was employed, while the DDT was used as a wettable powder.

In 1948 there were 1,034 cases of malaria reported by the dispensaries in the Jerte River Valley. In 1949 this had dropped to 400 and by 1950 to 80 cases. It should be pointed out that there was a spontaneous recession prior to the use of DDT. For instance, there were 1,440 cases reported in 1947. A more spectacular reduction is apparent when statistics for the entire province are considered. In 1946 and 1947 there were, respectively, 11,980 and 10,675 cases of malaria reported from the province of Caceres. This decreased to 9,174 in 1948 and by 1949 only 5,678 cases were recorded. This number had decreased to 2,881 in 1950 although it must be remembered that an insecticide program was carried out in the latter two years in Caceres Province. The campaign was extended to other provinces including Badajoz and Sevilla.

In an earlier report, CLAVERO and ROMEO VIAMONTE [100] described promising results in the reduction of anophelism and malaria in a rice-growing area 6,000 hectares in extent. Both DDT and BHC were used as a residual spray against the vector, *A. m. labranchiae*.

Portugal

The principal vector of malaria in Portugal is *A. atroparvus*, a species that is predominantly zoophilic. The annual incidence of malaria based on reported cases is usually between 50,000 and 70,000. The mortality during the period 1940 to 1947 ranged between 83 and 765 deaths per year. Portugal relies on modern insecticides for malaria control and has used DDT residual sprays in part of the endemic areas and benzene hexachloride in another. The program was initiated in 1948, and the total number of persons protected in 1949 was 33,776. The DDT was used at a dosage of 2 g/m^2 of treated surface, and BHC at a dosage of 1.875 g/m^2. A statistical appraisal of the effect of the program on the incidence of malaria is, unfortunately, not available.

Netherlands

Malaria in the Netherlands is endemic in the provinces bordering on the North Sea, North Holland, Friesland, and Zeeland, and occurs in epidemic waves every twenty years or so (HACKETT [241]). The only vector is *A. m.*

atroparvus, and nearly all malaria is *P. vivax* (SWELLENGREBEL [567]). The vector breeds in tremendous numbers in slightly brackish waters having a salinity of about 0·8% or less (HACKETT [241]). During World War II malaria incidence rose, and between 1941 and 1945 it doubled as a result of the flooding of the malarious zone with sea water by the enemy.

SWELLENGREBEL and DEBUCK [568] attributed the persistence of malaria in the prosperous, highly developed, cattle-raising area of the country to two factors. First, during the winter, *A. atroparvus* retreated indoors into warm shelters where food was plentiful, and many of them overwintered in houses, feeding on the inhabitants. Thus, if some member of the family was a carrier, soon all members would have malaria. Second, malaria was not evenly distributed, infective anophelines being most numerous in certain houses where cases appeared year after year. These houses belonged to families with large numbers of children who are the most persistent carriers. Since overwintering mosquitoes did not remain infective until spring and since the continuity of infection is dependent on carriers, the birth rate — the Dutch one being the highest in Europe — seemed to be a factor in *atroparvus*-borne malaria.

In North Holland, primary attacks occur in the late spring and early summer, declining generally in August. At the same time, infected anophelines are extremely rare in spring and early summer and are generally found from late August to October when they reach their maximum. Therefore, it is probable that persons are bitten at this time but show no symptoms of malaria until the following spring or summer. Occasionally, however, a clinical case is seen in the fall (SWELLENGREBEL [567]).

While malaria has been known in the Netherlands for many years, in the village of Wormerveer it has been observed and recorded for about fifty years (SWELLENGREBEL [567]). A study of the available records shows that there were peaks in malaria prevalence in 1901, 1922, and 1946, at intervals of 21 to 24 years. The typical epidemic pattern seems to be a slow rise in malaria incidence over a period of about three years. While only a small proportion of the total population is affected, the figure at the peak of the epidemic may be ten to thirty times that when the incidence is lowest. The periodicity of malaria outbreaks is strictly local. Neighboring villages experience epidemics in different years, as did Uitgeest which is only four miles from Wormerveer but had a malaria epidemic in 1936.

Throughout the years a variety of malaria control measures have been used in North Holland. Larviciding is expensive since the mosquito breeds in the drainage system which is vital to the agricultural life of the country. In 1920, the attack was directed at hibernating adults; but as the solution sprayed was lysol, this was not very effective. Starting in 1926, pyrethrum sprays were used in houses and stables, pigsties, etc. The populace killed many *C. pipiens* during the summer but ceased spraying by mid-August when malaria vectors became abundant. The responsibility for carrying out this measure was therefore transferred from the householder to an official body in 1936. Emphasis was placed on spraying houses where recent cases of malaria had occurred.

Domestic spraying with DDT was introduced in 1945. Only in one village was spraying done throughout; in others, pinpoint spraying, as used in the pyrethrum program, continued. DDT as a 5 per cent solution in purified kerosene was applied to walls and ceilings at the rate of 40 ml/m². As of 1950, the reduction in malaria incidence in sprayed villages was more marked than that in untreated villages and towns (Table 5).

In a psychiatric institution at Franeker, Friesland Province, studies of the effect of DDT residual spraying on mosquitoes and flies were started in 1947 (FIRTOS and DE JONG [184]). Several rooms were treated with 3 to 5 per cent solutions of DDT in kerosene at the rates of 1 to 2½ g of DDT per square

Table 5

Effects of one Application of DDT in 1947 on
Percentage of Malaria Patients in North Holland[1])

Year	Untreated towns	Towns where pinpoint spraying done	Area where spraying done throughout[2])
1943	0·6	0·2	0·5
1944	0·5	1·1	1·0
1945	0·6	4·1	3·0
1946	1·2	12·8	8·1
1947	1·4	3·7	1·8
1948	3·4	0·5	0·12
1949	0·3	0·1	0·05

[1]) Adapted from SWELLENGREBEL [567].
[2]) One area only.

meter of wall and ceiling surface. During 1947, 1948, and 1949, counts of living and dead insects were made weekly between April and November. In rooms where 1 and 2 g of DDT had been applied, there was a 100 per cent residual effect from 12 to 14 months after spraying; where 2 to 2½ g of DDT were applied, this effect lasted 14 to 18 months.

During this period, in addition to these studies, a spray program with DDT was undertaken in an attempt to reduce the number of cases of malaria in the institution. Because of the proximity of the town and its inhabitants to the institution, house-to-house spraying was done there also. The DDT was applied starting just before September 1 each year. Mosquito and fly populations were both greatly reduced; in fact, no mosquitoes engorged with human blood were found. Malaria cases among patients at the institution dropped from 38·7% in 1946 to 9·7 in 1947, to 3·1 in 1948, and to 0 in 1949. It was felt that this was due primarily to the DDT residual spraying.

In the course of their work the authors noted a change in the behavioristic pattern of some of the mosquitoes. This was first observed about 14 months after the initial spraying. When the mosquitoes came in contact with old residues, they became restless; this restlessness was identified as the first sign of incipient DDT intoxication. This response resulted in a change in the settling habits of the mosquitoes and in a reversal of the normal reaction to light; in addition, they were impelled to fly away from the sprayed area and, therefore, were sometimes killed and other times not, depending on how much DDT they had absorbed. This was not believed to be a truly repellent action since some contact with the insecticide was necessary to stimulate this response.

Malaria control with DDT was undertaken in other areas of Friesland also. While Franeker was a large malaria focus, malaria existed throughout the province. In 1946, between 2·4 and 2·7$^0/_{00}$ of the population suffered from malaria (DE JONG [138]). In 1947, these figures were 0·72 and 0·78$^0/_{00}$ (DE JONG [137]). DDT spraying was done on an experimental basis in two towns other than Franeker that year. In Harlingen, the walls and ceilings of horse stables were sprayed in July with a 3 per cent DDT solution in kerosene at the rate of about 1 g of DDT per square meter. In Bolsward, similar operations with an insecticide containing 5% of DDT, applied at the rate of about 3 g of DDT per square meter, were instituted. In 1948, the number of cases per 1,000 inhabitants fell to 0·38 (DE JONG [136]). This decreased further in 1949 to 0·06$^0/_{00}$, and of these over one third were cases imported from Indonesia (DE JONG [135]). In 1950, there were no autochthonous malaria cases in Friesland (DE JONG [134]); on the other hand, the number of cases of imported malaria was considerably higher than in 1949. While the spraying operations of the Health Service were not directed against malaria mosquitoes alone but against insects in general, the near disappearance of malaria from Friesland provinces must be attributed to the effects of DDT.

Yugoslavia

Malaria is not a reportable disease in Yugoslavia; therefore, its incidence is probably much greater than available figures indicate. Free antimalaria centers operate, and as late as 1947 they reported 78,000 cases. There was a spontaneous decline in the disease during the next few years, but the drastic reduction to 8,562 recorded cases in 1949 no doubt can be attributed in large measure to the use of DDT residual spraying.

The vectors in Yugoslavia are *Anopheles labranchiae, A. sacharovi,* and *A. superpictus.* The DDT residual control program was initiated in the spring of 1947. In addition to the treatment of residences, larvicide measures were also carried out, particularly around municipalities where residual spraying would be too expensive. The usual dosage employed in residual spraying was 1 g/m^2 of surface treated. House spraying was limited to the ceiling and a band about 1 m below. According to reports received by PAMPANA [424], the campaign during its first three years was sufficient to stop transmission of malaria. The autumnal peak, which corresponds to new infections, has been prevented. New infections seem to be confined to hypoendemic zones which, due to the low endemicity of the disease, were not treated with insecticides.

Greece

A history of malaria would draw heavily upon Greece. As with Italy, the disease has been an ever-present factor in the undulating career of the nation. According to Livadas and Belios [331], the first clinical and epidemiological observations of the disease appear in the works of ancient Greek writers and physicians where also are found such terms as $\tau\varrho\acute{\iota}\tau\alpha\tau o\varsigma$ (tertian), $\tau\epsilon\tau\alpha\varrho\tau\alpha\tau o\varsigma$ (quartan), $\sigma\pi\lambda\eta\nu o\mu\epsilon\gamma\alpha\lambda\iota\alpha$ (splenomegaly), and $\dot\epsilon\lambda\acute\omega\delta\eta\varsigma$ $\varkappa\alpha\chi\epsilon\xi\acute\iota\alpha$ (malarial cachexia).

Greece has several effective malaria vectors. In order of importance they are: *A. sacharovi*, *A. superpictus*, and *A. maculipennis* (var. *typicus* and *subalpinus*). *A. sacharovi* breeds chiefly in seashore swamps, *A. superpictus* in numerous running streams throughout the country, and *A. maculipennis* in swamps and miscellaneous waters. The latter species is more zoophilic than the former two and thus of much less importance. Barber and Rice [35] found the sporozoite index of the vectors as follows: *A. sacharovi* 0·93 to 1·22%, *A. superpictus* 0·45 to 1·48%, and *A. maculipennis* 0·07 to 0·13%. Malaria transmission occurs from about the end of May until mid-October. There are three malaria parasites in Greece: *Plasmodium falciparum*, *P. vivax*, and *P. malariae*. A survey among school children in the autumns of 1930–38 showed 49% of infections to be *P. falciparum*, 27% *P. vivax*, 22% *P. malariae*, and 2% mixed (Livadas *et al.* [335]).

Malaria is not a notifiable disease in Greece, but records from various clinics and health establishments, particularly since 1937, are sufficient to allow a reasonable estimation of morbidity and to determine trends in endemicity.

Afendoulis [2], describing the general malaria epidemic of 1886 on the Island of Salamis where 4,500 of the 5,000 inhabitants were affected, states: 'No province of Greece remained unaffected by the malaria fever, no corner remained undug or unploughed for graves.'

Greece was for many years the most malarious country of Europe. Sir Ronald Ross [480] stated: 'I have never seen anything worse in the most malarious parts of India and nothing as bad in Africa, where in all probability the natives are endowed with greater immunity.'

Greece has waged an ever-increasing battle against malaria throughout this century and especially since 1937 when the malaria control service was organized. Control measures consisted of antilarval attacks, both by chemicals, employing the older insecticides such as paris green and oil, and by drainage. House screening was also promoted and quinine was used, although not in a systematic program.

Coincident with these efforts, the general trend of the disease was downward, but even so malaria remained a potent deterrent to the social and economic development of this country.

As recent as the prewar decade, 1930–40, Livadas *et al.* [335] and Livadas [326] summed up the situation as follows:

1. The population of endemic areas amounts to approximately 5,000,000 or 68% of the population...

2. Malaria mortality during 1930–37 amounted to 72·2 deaths per 100,000 population. Malaria is the 9th death cause, regardless of age and 2nd in childhood, after pneumonia.

3. The number of persons yearly attacked by malaria ranges approximately between 1 and 2 million and the number of working days lost between 20 and 40 million a year.

4. Spleen and parasite rates of school age children range between 25 and 100% and 10 and 50% respectively.

5. Quinine annually consumed amounts to 30,000 kg.

6. The direct loss from malaria every year is estimated at 3,875,000,000 prewar drachmas, or 27,678,500 dollars (1940 valuation).

This situation was bad enough but it was to grow worse, before better. During the war years (1941–44) malaria control activities were badly disrupted and the incidence of malaria began to increase. It reached a level higher than had occurred since establishment of the Service, and in 1942 it reached the highest peak in recent history (PAMPANA [425]).

At this point, the new insecticide, DDT, entered upon the scene. MANDEKOS [356] conducted both laboratory and field experiments in 1943 and 1944 against anopheline and *Culex* larvae and adults. A 5 per cent DDT (Neocid) spray used at the rate of 1 g/m² freed bedrooms and stables from mosquitoes for five weeks. A DDT emulsion applied at the rate of 0·007 g/m² of water surface effected good initial kills, but needed repeating after one week.

During 1945 malaria control activities returned to prewar levels, and DDT was used, beginning in the early summer of that year, as a residual spray in three groups of villages. Treated buildings were usually negative for anopheline mosquitoes, and the spleen indices dropped as much as three fifths in some areas. The cost of the method was less than larval control with paris green, and in general the results were so promising that a full-scale program was initiated in 1946 (LIVADAS *et al.* [333]).

The extended malaria control program, initiated in 1946, was based exclusively on the use of DDT. Neither drainage projects nor dispensaries for the issuance of drugs, which were suspended during the war, were re-established.

DDT was used both as a residual house spray and as a larvicide. All houses and outbuildings were treated once a year with DDT at the rate of 1·8 g of DDT per square meter. The spray was applied with pressure type hand sprayers fitted with fan-spray nozzles. During the first year, 4 to 5 per cent DDT oil solution was used, but in the following years a 25 per cent emulsifiable concentrate, diluted to 5% with water, was employed. In 1948 a small amount of DDT wettable powder was tried, but its use was abandoned due to its unsightliness on walls and its easy removal by cleaning operations.

Table 6 (LIVADAS *et al.* [335]) is a record of residual spraying operations of the four-year period 1946 to 1949.

In ground larviciding, a 25 per cent emulsifiable concentrate, diluted 1:1000, was used. This was applied at the rate of 0·02 g/m^2 at cycles varying from 10 to 15 days. From 92 to 147 local larval control programs were in operation during the four-year period, mostly to protect urban areas.

The population in the larvicide areas varied from about 1¼ to over 1½ million people during the respective years.

Airplane larviciding was done with Stearman airplanes applying DDT in the form of a thermal aerosol at the rate of 0·012 g/m^2 of water surface. The total estimated treated swamp area varied from about 1400 to 2200 km^2 during the years 1946–48, and was slightly over 300 in 1949. During 1946 an average of 3·3 sprayings was made. The number increased to 6·2 in 1947, was

Table 6

Technical and Financial Record of Residual Spray Program for 1946 to 1949

Item No.	Description (Items and representative rates)		1946	1947	1948	1949[1]
1	Villages and towns sprayed . .	No.	4,139	5,780	4,798	5,410
2	Houses sprayed	No.	487,300	663,600	678,304	689,124
3	Total population of sprayed villages	No.	2,852,000	3,490,000	3,530,024	3,981,417
4	Total quantity of final solution or emulsion used	l	3,602,700	6,232,900	5,883,809	6,030,934
5	Average quantity of pure DDT used per capita	g	51	78	81	76
6	Foreman and spray-man days spent	No.	67,885	102,354	99,214	96,898
7	Average rooms sprayed per man day	No.	42	40	42·4	45·7
8	Average cost per protected person	$	–	0·31	0·31	–

[1]) Figures collected by November 10, 1949.

5·8 in 1948, and 7·3 in 1949. The spraying cycle varied from about 23 to 49 days during the spraying seasons of the four years.

The total cost of the program in 1947 was $1,444,000 and in 1948 was $1,495,585. The annual cost of the program is approximately the amount saved each year by the reduced consumption of antimalaria drugs. The amount of pure DDT (100%) used was 187·5 in 1946, 333·5 in 1947, and 364·2 in 1948 (LIVADAS *et al.* [335]). Against this cost it has been calculated that approximately 19,000 human lives were saved during the period 1946–49 (based on mortality estimates for the period 1926–37, i.e., 5,032). Sickness avoided during the same period was estimated to be 80,000,000 work days (PAMPANA [424]). According to HEDLEY [258], farm production in some areas has increased almost 40% as a result of the conquest of malaria.

After the experimental program of 1945, LIVADAS and ISSARIS [334] noted a virtual disappearance of mosquitoes from treated houses, and this condition continued during subsequent operations (LIVADAS and BELIOS [332]; LIVADAS [329]). It was not until May of 1949 that regular catching stations were established, which consisted of untreated quarters in sprayed areas in Attica. Systematic examinations of anopheline breeding areas were also instigated at this time (LIVADAS [328]).

As a result of entomologic inspections, no *A. sacharovi* or *A. maculipennis* adults or larvae were found in the Northern Sector except for 12 *A. sacharovi*

Table 7

Comparison of Parasite Indices of School Age
Children (6–13 Years) by Homogeneous Area Groups

Groups of areas	Number of areas and total population	Approximate average parasite index % [1] 1933–1945	Parasite index during DDT program years %		
			1946	1947	1948
Group 1	84 174,817	16·36 (39,333)	1·67 (6,255)	0·24 (7,428)	0·10 (6,735)
Group 2	17 25,232	19·94 (5,215)	0·63 (1,102)	—	0·09 (1,053)
Group 3	44 112,831	17·52 (17,924)	—	0·18 (4,403)	0·15 (3,947)
Group 4	24 26,269	22·10 (3,036)	—	—	0·32 (1,258)
Group 5	18 11,504	—	2·64 (1,250)	0·41 (1,196)	0·17 (1,198)
Group 6	3 2,072	—	5·67 (141)	—	1·25 (159)
Totals		17·23 (65,508)	1·74 (8,748)	0·23 (13,027)	0·15 (14,350)
Number of areas		169	122	146	190
Population		339,149	213,625	299,152	352,725

[1] Number of persons examined in parenthesis.

females found in one treated barn filled with hay. In the Eastern Sector of Attica a few *A. sacharovi* adults were found, but these disappeared during the season of their normal highest potential. *A. sacharovi* and *A. maculipennis* larvae occurred in the Eastern Sector throughout the survey. Only two *A. superpictus* adults were found in the Southern Sector, and these in an unsprayed tunnel. Larvae of *A. superpictus* were found in small numbers in all three areas. It was noted that *A. claviger* had a tendency to occupy the areas abandoned by *A. superpictus*.

MANDEKOS and DAMKAS [358] and MANDEKOS [355] stated that *A. sacharovi*, the most important malaria vector in Greece, could be eradicated by the use of DDT. They believed, however, that the eradication of *A. maculipennis* and *A. superpictus* would be difficult if not impossible because of their zoophilic habits and wide distribution. In areas where the latter two species were found, it was recommended that control be carried on over a long period with reduced dosages of DDT, in order to keep the vector population below the transmission threshold.

Comparative epidemiologic data are the key to the efficacy of any disease control program. The lack of a uniform reporting system on malaria in Greece precludes an epidemiological analysis of the precise changes which occurred after the institution of the DDT control program in 1946. The intensification of malaria control activities at that time included, however, an increase in mortality and morbidity surveys by the Malaria Service and by other groups, which appear to have produced sufficient data to permit general conclusions as to the efficacy of the program.

LIVADAS *et al.* [335] and LIVADAS [326] have summarized the results of three years of work, which are presented in Table 7.

There was a total of 13,861 children examined to March 8, 1949, of which seven, or 0·05%, were positive, indicating a continuing downward trend.

An examination of 558 infant blood smears from 57 villages in the autumn of 1946 revealed 0·18% positives. In the autumn of 1947, 823 blood smears examined from 70 villages, were all negative. This was likewise the case in 1948 when 1,125 infant smears from 136 villages were all negative. This condition existed to the last report in March 1949.

Examinations made in the untreated village of Rodonia showed a parasite index of 24% among school children in 1945 and this had risen to 33% in 1946, at which time the infant index was 9%. Two comparable villages, Zelifton and Kastri-Paliouri, which were treated in 1946 showed a parasite index of only 8 and 2·7%, respectively. An outbreak of malaria occurred in August 1947 in a small settlement in Attica which had not been treated during that year.

In all treated villages of Attica, the parasite index was nil. In a treated village from which the population migrated to low farm lands during the summer, the parasite index remained high.

Spleen indices of school children made during 1946–48 in treated areas showed a progressive decline of one half to over two thirds of the prewar incidence. There was also a reduction in spleen volume, IV degree splenomegaly had practically vanished, and classifications II and III were rare.

There was a marked decline in the number of cases of malaria incidence in the Army from 1946 to 1949 even though the number of men was progressively increasing.

Mortality statistics from malaria in Greece are very poor, but LIVADAS [326] stated that deaths from malaria had been reduced to a point nearing zero.

Additional evidence of malaria reduction was the decline in the consumption of quinine. From an average annual total of 30,000 kg used there was a drop to 3,930 in 1947 and 2,320 in 1948. LIVADAS [326] stated that most of the reduced quantity was consumed for other than anti-malaria purposes. He also believes that the consumption of other antimalaria drugs was very limited.

Probably the most recent information on the results of the malaria campaign in Greece is by MARK and BROWN [363]. Figures 4 and 5, showing the malaria deaths as reported from towns with over 5,000 population and the malaria parasitic index in infants and children, illustrate further that malaria has, at least for the present, been conquered. One of the difficulties in corre-

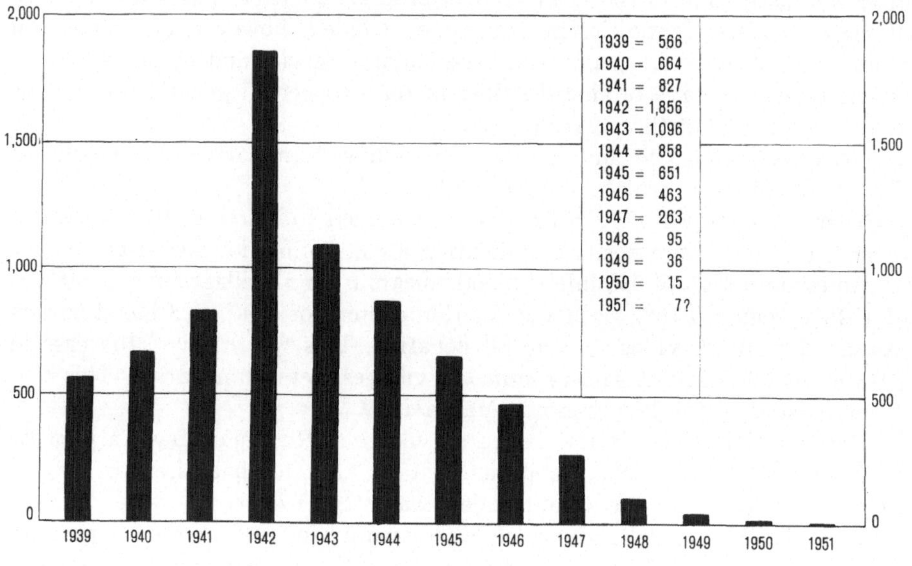

Figure 4

Malaria deaths in Greece as reported from towns over 5,000 population.

lating the epidemiological data on the downward trend of malaria is occasioned by the fact that consistent reporting from the same area is infrequent. The type of reporting also varies considerably, one worker giving information on death rates, another on parasite indices, another one on spleen rates, and often from either overlapping or widely separated areas.

All of the foregoing evidence presents a striking example of the efficacy of DDT in malaria control, particularly when compared to the 1933–1945 period. If, however, the epidemiological data available in Macedonia and Thrace for the period following the 1942 epidemic are examined, there is noted a definite and sharp decline in malaria incidence during the two years preceding initiation of the DDT program. FOY *et al.* [191] working in Macedonia and Thrace showed that a major reduction in malaria incidence occurred several years prior to the use of DDT. Malaria, which had constituted 28·4% of total admissions to two large hospitals in Salonika in 1942, had dropped to 13·7% in 1943, to 8·3% in 1944, to 2·3% in 1945, and to only 1·7% in 1946. There was likewise a similar reduction in malaria mortality from 1·6% of total admissions in 1942 to 0·01% in 1945. The mortality was 0·5% among the one- to five-year age group in 1945, with no deaths for the age group beyond five years.

There was a great variation in the parasite and spleen indices in the fifteen areas surveyed, and a comparable variation from season to season, with the lower indices occurring during the winter. In the winter of 1946, spleens of

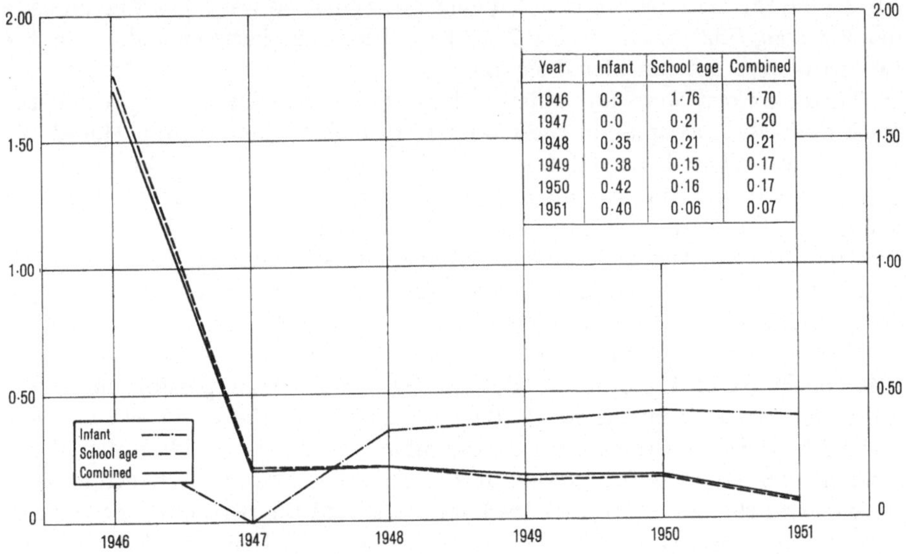

Year	Infant	School age	Combined
1946	0·3	1·76	1·70
1947	0·0	0·21	0·20
1948	0·35	0·21	0·21
1949	0·38	0·15	0·17
1950	0·42	0·16	0·17
1951	0·40	0·06	0·07

Figure 5

Malaria parasitic index shown as percent positive blood slides
for infants and Children of school age, Greece.

11,905 school children (4–14 years old) were examined, together with 7,331 bloods. The parasite-spleen ratio was observed to have fallen from its previous usual figure of 1:2–1:4 to a low of 1:30–1:70, which indicated a low transmission rate.

The geographic and seasonal variations encountered in parasite and spleen indices did not appear to be correlated with DDT treatment. Children from Peponia in the Serres region which had been treated with DDT had a spleen index of 82·4% and a blood index of 4·0%, whereas those from Skitholithos which had not been treated with DDT had a spleen index of 94·4% and no positive bloods in 36 examined.

Bloods were taken from 2,207 infants under one year of age, who had lived through not more than one transmission season, of which 21 were positive, giving a parasite index of 0·9%. One of the highest indices encountered was in the village of Filippi which had been properly treated with DDT. A total of 1,528 spleens were palpated and 913 children's bloods and 311 infants' bloods were taken in 18 villages in the Evros region. Twelve of the villages had not been treated with DDT or had been treated at such a time that it was not effective. Two positive infants (16%) were found in the 12 untreated villages and one positive (16%) in the six treated villages.

In the treated areas of Macedonia and Thrace the mean parasite rate among 4,637 children was 1·2% and the mean spleen index among 7,364 children was 28·2%. The mean parasite index among 1,379 infants was 1·1%. In untreated areas, the mean parasite index among 2,909 school children was 0·67% and the mean spleen index among 3,675 children was 23%. The parasite index among 732 infants was 0·7%. The differences between indices of the two areas were not considered significant.

These workers were not able to draw concise conclusions regarding the place that DDT occupied in the reduction of malaria incidence in Macedonia and Thrace. They believed, however, that it played a very limited part. *A. superpictus* is an important vector in Macedonia and Thrace. Its breeding habits, zoophilic tendencies, the inaccessibility of the endemic areas, the poor construction of houses and stables, and the habits of peasants of sleeping out of doors were factors considered as making control by insecticides none too easy a problem.

MANDEKOS *et al.* [359] working in Macedonia during the period 1934 to 1946 concluded, on the other hand, that DDT would make possible the eradication of malaria from the most affected countries of the world within a few years. These workers reported a considerable drop in the number of patients reporting for blood examinations at the Salonika Malaria Laboratory concurrent with the use of DDT in 1945 and 1946, and the percentage of positive cases likewise decreased. Examination of school children revealed that the decline of the parasite index was more conspicuous in 1945 and 1946 when DDT residual spray was first generally used.

It is obvious that there is a difference of opinion regarding the efficacy of DDT in the control of malaria in Greece, but it is interesting to note that FOY *et al.* [191] and MANDEKOS *et al.* [359], although both groups worked in Macedonia, did not survey the same villages. Of the 32 villages in Greek Macedonia used for parasite surveys by MANDEKOS, only two were in common with the fifteen surveyed by FOY. In both instances the number of positives in many small areas were often too little to establish a reliable percentage index. There is much logic in questioning the value of DDT in a situation such as existed in Greece, where the malaria incidence was rapidly declining prior to the initiation of the DDT program. This same phenomenon existed in the United States, in much of South America, and in other places. It is generally recognized that malaria often appears in cycles, and in Greek Macedonia and Thrace this has been stated to be every two to five years (LIVADAS *et al.* [335]). Based on information available from most countries including Greece, where

DDT has been widely and properly employed, the pre-DDT downward trends are rarely so steep nor reach such low levels as occur following the use of DDT. Subsequent to the controversy in question, malaria in Greece as a whole has continued downward to a point where it is no longer an important problem. The use of DDT in certain other countries, where there was a high incidence of malaria coincident with the initiation of DDT spraying activities, was followed by a dramatic drop. This would appear to be further evidence that in those countries where the disease was on the decline, the trend was considerably accentuated by the use of DDT insecticides. We would guess that Greece is no exception to this generalization.

Malaria in Other European Countries

Following World War II, countries traditionally devoid of malaria had cases among returning veterans, but in general these did not create a serious problem in producing an epidemic of malaria.

Bulgaria, Hungary, Roumania, and European Russia, as well as most other European countries where malaria occurs, have a control program utilizing modern insecticides including DDT. There are, however, little data available on the results obtained, but the fragmentary information at hand indicates that the results have been relatively consistent with those in other European countries.

Asia

Turkey

Turkey has been malarious since antiquity, particularly in the coastal areas where now much of the country's estimated 22,000,000 population is concentrated. Malaria has played a leading role in the migration to these areas from the West and Southwest of Turkey where many ancient cities have disappeared or lie in ruins as a result of malaria, earthquake, and decadence. As the Ottoman Empire began to shrink, Turks returning from the Balkans and tropical areas in the Near East introduced new strains of malaria to susceptible local populations, causing widespread epidemics out of proportion to all previous experience (KRATZ et al. [316]). Malaria is today a disease of villages rather than cities in Turkey. At Antalya, on the southern coast, the spleen rate about 1923 was around 86%, but the disease has subsequently been largely conquered by the very effective control operations.

Turkey can be divided into four topographic zones. The coastal plain on the north, west, and southwest comprises about 10% of the total land area and is the most densely populated. Rainfall is plentiful throughout. Malaria is more prevalent on the Aegean and Mediterranean coasts. Water from the mountains, flood plains of rivers entering the sea, and outcroping of the water

table provide ample breeding places, and in southern areas breeding is continuous.

Immediately behind the coastal plains lies a saucer-shaped plateau, 1,600 to 5,000 ft in altitude, which occupies approximately one half of the land area of Turkey. This region — Anatolia — is the heart as well as the bread basket of the country. The hot and dry summers limit malaria to villages adjacent to water courses or to the numerous brackish-water marshes or lakes. Breeding occurs only from June to October.

The Anatolian plateau ends abruptly at the eastern end where the mountains become progressively higher, rising to Mt. Ararat (16,946 ft) on the Russo-Iranian border. Because of the rugged country, towns are forced to locate in narrow strips of land in river valleys. The winters are very cold, and the short summers limit mosquito breeding so that only along the Russian frontier where the land falls away rapidly can malaria be a threat.

The foothill region in the Southeast bordering on Syria, Iraq, and Iran is the most undeveloped in Turkey and shows high spleen rates. The peak rainfall is in the winter and accounts for about one half of the annual precipitation. The primitive living conditions and the long summers create a serious malaria problem.

The distribution ranges of two major malaria vectors overlap in Turkey. The country lies in the western portion of the range of the more Asiatic *A. superpictus* and in the eastern portion of the range of the European *A. sacharovi*. In addition the seasonal prevalence of these two species complement each other, *sacharovi* populations being greatest from spring to midsummer and *A. superpictus* becoming more abundant from midsummer through fall. These two species are the principal vectors and the success of any antimalaria program rests on their control. *A. maculipennis* is the most important secondary vector.

Certain native practices are conducive to malaria transmission, thus adding to the problem of malaria control. Irrigation is practiced extensively and ranges from continuously inundated rice fields to nighttime watering of cotton fields. The danger from rice fields is somewhat diminished by a Turkish law which requires that all rice fields nearer than two miles to a village be dried for 48 hours every 10 days. However, this is the only crop so regulated. Generally there are few fences in Turkey, and this, plus the custom of night irrigation, necessitate night crop-watching against both animal and human marauders. The guard — and frequently his family — being the only warm-blooded animals in the vicinity, are abundantly bitten by capable vectors. Like many other countries in the Near East, many people, to escape the heat, sleep on rooftops with no protection from mosquitoes. In addition, transient farm labor and migratory herdsmen serve as efficient reservoirs of malaria, transporting the disease from area to area.

Malaria had become such a burden in the years following World War I and during the early years of the republic that by 1925 control activities were begun. These were in continuous operation until World War II with fair progress. The wartime economic crisis caused a breakdown in the program thus bringing widespread epidemics to the Aegean and Mediterranean regions. In 1944, specific control areas were set up but these were virtually ineffective in the face of wartime shortages. In 1946, a separate Directorate of Malaria Control was established by the National Assembly with broad authority to control malaria. New legislation gave control workers virtual police powers.

After World War II, the old systems of larviciding, drug therapy, and prophylaxis continued to be used and the change to DDT residual spray techniques was slow. The growth of this method of malaria control in Turkey is exemplified by the progressive annual increase in the amount of DDT available, as shown in Table 8. In the early years when amounts of DDT were limited, it was used only in special circumstances. Full effectiveness of the spray was not at first realized since it was used largely as a DDT-xylene-triton emulsion or as a DDT-fuel oil solution on highly absorptive adobe walls. Since 1951, however, 75 per cent water-wettable DDT has been used for most rural home spraying activities. The 30 per cent DDT-xylene-triton emulsifiable concentrate continues to be used for residual spraying of non-porous surfaces, for other surfaces where spotting by water-wettable DDT is objected to, and for larviciding.

Table 8

Malariometric Data for Turkey
Correlated with the Use of DDT in the Control of Malaria Vectors[1])

Year	DDT available metric tons	Population in control areas	Per cent examined	Per cent with positive spleens	Per cent reduction
1946	2	6,032,573	78·1	25·9	
1947	18	6,036,073	73·9	19·8	24
1948	62	6,403,475	70·3	14·7	26
1949	99	5,398,767	80·1	10·5	28
1950	110	5,307,420	85·6	7·1	30
1951	184	5,717,394	90·9	4·3	54
1952	936[2])	5,983,502	75·5	2·1	40

[1]) From KRATZ *et al.* [316].
[2]) This proportionately larger amount includes a part of the 1951 supply of DDT which arrived too late to be used that year.

Examinations for malaria in rural populations have been carried out since 1946. The percentage of the population examined in the control areas each year has ranged from about 70 to 90%. Prior to 1951, only spleen examinations were done routinely, giving a higher number of cases than actually existed. The above table gives the results of malaria examinations since 1946, which shows a decline in the disease paralleling the increased use of DDT.

The control of malaria has also had a favorable effect on birth and death rates in Turkey (AMERICAN MEDICAL ASSOCIATION [13]). In 1943, the birth rate in the malarious region was 29⁰/₀₀ and the death rate 19·68. By 1945, the death rate had dropped to 12·37; and by 1948, the birth rate was 30·5 and the death rate 11·45.

In Turkey, DDT residual spraying at the rate of 200 mg/ft² has broken the chain of infection. There are no deaths now from malaria and the index in malarious regions has dropped from 30·5 to 1·4% [586]. In addition, DDT has been responsible for the control of many other insect household pests, thus making the program very popular. Where DDT-resistant flies have developed in urban areas, a chlordane-DDT-xylene-triton mixture has been used. The only adverse effects noted from the use of DDT has been a higher mortality among silkworms. Where this is a major industry, special precautions are now taken to preclude injury.

Cyprus

Cyprus, an island in the Mediterranean Sea, is about 40 miles distant from Asia Minor; it is approximately 60 miles wide by 140 miles long and contains a population of about 450,000. The average annual rainfall is 18 in. As early as 1394 it was noted by NAMIK KEMAL, a Turkish poet, that 'fever and bad air' were a source of much trouble in Cyprus (SHELLEY and AZIZ [507]). Following the British occupation of the island in 1878, troops suffered severely from malaria and in the year 1900, 470 of 503 patients examined were positive for blood parasites. These were divided equally among *P. falciparum* and *P. vivax*, with an occasional *P. malariae* being present. Sir RONALD ROSS visited the island in 1913 and found that 25·4% of the children had splenic enlargements. Upon his recommendation, drainage was undertaken with a consequent improvement in malaria endemicity. AZIZ [27] reported that blood parasite indexes reached 70% in some villages before instigation of the *Anopheles* eradication program.

There are three vectors of malaria in Cyprus, *A. superpictus*, *A. sacharovi* and *A. bifurcatus*. In the case of *superpictus*, a sporozoite rate of 7·8% has been found which exceeds that found in *A. gambiae* in certain malarious regions of Africa (AZIZ [27]; SHELLEY and AZIZ [507]). A DDT spraying campaign was started in April 1946 to eradicate anopheline vectors from the island. An area of 500 square miles, with 34,000 inhabitants was chosen for the initial attack. Unlike most other campaigns, this one was directed almost entirely at the immature or larval stage of the vectors. AZIZ [28] doubted that residual spraying would be effective to control or eradicate malaria in Cyprus because so many of the people, such as foresters and shepherds, sleep in the open and consequently contract malaria outside their places of abode. The larvicide used consisted principally of 3 to 4 per cent solutions of DDT in gas oil (SHELLEY and AZIZ [507]). 'Malariol', Paris green, and fuel oil were used in some cases. During the winter months adult anophelines were searched for and attacked with DDT spray in their hibernating places.

Following the first year of the program malaria diminished and anopheline and other household pests practically disappeared. In 1947 the work was extended to about half of the island, which included an area of about 2,000 square miles. Following extensive larviciding during the breeding season, hibernating places of adults were again attacked with DDT residual spray. The campaign in 1947 resulted in a further drastic reduction in malaria and its vectors (AZIZ [29], [27]). The 1947–48 campaign was island-wide and in 1949 it was reported (SHELLEY and AZIZ [507]) that no adults had been discovered in the entire island for 13 weeks, and it was anticipated that the eradication campaign proper could be brought to an end within 2 months.

Although the malaria incidence had been declining somewhat since 1944, the DDT campaign achieved spectacular results. Two years previous to the initiation of the campaign, 7,686 cases of malaria were reported, and this was reduced to 406 in 1948, of which only 3 were considered to be recent infections. Splenomegaly and parasitaemia were found in 31·4 and 51·9%, respectively, of school children examined in 1944, but this had declined to 10·6 and 1·3%, respectively, by 1948. Of 655 infants (under 2 years) residing in formerly highly malarious villages, all were negative for blood parasites in 1948 (SHELLEY and AZIZ [507]).

The campaign in Cyprus is of especial significance since it demonstrates the feasibility of malaria eradication by the use of larvicides. Such a scheme is worthy of consideration in those areas where house treatments would be expensive and time-consuming, or where many of the residents sleep out of doors.

The Levant

While the countries of the Levant (Lebanon, Israel, Syria, and Jordan) are defined politically, geographically they are a single unit consisting of a narrow coastal plain, a coastal range of mountains rising in places to a height of 9,000 ft, a rift valley which varies in altitude from +3,000 to −1,300 ft, an inland mountain range, and east of this a vast expanse of desert (HACKETT [241]). Rainfall decreases from west to east.

A. sacharovi is the most important malaria vector of the region and is well adapted to the hot dry climate (HACKETT and MOSHKOVSKI [243]). On the coast the vector is more tolerant to salt and breeds in marshes and lagoons, but as it proceeds eastward, it becomes more and more of a fresh-water form until in eastern Israel (Palestine) and beyond it breeds only in standing fresh water and inland marshes. This mosquito does not breed in rivers, but throughout almost its entire range it is accompanied by *A. superpictus*, a stream-breeder. *A. claviger* breeds in cisterns or other water stored underground and is one of the few urban vectors. Around oases in the eastern desert, *A. sergenti* is responsible for 'oasis fever'.

Malaria in the Levant is the most important and the most prevalent disease (HACKETT [241]). Its incidence is relatively light in regions where there are practically no streams or marshes such as coastal Syria and Lebanon; but elsewhere it prevails on the coast, in the rift valley, in the hills, and in the oases. Malaria was once highly endemic in the big cities of Israel, Lebanon, and Syria. During the occupation of this area in World War I, army officials either closed or oiled all cisterns. These measures were continued after the war and in Jerusalem, for instance, the spleen rate dropped from 44% in 1919 to 2% in 1940 (HACKETT and MOSHKOVSKI [243]).

A number of studies have been made in Lebanon to compare the values of chloroquine and DDT as malaria preventives. Two villages, Saideh located in the Bekaa plain 18 km southwest of Baalbeck and Azouniyeh lying in the Lebanon mountains south of the Beirut–Damascus road, had long suffered from hyperendemic malaria (BERBERIAN and DENNIS [42]). In 1946, chloroquine was administered to the inhabitants, and a malariometric survey made prior to the spring of 1947, when before the seasonal appearance of anophelines, DDT residual spray was done and repeated in midsummer. The village of Amik on the Bekaa plain, also in a hyperendemic malaria zone, was selected to study the effect of DDT residual spray alone on spleen indexes (BERBERIAN and DENNIS [41]). Results were compared with those obtained in Saideh after drug prophylaxis alone. In Amik the splenic index was reduced from 70 to 50% over a period of eight months while in Saideh the reduction was from 59 to 6%. From these field experiments, it was concluded that weekly medication with chloroquine combined with adequate DDT antimosquito measures is the procedure of choice in eradicating malaria from a community, since it eliminates both human and insect reservoirs of the disease.

Malaria incidence in the Israel portion of Palestine in the years immediately following World War II was as follows (MER [371]):

1945	1,207
1946	466
1947	226
1948	1,227

The sudden rise in the number of cases in 1948 came with the addition of 150,000 persons to a relatively static population numbering some 600,000. In addition, about 500 cases can be added to the 1948 figure as a result of long latent *vivax* infections, which did not manifest themselves until the spring of 1949 after the discontinuance of proguanil prophylaxis. Malaria control in Israel in 1945 included the classic methods, larviciding, screening, bed nets, and repellents. Where malaria prevalence was high, pyrethrum spray was used against adults. In this year, the first experiments were made with DDT as a residual spray. In 1946 and 1947, DDT residual spray, in addition to antilarval work, was carried out in hyperendemic areas of all Jewish and Arab villages where the vector is *A. sacharovi*. In other malarious areas all Jewish villages and some Arab ones were treated. In 1948, spraying continued in all villages, some of which due to the political situation were in ruins. Many inhabitants slept in the open, in tents, or in dugouts. These shelters were sprayed with DDT, and repellents were also used. Breeding areas were oiled with DDT in kerosene. Proguanil was administered, but only after the appearance of the first case in a community. More and more people availed themselves of this until by fall about 80% of the population was taking proguanil, 0·2 g biweekly. In the DDT residual spray program, a 5 per cent solution of DDT in kerosene, to which 0·5% of sesame oil was sometimes added, was applied to the walls and ceilings of buildings at the rate of 1 g of technical grade DDT per square meter. In 1946 and 1947, spraying in *A. sacharovi* areas was done three times a year, in February, May, and September; for *A. superpictus* and *A. sergenti* spraying was done twice each year, late spring and early fall. It was noted that in hot weather two months after spraying, the number of anophelines rose; therefore, in 1948 the spraying schedules were altered so that treatment was done every 4 to 6 weeks with a subsequent reduction in anophelines. Breeding of *A. sacharovi* was considerably reduced as a consequence of residual spraying, but *A. superpictus* and *A. sergenti* seemed unaffected. In 1948, an essential difference in anopheline behavior was noticed. Whereas in years before, biting by anophelines out of doors was minor, in 1948 it was necessary to use repellents to protect against them. The significance of this phenomenon is yet to be determined.

Iran

Of the estimated 16 million population of Iran, some 11 million live in about 43 thousand small villages averaging 50 families or 250 persons per village (ZIONY [668]). Two fifths of these villages lie in areas where malaria has been intensely endemic and another fifth where malaria has occurred intermittently. Malaria in Iran has constituted a substantial hindrance to the economic development of the country. Any reduction of this disease would improve the standard of living, particularly through increased agricultural production.

Malaria is reported from all 10 Iranian provinces but is most prevalent in those bordering the Caspian Sea and in those with the most land under irrigation. In the Caspian provinces malaria mosquitoes breed both in natural swamps and in rice fields.

The anophelines which are probably of most significance in malaria transmission in Iran are *A. superpictus*, *A. sacharovi*, *A. stephensi*, and *A. maculipennis*. In northern Iran the malaria season extends from May to October, July being the peak month; in the south the season starts earlier and continues later in the year. New cases and relapses have totaled as much as 3 million or more per year (ZIONY [668]).

During World War II, the US Army started malaria control at its base in Khorramabad. In 1945, DDT was introduced to combat typhus and when the Army departed, this was left for use in malaria control. Iranians who had been trained at Khorramabad got antimalaria operations under way at Isfahan and Shiraz. These were extended in 1946 to Sanandaj, Kermanshah, and Bushire, and in 1947 to Khorramshahr, Ahwaz, Palisht area, and Parchin.

Starting in 1948, the Near East Foundation undertook a DDT spray program in the Caspian region (NEAR EAST FOUNDATION [407]). A spray program started in 1949 in Baluchistan province resulted in the eradication of malaria mosquitoes and other disease-bearing insects from 3 cities and 51 villages of the province. The Anglo-Iranian Oil Company has carried on malaria control in company areas for a number of years (ANDERSON [14]). At the large refinery at Abadan in 1944 and 1945 they cooperated with the armed forces in a fairly extensive antilarval campaign. After the troops moved out at the end of 1945, malaria incidence rose. In 1947, the company obtained DDT and started a residual spray program, and the high malaria incidence of 1946 was markedly reduced. Where long-standing antilarval campaigns had been operated in the field areas, an apparently irreducible minimum in malaria infection had been reached, but in 1947, DDT was added as a residual spray and malaria has declined sharply.

In July 1949, the Majlis, the Iranian Parliament, enacted a bill for a seven-year plan for the economic improvement and development of Iran. Of the money allocated for operations under this plan, $7 \cdot 2\%$ was designated for public health. Antimalaria activities were outlined as a 4-year program, the first year (1949) to be a demonstration project and the next 3 years, large-scale operations designed to give maximum coverage by the end of 1952.

Tests with DDT sprays were conducted in 1949 in 3 comparable villages near Varamin (DOW [149]). One village was sprayed with a 5 per cent solution of DDT dissolved in kerosene at the rate of $0 \cdot 2$ g of DDT per square meter, a second one with a water suspension of 50 per cent wettable powder applied at the same rate, and the third left unsprayed. Prior to the treatment, the average number of female *A. superpictus* found per inspection station was $21 \cdot 0$, $22 \cdot 4$, and $33 \cdot 5$ for each of the 3 villages. After one round of spraying these figures averaged $0 \cdot 27$, $0 \cdot 0$, and $26 \cdot 9$, respectively, over a 6-week period.

Since the spring of 1950, a World Health Organization advisory malaria control unit has been assisting the Iranian Government in its national program of malaria control (WORLD HEALTH ORGANIZATION [660]). That year $70,000,000$ m^2 of surface in $404,690$ houses were sprayed with DDT, protecting some $700,000$ persons (WORLD HEALTH ORGANIZATION [655]). Entomologic surveys made in 1950 revealed that mosquito trap catches were reduced by between 76 and 91% during the first five days following the treatment of buildings with DDT and continued to decline thereafter (GARRETT-JONES [215]).

Everywhere in Iran where DDT spraying has been done the incidence of malaria has declined abruptly. In one village a spleen rate in children of 89% established by the World Health Organization in 1949 dropped to 50% by the next spring (WORLD HEALTH ORGANIZATION [646]). In 1951, only 21% of children had enlarged spleens and most of the youngest ones born since DDT spraying was begun showed no signs of infection.

The malaria program in Iran is now (1953) sponsored jointly by the Iranian Ministry of Health, the Institute of Malariology of the University of Tehran

School of Medical Sciences, the World Health Organization, and the US Foreign Operations Administration (JOHNSON [298]). In 1951, during the early part of the program, 20 t of DDT 75 per cent wettable powder had to be airlifted to Tehran to prevent collapse of the program. During 1952, the houses in approximately one third of all the villages of Iran were sprayed with DDT at the rate of 2·3 g/m², protecting about 4,000,000 of the estimated 5,000,000 persons who live in malarious areas. These operations required 1,490 t of DDT. Between March 17 and October 25, 1953, a total of 4,244,000 people were protected by spraying approximately 16,000 villages. A total of 1,650 t of DDT were used at the rate of 2 g of technical DDT per square meter. During the 1954 campaign, it was estimated that 1,800 t of DDT would be needed.

The results from the spraying are striking. Physicians in the Caspian Sea area, for instance, each formerly saw as many as 100 malaria patients daily; now they see 10 to 15 per season. Agricultural productivity per individual has also been greatly enhanced.

Flies that are resistant to DDT have been encountered, and this is believed to be the result of DDT spraying prior to 1952. No DDT resistance has been observed in the mosquito population to date.

Afghanistan

In 1949 a field experiment on malaria control using DDT indoor residual sprays was carried out in Laghman district of the Eastern Province of Afghanistan (RAO [466]). This district is mountainous with several peaks rising to over 10,000 ft above sea level. The hills are rocky and practically devoid of vegetation, but the valleys which lie about 2,000 ft above sea level are well irrigated and are intensely cultivated, the chief crops being rice in summer and wheat in winter. Wheat being a dry crop, the fields are irrigated only twice to wet them, but from June to October, during the rice-growing season, the fields retain standing water. The two rivers crossing the district have rocky bottoms and provide good breeding places at all times for the vector anophelines, *A. culicifacies* and *A. superpictus*. The climate is dry and the little rain that falls does so mainly in the winter.

Prior to the advent of DDT, malaria was hyperendemic in the area, the season of transmission extending from May to October. The infant parasite rate was 70%. DDT as an indoor residual spray was applied in August 1949 in 13 villages with a total population of 14,000. DDT emulsions and water-wettable powders both were used at the general rate of 200 mg/ft², and anophelines remained under control even 12 weeks after spraying. No differences were observed between the formulations. At the end of the season the average cumulative spleen rate in sprayed villages was 62·0% and in unsprayed ones 89·1%. Infant parasite rates were 8·3% in sprayed villages and 70·0% in unsprayed ones.

In 1952, as a result of the successful World Health Organization-assisted project in North Afghanistan antimalaria campaigns were initiated in ten new areas, protecting a population of some 385,000, where the disease was a serious deterrent to economic development (WORLD HEALTH ORGANIZATION [647]). Spraying was carried out in 210 villages over an area of 820 square miles and effectively protected a population of nearly 150,000.

After four years of intensive antimalaria operations undertaken by the Government with the assistance of the World Health Organization and the United Nations International Children's Emergency Fund, malaria was controlled among approximately two thirds of the total malarious population of the country (WORLD HEALTH ORGANIZATION [637]). It was planned to expand activities to new areas during 1954 and 1955 and this was expected virtually to rid the country of malaria. Striking economic benefits of malaria control were already evident. In the town of Pulikhumri, where the total population had been 5,000 and where the output of textile mills located there had been seriously affected, improvement in health conditions resulted in an increase in the population to 20,000 and in the doubling of production at the mills. In agricultural areas such as Kunduz and Khanabad districts of Kataghan Province, large tracts of land have been brought under cultivation since antimalarial activities started in 1950 and the annual yields of rice and cotton crops have almost doubled, due primarily, it is thought, to the decline in malaria prevalence.

Pakistan

Baluchistan: Malaria in the area around Quetta, the administrative center of Baluchistan which lies on a high plain at about 5,000 ft altitude, is highly endemic and is transmitted by *A. superpictus*, *A. culicifacies*, and *A. stephensi* (AFRIDI and BHATIA [3]). These mosquitoes favor springs, irrigation channels, rivers, and perennial streams as breeding habitats. The transmission season is relatively short, mid-July to mid-September.

In 1946, a trial control program was activated to determine whether malaria in a rural population could be controlled by periodic spraying with DDT to kill adult mosquitoes, whether this control measure could be conducted at a reasonable cost, and whether spraying operations could be successful where purdah (seclusion of women) is practiced. Earlier experiments in this region (PURI and BHATIA [456]) had shown that DDT in a kerosene solution sprayed at the rate of 50 mg of DDT per square foot was effective against anophelines for nearly ten weeks, which would cover adequately the transmission season. While emulsions and suspensions sprayed at the rate of 25 mg of DDT per square foot reduced substantially both the numbers of vectors and their infection rate, an average dosage of 50 mg of DDT per square foot applied annually was found to be a more effective control measure. However, in the control scheme which predated the initiation of a provincial malaria organization, DDT emulsion was applied twice during the malaria season at the rate of 25 to 38 mg of DDT per square foot; doorways, windows, and other openings being treated more liberally (AFRIDI and BHATIA [3]). The first round of spraying was done in late July after the start of the malaria transmission season, and the second round in late August.

Both indoor and outdoor spraying were tried, but the indoor procedure proved more economical both in the amount of DDT used and in the time necessary to apply it. Spleen and parasite rates fell significantly in two of the three sprayed areas. In the third sprayed area both these indices rose, but it was not believed that this denoted a failure in the effectiveness of DDT but rather that this area, being the only one of the three immediately adjacent to an unprotected area, was subject to introduction of malaria by imported vectors and infected persons.

East Bengal: In May 1949, a joint WHO/UNICEF Malaria Control Demonstration Team started operating in a densely populated rural area of East Bengal with a population of 13,000 (WORLD HEALTH ORGANIZATION [648]). In this area malaria is transmitted by *Anopheles philippinensis* which is prevalent in great numbers from April to July and again from October to mid-November, the transmission season coinciding with that of anopheline prevalence. Spleen rates are high both in adults, 48%, and in children, 70%, with lower parasite rates, 10 and 24%, respectively. DDT as an imagocide was employed as a suspension at a dosage of approximately 200 mg/ft² once each year with the area of control being extended annually (WORLD HEALTH ORGANIZATION [648]). A post-spraying survey revealed that spleen rates in the treated villages dropped to less than one third of those prevailing in June 1949, the figures before and after spraying being 74·5 and 21·2%, respectively (WORLD HEALTH ORGANIZATION [661]). On the other hand, in the unsprayed villages spleen rates rose from 74·5% in June 1949 to 81·1% in January 1950. The differences in parasite rates were not as spectacular, but significant, being 5·78% in sprayed areas as compared to 18·2% in unsprayed ones. Entomologic observations revealed that DDT remained effective for six to seven months after application. This was true even after renovation and re-plastering of most of the houses, apparently because the DDT residual on unmolested walls and furniture was sufficient to protect against mosquitoes. Although the accumulation of adequate statistical data is not complete, it was predicted that the agricultural output had been increased materially, due to the increase in available labor as a result of the elimination of malaria (WORLD HEALTH ORGANIZATION [661]).

The presence of the WHO Malaria Control Demonstration Team in 1949 and 1950 led to the formation in East Pakistan of two national teams, assisted by UNICEF (NASIR-UD-DIN [406]). These teams worked in Parashuram, where the vector is *A. minimus* and the population protected was 61,000; and in Gabtali where the vector is *A. philippinensis* and the population 23,000. *A. philippinensis* is the malaria vector on the higher plains of East Bengal, while *A. minimus* is responsible for malaria transmission in the hills and foothill regions (QURAISHI *et al.* [458]). One round of spraying was completed in 1950 using DDT wettable powder on houses and animal shelters at an average dose of 187 mg/ft² in the first area and 107 mg in the second (Nasir-Ud-Din [406]). In the sprayed areas there were significant reductions in the spleen, general parasite, and infant parasite rates, while in the unsprayed areas there were seasonal increases in these rates. Vector anophelines disappeared from the treated areas and tests showed that the sprayed surfaces remained effective for 13 months in the first zone and for 9 in the second. It was recommended, therefore, that spraying be done on an annual basis. It appears this will be sufficient to break the chain of transmission.

India

India has played an important role in the history of malaria. It was there that Sir Ronald Ross made the fundamental discovery that anopheline mosquitoes transmit malaria, and subsequently many famous malariologists contributed to the knowledge and decline of the disease by their work in India.

Geographically, India has four general types of terrain, a low coastal belt around all peninsular India, vast plains which occupy most of central India from the Punjab in the north to Mysore State in the south, the foothills which border this great central plateau on both east and west, and the mountains where altitudes exceed 5,000 ft above sea level. There is a direct correlation between these regions and the incidence of both malaria and vector anophelines. The western coast of peninsular India, including the rice-growing lowlands where *A. fluviatilis* is the principal vector, is almost free from malaria; spleen rates are 5% or less (Covell [115]). In northwestern coastal areas, where *A. stephensi* is the chief carrier, the amount of malaria is usually moderate. On the eastern seaboard there is a considerable amount of endemic malaria around Madras City due to profuse breeding of *A. culicifacies*. Further up the coast, *A. stephensi* is responsible for occasional localized epidemics. In the north, *A. sundaicus*, a brackish-water breeder commonly native to the estuarine parts of Bengal, is invading new areas. It has established itself around Calcutta and in certain parts of Orissa (Covell and Pritam Singh [118]), causing severe outbreaks in many localities.

Malaria in the northwest and peninsular plains is markedly seasonal with a low or moderate endemicity. In wet years the number of infections is greater during the spring or early summer, but in dry years outbreaks occur for the most part in the fall. Within this tract localized areas of high or hyperendemicity may be found which are frequently the result of poor irrigation management. Epidemics may occur, which in mortality can at times be compared with the more spectacular epidemic diseases, cholera, plague, and smallpox. These occur about every eight years, generally in a season of heavy monsoon rainfall accompanied by river flooding after one or more years of insufficient precipitation. The effect of these epidemics is compounded by the lack of communal immunity, as evidenced by low spleen rates in the year just preceding an epidemic. These rates rise to 80 or 90% in the epidemic year and then gradually regress. These outbreaks are directly correlated with the population of *A. culicifacies*, the vector concerned. They occur when climatic conditions favor both the production of enormous numbers of the species and the longevity of the individual insects.

Malaria conditions in the eastern plains are quite different. In areas subject to unobstructed and extensive flooding, such as deltas, malaria is generally of low or moderate intensity. Regions where rivers have become sluggish, due to the deposition of silt in their upper reaches or to the construction of embankments to prevent flooding, are usually highly malarious. In striking contrast with conditions in the northwest, there is little malaria where the subsoil water level is very high; but where it is low, the reverse is true. *A. philippinensis* is associated with such areas and breeds in ponds, pools, and ditches containing submerged vegetation (COVELL [116]).

The foothills throughout India are almost invariably characterized by the prevalence of hyperendemic malaria, associated with blackwater fever and pernicious types of *falciparum* malaria. The aboriginal tribes inhabiting this area suffer malaria infection from childhood and must acquire a very considerable degree of immunity, for adults often show very little evidence of its effects. Nonimmune immigrants, however, are liable to be stricken with intensely severe forms of the disease. Throughout this region there are widespread tracts of fertile country which have remained undeveloped because of the ravages of malaria; even industrial and engineering projects have been hampered or abandoned on this account. The principal vectors of the foothill regions are *A. minimus* in the northeast and *A. fluviatilis* in the south and west and also in the Northwest Tarai in Uttar Pradesh. Both these species breed in stream beds and seepages. They are domestic and have a marked preference for human blood. Malaria incidence in the foothills rises twice during the year, just before and again following the southwest monsoon rains. These wash out breeding places and, when heavy enough, may halt transmission for several weeks. In the north there is a malaria-free period of several months during the winter, but in the south transmission occurs throughout the year.

In areas of India where the altitude exceeds 5,000 ft, malaria is believed to be absent. The vector, *A. superpictus*, which transmits malaria in the adjacent high regions of Pakistan, does not penetrate to the northwest areas of India.

Malaria in India is restricted, for the most part, to rural areas. Only one vector, *A. stephensi*, has been able to adapt itself to urban conditions. In certain of the large cities, Calcutta, Bombay, Bangalore, and others, it causes outbreaks which are often extremely localized. While control is comparatively easy, it is seldom enforced and therefore malaria still occurs far too commonly in many of the great cities of India.

Malaria in the Adaman and Nicobar Islands is transmitted only by *A. sundaicus* and is therefore limited to areas in the vicinity of salt-water swamps or where embankments have been built to protect rice fields from the effects of tides. Villages near such localities are always highly malarious.

Nonimmune immigrants are subject to cerebral malaria and blackwater fever. A high proportion of the infections are with *P. malariae;* May to September is the period of highest incidence and also of maximum rainfall (CHRISTOPHERS [99]; COVELL [117]).

Malaria control in India has long been a difficult problem. In the most malarious regions the inhabitants live in villages composed of small dwellings, often scattered over wide areas. The *per capita* income is very low, which makes local support of antimalaria programs almost impossible. The people have little comprehension of the problem and consequently are apathetic.

Up until about 1930 oiling of breeding places was the main method employed (RAO [467]). Oil is costly and its transport and storage is difficult. When Paris green was introduced, it was anticipated that this would be a cheaper and easier measure. However, difficulties were encountered in obtaining and storing the diluent. Spraying had to be done frequently; the technique of application required skill and patience, so only where workers were enthusiastic were antilarval measures successful. The cost was such that work was necessarily limited to industrial and military requirements while vast rural areas where malaria was the greatest scourge remained unprotected. So-called permanent control measures — drainage, filling, etc. — were considered but were not adopted on a wide scale because of excessive initial costs.

Spraying with pyrethrum to control adult mosquitoes was tried with some success, but the lack of residual qualities made frequent treatment necessary, and this was a limiting factor because of costs involved.

DDT was first employed in India on a practical scale by civilian organizations in 1945 (SINGH [523]). Due to the differences in bionomics of the vector species of anophelines in India, each region, to make the best use of the insecticide, had to analyze its individual problem and organize its antimalaria program accordingly.

In 1953, a malaria control program to protect approximately 75 million people in India was launched by the Technical Cooperation Administration of the United States in cooperation with the Government of India. This was to be in full operation by March 1954 (ANONYM [22]).

According to the best available statistics, there are 100 million cases and 1 million deaths per year from malaria in India. The cost of this disease to the country is unknown. It could be approximated, however, by using figures determined for a representative locality. In the Tarai region of Uttar Pradesh in the foothills of the Himalayas, a World Health Organization malaria program has resulted in a 35 per cent increase in land under cultivation, a similar percentage of increase in grainfood production, and more than 100% increase in grain, edible oil, and certain other industrial undertakings.

Assam: Starting in 1945 and continuing through the 1949 monsoon season, a malaria control study, using both DDT and 'Gammexane' residual sprays, was conducted in a hyperendemic area of Upper Assam (BERTRAM [48]). This region is characterized by high rainfall and is traversed by a number of rivers and streams. The mosquito infestation was heavy and due to the breeding habits of *A. minimus*, the malaria vector which breeds in flowing water, biological control measures could not be successfully carried out. Both DDT and 'Gammexane' were applied either as a five per cent or a one per cent emulsion or as a five per cent or a one per cent solution in grade 2 kerosene. Experience showed that a one per cent DDT emulsion or suspension sprayed at the rate of 1 gal/1,000 ft² every two months would give adequate protection against *A. minimus*. A similar 'Gammexane' emulsion sprayed at the same rate proved effective though not with the same degree of certainty as the DDT.

In one area, however, after four complete treatments during the season with one per cent DDT emulsion, followed the next season by treatment with a one per cent DDT suspension at bimonthly intervals, only small reductions in the general malaria and spleen rates in children between the ages of two and ten were observed. Investigations showed that, while there was no question of the lethal effects of DDT, this compound had an excitatory effect also and a limited repellency, which often prevented mosquitoes from getting toxic doses of the insecticide. Even at a dosage of 215 mg of DDT per square foot about 20% of the mosquitoes survived. While this excitation factor also prevented many of the mosquitoes from feeding in sprayed areas, it did permit the survival of previously infected vectors. No excitation was observed with mosquitoes entering 'Gammexane'-sprayed buildings. It was concluded, therefore, that in Upper Assam, where *A. minimus* was the malaria vector, 'Gammexane' was the insecticide of choice for malaria control.

Concurrently with the above program, a study to assay the effects of DDT and 'Gammexane' on populations of mosquito larvae was conducted (BERTRAM [49]). The insecticidal preparations used were five per cent DDT in malariol, and two per cent 'Gammexane' liquid concentrate L. G. 140 in malariol; in every case the mixture was applied at one point only. The amount of these compounds necessary for the control of mosquito breeding along a 100-yd stretch of a small stream about 2 ft wide and 6 in. deep was reported to be three fluid drachms (approximately 10·65 ml); for a river 35 ft wide the amount to cover an area 1,500 yd in length is 2 gal.

United Provinces: The Tarai region, United Provinces, lies in the foothills of the Himalayas in northern India. The country in general is a tract of forests and swamps with scattered patches of cultivated land (SRIVASTAVA [549]). The northern half consists chiefly of jungle or savannahs of grass and reeds. Springs are numerous and give rise to a number of sluggish winding streams with ill-defined channels. The annual rainfall varies from 47 in. to 65 in.; July and August are the wettest months and November and December the driest. The soil in the region is highly fertile, but the district has been partially abandoned by its inhabitants as the high incidence of malaria has made agricultural work almost impossible (WORLD HEALTH ORGANIZATION [665]).

The streams have been used since earliest times for irrigation, and many simple dams have been constructed (SRIVASTAVA [549]). This system converted arable land into immense swamps which ultimately affected the climate. Day and night temperatures vary greatly, as much as 30° F (SRIVASTAVA and CHAND [548]), and the excessive humidity, between 60 and 90%, combines with the severe heat of summer to render the climate distinctly enervating. These conditions of temperature and general saturation provide an ideal environment for both the mosquito and the malaria parasite, and it is therefore not surprising that malaria is hyperendemic in the area. *P. vivax* is the predominant species of malaria throughout most of the Tarai; however, on the eastern border there is a preponderance of *P. falciparum* (SRIVASTAVA and CHAND [548]). September to December is the period of highest malaria incidence; however, transmission does occur throughout the year. The chief vector is *A. fluviatilis* except for the months of July, August, and September when *A. culicifacies* is responsible (LIVADAS [325]).

Efforts to colonize this region in the past have failed because of the excessive paludism, but development of the Tarai was again considered after World War II when the Government was faced with the problem of locating returned servicemen and displaced persons from Pakistan, and with an acute food shortage. One antimalaria unit was established in this area on September 1, 1947 (SRIVASTAVA [549]). Surveys carried out in the zone selected for this preliminary control effort (Kichha, Naini Tal District) revealed spleen rates varying generally from 50 to 100%, and malaria parasite rates ranging from 0 to 96·5%, the averages among surveyed villages where parasites were observed being 24·5% and among all villages surveyed 7·3%. In planning the control measures, it was decided to use both DDT residual spray and chlorguanide prophylaxis and treatment. DDT was initially applied at intervals of six weeks to all permanent huts, bungalows, and other structures, as a 2·5 per cent emulsion prepared with gum acacia and gelatine. In later applications a 1·25 per cent suspension was used at the rate of 50 mg of DDT per square foot. These measures, combined with the administration of 100-mg tablets of chlorguanide, proved effective. This conclusion is supported by the fact that the developmental scheme proceeded without hindrance from malaria; in fact, new settlers remained free from the disease and the number of fever and malaria cases among residents declined markedly.

After this successful beginning, DDT spraying, using the schedule established at Kichha, was initiated in the Western Tarai in 1949, and malaria transmission was definitely stopped (LIVADAS [325]). Spraying was started one year later in the Eastern Tarai with a remarkable reduction in malaria. At this time the World Health Organization provided medical teams to assist in the program, and the Indian Government furnished the necessary labor and provided physicians and nurses to work alongside the international teams (WORLD HEALTH ORGANIZATION [638]). In addition, the United Nations International Children's Emergency Fund sent transport, equipment, and materials. In 1950 in the Tarai, 778 children were examined in 28 villages and 548 (or 70·4%) had enlarged spleens. Of 187 blood smears examined from children under two years of age, 49% were positive for malaria parasites

(WORLD HEALTH ORGANIZATION [663]). It was estimated that no fewer than 77 children out of 100 were infected (WORLD HEALTH ORGANIZATION [636]). Systematic spraying with DDT was continued, and in 1951 only 3 children out of every 100 had malaria. Before the spraying operations were undertaken, microscopic examinations of blood samples from thousands of infants revealed that from 50 to 75% of them contracted malaria during their first year of life. After the first year of DDT spraying, these figures fell to 2·6%, and after the second year, essentially no babies were becoming infected (WORLD HEALTH ORGANIZATION [649]). By 1952, almost 100 per cent protection against malaria was being given to inhabitants of 1,300 villages. Today, over an area of 1,750 square miles, the scourge of malaria is under control (WORLD HEALTH ORGANIZATION [636]). Since the beginning of the WHO participation in the program, the population has almost doubled; in addition, there has been a 36 per cent increase in land under cultivation, an equal percentage of increase in grain-food production, and more than 100% in grain, edible oil, and other industrial undertakings (ANONYM [22]).

To assist in the proposed industrial and agricultural development of the Tarai, the Government decided to construct a hydroelectric plant which could generate cheap power (SRIVASTAVA and CHAND [548]). This is being achieved by diverting the Sarda Canal which lies at the eastern end of Naini Tal District in the Tarai region. At the time construction started, the working season in the area was limited to seven months of the year because of malaria. The antimalaria scheme went into effect at the site in 1948; soon thereafter laborers were able to work all twelve months and have continued to do so since that time. Systematic indoor residual spraying with DDT was done every six weeks using a 2·5 per cent suspension of DDT at the rate of 50 mg of DDT per square foot in initial and 25 mg in subsequent applications. Spleen rates fell from 12·3 in 1948 to 8·5 in 1950, and malaria rates from 9·8 to 7·8. Spleen, parasite, and malaria morbidity rates have continued to remain low among the resident population.

Apart from the Tarai, DDT has also been used in other parts of the State in malaria control operations. DDT spraying was started in the Ganga Khadar tract in Meerut District in December 1947 (SRIVASTAVA et al. [547]). This is a river tract comprised of patchy cultivation, interspersed with tall grasses, thorny bushes, and date palm trees, and a large number of ponds, lakes, and nullahs covered with grasses and weeds which form ideal breeding places for mosquitoes. In 1948 and 1949, DDT as a water suspension prepared with gum acacia and gelatine was used. For the initial spraying, a 2·5 per cent water suspension was applied which was changed in subsequent sprayings to a 1·25 per cent suspension. In 1950, however, a 2·5 per cent DDT water suspension made locally as above with a 50 per cent water-wettable powder was employed, supplemented in 1951 and 1952 with a 2·5 per cent DDT emulsion with aromax and soft soap. As a result of these measures, the spleen and parasite rates decreased from 14·5 and 9·4%, respectively, in 1948 to 3·2 and 1·5% in 1952. The population increased from 6,000 in 1948 to 18,000 in 1952. This region which was notoriously unhealthful because of malaria is now humming with activity. Jungles and swamps have been cleared, making available for cultivation vast areas of hitherto uncultivated land. Small cottage industries have grown up and the economic condition of the people has improved.

Central Provinces: Early in 1948 the Government of Central Provinces and Berar suggested a cooperative malaria control program between the government and certain local organizations, with the former supplying the DDT insecticides, spraying equipment, and technical supervision, and the latter the labor (SUBRAMANIAN *et al.* [563]). That year this suggestion was carried out on a trial basis in a limited area in the western part of the State, a foothills region comprising 34 villages with a total population of 15,000. Malaria was intensely hyperendemic there, with both *A. culicifacies* and *A. fluviatilis* vectoring the disease and showing infection rates of 0·21 and 1·28%, respectively (SUBRAMANIAN and DIXIT [564]). Transmission occurs principally from September to February. Three rounds of spraying at two-month intervals were completed between July and December 1948. For the first two rounds the dosage of DDT applied was 50 mg/ft^2 which was reduced to 25 mg in the third round. All human, animal, and mixed dwellings were sprayed. In July, the anopheline density per man hour of catching was 30, but by late September this had fallen to practically zero, where it remained for the remainder of the malaria transmission season. Spleen rates in sprayed villages fell from about 92 to 85% as compared to the rate in an unsprayed village which rose from 81 to 87%. At the end of the transmission period, parasite rates in a sprayed village were 10% as compared to 75% in an un-sprayed one. The parasite rate in infants of a sprayed village was 8% and of an unsprayed one 40%. These results after only six months of malaria control activity led to requests from neighboring zones for similar cooperative pro-grams.

In 1949, malaria control in a rural area in the southeastern part of Central Provinces was started (SUBRAMANIAN and VAID [562]). That year all the houses in Kanker Town, a town with a population of 4231, were sprayed with DDT twice during the malaria season, in August and October. The dosage used was 56 mg/ft^2. This was repeated in 1950 and the program was extended the following year to include 29 villages surrounding Kanker Town, affording protection to about 11,500 persons. In 1951, benzene hexachloride was substituted for DDT. It was applied at the rate of 11 mg of gamma isomer per square foot twice during the season at an interval of six weeks. In Kanker Town the spleen rate in children dropped from 25% in August 1949 to 5% in May 1952. In the neighboring area the rate dropped only 3% after one season of control. A comparison of infant parasite rates at the height of the 1951–52 season revealed a rate of 6% in sprayed areas as against 60% in nearby unsprayed villages. Some rise in these rates at the end of the season indicated that, in future operations with BHC, spraying would have to be done more often at perhaps an increased dosage.

Delhi State: The scheme for malaria control in Delhi State was activated in July 1947 (SINGH and SINGH (526]).

The incidence of malaria in the rural sections of the State varies considerably in different localities, even from village to village, depending upon the presence or absence of breeding places for *A. culicifacies* (AFRIDI and SINGH [4]). However, a uniformly high degree of malaria prevails in villages situated in the canal irrigation sector, and in those exposed to the flooding of the River Jumma or of Najafgarh Lake.

DDT residual spraying directed against adult mosquitoes was selected for the rural areas since larviciding was both costly and difficult to supervise adequately.

In order to conserve both material and man power; villages with spleen rates below 5% were excluded (except in a three-mile zone surrounding urban Delhi); those with rates of 5 to 20% received one spraying, while those above 20% were sprayed twice during the season. The spraying schedule was coordinated with the incidence of infective mosquitoes which in Delhi begins in mid-July and reaches its peak in mid-September. The villages which required two sprayings received the first between early July and early August, and the second during September. In the intervening period, spray teams treated the villages needing only a single spray. Since the effect of each application of DDT at the intended dosage, 50 mg/ft^2 (i.e., 2 ml of 2·5 per cent DDT emulsion per square foot of surface area), was found to last at least six weeks in Delhi (PURI and KRISHNASWAMI [457]), this spray schedule provided protection through the peak period of infection in mosquitoes. The spray was applied to the interiors of all human, animal, and mixed dwellings, and of verandas.

In addition to the residual indoor spraying, antilarval measures were carried out in villages within the three-mile zone around urban Delhi.

At the end of the second year of rural malaria control in Delhi State (1948), the prevalence of *A. culicifacies* in sprayed structures had been reduced 98·5%. Surveys in certain selected villages revealed that the infant parasite rate in a sprayed village was nil and in an unsprayed one 20%. The incidence of malaria in the unsprayed village was sixteen times that in the sprayed one. Climatic conditions in 1948 favored an epidemic of malaria in Delhi, but the incidence was found to be lower in that year than in the preceding four years.

In 1951 the incidence of malaria per 1,000 inhabitants was 2·3 in urban areas of Delhi, a reduction from 30·9 in 1945. In rural areas, the total number of malaria cases in dispensaries decreased from 18,344 in 1945 to 3,725 in 1951 (SINGH [522]).

Punjab State: Field studies on the comparative effectiveness of DDT and BHC were carried out during 1950 in a group of seven villages near Palwal, District Gurgaon, Punjab (SINGH *et al.* [525]). Spleen rates in this area ranged from 34 to 54%, and *A. culicifacies* was the principal vector. Transmission occurs from July through November, and it was during this period that these experiments were conducted. DDT was used both as an emulsion and as a suspension; BHC was used as a suspension; and DDT and BHC in combination was used as a suspension (equal parts of 2·5 per cent DDT suspension and 0·5 per cent BHC gamma isomer suspension). All structures, both human dwellings and cattle sheds, were sprayed. On the basis of entomologic data, chemical estimation of insecticides from scrapings of treated surfaces, and malariometric information, it was determined that, irrespective of the dosage used, DDT suspensions were more effective than emulsions on mud-plastered walls, and massive applications of either DDT emulsion or suspension did not prolong the residual effects. DDT at 50 mg/ft^2 and BHC at 10 mg of gamma isomer per square foot gave equal protection for about six weeks, and DDT and BHC in a combined spray (25 mg DDT and 5 mg BHC gamma isomer per square foot) was not only effective over a longer period of time (8 weeks) but was more efficient in suppressing mosquito densities than either insecticide alone.

Orissa State: In the early 1940's, efforts to control the malaria vector anophelines, *A. fluviatilis*, *A. varuna*, and *A. minimus*, in the Jeypore Hills of Orissa State, by the spray-killing of adult mosquitoes once a week with pyrethrum were not very successful (RAO [467]). Residual spraying with DDT was undertaken in 1945. In the second year of spraying, 1946, there was a general reduction in spleen rates, and the density of vectors and their infection rates were reduced significantly (SENIOR WHITE and GHOSH [505]). The average vector infection rate for all three species was 1·8% in a comparison village as against an average of 0·4 in areas treated with DDT several times, at 2- to 6-month intervals, during this period. In the treated areas, however, only *A. fluviatilis* were found infected. The rate of application of DDT was increased in January 1946 from one quart of 5 per cent solution per 1,000 ft^2 to one quart per 600 ft^2.

In 1949, a malaria team began operating in this area under the technical supervision of the World Health Organization and with supplies furnished by the United Nations International Children's Emergency Fund (WORLD HEALTH ORGANIZATION [662]). The primary responsibilities of this team were to demonstrate modern malaria control methods and to train local personnel. During 1949, the team's operations covered an area of 880 km^2, affording protection to some 61,000 persons. Houses inspected nine months after spraying done in July and August 1949, were found free of vector anophelines.

At the end of two years, the program sponsored by the World Health Organization-United Nations International Children's Emergency Fund was turned over to the Orissa Government which continued the malaria control operations (WORLD HEALTH ORGANIZATION [656]). A summary of WHO-UNICEF activities reveals that in the demonstration area spleen rates were reduced 15 to 20% with a 30 per cent reduction in the average size of the spleen; the parasite rate decreased by about 50%; and the monthly malaria morbidity rate fell from a range of 18 to 20°/$_{00}$ in 1950 to 1·5 to 3·7°/$_{00}$ at the end of 1951. These results are surprising when it is noted that in this area the local custom is to replaster all houses twice a year and to rethatch roofs once a year (SINGH [522]). However, in order to obviate any difficulties which might arise from these customs, the spray schedule was modified. The area was divided into three zones as follows: In the first zone, the technique of one spraying per year before the rainy season (June–July) at the rate of 200 mg/ft² was retained; in the second zone, spray was applied at the rate of 100 mg/ft² twice a year, once in early February following the replastering of the houses and again in June–July before the rains; and in the third zone, three spray applications were made at the rate of 60 mg/ft², the first in early February, the second in June–July, and the third at the end of October following the second replastering of the houses. It was observed that for all practical purposes DDT applied at the rate of 100 mg/ft² gave the same results as when applied at 200 mg/ft². Results of the 60-mg applications were not given.

Madras State: In 1947 the Malaria Institute of India undertook a pilot scheme for malaria control in a betelnut-growing area in South Kanara District, Madras State (RAMAKRISHNAN *et al.* [464]). A preliminary survey was completed in June and sixteen villages, in an area of about 250 square miles, were selected for the test.

Malaria incidence in this District varies. The western part adjacent to the coastal region has only scattered endemic foci of malaria. In the foothills of the Western Ghats, which form the eastern border of this area, malaria is hyperendemic and it is here, along the upper reaches of the three rivers which traverse the sector, that betelnut cultivation is widespread. From 37 to 57% of the high death rate in this region has been attributed to malaria. *A. fluviatilis* is the sole vector with a high infection rate of 8·7%. It enters houses at night to feed, so that the best attack against it is indoor residual spraying. The larvae breed in streams, river edges, irrigation channels, and paddy fields. There is no part of the year intrinsically unsuitable for transmission, the only limiting factor being the lack of breeding places during the monsoon months when all watercourses are frequently flushed (VISWANATHAN AND RAO [613]). January to June is the period of maximum transmission.

DDT both as a suspension and as an emulsion was applied to the interior surfaces of dwellings and cowsheds at the rate of 50 mg/ft². On the basis of epidemiologic and entomologic data, it was determined that malaria could be controlled by the DDT spraying of houses and cowsheds. The number of anophelines collected in biweekly half-hour periods from catching stations in dwellings and cowsheds during the season of 1947–48 was 308 in sprayed areas and 1,317 in unsprayed ones. Of these, however, only an extremely small proportion were *A. fluviatilis*, the vector mosquito. The spleen and parasite rates among children in the sprayed area were reduced in all cases whereas in the unsprayed area these rose as the transmission season advanced (Table 9).

In Wynaad, Malabar District, malaria is hyperendemic with a few relatively healthy areas (VEDAMANIKKAM [601]). In general, the vector *A. fluviatilis* breeds in the streams and channels which run through the many valleys of the region. Most dwellings and cattle sheds are located on the slopes rising

304   S. W. Simmons

Table 9
*Spleen and Parasite Rates in Selected Villages,
South Kanara District, Madras State, India*

	Spleen rates		Parasite rates	
	Sprayed villages	Unsprayed villages	Sprayed villages	Unsprayed villages
June–July 1947	70·0	—	—	—
January–February 1948 .	64·0	63·0	5·0	4·6
April 1948	55·5	70·3	2·3	3·6
June 1948.	52·6	76·0	2·1	19·3

from these valleys, and it is here that the vectors come to feed and rest. Prior to 1946 only antilarval control measures were used, but during the season of 1946–47 spray-killing of adults with pyrethrum extract was done. This method was gradually replaced by DDT indoor residual spraying until by May 1947 it was the only one employed. DDT was applied as a 5 per cent emulsion at the rate of 50 mg/ft^2. At the end of two seasons, the incidence of *A. fluviatilis* adults became almost nil with a corresponding marked decrease in the numbers of larvae.

In 1952 there were 36 DDT malaria control schemes operating in Madras State, covering an area of about 10,000 square miles and protecting some 2,300,000 people from infection at an approximate annual cost of $420,000. Eleven of the schemes were in conjunction with new land colonization and development.

Mysore State: Malaria is the most important health problem in Mysore State (RAO [468]). Except for a hilly area along the western border the country is gently rolling. In the hills the annual malaria incidence is about 27%, and in the irrigated tracts of the remainder of the State it is about 8%. It has been estimated that every year, directly or indirectly, deaths due to this disease number 40,000, which is approximately 36% of the total mortality. Epidemic malaria occurs only when unseasonal and heavy rainfall favors a large output of the vector anophelines in areas which normally are very dry. The vectors are *A. fluviatilis, A. culicifacies,* and *A. stephensi.* The period of natural transmission of malaria ranges from three to eight months. In the hilly areas it lasts from October to June while in the rolling country it continues throughout the year, the months between October and April being the more favorable ones.

Control measures against both larval and adult anophelines using Paris green and pyrethrum, as well as naturalistic methods, such as shading, cleaning, weeding, and sluicing, have been practiced since the 1930's with only minimum success. When the results obtained with DDT during World War II demonstrated that Indian rural malaria could be attacked effectively, an experimental program was organized in 1946–47 in the foothills area to determine if this method would be compatible with the economic limitations of the local government. After only one round of DDT spraying, the reduction in malaria morbidity and anopheline prevalence was such that health authorities were pressured for a permanent DDT program. Since October 1948, spraying with 2·5 per cent DDT emulsion has been part of the comprehensive medical and public health program in this area. The success of the work created an increasing demand for the extension of the program to all affected areas in the State. This has been done as personnel, trained in malaria control methods, have become available. Malaria control was incorporated into the functions of the 118 health units, and protected a population of some 850,000 people. With the advent of the National Malaria Control Program, however, the protection was extended in 1953 to 5 million people (SINGH [522]).

Coorg State: Coorg is a small state situated on the top of the Western Ghats between Mysore and Madras States. It is a very heavily forested region covering some 1,600 square miles (SINGH and KARIAPA [524]). Rainfall is intense, starting about the middle of June and lasting through October. Malaria is sufficiently prevalent throughout the State to class Coorg as a hyperendemic area. *A. fluviatilis*, the vectoring mosquito, is present in large numbers but is infrequently recovered in houses. In early 1947 the spleen rate was 58·2% and the malaria case rate 38·3%. DDT, as a 5 per cent solution applied at the rate of 2·5 ml/ft^2, was employed as an indoor residual spray for the first time in 1947 in Mercara, the capital. In 1948 and subsequently, DDT suspensions were used and the program expanded to the rest of the State. The results were most satisfactory. In 1951 case rates had fallen to 5·0%, and spleen rates were below 5% (SINGH [523]). By 1953 the spleen rate for the whole State was 1·5% and the malaria case rate was 1·6% (SINGH [522]). One hundred forty-two square miles of land had been cultivated in areas which were hyperendemic in 1946.

Bombay State: The Bombay State Malaria Organization was established in 1942 (VISWANATHAN [610]). While it was expected eventually that the activities of the organization would be extended to the entire State, for the first few years its scope of work was restricted to Kanara District in the southern part of the State where malaria prevalence was the highest. With the advent of DDT the control area was enlarged, first to include Kanara and Dharwar Districts, but later to add others until by 1950–51 control measures were under way in ten districts and surveys were being completed in six more. Surveys were to be completed for the entire State by the end of 1952.

Kanara District can be divided into two physiographic regions, a narrow coastal strip and a broad plateau of 2,000 ft altitude. The coastal strip area is more or less healthful, having a spleen rate of only 10% except where hills reach the sea. The plateau can again be divided into two regions, a western portion where rainfall is heavy and malaria transmission commences about two months after the monsoon and extends from December to June, and the eastern half where rainfall is much less and malaria transmission occurs soon after the first of the monsoon showers in June and lasts through November.

More than 50% of the population of Kanara District lives on the coast. The prevalence of extreme ill health in the plateau where malaria is hyperendemic has resulted in the decimation of the population in that area both by the excess of deaths over births and by emigration.

Of the 23 species of anophelines identified in this district, only one, *A. fluviatilis*, has been found naturally infected with human plasmodia. Streams and channels with marginal vegetation, rice fields, seepage pools, and shallow field wells constitute the principal habitats of this mosquito. In this region this species is predominantly anthropophilic and while it transmits malaria indoors, it lives for the most part outdoors. Thus, in areas where the natural infection rate is very high, a good number of vectors can be observed resting indoors during the daytime in human habitations, but a much greater number take shelter in the wild by day. In areas where the vectorial capacity of this species is low, most specimens are recovered from animal shelters.

A parasite survey in 1942 showed a 21·6 per cent plasmodia infection of children examined. The relative proportions of *P. malariae*, *P. falciparum*, and *P. vivax* were 58·2, 20·1, and 27·7%, respectively. This high incidence of *malariae* was unusual, for in later surveys only 2 to 10% of total plasmodia prevalence could be attributed to that species.

Dharwar District lies immediately to the east of Kanara District. Topographically the western portion is an undulating plateau which is a continuation of that described for eastern Kanara. This is the malarious section of the district with spleen rates ranging from 20 to 100%. Rice is cultivated extensively in large sections of this zone. In the rest of the district, rice cultivation is limited and malaria is found only in villages close to irrigation water storage tanks and perennial streams. This latter region is more liable to epidemic exacerbations. Transmission takes place from July through November (VIS-WANATHAN and RAO [613]).

A. fluviatilis is the vector in the western zone, but in the east *A. culicifacies* has been found naturally infected. Adults of this species are believed to be predominantly zoophilic and during the months of transmission may be found resting indoors during the day in all types of dwellings, human, mixed, or animal.

The first shipment of DDT was received in September 1945 to be used for field trials in civilian areas (VISWANATHAN [610]). Three sets of villages, each representing one of the three epidemiologic groups in Kanara District, were selected for the pilot experiment. A 5 per cent solution of DDT in kerosene or light diesel oil was used. It was determined by entomologic evaluation that three rounds of spraying at intervals of two months during the transmission season would be effective against *A. fluviatilis*. The average dosage of DDT used per square foot was 56 mg.

In treated areas the prevalence of *A. fluviatilis* was reduced from 2·8 per 10 man hours of mosquito catching to 0·08, as compared with the control group where the density was 2·4 per 10 man hours. The spleen rate among children in the sprayed zone was 44% while among the comparison group it was 55%. Similarly the parasite rate was 16·3 and 21·5 in the two groups.

The experiment was so successful that before the statistical data had been evaluated a scheme was drawn up for extensive control of rural malaria in Dharwar and Kanara Districts involving the use of DDT as an indoor residual spray. The operational techniques established by the field trials have been used with only minor variations since the start of the scheme.

The striking reduction in the population of malaria vector mosquitoes seen during the experimental period was maintained throughout. In the sprayed villages of Kanara in 1948–49, *A. fluviatilis* was recovered in human and mixed dwellings during the month of September only; in fact, this catch represented only 3 specimens, the total for the year (VISWANATHAN and RAO [612]). In contrast, the incidence of *A. fluviatilis* in unsprayed villages was much higher in the monsoon and post-monsoon months of 1948 than in previous years. *A. culicifacies* prevailed at negligible levels in sprayed houses

Table 10

*Spleen and Parasite Rates in DDT Treated and Untreated Villages.
Kanara and Dharwar Districts, Bombay State, India, 1942–1949[1])*

| | Kanara | | | | Dharwar | | | |
| | Spleen rates | | Parasite rates | | Spleen rates | | Parasite rates | |
	Sprayed villages	Unsprayed villages	Sprayed villages	Unsprayed villages	Sprayed villages	Unsprayed villages	Sprayed villages	Unsprayed villages
1942–1945		70·0		20·0		[2])		[2])
1946–1947	14·4	72·2	3·8	14·6	19·7	28·3	4·3	7·5
1947–1948	11·6	47·1	2·7	19·5	10·6	25·1	0·9	5·2
1948–1949	7·0	52·9	2·1	21·6	7·8	18·0	1·1	3·2

[1]) After VISWANATHAN [610].
[2]) Unknown.

while in unsprayed villages their densities were well above the critical limit throughout the transmission season.

Spleen rates dropped in Kanara District from a pre-DDT figure of 70 to 7% in 1949, and parasite rates dropped from 20 to 2% (Table 10 and Figure 6). In Dharwar District spleen rates dropped from over 25 to 7·8%, and parasite rates from about 7·5 to 1·1%. In unsprayed villages the rates remained high.

The malaria death rates in the treated areas show a most pronounced reduction from expected rates as compared with the rest of the province (Table 11) (VISWANATHAN [611]). While the reliability of classification of causes of death might be questioned, a static over-all malaria death rate in the rest of the province from 1936 to 1948, compared with a descending one in the treated areas, might be taken as adequate evidence of the effect of malaria control measures on the reduction in malaria death rates. This was not due to any improved methods of treatment, but almost entirely to reduced

S. W. Simmons

Table 11

Predicted and Observed Malaria Death Rates for Three Years 1946 to 1948.
Kanara and Dharwar Districts, Bombay State[1])

Year	Treated				Untreated	
	Kanara		Dharwar		Rest of Province	
	Predicted	Observed	Predicted	Observed	Predicted	Observed
1946	2·02	1·29	3·62	1·23	1·59	1·52
1947	2·00	1·33	3·86	1·19	1·62	1·48
1948	1·98	0·80	4·09	0·85	1·65	1·51

[1]) After VISWANATHAN [611].

morbidity as evidenced by case figures of those treated at various dispensaries in the control areas, which in 1948 declined by more than 50,000. As dispensaries see only about one sixth of the population, it can be estimated that there was a reduction of about 300,000 cases of malaria per year in the treated area.

In 1948 field trials were instituted and continued through the 1949–50 season in Thana District, Bombay State, to test the comparative efficacy of DDT and BHC for malaria control (VISWANATHAN *et al.* [614]). In areas where *A. culicifacies* and possibly *A. fluviatilis* are vectors, both insecticides are

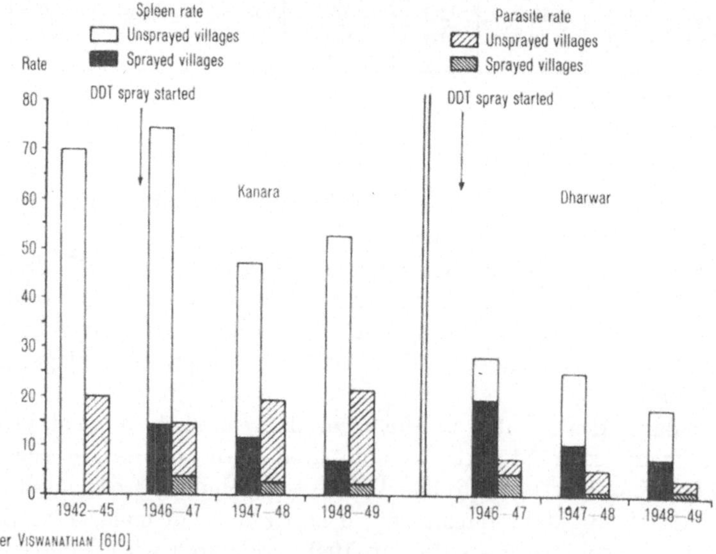

Figure 6

Spleen and parasite rates in DDT treated and untreated villages Kanara and Dharwar districts, Bombay state, India, 1942–1949.

capable of producing good malaria control. It appears, however, that DDT
is the preferred insecticide for most Indian anophelines, even though BHC can
be used successfully.

In 1951, 5·4 million people or nearly one seventh of the population of
Bombay State were directly protected from malaria infection at a cost of
about $567,000, or 14 cents per capita (SINGH [523]). In 1952 a total of 6·8 million
people were protected at about the same cost (VISWANATHAN [609]). In each
of these years the total population that benefited, including those who were
indirectly affected, was 11·7 million. In 1953, malaria control was extended
to all the malarious parts of the State, with a total population of 19 to 20
million directly benefiting and the entire State population of 36 million
deriving indirect benefits. It is estimated that in each of the years 1951 and
1952 about a million cases of malaria were prevented in the State and in
1953 about 1·7 million cases were prevented (VISWANATHAN [609]).

Ceylon

Malaria has been prevalent in Ceylon for many centuries. DUNN [159] refers
to the work of BRIERCLIFFE [66] in which the latter states that the Government Archives show a map that the Dutch published in 1638 in which the
Southern Province was marked with a statement that it had been depopulated
and deserted 300 years previously by 'fever sickness' and, also, the Northern
Province had been 'depopulated by sickness'.

A rather extensive survey of the endemicity of malaria among children
was made by CARTER in 1921–22. According to DUNN [159] this survey revealed
a spleen rate of 13·6% and a parasite rate of 13·5%.

Ceylon can be divided into three climatic zones characterized by differences
in the rainfall. In the dry zone (rainfall less than 20 in. per year) malaria is
hyperendemic; whereas, in the wet zone (rainfall over 40 in.) and in the dry
zone (rainfall 20 to 40 in.) malaria is not as prevalent (SIVALINGAM and
RUSTOMJEE [527]).

A characteristic of malaria incidence in Ceylon is the regional epidemics
which occur every five to seven years. These epidemics can be of great severity,
the one of 1934–35, for instance, having been responsible for approximately
87,000 deaths in the space of seven months. It is estimated that there were
at least 1,500,000 cases of malaria during the outbreak (DUNN [159]). Regional
outbreaks always occur in periods of drought caused by the failure of the monsoon rains. In years when rainfall is normal or excessive and the rivers and
streams run full, breeding places for *A. culicifacies*, the only mosquito incriminated as a vector in Ceylon, are few. In years of drought, however, the waterways
become sluggish, leaving many pools in which this species can breed profusely.

Malaria control in Ceylon has been practiced since about 1920 (RAJEN-
DRAM and JAYEWICKREME [463]). Early larviciding efforts consisted chiefly
of oiling, although some channeling and clearing were done. Spray killing
of adults with pyrethrum was also tried. Following the disastrous epidemic

of the mid-thirties, a new malaria control scheme was organized which applied the above measures but on a more permanent basis.

Ceylon was the first country in Asia — and one of the first in the world — to undertake a national antimalaria campaign, using residual DDT, to control both epidemic and endemic malaria (PAMPANA [424]). The first two units got underway in late 1945. Implementation proceeded as material became available so that by 1946, all of the paludic region of the island was included in the campaign plan.

At the start technically pure DDT in a 5 per cent solution in kerosene was sprayed at the approximate dosage of 120 mg/ft² (RAJENDRAM and JAYEWICKREME [463]). The next year a 40 per cent emulsion of DDT in xylene with Triton X-100, diluted in the field to make a 5 per cent emulsion, was employed. From about the end of 1948, water-wettable DDT powders replaced all DDT formulations. At the same time, the dosage was reduced to about 100 mg/ft². Since October 1949, the standard dosage employed has been 50 mg/ft². Spraying is done at intervals of from six to ten weeks depending on the endemicity of the disease and the type of housing. In a few areas, BHC (Gammexane P.520 — a wettable powder) at a dosage of 10–11 mg/ft² has been sprayed at intervals of six weeks.

Subsequent to the residual spraying of houses in the epidemic zone, the number of *A. culicifacies* larvae recovered per 100 dips fell from about 4·0 to 0·28 in 1947 and 1948, and was 0·45 in 1949. Similarly, the catching rate per hour of adults in dwellings has gone from 0·48 in 1945 to 0 in 1947 and 0·002 in 1948. In 1948, climatic conditions were favorable for a malaria epidemic, but it is believed that the malaria control efforts were responsible for averting an outbreak. The success of the campaign is best depicted by the decline in death rates from 1946 to 1949. Infant mortality fell from 141 to 87⁰/₀₀, maternal mortality from 15·5 to 6·5 per 1,000 live births, and the national death rate from 22·7 to 12·6⁰/₀₀ per year (WORLD HEALTH ORGANIZATION [639]). Malaria death rates dropped from a pre-DDT program average of 0·8 to 1·8⁰/₀₀ to 0·2 in 1952 (WORLD HEALTH ORGANIZATION [640]). Comparable reductions were observed in morbidity and in spleen and parasite rates.

As a result of the malaria control campaign, Ceylon has reclaimed and brought under irrigation 206 square miles of formerly uninhabitable malarious jungle providing land for some 91,000 people (WORLD HEALTH ORGANIZATION [640]). This is a monumental contribution to the social and economic well-being of the country.

Burma

Malaria in Burma affects half the population and is found in three quarters of the country (BURMA [78]). It is a severe hindrance to economic development, and its control would mean new health and wealth for the country.

Malaria is intense in the northwest coastal plain and hill regions of Burma (Arakan), but its distribution is uneven, the spleen rates of neighboring villages often differing markedly (MACAN [342]; JONES [301]). *P. vivax* is the most common malaria parasite, with only slightly less *P. falciparum* and small amounts of *P. malariae.* Breeding areas for vector mosquitoes are extensive in ill-drained valley bottoms in the foothills. However, where this land has been cleared and terraced for rice cultivation, non-vector anophelines flourish and malaria transmission is reduced. On the coast the reverse of this is true. The greater part of this region was malaria-free mangrove forest until the land was reclaimed by the building of sea walls; this has resulted in seepage of sea water and the establishment of brackish pools ideally suited for the breeding of *A. sundaicus.* This species, *A. minimus,* and *A. leucosphyrus* have been incriminated as vectors in this zone (DHALIWAL [143]). Transmission is known to occur in March and April, during the rainy season which extends from June until late September, or into October. In the more central valleys of the country, malaria incidence is much the same (MACAN [343]), spleen rates averaging 80%. Transmission by *A. minimus* and *A. leucosphyrus* takes place throughout the year but is most intense from May to December. In the Inle Lake region of Shan States, eastern Burma, spleen rates average about 70% (MOFFAT *et al.* [384]). At Lashio, Shan States, in a spot spleen survey, a spleen rate of 94% was found (WORLD HEALTH ORGANIZATION [650]).

Antilarval and anti-adult measures employed in Arakan prior to the advent of DDT were not very effective. The initiation of DDT residual spraying on a large scale brought hospital admission rates in this area down from an average of 77 per thousand persons per month in November 1943 to $9.8^o/_{oo}$ per month in November 1944 (DHALIWAL [143]).

In October 1951, a World Health Organization malaria control demonstration team arrived in Lashio to initiate malaria control operations (WORLD HEALTH ORGANIZATION [650]). By 1952, the program covered an area of 800 square miles (WORLD HEALTH ORGANIZATION [641]). In a zone where houses were treated with 2 g of DDT per square meter, malaria incidence among children and adults was reduced 80%. The infant parasite rate dropped within a year from 35.2% to 0, whereas in an unsprayed check area, this rate, during the same period of time, increased from 41.2 to 46.0% (WORLD HEALTH ORGANIZATION [642]).

A malaria control program for the Union of Burma, based on the experience gained by the World Health Organization and the Economic Cooperation Administration aided projects, was proposed in 1953 to extend within five years over all areas where malaria is a serious problem (BURMA [78]). It was planned to effect control by the application of residual insecticide in the living quarters of all residences, and other buildings where necessary, at the rate of 200 mg/ft^2.

Thailand

Malaria is the greatest single cause of sickness and death in Thailand (FLEGEL *et al.* [185]). The population of Thailand is approximately 18,000,000, of which about 5,000,000 live in highly malarious areas (EJERCITO *et al.* [169]).

More than 50,000 malaria deaths are reported annually, an estimated 20% of deaths from all causes. Cases are placed at upwards of 3,000,000 per year.

Malaria is widespread throughout Thailand. High endemicity with spleen rates of 25 to 50% and parasite rates exceeding 15% is general over large areas; and hyperendemicity with spleen rates exceeding 50% and frequently over 75%, and with parasite rates reaching more than 50%, is common in the most malarious areas (KOSOL and GRIFFITH [314]). The most highly malarious areas are in the northern fertile lands where the disease affects agricultural output sufficiently to be one of the more important factors in reducing the economic productivity of the country. Thus, for medical and economic reasons, the Thai government regards the extension of malaria control activities as both essential and urgent.

A. minimus, a flowing-water breeder, is the only species which has been found naturally infected with malaria parasites in Thailand (THAILAND [577]). While some malaria transmission goes on all year round, most of it occurs coincidently with the rainy season, June through October, declining during the cool season, November through January, and reaching its lowest level in the hot season, March through May. These seasons become less distinct from north to south in Thailand so that the seasons of transmission are also less distinct.

Organized malaria control began in Thailand in 1930; the present Division of Malaria Control in the Ministry of Public Health was established in 1943. In 1949, a World Health Organization/United Nations International Children's Emergency Fund malaria control demonstration project was initiated in Chiengmai, northern Thailand (WORLD HEALTH ORGANIZATION [651]). Populations of 40,000 the first year increasing to 175,000 by 1951 were protected with DDT indoor residual spray applied at the rate of 200 mg/ft^2. As indicated by entomologic and malariometric data, this program was highly successful. Malaria transmission was interrupted and the vector disappeared to a considerable extent from the treated area. In a group of seven villages sprayed in 1950 and left unsprayed in 1951, only one adult and one larva were recovered during the entire 1951 season. Morbidity dropped from 50·5$^0/_{00}$ in 1949 to 6·5 in 1950 and to 2·1 in 1951; this figure was 42$^0/_{00}$ in untreated areas. These outstanding results led the Thai government to establish a malaria control program which would protect a population of 1,400,000 in 1952, increasing to 3,500,000 in 1954, and to 5,000,000 no later than 1957.

Through the malaria control program in Thailand, it is hoped to eliminate not only malaria but also the vector which is at present very susceptible to DDT. Residual spraying in houses and other buildings occupied by humans is done during the hot season each year, March–May, with some necessary exceptions in south Thailand, using a 5 per cent DDT suspension formulated from 75 per cent water wettable powders. DDT emulsion is employed in limited areas. Throughout the rest of the year, pre- and post-spray entomologic and malariometric surveys are made to evaluate the effects of the program for future guidance. The 1952 survey revealed that the cumulative percentages of houses with *A. minimus* were 40·71 in uncontrolled areas, and 0·20, 0·08, and none in areas of one, two, or three years of control, respectively (KOSOL and GRIFFITH [314]). In the same areas, cummulative spleen rates among children were 50·72%, and 17·16, 6·93, and 5·96%, and the cumulative parasite rates were 23·77%, and 7·42, 1·21, and 1·32%, respectively. The cumulative infant parasite rates were 17·91% in uncontrolled areas; 3·84% among infants in sprayed areas born before the first spraying; and 0·54, 0·47, and 0·00% among those born after one, two, or three years, respectively, of spraying for malaria control. A survey in northern Thailand prior to the malaria control program established that each malaria case incapacitated its victim about 7·6 days (WORLD HEALTH ORGANIZATION [643]). In DDT-sprayed areas, about 50,000 malaria cases were prevented. Since about one half of these cases involved workers, an estimated saving of 25,000 work-weeks, or 175,000 man-days was realized. There was no evidence at that time of resistance to DDT in the vector. It was believed that spraying could be omitted in demonstrably controlled areas after about three years of operation while still maintaining the *status quo*. If no new areas were added to the program beyond the planned 5,000,000 population, it was expected that the continued omission of these old areas would terminate the extended house spraying campaign no later than 1957. The program would then consist principally of surveillance and prevention activities.

Malayan Federation

The Malayan Federation covers an area of about 56,000 square miles and has a population of about 5,500,000 (SIMMONS *et al.* [514]). The Malay Peninsula is divided by a range of mountains into two main sections, a smaller western one characterized by a belt of mangrove trees many miles deep and an eastern one which is kept free of these trees by the force of the northeastern monsoon and is broken by well-forested rocky headlands. Seventy-two percent of the peninsula is covered with heavy forest. The climate is moist with an average annual rainfall of 100 in.; on the eastern coast this has amounted to as much as 235 in. The rains come in March, April, May, September, October, and November.

Twenty-five species of anophelines have been reported from the Malayan Federation. Of these, only five are known to play a significant role in the transmission of malaria. The most important vector is *A. maculatus*. It is the scourge of the rubber estates and is found in the hilly lands from sea level to 5,000 ft wherever there has been sufficient clearing to expose springs, seepages, and small streams to sunlight. On the coastal plains there are three vectors which breed in still and more or less shaded waters. These are *A. umbrosus*, *A. letifer*, and *A. barbirostris*. The first of these prefers acid waters although it is not confined to them. *A. umbrosus* is at home in the deep shade of virgin swampy jungle, while *A. letifer* prefers the lighter shade of cultivated and semi-cleared land and is therefore more in contact with man. *A. barbirostris* will enter the landward edge of the brackish water zone avoided by

A. umbrosus and *A. letifer* and can maintain a high endemic rate of malaria though it seldom causes severe epidemics. In the brackish water zone along the coast, the important species is *A. sundaicus*, one of the more important vectors of the world. It breeds prolifically in still sunlit brackish waters, especially if there is floating green algae. Transmission may occur at any time of the year, but it is enhanced in the hilly areas from March to July and in the plains and the brackish water zones around the coast from September to November (MALAYA [351]).

Malaria is the chief menace to the health and well-being of the people of the Malayan Federation. The disease is hyperendemic and exists in all but a few parts of the area. It occurs throughout the year with a slight decrease in incidence during February and March followed by a gradual increase during the summer and fall months. Since the last century, the Malayan health department and its predecessors have done much toward the control of mosquitoes and malaria which has reduced the incidence of the disease, especially around the cities and on some estates (MALAYA [352]).

The prewar systems of malaria control were the classical ones, aimed either at mosquito larvae (ditching, filling, oiling, etc.) or at the parasites (quinine and atabrin). During the war, actually during the Japanese occupation, residual spray for the destruction of mosquito adults was started. In June 1946, indoor residual spraying was conducted in an entire hyperendemic village with 5 per cent DDT in kero III, at the rate of 4 pints of DDT solution per 1,000 ft^2 of area treated. This resulted in a 25 per cent reduction in the spleen rate (NAIR [405]). Other investigations (NAIR [403], [404], [402]; WALLACE [616]) showed the potential value of this compound as a malaria control tool in rural Malaya. Late in 1948, the Institute for Medical Research started on a three-year experiment to test the efficacy of house spraying with DDT and BHC (gammexane) for malaria control in Malay settlements where *A. maculatus* was thought to be the vector (MALAYA [353]). Four valleys with a population of about 4,000 were chosen because of their comparatively high incidence of malaria, their topographic similarity, and their relative isolation. In one valley, the houses were sprayed with DDT at the rate of 200 mg/ft^2 and in another, with BHC; in a third proguanil prophylaxis was tested; and the fourth served as a control area. From entomologic studies it was determined that although the DDT remained effective against *A. maculatus* for nearly six months, this species could not be eradicated by house spraying with residual insecticides. *A. umbrosus* was also susceptible to DDT, the death rate within 24 hours of exposure being 70 to 80% for 6 to 7 weeks after treatment. For the first three weeks after DDT spraying, the kill among *A. sundaicus* and *A. barbirostris* was high, 80 to 100%; but this fell sharply to well under 50% when the DDT deposits were 4 to 9 weeks old. *A. letifer* was apparently only slightly susceptible to DDT, the death rate during the six weeks following DDT application averaging only 31% and never exceeding 50%. Studies on larviciding revealed that DDT as an emulsion was about as effective a larvicide as DDT in oil and was much cheaper. Malaria surveys showed that the insecticides were beginning to reduce the amount of malaria infections. Comparisons of malaria incidence before and after DDT spraying and in the control area are shown in Table 12.

At the end of the experiment in 1952, it appeared that DDT house spraying was the most suitable method for rural malaria control. In fact, before the end of the experiment, all four areas had been sprayed with DDT emulsion (MALAYA [351]). The results were not considered to warrant recommending a country-wide spraying campaign; however, routine DDT spraying has subsequently been started in villages and rural areas throughout the country. The success of antilarval methods around urban communities renders unlikely

Table 12

Comparison of Malaria Incidence in a Treated Area Before and After DDT Spraying and in an Untreated Area. Malay Federation

	Sprayed area		Control area	
	Before DDT	After DDT	Before Spray[1]	After Spray[1]
Infant infection rates	10·3	1·9	4·8	24·1
Parasite rates, children 12 and under .	29·8	12·0	23·0	24·0
Spleen rates, children 12 and under . .	58·3	36·0	52·3	51·0

[1] Spraying not done in this zone.

a change to residual spraying in those areas. It is the plan of the government of Malaya to continue its present policy of protecting from malaria infection as many people as is economically possible, and DDT will play a large part in these plans.

Associated States of Indochina

Malaria is responsible for most of the recorded, and doubtless much of the unrecorded, illness in the three associated states of Indochina, Laos, Cambodia, and Vietnam (SIMMONS *et al.* [514]). The unrest in this area since 1940 has resulted in a continuous migration from the disturbed rural areas to the comparatively secure large cities, greatly increasing the potentialities of malaria outbreaks in these cities. The highest incidences of malaria are reached on the plateaux and in the foothills of the mountains up to 1,500 ft where the stream-breeding *A. minimus* is the chief vector (PRATT [454]). In the low delta rice-growing regions in North (Tonkin) Vietnam around Hanoi and in South (Cochin China) Vietnam near Saigon where *A. hyrcanus sinensis* often breeds in tremendous numbers, a low-grade malaria with occasional explosive outbreaks occurs. In addition to these two species of mosquitoes, *A. sundaicus* is credited with the transmission of malaria in isolated areas along the coast of Vietnam while *A. culicifacies* is a minor vector in certain mountainous areas of Laos (BAKER [30]). The incidences of malaria are directly related to the two rainy seasons in this region [596]. There are four peaks, one at the beginning and one at the end of each rainy period with a high average during the rains.

In 1950 an ECA (Economic Cooperation Administration, USA) mission was assigned to Indochina to assist in the organization of a joint ECA-Vietnamese health department malaria control program, part of the Vietnam ten-year program for health. At the time the program was initiated, blood and spleen rates in urban areas varied from 0 to as high as 17%; in rural areas, however, the picture was quite different with spleen rates of 80 to 90% (PRATT [454]). Operations were first started in the Hanoi area of North Vietnam and the Saigon area of South Vietnam during September and October 1950, using 75 per cent DDT water-wettable powder as a 5 per cent DDT suspension. Up to October 1, 1950, a total of about 4,000 lb of DDT had been used to spray 2,500 houses (LYMAN [341]). This rate per application is higher than that generally used due to the nature of the housing units. These units consist of a walled compound containing the family living quarters, and often the working quarters, as well as shelters for domestic animals. Blood slides taken at that time averaged 11·5% positive for malaria, with *P. falciparum* responsible for about twice as many infections as *P. vivax*. By August 1951, 80,054 lb of DDT had been applied to 114,546 houses (BLUM *et al.* [57]). The budget for 1952 called for 250 t of DDT to be purchased.

It is anticipated that this malaria control program will be expanded to all parts of the country protecting some 28,000,000 people in an area of 285,714 square miles (740,000 km²). Each of the three governments has established a malaria control organization for this purpose.

Republic of China, Taiwan (Formosa)

Malaria is the most important endemic disease in Taiwan [572]. Almost all of the rural population of 5,300,000, or 70% of the total population of the island, is subject to infection. This has hampered agricultural and industrial development to a considerable extent. The chief vector is *A. minimus* [98]. Transmission takes place principally in May and June in Central Taiwan, in July and August in Northern and Eastern Taiwan, and in October and November in Southern Taiwan.

A four-year island-wide plan for eradication of malaria was proposed in 1951, based on the use of water-dispersible DDT in a 5 per cent suspension applied at the rate of 2 g of the technical material per square meter of surface treated. The total cost for this program was estimated at less than $2,000,000, or less than 14 C. per person per year [572].

During the first two years of the island-wide operations (1952 and 1953) priority was given to the highly endemic foothill areas [98]. The malarious coal mining section of Northern Taiwan was also included in the 1953 campaign. The less accessible mountain regions, which may have equal or slightly higher malaria rates, were scheduled for inclusion in 1954 and 1955.

Activities which protected 156,000 people in 1952 were expected to protect 5,000,000 by 1954. In the first year of operations the average number of anophelines collected per house remained less than one in completely or selectively sprayed areas while this number was never less than two and during the transmission season reached a peak of 266 per house in unsprayed areas.

Philippine Islands

Malaria has long been a problem in the Philippine Islands, ranking as the primary cause of morbidity and among the top five causes of mortality (EJERCITO *et al.* [168]). Prior to World War II, malaria was responsible for 10,000 to 20,000 deaths and for at least 2,000,000 cases annually in a population of about 13,000,000 (RUSSELL [488]). This number increased during the war years to an estimated 4,000,000 cases annually [593]. Since then there has been a gradual decrease, there being 10,558 reported deaths in 1948 and 7,170 in 1952 among a population of approximately 20,000,000, of which about one third live in malarious areas [442]. In over half of the provinces of the Philippines, malaria is hyperendemic in one or more areas. It is a rural disease and is a great deterrent to agricultural and economic development in the Islands. It is estimated to cause losses each year of between 100,000,000 and 750,000,000 pesos ($50,000,000 and $375,000,000).

A. minimus flavirostris is the principal vector with *A. mangyanus* and *A. maculatus* being secondary vectors [593]. All three species are stream breeders. In areas where there are distinct wet and dry seasons, transmission peaks occur at the beginning and the end of the dry season; in other areas, transmission is perennial (RUSSELL [488]). Most of the malaria in the Philippines is *P. vivax* and *P. falciparum*, the former accounting for about two thirds of the infections.

Since 1906, efforts have been made to control malaria in the Philippines using drugs, naturalistic measures, biologic methods, or combinations of these, with little success and this primarily in military and hacienda communities only (RUSSELL [488]). Even these comparatively limited activities were halted during World War II. They were revived immediately after the end of hostilities, but it was the advent of DDT that gave the first real hope for a rapid nationwide reduction in the prevalence of malaria. Preliminary experiments using 5 per cent DDT emulsion applied at the rate of 200 to 350 mg/ft² to 1,750 houses and other shelters in three areas of the provinces of Laguna, Negros Occidental, and Palawan showed that within one year parasite rates in both adults and children were reduced 6 to 85% (SMITH and DY [531]). Since this was a small operation, it was decided that before the use of DDT residual sprays could be incorporated into a nationwide malaria control program in the Philippines, a field study on a large scale would be necessary. Therefore, a pilot project was organized with the aid of the World

Health Organization and the United States in a highly malarious sector of Mindoro (SAMBASIVAN *et al.* [497]). The population numbered some 25,000 and there were about 5,000 houses to be sprayed. The spleen and infant parasite rates immediately before spraying were 77% and 32% respectively, as compared to 57% and 33% in an adjacent comparison area. Spraying was done using a 75 per cent DDT water-dispersible powder as a 5 per cent DDT water suspension at the rate of approximately 200 mg/ft^2 applied to houses and outbuildings. Observations made eleven months after spraying revealed that malaria transmission was effectively controlled, that the spleen rate had dropped to 32%, that the incidence of parasite positive fever cases had decreased significantly, that the densities of the vector in sprayed houses were at or near zero, and that the numbers of larvae and of adult vectors in outdoor resting places were lowered substantially (EJERCITO *et al.* [168]). It was concluded that malaria-carrying *A. minimus flavirostris* could be controlled by one yearly application of DDT at 200 mg/ft^2 to the inside walls of houses. This conclusion has provided the basis for a nationwide program of DDT residual spraying.

The objective of the plan for malaria control throughout the Philippines devised by the staff of the Malaria Control Division and the Technical Assistants of the US Operations Mission to the Philippines and entitled *A Six Year Philippine-American Plan for Malaria Control in the Philippines* is to reduce malaria to a level where it will no longer constitute a public health problem. This is to be accomplished by applying DDT residual spray to houses in malarious regions once each year for a minimum of three consecutive years. Concomitantly, a long-range program of more permanent naturalistic control measures will be developed which will aid the Philippine Government in keeping malaria under control and possibly in eradicating the disease [442].

By the end of the 1953 spraying season (the first season's operations), 215,000 houses had been sprayed (EJERCITO *et al.* [168]). Initial appraisals indicate good results which conform with those obtained in the Mindoro Pilot Project. It is estimated that between one and two thousand lives were saved and 300,000 cases of illness were prevented as a result of the 1953 operations. In 1954, the program will be expanded to a nationwide operation involving over 1,000,000 house-sprayings, bringing protection from malaria to over 5,000,000 people. The total number of houses to be sprayed during the six-year period is estimated at 1,250,000, requiring some 4,000,000 house-sprayings.

Indonesia

The republic of Indonesia is made up of three large islands — Sumatra, Java, and Celebes — and parts of two more — most of Borneo and the western half of New Guinea — plus some 3,000 small islands such as Bali, Lombok, Soembawa, Soemba, Flores, Timor, and the Moluccas (Spice Islands). The entire area of Indonesia is about 753,000 square miles (SIMMONS *et al.* [514]).

The population is estimated at between 70 and 80 millions, of which some 30 million live in hyperendemic malarious regions [286].

Malaria is the most important disease in Indonesia, causing more deaths than any other and limiting the productive capacity of the country (KETTERER [306]). Generally the greatest number of cases is reported following the rainy season; however, acute outbreaks can occur at any time.

In certain coastal areas of Java, the Sunda Islands, Sumatra, Celebes, Nias Island, and Buton Island, *A. sundaicus* is a very efficient vector (SIMMONS *et al.* [514]). Where rice cultivation is done — Java, Sumatra, Soemba, and the Minahassa Peninsula of Celebes — malaria is transmitted by *A. aconitus*. At altitudes of 4,000 to 5,000 ft (in the interior of Java), *A. minimus minimus*, *A. maculatus maculatus*, *A. sundaicus*, and *A. aconitus* are the recognized vectors. Both along the coast and in the interior of Borneo and Bangka Island, malaria is transmitted by *A. umbrosus*. In Celebes, *A. subpictus subpictus* and *A. sundaicus* are incriminated as vectors, while *A. hyrcanus nigerrimus* is the vector in the marshlands of southern Sumatra. In eastern Indonesia and in most of the coast and interior of New Guinea, the vectors are *A. punctulatus punctulatus*, *A. punctulatus moluccensis*, and *A. bancrofti*.

These areas are those where malaria incidence is high; however, there are only a very few regions of Indonesia where it is low. In coastal areas where the incidence is apt to be high, the spleen index is frequently 90% or more (JENNEY [293]). Indices of 80% have been found in many areas (KETTERER [306]). *P. vivax* and *P. falciparum* occur in about equal proportions with some *P. malariae* seen irregularly.

Soon after World War II, private interests became concerned with the necessity for malaria control. During 1947 at the time of the reopening of oil fields on Sumatra, a full-scale malaria control program was initiated by a petroleum company in company areas and villages using DDT for both larviciding and residual spraying (AVERETT [25]). It is believed that this was the first program of this type on Sumatra. Malaria incidence at that time among both the indigenous population and the former (prewar) employees returning to work was between 40 and 50%. Within 12 to 18 months, this was reduced to 10% (by laboratory check), and by the end of 24 months the rates were less than 5% in the majority of the employee groups. Continuous survey has shown a marked reduction in both the total number of adult and larval mosquitoes recovered and in the number of adults found infected.

Postwar efforts in malaria control by the government of Indonesia started with field experiments in 1947–48 to study the effects of DDT house spraying on *A. aconitus*, the *A. punctulatus* group, and *A. sundaicus* (SOEPARMO and STOKER [535]). The results of these trials were excellent (SWELLENGREBEL and LODENS [569]; BONNE-WEPSTER and SWELLENGREBEL [60]); therefore, the Malaria Section of the Ministry of Health of Indonesia decided to initiate studies on the effects of DDT house spraying on malaria incidence prior to its use over large areas of the country. In one of the experimental villages, Marunda in Java, after DDT spraying in the fall of 1949 and in the spring of 1950 at the rate of 2 g/m², malaria infection among babies born during that time was absent and malaria incidence in children aged one to two years was reduced markedly (VAN THIEL and WINOTO [600]). In a nearby control village, malaria conditions remained unchanged.

In 1951, when supplies of DDT and sprayers became more plentiful, spraying operations on a somewhat larger scale were started using both wettable powders and emulsions for control of adult and larval mosquitoes (JENNEY et al. [294]). The first project was at Metro, Sumatra, a center for the transmigration efforts of the Indonesian government to relieve the overcrowded conditions in Java (SOEPARMO and STOKER [535]). In 1951, about 80,000 people were protected from malaria there. That same year two other projects were activated, one in an almost abandoned rice-cultivation zone and the other on a small government rubber estate. In 1952, these activities were expanded so that there was at least one antimalaria spray program in each of the ten provinces of Indonesia protecting some 1,900,000 persons. The entire control area is covered once each year with DDT wettable powder at the dosage of 40 ml of 5 per cent DDT suspension in water per square meter. It was planned to extend the campaign to give protection to 5,000,000 inhabitants in 1953 and to 30,000,000 by 1961 [286]. The annual budget for 1952 was $87,719.30 and the ultimate budget for 1960 is estimated at $701,754.39. There is a great deal of interest in the construction of a plant for the manufacture of DDT which, it is thought, would effect some savings.

In April, May, and June 1952, DDT spraying was done in a rural area around Tjulatjap, Java, under the guidance of a World Health Organization malaria control demonstration team. A total of 56,884 persons were protected against malaria in that area (SOEPARMO and STOKER [535]).

The Indonesian malaria control program is expected to produce beneficial effects on several phases of the country's economy, particularly in the production of rice and such export products as rubber and palm oil, in the establishment of transmigration projects, and in the sea and inland fishing industry (KETTERER [306]).

After two years of spraying with DDT for malaria control at Tandjung Priok, the harbor area for Djakarta, transmission had not been reduced appreciably (LAIRD [318]). The infant parasite rate which in 1952 was 18·8% dropped only to 16·5% in 1954. This port was being developed and enlarged and it was believed that the resultant increase in the population might have played a part in this lack of effectiveness of DDT. However, from a series of biological tests it was soon evident that *A. sundaicus* had developed resistance to DDT. Dieldrin has been substituted in this area and from early observations seems to be effective. The appearance of vector resistance to DDT may upset malaria control plans of not only Indonesia but other nations as well. The susceptibility of DDT-resistant mosquitoes to dieldrin or other chlorinated hydrocarbon insecticides is of considerable immediate importance, but it remains to be seen if mosquitoes follow the pattern of the house fly and acquire resistance to these chemicals as well as to DDT.

Other Areas in the South Pacific

At the military base on Espiritu Santo, New Hebrides, and adjacent small islands during World War II, DDT was used in 1944 to control *A. farauti*, the vector of malaria and filariasis (YUST [666]). Later that same year native huts and plantation buildings in and near the previously treated area were sprayed with 5 per cent DDT in kerosene at the rate of 1 gal/1,000 ft². This dosage kept these buildings free of the vector for 3 months. Breeding sites within a one-mile radius of the camp were sprayed weekly with a 5 per cent solution of DDT in No. 2 diesel oil at the rate of about 2 quarts per acre. When American troops first occupied this area early in 1942, the malaria incidence in some units was 60 % within 60 days (SAPERO [499]). One year later, following the instigation of control measures, malaria was no longer a problem. By 1944, and with the appearance of DDT, the occurrence of one case was a matter for concern.

In the village of Papua, New Guinea, the number of malaria vectors, *A. p. moluccensis* and *A. subpictus*, was reduced to less than 5% of the number found in neighboring villages, through the treatment of all hut walls with DDT at the rate of 100 mg/ft² over a period of 4 months (BANG *et al.* [31]). The parasite and spleen indexes in a neighboring uncontrolled village increased while in the treated village the parasite index was slightly decreased and the spleen index stayed the same. Relapses of *P. falciparum* over this period could have masked the full effects of the DDT treatment.

Africa

Modern methods of malaria control have made huge inroads into the prevalence of the disease in many places throughout the world, but in the 'dark continent', the light, in this respect, has hardly begun to illuminate the extensive domain of the monarch of diseases.

Malaria affects the greater part of Africa, and except for a relatively small proportion, where modern control programs are underway, the malady rages unabated. In general, the status remains about the same, varying from hypo- to holo-endemicity (RUSSELL [487]). While there has been a spontaneous decline in malaria in many countries, an increase has occurred in some parts of Africa. In Nigeria, for example, the number of cases among the indigenous population increased from 84,401 in 1944 to 123,265 in 1948, and the case fatality rate per 1,000 inhabitants rose from 1·56 to 1·95 over the same period, based on such epidemiologic data as was available (BRUCE-CHWATT *et. al.* [75]).

Due to the lack of adequate health organizations in most of Africa, epidemiologic data are sparse so that the actual incidence of malaria is at best an estimate; likewise, the economic and social impact of the disease, although great, can only be guessed at.

A. gambiae, perhaps the most vicious of all malaria vectors, is the principal vector in Africa. A sporozoite rate of 38·2 has been recorded for this species in French West Africa. *A. funestus* is second in importance to *gambiae*. DeMEIL-

LON [140] divides the vectors into two main groups: (1) '*Anopheles gambiae gambiae, A. funestus funestus* [and *funestus* var. *imerinensis* (Madagascar only)] are the principal vectors over most of Africa south of the Sahara. These are mostly endophilic species and both may be responsible for endemic malaria, but *A. gambiae* is the most important vector of epidemic malaria; (2) *A. brunnipes* (Leopoldville), *gambiae* var. *melas* (French Guinea, Gold Coast, Ivory Coast, Nigeria, Sierra Leone), *hancocki* (Belgian Congo, Nigeria, Uganda), *hargreavesi* (Southern Nigeria), *moucheti moucheti* (Belgian Congo, Uganda), *nili* (Belgian Congo, Nigeria, Sierra Leone), *pharoensis* (French West Africa), and *rufipes* (French West Africa, ? Sudan) — is of importance in the spread of malaria only in some restricted localities where the primary vectors are absent or rare. Elsewhere they are harmless and mainly exophilic.' A third group of short-lived exophilic species are unimportant as malaria vectors.

A. gambiae is anthropophilic and endophilic over most of its range, but this species shows a high degree of behavioristic variations in different areas. At high altitudes it may be rare, zoophilic, and relatively harmless; or it may, on the other hand, be an important vector under such conditions. In some areas (Kenya), *gambiae* was reported not to rest outside during the day except just after emergence, but in other localities it attacks man out-of-doors. In the forest of Uganda, *gambiae* bites readily during the day, but elsewhere its habits are usually nocturnal.

Climatic conditions certainly influence the habits of this vector. The transmission period and its intensity are governed to a large extent by rainfall and temperatures. In general, *gambiae* is most active during wet seasons. Breeding density seems, however, to be related more to tides than to rainfall. Its flight range, with the prevailing wind, has been shown to be as much as 4 miles (6·4 km). The geographic behavioristic variations of this vector are factors complicating the establishment of uniform and effective control procedures.

A. funestus is both anthropophilic and endophilic and is a weaker flier. In the absence of prevailing wind, 800 m seems to be about the usual maximum limit, although with prevailing winds a flight distance of 8 km has been recorded. This species is an important vector in some areas but definitely takes second place as an African malaria vector. It prefers streams and shade as breeding sites but does breed in other locations. A sporozoite rate of 33·0 was recorded for *A. funestus* from Transvaal.

In much of Equatorial Africa, transmission of malaria occurs throughout the year but diminishes progressively as latitude increases. The transmission period becomes shorter as altitude increases. The disease may be endemic at 1,800 m altitude and, at least in Kenya, does not disappear until 2,700 m is reached. However, *A. gambiae* is, in general, a lowland species (BRUCE-CHWATT *et. al.* [75]).

Plasmodium falciparum is the principal malaria parasite in Equatorial Africa, although *P. vivax* and *P. malariae* are present in certain localities. *P. ovale* has also been found in both East and West Africa.

One of the most interesting, and controversial, problems to arise as a result of recent work in Africa has been with reference to the efficacy of DDT as a residual spray against *A. gambiae*. MUIRHEAD-THOMSON [400], [401] tested the DDT-kerosene residual sprays against *A. gambiae* and *A. melas* in Nigeria. He concluded that these species entered treated native huts, fed, and escaped without showing appreciable mortalities. Further substantiation of his conclusions was the finding of blood-fed gravid females in outdoor resting places in a village where all houses were treated with DDT. The apparent elimination of mosquitoes from treated buildings was attributed to a shift in the population from inside to outside resting places, due to the irritating, but not lethal effects of the DDT. The sharp decline of vectors in treated houses was not considered to evidence mosquito reduction or effective control.

In later tests in Tanganyika, MUIRHEAD-THOMSON [399] treated huts with water-dispersible DDT at the rate of 400 mg/ft². He stated that the number of *gambiae* found dead in the treated hut was only 1% of those that escaped alive after feeding, and of those that escaped, 98% were alive after 24 hours and at least 80% after 48 hours. Similar huts treated with BHC proved completely lethal to all *gambia* that entered for 13 weeks subsequent to treatment.

WILKINSON [628] conducted tests in Uganda to determine the fate of *A. gambiae* and *A. melas* which entered treated mud and wood huts. With DDT-treated wood and fiber huts, the over-all kills were 97% with *gambiae* and 100% with *A. funestus*. The treated mud huts killed 62 and 78% respectively. BHC-treated wood and fiber huts produced about the same kills as the DDT. On mud walls, BHC was superior to DDT, producing 98% kills of *gambiae* and 93 of *funestus*.

HOCKING [272] conducted tests in Kenya on the efficacy of DDT residuals against *A. gambiae* and *A. funestus*. He concluded that these species tend to leave lightly treated buildings, but that with a dosage of 200 mg of DDT per square foot the final mortality would approach 100% and that the toxicity on most surfaces would last from four to six months. HOCKING [271] stated that serious differences between the DDT wettable powder used by MUIRHEAD-THOMSON and that used by WILKINSON have come to light. The material used by MUIRHEAD-THOMSON had large particle sizes which would produce sedimentation, both before and during application, so that the actual dosage of DDT on the walls might be much less than estimated. No chemical determinations were made to check this possibility. HOCKING, therefore, believes that MUIRHEAD-THOMSON's failure might have been due to low dosage or poor formulation.

KARTMAN and DASILVEIRA [304] in laboratory tests in Dakar, French West Africa, found that all *A. gambiae* females exposed to DDT residue for only 30 seconds died between 10 and 24 hours after exposure. Those exposed for 60 seconds were dead within 9 hours. The data indicated that extremely short contact with DDT inhibited most *A. gambiae* females from taking a blood meal at between 1 and 2 hours subsequent to exposure. These tests were conducted in glass tubes of a size to insure constant exposure. It appears, therefore, that the results could not be transposed to field conditions.

DAVIDSON [123] in Taveta, Kenya, constructed mud and thatch huts fitted with window traps to further check the work of MUIRHEAD-THOMSON. These were treated with 0·15 g BHC per square foot, 0·1 g water-wettable DDT per square foot, and 0·3 g oil-bound DDT per square foot, respectively. For the first month, most of the mosquitoes, *A. gambiae* and *A. funestus* were found dead on the floor of the BHC treated hut. In the DDT treated huts, most of the mosquitoes were in the window traps, but 80% or more died within 24 hours. The water-wettable formulation was less effective than the oil solution. Neither DDT nor BHC was repellent, but the contact time before irritation and flying were induced was shorter with DDT, and this time was insufficient to produce significant kills of female *A. gambiae*.

The author states that since no chemical tests were made, MUIRHEAD-THOMSON may have had a lower concentration than the 0·1 g of DDT per square foot used in his own experiments. It was also postulated that DDT formulations containing very small crystals might prove to be as effective as BHC initially, and be longer lasting.

In some areas of Africa where *A. gambiae* is the chief vector, the reduction in malaria morbidity, following DDT residual spraying, lends credence to the work of those claiming good results by this method.

At the World Health Organization Malaria Conference in Equatorial Africa [657], it was stated; 'The conference, after studying the data available, is of the opinion that it is possible to bring about a very large reduction in the transmission of *gambiae* malaria by the application of residual insecticides in human habitations and other mosquito shelters, using either: (1) the gamma-isomer of benzene hexachloride at a dosage of not less than 10 mg/ft² (about 0·11 g/m²) every three months, or (2) DDT at a dosage of 200 mg/ft² (about 2·2 g/m²) of the *p,p'*-product every six months.' It further stated: 'The conference is also of the opinion that it is possible to prevent transmission of *gambiae* malaria completely (1) in some highland areas above 6,000 ft (1,830 m), and (2) in areas where transmission is confined to relatively short seasons.' It is a fact, however, that a certain proportion of some *Anopheles* do escape from DDT-treated buildings without obtaining a lethal dose of the insecticide. The proportion of such escapes could be relatively high under some circumstances, i. e., poor formulations of the insecticides, unusually absorptive walls, old deposits, etc. At any rate, only more and larger-scale control programs in Africa can definitely settle the question. Such programs have not been attempted in appreciable numbers; however, there are a few exceptions where an appraisal of the efficacy of this procedure is, in some instances, possible.

Egypt

There is a great story of malaria control from Egypt concerning the eradication of *A. gambiae*. The fight started with the invasion of upper Egypt in 1942 by *gambiae* from Wadi Halfa (SHOUSHA [508]). Paris green larviciding was the method of choice, but Malariol was also used, and pyrethrum was employed for disinsectization of houses, vehicles, and vessels.

DDT was available only in small amounts and was used in both 5 and 10 per cent solutions to spray-paint railway passenger wagons, airplanes, and boats at the rate of 1 or 2 g/m². A total of only 4,773 l was used. The DDT was of great value for the treatment of passenger wagons because of its long-lasting residual qualities, which greatly decreased the chances of *gambiae* being transported from one area to another.

The last *gambiae* was found on February 19, 1945 during which year there were 28 primary cases of malaria in the *gambiae* area, as compared to 32,823 in 1944.

Morocco

GAUD and MECHALI [216] reported that DDT residual house spraying was singularly unsuccessful in Morocco, in that suppression of mosquito populations seldom lasted as long as two months. Various DDT formulations were used, the best results being obtained with a spray made from 50 per cent water-wettable powder. *Anopheles* mosquitoes reappeared in buildings within 15 days after they had been treated with a DDT-kerosene solution at the rate of 2 g/m². A DDT-water suspension applied at the rate of 1 g/m² lengthened the mosquito-free period to 25 days (GAUD *et al.* [217]).

The authors cite several possible reasons for the failure of DDT. The very hot dry climate might have been a factor, and the nomadic character of the people makes it impossible to protect them from malaria by antimosquito measures alone. Drugs were considered to be very important in the suppression of epidemics, which these workers believed might be started by exotic species such as *A. turkhudi* (*hispaniola*) rather than by *A. maculipennis*, the vector usually responsible for endemic malaria.

The number and extent of vector breeding places are often small compared with the large number of houses that would have to be treated. For economic reasons, therefore, it was considered that the continued use of antilarval methods would be required. MECHALI [370] stated that as a result of such measures, all large centers of population, comprising more than one fifth of the total population, were free from malaria. Management of irrigation and reclamation projects was considered to be a potent factor in suppression of vector populations. Anti-adult measures were considered to be more practical in rural areas, but in sparsely settled and relatively inaccessible territory the method was considered to be too costly, necessitating suppressive medication practices.

Liberia

A report in June 1954 of the malaria control activities in Liberia showed that 52,791 people had been protected to date by the malaria-insecticide program in that country. The activities reported were in Monrovia, Bushrod Island, Sinkor, Brewerville, Klay, White Plains, and Harper. A total of 242 square miles had been covered [589]. The main program, which consisted of residual spraying, was carried out with dieldrin. This was used at 0·3 and 0·5 per cent dieldrin wettable powder and 0·3 per cent dieldrin emulsion [590]. Some DDT spraying was employed, and on celotex and mud walls in the interior of the country it was reported to have maintained a residual for 6 months, although along the coast it lasted only about 1 month (BURTON [81]). DDT was used as a larvicide at a strength of 1·25% in oil, repeated at weekly or biweekly intervals.

Malariometric data are meager but indicate considerable protection of the populace by the activities under way.

Nigeria

Residual spraying with DDT is used in Nigeria only on a small scale by institutions, industrial concerns, and private individuals. The use of DDT as a larvicide by public health authorities was increasing, however, in 1950–52 (BRUCE-CHWATT [73]).

The only organized residual spray program seems to be the Ilaro Experimental Malaria Eradication Scheme, instigated by a suggestion of the World Health Organization Expert Committee on Malaria. Benzene hexachloride, and not DDT, is used in this program and initial spraying was in March 1950. Ilaro in South Western Nigeria has a population of 12,000 inhabitants and 2,300 houses, and is considered to represent hyperendemic malaria conditions in Nigeria. The results are not complete but the program has reduced the number of *A. funestus* adults by about 99%, and larval breeding has been reduced threefold. Substantial reductions of *A. gambiae* adults and larvae also occurred. Malariometric indices have declined slowly; however, a pronounced parasite reduction among infants has occurred, and malaria morbidity, recorded at the Ilaro dispensary, has decreased (BRUCE CHWATT [74]).

Kenya

GARNHAM [209], [210] has conducted an experiment on the use of DDT residual sprays for the control of malaria in the Highlands of Western Kenya where malaria has been on the increase since at least 1940. The program was started in early 1946 in a 60-square-mile area of native reserve between Kericho and Sotik. The area contained about 2,500 huts and a population of approximately 10,000 people. An adjoining area was selected as a check, and parasite rates were determined in 300 people in both areas prior to initiation of the program. Parasite rates were 7% in the area to be treated and 8% in the untreated area. The terrain and lack of roads made treatment a difficult task. The insecticide in solution had to be carried into the mountains on donkeys. The general physical difficulties were such that a single spray team could not treat more than 20 huts per day. The formulation used was 5 per cent DDT in kerosene and the dosage intended to be applied was 200 mg of DDT per square foot. Chemical analysis showed that there was a great variation in the amount of DDT applied to the surfaces and actually an average deposit of about 300 mg was obtained. Three treatments were applied, the first in May just before an epidemic of malaria, the second in August, and the third in December. Much of the area was at an altitude of about 6,000 ft where the transmission season is short, and it was stated that one treatment per year would suffice.

The actual vector density at these altitudes is not great, but small numbers of *A. gambiae* were sufficient to produce new infections in at least one third of the population.

Table 13

Malaria Rates in Different Age Groups in DDT Area and in Control Area
Post Epidemic Figures (Western Kenya)[1]

	Rates all ages %	Ages			
		Under 1 year	1–10 years	11–15 years	Adults
		Parasite rate (%)			
DDT	16	0	26	41	12
control	36	10	44	50	32
		Crescent rate (%)			
DDT	4	0	6	3	3
control	8	0	16	5	6
		Spleen rate (%)			
DDT	13	0	16	21	12
control	17	10	20	23	16

[1] From GARNHAM [211].

Children under 1 year numbered only 17 and 10, respectively, in the DDT and control areas and rates are probably insignificant.

Entomologic inspections were made from June to November following treatment and only five *A. gambiae* were caught in huts in the treated area, while 545 were caught in the same number of huts in the adjacent untreated area. The question as to whether this difference was due to *gambiae* leaving the huts because of irritation by the DDT before they had obtained a lethal dose was not considered to be an important malaria factor in this area, the reason being that in the Highlands the temperature outside of the huts is too low for development of the parasite.

The epidemic of 1946 in the area swept all parts of the district except where DDT had been used, and there was an increase in the parasite rate even in the treated area. Prior to the epidemic in the untreated area, the parasite rate was 8%, but this rose to 36%. In the treated area, it rose from 7 to 16% which suggested that the transmission was only one half as great in this area.

It was believed that more conclusive results would have been obtained had the area been larger than 60 square miles. It was probable that those who contracted malaria in 1946 did so largely while outside the treated zone. Older children were more prone to infection (Table 13), probably because they often stay outside until near midnight, whereas the adults and infants remain inside the huts (GARNHAM [211]).

Uganda

The incidence of malaria in Uganda is often quite high, and has recently been reported on the increase in some areas: Kigezi and adjacent Ruanda Territory. Spleen rates in 1947 in Kigezi were 32% for infants, 82 for children 6–10 years old, and 73 for adults over 20 years old. *A. gambiae* and *A. funestus* are the principal vectors; *A. christyi* has recently been found to transmit malaria in at least the southwest portion of the country (GARNHAM *et al.* [214]).

A preliminary report of a small-scale DDT residual treatment of native huts near Jinja was made by the Colonial Insecticides, Fungicides and Herbicides Committee [5]. Thirteen hundred native huts were treated internally five times at six months' intervals with four per cent (p,p') DDT in Diesoline at the rate of 200 mg/ft². Very little effect was observed on the parasite and spleen rates.

A rural area of about 60 square miles near Kasanje was treated to compare the effectiveness of DDT and BHC. Six different treatments were made over a 2½-year period, but the areas chosen were too small for conclusive results.

The vector populations in treated huts were reduced as much as 98–100%, and the parasite rate reduction was between 57 and 22%. The parasite rate over the whole area was reduced from 38 to 20% and the spleen rate from 52 to 14%.

It was apparent from those experiments that the mud walls of the huts absorbed the spray to a serious extent. DDT wettable powder was absorbed less, and the third experimental spraying employed this formulation.

Special semi-field studies revealed that over 60% of mosquitoes entering untreated huts before 10:00 p. m. remained until dawn, and over 70% that entered during the night remained for two hours or more. Preliminary results did not confirm the work of MUIRHEAD-THOMSON [400] in Nigeria to the effect that nearly all of the mosquitoes (*A. gambiae* and *A. melas*) which entered treated huts were able to feed and escape unharmed.

Tanganyika

A DDT air spray larvicide experiment giving inconclusive preliminary results was reported by WILSON [630]. The work was conducted in Dar es Salaam during the rainy season of April and May 1951. The spray used was 5 per cent

DDT in Malariol H. S. and was applied initially at a dosage of one quart, 0·125 lb of DDT, per acre; later being increased to two quarts per acre. Recovery of DDT at ground level was 75–80%, but only 30% of the oil was recovered. A swath width of 55 yd was employed, and the intervals between treatments varied from 7 to 11 days.

Houses were selected in both treated and nontreated areas as catching stations for the appraisal of results. Only *A. gambiae* was considered since it is the principal vector during the rainy season, when these tests were conducted. There was a significant difference in the population levels in the treated and extralimited zones during only one week of the spray period. The normal suspected rise in *gambiae* density during the long rainy season was substantially smaller than in previous rises. It was not evident whether this was due to the air spray or to the normal antimalaria activities of the municipal control organization.

Evidence did not indicate that the treated areas were being populated by migration from the extralimited zones, and it was indicated in preliminary tests that sufficient DDT was being deposited at ground level to effect complete control.

The tests were on a small scale, the spraying cycle varied in some instances perhaps sufficiently to permit emergence of adults, rain impaired the efficiency of the treatments and in some areas the test was complicated by other mosquito control activities. These and other factors prompted the attitude that the results of the experiment could not be considered other than preliminary, and in no way conclusive.

Belgian Congo — Katanga

Some form of malaria control has been a part of the responsibilities of the Health Service in Katanga since about 1925. Large-scale drainage undertaken mainly for soil reclamation purposes rather than malaria control was the principal control procedure. Less extensive work was carried out against *Anopheles* breeding in some areas, consisting of periodic inspections, oil larviciding, and some small-scale drainage. In a few places such as Matadi, where mosquito control was undertaken principally to prevent a recurrence of the yellow fever epidemic of 1928–29, good malaria control was obtained. In general, however, malaria control in Katanga in the pre-DDT period was spotty and only partially effective.

In 1947 the first DDT house spraying campaign was undertaken in Katanga (VINCKE [607]). The program was initiated at Elisabethville and later extended to Manono, Albertville, Kasenga, Bukama, Kongolo and Kamina. The 'Union Minière du Haut Katanga' initiated treatment in the town of Jadotville in 1947 and gradually extended the treatment to all its centers in 1948. The organization in Katanga was modeled after the work in Italy and Sardinia. DDT was applied at the rate of 2 g/m², and emulsions were used for permanent structures and suspensions for temporary ones. The campaign, according to

Vincke, was not conducted with the thoroughness desired due to the lack of trained personnel.

A. funestus and *A. gambiae* are the two malaria vectors in the area, and both species are normally highly infected. Following DDT treatment, *Anopheles* almost disappeared from treated structures. Anopheline density was determined by means of captures in sprayed and unsprayed houses and in human-baited traps. In the European towns, one house per block was chosen in advance of treatment as a control house, and examinations for *Anopheles* were made throughout the year. After treatment, only 4 *A. funestus* were found in 3,017 visits, or 0·13% as against 24% prior to treatment. No *gambiae* were found after treatment, whereas the pretreatment rate was 25%. The 1949 and 1950 results confirm this condition. Counts of mosquitoes made on man during the night revealed a similar dramatic reduction in the *Anopheles* population. At Albertville, the rate per 100 visits to human-baited traps dropped from 503 *A. funestus* in 1945 to 6·3 in 1948. No catches were recorded in 1950. The rate per 100 trap visits for *A. gambiae* dropped from 45 in 1945 to 0 in 1948. During May 1950, a rate of 4·87 was recorded.

As would be expected, the malaria incidence declined following the effective control of the vector. The average malaria infection rate of mosquitoes from January 1944 to January 1946 was 2·8% for 6,626 *A. funestus* dissected, and 4·6% for 1,603 *gambiae*. The infection rate in the rainy season was as high as 9·84% for *funestus* and 12% for *gambiae*. During July, August, and September the infection rate is normally very low and often approaches zero. From the period December 1947 to November 1948, the sporozoite rate was 0·22% among 450 *funestus* examined and no infection was found in 214 *gambiae*. The only infected mosquito was captured at the extreme edge of the zone and could have originated from outside the protected area.

A critical evaluation of human incidence is difficult from the information available, but the low infection rate among mosquitoes accompanied a decline in human malaria. Examination of children at native schools at Elisabethville during the period October 1947 to December 1949 revealed a striking decrease in the number of positive cases. The plasmodium rate decreased from 37·03 to 7·09% among native school children 4 to 8 years old during this period. In the 8- to 12-year group the decrease was from 43·02 to 17·14%; the 12- to 16-year group from 40·73 to 4·47%; and over 16 years from 34·35 to 3·33%. The spleen rate at Manono among children 0 to 4 years of age decreased from 36·36% in June 1948 to 3·87% in November 1949. A similar decrease occurred in the age groups, 4 to 8 and 8 to 12. Spleen rates in control areas were significantly higher, reaching 59·67% in the 0- to 8-year bracket and 42·85% in the 8- to 12-year group.

Following the DDT campaign, there was a decided drop in the number of working days lost among Europeans because of malaria. In 1949, the figure for Elisabethville was 6·8% of the lowest figure during the six years preceding DDT spraying, and for all sites combined it was 25·23% of the lowest figure for the same period (Vincke [607]).

Child mortality declined and in 1949 was only 23% of the lowest figure registered since 1937. It is abundantly clear that DDT spraying completely revolutionized malaria control in Katanga and has resulted in a greatly improved state of health for both natives and Europeans, in addition to contributing to the economic well-being of the area.

Ruanda-Urundi

Ruanda-Urundi, a Belgian protectorate just east of the Belgian Congo, is a hilly, plateau type of country interspersed with swampy valleys. The altitude averages 5,500 ft, and the population density ranges from 1,030 to 1,300 per square mile.

There are three *Anopheles* vectors present, *A. funestus*, *A. christyi*, and *A. gambiae*. *A. funestus* is the chief vector and *gambiae* is rare. A sporozoite rate of 6–18% has been recorded for *A. funestus*. Trials with DDT residual spray formulations were instituted by JADIN *et al.* [289] using a water-wettable DDT suspension applied to inside resting surfaces of buildings at the rate of 200 mg/ft². The internal surfaces of some buildings were also treated with DDT dust. Both the spray and dust effected a reduction in the vector population. Following these tests a large-scale program was instituted using the DDT wettable powder spray at a dosage of 200 mg of DDT per square foot. An area of 25 km in diameter around Astrida was treated. A total of 1,443 kg of wettable powder was used to treat 22,446 huts, housing a population of 37,736 people. In a later program, 33,220 huts were treated involving 50,295 people. In addition to the residual spraying, some larviciding was performed with 5–10 per cent DDT dust, during the dry season, in breeding areas in proximity to concentrations of huts.

As a result of the program in Ruanda-Urundi anopheline vectors were reported to have disappeared from the mountainous region of Astrida, and the parasite index, based on examination of 23,028 thick blood films, dropped from 51·13 to 24·06, a decrease of 52·92%.

Southern Rhodesia

Malaria is seasonal in Rhodesia and is transmitted by both *A. funestus* and *A. gambiae*. The former is mainly responsible for maintenance of endemicity and the latter for seasonal exacerbations and epidemics. The Zambesi and Mazoe Valleys are particularly malarious, but areas over 4,000 ft in altitude are substantially free of the disease (WILSON [631]). The country has a population of about 2,000,000 people. The first experimental use of modern insecticides against malaria and blackwater fever was in 1946 and the results obtained were sufficiently encouraging to justify a large-scale program. This was initiated in the Mazoe Valley in 1949 and consisted entirely of the application of residual spray to buildings. No anti-larval measures were employed, no organized prophylactic campaign was undertaken, nor were educational activities conducted (Rosen [479]). The area selected contained 1,900 square miles encompassing farms, small mines, small urban settlements, and native reserves.

The first treatment was made with DDT exclusively and the second, three months later, with BHC wettable powder at the rate of 46 mg of gamma isomer per square foot. In some instances, Europeans objected to the BHC, in which cases DDT was employed. A total of 66,712 habitations were treated, in addition to 6,122 rooms in European dwellings during the first two spray applications.

A malariometric evaluation of the program is yet to be made; however the parasite rate after the first year of treatment was 40.6% compared with 77.3% in adjacent untreated areas. Malaria admissions to European Hospitals dropped from a nine-year average of 78 cases annually to 55 cases in 1949, and 2 in 1950. Farmers reported a marked decrease in African sickness and in absenteeism from work.

Plans are to encourage municipalities, industrial, mining and agricultural groups to initiate residual spray programs to protect their communities, and a second line of attack envisages the inauguration of a ten-year plan to treat with residual insecticides all native reserves of Southern Rhodesia. Such a program would involve some 3,600 square miles with a population of 174,000 (Blair [53], Pampana [424]).

If preliminary results are an indication of final accomplishments, it can be assumed that such a campaign will open large areas of fertile lowlands to agricultural development.

One interesting feature of the Southern Rhodesian program is the year-round employment of operating personnel. The period from October to April is devoted to malaria control, and the period from May to September to the control of bilharziasis. This scheme is worthy of consideration in other countries which have summer and winter health problems that can be handled by such personnel.

Union of South Africa — Transvaal

It has been said that the Transvaal is a malaria-free island in the Union of South Africa, and the control of the disease is like adding a fifth province to the Union (Olivier [419]). Transvaal contains about 58,000 square miles (150,000 km^2) of malarious territory, with a population of about 1,350,000 of which only about 100,000 are Europeans. About 98% of the infections in North Transvaal before 1944 were due to *P. falciparum*. *A. gambiae* is the vector of summer epidemics, with *A. funestus* being confined to certain endemic zones.

A malaria research station was established in 1931 in a hyperendemic area in Transvaal. A small staff attempted to control malaria in limited areas by educating the populace concerning the disease, including information on housing, screening, and insecticidal spraying. It was shown that the disease could be controlled since in one hyperendemic settlement the incidence of malaria was reduced from 51 to 7%. In spite of this work, however, the overall malaria situation in Transvaal was little changed, and two epidemics ravaged the area in 1939 and 1943, respectively (ANNECKE [21]).

A malaria campaign using DDT insecticide was initiated in 1945 and covered an area of about 42,000 square miles. Both anti-larvae measures and residual house spraying were practiced. The residual spray program encompasses an area of 17,450 square miles (45,000 km²), and in the summer of 1948–49 a total of 350,000 houses were treated (PAMPANA [424]).

In 1950, according to ANNECKE, the following program was in effect:

(a) Spraying of all European dwellings with 5 per cent DDT-kerosene every two to three months, aiming at an even coverage of 100 mg/ft².

(b) Spraying of all native dwellings every two to three months with 50 per cent DDT wettable powder diluted to 5%, aiming at an even coverage of 100 mg/ft².

(c) Oiling of all breeding places in summer every fortnight with 27½ per cent DDT emulsion; diluted with any local water 1:300.

(d) Oiling of *A. gambiae* winter nurseries in winter with 1:300 emulsion every month.

DDT larvicides are applied in the winter when the vector population is low, mainly to protect the densely populated areas or areas of high endemicity where residual spraying would be expensive. The winter larviciding results in a delay of the summer vector population; thus a single application of DDT residual spray may suffice for the entire season (OLIVIER [419]). Larviciding should, according to OLIVIER, constitute the bulwark of the well-directed malaria control campaign.

The malaria research and control station at Tzaneen, Transvaal, reported on the incidence of malaria from examinations made at that station during the period, 1939–49, as shown in Table 14. After the introduction of control in 1944–45, spleen rates dropped from a maximum of 93% in the age group 2–5 years to 16% in the age group 11–15 years. Spleen, parasite, and crescent rates among native children, 0–2 years of age, which were those born after the initiation of malaria control, were extremely low compared with the precontrol period, indicating that little active transmission was occurring. The decrease in spleen rates in this age group was from 62% in the summer of 1931 to 12% in the summer of 1950. Parasite rates had a corresponding decline (ANNECKE [21]). Admissions to Transvaal hospitals for malaria treatment dropped from 2,070 in 1942–43 to 1,177 in 1944–45 to 263 in 1949–50. Reported cases of blackwater fever declined from 23 in 1932 to 0 for the years 1947–50, and no deaths have been reported since 1945.

Coincident with the DDT malaria control program, the use of antimalaria drugs, whose use parallels that of malaria prevalence, rapidly declined. The experience in Transvaal leaves no question as to the value of vector control, as compared with drug control in human hosts, for malaria control (ANNECKE [21]).

334   S. W. Simmons

Although drugs are still made available to the public, consumption of quinine (gr. 5) dropped from 1,849,000 in 1943–44 to 217,000 in 1948–49 and of atabrin (0·1 G) from 41,450 in 1944–45 to 9,000 in 1948–49.

ANNECKE [21] discussed the cost of larviciding operations versus adult control practices in the Transvaal. Areas with a high human population density, limited mosquito breeding in accessible places, and with plentiful and cheap labor lend themselves to larviciding practices; whereas, a low human population density in an area of numerous inaccessible breeding places and limited expensive labor, usually is more suitable to adult control programs. In the Transvaal, there was little difference in the results obtained, but control work

Table 14

The Incidence of Malaria Determined by Positive Blood Smears Examined at the Malaria Research and Control Station, Tzaneen, Transvaal, 1939–49

Year	Total examined by malaria staff	No. positive
1939–40	10,312	4,494
1940–41	6,544	4,287
1941–42	9,514	4,498
1942–43	5,994	2,396
1943–44	6,713	3,980
1944–45	5,151	1,831
1945–46	2,938	1,263
1946–47	1,898	348
1947–48	786	441
1948–49	945	128

over lightly populated areas was usually more economical with residual sprays than with larvicides. The per capita cost of the program during 1948–49 was US 45 C. (OLIVIER [419]).

The economic and social aspects of malaria control are exemplified in the changes that occurred following the instigation of the program in the Transvaal. Prior to 1944 the plight of the farmers was one of serious economic instability which resulted in migration from malarious areas, and a drastic decline in farm prices. In the Tuinplaats area, the maximum prices for land in 1944 were from 5 to 6 £ per morgen for well improved farms. Prices were showing declines chiefly due to the presence of malaria. After 4 years of malaria control, the prices of the same farms rose from 5 to 6 £ to 22 £ per morgen, and production increased by over 400%. Post offices, banks and other businesses showed an increase in business of 400 to 500% over the pre-malaria control period (ANNECKE [21]). In general the Transvaal has experienced an economic revolution coincident with malaria control and provides an additional example of the value of disease control in the conservation of human resources.

Swaziland

Malaria in Swaziland is predominantly due to *P. falciparum*, and the principal vector is *A. gambiae*. The country includes hyperendemic, endemic, and intermediate zones, depending upon the altitude. DDT was first used, on a small scale, in 1948–49, and the promising results obtained stimulated larger-scale operations in 1950. In that year, 2,500 square miles of territory were covered and 35,000 people protected. DDT was used as a 3·5 per cent emulsion and as a 5 per cent water-wettable powder. The average dose was 0·43 gal per hut treated.

The program resulted in a satisfactory reduction of the vector, *A. gambiae*. Parasite surveys were made prior to treatment and at monthly intervals during the program, in both treated and untreated areas. The parasite rates were consistently less in the treated than the untreated zones, except in those non-hyperendemic areas where the rates remained at about their preseason level, but without the rise which occurred in comparable untreated territory. The treated hyperendemic areas showed an appreciable rise in parasite rates, but this did not reach the level attained where no DDT treatment was applied (MASTBAUM [364]).

Madagascar

The use of DDT for malaria control in Madagascar began with a house spraying program in Tananarive in September 1949. This was supplemented by the employment of the usual anti-larval measures and the prophylactic use of antimalarial drugs. The latter acted as a 'stop gap' filling for the holes in the chemical 'mosquito netting' (MERCIER [372]). The program continued on a biannual cycle throughout 1950, a total of 37,021 buildings being treated from January to December 1950; protecting a native and European population of 179,789 persons. The DDT was applied as a 5 per cent solution in kerosene and as a 3 per cent suspension of water-wettable powder.

There was a near disappearance of domestic insect parasites, and a limited survival of the malaria vectors, *A. funestus* and *A. gambiae*, as well as culicine mosquitoes.

BERNARD [43] reported that more than 71,000 buildings were treated in Madagascar during the period, September 1949 to March 1950. Collections made in 2,227 houses 4 to 6 months after spraying resulted in the recovery of 2,302 mosquitoes, only 169 of which were alive, including 18 *Anopheles*. The same worker, BERNARD [44], presented data from blood smear examinations, ranging in numbers from 25,000 to 43,000 annually, in which the positives declined from 19·17% in 1948 to 6·56% in 1949 and 3·03% in 1950. The first decline coincided with the introduction of nivaquine in malaria prophylaxis, but the record was attributed to insecticide spraying.

As a result of the work in Tananarive, MERCIER [372] reported a reduced parasitemia and malaria morbidity rate, resulting in a decrease in absentee workers and in reduced hospital admittance. After the house spraying campaign, the usual annual malaria epidemic, from December to May, did not occur. Prior to the use of DDT, however, the annual epidemics had been partially depressed by the use of prophylactic drugs among pre-school and school children (Table 15).

Table 15

Native Malaria Mortality, Tananarive, Madagascar, 1946–50[1])

	1946	1947	1948	1949	1950
Malaria mortality (cases)	843	632	619	445	189
Native population	143,202	145,295	150,372	154,049	159,595
Rate per 100,000 persons	588·67	434·98	411·64	288·86	118·42

[1]) From MERCIER [372].

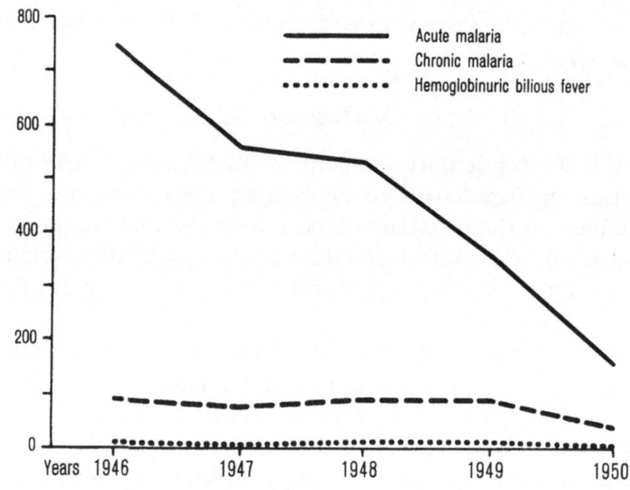

Figure 7

City of Tananarive, Madagascar: Native malaria mortality since 1946.

Although deaths from acute malaria decreased sharply in 1949–50, chronic forms of the disease, as would be expected, showed a slower decline (Figure 7).

The program not only exerted a pronounced influence on the total death rate but produced a shift in the principal causes of mortality among the native population. Malaria was the leading cause of death in 1946–47, but shifted, by a small margin, to second place in 1948. In 1949 there were more deaths from respiratory diseases, and in 1950 digestive and circulatory diseases also produced more deaths. The relation of reported malaria deaths to total deaths (excluding still-births) was 21·29% in 1946, 20·06 in 1947, 18·76 in 1948,

16·76 in 1949, and only 8·16% in 1950. The use of DDT coincides with the sharpest decline in the relative position of malaria as a cause of death.

Malaria deaths occur mainly in the 1- to 15-year group, where the mortality was reduced about 78% among the native population between 1946 and 1950. The death rate from all causes among ages 0 to 15 years decreased 36·35% over the same period. Most of the deaths of infants under one year were caused by pulmonary and intestinal diseases, but even these declined by 31·44% over the five-year period. Since decreased infant mortality, due probably to fly-borne diseases, has been observed following DDT residual spray campaigns, one wonders if this was not a factor in Tananarive, although it was not mentioned as such.

BERNARD [44] stated that the demographic situation of Madagascar remained practically stationary up to 1947, but since that time the increase in population has doubled each year. What is most significant, the results of the antimalaria spraying, which began in September 1949, had not at this time clearly shown their full effects on the demographic picture. Malaria has not been as great a peril to the European as to the native population of Madagascar, particularly as a cause of death. The vital statistics of this group are more accurate, and although the amount of malaria is relatively small, it is significant that in 1950, following the use of DDT, acute malaria cases were only one third as prevalent as in 1947–48 and there were only half as many as in 1949. The disease was the main cause of death among Europeans in 1946 but dropped to fourth place in 1950. In 1946, it accounted for 15·42% of the total mortality, but in 1950 only 6·21% of deaths, recorded by the municipal Bureau of Hygiene, were due to malaria.

In summation, the most important decrease in malaria incidence, particularly among the native population, occurred in 1950 when there was a decline in malaria mortality among children (0–15 years) of 61·24%, and a decline among adults was 45·79%.

Since the beginning of the insecticide program, the usual annual cyclic increase of malaria has been replaced by a decrease. MERCIER [372] believes, however, that it will be impossible to suppress the disease further, but states that the prevailing low incidence can be maintained by a continuation of the insecticidal program.

Mauritius

Small-scale experiments with a 5 per cent DDT-kerosene residual spray were conducted in Mauritius in 1945 by TONKING et. al. [580]. A pronounced reduction of *A. funestus* and *A. gambiae* occurred in treated dwellings. In 1946 the experiment was enlarged to cover a hyperendemic area of 16 square miles on the West Coast of the Island. In this area, *A. funestus* outnumbered *A. gambiae* 36 to 1 in the winter and 98 to 1 in the summer. A DDT-kerosene solution was applied three times at 12 to 14-week intervals at the rate of 178, 112, and 150 mg of DDT per square foot respectively. The interiors of all human and animal dwellings were treated. *A. funestus* populations which varied from 6 to 839 in huts before treatment were reduced to nil, while in an adjacent untreated area, the population averaged 202 per hut. It was concluded that the spraying lowered the parasite index of the general population by two thirds and of the children by one half in the treated area (TONKING and GÉBERT [579]).

Table 16

Results of Residual Spraying on the Population of Specified Mosquitoes in Mauritius[1])

Number of houses treated	Anopheles gambiae	Anopheles funestus	Aedes aegypti
Before Spraying 7,767	2,486	31,196	1,716
After Spraying 76,246	703	474	1
Reduction (%)	97·1	99·9	99·9

[1]) From Dowling, [151].

Table 17

Spleen and Parasite Rates in Children in Mauritius

Surveys	Number of children examined	Rates for whole island	
		Spleen rate %	Parasite rate %
First survey, 1948 (before spraying)	3,585	34·8	9·5
Second survey, 1949 (after initial spraying)	12,105	15·3	2·4
Third survey, 1950 (after 2 sprayings)	14,526	2·8	0·36

A malaria research team was appointed in November 1948 to direct the campaign to eradicate both malaria and its vectors from the Island. The basis for the program was the spraying of the interior of all buildings with DDT or BHC. The Island was divided into four principal zones, a central untreated zone where the vectors had already been eradicated and three coastal zones where the disease was prevalent. DDT was used, both as a kerosene solution and as a water-wettable powder, and applied at the rate of about 200 mg/ft². BHC was used as a water-wettable powder suspension applied at the rate of about 24 mg of gamma isomer per square foot. Spraying was initiated in January 1949 and the first spraying cycle was completed in May 1949. A total of 75,829 houses was treated, protecting 324,191 people at a cost of 1.19 rupees

(1 rupee = US 21 C.) *per capita*. A second treatment was applied in August 1949 at a cost of 0.77 rupees *per capita*. After 1949, the use of BHC was abandoned and DDT used exclusively (DOWLING [152], [153]; PAMPANA [424]).

Entomologic inspections for adult mosquitoes were made regularly in about 5,500 houses scattered throughout the Island. There was a reduction of 97·1% for *A. gambiae* and 99·9 for *A. funestus* and *A. aegypti*. Table 16 illustrates the entomological results of the program as reported by DOWLING [154].

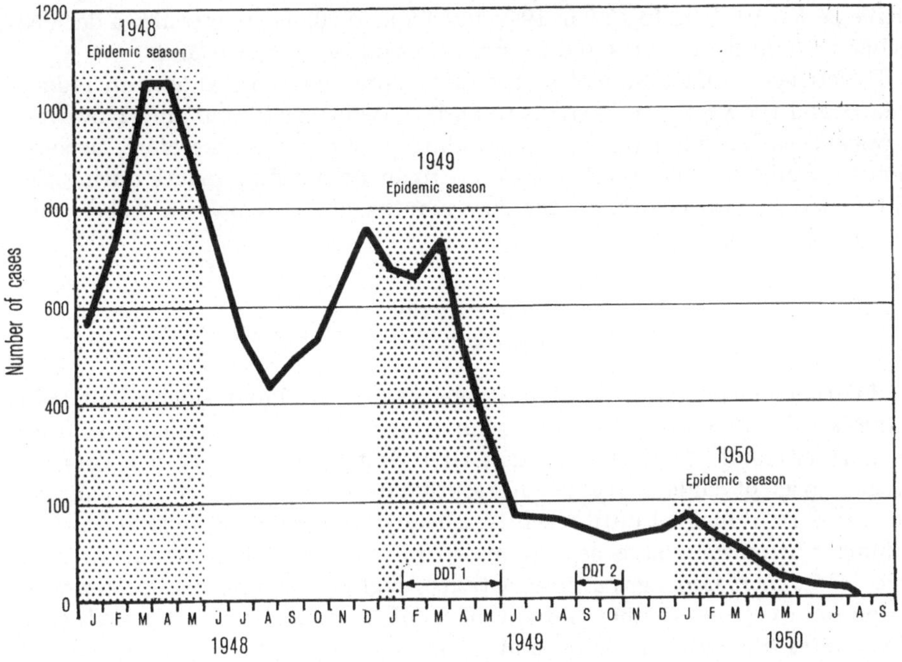

Figure 8

Cases of malaria notified by dispensaries and hospitals. Flacq, Mauritius, 1948–1950.

A malariometric survey was conducted prior to spraying in which spleen and blood examinations were made of children of all ages from birth to 15 years. A second survey was conducted after the initial spraying in 1949 and a third in 1950. Table 17 gives the results of these surveys.

There was a reduction in total spleen rates of children examined from 34·8 during the first survey to 2·8 during the third survey. There was likewise a reduction in Class II, III, and IV spleens. A survey of 6,361 infants born subsequent to the completion of the first spraying revealed 17 positive blood slides, or a parasite rate of 0·27%. The rate among 704 children over one year of age was 0·28%.

Coincident with the general reduction of malaria, the proportion of *P. fal-*

ciparum infections in school children decreased. *P. vivax* was the predominant parasite in all surveys with *P. malariae* about as prevalent as *falciparum*.

Figure 8 illustrates the downward trend of malaria based on reports from dispensaries and hospitals in Flacq from 1948 to 1950.

The total number of malaria cases reported by hospitals and dispensaries during the months of January to June 1950 showed a reduction of 86% as compared with 1948. DOWLING [151] believes that many of the reported cases were not malaria since only 3% of 1,760 bloods examined were positive. The malaria death rates per 10,000 population declined from a mean of 32·2 ± 8·07 for the period 1934–48 to 12·9 in 1949 and 4·9 in 1950. There was also a decided decline in total death rates and in infant mortality for the Island.

Experience to date indicates that DDT spraying alone is able to reduce malaria and the vector *A. funestus* to an inconsequential minimum. However, *A. gambiae* is less vulnerable. A larviciding program was, therefore, decided upon to eliminate this vector from its permanent breeding places during the winter season, particularly in the dry months of September, October, and November (DOWLING [151]).

North America
United States

Malaria incidence in the United States has been declining since about 1875 (BARBER [33]; BOYD [61]; ACKERKNECHT [1]). The disease receded first from the northern and, to a great extent, the western sections of the country leaving areas of endemicity in the southeastern quadrant of the country where malaria remained until recently a heavy financial and physical burden. A number of reasons have been proposed for this recession: urbanization of large areas, increased agricultural activity, higher incomes, and better protection from vectors. Reduction in numbers of the vector anophelines, *A. quadrimaculatus*, the eastern vector, and *A. freeborni*, the western vector, did not apparently play a deciding role. These mosquitoes continued to be numerous throughout formerly malarious areas.

Malaria has been an important cause of disability and unproductiveness to the country, and as late as 1938 the economic loss attributed to the disease was estimated to be one half billion dollars (WILLIAMS [629]). Apparently the first suggestion for the eradication of malaria in the United States was made by HOFFMAN in 1915 [275]. From this time efforts to control the disease became better organized and coordinated. Experience gained in World War I in malaria control with screening and larviciding, in addition to more expensive drainage, led to the widespread use of these measures. In 1921, the larvicidal potentialities of Paris green were discovered and gave added impetus to control efforts (BARBER and HAYNE [34]). During the depression years of the early 1930's, labor from relief organizations was used widely for malaria control (ANDREWS [17]). The passage of the Social Security Act in 1935 and its extension in 1939 made possible the addition of malaria survey and control personnel to State health departments and stimulated greater interest among the States in this activity (ANDREWS [16]). The Tennessee Valley Authority, established for the development of the Tennessee river system, did much through its own operations or those under its supervision, to reduce malaria morbidity over a large area of former high endemicity. All of these factors contributed to the over-all advancement of environmental malaria control in the South until the end of 1941.

Mobilization for World War II took from the States many trained personnel for use in military malaria control. In addition, large training centers were constructed, particularly in the malarious Southeast where the climate favored year-round outdoor activities. This placed an additional burden on already understaffed State health departments, and the Federal government had to assume the direction and coordination of malaria control activities around extra-cantonment military areas. For this, the US Public Health Service opened an office for Malaria Control in War Areas. Through this office, operations were carried out between 1942 and 1945 near some 2,200 localities in 19 States at a cost of $ 25,000,000 (ANDREWS [15]). These operations were generally credited with having prevented malaria from interfering with the war effort.

DDT first came to the US in 1942 and in the following years much research was done to establish proper procedures for its use in insect control. It was first employed on military reservations, but by 1945 was made available for civilian use. In this year, an extended malaria control program was inaugurated by the United States Public Health Service in cooperation with the various State health departments to spray DDT on the interior walls and ceilings of rural homes and privies. Counties were selected for treatment where relatively high malaria mortality — 10 per 100,000 — was experienced during the five years preceding World War II. This was continued for two years during which time nearly 2,500,000 house-sprayings were done in 315 counties at a cost of 11·5 million dollars.

The decline in malaria morbidity and mortality had been steady since the last epidemic years, 1933–36, except for the latter years of World War II. A temporary rise at this time was due to the repatriation of thousands of service personnel from malarious areas overseas. The DDT spraying program no doubt contributed to the control of the disease and no epidemics resulted from the introduction of foreign malaria, which experiments demonstrated could be transmitted by local anophelines. These factors, plus greater awareness of malaria by the people and the successful extermination of *A. gambiae* and *A. aegypti* from parts of Brazil (SOPER and WILSON [544]), seemed to favor the success of an eradication program; and on July 1, 1947, the National Malaria Eradication Program, a cooperative undertaking of State and Federal health organizations, was launched (ANDREWS and GILBERTSON [18]). The purpose of this program was to reduce concurrently the numbers of malaria parasites and vectors to the point where malaria transmission would not occur. As a supplementary phase of the program, surveillance-and-prevention teams were assigned to State health departments to prevent reestablishment of endemicity. Operations of the Eradication Program were those developed on the extended program, utilizing DDT.

The general procedure was to apply 5 per cent DDT in a Triton-xylene-water emulsion to interior surfaces of homes and privies at the approximate rate of 200 mg of DDT per square foot (COMMUNICABLE DISEASE CENTER [106]; STIERLI *et al.* [558]; SIMMONS [517]; QUARTERMAN [460]). This was usually done twice per season although in some areas this was reduced to one annual treatment. At first, spraying was done only in counties with a 1938–42 death rate of 5 per 100,000; but as malaria mortality was eliminated in many coun-

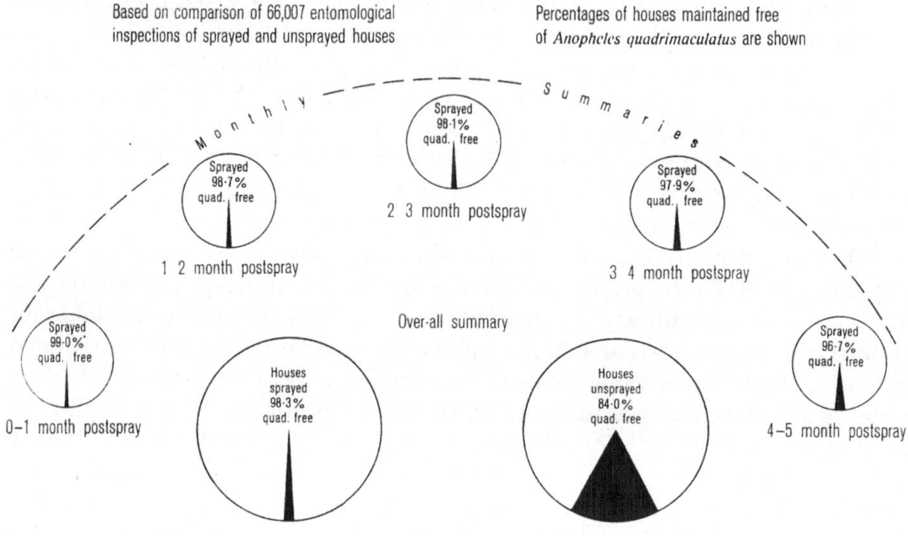

Figure 9

Communicable disease center malaria eradication program. Five-year summary of entomological evaluation of malaria eradication program in 13 southeastern states 1945–1949, inclusive.

ties, this was reduced until by 1949 counties with the above death rate plus a rate of 1 per 100,000 from 1943–46, counties with a rate of 4 or more per 100,000 for this period irrespective of previous rates, and rural homes within one mile of malaria cases were included (BRADLEY and LYMAN [65]).

Malaria control activities were the most widespread in 1948 when DDT residual spray was applied to 1,364,950 dwellings in 360 counties of 13 States (ANDREWS [15]). Entomologic surveys showed that 97·2% of sprayed houses and 83·3% of unsprayed ones were free of *A. quadrimaculatus*, an indicated percentage of control of 83·2 (BRADLEY and LYMAN [64]). Entomologic evaluation of the National Malaria Eradication Program over a period of five years indicated the high percentage of houses maintained free of *A. quadri-maculatus* (Figure 9). Reported malaria morbidity and mortality which had totalled 58,781 and 861 respectively in 1942, declined to 16,203 and 214 in 1947 and to 9,606 and 170 in 1948 (ANDREWS and GRANT [20]). The apparent success of the Eradication Program led to reductions in activity, and by 1950, Federal support of malaria control in six States on the outer edge of the malarious area was withdrawn. In 1951, United States Public Health Service participation in malaria control operations was stopped. However, support was given to State health departments in enhancing their surveillance and prevention programs aimed at maintaining the *status quo* and at completing the eradication of endemic malaria from the United States. Only 2,184 cases — almost three fourths of these were from one State which did not require identification of patient and physician for each case — were reported in 1950. The National Malaria Society in 1950 established criteria to determine when malaria ceases to be an endemic disease in the United States (HINMAN [267]). The eradication goal seemed to be in sight. However, in that year, US troops entered the Korean conflict and by 1951 great numbers of service-men infected with Korean malaria were being repatriated (ANDREWS and GRANT [20]). This strain of malaria exhibited the same long incubation period characteristic of other temperate zone malarias and, therefore, many primary attacks were experienced after troops had returned to the United States. The number of reported malaria cases for 1951 rose to 5,600, an incomplete figure, since over 12,000 malaria attacks which occurred among Army personnel were not reported to the National Office of Vital Statistics. Early in 1952, mass prophylaxis with primaquine of all troops returning from Korea by ship was initiated. This did not go into full effect until June 1952, and malaria morbidity again rose that year to 7,023. The first outbreak of malaria in the United States since 1945 also occurred in 1952 (ANDREWS *et al.* [19]). A Korean veteran on a weekend camping trip near a Campfire Girls' summer camp experienced a relapse of *P. vivax* malaria. As a result, 35 girls came down with malaria, 9 in August and early September 1952 and 26 between February and August 1953 with indicated incubation periods of 199 to 308 or more days. There was no evidence of transmission at the camp in 1953.

Total reported malaria morbidity for 1953 was 1,418[1]) from both civilian and military sources. From this figure, it is evident that malaria incidence is again declining. Even though widespread insecticidal operations against vector mosquitoes have been discontinued, there is no evidence that malaria endemicity has been reestablished in spite of the influx of Korean malaria. It is believed that careful surveillance and prompt institution of antimalarial measures where a case has been reported have been instrumental in the success of the National Malaria Eradication Program in the United States.

Mexico

Mexico includes an area of about 2,000,000 km² and has a population of approximately 25 million. Malaria is widespread throughout the country and

---

[1]) Preliminary figure.

is a leading cause of sickness and death. It occurs in all the States of Mexico from sea level to altitudes of 7,000 ft. All the southern and seaboard States contain innumerable low-lying areas in which temperature, rainfall, and humidity are very favorable for the perpetuation of endemic malaria. Pernicious and fatal cases of malaria are frequent. The States in north and central Mexico also have endemic malaria but cases are fewer in number and are less severe. The disease is transmitted in the hot, moist low-lying coastal areas by *A. albimanus*, in both low-lying and elevated areas by *A. pseudopunctipennis*, the most prevalent species, and in areas above 5,000 ft by *A. aztecus* (MEXICO [373]; PAMPANA [424]).

Malaria morbidity in Mexico has been estimated at 2,000,000 cases per year (PELAEZ [432]). In 1946 there were 25,124 deaths from the disease or a rate of 110·3 per 100,000 population [427]. In Tabasco, one of the more heavily infected States, the case rate has varied between 3,500 and 4,500 per 100,000 population, with an average death rate of 140·4. During the period 1939–45, other States reported even more impressive mortality rates, 557·4 in Oaxaca and 309·9 in Chiapas (MEXICO [374]). The majority of infections are *P. vivax*, with *P. falciparum* occurring in lesser amounts (PAMPANA [424]).

During the mosquito season of 1945, an experiment was launched in Mexico in the State of Morelos to determine whether spraying the interior walls and ceilings of buildings with DDT would protect the inhabitants of that area from malaria (GAHAN and PAYNE [204]). Two small villages, Temixco and Santa Ines, were selected and were treated between April 2 and May 14. Data accumulated in twelve series of observations made between May 15 and October 18 showed that the number of adult mosquitoes resting in houses in the afternoon was reduced more than 99%. In San Jose, it was indicated that DDT residues were responsible for reducing the number of larvae 85% or more over a four-month period, when compared with the larval population of a comparable town. In subsequent surveys in 1946, 1947, and 1948 following annual applications of DDT, the number of *Anopheles* adults found per house averaged 0·02 in Temixco and 0·5 in Acatlipa. The figure for Acatlipa was 92 in 1945 when the village was left untreated (GAHAN *et al.* [202]). The concomitant reduction in larvae also continued. Laboratory tests proved that after four annual sprayings there was no resistance to DDT among *A. pseudopunctipennis*. Five years of spraying in these towns resulted in a reduction in malaria incidence estimated to be between 75 and 90% (DOWNS *et al.* [157]). Only a few autochthonous cases of malaria have been seen in the resident population during this time. DDT residual deposits applied to adobe walls and thatched roofs remained active against the vector, *A. pseudopunctipennis*, for one year (DOWNS and BORDAS [154]). An irritant factor was noticed among exposed mosquitoes, 35·1% of them leaving treated huts. Mortality among these, however, was twice that of unexposed mosquitoes. Gammexane also remained effective for one year although toward the end of that year its effects were noticeably diminished. The irritant factor was much less with this compound, only 20·1% of mosquitoes leaving treated houses; but mortality was more than triple that of mosquitoes not exposed. The DDT-irritated mosquitoes tended to fly towards light, while the gammexane-affected ones flew around restlessly and aimlessly.

An antimalarial campaign was started in the port of Veracruz in 1947 (ELORDUY [172]). That year Paris green larvicide only was used; in 1948, DDT both as a larvicide and as a residual house spray was substituted. This was repeated annually for three years with excellent results as shown in Table 18.

In addition to the reduction in the number of houses infected with anophelines, the number of vectors per house was also greatly reduced. The average number of anophelines per house in 1947 was 29·60; in 1948 1·67; in 1949 1·04; and in 1950 0·01. *Aedes* species were eliminated from houses after the first year of DDT treatments in 1947.

Table 18

Malaria Morbidity and Mortality and Anophelism in Puerto de Veracruz 1941–47

Malaria cases and deaths			Houses infected with anophelines	
Year	Cases	Deaths	Year	Percentage
1941	1,331	24		
1942	1,131	25		
1943	569	16		
1944	612	11		
1945	378	30		
1946	368	26		
1947	349	21	1947	2·54
1948	195	9	1948	0·53
1949	32	5	1949	0·37
1950	34	0	1950	0·01

A malaria control program using a DDT suspension prepared from 50 per cent water-wettable powder was applied at the rate of 200 mg/ft^2 in the Distrito Federal, where both the endemicity of *P. vivax* and the density of the vector, *A. aztecus*, were very low. The result was a considerable decrease of splenic indexes and a reduction of the parasitic index to zero (DOWNS *et al.* [156]; DOWNS and BORDAS [153]).

In Lower California, which is a semidesert region, *A. pseudopunctipennis* must seek favorable microclimatic conditions inside houses (DOWNS *et al.* [155]). Therefore, very effective control of malaria has been achieved through DDT residual spraying of dwellings in this area.

These various projects and many others not discussed have been combined into a wide-scale national campaign against malaria. By 1948, DDT was being used in 13 States, protecting some 375,000 people (PAMPANA [424]). All States having high malaria morbidity rates are now using DDT as a residual spray (MEXICO [373]).

Puerto Rico

Puerto Rico is a 100-mile-long island in the Caribbean Sea, with the Cordilleras mountain range extending its full length. The highest peaks range from 2,000 to almost 4,400 ft. The lowlands consist of a narrow belt not more than 5 miles in width along the coast. The annual mean temperature during January is about 73° and during August about 79°. The rainfall varies considerably in different parts of the Island, exceeding 200 in. in certain mountainous sections and being below 21 in. in certain portions along the southern coast. The average annual rainfall is about 71 in.

Malaria has long been a serious health problem in Puerto Rico, particularly along the coast, although no part of the Island is free from the disease. Malaria has in the past accounted for the death of 1,000 or even 2,000 people annually. The disease was considered of sufficient importance during World War II that, during 1942 and 1943, over a million dollars was spent by the various insular and federal government agencies to control this disease. *A. albimanus* is the principal vector and control efforts are based on the suppression of this mosquito. *A. albimanus* breeds most abundantly in coastal areas and will breed in brackish water containing as much as 50% sea water. Prior to 1944, drainage and larviciding with Paris green or diesel oil were the chief measures of malaria control. However, these did not prevent the seasonal rise in *albimanus* populations. In 1944, the United States Public Health Service, in cooperation with the Insular Health Department and the School of Tropical Medicine, began practical experiments with DDT spray for the control of *A. albimanus*. Humacao Playa, on the east coast, was chosen for treatment, while Loiza Aldea, some 30 miles distant on the northeast coast, was used as an untreated check village. The DDT was applied as a residual to the inside walls of dwellings. The first spraying used 309 mg/ft² treated and the second 147 mg. The indices of *albimanus* in each village were obtained by animal bait and light trap collections made throughout the year. There was no reduction in the *albimanus* population based on trap collections in either the treated or untreated town. There was, however, a reduction in malaria incidence in the untreated town of Loiza Aldea. The malaria parasite rate followed closely the normal cycle for Puerto Rico. In November 1944, the rainy season, 4·7% positives were found, and this decreased to 1·5% during the dry season of March and April, 1945. There was, however, an increase to 3·8% during the rainy season of November 1945. In Humacao, which had been treated twice with DDT during the period November 1944 to October 1945, there was a decline in positive slides from 5·8% in October 1944 to 2·8% in March 1945. Most significantly, there were only 0·91% positives during the rainy season of October 1945 (STEPHENS and PRATT [551]). In 1945, larviciding was instituted with 0·5% emulsion of DDT in diesel oil containing an emulsifier (STAGE and PRATT [550]). These relatively small-scale tests definitely indicated the efficacy of DDT in the control of malaria transmitted by *A. albimanus*.

In 1944, the US Army initiated airplane spraying using 5 to 10 per cent

DDT in diesel oil, and within a short time, this reduced the counts of *albimanus* in the area. The US Navy began similar operations in 1945, and both the Army and Navy utilized airplanes for malaria control in 1946. Planes were equipped with thermal aerosol apparatus for treating large areas.

Following the encouraging results obtained with DDT, the Puerto Rico Department of Public Health began an extensive program of malaria control covering much of the coastal area of the Island. Surveys following the use of DDT indicated significant reductions in the malaria rate. The use of DDT gave such promise for reducing malaria, particularly in rural areas where drainage and larviciding are not economical, that consideration was given to a campaign for the eradication of *albimanus* from the Island.

Table 19
Reported Cases of Malaria in Puerto Rico, 1951–53

	Years		
	1951	1952	1953
Cases reported	88	128	28
Cases reported from civilian sources	84	18	10
Cases reported from military sources	4	110	18
Deaths reported	33	13	2
Imported cases other than military	0	0	3
Confirmed indigenous cases (primary or relapses)	34	5	5

The value of drainage, particularly of the lowlands on the Central Constancia area, southeast of Ponce, should not be underestimated. Drainage was provided to dense breeding areas both by pumping and gravity-flow systems. This not only reduced the production of malaria vectors but added to the agricultural value of the land (NEILL [413]).

Indigenous malaria has been rare in Puerto Rico since 1949. Table 19 shows details of reported cases during 1951–53 (PONS [453]).

None of the deaths were confirmed as being due to malaria; in fact, those investigated were found to be due to other causes.

Central America

Guatemala

Malaria is found throughout Guatemala although its prevalence varies with the differences in altitude and in the amount of yearly rainfall (FAUST [178], [179]). The disease is endemic in the coastal areas and much lighter in the central plateau and in the mountains.

In July 1950, with aid from the World Health Organization and the United Nations International Children's Emergency Fund, an insect-control program was started in Guatemala providing for DDT to be sprayed at least twice per year throughout the whole endemic malaria area [652]. By July 1951, from 60 to 75% of the planned work had been done. During the first seven months of 1951, DDT at the rate of $2 \cdot 27$ g/m² was applied in 1,500 communities, including some 123,000 houses, and protecting a population numbering 630,000. Although a complete epidemiological appraisal has not been made, the program resulted in a reduction in the number of malaria cases, and also in an increase in the number of *Ae. aegypti*-free areas.

Nicaragua

Malaria in Nicaragua is endemic along the entire Caribbean coast, the Northwestern Pacific coast, and around Lake Nicaragua and Lake Managua (FAUST [178], [179]). This disease in 1943 was the principal cause of death, the mortality rate being $410 \cdot 4$ per 100,000 for the entire population but for children under one year of age it was 2,440. In the endemic area, some 70 to 80% of the population probably had malaria, and it was stated that the disease constituted 99% of all febrile conditions.

In 1951, the second DDT-spraying cycle was completed under the insect-control program carried out by the Government with the assistance of the World Health Organization and the United Nations International Children's Emergency Fund [653]. Two thousand communities with 115,000 houses were sprayed with DDT at the rate of 2 g/m², protecting 600,000 persons. In the years between 1948 and 1950, many children with malaria came to the clinics in Managua; however, during 1951 only rare cases of malaria were reported. Due to a reduced demand, fewer antimalarial drugs have been imported since DDT spraying was initiated.

Costa Rica

The malarious zone of Costa Rica occupies over 60% of the country's total area and is inhabited by approximately 40% of the population (CHAVARRIA and ARGUEDAS [97]). There is $1 \cdot 5$ km² in the malarious zone for each square kilometer in the nonmalarious one; in contrast, there are $184 \cdot 2$ persons in the nonpaludic region for each 100 in the paludic.

In 1942, malaria was the second cause of mortality, being responsible for 9% of total deaths. DDT residual spraying was first initiated in Costa Rica in 1946 by the Banana Company of Costa Rica around company installations. This resulted in an immediate decline in the number of hospitalizations for malaria. This decline continued to 1950 when the Government of Costa Rica with the aid of UNICEF started DDT residual spraying for populations not included in the Banana Company's activities. This was followed by an acceleration of the decrease in the incidence of malaria. In 1952, malaria as a cause of death ranked eleventh, only 1·6% of the total mortality being ascribed to this disease. In the eleven years between 1942 and 1952, hospital admissions for malaria fell 91·2%. It is interesting to note that while hospitalization for malaria was at its lowest in 1952, total hospital admissions were at their highest.

Panama

In Panama and the Canal Zone, *A. albimanus* is the principal vector of malaria (TRAPIDO [583]). This species is not a domestic one and does not remain long in houses, returning to the jungle or to breeding places soon after feeding or early in the morning. The use of residual insecticides in homes, therefore, appeared to be the best method of destroying that segment of the vector population of most importance, comprised of mosquitoes seeking human blood meals.

The first test with DDT in Panama, in April 1944, suggested that it might be of great value in reducing malaria rates where *A. albimanus* was the vector (LINDQUIST and McDUFFIE [321]). The area on the middle Chagres River selected for these tests was one where the Gorgas Memorial Laboratory had for many years been conducting extensive studies on malaria and *Anopheles*. The initial thick blood film survey made by these workers in 1929 revealed a malaria positive index of 45·6% [233]. Quinine treatment was started in 1930 and by 1934 this rate had dropped to 21·6%. From 1935 through 1947, atabrin and plasmoquine were administered to all persons with positive blood films and the rate fell to 10·1%. Other measures used prior to 1944 in addition to drug therapy included use of herbicides to kill aquatic vegetation in breeding areas and the application of Paris green to the rivers near towns every week or 10 days.

In the summer of 1944, a field experiment was started in this area to study the effectiveness of DDT residual spray in native dwellings as a means of controlling malaria and its vector (TRAPIDO [583]). A 5 per cent solution of DDT in kerosene was sprayed at 4-month intervals on both the inner and outer walls of houses in one village. From this experiment it was learned: (1) that the number of mosquitoes visiting houses are greatly reduced; (2) that there is a marked reduction in the numbers of engorged mosquitoes recovered from treated dwellings; and (3) that the survival rate among engorged mosquitoes is low for three months after treatment (TRAPIDO [583]). Beginning in 1948, DDT spraying was carried out in all of the Chagres River villages three times a year [233]. In addition, this compound was used as a larvicide on the rivers, replacing Paris green. The average monthly malaria rates dropped to 3·6 in 1948, to 2·2 in 1949, to 1·06 in 1950, and in 1951 was less than 1%. During the first four months of 1952, following the twentieth spraying of one village and the thirteenth of another, two of the effects of DDT spraying noted above were no longer apparent, only the reduction in engorged mosquitoes remaining the same (TRAPIDO [581]). This was thought to be the result of the selection of an *albimanus* population which was either physiologically resistant

or hyperirritable to DDT or both. There has fortunately been no rise in malaria
rates as a result of this development.

In another experiment in 1946, 5 per cent DDT in diesel oil was applied
by airplane over one seaboard town in Panama, and in a second town the
interior and some exteriors of houses were treated with 5 per cent DDT in
kerosene. A third town was selected for comparison purposes in which no
control procedures were carried out (ELMENDORF [171]). A virtual elimination
of mosquito larvae and adults followed the airplane spraying, but a reduction
in malaria incidence did not become evident until twelve months later when
it was found to be approximately one fifth of that in the untreated comparison
town (ELMENDORF [170]). Twelve months after the house spraying at the rate
of 300–400 mg of DDT per square foot, malaria incidence had dropped from
92·1% to 15·6%.

South America

Colombia

Malaria in Colombia, which is populated by 6,800,000 inhabitants, occurs
in an area from sea level to an altitude of 1,700 m. The disease is more intense
in the llanos or savannas east of the Andes, bordered on the north and east
by Venezuela and on the south essentially by the River Guaviare and the
Macarena Range of mountains. The elevation in this area is generally less than
500 m. The rainfall is usually above 60 in. per year, and it may be as high
as 175 in. in some areas. The heaviest precipitation occurs during the months
of April to November. It has been estimated that 700,000 persons suffer from
malaria annually, and the disease represents a considerable obstacle to the
economic development of the country.

The chief vector is *A. darlingi* which is apparently not always associated
with humans in this region. Various other species of *Anopheles* mosquitoes
occur in the area and in some cases are found in appreciable numbers in
dwellings. Laboratory examinations of the various species are not adequate
to clearly establish whether they are infected with malaria. The malaria indices
in humans varies considerably with the highest rates in the areas frequented
by *A. darlingi*. Spleen rates in various areas range from four up to 51% in the
general population, with a parasite rate of as high as 40%. Among approximately 7,000 people examined, roughly 50% of those in which blood smears
were positive for malaria parasites were infected with *P. falciparum;* 46%
were *P. vivax;* and 4% were *P. malariae*. Seasonal fluctuation can be excluded
from these findings since the respective malaria surveys were made at the
same time of the year. However, the studies did not cover a long enough
period to rule out cyclical variations. The suggestion that malaria peaks tend
to occur every five years in the northern part of South America is consistent
with the observations in the eastern part of Colombia. A spontaneous reduction
of malaria without the use of DDT occurred from 1946 to 1950 (RENJIFO and
DEZULUETA [472]).

DDT was first used in the Villavicencio area in 1948 and this was later extended to several other areas. In 1950, the population protected by DDT was approximately 31,000 inhabitants. RENJIFO and DEZULUETA [472] reported that during 1948 and 1949 technical grade DDT was sprayed at the rate of 1 g/m^2 of treated surface. It is presumed that they refer to an emulsion or solution for they later state that in 1950 a DDT wettable powder was used at the rate of 2 g of powder per square meter. They found the residual from the wettable powder to be considerably better on rough surfaces than that of the technical grade DDT, lasting five months as against two months with the technical grade material. There was a marked reduction in the *darlingi* population and also a marked reduction or complete disappearance of certain other

Table 20

Human Malaria Parasite and Spleen Rates (%) Before and After the Use of DDT in Colombia (Preliminary Results After MIRANDA [376]

Locality	Pre-DDT	Post-DDT
Quibdo, Choco		
Parasite rate	18	6
Spleen rate	31	8
Lloro, Choco		
Parasite rate	50	6
Spleen rate	79	18
Cucuta-Santander sector		
Parasite rate	15	11
Spleen rate	48	21
Bucaramanga-Wilches sector		
Parasite rate	16	7
Spleen rate	48	18

species, some of which were seldom caught in human dwellings in this area prior to the use of DDT. This reduction in non-house-frequenting mosquitoes is difficult to explain. Two species (*A. benarrochi* and *A. triannulatus*) actually increased in abundance at this time. The malaria parasite index among the general human population in two of the treated areas was reduced to zero in the summer of 1950, as compared with 14 to 25% in the same months of 1946. The spleen index was also greatly reduced.

In 1952 the Government of Colombia instituted a large-scale vector control program aimed principally at malaria and yellow fever vectors. At the end of the first year's operations protection had been given to 1,225,000 persons. Initially the program was chiefly in the Departments of Atlantico, Magdalena, and Bolivar. Between April 1952 and February 1953, a total of 205,000 houses in 1,777 communities had been sprayed with DDT [644]. The results have been reported as very good although only limited statistical data are as yet available.

Table 20 is an example of preliminary results obtained (MIRANDA-FRANCO [376]).

Two of the most important ports of Colombia, Barranquilla and Cartagena, have been freed of *Ae. aegypti*, apparently as a result of the spray program.

Venezuela

Venezuela may be divided into three geographic regions, the Costa-Cordillera with the coast and the Andean mountains and valleys, the llanos including the flatlands between the foothills of the southern slope of the mountains to the Orinoco River, and the Guayana covering the valleys and mountains south of the Orinoco River. Malaria was once the most important health problem in Venezuela (GABALDON [195]). While the disease existed in all the geographic areas, it was most severe in the Llanos where spleen indices sometimes rose to 100% (PAMPANA [424]). No other disease, including influenza, caused a higher mortality than epidemic malaria in Venezuela during the period 1905–1945 (GABALDON [199]).

Since 1936, the Malaria Division in the Ministry of Health and Social Welfare has been conducting research and training regarding malaria and its control (GABALDON [195]). Before 1945, malaria control consisted mainly of nationwide free drug distribution, large drainage and filling schemes around some towns, and the' use of Paris green larvicides and imagocides (mostly pyrethrum), particularly in epidemic areas (GABALDON and BERTI [200]). *A. albimanus*- and *A. darlingi*-transmitted malaria were successfully controlled by the drainage, *darlingi*-transmitted malaria being eliminated for the first time from some towns (BERTI [47]).

In 1945, DDT spraying was incorporated into the malaria control program, resulting in a sharp reduction in malaria prevalence. There were no preliminary trials of this new insecticide; it was used from the beginning throughout the entire malarious zone which in Venezuela comprised any populated area where the two principal mosquito vectors, *A. albimanus* and *A. darlingi*, or malaria parasites were found. Neighboring communities with similar topography were also treated. DDT was applied to all houses and verandas, to stables, privies, and other outbuildings, and even to the under surfaces of eaves and furniture (PAMPANA [424]). Only the peripheries of larger towns and cities were treated. In the beginning, DDT was applied at the rate of 1 g/m² every three months and then every four months. Since 1948, a rate of 2 g/m² every six months has been the standard procedure. Since malaria transmission is uninterrupted throughout the year in Venezuela, there being only a slight decrease during the dry season, operations are continuous. DDT is usually employed as an aqueous suspension since rural houses are mostly of adobe construction with absorptive walls. In the relatively few houses with painted walls, solutions in kerosene are applied (GABALDON [196]). By 1950, some 70% of the houses in the malarious zone were being protected twice a year (GABALDON [195]), and by 1953 protection was 85% (BERTI [46]). The remaining 15% were in very isolated or in marginal zones whence malaria had disappeared as a result of the campaign in the more malarious ones.

Malaria mortality for the entire country which averaged 110·0 per 100,000 persons for the period 1941–45, fell to 16·2 in 1948, 8·9 in 1949, 8·0 in 1950, 5·3 in 1951, 2·0 in 1952, and 1·9 in 1953 (BERTI [46]). In one section of Venezuela including the central portions of the Costa-Cordillera and llanos regions, malaria has been eradicated (GABALDON and BERTI [200]). The population of this area numbers 2,430,986, or 49% of the total population of the country. Prior to the residual spray program, highly endemic or epidemic malaria had been present in parts of the sector. Epidemiologic investigations revealed that in one area of 20,000 km², no primary indigenous cases of malaria had been found for three years, and in another area of 160,000 km² there had been none for one year. *A. darlingi* also seems to be eliminated from the section. The densities of other anophelines, however, including certain secondary vectors have remained high, showing that with residual spraying malaria eradication can be attained without vector eradication. It is anticipated that in the near future malaria will be eradicated throughout the country with the exception perhaps of two small areas: about 3·4% of the previously malarious zone, where outdoor transmission continues, and certain districts where nomadic and uncivilized Indian tribes are found.

Trinidad

There are two important vectors of malaria in Trinidad, *A. bellator* and *A. aquasalis* (GABALDON [197]). In the area of high rainfall, *bellator*-transmitted malaria is found; in other regions, particularly along the coast, the disease is transmitted by *aquasalis*.

From 1945 to 1948 a campaign to eradicate *A. aquasalis* was carried out in Trinidad (GILLETTE [223]). Some 30,000 houses in seven areas were treated once or twice a year with 5% DDT in kerosene. In those areas where malariometric data obtained prior to residual spraying were available, the effectiveness of the program was proved by a reduction in malaria incidence.

British Guiana

British Guiana just north of the equator on the northern coast of South America has a population of about 425,000, largely of West African and East Indian descent (SYMES [570]; GIGLIOLI [220]). About nine tenths of the population live in the coastal plantation belt, a narrow fringe only 5 to 10 miles deep and 280 miles long, much of which is below sea level. It is drained by an extensive canal system and protected by a sea wall. Sugar cane and rice, the main crops, are irrigated by more canals bringing water from inland catchment areas. There are also large swamplands where cattle graze sometimes up to their necks in water. The rest of the country may be divided into three more regions topographically: the peneplain of the interior, the backlands of the coastal plain, consisting of a long series of flat-topped, uniform sand hills averaging 150 ft in height but rising gradually to 400 ft in the west; the rolling grassy savannas of the southwest, between 300 and 450 ft above sea level; and the western highlands where the ground rises rapidly and progressively with a series of cliff-like escarpments approximately 1,200 to 2,000 ft high (GIGLIOLI [221]). About 50,000 people live in the interior in widely scattered small settlements along the lower tidal reaches of the rivers, in mining and timber camps on the plateau, and on the savannas (GIGLIOLI [220]).

Until the advent of DDT, malaria was endemic, hyperendemic, or epidemic throughout the country and was chiefly responsible for retarding the economic development of the Colony and the natural increase of its population (GIGLIOLI [221]). Spleen and parasite rates of 50 to 80% were common. The wetness of the coastal belt made that area a particularly difficult problem. In the interior, widespread uncontrollable seasonal flooding gave rise to a similar problem. Breeding conditions were ideal throughout the country for the production of the only vector of malaria in British Guiana, *A. darlingi*, an anthropophilic species which enters houses in large numbers to feed and rest.

From 1933 to 1945, a number of research studies on malaria epidemiology and mosquito biology were sponsored by the Colonial Government, the British Guiana Sugar Producers' Association, and The Rockefeller Foundation (GIGLIOLI [220]). Much valuable data on the malaria problem and on the ecology of the indigenous anophelines were obtained, but no successful control measures applicable to the peculiar conditions in British Guiana were evolved.

This information paved the way, however, for the rapid and appropriate use of DDT in residual spray experiments in 1945 when this insecticide became available (GIGLIOLI [220]). These activities expanded rapidly; in 1946, densely populated districts on the Demerara and Berbice River estuaries, with a rural population of about 60,000, were included. The success of these preliminary trials led to the initiation on January 1,1947, of a colony-wide DDT residual spray program under the joint cooperation of the Mosquito Control Service of the Medical Department and the sugar industry. By December 1948, 90% of the population was protected; 95% by the end of 1949; and approximately 98% by the end of 1950, including some of the population in the more remote and almost inaccessible interior.

In British Guiana, houses are constructed of a variety of materials. On the coast, they are wooden structures, for the most part, and DDT as a 5 per cent solution in kerosene gives an adequate residual deposit (GIGLIOLI [220]). In rural areas of the coastal belt, some mud-and-wattle houses are found and to these a 10 per cent aqueous suspension of DDT made from a 50 per cent wettable powder was applied. In the interior, houses are built of wood, split palm trunks, palm leaves, tree bark, mud-and-wattle, or sun-dried bricks; but because of transportation difficulties, only DDT emulsions were used. The residual dosage was 150 mg of DDT per square foot. All interior surfaces of the houses were treated to the height of the reach of a three-foot lance; the under and back surfaces of all furnishings were also treated. In rural districts, all houses were sprayed; in cities, this was limited only to those on the outskirts. The interval between spraying has been modified as experience indicated. In 1946–47, it was eight months; in 1948–49 twelve months; and in 1950 eighteen months. DDT spraying has been made compulsory in certain areas. Throughout this program, which started in 1945, no antilarval measures were conducted.

Houses sprayed with DDT became negative immediately for *A. darlingi*, and within 2 to 4 weeks larvae of this species also disappeared from breeding places in the vicinity, although larvae of other anophelines continued to be present (GIGLIOLI [220]). Thus, residual DDT not only eliminated adult *A. darlingi* from houses but caused oviposition to cease, indicating that most mosquitoes of this species entered houses and made lethal contact with DDT. It was reported, therefore, that in British Guiana, DDT had no repellent effect on *A. darlingi*. This formerly house-frequenting anopheline has completely disappeared from all treated areas. No specimens were recovered during three years of search in these areas, and it seems safe to say that the species has been eradicated from the inhabited parts of British Guiana.

This cannot be said yet of malaria. Old infections continue to relapse and new infections are contracted outside of the control area and imported. However, neither endemic nor epidemic malaria any longer exists in the Colony. Parasitologically positive cases are rare. Spleen rates ranged from 10 to 25% in the second year of the program (GIGLIOLI [221]), and in 1950 was 6% (GIGLIOLI [220]).

It is felt that this change in the local malaria situation has been directly brought about by the use of DDT residual spray (GIGLIOLI [220]). The decline in malaria incidence has been consistent in spite of the fact that in the three years, 1948, 1949, and 1950, rainfall was excessive. The maintenance of this status will continue a problem, as British Guiana is surrounded on three sides by regions where *A. darlingi* continue to flourish. Ideal breeding conditions for *A. darlingi* still exist in the Colony. Present control activities are based on the premise that the vector anopheline cannot re-enter the country except along pathways adaptable to its ecologic necessities. Therefore, spraying is continued in houses on the banks of rivers and their estuaries, and in the eastern coastland which is subject to *A. darlingi* invasion. This is integrated with existing natural barriers, such as arid atmospheric conditions, and salt water containing over 150 mg of NaCl per liter or acid water of a pH below 4.5 which is not favorable for breeding. It is estimated that *A. darlingi* and malaria control can be maintained throughout the densely inhabited coastlands of British Guiana by routine DDT residual spraying at eighteen-month intervals of one fourth to one third of all houses.

French Guiana

French Guiana includes an area of 86,000 km² near the equator with scarcely any coastal plains. Most of the area is a continuation of the Amazon forest.

                                   S. W. Simmons

Its climate is warm and humid, the temperature averaging 79° F with little variation during the year.

The population is less than 30,000 with most of the inhabitants residing in the coastal zone, 11,000 of them in Cayenne.

Malaria occurs throughout the country. *P. falciparum* constitutes approximately 80% of the infections, *vivax* accounting for most of the rest. *P. malariae* is very rare. The most important vector is *A. darlingi* but there are 20 other species of *Anopheles*, two of which, *aquasalis* and *pessoai*, are considered to serve as vectors. *A. triannulatus* has been infected in the laboratory but is not particularly anthropophilic and is considered definitely secondary as a vector. The principal vector, *A. darlingi*, has been found naturally infected with malaria to the extent of 1·5%. It is believed that this species cannot be eradicated from French Guiana because of the great reservoir maintained in the tropical forests of the country. *A. darlingi* scarcely exists in Cayenne where *A. aquasalis* is the principal *Anopheles*.

DDT was first used in 1949 when it was directed primarily against the yellow fever mosquito in Cayenne. Five per cent DDT in oil solution was applied as a residual spray. A total of 3,707 houses out of a total of approximately 4,400 in Cayenne were treated during this year (FLOCH [189]). Twelve thousand kilograms of DDT were used (ALVARADO [8]).

In 1950, Prefectoral Order No. 156/C dated March 21, 1950, made DDT indoor residual spraying obligatory in Cayenne. Again the yellow fever mosquito was the primary object, but *A. darlingi* was also included and the campaign was extended to the entire inhabited coastal zone plus two interior centers. Twenty-eight hundred kilograms of DDT were used and 23,000 people were protected. The cost per person protected in 1950 was US $0.54. This amounted to US $1.83 per house or US $0.0082/m². 70% of this cost was for DDT and solvents, and only 18% for labor (ALVARADO [8]). The campaign was considered very satisfactory against *A. darlingi* in the treated zones but less so against *A. aquasalis* because of its less anthropophilic tendencies (FLOCH [189]).

The data on malaria in French Guiana has indicated a marked seasonal variation with the peak occurring usually in September (FLOCH [187]). During the years 1946 to 1948 there was an average of 778 cases of malaria confirmed by blood examinations at the Institut Pasteur de Cayenne. In 1949 when the use of DDT was directed primarily against *A. aegypti* in Cayenne, there were 595 cases of malaria confirmed for the entire country. In 1950 when the campaign was extended to include most of the inhabited portion of the country, the number of confirmed cases dropped to 160.

It is notable that in the year 1950 more cases occurred during the first six months of the year than occurred during the last six months which normally has the greater preponderance of cases. Actually, January of 1950 had more cases than any two other months combined. In 1951, during the first six months, only 12 malaria cases had been confirmed (FLOCH [187], [188]). Hospitalization records, though less significant, show a parallel trend.

In view of the results to date it appears appropriate to repeat the conclusion of FLOCH [187]: «Une ère nouvelle a commencé en Guyane française avec les pulvérisations résiduelles de DDT, en mai 1949.»

Brazil

Brazil covers a total area of about 3,385,000 square miles which may be divided administratively and geographically into five major regions, the northern, northeastern, eastern, southern, and central western zones (PINOTTI [448]). At the end of 1948 the population of Brazil was estimated at 48,900,000. The climate falls into three types: equatorial, which is hot and humid with abundant rains, as in the Amazon region; tropical, which is less humid than the equatorial and with variable amounts of rainfall; and subtropical or temperate climate such as is found in Southern Brazil and in some or the mountainous zones.

Malaria occurs in all 20 States, in the Federal District, and in the 4 Federal Territories of Brazil. In certain States of the northern and northeastern regions, malaria is found everywhere. In others of the eastern and southern regions, a few counties may be found free of the disease; and in some States in the southernmost and in the extreme northeast regions of Brazil, malaria occurs in only a few counties. In 1948, the National Malaria Service reported malaria in about 65% of the 1,780 counties of the country. Endemic malaria rates vary greatly from county to county. In certain areas, malaria has appeared in epidemics. In 1940, approximately 70% of the population (17,000,000 people) in the malarious regions of Brazil was exposed to endemic or epidemic malaria.

The principal vectors are *A. darlingi*, which is responsible for most of the malaria found in Brazil, and *A. aquasalis*, *A. albitarsis*, *A. cruzii*, and *A. bellator*. Surveys have showed that spleen indices were generally higher in areas where *A. darlingi* was the vector, either alone or in combination with *A. aquasalis* and *A. albitarsis*, than in areas where these two species were found alone or together. Where malaria was transmitted by the *Kerteszia* species (*A. cruzii* and *A. bellator*), which breed in arboreal bromeliads, the spleen indices were similar to those found in areas where *A. darlingi* was the vector. The highest malaria incidence has been found in the São Francisco River valley of the Eastern region, in the coastal lowlands of the State of Rio de Janeiro, and in certain other areas located along the seacoast of various states. The percentage of mosquitoes found inside houses is often not very high, but nevertheless most of the malaria transmitted by them is contracted indoors (PINOTTI [449]). Seasonal variations in the incidence of malaria are influenced by rainfall, the rise in malaria incidence following the rise in rainfall by about two months (PINOTTI [448]). Only in the southern temperate zone does temperature appreciably affect the seasonal occurrence of the disease. In all regions except the southern, some transmission occurs throughout the year with only very occasional brief interruptions.

Malaria control was first attempted seriously in Brazil at about the beginning of the century. CHAGAS was successful in controlling malaria by killing mosquitoes inside houses (GABALDON [198]). In the 1920's, malaria control activities included principally antilarval measures (BOYD [62]). Drainage of lowlands, supplemented by the use of native fish and of larvicides where inspection indicated this was necessary, was done within a radius of 1 km of the population to be protected. Some quinine was distributed in connection with this program. Following these control operations, there were fewer fever cases and the number of mosquitoes was reduced. In 1930, it was discovered that *A. gambiae*, the most notorious of the African malaria vectors, had been imported to Natal in Northeast Brazil (SOPER and WILSON [543], [544]). An organized control campaign in 1931 apparently resulted in the eradication of *gambiae* from Natal, but the vector had already established itself in the interior of Rio Grande do Norte. From 1932 to 1937 very little was done to eject the vector; but in 1938 a severe malaria epidemic struck in Northeast Brazil, and this stimulated the organization of the Anti-Malaria Service. This Service fused in 1939 with the Malaria Service of the Northeast, an organization formed for the purpose of combatting *A. gambiae* and supported by the Ministry of Health of Brazil and the Rockefeller Foundation. The principal method of attack was the use of insecticides; Paris green was employed as a larvicide and pyrethrum spray as an adulticide. Progress at first was slow; but by 1940, large areas could be left untreated with no danger of reinfestation by *A. gambiae*, and after January 1941 all control measures were suspended since apparently this species had been eliminated.

In 1941, the National Malaria Service was organized (PINOTTI [448]), and from that time until 1946, the antimalaria campaign was directed largely against the aquatic phases of the vectors and was limited, in the main, to State capitals and principal cities. Many permanent drainage channels were constructed and larvicides, especially Paris green, were applied to breeding areas. Insecticides containing pyrethrum were used occasionally to spray houses as a protection to troops on military bases during World War II. In rural areas, the free distribution of drugs was the chief means of relieving these populations from the suffering of malaria.

In 1945 the National Malaria Service began experiments with DDT. A total of 2,673 houses were sprayed that year. The favorable results from these experiments led in 1946 to the treatment of 6,419 houses in 50 small localities. Control of malaria transmitted by *A. darlingi* was excellent, as was the reduction in populations of *A. aquasalis* and *A. albitarsis*. In areas of bromeliad malaria it was shown that this technique also gave satisfactory results (BUSTAMANTE and FERREIRA [87]).

In the south of Brazil, it has been customary to control bromeliad malaria by the manual removal of bromeliads and by deforestation. While this has proved successful in several important towns, it is not practicable for rural areas. Therefore, in 1948 an experiment was conducted in such an area to determine the effectiveness of a 5 per cent aqueous suspension of wettable

DDT powder applied from a helicopter at the rate of 546 g of DDT per 10,000 m² (RACHOU *et al.* [461]). A large reduction in the numbers of *A. cruzii* and *A. bellator* was observed, but reinfestation of the area took place rapidly, especially if the spraying was done during the period of the heavy rains. Therefore, it was concluded that in regions where bromeliads prevailed, the destruction of these plants must continue if malaria is to be controlled.

The first extensive DDT house spraying programs carried out by the National Malaria Service were started early in 1947 (PINOTTI [448]). A total of 41,339 houses in the São Francisco River valley and 130,419 houses in the coastal lowlands of the State of Rio de Janeiro were sprayed. The program was

Table 21

Growth of the Malaria Control Program in Brazil From 1945 to 1949[1]

Year	Appropriations received, all sources (in cruzliros[2])	DDT used lb	Number of houses sprayed	Populations protected
1945	48,430,230		2,673	5,572,014
1946	84,900,310	7,717·5	6,419	4,154,721
1947	112,130,970	207,270·0	186,189	6,933,399
1948	137,797,680	1,161,452·9	968,611	11,744,847
1949	213,029,470	2,754,794·7	2,364,279	17,763,179

[1]) After PINOTTI [448].
[2]) One cruzliro equals 5 cents US (approximately).

rapidly expended in 1948 and 1949, and by 1950 it covered most of the malarious areas of Brazil, at a cost of about US $3.50 per house treated. This expansion is shown in Table 21. In the area first put under treatment, spraying continued twice a year; in other States, because of the enormous size of the total area which had to be treated, only one spraying per year could be done. This was scheduled for two months before the annual rise in the incidence of the disease. Malaria has not been eliminated from these areas, but gross transmission has been stopped. In 1950, only 12% of the houses included in the program were sprayed twice a year.

DDT is used most frequently as a water emulsion prepared from a 30 per cent emulsion concentrate. A mixture of xylene-toluene-benzene is the solvent and a monoglycerate made from castor oil and glycerine is the emulsifying agent. Where houses are made of straw or clay, suspensions of wettable DDT powder are used. During 1949, about 80% of the DDT used was in the form of an emulsion, and 20% as a suspension of wettable powder. These mixtures are applied at rates varying between 1·5 and 2·0 g/m² to all interior wall surfaces and to ceilings, where *A. cruzii* and *A. bellator* are the vectors.

The results of the antimalarial program in Brazil, which uses both house-spraying with DDT and the destruction of bromeliads as the methods of control, have been excellent (PINOTTI [449]). Malaria morbidity has decreased as shown by cases reported from representative areas as follows.

1945	33,945
1946	26,295
1947	27,433
1948	17,136
1949	4,073

In another instance cases decreased from 14,198 in 1947 to 1,192 in 1949, in eleven localities where *A. aquasalis*, *A. darlingi* and *A. albitarsis* were the vectors, and where DDT had been used twice per year (PINOTTI [448]). Comparable reductions in vector mosquito populations were noted especially inside sprayed houses. In certain other parts of South America *A. darlingi* has been eradicated from large areas through the use of DDT, but this has not been true in Brazil (BUSTAMANTE [86]). Specimens, both larval and adult, are still being found in sprayed areas on the coast, and in the hinterlands as well.

As of 1950, about 20% of all houses in the malarious areas of Brazil had not yet been included in the malaria control program. The majority of these are in isolated areas; and while it may be some time before these are reached, it is the purpose of the National Malaria Service to spray everywhere that malaria exists.

Brazil is a large country in a tropical region where malaria is widespread and difficult to control. The accomplishments which have been made are a tribute to those engaged in malaria control work, and they also reflect the efficacy of DDT as a principal tool of attack against this disease.

Ecuador

Malaria occurs in almost all of Ecuador up to 8,500 ft of altitude. The disease is transmitted by *A. albimanus* on the coast and by *A. pseudopunctipennis* in the valleys of the Andes provinces (MONTALVAN [386]). It has been a serious obstacle to the economic and agricultural development of the country (PAMPANA [424]).

In 1948, the Malariology Section of the National Institute of Hygiene organized a control program using residual insecticides, with the ultimate aim of spraying all houses in the malarious zone. In preliminary trials made from May to July 1949, 10,000 houses were treated with DDT. A large-scale campaign followed in November of that year. Up to March 1950, 101,613 homes had been sprayed, giving protection to some 800,000 persons. DDT was applied at the rate of 2 g/m², generally as an aqueous suspension, infrequently as a solution. In most of the coastal provinces, one spraying per year is believed sufficient; but in other areas and in the Andes provinces, 2 sprayings a year are recommended (MONTALVAN [386]). Benzene hexachloride was used as an aqueous suspension at the rate of 130 mg of gamma isomer per square meter in a less extensive area where the concurrent elimination of *Triatominae* was desired (PAMPANA [424]). After March 1950, not only dwelling houses but all outbuildings were also treated.

In 1951 only 8 positive bloods were found in an examination of 2,519 smears, and not a single case of malaria was reported from various coastal zones which formerly had been highly malarious. By 1952 some 200,000 houses had been sprayed, protecting a population of more than one million. It could be announced at this time that malaria had been eradicated from most of a region comprising about 95% of the total area of the country (AYER [26]).

Peru

Approximately 60% of Peru is affected by endemic malaria. The area involved is about 800,000 km^2 containing a population of approximately three and a half million people, or 40% of the total population of the country (ALVARADO [8]).

The vector in the coastal region where malaria is a serious problem is *A. pseudopunctipennis*. In the more northern region, *A. albimanus* is the vector though *A. punctimaculatus* has also been found infected (PAMPANA [424]).

In 1946, before the first experiments with DDT in Peru, there were approximately 58,000 cases of malaria reported to the Departamento de Malaria. This was a marked reduction from the 81,000 reported in 1945 and represents a downward trend which has continued, with 51,000 cases in 1947, 40,000 in 1948, and 30,000 in 1949.

Laboratory examination of blood smears during these years showed some reduction from the 54% positive in 1945 but the positives remained at approximately 40% during 1946, 1947, and 1948. The total number of bloods examined was greatly reduced in 1949 (ALVARADO [8]).

In 1947, a DDT spray experiment was conducted in the Mala Valley using a kerosene solution of the insecticide applied at the rate of 2 g/m^2 to all houses in the valley. A total of 590,000 m^2 was treated and over 8,000 people were protected. The results measured in terms of the parasite index of the population as a whole were very striking. The parasite indices during the malaria season (June–July) from 1942 to 1946 varied between 11% and 28.9% with a peak in 1943 and a gradual reduction from then to 1946. Following the application of DDT in January, February, and March of 1947, the index dropped to 0.5% in July of 1947. The reduction occured in all three species of malaria but was most striking in the case of *P. vivax* which is the most prevalent parasite. No *falciparum* malaria was present in July of 1947 and only a trace of *malariae*. A very great reduction in the number of *A. pseudopunctipennis* was noted in larval foci during the season when they were normally most abundant (CORRADETTI [114]).

In 1947, an experiment was also conducted in the valley of Santa Eulalia. This experiment was directed against both bartonellosis and malaria transmitted by *A. pseudopunctipennis*. Several towns and adjoining suburban residences were treated with DDT at a rate varying from 2 to 3 g/m². Of nearly 700 blood smears examined in children under 12 years of age following the treatment, only six cases of *P. vivax* and 4 cases of *P. falciparum* were detected (CORRADETTI [111]).

The campaign was gradually extended to the malaria-infested valleys of the coastal region and into the foothills of the western spur of the Andes to an altitude of 1,700 m. The valleys of the Chilean frontier had priority in accordance with an agreement reached between the two countries (PAMPANA [424]).

By 1949, 40% of the population in the endemic malaria zone was protected with DDT, making use of nearly 72,000 kg applied primarily as a solution in kerosene. Ninety-five thousand houses were sprayed once a year in this program at a cost of US $1.79 per house, $0.44 per person protected, and $0.007 per square meter treated. Roughly, 43½% of the cost was for labor.

Parasitemia among school age children in the endemic zone dropped from 7·3% positive in 1945 and 5·6% positive in 1946 to 0·9% positive in 1947 and 0·1% in 1949, the last figure being based upon over 31,000 examinations. There was a striking reduction in the proportion of *falciparum* malaria among children from approximately 20% of the positive slides through 1948, to 5·3% in 1949 (ALVARADO [8]).

In the valley of Canete, the number of cases of malaria confirmed by positive blood smears was reduced from 3,750 in 1944 to 0 in 1949. It was stated in the Health Appraisal Forms presented by the official delegations to the 13th Pan-American Conference in the Dominican Republic in 1950 that generally similar results were obtained in all of the valleys of Peru following the application of the DDT residual treatments [428].

Bolivia

About 70% of the area of Bolivia is malarious (PAMPANA [424]). Of the total population of 4 million, approximately 1·7 million live in the endemic zone and 200,000 in the epidemic zone. Malaria is unreported from only one of the nine departments of the republic. In 1946, one sector of Bolivia with a population of 180,000 had an estimated malaria morbidity rate of 21% and a malaria mortality rate of 2%. There are two vectors of malaria in Bolivia, *A. darlingi* which is found in the Amazon region and *A. pseudopunctipennis* which occurs in the Andes region, and transmission occurs from March to May.

DDT residual spray was first used for malaria control in Bolivia in 1946. It was applied as a solution in oil twice annually at the rate of 2 g/m². The initial house spraying was done in the town of Guayaramerin, on the Brazilian border, and produced quick results. There were fewer adult mosquitoes and fewer cases of malaria (HART *et al.* [254]). This activity was soon extended to other parts of the country and as a result of six treatments at six-month intervals, spleen rates in DDT-treated areas dropped from an average of 49·4% in 1947 to 12·6% in 1949 and parasite rates decreased from 18·4% to 1·5%.

Chile

Malaria has apparently been eradicated from Chile, no autochthonous case having been reported since April 1945. The last published report of adult anopheline vectors was a collection of five *A. pseudopunctipennis* in March 1948 in a small and remote oasis (Suca) where antimosquito work was not begun until the summer of 1947 (NEGHME *et al.* [411]). A few larvae of *A. pseudopunctipennis* continue to be found each year, however, in the most northern valleys, near the Peruvian border (NEGHME [408]).

The malaria control campaign was initiated in July 1937 in the port city of Arica. In the first stages, the antianopheline measures were almost entirede directed against larvae and were mostly engineering procedures, but includly oiling and the use of Gambusia fish, depending upon the topographic conditions. Antimalarial drugs were also used in this stage.

The traditionally malarious zones of Chile are valleys and oases of Tarapaca Province ranging from 18° to 21·5° South. In the endemic zones the splenic index was as high as 97% in the general population and 100% among infants in the Camarones Valley in March 1941. Other typical values for the summer season ranged from 39·5% for the general population and 20·4% for infants in the Vitor Valley during February 1940, to 80·6% and 61% for the general population and infants, respectively, for the same locality in February of 1942. The parasite index in March 1941 varied from 20% in the Lluta Valley to as high as 40% in Camarones. *P. vivax* was the predominant species, *P. falciparum* was roughly half as common, and *P. malariae* maintained a rather low level. The peak of cases occurred in the late summer or early fall (around April), and there was little transmission in the winter and early spring months of September and October.

In October 1944, DDT was introduced into the program. At first it was used at a concentration of 0·1 or 0·2% and resulted in a residual effect lasting about two or three months. Finally, by the end of 1946, the concentration of DDT was raised to 1 or 2% and the period between treatments was increased to six months. Kerosene, water-miscible petroleum oils, and xylene containing an emulsifier were used as solvents, and wall dosages were 1 to 2 g/m² treated. As a larvicide, a concentration of 0·01% DDT in No. 2 diesel oil, petroleum oil, or xylene was used. In the final efforts to eradicate anophelines, domestic animals were sprayed with 0·5% DDT and even airplanes, automobiles and trains, especially those coming from endemic zones of neighboring countries, were treated with DDT (NOÉ *et al.* [415]).

Since the last autochthonous cases of malaria occurred in April 1945, only six months after the introduction of DDT, it is quite apparent that the campaign against malaria had been notably successful even without DDT, but DDT probably hastened the complete reported elimination of the vector.

Argentina

Although malaria has not been eradicated from Argentina, it has ceased to be a major public health problem in that country and a large portion of the credit belongs to DDT (ALVARADO *et al.* [11]).

That the malaria program is seriously designed for total eradication of the disease is evidenced by the stringent law that requires reporting of all cases of malaria within twenty-four hours of diagnosis, and provides severe penalties not only for failure to report cases but also against any patient that refuses to permit the taking of a blood sample for laboratory confirmation of the diagnosis (ALVARADO [9]).

The National Malaria Service of the Argentine Republic was established in 1937. At that time, the annual morbidity from malaria was about 300,000 with a population of 1,000,000 in the endemic zone. The cases occurred in two distinct areas. The first, in the northwest, extends from the border of Bolivia to the 32nd parallel South involving the fertile and productive valleys and plains of the Andean foothills. This area includes about 120,000 km² where malaria is usually endemic. The vector is *A. pseudopunctipennis*.

The second zone, consisting of about 100,000 km², is the littoral northeast where epidemics of malaria have occured sporadically; presumably originating from the endemic zones of the neighboring countries of Paraguay and Brazil. In this zone two vectors exist: *A. albitarsis*, an indigenous species, and *A. darlingi*, brought in by floods in the Upper Parana River.

Until the advent of DDT in 1946, the antimalarial campaign in Argentina was entirely directed against larvae of the vectors, and the cost was based on the area to be controlled and the nature of the breeding areas rather than on the size of the human population protected.

The DDT residual spray program was initiated in September 1946, using solutions, emulsions, and suspensions applied at the rate of 2 g of DDT per square meter. During the first year, 40% of the population of the malarious zone were protected, and this was extended to 80% during the second year (ALVARADO and COLL [10]). At this time, epidemiologic studies were intensified and the origin of each case was studied as to the date of onset, date upon which each house was sprayed, and the formulation used. Blood samples were taken from the patient and also from other members of the family and neighbors. Entomologic inspections were made of the house and of those of the neighbors.

A Surveillance Service was established to control rigorously and study exhaustively the epidemiology of all cases confirmed or suspected.

In 1949 there were 115,449 kg of DDT used, about 90% as a 5 per cent solution and 10% as a 5 per cent suspension (ALVARADO [8]). During the first half of 1949, the period of transmission, 2,785 cases of malaria were reported, and of these 134 were new infections. Seventy-six, or 57%, of these new cases were from houses which had not been sprayed with DDT. Thirty-eight, or 28%, were from houses which had been treated over four months previously, or for various other reasons did not have an effective DDT residue. In 20 cases, equal to 15%, the houses, or at least the bedrooms, had been well sprayed. In part of the cases, there was evidence that the disease had been contracted while the patient was temporarily living elsewhere. It therefore appears probable that in at least part of the 20 cases, transmission was outside the treated houses (Alvarado *et al.* [11]).

The cost of spraying in 1947 was 0·05 pesos/m² in compact urban areas and 0·07 pesos/m² in rural zones (5 pesos = US $1.00). By the first of April 1947, over 151,500 houses had been treated which was the equivalent of about 80,000 houses per cycle of three months. Since the program was operational and not experimental, adequate comparative statistics are not available; but it appears clear that the DDT program reduced the number of anophelines in houses below the critical level of significant transmission, thus reducing the number of primary malaria infections. For example, the results of laboratory examinations of 13,791 blood smears in the Province of Tucuman in 1946 showed 5,105 positives, of which 15·8% were *falciparum* malaria. In 1947, an examination of 5,621 smears showed 830 positives of which 6·6% were *falciparum* (Alvarado and Coll [10]).

In the zone where malaria is transmitted only by *A. pseudopunctipennis*, reported incidence of the disease dropped from 129,248 in 1945 to 5,324 in 1949 (Soper [537]). Blood examinations in 1945 were 47·5% positive; whereas, in 1949, this figure was 6·2%. Only 232 new malaria cases were found in the endemic zone in 1949. A survey of school children in 1945 yielded 10·9% positive bloods compared to a blood positive percentage of 0·1 in 1949. In preschool children this drop was even more dramatic, from 8·9% positive in 1945 to 0·05% positive in 1949.

It does appear that endemic malaria has ceased to be a significant health and social problem in Argentina.

Filariasis

Filariasis, a disease produced by the nematodes, *Wuchereria bancrofti* and *W. malayi*, was known, according to Wilson and Reid [633], by the ancient Hindus and certain Persian physicians. It occurs indigenously in almost every tropical and subtropical country, and Stoll [559] calculated the number of human infections in 1947 as 189,000,000. Since Manson's discovery in 1878 that the mosquito, *Culex quinquefasciatus*, acted as an intermediate host for the parasite, there have been at least 60 other species and varieties of mosquitoes incriminated, involving 5 genera, in many different countries. The fact that there are many vectors with varied habits and wide distribution means that chemical control measures will vary in their effectiveness. For example, better disease control could be expected in an area where a house-frequenting mosquito such as *Aedes aegypti* is the vector than in localities where *Mansonia*, a mosquito difficult to reach with insecticides, is involved.

It has been estimated that, owing to the slow maturation and longevity of the adult filaria in the human host, and the chronicity and permanent disfiguration of elephantiasic manifestation, 10–15 years will be required to produce clear-cut clinical evidence of reduced incidence of the disease (GIGLIOLI [221]).

The most clear-cut results on filariasis control appear to be those reported by BROWN and WILLIAMS [68] from the experiment on St. Croix, Virgin Islands.

Table 22

Filaria Infections (W. bancrofti) in Children of St. Croix, Virgin Islands, Before and After the DDT Spray Program

Age years	Number examined		Percent positive	
	1946 Pre-DDT	1948 Post-DDT	1946 Pre-DDT	1948 Post-DDT
3	16	2	0	0
4	17	9	0	0
5	46	12	0	0
6	145	67	6·2	2·9
7	145	85	8·9	3·5
8	151	104	11·9	6·7
9	149	118	11·4	9·3
10	177	101	14·4	13·8
11	142	116	14·1	12·9
12	139	101	15·9	17·8
13	106	85	25·5	25·8
14	78	106	16·7	16·0
Total	1,311	906	13·3	10·6

This program was initiated in October 1946 as an island-wide DDT residual spray program to control *C. quinquefasciatus*, the principal filariasis vector of the area. A 5 per cent DDT emulsion spray was applied to the interior of homes and privies at the rate of 200 mg of DDT per square foot of surface area. Four treatments were applied from October 1946 to May 1948. Each over-all treatment involved from 2,530 to 2,934 houses, and a total of 11,078 premises were treated.

The work did not extend over a sufficiently long period for a proper appraisal of the value of DDT in filariasis control, but the results over the period involved were very encouraging. The population of *C. quinquefasciatus* was reduced approximately 50% in the houses, and the number of infested houses was reduced by 57%. *Ae. aegypti* was completely eliminated from treated houses. There was a 50 per cent reduction in *C. quinquefasciatus* infested with

W. bancrofti developed beyond the ex-sheathing stage. After the spray program not one infective-stage larva was found, whereas 0·40% of all *C. quinquefasciatus* examined prior to DDT spraying harbored this stage. The infection rate in school children dropped from 13·3 to 10·6% during the spray program (Table 22) and the average microfilaria count fell from 74·1 per 0·04 ml of blood to 45·8 (Table 23.)

Table 23

Microfilaria Counts of School Children Before and After the DDT Spray Program, St. Croix, Virgin Islands

Age years	1946 — Prespray			1948 — Postspray		
	Number infected	Total count	Average count	Number infected	Total count	Average count
6	9	709	78·7	2	98	49·0
7	12	1,159	96·5	2	38	19·0
8	18	936	52·0	6	432	72·0
9	17	1,486	87·4	10	229	22·9
10	25	2,215	88·6	12	401	33·4
11	21	2,069	98·5	13	901	69·3
12	22	1,242	56·4	18	615	34·1
13	27	1,658	61·4	17	1,356	79·7
14	13	683	52·5	16	333	20·8
Total	164	12,157	74·1	96	4,403	45·8

GIGLIOLI [221] showed that *Anopheles darlingi*, one of the vectors in British Guiana, could be eliminated by DDT residual spraying. Results were not as spectacular with *C. quinquefasciatus*. DDT residual sprays applied at the rate of 100 mg/ft^2 on a six- to eight-month cycle, combined with the spraying of pit latrines on a three-month cycle, drastically reduced the *Culex* population in bedrooms where presumably most of the transmission occurred. A total of 818 room visits were required to collect 1,695 mosquitoes in Lodge Village, whereas before DDT treatment 10 to 20 rooms would have produced this number. Such drastic reduction of vectors in bedrooms will no doubt be reflected in a reduced disease incidence.

C. quinquefasciatus that were dissected immediately after capture showed an infection rate of 1·05% whereas the rate was 1·29% before DDT was used. The post-DDT dissection revealed no mature larvae which would render the mosquito infectious, but pre-DDT dissections showed 0·25% of *C. quinquefasciatus* infectious on the day of capture. Dissections of *C. quinquefasciatus* females which survived captivity for 7 days showed 10·8% infected with filariae and 1·89% actually infective. Prior to DDT treatment these figuree were 7·27 and 1·24%, respectively. This high positive delayed infection rate was considered to be of little significance, however, since the death rate of female mosquitoes used in the tests was unduly high. A mortality of 24% was observed among female mosquitoes taken from DDT-treated houses and held in DDT-free environment. It was reasoned that the chances of survival in the field where contact with DDT was a likelihood was much less, and the chances of field survival and of the successful incubation of microfilariae to the infectious stage was much reduced.

An appraisal of GIGLIOLI's work in British Guiana indicates that the DDT spraying, because of the greatly reduced vector populations in treated homes, and the increased mortality of vectors in the treated area, and the reduced positive infectiveness and infection rate among 760 female *C. quinquefasciatus* dissected, resulted in considerable control of filaria transmission in Lodge Village, the area under DDT treatment.

The work of DeCAIRES [129] in British Guiana showed that even though the vector of filariasis is a house-frequenting species, it may not necessarily be controlled by DDT residual sprays due to natural or acquired resistance. His work demonstrated a natural resistance in *C. quinquefasciatus* of that area to DDT on initial exposure and an increase of this resistance by subsequent exposures. This worker stated that species eradication could not be obtained by the use of residual house spraying with DDT, benzene hexachloride, or chlordane. It was found, however, that the control of DDT resistant *Culex* could be obtained by switching to one of the other chlorinated hydrocarbon insecticides. If resistance developed to these materials, a reversion to the use of DDT gave good results. This is not, however, a consistent phenomenon, particularly with other species of insects such as flies. *Ae. aegypti* and *A. darlingi* are sufficiently susceptible that eradication is possible under favorable conditions.

BEYE *et al.* [51] in a preliminary report of work conducted on Tahiti and Maiao in the Society Islands, demonstrated a reduction in the percentages of *Ae. pseudoscutellaris* infected with microfilariae of *W. bancrofti* following the use of 5 per cent DDT applied as a residual spray to the interior of homes and sheds each three months. The exterior of buildings and surrounding shrubs were also sprayed with a 1 per cent DDT emulsion. One area where only DDT was used showed 6·5% of dissected specimens positive prior to the experiment and 2% positive at the close of the tests one year later.

There was a drastic reduction of *C. quinquefasciatus* in resting stations, but the control of *Ae. pseudoscutellaris*, which comprised practically the entire mosquito population of biting stations, was far less dramatic.

Recent work by JACHOWSKI and OTTO [287] in America Samoa indicated that DDT residual spraying of houses could not be expected to control filariasis caused by *W. bancrofti* where *Ae. pseudoscutellaris* was the vector. These workers showed the principal habitat of this species to be in the brush rather than in the open villages. The vector infection rate in the villages was somewhat higher than in the brush, but the preponderance of mosquitoes in the brush created a higher index of transmission in that environment. It was maintained that practically all human infections are contracted by natives while they are in the brush and that transmission in villages, particularly open villages, is insignificant. They referred to the work by POYNTON and HODGKIN which showed that *W. malayi* in the Federated Malay States is transmitted in the plantation areas or along trails by *Mansonia*, and they believe that a re-examination of the usual concept of household transmission should be made.

It will be a number of years before epidemiologic data are available which will permit a critical evaluation of the efficacy of DDT in filariasis control. It appears, however, that the effectiveness of DDT in the suppression of this disease will depend upon the habits of the vector and their susceptibility to DDT. In the case of most house-frequenting species, good results can be expected; but where natural resistance exists, initial treatments will sometimes not effect satisfactory control of the vector. In instances where the vector is a brush-dwelling species, and where consequently the disease is contracted out-of-doors, new techniques will have to be developed if DDT is to be a potent weapon in the attack.

Most evidence on the role of DDT in the control of filariasis at present is by inference, based on control of the vectors. It is most probable that later, in many instances, this will be translated into actual disease suppression.

Dengue Fever

Dengue fever, sometimes known as breakbone fever, dandy fever, and by other local names, is a virus infection transmitted principally by *Aedes aegypti*, although *Ae. albopictus* and probably other species may at times be involved. This disabling but nonfatal disease is prevalent throughout many tropical and subtropical regions of the world, and often appears in epidemic or pandemic form. It is estimated that some 500,000 to 600,000 cases occurred in Texas alone during the epidemic of 1922 (CHANDLER and RICE [96]). In the Grecian epidemic of 1928, 239,000 cases occurred up to September of that year (MANSON-BAHR [354]). A total of 82,392 cases were reported in US Army troops during the period, January 1942 to August 1945, and the annual hospital admission rate was 3·7 per 1,000 (SIMMONS [513]).

SIMMONS [513] cites a spectacular example of the effectiveness of DDT in the control of dengue in Saipan in August 1944. An extensive epidemic was terminated by spraying the occupied areas with an oil solution of DDT from airplanes.

HOLDSWORTH [277] was able to protect a group of US Navy personnel quartered in the center of Hankow, China, while the city was being ravaged by an epidemic of dengue in 1945. It was estimated that 90% of the Chinese population contracted dengue sometime during the epidemic. The vector was *Ae. albopictus*, and due to the short flight range of this mosquito the application of DDT residual spray to buildings within one block of the locale to be protected, coupled with the spraying of breeding places, resulted in the elimination of the vector from the protected zone.

Most of the first contingent of personnel to enter Hankow contracted the disease. After control was initiated, however, there were only two cases of dengue among Navy personnel and these were contracted by dengue control crew members working in an infested cellar without the use of repellents. The author states that the military value of this work is the demonstration that men can be taken into the midst of a dengue epidemic and be protected from the disease by creating a dengue-free area through the use of DDT.

In the Hawaiian epidemic of 1943–44, insecticides were employed along with other methods to check the disease (GILBERTSON [222]). If DDT had been readily available, it no doubt would have played a vital part, since its high effectiveness has subsequently been demonstrated against the vector.

The Surgeon General's office of the US Army [587] recommended the use of aircraft for large-scale aerial control of mosquitoes in the prevention of dengue fever. Oil sprays containing DDT were used with dramatic success to bring dengue epidemics under control. This method was considered to be an effective weapon in the face of outbreaks of dengue among troops since it was immediately effective in destroying infected adult mosquito populations as well as the larval forms. Adequate disinsectization of aircraft and ships was recommended as a necessary measure to prevent the transporting of dengue vectors from infected to dengue-free areas. Insecticide aerosols were used for this purpose supplemented by the larviciding of possible breeding places on shipboard.

Vector control has been demonstrated to be an effective means of controlling dengue; thus the efficiency of DDT against dengue vectors is of the highest significance. UPHOLT *et al.* [598] demonstrated the superiority of DDT as a larvicide for *Ae. aegypti* in Savannah, Georgia. The insecticide was shown to be highly effective at a concentration of 0·25 ppm, and the treatment of glass, wood, rubber, and metal containers was effective for six months or longer when exposed to both summer and winter conditions.

Yellow Fever

Yellow fever is an acute febrile disease caused by a filterable virus that is transmitted from man to man by the mosquito, *Aedes aegypti*. The disease is not mentioned in ancient medical literature nor is a disease described that can be identified as yellow fever. STRODE [560] believes that yellow fever did not exist in ancient civilization. The disease was initially identified as an entity in the seventeenth century, first in America and later in Africa. Yellow fever still exists on these two continents, and it is endemic in Africa over an area extending from the West Coast south of the Sahara through the Belgian Congo into Northern Rhodesia, Nyasaland, Uganda, Kenya, and Eritrea. Endemic areas include much of South America, particularly in the rain forests of the Orinoco, Magdalena, Altrato, and Amazon watersheds, the southern coast of Bahia in Brazil, northern Argentina, and Ecuador. Very recently yellow fever has advanced westward and northward into Panama, Costa Rica, and Nicaragua (MACKIE *et al.* [348]). Since there are few early writings from the traditional yellow fever areas, it is possible that the disease was in existence long before it was known to the outside world. Yellow fever, often referred to as 'yellow jack', was one of the most devastating diseases for some 200 years throughout the tropical and subtropical regions of America and even in northern North America, England, Italy, and other parts of Europe where it was introduced. It has been estimated that the epidemic of the Mississippi Valley in 1878 caused 13,000 deaths (STRODE [560]). The last epidemic in the United States occurred in 1905 in New Orleans and other southern ports, resulting in 5,000 cases, with 1,000 deaths. Recent cases have been diagnosed in Argentina in 1948; Bolivia, 1950; Ecuador, 1951; Brazil, 1951; and Panama, 1950. In each instance, however, modern methods of control, which in many cases included the use of DDT, prevented an outbreak of the disease (DECAIRES [130]).

Dr. CARLOS J. FINLAY of Havana, Cuba, read a paper before the Royal Academy of Cuba on August 14, 1881, in which he proposed the theory of transmission of yellow fever by mosquitoes (FINLAY [183]). The Yellow Fever Commission of the US Army confirmed this hypothesis by experimentation in Cuba in 1900–1901. Major WALTER REED, Dr. JAMES CARROLL, Dr. JESSE W. LAZEAR, and Dr. ARISTIDES AGRAMONTE published several papers on the etiology and transmission of the disease as a result of their work (STRODE [560]).

The classical concept of the epidemiology of yellow fever received a sharp blow when jungle yellow fever was identified. In 1907, Dr. ROBERTO FRANCO and his associates investigated an epidemic in Muzo, Colombia, which was described as yellow fever associated with a fever of spirochetal origin. They stated, 'The yellow fever has certain peculiarities from an etiologic point of view; (a) it is contracted in the forest and not in the neighborhood of the houses; (b) it is transmitted by *Stegomyia calopus* and probably also by other culicines; (c) inoculation takes place during the daylight hours, which are spent by the workers in places where the transmitting mosquitoes predominate (STRODE [560]). In 1932, SOPER *et al.* [542] reported an epidemic of rural yellow fever in Brazil in which *Ae. aegypti* played no part, the transmission being apparently due to other species of mosquitoes. Infections were carried to man from jungle animals, principally primates, by mosquito vectors.

Since no significant epidemics of yellow fever have occurred subsequent to the advent of DDT, there is a paucity of information on actual disease control with this insecticide. There is, however, voluminous literature on the efficiency of DDT in the control of the vector, *Ae. aegypti*. In the Americas, this mosquito is so domesticated that it is found only in proximity to man, and the larvae breed only in artificial containers. Due to these habits and its sensitivity to DDT, the insecticidal control of this species has been very successful. In Africa, however, *Ae. aegypti* is more sylvan; it breeds in places such as tree holes and lives in forests as well as about habitations. The tech-

S. W. Simmons

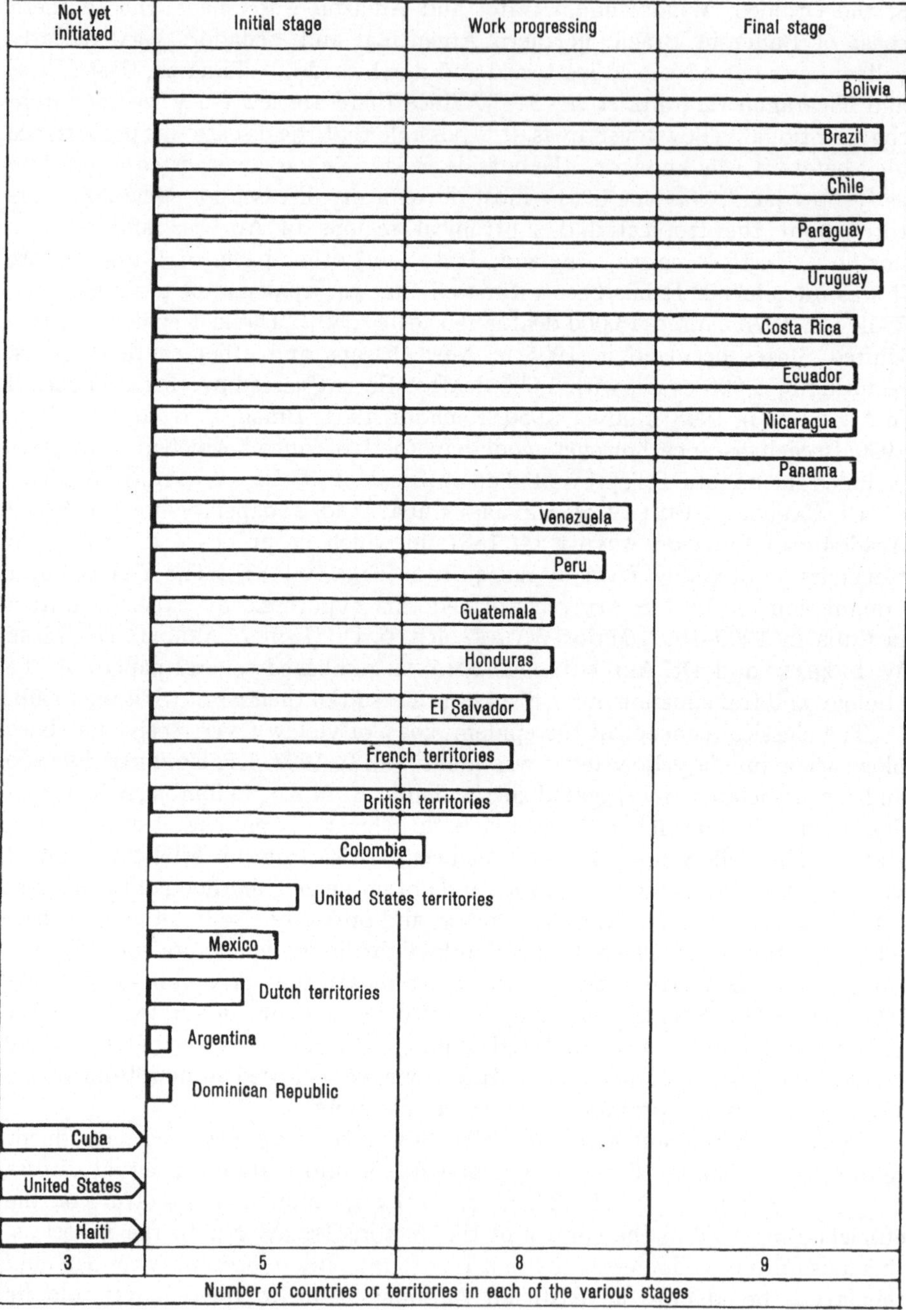

Figure 10

Aedes aegypti eradication in the Americas. Status of the campaign in the various countries and territories, June 1952.

niques for control of the African *aegypti*, therefore, have to be adapted to the local conditions and are not so simple as the procedures used in the Americas.

Ae. aegypti eradication programs have been under way in some South American countries for many years, and the advent of DDT has stimulated other countries to undertake similar activities. Figure 10 shows the status of *Ae. aegypti* eradication work in the Americas in June 1952 when 22 countries or territories had programs (SEVERO [506]).

DDT is used for the control of *Ae. aegypti* in several different ways:

(1) as a residual spray in dwellings in the same manner as used for malaria control;

(2) applied to the outside and inside of water deposits with or without water and to walls and other resting places nearby;

(3) as a larvicide: (a) in water tanks or receptacles with or without water; (b) in central municipal water supplies.

(4) for the disinsectization of aircraft and vehicular equipment entering yellow fever-free areas from infected zones.

The technique of residual spraying is the same as that used for malaria control. These procedures have proved to be quite effective in many areas, and eradication has been achieved in some instances (DECAIRES [130]). The application of DDT according to the second or 'perifocal' method is less expensive than residual house spraying and is preferred by some countries. In Brazil, it is considered the most practical and effective method (ROUANET [481]). DDT has proved to be an ideal larvicide for *aegypti* control. UPHOLT *et al.* [598] found that it was effective at dosages of 0·25 ppm. The treatment of water containers, whether they be glass, wood, rubber or metal, resulted in a residual action which was effective against *Ae. aegypti* larvae for 6 months or longer. The insecticide is effective in emulsifiable solution or as wettable powder which can be dusted into wet or dry containers. One favorite way of applying DDT to water is in alcoholic solution. A 2 per cent solution has been recommended by SASSE [500]. This is used at the rate of 1 ml/6 l of water repeated every 4 to 5 weeks.

Several South and Central American countries, including Chile and El Salvador, have applied DDT at the rate of 1 ppm to the reservoirs containing the city water supply. Application was made in the form of an aqueous suspension (NEGHME *et al.* [409]; RODRIQUEZ [476]).

Since DDT became available, it has been used in practically all formulae which have been devised for airplane and ship disinsectization. A program of disinsectization of carriers in international traffic, associated with good airport sanitation, including frequent inspections for *Ae. aegypti*, is a safeguard against the establishment of yellow fever in free areas. Further details on aircraft and ship disinsectization will be found in the section on that subject.

The campaign against *Ae. aegypti* really began in 1901 after FINLAY's mosquito transmission theory of yellow fever was proved by the US Army Yellow Fever Commission working in Havana, Cuba. By 1910, many workers believed that the yellow-fever problem was solved since the disease had disappeared from most urban centers. In 1928, however, yellow fever again appeared in Rio de Janeiro and parts of Bolivia and Colombia. Shortly after this (1932), SOPER and co-workers proved the existence of sylvatic yellow fever, and the fight against *aegypti* was waged with renewed vigor. The present campaign for eradication of *aegypti* in the Americas was given great impetus, not only because of the availability of an effective weapon in DDT, but also because of the support by various international health organizations. In 1947 in Buenos Aires, the Directing Council of the Pan American Sanitary Bureau entrusted to that body the 'solution of the continental problem of urban yellow fever, based fundamentally on the eradication of *Ae. aegypti*, without prejudices to other measures which regional circumstances may indicate...' (SEVERO [506]). In addition to the Pan American Sanitary Bureau, the World Health Organization, the United Nations International Children's Emergency Fund, the Rockefeller Foundation, and the Institute of Inter-American Affairs have given much assistance to the *Ae. aegypti* eradication program, as well as to the suppression of other arthropod disease vectors.

DDT, not only when employed directly for *Ae. aegypti* control, but also when used for malaria control, aided in the early success of existing *aegypti* eradication programs in several countries. *Ae. aegypti* is reported to be eradicated from Bolivia, Brazil, Costa Rica, Ecuador, Panama, Paraguay, and Uruguay. Programs for eradication are well under way in Chile, Peru, Venezuela, and some Central American republics. New programs have been initiated in Mexico and in several French, British and United States territories in the Caribbean area (PAN AMERICAN SANITARY BUREAU [426]).

The efficacy of an *Ae. aegypti* eradication program is dependent a great deal upon the circumstances encountered in each country. In temperate zones, where yellow fever is not endemic, and where cases have occurred only when the virus was introduced, there is less need for a complete *aegypti* eradication program. In subtropical and tropical countries where sylvatic yellow fever exists and where the disease is endemic in man, the eradication of *Ae. aegypti* becomes a problem of some concern, and the amount of money and effort that can justifiably be expended may be considerable.

The cost of an eradication program naturally varies a great deal from one country to another. In the Caribbean area DECAIRES [130] gave a figure on the cost of residual DDT spraying of $ 0.75 (US) per house in Puerto Rico. He also gave a method for estimating the cost of *Ae. aegypti* eradication programs in the Caribbean area, by which the approximate cost of a 2-year campaign was calculated to be $ 4,500 B.W.I. (B.W.I. $ 1.00 = US $ 0.59) for each 10,000 of population. In countries where labor is more expensive and where the population is less dense, this cost will, of course, increase.

There are probably few arthropod disease vectors that are more amenable to species sanitation than *Ae. aegypti*, both because of its domesticated breeding habits and its susceptibility to insecticides. The freeing of all international airports from *Ae. aegypti* and the assurance of continued protection by constant inspection is a necessity. Insecticidal treatment, not only of planes but also of boats from yellow fever zones, and in the case of vector-free areas, also from zones infested with *Ae. aegypti*, is an important aspect of the campaign against yellow fever. International quarantine regulations probably play a greater role in preventing the spread of *Ae. aegypti* than is the case with most other arthropod disease vectors.

Arthropod-Borne Virus Encephalitides

The arthropod-borne virus encephalitides (HAMMON [245]) include four diseases encountered in the Western Hemisphere, eastern, western, and Venezuelan equine encephalitis, and St. Louis encephalitis; three from the Orient and Pacific areas, Japanese B encephalitis, Russian Far Eastern tick-borne encephalitis, and louping ill (OLITSKY and CASALS [418]); and one from Africa, Mengo encephalitis found in Uganda (DICK [145]); DICK *et. al.* [146]). As far as is known, all of them are transmitted by insects or acari and all of them are caused by filtrable viruses (OLITSKY and CASALS [418]). The natural history of these agents indicates that man is only an accidental host in a complicated infection chain based on completely inapparent infections in nature (FOTHERGILL *et al.* [190]; HODES *et al.* [273]; HOWITT [282]; MEYER *et al.* [375]; SABIN [490]; SABIN *et al.* [491]; WEBSTER and FITE [621]; WEBSTER and WRIGHT [622]). The virus cycle involves birds as hosts and mosquitoes as vectors, with arthropods, Aves, or some unknown animals serving as long-term reservoirs (DAVIS [128]; GILYARD [224]; HAMMON [245], [246]; HAMMON and REEVES [248], [249]; HAMMON *et al.* [250]; HOWITT *et al.* [280], [281]; JENKINS [292]; KITSELMAN and GRUNDMANN [310]; NORRIS [416]; REEVES *et al.* [470]; SMITH *et al.* [533]; SMITH *et al.* [532]; SULKIN [565]; SULKIN and ISUMI [566]; TEN BROECK [575], [576]). This inapparent infection in nature can become a public health problem when the causative agent is transmitted to susceptible human hosts. Eastern equine, Japanese B, and Far Eastern tick-borne encephalitis result in a high proportion of severe and fatal cases, while western and Venezuelan equine, Mengo, and St. Louis encephalitis are more benign illnesses, at least in man.

Table 24 adapted from KUMM [317] gives the species of arthropods found naturally infected with encephalitis viruses.

Knowledge regarding encephalitis to date is insufficient to justify the control of this disease by attacking the widespread avian hosts or their mite ectoparasites. Therefore, the method of choice for the control of the arthropod-borne virus encephalitides would seem to be the reduction in endemic — and epidemic — areas of the numbers of arthropods which may transmit the virus to man. The lack of knowledge of the vectors in different regions, however, greatly reduces the value of this procedure as a general practice.

Studies in Kern County, California (REEVES *et al.* [471]), have established that the infection chain for western equine encephalitis in that area is mosquito–bird–mosquito–bird, etc., with the mosquito *Culex tarsalis* occasionally biting and infecting man, horses, and other animals. For some time antilarval efforts had kept the mosquito population just at, or below, the pest threshold for man; but during this period infections with encephalitis among mosquitoes, fowl, horses, and man continued to occur; therefore, it was felt that if a sufficient proportion of the infected and infective mosquitoes could be killed through the

Table 24

Arthropods That Have Been Found Naturally Infected With Encephalitis Virus

Species	St. Louis	Western Equine	Eastern Equine	Venezuelan Equine	Japanese B	Far Eastern Tick-borne	Louping ill	Mengo
Aedes dorsalis	×	×						
Aedes taeniorhynchus				×				
Anopheles freeborni		×						
Culex pipiens pallens					×			
Culex pipiens pipiens	×	×						
Culex restuans		×						
Culex stigmatosoma		×						
Culex tarsalis	×	×						
Culex tritaeniorhynchus					×			
Culiseta inornata		×						
Taeniorhynchus fuscopennatus								×
Mansonia perturbans			×					
Mansonia titillans				×				
Dermanyssus gallinae	×	×	×					
Eomenacanthus stramineus			×					
Ixodes persulcatus						×		
Ixodes ricinus							×	
Liponyssus bursa		×						
Liponyssus sylviarum		×						
Triatoma sanguisuga		×						

use of residual DDT deposits in their known domestic resting places, the infection chain might be broken. Accordingly, the inside walls of all domestic fowl shelters in the experimental plot were sprayed twice with DDT at an average rate of 200 mg of DDT per square foot. The first application was made in late April using a 3·7 per cent DDT solution in No. 2 diesel oil. Because much of this solution when sprayed on unpainted dry wood was carried below the surface leaving virtually no residual surface deposit, a second spraying was done in late June with 5 per cent DDT in a xylene-water emulsion. Complete coverage of each shelter was necessary as any surfaces left unsprayed harbored mosquitoes. Limited larval control measures were also carried out. Analysis of these studies showed that western equine virus infection rates remained high in *C. tarsalis* collected in the DDT-sprayed area (HAMMON and REEVES [247]). *C. tarsalis* continued to feed on chickens in DDT-sprayed shelters and apparently escaped, at least temporarily, the lethal effects of this insecticide. DDT residual spraying did not eliminate equine encephalomyelitis, for one case occurred in the center of the zone under control.

In the summer of 1945, an outbreak of Japanese B encephalitis occurred among the natives of Okinawa and neighboring islands, Heanza and Hamahika (MOSHER [395]). The incidence of the disease was 4·1 per 10,000 on Okinawa and 44·7 per 10,000 on Heanza and Hamahika; in both areas the mortality rate was about 30%. The incidence was greater in the northern mountainous wooded areas of Okinawa where DDT mosquito control measures were instituted late or were not effective on account of the terrain, which was treated by power sprayers and airplanes. The epidemic on the island of Heanza subsided abruptly two to three weeks after the start of intensive antimosquito measures using DDT.

In the Missouri River and the Arkansas-Red-White River basin areas of central United States there are, at the present time, programs to extend greatly the supply of water in these regions, converting dry farming to irrigated farming areas (COCKBURN *et al.* [102]; ROWE [482]; SMITH [530]). This increase in surface water, if not properly managed, could result in an increase in the numbers of mosquitoes, some of which are vectors of western equine and St. Louis encephalitis, both known to be endemic in the area (REEVES [469]). Because of this potential for increasing the incidence of these diseases, appropriate preimpoundment control measures and methods of water management are being incorporated into this program during the developmental period to prevent the creation of mosquito-breeding foci. In postimpoundment areas, antimosquito activities include a limited amount of larviciding with DDT, particularly in recreational areas.

There is a group of encephalitis-like viruses — West Nile virus, Bwamba fever, Semliki Forest virus, and Bunyamwera virus — found in Uganda which may be transmitted by arthropods (VAN ROOYEN and RHODES [599]). Murray Valley encephalitis of Australia is suspected of being carried by a mosquito, *C. annulirostris* (BURNET [79]). However, since no arthropods have definitely been incriminated as transmitters of these diseases, there have been no organized control operations directed against possible vectors.

The value of DDT and other insecticides as standard weapons for the control of the encephalitides can be determined only after more is known of the vectors and their ecological relationship to the viruses in nature.

3.

FLY-BORNE DISEASES

Diarrheal Diseases

That flies play a part in the dissemination of enteric infections has been an accepted thesis for years. Even before the existence of scientific data, flies were suspected of disseminating enteric infections because of their association with human excrement and filth. The increased prevalence of such diseases as dysentery, diarrhea, and typhoid fever, coincident with the seasonal prevalence of flies, led many to suspect that flies were involved in the transmission of these diseases. Early attempts to obtain scientific corroboration consisted principally of culturing organisms found on the bodies and in the voidings of house flies. Such work resulted in the detection of a large variety of viable pathogenic microorganisms associated with flies, as circumstantial evidence had suggested.

The presence of viable disease organisms on flies indicated that flies probably were vectors of human enteric infections but data were not sufficient to establish this. The most substantial proof of the practical significance of flies in transmission of pathogenic microorganisms could be obtained from comparison of enteric disease incidence in areas that were comparable except that the fly population was high in one and relatively low in the other. Such a study was possible only after the discovery of DDT. Before that time no insecticide was available that would essentially eliminate a fly population, and sanitational measures were rarely practiced to the extent necessary for the required reduction in fly densities. DDT was a tool that could be used to ascertain the incidence of enteric diseases in the near absence of flies and thus help establish the role of flies in the dissemination of such diseases.

United States

A comprehensive study of the problem was made by WATT and LINDSAY [619]. These workers selected two groups of small towns in the Lower Rio Grande Valley of Texas with similar environmental conditions. Group A towns consisted of five villages with a total population in 1946 of 46,100, of which 28,800 were Latin-Americans. Group B towns were four in number with a total population of 32,800, of which 25,600 were Latin-Americans.

In January 1946 fly control by DDT residual spraying was initiated in group A towns, and group B towns were left untreated as a control. At the outset a fixed schedule of applying residual sprays every six weeks was tried, but more frequent applications were found necessary in certain problem areas. Subsequent re-treatment was based upon the finding of one or more high counts of ten or more flies on the Scudder grill (SCUDDER [503]), irrespective of the interval since last treated. A comparison of the fly populations in treated and untreated towns is shown in Figure 11.

To determine the prevalence of enteric infections, anal swabs (HARDY *et al.* [251]) were made from Latin-American children under 10 years of age and cultured for the *Shigella* and *Salmonella* groups of organisms. Representative samples were taken from children residing in comparable blocks in each town,

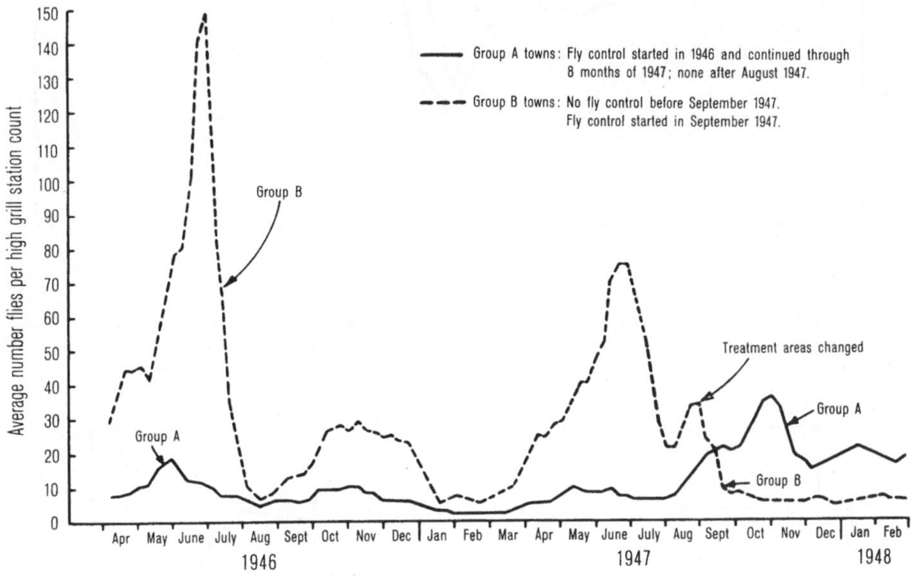

Figure 11

Three-week moving average of high grill index of total flies.

and the sampling frequency was established to provide 1,300 cultures per month divided equally between the two study areas.

The results obtained showed the amount of infection, disease, and death due to dysentery in the DDT-treated towns to be significantly lower than in the nontreated ones. There was a much greater difference between incidences of *Shigella* infections than between *Salmonella* infections, which were not greatly affected. Figure 12 shows graphically the effect of fly control on the attack rate of diarrheal diseases in children under 10 years of age, and Figure 13 shows the difference in new *Shigella* infections for a similar group.

In September 1947 treatment was stopped in the group A towns and initiated in the group B towns. Almost immediately the total fly population in group B towns dropped for the first time below that in group A towns; the population

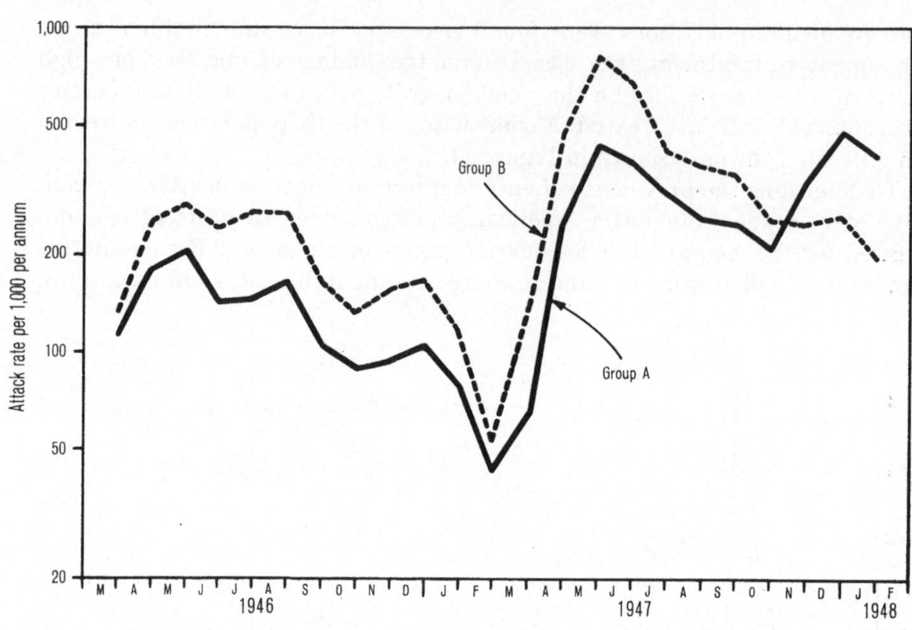

Figure 12

Attack rate per 1,000 per annum of reported diarrheal disease in children under 10 years of age, figured on a 2-month moving average.

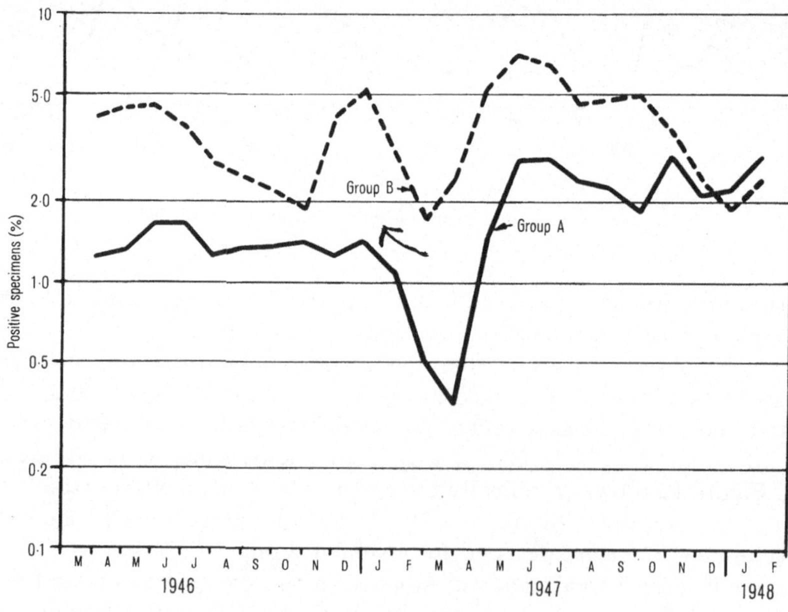

Figure 13

Percentage of children under 10 years of age, figured on a 2-month moving average.

in group A towns rose to a level higher than had existed since beginning of the experiment. The most significant occurence was a comparable switch in the incidence of diarrhea infections, disease, and deaths. Within a period of a few months, attack rates, new *Shigella* infections, and the death rate were reduced in the previously nontreated high-incidence group, and a comparable rise occurred in group A towns where treatment had been stopped. It was established, therefore, that a definite and significant correlation existed between the

Table 25

Reported Deaths Under 2 Years of Age and Death Rates per 1,000 per Annum Under 2 Years in Latin-American Children of Group A and Group B Towns for Selected Causes

| | I Diarrhea and dysentery | | | | II Unknown cause | | | |
| | Group A | | Group B | | Group A | | Group B | |
Quarters	Number of deaths	Rate[1]	Number of deaths	Rate[1]	Number of deaths	Rate[1]	Number of deaths	Rate[1]
1945								
March–May	26	47·2	20	39·8	4	7·3	7	14·0
June–August	26	46·8	25	49·2	10	17·9	8	15·8
September–November	5	8·8	8	15·6	9	15·9	8	15·6
1946								
December–February[2]	10	17·4	15	29·0	12	20·8	7	13·5
March–May	8	13·7	21	40·4	0		9	17·2
June–August	15	25·3	19	36·0	8	13·5	10	19·0
September–November	0		4	7·5	4	6·6	5	9·4
1947								
December–February .	5	8·2	6	11·2	4	6·6	5	9·3
March –May	8	12·9	19	35·0	6	9·7	9	16·6
June–August	23	36·6	30	54·8	5	8·0	9	16·4
September–November[3]	13	20·4	11	19·9	8	12·6	4	7·2
1948								
December–February .	17	26·3	1	1·8	19	29·4	8	14·4

[1] Rate per 1,000 per annum under 2 years of age.
[2] Fly control started this quarter in group A towns. No fly control in group B towns.
[3] Fly control started September 1947 in group B towns. No fly control in group A towns.

fly population and the incidence of diarrhea. The work demonstrated that diarhea, at least that caused by *Shigella* infections, might be expected to be reduced in municipalities where fly control is accomplished. Where flies have not developed a resistance to DDT, as was the case in the experiment reported, DDT is the most effective chemical for control of fly-borne diarrheal diseases.

Since the reported diarrhea death rate was greatest in Latin-Americans under two years of age, a comparison was made of the mortality among this group in the two areas. Table 25 shows that a significant reduction in diarrheal deaths occurred in the group of towns treated with DDT. There was also a reduction in death from unknown causes, and no doubt this included diarrheal and dysentery cases.

Later studies by LINDSAY *et al.* [324] in an area of moderate morbidity confirmed the earlier findings. Six towns in Thomas County, Georgia, were selected for study, and DDT residual and space spraying were initiated in three of them in July 1949. The Scudder grill technique was employed to measure fly prevalence, and a count not exceeding three flies per grill, as determined from the average of the five highest counts, was designated as the criterion of

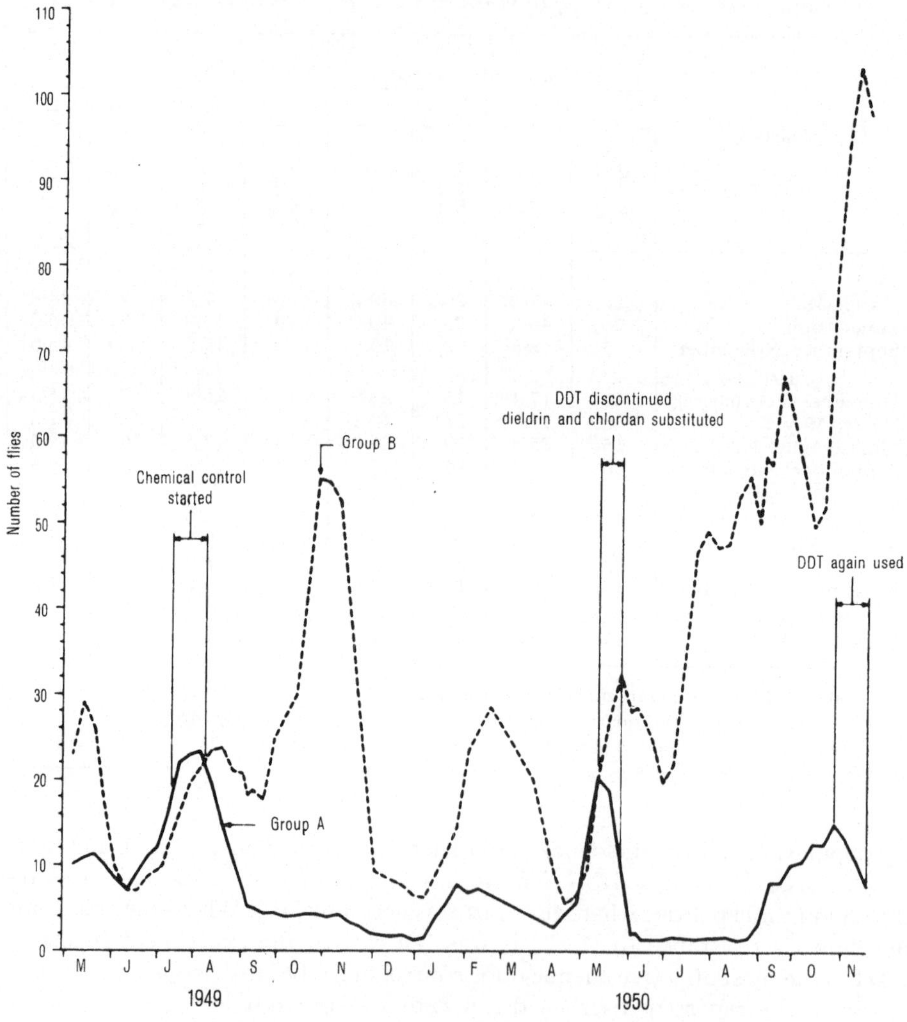

Figure 14

3-week moving average 3rd high-fly count, all species combined.

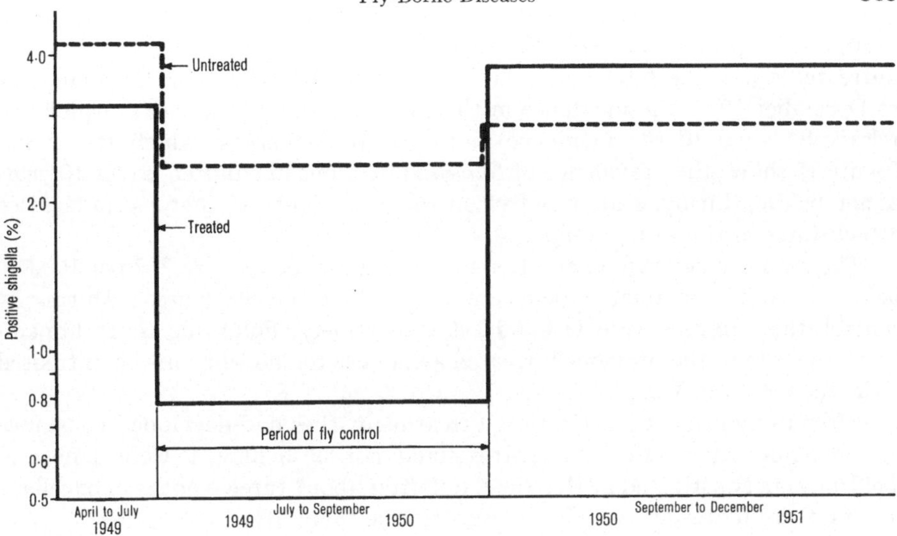

Figure 15

Percent positive Shigella in children under 10 years of age before, during, and after fly control.

effective control. House flies, *Musca domestica,* were predominant, outnumbering all other species combined. The relative population of all species of flies in treated and untreated towns is shown in Figure 14.

Before fly control, children under 10 years of age in the areas to be treated showed a 3·2 per cent positive infection rate for *Shigella.* The rate was 4·2% in the areas to be left untreated. During fly control from September 1949 to August

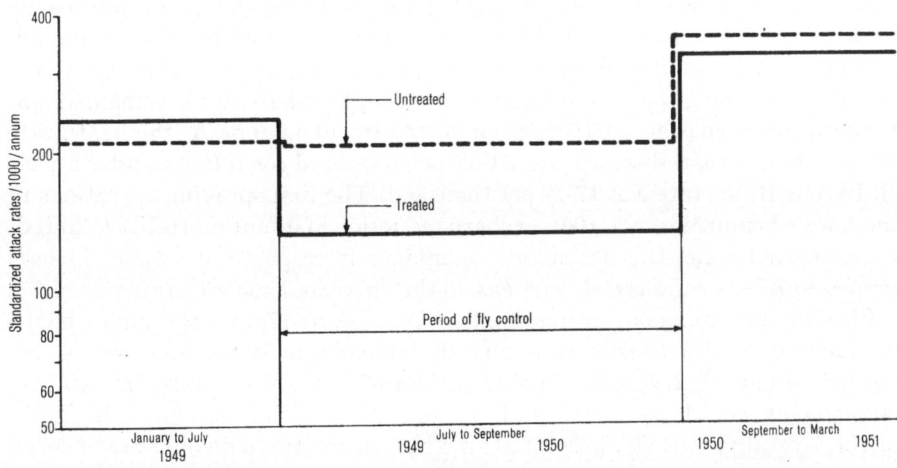

Figure 16

Standardized attack rates for diarrhea in children under 10 years of age before, during, and after fly control.

1950, the *Shigella* incidence in the treated areas dropped to 0.8%, while in the untreated areas, the prevalence was 2.3%. After fly control, September 1950 to December 1951, the incidence in the treated areas rose to 2.4%, which was relatively equal to the incidence in the untreated areas, which was 2.8%. Figure 15 shows the prevalence of *Shigella* infections in children under 10 years of age before, during, and after fly control, and Figure 16 shows standardized attack rates in the same group.

The crude attack rate in children under 10 years of age was 207 per 100,000 before fly control in areas to be treated, and 208 in the check areas. During fly control, these figures were 113 and 195 respectively. Following control operations, the rate in the previously treated areas rose to 258 while in the untreated areas the rate was 335.

After 10 months of DDT residual treatment, flies had developed resistance to the point where effective control could not be achieved. Chlordane and dieldrin were substituted at this time, but after about three months of excellent results, these materials were even less effective than DDT.

An evaluation by HEMPHILL [259] showed that trends of mortality from diarrheal diseases in the United States were downward for the period 1933 to 1946 inclusive. The sharpest drop occurred in 1946, and he attributes this in part to the widespread use of DDT and better medical care.

Italy

Work in Italy has demonstrated the efficacy of DDT spraying in the reduction of gastrointestinal diseases among infants. CORBO [107, 108] made an epidemiological appraisal of gastrointestinal morbidity and mortality among infants in Latina Province where DDT residual house treatment was begun in 1946. Two zones were selected for study. Zone A, consisting of 19 communities inhabited by approximately 200,000 people, was treated with DDT and zone B, comprising 14 communities with 60,000 inhabitants, remained untreated for comparison. In 1945, prior to treatment in zone A, the death rate from gastrointestinal diseases was 31.33 per thousand for infants under a year old. In zone B, the rate was 17.79 per thousand. The first spraying operations in zone A were begun March 5, 1946. A sharp reduction in infant mortality followed. In the treated zone, the death rate of infants from gastrointestinal diseases dropped to 7.74 per thousand; whereas, in the untreated zone the rate was 11.71.

Flies in this area became resistant to DDT by 1947, and the infant death rate due to diarrheal diseases consequently increased to 19 per thousand. Octa-Klor (chlordane, 1, 2, 4, 5, 6, 7, 8, 8-Octachloro-3a, 4, 7, 7a-tetrahydro-4, 7-methanoindane) was added to DDT in 1948, and resistant flies were brought under control. Mortality was again reduced; the rate in the treated zone was 11.59 as compared to 13.27 in the untreated area. In 1949 the rate was further decreased to 3.36 in the treated zone as compared to 11.40 in the untreated zone. By 1950 flies had become resistant to the combined DDT-Octa-Klor formulation, and the infant death rate due to gastrointestinal infection rose to 8·10 in the treated area; beginning in August 1950, there was a comparable increase in morbidity.

During 1950, the combination DDT-Octa-Klor residual spray was used in zone B. Since this zone had not been treated previously, flies were not resistant to the insecticide and good control was obtained. There was a drastic reduction in the infant death rate from gastrointestinal diseases, from 11·40 per 1,000 in 1949 to 5·16 per 1,000 in 1950.

This study showed that infant mortality caused by gastrointestinal diseases was always much lower in sprayed than in unsprayed areas if the insecticide was effective in killing most of the flies, and that where resistant flies became prevalent there was an increase in the incidence of gastrointestinal infections among infants. These results led to the conclusion that flies played a major role in Latina in the dissemination of epidemic summer outbreaks of diarrhea, which could be greatly reduced by good fly control.

SPANEDDA and MARCHINU [546] made a preliminary evaluation of the effect of DDT residual treatment of buildings in Sardinia for mosquito control on the incidence of enteric infections. Spraying by ERLAAS began in November 1946 with 5 per cent DDT in oil or other solvents. The secondary effect was the almost total absence of flies, and in 1947 there was a decrease in typhoid, paratyphoid, bacillary dysentery, and infant enteritis. The effect of DDT on the fly population was suspected to have influenced this decrease in enteric infections.

Bacillary dysentery was not an important disease in Sardinia, and infant enteritis was very poorly reported, so that the evaluation of these diseases appeared to be somewhat speculative at best.

A later study by SPANEDDA [545] indicated, however, that the insecticidal treatment did not result in any noticeable decrease in the incidence of typhoid and paratyphoid infections. It was concluded that these diseases in Sardinia were spread by direct contagion rather than by insects.

India

In connection with a study of the effects of malaria control on the other diseases in Kanara and Dharwar Districts of Bombay Province, VISWANATHAN [611] obtained data that showed the effects of DDT spraying in reducing diarrhea and dysentery in the Dharwar District. During the ten-year period from 1936 to 1946, there was a general, but not significant, rise in the incidence of diarrhea and dysentery in Dharwar and a fall in the rest of the Province. DDT spraying was initiated in 1945, and the death rates from diarrhea and dysentery showed a significant decline in 1947 in Dharwar District but not in Kanara. An insignificant proportion of *Anopheles fluviatilis*, the vector of malaria, rested in cattle sheds during the day and these structures were not treated. Cattle sheds were sprayed, however, in the Dharwar District and a significant decrease in diarrhea and dysentery death rates followed.

The DDT spraying program was designed specifically for mosquito control, but the author raises the question as to whether inclusion of the treatment of cattle sheds contributed to the reduction of the house fly population and consequently to the diarrhea and dysentery incidence in Dharwar District. He believed that there is a possibility that modification of the malaria program to include the treatment of certain outbuildings might yield collateral benefits in the control of other insect-borne diseases such as diarrhea and dysentery.

Greece

The DDT antimalaria campaign in Greece, as in other countries, resulted in a drastic reduction in the house fly population. VINE [608] stated that, 'To have seen a house fly in Athens last summer was something to comment upon.' The officially reported cases of dysentery for Greece for the summer of 1945, before DDT spraying, as compared to 1946 were, respectively: July 268:151; August 581:85; September 510:2; and October 152:6. This spectacular decline of dysentery following the use of DDT is certainly indicative of considerable fly transmission and of the efficacy of DDT in controlling the disease.

Finland

The role of flies in the dissemination of diarrhea apparently varies greatly in different localities. In Finland, ILPPÖ et al. [285] concluded that flies play a negligible part in the spread of diarrhea in that country. This was determined by comparing the incidence of infant diarrhea in a DDT-treated town with that in a comparable untreated one. In the treated area 11·3% of the infants had diarrhea while in the untreated town 13·9% had the disease.

Throughout the literature on the control of malaria by DDT spraying, mention is made of the reduction obtained in diarrhea and dysentery morbidity and mortality. Oftentimes there is a lack of sufficient data to substantiate adequately such a hypothesis, but the frequency of the observations, together with the data that have been obtained from two or three well-planned programs, is certainly strong evidence of the validity of these statements. There is little doubt that the significance of flies in the dissemination of enteric diseases varies greatly from one region to another because of differences in environmental sanitation practices, but there appears to be no doubt that good fly control contributes to a reduced incidence of certain enteric infections, particularly diarrhea and dysentery. Data on the effects of DDT spraying on the incidence of amebiasis and typhoid fever are not available.

Sir MALCOLM WATSON stated that 'when we cleared malaria out of the Malay States, we closed our dysentery ward' (SOPER [538]). The explanation given was that the chief cause of dysentery was something that lowers the vitality, and in Malaya this was malaria. Thus, insecticides may have an indirect, as well as a direct, value in the control of dysentery at least in some parts of the world.

Cholera

Cholera, an enteric disease whose etiological agent, *Vibrio cholera*, is passed in viable form in the feces of infected persons most certainly is transmitted to some extent by flies. The effect of fly control upon the suppression of a cholera epidemic apparently has not been adequately evaluated, but fly control has been shown to reduce other bacterial enteric infections (WATT and LINDSAY [619]).

In the 1947 cholera epidemic in Egypt one of the first problems undertaken by the Ministry of Health, following the outbreak in the latter part of September, was the control of flies. DDT was used for this purpose, along with sanitation; quarantine measures and vaccination were also employed (SHOUSHA [508]; OUCHTERLONY [422]). Principal attention was given to immediate killing of the adult fly population, and DDT was applied by four different methods as follows: (1) as an aerial spray from planes; (2) as a residual spray in buildings; (3) as a ground thermal spray; and (4) as a flit-type spray for the disinsectization of vehicles.

Planes began spraying operations on October 22 at villages in Menufia Province, and later spraying was carried on in Cairo and Alexandria. The residual program was carried out in specific infected sites such as homes of cholera victims, hospitals, and isolation shelters. In some instances entire infected villages and neighboring noninfected villages were treated.

Fogging with thermal sprays was carried out principally in infected villages and cholera hospitals in conjunction with the residual spray program.

The car-spraying program was carried out at specified traffic control points around the city of Cairo in an attempt to prevent introduction of infected flies into the city (EGYPT [166]). The program resulted in a reduction in the fly population in the treated areas. SHOUSHA [508] stated that the reduction ranged from 24·4 to 72% with an average kill of 48%. These figures were obtained by exposing flies in the treated areas; however, untreated villages were also used as a means of evaluating the effectiveness of the spray program. Twenty thousand cases of cholera, with a 50 per cent mortality, occurred before cessation of the epidemic. No adequate evaluation of the effects of the anti-fly campaign in the course of the epidemic was made.

Conjunctivitis

No epidemiological information seems to exist on the control of epidemic conjunctivitis with insecticides, but since it is disseminated to some extent by various species of small flies, it is a malady which would lend itself to insecticidal attack. BERBERIAN [40] said that the residual DDT spraying for malaria control in several villages in Lebanon resulted in the destruction of insects responsible for dissemination of conjunctivitis.

The 'eye fly' *Siphunculina funicola* of India, Ceylon, and Java is reported to be a vector of conjunctivitis (MANSON-BAHR [354]), and *Hippelates pusio* is a vector of undetermined magnitude in the southern United States.

Experimental studies by the US Public Health Service [591] indicated that *H. pusio* breeding in fields could be controlled by the application of 3 lb of aldrin (1,2,3,4,10,10-Hexachloro-1,4,4a,5,8,8a-hexahydro-1,4-endo,exo-5,8-dimethano-naphtalene) or heptachlor (1,4,5,6,7,8,8-Heptachloro-3a,4,7,7a-tetrahydro-4,7-methanoindane) per acre as a larvicide. The toxicants were applied as granular dust from a plane and turned under the same day. If the material is left on the surface, its residual is lost after two months. DDT applied at the rate of 16 lb/acre was ineffective for gnat control.

Phlebotomus Fever

Phlebotomus fever is also known as sand-fly fever, pappataci fever, three-day fever, Bessarabia fever, 'Hundfieber', and by other names. It is a virus infection which has been known for perhaps a century, and its transmission by *Phlebotomus* was established in 1908. Phlebotomus fever is endemic in the Mediterranean area, East Africa, India, portions of China, Ceylon, and in parts of South America.

During World War II there were 12,228 recorded cases of phlebotomus fever among American troops (SIMMONS [513]). At the present time, *Phlebotomus papatasii* is the only proven insect vector, although the disease does occur where *P. papatasii* has not been found (MATHESON [365]).

HERTIG and FISHER [264] conducted tests in Palestine in 1944 on the effect of DDT residual sprays against *Phlebotomus*. The treatment was very effective against *P. papatasii* for a 24-day observation period, after which the sand-fly season had come to an end.

JACUSIEL [288] continued tests with DDT residual treatments in Palestine. A high degree of protection was obtained at Rosh Pina in 1945 against *P. papatasii*, *P. major* and *P. chinensis* by the spraying of rooms with a DDT-kerosene solution at the rate of 1 or 2 g of DDT per square meter of area treated. Fewer sand flies than formerly were present in untreated rooms which indicated a general reduction in the population of the area due to the treatment. An attempt to protect houses by a DDT treated barrier strip within a radius of 50 m failed. HERTIG and FAIRCHILD [263], however, reported a marked reduction in the sand-fly population of houses by the use of this technique in Peru.

All military installations in Palestine were treated with DDT during the summer of 1945. Many complicating factors made a specific epidemiological appraisal of the program impossible; however, the sand-fly fever incidence in 1945 was much lower than previously even though nonimmune replacement troops were more numerous in 1945 than in 1944. In one British Royal Air Force station, the effect of DDT spraying on the disease incidence was conclusive. This station had a continuous supply of nonimmune personnel, was continually heavily infested with *Phlebotomus*, and showed a high incidence of sand-fly fever. Heavy infestations of *Phlebotomus* and large numbers of fever cases began to appear in May and early June, 1945. A large-scale application of DDT was made in mid-June followed by reapplications every six weeks until October. The incidence of sand-fly fever fell abruptly at the end of June and remained negligible throughout the usual peak season of the disease.

SEMPLE [504] describes an epidemic of phlebotomus fever in Malta in 1944 where the incidence rate among British military personnel in shore establishments was 349 per 1,000. A similar epidemic began in 1945, but the use of a 5 per cent DDT residual spray at the rate of 56 mg/ft^2 was followed by a reduction in the disease incidence. Spraying was not initiated until the latter part of June at which time the epidemic was at its height, but it was credited with the practical elimination of sand flies and within one month the weekly case rate had dropped from near 13 to about 3 per 1,000 and the rate for the entire epidemic was only 136. A total of 1,365 cases were recorded among shore personnel, with an average weekly strength of 9,988 men, as contrasted to 2,795 cases at a strength of 8,010 men during the 1944 epidemic.

Due to the shortage of sand-fly nets, DDT impregnated mosquito nets were tried and gave good protection from sand-fly bites and at the same time allowed more air circulation with consequent increased sleeping comfort.

HERTIG [260] reported the results of an extensive survey on the effect of DDT campaigns, mostly for malaria control, on the density of *Phlebotomus* in Greece and Italy. He concluded that DDT residual spraying gave immediate and virtually complete protection from sand flies indoors. Based on the investigations, it was concluded that an annual treatment of houses alone, preferably before the sand-fly season, would eventually reduce *Phlebotomus* almost to the vanishing point.

The use of DDT for malaria control in Lebanon resulted in the destruction of insect vectors of sand-fly fever, as well as malaria, dengue, conjunctivitis, and enteric diseases, according to BERBERIAN [40].

Bartonellosis

HERTIG and FAIRCHILD [263] conducted a DDT residual spray program for sand-fly (*Phlebotomus verrucarum*) and bartonellosis control at two construction projects in Peru. Austisha, a hydroelectric project about 50 miles from Lima, housed 800 persons. The sand-fly population was relatively low, but workmen complained of being bitten; and bartonellosis had been serious in the early years of the project. A 5 per cent DDT-kerosene residual spray was applied at the rate of about 1·5 gal/1,000 ft^2. The first treatment was applied in July 1945 to outside structures such as pigpens and retaining walls, which served as resting places for sand flies, and this resulted in a marked reduction of the population. Subsequent treatments were applied in December 1945 and March 1946 and included, in addition to the outside structures, the inside and outside of dwellings as well. Treated retaining walls were still free of sand flies one year after treatment; however, people were still being bitten.

The Cañon del Pato, in the Santa River Valley about 250 miles north of Lima, housed 1,500 workmen and their families in seven or eight camps. Most of the houses were of whitewashed adobe, and bartonellosis was a problem. One camp had been abandoned on account of the disease, there having been several fatal cases. The camp of Los Cedros was treated in July 1945 with a 5 per cent DDT spray applied to loose stones and boulders within a radius of 50 yd of the camp. After spraying, semi-weekly inspections for sand flies were negative. Additional treatments were made in October and November 1945, and January, March, April, and May 1946. After the first three treatments, the inner walls of the houses were treated. In the period from March through July 1945, the camp had 11 cases of bartonellosis. Only three later cases were known in this camp; at least one and perhaps all were contracted elsewhere. More extensive spraying began in 1946, including all camp buildings and being extended to include the town of Huallanca. In the period from April 1946 until the investigators' report in 1948, only one or two cases of bartonellosis came to the attention of the Chief Medical Officer, these being contracted outside the treated area.

The workers summarize their results as follows:

'1. The results furnish additional support for the effectiveness of house spraying in protecting persons indoors.

2. Treatment of stone walls (the principal outdoor shelters and breeding places) produced marked reduction of sand flies.

3. Treatment of stone walls combined with house spraying reduced sand flies to an extremely low level. This effect still persisted after 12 to 19 months.

4. The results were sharply localized within the sprayed areas, sand flies occurring in normal abundance in houses or caves 75 to 200 yd distant.

5. Practical control programs in camps of two large construction projects gave an extremely high degree of sand-fly control, followed by the virtual cessation of new cases of cutaneous leishmaniasis or bartonellosis.

6. Analysis of the results in terms of the habits and life history of *Phlebotomus* supports the possibility of achieving practical control by methods applicable to many of the *Phlebotomus* regions of the world.

(a) Their flight habits make sand flies vulnerable to residual DDT throughout their adult life.

(b) The long lifecycle delays the recovery of a depleted sand-fly population.'

CORRADETTI [114] reported complete protection of workers from bartonellosis in a construction project in the Valley of Santa Eulalia of Peru by mean DDT residual spraying of dwellings and huts and around houses, at the rate of 2·5 g/m² of surface treated.

Leishmaniasis

HERTIG and FAIRCHILD [263] obtained good control of the vector, *Phlebotomus verrucarum*, in two construction projects in Peru by the use of a 5 per cent DDT residual spray. Initial attention was given to outside resting places such as pigpens and stone retaining walls, which are both resting and breeding places. Subsequently entire houses were usually treated, and the program resulted in almost complete cessation of cutaneous leishmaniasis in the camps.

HERTIG [261] made an evaluation of the effects of the extensive DDT-malaria campaign on *Phlebotomus* control in Athens and Crete in Greece during 1948. The inner walls and ceilings of houses, stables and other outbuildings were treated at the rate of 2·0 g of DDT per square meter. Kerosene solutions of DDT were initially employed but subsequent treatments were mainly with water emulsions. Untreated places supported normal sand-fly populations, but sprayed buildings were uniformly negative for *Phlebotomus* of any species.

A sharp decline in the incidence of oriental sore in Canea (Crete), with the development of few new cases during the two years preceding the study, coincided with the introduction of DDT in 1946. The marked decline of kala-azar which occurred prior to the use of DDT was attributed to the destruction of infected dogs and to the reduction of the general dog population of Canea.

In the Prefecture of Kavalla there was an annual average of 11 cases of leishmaniasis during the decade 1930–40, 15 in 1945, 13 in 1946, 4 in 1947, 4 in 1948 and none in 1949 (January to July). The decline of the disease coincided with the use of DDT for malaria control, the sharpest decline occurring in 1947 following the first year of the malaria control program (LIVADAS [330], quoted from HOURMOUGIADIS).

CORRADETTI [113], [112], [109] working in the Province of Teramo, Italy, demonstrated the effectiveness of DDT residual sprays in reducing the prevalence of oriental sore. This work was in an area where two thirds of the population was rural and one third urban. All houses and stables within the territory of four towns of the Province were sprayed during the summer of 1948 with 2 g of DDT per square meter of surface treated. There were three species of *Phlebotomus* present: *P. papatasii*, *P. perniciosus*, and *P. perfiliewi*, of which the latter was the most abundant and considered to be the principal vector. Following the initial treatment there was an observed complete disappearance of *Phlebotomus* inside houses and stables in the treated area, with the exception of one premises. The untreated test zone simultaneously contained large numbers of the insects. During the summer of 1949 a second application of DDT was made following which no *Phlebotomus* were detected either indoors or outdoors in the treated zone.

In a pretreatment survey 28,599 persons were examined, of which 2·9% had active lesions and 20·8% had scars; thus 23·7% of the population either showed active lesions or scars of oriental sore. The DDT treatment essentially stopped transmission, reducing the number of new cases to 7 per 1,000 following the first treatment and to 1·3 following the second application (Figure 17). The untreated zones showed 34·7 and 40·4 new cases per 1,000 respectively in parallel surveys. The reduction of the disease below 1·3 per 1,000 in the area would probably not be possible without the treatment of the entire endemic area, which would eliminate contraction of the disease from nearby untreated zones. The work of CORRADETTI seems to demonstrate adequately the effectiveness of DDT in the suppression of this disease at the very nominal cost of 25 cents/person/year. CORRADETTI [110] also called attention to the reduction of leishmaniasis in Sardinia, Sicily, Calabria, and Puglie as a result of the DDT malaria control programs of those areas.

AMALFITANO *et al.* [12] made a survey of visceral leishmaniasis in the Fondi plains of Italy in 1948. A significant number of cases were uncovered and it was the opinion of the workers that an epidemic had been averted by the timely use of DDT in the area.

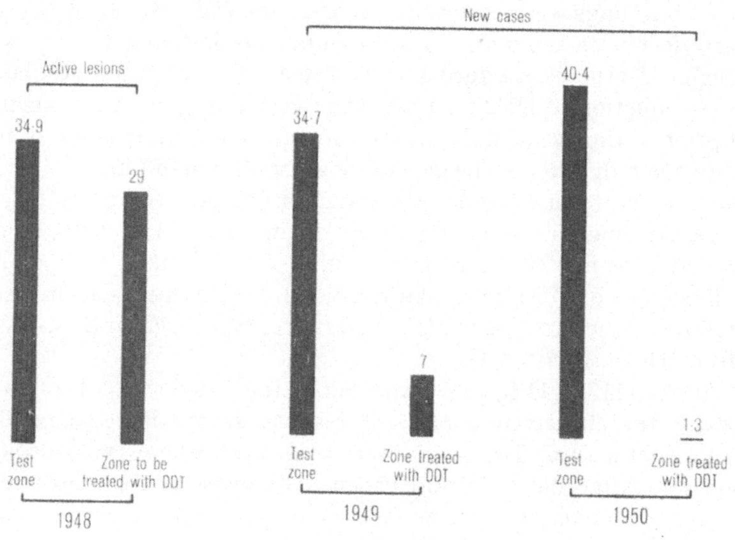

Figure 17

Incidence of oriental sore (cases per thousand persons) before and after DDT. Teramo Province, Italy.

D'ALESSANDRO *et al.* [121] reported an outbreak of visceral leishmaniasis in Palermo in 1944/45. *P. papatasii* was believed to be the vector, and by pasting broad strips of DDT-treated paper along the walls near the beds, they were eliminated from the rooms. The specific effect of vector proofing the homes on the course of the disease incidence was not reported, but a more recent communication (HERTIG and FAIRCHILD [262]) was to the effect that new cases of leishmaniasis within the 200 km² treated area had been reduced to straggling cases. The rate of infection had dropped to a level only one four-hundredths of that in the surrounding untreated area. HERTIG believes that large-scale control measures against leishmaniasis in many parts of the world could be undertaken with complete confidence of effective results.

A study was undertaken by a malaria control team in East Pakistan (Bengal) to determine the incidence of kala-azar and of the possible effect on its transmission by the residual spraying of DDT for malaria control (GRAMICCIA and SACCA [238]). Spleen examination and blood test of children below 15 years of age were made before and after spraying of the area with DDT. The test was in the Mymensingh area, involving 42 square miles and a total population of 45,000 people.

For the purpose of the test, operations were divided into two geographical sections. In one of these, all inside surfaces of every structure were first sprayed with DDT at the rate of 2 g/m² in June 1949. The second section was left unsprayed as a check area but was eventually treated for the first time in May–June 1950, with the same dosage of DDT. During the first survey from January to June 1950, 6,108 children below age 15 were examined, and 510, or 8·35%, gave a positive reaction to the formaldehyde test for kala-azar. In the second survey, October 1950 to January 1951, of 899 children who lived in the area which had been sprayed since 1949 and who were negative on the first examination, only 1 gave a positive reaction, and this individual had been outside the treated area for 2 months. Of 2,153 children living in the area sprayed for the first time during May to June 1950, 40, or 1·85%, gave a positive kala-azar test. This suggested that transmission had occurred to some extent since the beginning of 1950 in the area that had been sprayed for the first time in May–June 1950, but transmission had not occurred, or at least only to a very limited extent, in the area which had been treated since 1949.

The *Phlebotomus* population in unsprayed areas increased considerably in the pre-monsoon months (April–May) and showed a minor increase after the monsoon (November). No specimens were captured in the premises treated with DDT except in one house which had new walls. The data from the Pakistan project are not sufficient to draw statistically significant conclusions but are an additional indication that DDT is of value in the control of leishmaniasis.

Yaws

MANSON-BAHR [355] deduced from historical records that yaws was of more ancient lineage than syphilis; however, the disease was first described by OVIEDO in the 16th century. It is a common disease in tropical Africa, Ceylon, the Pacific Islands, Papua, the East Indies, the Malay States, and the West Indies.

Yaws is a contagious disease caused by *Treponema pertenue* and can be transmitted by direct contact. It is generally considered, however, that various flies play an important role in its dissemination. Minute flies of the genus *Siphunculina* are suspected of spreading the disease in the Kampongs of Malaya (FIELD [182]). *Hippelates papillipes* has been suspected of transmitting the disease in Jamaica.

ROCK [475] investigated a virulent type of yaws relapse in 4 villages of Guam during 1944 and 1945, in which 1,030 children were treated with penicillin and bismuth subsalicylate. Unidentified small bluish flies were observed persistently feeding on serum-filled ulcers of children at clinics, and these flies apparently followed the children to and from school. DDT was sprayed on the interior of houses, beginning 1 week prior to the first treatment clinic and for the 3 following weeks. The Quonset dispensary and schoolhouse were likewise treated on each morning of the treatment clinic. Spraying was repeated at 4 monthly intervals until finally interrupted by a typhoon.

There was a definite correlation between the relapse rate and the fly population. The total relapse rate for the best sanitated village with practically no flies was 4·7%; whereas, in the poorest sanitated village with a high fly population, the relapse rate was 36·1%. Villages intermediate in sanitary improvements also had intermediate relapse rates. Although adequate epidemiological evidence is lacking on the control of yaws with DDT, the potential for an insecticidal approach to the control of a portion of the transmission of this disease appears bright.

Trypanosomiasis (African Sleeping Sickness)

According to MATHESON [365], African human trypanosomiasis has been known for almost two centuries. However, it has only been about a half century since its cause was elucidated and since the tsetse fly was found to be responsible for its dissemination. The disease is prevalent in a considerable portion of Equatorial Africa. MANSON-BAHR [354] gives its distribution as follows: French West Africa, on the Gambia, in Sierra Leone, the Gold Coast, Nigeria, Cameroons, the Southern Sudan, and Uganda. The main stronghold, however, is along the Congo and its branches. It has been found in Zululand of the Union of South Africa. Rhodesian sleeping sickness is confined mainly to East Africa.

The etiological agents involved are *Trypanosoma gambiense* and *T. rhodesiense*, and the principal vectors are *Glossina palpalis*, *G. morsitans*, and *G. tachinoides*. Other species have been shown experimentally to be capable of transmitting the disease. *G. palpalis* is the chief transmitter of Gambian sleeping sickness although *G. tachinoides* is also involved. Rhodesian sleeping sickness is transmitted mainly by *G. morsitans*.

Trypanosomiasis is a serious public health and economic problem in large portions of Africa, and the disease has not yielded readily to control. The advent of new insecticides during the last decade has given some promise of assistance, but a full evaluation of their efficacy has not been made.

DDT was tried in the area control of the vector in 1945/46 by DuToit [160] after laboratory tests had established a high susceptibility of the vector. The area treated covered 30 to 35 square miles of the Mkuzi Game Reserve lying east of the Ubombo mountains and south of the Mkuzi River in the Union of South Africa. Treatments were made by plane, employing a 5 per cent DDT solution in furnace oil. Three treatments were applied between November 30, 1945, and January 15, 1946, at the rate of 0·5 lb of DDT per acre per treatment.

The population of *G. pallidipes*, the principal fly of the area, showed a steady increase during the month of November, prior to spraying. The week prior to treatment, a total of 7,053 specimens were taken from 230 Harris traps in the high fly density area. In the low fly density area, 1,089 flies were captured in 61 traps.

At the end of the week of the first spraying the *G. pallidipes* population of the high density area dropped to 37·5% of the previous week. After the second treatment the figure reached 19% and after the third treatment, 14%. By the end of January the population was only 7% of the pretreatment numbers. A similar decrease was also noted in the low fly density area. The workers point out that the dry period of the treatments might have been responsible for a portion of the decrease after November.

The dense canopy of vegetation was a definite barrier to proper penetration by the spray, and treatment was recommended during the winter months during maximum leaf drop. In order to get through the canopy, it was decided to try the aerosol method of dispersal so that the smaller droplets would better penetrate the dense vegetable growth.

In August 1946, a DDT aerosol generated by metering a DDT solution into the exhaust of the plane was applied in the Mkuzi Reserve. Approximately 40 square miles were treated in some 10 hours of flying time. The calculated dosage of DDT was 0·32 lb/acre.

A comparatively high density of *G. pallidipes* was present in the western section of the reserve at the time of treatment, and caged flies were exposed in various parts of the test area, including extremely dense vegetation. In two instances, the mortality of the caged specimens was 85 and 100% respectively. After these initial tests, the workers were unable to find sufficient flies in nature for further cage exposures. DuToit believed that six treatments of the area at intervals of 14 days held great promise for excellent control. This schedule would take into account the life cycle of the fly so that the unhatched pupae would have an opportunity to hatch during the period and the emergent adults to be destroyed. Du Toit [161] does not, however, overlook other methods of control and states that the use of the newer insecticides in conjunction with bush clearing and game control appears to offer more hope of success.

Symes *et al.* [571] conducted field experiments with DDT against *G. palpalis* in Uganda. Hand spraying of the vegetation on 1·7 acres of a 30 acre island in Lake Victoria, to produce a 100 mg/ft^2 deposit of DDT from a 5 per cent solution in oil, resulted in approximately an 80 per cent reduction in *G. palpalis* for about a week, followed by a sudden increase in population. In an attempt to maintain a toxic deposit over the maximum pupal period of the fly, four similar treatments of selected areas along the coast and in the center of Mbirubuziba Island were subsequently applied at about 2-week intervals. A total area of about four acres were treated, resulting in good control for at least 5 months. Steyn [557] stated that dust impregnated with 4 per cent DDT applied from an airplane had no effect on the fly on Nkuzi Island.

Four treatments were applied in a two-month period to a small percentage of the vegetation in a 130-acre area of the tip of a small peninsula into Lake Victoria. Two applications were made with 5 per cent DDT emulsions and two with 5 per cent DDT solutions in oil. Sample papers fixed to leaves during the third spraying collected from 20 to 87 mg of DDT per square foot, depending upon their position in the vegetation. Heavy rains during operations prevented

396   S. W. Simmons

treatments at regular intervals, and probably affected, to some extent, fly activity. A reduction in the population of *G. palpalis* of about 98% was attained, however, over the entire area.

After the final spraying, natives began to move cattle into the area from fly infested brush, which was the apparent cause of flies appearing in fair numbers at two catching stations one week after the terminal treatment.

Formulations of DDT used were absorbed by the vegetation which decreased its insecticidal activity, and deposits on hairy leaves gave very poor fly kills in laboratory tests. It was concluded that in spite of the control obtained, the laborious hand application of DDT was not likely to eliminate *G. palpalis*.

Verhaeghe [602] conducted two tests with DDT smoke bombs against *G. palpalis* by discharging the bombs in infested areas at daybreak, at which time dispersal of the smoke was good in all directions, both in vegetated and cleared areas. Judging from the post-treatment population of flies he could detect no effects of the treatment. Only three bombs were used in one test and four in the other.

Pires [450], working in Portugese East Africa, attempted to control *G. brevipalpis*, first by ridding the area of host hippopotamus, and second, by a campaign of dipping cattle in a mixture of DDT and BHC with the intention of killing tsetse that lighted on them. No results of the tests were given.

Wilson [631] fed two cows DDT powder containing 83% of *para-para* isomer at the rate of 0·5 g/kg and 0·25 g/kg, respectively. He fed 18 *G. palpalis* on each animal 18 hours after dosing and two engorged males were paralyzed after three hours and died after eight. Two additional males and one female were found dead after 24 hours, but the remaining flies were apparently not affected. Flies fed on the animals 24 and 48 hours after dosing were likewise not affected. Wilson concluded that the experiment had little practical value. Similar tests with benzene hexachloride gave more promising results.

De Kock *et al.* [139] states that the government departments have combined their efforts with the farmers in an attempt to stamp out tsetse flies in the Union of South Africa. The methods include brush clearing, elimination of host animals, and the use of insecticides such as DDT. It is generally conceded by most authorities that the use of insecticides should be supplemented, where at all possible and feasible, with appropriate brush clearing and control of host animals of the area.

Onchocerciasis

Onchocerciasis is a filarial infection caused principally by *Onchocera volvulus*, although at least two additional species, *O. colcutiens* and *Dipetalonoma streptocerca*, have been incriminated. The disease is present in portions of Africa, in Central and South America, and Mexico. Transmission is by several species of blackflies of the genus *Simulium*.

FAIRCHILD and BARREDA [174] obtained complete eradication of *Simulium* larvae from swift flowing mountain streams in Guatemala for distances up to 10 km by the use of DDT as a larvicide. The chemical was applied as a 4 per cent xylene-water emulsion to effect a concentration of 1 part DDT to 10 million parts of water. Tests with DDT powder made into a suspension with water by the aid of a wetting agent gave equally satisfactory results. *S. metallicum* and *S. ochraceum* were considered the major vectors of onchocerciasis in the area although larvae of *S. callidum, S. mexicanum* and *S. exiguum* and several other unidentified species were present in the area treated.

STEWARD [556], in laboratory tests, found that DDT was lethal in a few hours to *Simuliidae* and at dilutions as high as 1 part in 4 to 8 million parts of water.

GARNHAM and MCMAHON [213], working in an onchocerciasis region of South Kavirondo, Kenya Colony, freed 65 square miles of area from *S. neavei.* This was accomplished in highly infested streams by adding DDT to the water. Preliminary applications were at the rate of 2 parts of DDT per million of water; however, this was later increased to 5 or more parts per million. A 30-minute treatment period was employed to insure exposure of the larvae for at least this time. A 10-day treatment cycle was used for the first 10 applications but this was later extended to 14 days and eventually to a month. The treatments extended over a 5-month period. Most aquatic life, other than the *Simulium*, had returned to the treated streams three months after the last treatment. In addition to *S. neavei, S. alcocki, S. lepidem* and *S. elgonensis* were also involved.

The very slow development of filaria in man precluded an epidemiological appraisal of the vector control work. A microfilaria index of 37% in children 4 to 8 years old was established as a base line for future surveys, which it is hoped will evaluate the effects of this work on actual disease suppression.

The experiences of GARNHAM and MCMAHON were so successful in controlling the aquatic forms of *Simulium* in South Kavirondo that the method was tried on a larger scale for the control of *Simuliidae* in the rivers in the Kodera District (GARNHAM [210]). DDT was introduced at 10- to 14-day intervals for 5 months. The last *S. neavei* was captured 2 months after treatment, and not a single fly was captured during the following year although frequent and intensive searches were made. The opinion was expressed that the fly had been permanently abolished from the area and that the effects on onchocerciasis would show up several years later; first, in children of age 4 to 6 who in two years' time should show a low microfilaria count instead of the 37% existing at the time of treatment.

A disadvantage of this method of control is the detrimental effect of the DDT upon fish, large numbers of which were killed by the treatment. The experimenters found that 'gammexane' used in a way similar to DDT, but in a dilution of 1 ppm, was satisfactory. Since 'gammexane' is less toxic to fish, it might perhaps offer an advantage for *Simulium* control in water where the fish population is of importance (MATHIS and QUARTERMAN [366]).

WANSON [617] working in the area around Leopoldville, Africa, used DDT both as an adulticide and larvicide against the *Simulium* vectors of onchocerciasis. The principal species was *S. damnosum* although *S. albivirgulatum*, *S. alcocki*, *S. unicornutum*, *S. cervicornutum*, *S. violacenum*, *S. ruficorne*, and *S. rutherfoardi* were also involved.

Benzene hexachloride was used as a larvicide by applying it upstream from breeding areas and good results were obtained. It was concluded, however, that the enormous quantities of emulsion needed to maintain an effective concentration over a sufficient period of time was not practical and an attack against the adults would be more feasible.

Initial tests against adult flies consisted of treating the banks of streams around Leopoldville for a length of 2 km and a depth of 100 m. Simultaneously, the grass or other vegetation over a perimeter of 800 m around the European quarters were treated. A water-wettable DDT powder was used and applied at the rate of 20 centigrams/m^2.

The treatments were a failure due to the long period required for applying the insecticide and to the rapid growth of new vegetation on which the insects rested without being exposed to DDT. The flight of *Simulium* from outside the treated areas was also an adverse factor.

The use of DDT fumigant bombs was likewise unsuccessful even when used at 10 or even 20 bombs/1,000 m^2.

The most successful control was obtained by the application of DDT by airplane to the reproduction and resting areas of the insect. In the early morning adults were found to be concentrated on vegetation near their breeding place, and could be destroyed by a DDT thermal generated aerosol applied by plane at the calculated rate of 26 mg of DDT per square meter.

The aerosol was generated from DDT dissolved in three parts toluene and seven parts gasoline by volume, and the flight altitude was from 10 to 15 m. There was a residual effect on the vegetation for 48 hours; however, this was not the case when benzene hexachloride was employed. Treatments were made daily either at dawn or an hour before sunset, and 90% of the dispersed particles were in the 50–100 μ size range.

The treatment of the wooded areas resulted in a reduction of the *Simulium* population of the inhabited quarters. Observations indicated, however, that migration of *Simulium* was occurring from larval beds across the river, and in October 1948, daily flights alternating between the breeding places of the two shores and the islands were instituted. This was carried on for 26 days but within the first two days Leopoldville West was practically freed from *Simulium*, and within four days the indigenous villages located within a perimeter of 35 km were also free. At the end of three weeks, the larval beds of the waters in the areas were negative. The treatment of all breeding and strategic resting areas of *Simulium* resulted in complete absence of the insect from both European and native quarters. In the islands before treatment, it was possible to capture monthly more than 2,000 females with an index of infectivity of 2·9%. After treatment, the population had so diminished that only 0 to 5 females could be captured per day, with a yearly average of one per day, and this was attributed to infiltration from untreated areas. The index of infectivity was reduced to 1·1%. The author stated that filaria transmission was broken by the destruction of the vector and that the problem of onchocerciasis for Leopoldville and Brazzaville seemed to be solved. The treatments neutralized the breeding beds at a distance from the city double the migratory range of the vector, and although daily treatments were made, it was stated that treatments on alternate days would probably exterminate the adults. For two years following the experiment, no new cases of onchocerciasis occurred among Europeans of the area.

4.

LOUSE-BORNE DISEASES

Epidemic or Louse-Borne Typhus Fever

There are few diseases that have had a more profound effect on the history of mankind than epidemic typhus fever. This louse-borne rickettsial disease has practically world-wide distribution, but has been more prevalent in Europe and Asia than elsewhere. Its prevalence is not restricted to war conditions, but it thrives best in such an environment and has probably destroyed more armies than all of the lethal devices created by man.

It is likely that typhus occurred in endemic form for centuries prior to epidemic manifestation and for centuries more in the epidemic form without diagnosis. According to ZINSSER [667], there is reason to believe that the earliest recorded severe European epidemic of typhus occurred in 1489 and 1490 when the soldiers of Ferdinand and Isabella brought the disease from Cyprus to Spain during their fight with the Moors for the possession of Granada. A disease in a monastery near Salerno some 400 years earlier is described in Cronica Cavense and is believed by ZINSSER to warrant the diagnosis of typhus fever.

A logical estimate of the number of victims of epidemic typhus is impossible, but even as recently as the period 1917–1923, it was estimated that there were over thirty million cases, with over three million deaths, within the confines of European Russia alone (MANSON-BAHR [354]).

Since the incrimination of the body louse in the transmission of typhus, attempts to control the vector have been foremost in the fight against this disease. Disinfestation of wearing apparel and bed clothing by heat and fumigation has been a part of most campaigns against epidemics, and a standard procedure with armies.

A really effective procedure for louse control, however, was not available until World War II when chemical disinfestation of the individual was perfected. Pyrethrum and Derris formulations were used first, but they were soon superseded by DDT, which could be applied rapidly and inexpensively to large numbers of clothed people, with the assurance of several weeks' protection from lice.

Swiss investigations (DOMENJOZ [147]) demonstrated the value of DDT as a lousicide, and as early as 1942 J. R. Geigy S.A. of Basel produced a 5 per cent powder known as 'Neocid' for this purpose. MOOSER [390], as a result of tests in 1941–1942, proposed DDT as a typhus prophylaxis.

The early work of DOMENJOZ on the impregnation of fabrics with DDT permitted the employment of this technique as a protective measure by armies of World War II. This worker continuously exposed lice to fabrics treated with various concentrations of DDT. Results of his tests are shown in Table 26.

DOMENJOZ also exposed lice to felt impregnated with DDT solutions while they were in contact with the skin, and a 1 per cent concentration killed all the lice in 24 to 36 hours. Concentrations of 0·1 and 1·0% stopped oviposition almost immediately, and 1 to 3 hours' exposure produced definite symptoms of poisoning of the lice. Feeding ceased with the first stages of actual poisoning, and in no instance did lice survive after the onset of definite symptoms. When 5 per cent DDT in talc was used, similar results were obtained and reproductive activities were disturbed almost immediately after exposure.

Table 26

Insecticidal Action of DDT in Impregnated Material. In Vitro Tests Using 30 Pediculus corporis (Reared Lice) for Each Impregnation Concentration [1])

Concentration of the impregnation solution	% of surviving lice after			
	24 h	30 h	48 h	96 h
1%	27	0	0	0
0·1%	30	10	0	0
0·01%	62	27	0	0
0·001%	70	33	13	0
Control (untreated)	93	90	80	15

[1]) From DOMENJOZ [147].

In America, BUSHLAND *et al.* [83] confirmed the effectiveness of DDT against the body louse, *Pediculus humanus corporis,* in December 1942 and proceeded to develop it for practical large-scale use on troops. It was shown that on grossly infested subjects, DDT powder, when applied to the patient, would provide almost complete protection for three weeks and would give good control for longer periods.

After experimenting with various concentrations of DDT louse powders, a 10 per cent concentration in pyrophyllite was recommended for use by the Armed Forces in 1943 (BUSHLAND *et al.* [85]).

It was found that all lice were down after six hours' exposure to cloth treated with this concentration, but that none were dead. The first death occurred after 8 hours, and all were dead after 20 hours. Freshly fed lice were more resistant than lice which had been starved for 14 hours or longer.

Exposure of lice which had been starved for 6 hours had no immediate effect upon their ability to feed, but none fed after 3½ hours' exposure, at which time 60% were down. It was evident from these experiments that lice would cease feeding prior to obtaining an immobilizing dose of DDT.

The impregnation of clothing by soaking in a DDT solution is often a more effective, longer-lasting, and more practical procedure than dusting. Garments impregnated with 0·5 per cent DDT solution were completely effective after three weekly washings; those impregnated with a 1·0 per cent solution withstood four weekly washings without losing effectiveness. A dose of DDT equivalent to 2% of the weight of the garment is recommended to give protection through six to eight washings (JONES et al. [300]). During World War II the Army of the United States impregnated underwear and bed rolls to a 2 per cent DDT content for issue to troops.

Although DDT has no ovicidal properties, its residual effect extends beyond the normal incubation time of the louse eggs, usually making a single treatment sufficient for any existing individual infestation.

In 1943, the efficacy of an organized louse control program in eliminating a typhus epidemic was demonstrated in a Mexican village, and thus began a new phase of the age-old battle against this disease. By the use of various lousicidal powders applied to the population, the louse infestation dropped from 64% to practically zero in two weeks (DAVIS et al. [127]).

Field investigations of louse powders were started by the United States of America Typhus Commission at Esbe Rameses, Algeria, in January 1943 and in Bisda, Egypt, in March of the same year (BAYNE-JONES [38]). The powder initially used was the MYL-pyrethrum formula, since DDT-louse powders were not available until May 1943. The MYL powder was developed by BUSHLAND et al. [84] and was recommended to the United States Armed Forces in September 1942. The MYL formula was: 0·2% of pyrethrins, 2% N-isobutylundecylenamide, 2% of 2,4-dinitroanisole, and 0·25% Phenol S in pyrophyllite. Some of the early techniques of mass application were with the MYL formula, but later, DDT was substituted.

The technique of dusting individuals without the removal of their clothes was worked out chiefly by the Rockefeller Foundation typhus team in Algeria in collaboration with the Army of the United States and the Pasteur Institute of Algiers (SOPER [536]). In October 1943, power dusting was demonstrated at Esbe Rameses and a large-scale dusting was carried out at Sakiet Meki in December 1943, which reduced the louse population and apparently stopped a threatened typhus epidemic.

DDT was available when the Rockefeller Foundation typhus team arrived in North Africa in June 1943, and under the leadership of these workers new techniques were developed and old ones refined for the mass dusting of individuals. The work was conducted in Algiers in cooperation with the Medical Section of the North African Theater of Operations, the Pasteur Institute of Algiers, the American Red Cross, the United States Department of State, and the Director of Health for Algeria. The tests were initiated at the *Maison Carrée Prison* near Algiers (SOPER et al. [540]). Both MYL and DDT powders were used and various types of commercial hand and power dusters were tested. It was suggested from this work that a population which has been thoroughly dusted twice may be considered safe from important outbreaks of typhus for a period of at least three months. The complete eradication of lice in either an institution or a community was indicated to be feasible, with the proper use of DDT or MYL powders.

A further demonstration of the techniques developed was made in a civilian population at the Village of L'Arba near Algiers in October 1943. In this demonstration, 9,376, or 66·8%, of the 14,030 Arab inhabitants of the commune were treated; 30·2% were dusted twice and 33·2% were not treated. MYL powder was the basic formulation employed, and it was concluded that the application of an insecticide without the removal of clothing gives results comparable with those observed after careful hand application. The treatments were quite effective in the control of lice but the absence of typhus precluded the obtaining of epidemiological data relative to disease control. Such information was, however, soon to be available from the epidemic which appeared in Italy, and the techniques evolved were soon to prove their worth in the arrest and elimination of the Naples epidemic of 1943 and 1944.

The Naples outbreak apparently was the first instance in history where an important war typhus epidemic was promptly arrested and terminated by the insecticidal control of the vector, or, for that matter, by any means.

According to WHEELER [624] the first cases of typhus in Italy were reported from Bari, across the peninsula from Naples, in March, and from Aversa, near Naples, in April 1943. The first authenticated case was reported in Naples in July 1943. By December 15, a total of 83 cases had been reported from Naples, the number rising to 371 by the end of the month. The United States of America Typhus Commission was placed in charge of the epidemic on January 3, 1944, and the use of louse powders was decided upon as the control method of choice. Dusting operations had been instituted on December 15, 1943. The technique employed consisted of blowing powder, by hand or power dusters, between the layers of clothing and between the innermost layer and the skin. To accomplish this, the duster nozzle was inserted first down the neck, then up the sleeves, around the waist line, and into the crotch area. Hair and headgear were also treated. Figures 18 and 19 illustrate the use of both hand and power dusters for the treatment of individuals for louse control. One to 1½ oz of powder were sufficient for the treatment of an individual.

Only about 10% of the dustings were done with DDT up to the end of 1943. Prior to that date the principal powders used were MYL and AL 63. The British AL 63 formula was a derris and naphthalene preparation. According to CAMERON et al. [92], all DDT dusting was done with 10 per cent DDT in talc. However, CHALKE [95] states that after January 1, 1944, the British troops and their civilian workers were protected only by MYL and AL 63 powders, that few were inoculated, yet that only two to three cases of typhus occurred among this large group.

The activities of the typhus control program were organized into four major sections (WHEELER [624]): (1) the case-finding section, (2) the contact

404   S. W. Simmons

delousing section, (3) the mass delousing section, and (4) the immunization section. The case-finding section, in addition to its other duties, dusted the immediate family contacts, as well as the patient or corpse, with DDT.

The contact delousing section received daily, from the case-finding section, cards which contained information as to where dusting should be done. Essen-

Figure 18

The use of a hand duster for the treatment of a soldier with DDT powder to prevent louse infestation. (By courtesy of the US Department of Agriculture.)

tially, the work consisted of a house-to-house campaign, including institutions, in an attempt to create a louse-free area around places where typhus cases originated. Where cases of typhus had been reported from an institution, thorough dusting of inmates, personnel, and beds was promptly carried out, and in some of the larger hospitals and prisons, a routine dusting schedule at 14-day intervals was employed.

A special 'ricovero' service was established to do DDT dusting in air raid shelters. All persons living within the shelters were treated, and where typhus was present, bedding was also dusted.

The mass delousing section was organized under the immediate jurisdiction of the typhus team of the Rockefeller Foundation for the purpose of delousing

Figure 19

Use of a power duster proved to be effective and fast for the treatment of a large number of individuals with DDT for louse control. (By courtesy of MANNING A. PRICE).

the civilian population of Naples. Dusting stations were located throughout the city in such places as churches, schools, hospitals, railway stations, and quarantine stations.

The first two stations opened on December 28, 1943; and by January 15, 33 stations were in operation. At this time, an average of 1,611 persons was dusted daily at each station and an average of 1 lb. of dust was used for each 21 persons treated (SOPER *et al.* [541]).

Since some typhus continued to occur in close proximity to dusting stations, it was evident that complete coverage was not being obtained. Selected block surveys covering over 100,000 persons revealed that 77% had been treated. Approximately 20% had been dusted once, 28% twice, and 29% more fre-

                     S. W. Simmons

quently. On February 6, workers began dusting persons in blocks having more than 5% louse infestation, or where less than 70% of the people had been treated, or where both conditions prevailed. If typhus had been reported, treatment was carried out if less than 90% of the inhabitants had been treated. Such a procedure insured practically complete coverage in highly dangerous areas and was a culmination to the thoroughness of the undertaking. The epidemic had begun to decline at the time of initiation of block dusting, which no doubt accentuated this trend.

A total of 3,265,786 dustings was carried out in Naples and vicinity, and the number of typhus cases was held to 1,914 (Table 27).

Table 27

Number of Typhus Cases and Individual Insecticide Dustings in the Naples Typhus Epidemic 1943–1944[1])

Cases		Number of dustings				Total
		Contacts	Special service	Ricoveri	Station	
Naples	1,404	270,315	56,725	112,294	2,214,122	2,653,456
Outside Naples	510	201,257	...	...	411,073	612,330
Total	1,914	471,572	56,725	112,294	2,625,195	3,265,786

[1]) From WHEELER [624].

SOPER *et al.* [541] have calculated the epidemic potential of the Naples outbreak using a potential transmission period of 18 days following onset of illness, unless the patient had died or been dusted in the meantime. For each case reported in Naples with onset between November 1, 1943, and April 30, 1944, the number of days up to 18 elapsing between onset and isolation, dusting, or death was counted. These periods of infectiousness, when accumulated on a chronological basis, constitute the epidemic potential of typhus with respect to the uninfected population. Figure 20 demonstrates the trend of these data, plotted on a semilogarithmic scale, compared with the trends of the number of persons treated daily by the contact and station dusting services. It may be seen that the epidemic potential began to fall about 18 days after contact dusting was begun, and underwent a possible increase in the rate of decline in about the same number of days after station dusting reached a high level.

By early February 1944, the typhus epidemic in Naples had been broken. Although immunization and quarantine were employed to some extent in the fight, it is generally conceded that delousing was the procedure which stopped a threatened major typhus epidemic, initiated under ideal conditions for maximum propagation. This is indeed an historic milestone in the control of a most important communicable disease by use of an insecticide. BAYNE-JONES [37] stated: 'The conclusion is warranted that for the first time in the history of typhus in wartime, epidemics of the disease were brought under control and stopped before they had run their previously customary course.'

Following the successful control of the Naples epidemic, DDT marched with the allied armies; and the record of its efficacy in typhus control was

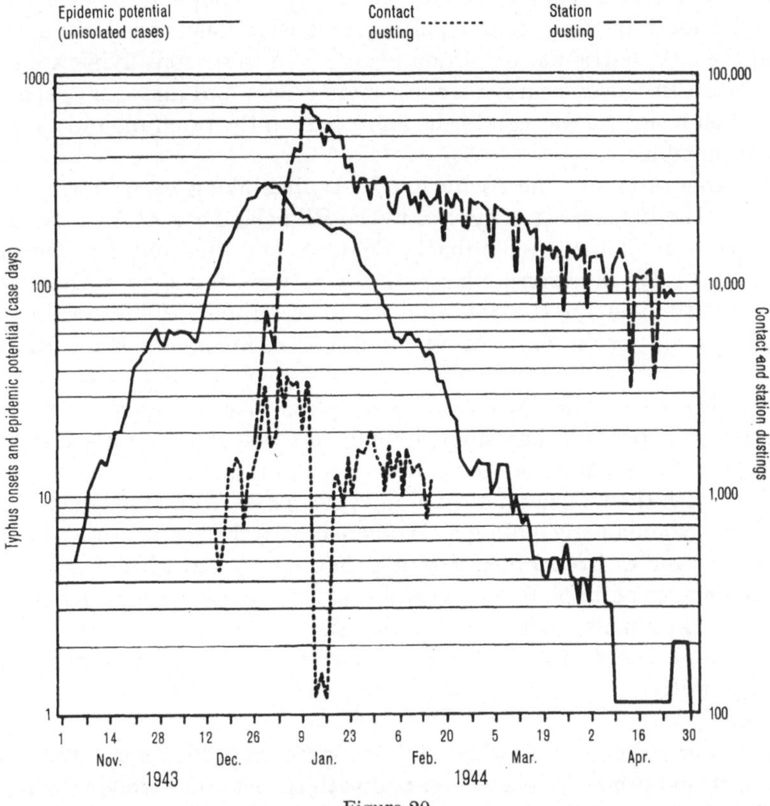

Figure 20

Typhus epidemic potential compared with the dusting activities, Naples, Italy (November 1, 1943, to April 30, 1944).

made in war-torn Europe under the most favorable conditions for typhus propagation.

A search of the records, by GORDON [231] and SNYDER [534], showed that typhus became established in Germany shortly after World War II began. When Germany was entered by the allied forces in the spring of 1945, typhus was encountered in many areas. The Army of the United States established a typhus fever control program which consisted of a system of case reporting, case finding, quarantine, immunization, and use of DDT for delousing of patients, of contacts, and, in some locations, of large numbers of the normal population, especially where a typhus epidemic threatened.

According to GORDON, delousing with DDT was the core of typhus fever control, the one reliance above all others. Military personnel were provided with a 2-oz can of DDT powder and were instructed to use it weekly on underclothing and inner surfaces of shirts and trousers. Where an outbreak occurred in a military installation, all personnel were systematically deloused. Where typhus was found in a civlian community having under 500 population, every individual was deloused; in larger municipalities, delousing was carried out in selected areas of the city, particular attention being given to persons living adjacent to typhus cases. In various camps for prisoners of war and displaced persons, delousing of all residents was established as a part of the program and carried out wherever possible.

Using this program, the Army fought typhus along its pathway into the Rhineland. The first report of typhus in the Rhineland was at Aachen, followed by reports from München-Gladbach, Cologne, and Hermülheim. During the first month of the invasion at least 31 towns were found to harbor cases of typhus; and to illustrate the potential for a catastrophic typhus outbreak, the following description by GORDON of conditions of foreigners in the Rhineland is given:

'The whole area seethed with foreign peoples, conscript laborers moving this way and that and in all directions, hoping to reach their homes, in search of food, seeking shelter. Most of the typhus was within this group and they carried the disease with them. They moved along the highways and in country lanes — now a dozen Roumanians pulling a cart loaded with their remaining belongings; here a little band of Frenchmen working their way toward France, there some Netherlanders, or perhaps Belgians; and everywhere, the varied nationalities of the East — Ukrainians, Poles, Czechs, Russians. They moved mostly on foot, halted, then gathered in great camps of sometimes 15,000 or more, extemporized, of primitive sanitation, crowded, and with all too little sense of order or cleanliness.

'These were the people where typhus predominated, more than a half million of them in the Rhineland, wearied with the war, undernourished, poorly clothed, and long inured to sanitary under-privilege and low level hygiene. Add to this shifting population the hundreds of released political prisoners, often heavily infected with typhus but happily far fewer in numbers; the German refugees, first moving ahead of our troops and then sifting back to their homes through the American lines. Rarely if ever has a situation existed so conducive to the spread of typhus.

'Typhus fever in a stable population is bad enough. It has demonstrated its potentialities in both war and peace. The Rhineland in those days of March, 1945, could scarcely be believed by those who saw it — it is beyond the appreciation of those who did not. It was Wild West, the hordes of Genghis Khan, the Klondike gold rush, and Napoleon's retreat from Moscow all rolled up into one. Such was the typhus problem in the Rhineland.'

The establishment on March 7, 1945, of the Remagen bridgehead necessitated immediate action to handle the typhus problem found in many places along the east bank of the Rhine and on into the inner Reich. A sanitary corridor was established along the east shore of the Rhine from the French-Swiss-German border to the Waal River and along the north side of the Waal to the North Sea. Delousing stations were established at ports of entry and crossing was permitted only after proof of recent disinfestation.

Typhus control in concentration camps presented especially difficult problems and was a challenge of the first order to the efficacy of DDT. This was particularly true in the Dachau Camp and the Belsen Camp. Upon liberation of the Dachau Camp by the US Seventh Army on May 1, 1945, an estimated 35,000 to 40,000 prisoners were found, and an undetermined number of typhus cases were encountered. The dusting of prisoners with DDT was started on May 3 and completed on May 8. Immunization was instituted as conditions permitted, but the principal reliance was on the delousing program. The number of cases was very small in the latter part of May and had practically disappeared in June, a contrast to the 100 to 150 new cases per day occurring before control measures were instigated. The confining of the epidemic to an estimated 4,000 cases, most of which occurred prior to the prophylactic effects of the delousing program, was attributed largely to DDT dusting. The coming of spring, the improvements in living conditions, and, of course, the use of immunization, played important roles, along with delousing, in the arresting and termination of the Dachau outbreak.

Apparently, the best estimate on the number of typhus cases within the American occupied territory of Germany was 16,506 patients from 518 localities. When the explosive epidemiological problems encountered, particularly in concentration camps, are taken into consideration, and when epidemics resulting from similar situations in pre-DDT days are considered, the real value of this material as a prophylactic for typhus fever is at once apparent.

The typhus epidemic at Belsen was of such proportion and was stopped so dramatically by the use of DDT that its consideration in some detail is justified. DAVIS [126] gave an excellent account of this epidemic, and most of the following discussion is from this source.

The Belsen Camp contained a variety of persons including Poles, Czechoslovakians, Hollanders, and other nationalities. Conscripted laborers from occupied countries and civilian political prisoners also helped to make up the population. As a result of the Russian offensive in the winter of 1944–1945, many eastern camps were evacuated and prisoners were brought to the overcrowded Belsen Camp. In January 1945, a trainload of Jews, heavily infested with lice and some having typhus, arrived from Hungary. By late January, the disease had spread throughout the camp. Delousing was carried on by the Germans, but reinfestation proceeded at a more rapid rate, since no delousing powder was available; and vaccine was not to be had. DAVIS estimated that there were 20,000 typhus cases in Belsen from January to April 1945. When the British

Army entered Camp 1 on April 15, 1945, it was estimated that there were 3,500 cases of typhus. The Military Government placed in charge of Camp 1 made an appraisal of the situation, established a field hygiene section which buried the dead, of which there were 10,000, erected latrines, improved the water supply, deloused and hospitalized the patients and, in general, made great improvements in the sanitary conditions. Delousing began in Camp 1 on April 22. Initial delousing was with the 5 per cent DDT preparation (British Insecticide AL63, Mark III), but after May 8 powder containing 10 per cent DDT was obtained and used.

The number of cases of typhus in Camp 2 with its 16,000 inmates was relatively few, and delousing permitted the removal of prisoners to other non-infected areas. On April 24, evacuation of Camp 2 was begun, and all inmates and their bedding were deloused before evacuation was begun.

From April 22 through May 25, a total of 77,695 inmates of the Belsen Camp were deloused, in addition to 131 British soldiers and 551 Hungarian soldiers and Polish girls. A total of 8,602 lb of DDT was used, or an average of 1 lb for each nine persons. An average of 113·9 persons were deloused per day by each worker. In summation, of the 61,000 living inmates in the various units of the Belsen Camp, approximately 3,500 cases of typhus existed at the time of liberation and 100% of the inmates were infested with lice. Within nine days of the initiation of the program, all inmates were dusted with DDT, and onset of typhus cases stopped abruptly 14 days after the first delousing. Twenty-five per cent of the inhabitants had some lice after one treatment, but only 3% after further treatment.

Without the preventive program, DAVIS states that the epidemic would have taken months to control, many cases of typhus would have been expected among troops, and louse-infested escapees from the prison would have spread the epidemic over wide areas, resulting in a prolonged, widespread, and costly epidemic. The fact that liberated internees left the Camp relatively free of lice cannot be overemphasized when it is considered that these people filtrated throughout Europe where conditions for the inauguration of a major typhus epidemic were most favorable. The benefits of the vast sanitary improvements, quarantine regulations, and immunization should not be underestimated; but the majority opinion, substantiated in most instances by sound epidemiological data, is that the DDT delousing program was the number-one prophylaxis which prevented a devastating European typhus epidemic.

In France, Belgium, and the Netherlands, although typhus was not reported through 1944, some delousing of certain groups of the civilian population was carried out as a preventive measure. This probably played a part in the prevention of outbreaks of typhus following the return of infested refugees to their homeland.

BOYER [63] reports 182 cases of typhus observed in Paris upon the repatriation of prisoners and deportees. He stated that in spite of delousing with DDT powder 15 of the 182 cases were among 167 persons repatriated from Germany. Transmission is credited to infested clothing not treated with DDT.

DDT dusting and vaccination are credited with the reduced typhus incidence in Turkey after 1944 (ERZIN [173]). The average cases per 100,000 population were 2·9 for the period 1935–1939. This rose during the war years to an incidence of 11·2 in 1940–1944 and 13·8 in 1945. The drop to 7·8 and 3·4 in 1946 and 1947, respectively, is coincident with the institution of the delousing and vaccination program.

When there is a migration of refugees from war-torn and disease-ridden countries, peripheral nations may expect to suffer from communicable diseases originating under the war-time environment of their neighbors.

Spain had an epidemic of typhus in 1941–1942 before DDT was available, and annual outbreaks occurred in some areas due to the influx of workers to the sugar cane factories. GIMENO DE SANDE [226] describes a successful campaign in the control of pediculosis in the endemic area of exanthematous typhus. The average cost of each disinfestation was much cheaper than the older hydrocyanic ampoule method, costing only about $ 0·046 per treatment.

GIMENO DE SANDE [225] describes the use of a 25 per cent DDT-benzol spray as a lousicide for the control of typhus in Motril. As many as possible of huts and other buildings were treated before being occupied. A single treatment was said to effect complete control, whereas three treatments of 10 per cent DDT powder were required to give the same results. The program resulted in the practical elimination of lice and typhus. In the year previous, there were 72 cases of the disease.

The 1941–1942 epidemic in Spain was the source of cases in Portugal as a result of the migration of gypsies. In 1943 and 1944, there were 181 cases of typhus registered in Portugal, and limiting the incidence of the disease to this number was brought about by a control program including the use of 10 per cent DDT powder at fixed and mobile anti-louse centers, together with disinfestation of houses, isolation of patients, sanitary inspection of travelers, and vaccination (DECARVALHO [131]).

According to epidemiological data from the World Health Organization [24], louse-borne typhus is endemic in Africa in: (1) French North Africa from Morocco to Tunisia, (2) Egypt, and (3) Union of South Africa. Some cases of typhus also occur in the highlands around the Great Lakes, and in Nigeria and Ethiopia. An African epidemic appeared and existed during the war, the peak being reached in 1942 when there were no less than 83,000 cases recorded in French North Africa from the three countries of Morocco, Algiers, and Tunisia. In Egypt, the number of cases increased and reached a peak of 40,000 in 1943. There was a postwar decline, and in 1948–1949 the number of cases reported from traditional African foci was the lowest on record. The use of DDT became increasingly common after 1945 and encouraged the hope that future epidemics would be prevented. Since 1946, DDT disinsectization has been carried out on a large scale in Morocco; nearly 224,000 disinsectizations were made in 1946 and 1,386,000 in 1947, and the regression of typhus is believed to be due to the increased use of the procedure. Beginning in 1946, the supply of DDT was sufficient in Tunisia to make mass disinsectizations possible in threatened areas and thus contributed to the rapid decline of the disease. Table 28 shows the number of cases of reported epidemic typhus in the period 1940 to 1949, and attention is called particularly to the rapid drop, beginning in 1946, which coincides not only with the end of World War II but with a plentiful supply of DDT.

In the Union of South Africa, 2,600 deaths from typhus were registered in 1943–1944 with a total of 5,623 reported cases from 83 districts which, of course, was far below the actual number. In spite of a lack of food and soap among the natives, abatement of the epidemic was rapid from 1945 to 1947, this regression being attributed to the use of DDT which was available at the end of 1944. Vaccination also was initiated in 1945 but its effect was not evaluated.

Table 28

Epidemic Typhus Cases Reported in Egypt, 1940–1949

	1940	1941	1942	1943	1944	1945	1946	1947	1948	1949
Lower Egypt	3,178	7,366	16,042	19,958	11,138	13,023	1,179	74	258	117
Upper Egypt	697	1,591	2,890	8,048	4,859	3,327	94	12	16	40
Cairo	364	168	2,244	8,751	1,758	1,254	142	49	31	16
Alexandria	117	170	521	1,473	413	422	24	14	9	11
Ismailia, Port Said, Suez and Damietta	25	28	244	1,723	246	202	94	20	8	3
Other Provinces	5	91	113	225	63	55	15	2	1	—
Total	4,416[1])	9,414	22,054	40,188	18,477	18,283	1,548	71	323	187

[1]) Corrected total.

MONTGOMERY and BUDDEN [388] reported on a typhus outbreak in Nigeria in 1945 from a native part of the town of Jos in northern Nigeria. 126 clinical cases were discovered, and it was found that the population was heavily infested with lice. DDT treatment was instigated, by which clothes were treated while worn; spare clothing and bedding were also treated and placed in dust bins in the sun for several hours to facilitate killing of the lice. Two weeks after the first campaign about 60% of the population still bore lice, a condition believed to be a result principally of contact with untreated people. After the second dusting campaign, no new cases of typhus were observed. A total of 80,000 persons and 1,300 residental compounds were treated. This appears to be an excellent example of the arrest and termination of a typhus epidemic by the use of DDT.

Typhus occurs almost every year among Africans in parts of eastern Transvaal and control was very difficult before the introduction of DDT. GEAR and MURRAY [218] stated that there is now a good prospect of eradicating the disease. They suggest that fabrics can be more effectively treated with a 5 per cent DDT-kerosene spray than with dry DDT which has a tendency to shake off.

The government of Afghanistan, assisted by the World Health Organization and the United Nations International Children's Emergency Fund, completely arrested a severe winter outbreak of typhus in Kabul and in Kandahar city and Province by the use of DDT dusting [23]. In 1952, 312,832 people, along with their bedding and clothing, were treated with DDT powder. A total of 44,000 lb of powder was used and 172 men and women workers were especially trained for the campaign. The Afghanistan Minister of Public Health noted the complete absence of typhus fever among the protected population of Kandahar city, a site of recent previous severe epidemics. There was an extremely small number of cases of typhus in 1952 in the treated areas. DDT certainly appears to have opened the curtain to a brighter future by its effectiveness in the typhus-infected areas of Afghanistan.

The typhus outbreak of 1940 in Yemen, Arabia, described by PETRIE [439] apparently is the first recorded instance of the disease in that country. The first definite case was noted in 1939 in San'a, the capital of Yemen, among hostages in the children's prison, with subsequent growth of the disease to epidemic proportions. In 1944, the establishment of insecticide dusting stations along the routes from Yemen to Aden apparently averted a catastrophic invasion of Aden by the disease. These dusting stations kept typhus outside their ring even when it was present on the periphery. The British AL63 formulation was used initially, but later, DDT powder was substituted. The DDT formulation was much more acceptable to the population, since the AL63 had an irritating effect. PETRIE states that wherever dusting was properly done it was effective, whether with AL63 or DDT. He describes the effectiveness of the lousicide program as follows: 'The Amiri and Lower Yafa' outbreaks make a useful contrast. In Dhala', the Amiri capital, as soon as the Amir understood the use of the dusting gun he used it himself, and saw that all who came to his palace were dusted. His example was followed by his people who came to appreciate the value of the dust for their lousiness if they did not understand its use as a typhus prophylactic. There were no secondary outbreaks in Dhala'. In Lower Yafa', the Sultan at his capital, El Quara, had at hand all the materials for prophylaxis, but he was too busy with his quat parties and did nothing. His brother sold the powder only to those who would pay, and so many people took typhus and many died.'

The use of vaccine was quite limited in the Arabian outbreak and as a consequence, major credit for the cessation of the disease would appear to be due to the use of lousicidal powders. By 1949, typhus apparently had been arrested or completely eliminated in south Arabia. Long-term prophylaxis is continuing, however, including the training of tribal dispensers in louse dusting techniques, and use of other anti-typhus measures such as vaccination and sanitation.

True to history, in Japan and Korea epidemic typhus fever followed in the wake of the war. The disease had been a considerable public health problem in Korea for many years prior to the war but had not been of major importance in Japan. Predisposing conditions, however, resulted in an epidemic in both countries during the season 1945–1946. These conditions were principally a shortage of housing, an overcrowded transportation system, inadequate general sanitation, the large-scale repatriation movement after hostilities, an ineffective public health department, and general apathy on the part of both physicians and other civilians alike (SCOVILLE [502]).

The control measures adopted by the occupation forces in both Japan and Korea, according to SCOVILLE, fell into three principal categories. The first was designed to prevent typhus in American military and civilian personnel. This program consisted of immunization, the supplying of each individual with a 2-oz can of DDT per month for dusting his clothes, the delousing with DDT of Japanese civilian employees of the Army and the American Red Cross, and of personnel of cabarets and dance halls frequented by American personnel. The billets of United States personnel were treated monthly with a DDT residual spray, and railroad cars utilized by Americans likewise were treated. The second measure was designed principally to prevent dissemination of typhus fever by repatriates. A port control program utilizing DDT powder was initiated at ten ports in Japan and four ports and three immigration stations in Korea. A team of 10 to 12 dusters and a supervisor, together with one or two individuals to fill the dust guns, processed from 400 to 600 individuals per hour. The third general procedure was designed to control typhus among Japanese and Korean civilians. Case-finding teams were created, and when typhus was found, the patient, as well as his family, was treated with DDT. In addition, the immediate neighborhood was searched by these teams for unreported cases. Dusting teams had the responsibility of delousing all persons, their clothing and bedding, in both focal and zonal areas, using DDT. The number of persons in a zone varied from 1,000 to 50,000. Special attention was given to jails, homes for vagrants, repatriation camps, and similar establishments, since they were often the origin of epidemics. Vagrants sleeping in railroad stations were also deloused at frequent intervals. All cases of typhus were removed to an isolation hospital, where possible, and quarantined for two weeks. The vaccination program also followed the focal and zonal types of administrative procedure.

The first outbreak of typhus in Japan was on the Island of Hokkaido. At the onset of the war, typhus was brought to the island by imported Korean laborers. During the first week in November 1945, 150 cases were reported, but the extensive use of DDT and typhus vaccine controlled the epidemic. All repatriates from Hokkaido were deloused with DDT at the ports. A port delousing program was put into effect at Hakodate, Uraga, Nagoya, Maizuru, Tanabe, Kure, Senzaki, Shimonoseki, Fukuoka, Sasebo, and Kagoshima (BLANTON and TANI [56]).

The outbreak in Osaka began in a city jail, and focal dusting with DDT was initiated. This was ineffective, however, and mass delousing was employed, an attempt being made to delouse the entire population of the city. These vigorous procedures finally resulted in control of the disease. The epidemic spread from the Osaka area to other cities, including Nagasaki and Tokyo. The Tokyo outbreak was extensive and the control procedures were ineffective, due, apparently, to incomplete case reporting and inadequate delousing techniques (SCOVILLE

[503].) The epidemic in Japan was a serious one but its limitation to some 30,000 cases with a fatality of 7 to 10% was surprising, considering the ideal conditions for typhus propagation (BLANTON and TANI [56]). The mass use of DDT powder was considered the outstanding control measure. To illustrate the magnitude and rapidity of the delousing program, 1,837,511 people were treated with DDT from February 16 to 19, 1946, and, according to BLANTON and TANI [56], an additional 1,306,360 persons were treated during the next 14 days, with a total of 201,757 lb of DDT being used. This mass dusting project was supervised by 600 enlisted men and 60 officers.

In Korea, the number of cases was less than in Japan. In 1945, foci of previous years were located and routine delousing and vaccination were instituted as prophylaxis measures. Such a program was effective, this being illustrated by the limiting of the outbreaks in Inchon and Seoul.

SCOVILLE [502] sums up the control program by stating that 'when control measures were faithfully used, a reduction of cases always occurred'.

This demonstrated effectiveness of DDT in the Japanese-Korean epidemic closes the chapter on epidemic typhus in World War II, a war in which only 64 cases of the disease were reported in the entire Army of the United States from January 1, 1942, to December 31, 1945, Incidence of typhus in other armies was likewise low, and the wartime experience with DDT demonstrated that no group of people need suffer from this disease in the future so long as there is available an effective lousicide for the treatment of the population in endemic areas.

Since World War II, however, experiences in Korea have shown that DDT is not a universally effective lousicide. HURLBUT et al. [284] reported the failure of 10 per cent DDT powder to control the body louse, *Pediculus humanus corporis*, among Korean military personnel during the winter and spring of 1951. Treatment was made without the removal of clothing, according to successful techniques developed during World War II; but despite weekly applications to all personnel, there was a simultaneous increase of infested persons. After five months of routine treatment, a random check showed infestations of 35·5, 49·2, 51·0, and 42·4% on successive weeks. In special tests, forty infested individuals were treated every third day for a period of 15 days, and 35 of them were still infested at the termination of the treatment. Another group of 40 men, 20 infested and 20 uninfested, were similarly treated and housed together, and after 15 days only five individuals were uninfested. Four separate samples of DDT were used, one of which had been tested against a standard laboratory strain of body lice, but none was effective in the control of the Korean infestation. Cloth impregnated with a 0·05 per cent solution of DDT is lethal to a standard laboratory strain of lice, but the Korean lice were reared on cloth impregnated with 0·1 per cent DDT through their complete life cycle and produced large numbers of viable eggs.

More recently reports have been made of resistant body lice in Egypt (BUSVINE [88]). Between the years 1946 and 1951, DDT powder was routinely applied to individuals for delousing purposes at 3- to 5-month intervals. In fact, most of the rural population was treated. In spite of such thorough treatments, however, some of the population continued to harbor lice, particularly during the latter portion of the period. Hurlbut *et al.* [283] initiated investigation to determine the extent of resistance as reported by Busvine. Efforts were made to establish colonies of field-collected lice on cloth treated with acetone solutions of DDT. Field-collected lice were propagated on cloth impregnated with 1:5,000 dilutions and, after two generations, colonies were established on 1:2,000 and maintained for four generations. The 1:1,000 dilution killed the lice from four localities in which it was unlikely that the specimens had been in contact with DDT. A colony could not be established on cloth treated with 1:1,000 dilutions, but a few individuals did survive to the adult stage. Tests with sleeves which had been dusted with 10 per cent DDT showed that after four days of wear, only 14% of the exposed lice were killed. Concentrations of DDT as low as 0·1 and 0·25% kill 100% of nonresistant lice in 24 hours. Although considerable resistance existed, DDT, was still a valuable weapon for louse control in Egypt. In one test an infestation of 56% was reduced to 9% four weeks after treatment while a control group showed no reduction.

Eddy and Selhime [165], working with the resistant strain of Korean lice, found that specimens fed on both rabbits and man and maintained without exposure to DDT lost most of their resistance. Where colonies were reared on cloth treated with 0·01% of DDT, resistance was maintained. This indicated that the resistance was a truly acquired resistance rather than a natural one. Toxaphene and lindane have proved to be effective against the resistant Korean strain.

These experiences place in jeopardy one of the most effective means of typhus prophylaxis ever devised, and only future work can determine whether the Korean experience will be sufficiently widespread to invalidate the use of DDT for typhus control. However, after having experienced the effectiveness of a residual insecticide in the control of typhus, it is hardly likely that the technique will fall into ill repute, but is more likely that newer and better materials will be developed, should DDT become completely ineffective.

In several villages in Mexico, Ortiz-Mariotte *et al.* [421] tested the efficacy of 5 per cent DDT powder in the control of lice. The DDT powder was applied to the clothing in contact with the skin, and in some instances village infestations were reduced from about 50% to slightly over 1%. In other instances, reductions were not this low but usually reduced the louse population to under 10%.

San Francisco Tlalnepantla, with a population of 671 people, had an epidemic of typhus beginning in September 1943 and reaching a peak during the first week of January 1944 when 22 cases appeared. DDT delousing began on January 4 and terminated on January 10. The first week following completion of treatment there were eleven new cases, the second week, four, and the third week, two, which terminated the epidemic.

An epidemic occurred in San Lorenzo Oyamel in the summer of 1943. The largest number of cases appeared in the latter part of June, at which time DDT treatment was initiated. A total of 739 of the 1,016 inhabitants was treated. Subsequent to dusting, only six new cases of typhus were detected, the last being on July 29. It was found that the elimination of lice from a village and the maintenance of a louse-free population was far more difficult than the temporary delousing necessary to stop a typhus epidemic.

A nationwide campaign was initiated in Mexico in 1951 to eradicate the typhus vector, *P. h. corporis*. A 10 per cent DDT dust was used alternately with other insecticide powder in order to prevent the development of resistant lice. The DDT was sold in envelopes containing 15 g of the powder, which was sufficient to delouse two articles of underwear. The price was 10 c in Mexican currency, and this allowed for a $3\frac{1}{2}$-c profit which was used for campaign activities. During this campaign it was found that laundry soap containing 5 per cent DDT was very effective in killing lice, and that soap with 3·2 per cent DDT was satisfactory. Soap that was one year old still retained 75% of the original DDT, a fact significant in manufacture and storage.

Mass treatment was carried out in communities which showed an infestation index above 30%, selective treatment of blocks was applied in communities with an index of 10 to 30%, and selective treatment of families was employed in communities with an index below 10%. All articles of clothing coming in contact with the body were treated with DDT, with particular attention being given to seams. Vaseline (petroleum jelly) containing 2 per cent DDT was used for head delousing.

The original infestation rate was 3·1 times that after treatment of the clothes with DDT powder, and 4·5 times that following the use of DDT soap. No reports of typhus incidence following the campaign are available, but, unless an appreciable degree of vector resistance develops, the outcome should be favorable (ORTIZ-MARIOTTE [420]).

There is relatively little information on the control of epidemic typhus in South America by the use of DDT. MONTOYA and OSEJO [389] carried out some very conclusive experiments on the use of DDT in the control of the body louse in two communities of Colombia: Imues and Yucuanquer. In September 1944, a total of 89% of individuals was found to be infested at Imues, while an 81·1 per cent infestation was found at Yucuanquer. The purpose of this study was primarily to determine the minimum frequency of application of DDT necessary to maintain a louse infestation index sufficiently low to render unlikely any appreciable transmission of typhus. A 10 per cent DDT powder in pyrophyllite was used initially; later the diluent used was commercial talc. Hand dusters were used exclusively in applying the powder to individuals, and house-to-house dusting was the technique employed. Reserve clothes, beds, mattresses, and bed clothing likewise were treated. The conclusions reached were that careful application of DDT at intervals of four months to at least 98% of the population should maintain the louse infestation rate at or below 5%. 5% was suggested by these workers to represent a reasonable limit below which a significant typhus outbreak might be precluded. In the community of Yucuanquer an attempt was made to determine the time interval required for a louse infestation to reach its original level after one treatment. Ten months after treatment, the infestation rate was approximately 60% of the pre-treatment index. This is illustrated in Figure 21.

VIEL and ROMERO [604] described the contrast between the ease with which an epidemic of typhus in south Chile was controlled with 10 per cent DDT powder and the elaborate organization needed to deal with an earlier epidemic prior to the availability of DDT.

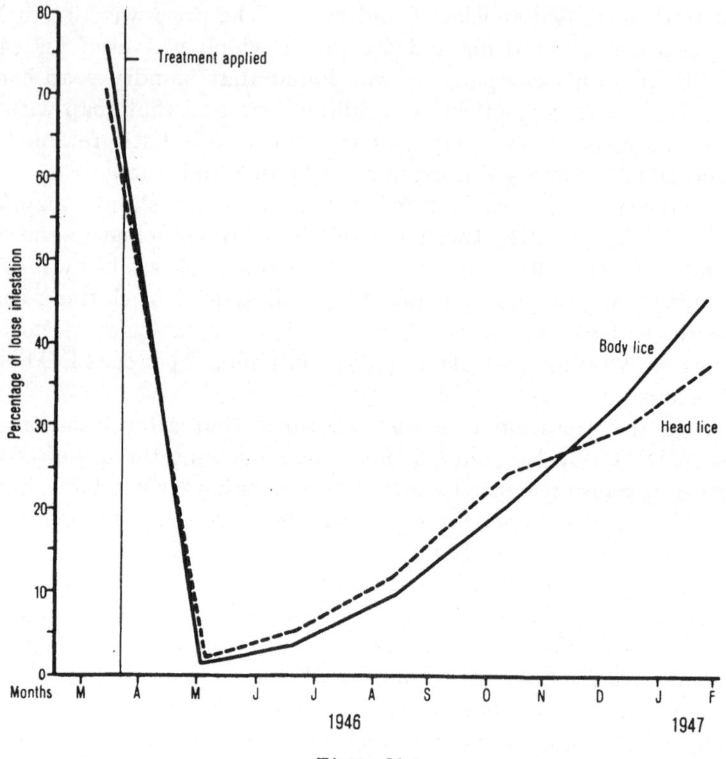

Figure 21

Variation of louse infestation at Yucuanquer, Colombia, after a single treatment with DDT-10 for body lice and phenyl-cellosolve lotion at 5% for head lice.

In 1946 the government of Guatemala, with the help of the Pan American Sanitary Bureau, initiated a nationwide typhus control program based on vaccination and the application of DDT to contacts of diagnosed cases (CABRERA *et al.* [91]). 70% of the population living in endemic zones were vaccinated, and all cases and contacts were dusted.

From June 1946 to December 1951, a total of 194,711 persons were treated with DDT and 1,538,126 articles of clothing were disinfected.

Table 29 shows the decrease in typhus morbidity and mortality over a nine-year period. The rapid decline of the disease following initiation of the control program is apparent.

Table 29
Typhus Incidence and Deaths, Guatemala, 1943–1951

Year	Cases	Deaths
1943	1,338	213
1944	2,144	381
1945	2,834	323
1946	1,043	135
1947	251	37
1948	69	9
1949	26	2
1950	10	2
1951	8	0
Total	7,723	1,102

Louse-Borne Relapsing Fever

As would be expected, the results of using DDT against louse-borne relapsing fever parallel in effectiveness those of epidemic typhus. GREAVES *et al.* [240] reported an interesting outbreak in Tunisia due to *Spirochaeta berbera*. Beginning with six cases in 1943, the epidemic grew to a total of 18,534 in 1944, and in the first four months of 1945 reached 23,221 cases. Mortality rates varied from 1·25% in one small village with 80 cases to 46·5% in another similar situation.

10 per cent DDT dust was used in one isolated Arab village wherein each member of the household was treated along with bedding and extra clothing. Ten days after treatment, neither lice nor nits were found on 75 people examined, and the disease completely disappeared within a fortnight after the treatment. In another test at Ferryville, 76·7% of 851 persons were dusted with DDT and 1,436 in an adjoining area were observed as controls. Table 30 clearly portrays the efficacy of DDT in arresting the epidemic.

The author states that one application of 10 per cent DDT powder, if applied thoroughly to the population, is almost completely effective in stopping a louse-borne epidemic of relapsing fever.

GARNHAM *et al.* [212] describes a serologically confirmed epidemic of louse-borne relapsing fever involving the hinterland of Mombasa and the Kenya Coast of Africa. There were approximately 2,000 cases, with a 40 per cent mortality in untreated cases. Prophylaxis measures included the treating of about 100,000 people with 5 per cent DDT powder, in addition to restrictions on funeral ceremonies and travel. The control measures rapidly terminated the epidemic.

Another account is given by GARNHAM [210] of the suppression of a relapsing fever epidemic on the Kenya Coast by the use of DDT as an anti-louse powder. The results are in general accord with those of other epidemics in that the treatment resulted in the suppression of the disease.

An epidemic of louse-borne relapsing fever in Morocco is described by SICAULT [510]. The dissemination was from Tripolitania beginning early in January 1945 and gradually invading the Protectorate. The number of cases rose from 162 in 1945 to a maximum of 5,027 in January 1946. The recorded cases in 1945 reached the total of 27,780, and there were 13,192 in the first

Table 30

Incidence of Relapsing Fever in the Bellevue Quarter of Ferryville, May 1945[1])

		Total people	No. deloused	Percent deloused	No. huts	No. sick	Percent sick	No. with positive blood smears
May 5–12	DDT	851	653	76·7	147	46	5·41	14
	Control	1,436	none	0·0	129	47	3·27	12
May 17–19	DDT					2	0·23	1
	Control					41	2·85	12
May 26–28	DDT					4	0·48	1
	Control					24	1·67	8

[1]) From GREAVES *et al.* [240].

three months in 1946. Principal prophylaxis measures involved the use of insecticidal powders containing DDT or Gammexane, and these measures were credited with stopping the disease. However, the distribution of food by the Government was credited with helping to check the severity of the outbreak.

In an epidemiological survey for 1946, PETRILLA [440], reporting from Budapest, noted a remarkable increase of typhus and relapsing fever. Outbreaks of relapsing fever occurred in practically the same areas as typhus. The incidence decreased with the extensive use of DDT powder, the establishment of new delousing stations, and general improvement in public health conditions.

BODMAN and STEWART [58] reported an epidemic of louse-borne relapsing fever in Abadan, Persia, which occurred between November 1945 and June 1946. The authors stated that the epidemic followed closely the degree of coldness of the weather until DDT was used in January, which caused a drop in the incidence of the disease. Lice were collected from 87·8% of the patients and *S. recurrentis* was isolated from the specimens.

Trench Fever

Trench fever, a rickettsial disease caused by *Rickettsia quintana*, was unknown until 1915, but during World War I it involved at least 1,000,000 men. It caused the loss of more man-days in the Armed Forces than did any other illness with the exception of influenza. Trench fever was believed to have come from Russia and is known to have occurred in England, France, Flanders, Salonica, Mesopotamia, Italy, Germany, and Austria. The disease seemed to have disappeared in the years between the two World Wars, but it was again recognized in epidemic form on the eastern European front in World War II, particularly in Jugoslavia and the Ukraine (WARREN [618]).

Trench fever is transmitted by the bite of the body louse, *Pediculus humanus corporis*. Once this insect has been infected, it continues to secrete the *Rickettsia* for the rest of its life but is not apparently affected itself by this agent.

Although to date there have apparently been no anti-louse campaigns directed specifically at the control of trench fever, any insecticide, such as DDT, which is effective against lice, in combination with any of several antibiotics useful in the treatment of rickettsial disease, should prevent this infection from ever appearing in epidemic form again.

5.

FLEA-BORNE DISEASES

Plague

Out of the mist of antiquity, plague emerged as a fully developed disease capable of striking terror and devastation among a population of defenseless people. In I Samuel IV, we find reference to the punishment of the Philistines after they had defeated the Hebrew Army: 'And He smote the men of the city, both small and great, and they had emerods in their secret parts', and 'the hand of God was very heavy there. And the men that died not were smitten with the emerods.' The most common translation is that emerods are 'swellings' or 'rounded eminences', similar to buboes of plague. Various authorities have interpreted this as evidence of an epidemic of bubonic plague.

One of the great pandemics of history was that of Justinian which began, apparently in Egypt, in 540 A. D. and raged until 590, devastating most of the civilized world. Towns were abandoned, food left unharvested, and general confusion reigned. As many as 10,000 people died per day in Constantinople alone, and the effect on the entire Roman world was devastating. The role which the pandemic played in the deterioration of the empire has probably been too lightly considered. There seems to be little doubt that the Justinian pandemic was principally bubonic plague.

The Crusades were probably more effectively impeded by epidemics than by armies. The 4th Crusade never reached Jerusalem because of an outbreak of bubonic plague which appeared soon after it left Constantinople.

Probably no event of Medieval history is more universally kown than the 'Black Death'. This pandemic of plague which occurred in 1348–1349 has been variously estimated to have destroyed one quarter of the population of Europe, or approximately 25,000,000 people. At times as many as two thirds of the population were afflicted, with most of them succumbing to the disease. Other epidemics, such as the one in Turkey in 1661 which finally spread to Amsterdam, Brussels, Flanders, and London, have played their part in the shaping of world history. Even the Swedes in 1708 found plague more formidable than the Russian Army and were rendered helpless by the disease (ZINSSER [667]).

The last great plague pandemic apparently originated in China, in Yumnan Province. It reached Canton in 1894, Calcutta and Bombay in 1896, and from there spread to most regions of the world, reaching San Francisco in 1900 (STRONG [561]).

At the present time plague exists in India, Ceylon, Burma, Siam, Indochina, Java, China, Manchuria, parts of Asiatic Russia, Hawaii, Ecuador, Peru, Bolivia, Argentina, Brazil, 15 Western States of the United States, Canada, Germany, the Azores, Madagascar, and many parts of Africa (SIMMONS and HAYES [519]). A great many species of fleas have been found to be either actual or potential vectors of plague. *Xenopsylla cheopis* and *Nosopsyllus fasciatus* occur in practically all places where domestic rats are found, and the former is considered to be the most important transmitter of the disease.

In the United States, 39 fleas have been infected with plague in the laboratory and 29 of these are capable vectors (WAYSON [620] and BURROUGHS [80]). Plague in the pneumonic form is also transmitted directly from man to man.

Essentially the use of DDT in plague control involves about the same techniques as those employed in the control of endemic typhus except that more intensification is usually advisable, and dusting of the population in an epidemic area, as well as buildings, is usually indicated. As soon as a case of plague is diagnosed, the institution of quarantine and DDT treatment should be immediate. Dwelling houses should be dusted, or sprayed with a 5 per cent DDT spray, and all clothing, bedding, and furniture, and all domestic animals should be dusted with a 10 per cent DDT powder. The area beneath the buildings, the attic, or any other place that rats are likely to frequent, and all outbuildings should receive a liberal treatment with 10 per cent DDT dust. The treatment should be applied to all buildings within the immediate vicinity of the foci. The treatment of occupants with 10 per cent DDT dust usually gives reasonable personal protection for at least one week.

If this focal type of operation is not sufficient to stop the disease, then area control must be carried out on much the same basis. Sometimes this can be confined to certain areas of the city where rats and fleas are most abundant; but in the case of smaller cities, it may be advisable to treat the entire population as well as the buildings. To determine the effectiveness of a campaign on the vector, the trapping of rats and the checking of them for fleas is a necessity.

GORDON and KNIES [232] believe that an oriental rat flea index above about 0·2 in a plague area calls for routine protection against human infestation. The index at which the chain of transmission will be broken, however, will vary considerably with reference to local conditions. Needless to say, any DDT control campaign should be conducted simultaneously with adequate quarantine measures, immunization, rat control, and general sanitary improvements. Although there are some instances which indicate that use of DDT, alone, is sufficient to quell an epidemic of human plague, the method should not be used as a substitute for general sanitation, and the employment of every other feasible method at hand in quelling the epidemic is advisable.

Plague appeared in Dakar in April 1944, and in this outbreak DDT was used for the first time to combat the disease. All quarters occupied by United States troops were dusted, as well as personnel and employed natives. In addition, an immunization and quarantine program was put into effect. Chemoprophylaxis was used to some extent, and a general rat clean-up was instituted (Gordon and Knies [232]). DDT dusting initially included all of the territory within a radius of one block of a plague case. It soon became evident, however, that focal dusting was not sufficient, and on October 24 a plan to treat the entire native area, the Medina, was undertaken, the work being completed on November 10. The native quarters were divided into zones and a cordon of gendarmes was placed around each zone to see that the workers were dusted as they left for work. After the working population had been treated, an innercordon was formed, and members of the dusting team completed dusting of the area by treating the natives who remained at home. It was estimated that 95% of the premises were treated by residual spray or powder, and in addition to homes, all public houses were likewise treated. A total of some 125,000 natives were disinsectized with DDT powder and after completion of the program, fixed stations were maintained where voluntary disinsectization was carried out. The campaign did not permit a clear evaluation of the effect of the dusting on the course of the epidemics, since the procedure was not initiated early enough to obtain an adequate prophylactic advantage. The DDT dusting, however, practically eliminated the fleas for several weeks after treatment of the premises. In dwellings where previously 200 to 300 fleas could be collected on a standard piece of fly paper in 5 minutes, the post-dusting examination was negative. Two weeks after completion of the dusting program only 7 houses of 316 examined were found to harbor fleas, and 5 of the 7 houses had been missed during the primary dusting. Prior to treatment, practically all houses were infested. It can be reasonably assumed from the drastic reduction of the vector that DDT had a very favorable effect on the course of the disease, since there was no pneumonic plague detected in the outbreak.

In Casablanca, plague appeared on July 20, 1945. American military authorities initiated control measures which included immunization, quarantine restrictions, and DDT dusting. People in the areas affected were treated with 10 per cent DDT powder; and buildings, vehicles, and ships were treated with DDT powder or DDT in kerosene. A rat campaign was initiated also, and the combined control measures limited the outbreak to three human cases, two natives and one European (Gordon and Knies [232]). It is considered that the attack on the flea is the desirable immediate objective in a plague control program, with rat control being a long-term consideration.

Davis [125] stated that 'localized recrudescenses of plague in Ngamiland were eradicated by dusting 6,000 huts once every four months with DDT'.

Schulz [501] reported on the control of the plague outbreak in Taranto, Italy, in 1945–1946. The principal measure utilized was a large-scale anti-rodent campaign with zinc phosphide and arsenious oxide. However, one of the first precautionary measures taken after appearance of the first case on September 6 was the spraying with DDT of the area where a cargo of imported rags was stored, and which were suspected as being the source of infection. A DDT zone two or three blocks deep, using both solution and powder, was laid down in the area surrounding the source of infection, and the barrier was renewed constantly throughout the campaign. All persons living in the zone were asked to visit dusting stations weekly for treatment. New cases which appeared were isolated, and their houses were sprayed with DDT and cleared of rodents. The last case appeared on November 29, 1945; and a total of 29 cases was recorded, with 15 deaths. Initially the control measures were carried out by the British Army since there was no effective local health organization, but later the responsibility was given to Italians working in collaboration with the United States Relief and Rehabilitation Administration.

VISWANATHAN [611] reported the complete absence of human cases of plague in the Kanara and Dharwar Districts of India following the use of DDT in the malaria control program during 1946–1948. Plague was present in rats, however, and in 1948 the adjoining Mysore State had a severe human epidemic of the disease.

Table 31

Results of DDT Residual Spraying for Plague Control in Bombay State, India

Method employed	Villages treated	Total population	Cases before treatment	Cases in next 3 days	Cases in days 4–10	Cases after 10 days
I	81	71,053	588	45	19	9
II	18	14,348	91	7	2	0
III	7	24,642	77	2	2	2

PATEL and RODDA [429] reported on the efficacy of DDT residual spraying in controlling plague in Bombay State. Three different methods were used in plague-infected villages: (1) all houses were sprayed, using a watery suspension or oil solution of DDT at the rate of 65–75 mg/ft^2; (2) in addition to the residual spraying, all rat burrows were dusted with 10 per cent DDT powder; and (3) only the infected locality was sprayed, and all burrows in the villages were dusted. The preceding table from the work of these authors shows the results obtained.

These workers came to the conclusion that the destruction of rats was not necessary or even advisable since residual spray was considered to be an adequate control measure when used alone. It was suggested, however, that a combination of residual spraying of houses and the dusting of burrows gave added protection.

One of the most dramatic instances of plague control with DDT is reported by POLLOCK [452] on plague in Haifa, Palestine. The first case of bubonic plague was admitted to the hospital in Haifa on June 26, 1947, and the second case on July 1. A program was started immediately which had as its first objective a breaking of the rat-flea-man chain by the use of DDT. Immediate measures were also directed against rats, and long-term measures were planned. However, the dramatic results obtained by the use of DDT were such that further control measures were not essential for arresting the epidemic. Conditions were especially favorable for an explosive epidemic, and 14 confirmed cases were isolated during the first eight days of July. The affected area was divided into sectors and each sector was treated with 5 per cent DDT residual spray in order to create a barrier between rat and man. The actual spraying started on July 2, the day after reporting of the second case.

Residual DDT treatments were applied to the houses and working areas of each suspected case, and to an ever-increasing circle around such localities. All residents were advised to use 10 per cent DDT dusting powder for their persons and a 5 per cent DDT residual spray in their homes. On July 4, dusting centers were set up in the areas and individuals were dusted with power-driven compressor blowers. These centers dusted 30,000 persons in the first 10 days. Even though the epizootic among rats was unabated, there was a sudden cessation of human cases seven days after initiation of the extensive DDT campaign, and the index of *X. cheopis* dropped from approximately 3 to less than 1 per rat within a three-day period.

In the early days of the campaign, only a limited amount of vaccine was available. This was conserved for persons being in direct contact with plague, and for staff members employed on the program. It was concluded, therefore, that DDT was principally responsible for arresting of the epidemic so that only 16 cases occurred in the city. The transporting of grain from Haifa carried the disease to workers of a flour mill 30 km distant, but an extensive DDT dusting and rat extermination program in the mill and the nearby village of Affula ended the outbreak. Further dissemination of the disease to outlying areas was guarded against by treating all lorries with 10 per cent DDT before they were allowed to leave the city. Stevedores and other workers were treated with DDT before they were allowed to board ships. After the completion of the main dusting program, focal dusting was continued in areas where plague-positive rats were found, and the last human case occurred on July 19 in an area which had been missed by the DDT residual spray operations. An attempt to evaluate the prophylactic value of sulphadiazine was precluded by the effectiveness of DDT in arresting the epidemic. It was concluded that the evidence produced from the Haifa campaign would indicate that DDT should be the first line of attack and that rat extermination would take a secondary place and should be planned on a long-term continuous basis.

An outbreak of plague in Calcutta was reported by AHMAD [6] in which disinfestation with DDT was one of the measures used, along with quarantine and immunization. The efficacy of the insecticide in the Calcutta outbreak does not, however, appear to have been clearly evaluated.

In 1949, an epidemic of plague occurred in Shimoga Town 46 miles from Sagar, Mysore, South India. At least 70 cases of plague were reported, with 27 deaths. Vaccination was given to approximately 6,000 people, but it is stated that DDT killed the rat flea, that the epidemic ceased one week after the spraying started, and that no cases of plague have been reported since (WORLD HEALTH ORGANIZATION [658]).

Pollitzer [451], reporting on plague in China, stated that DDT had been found effective for plague prevention. Prophylactic administration of sulpha drugs and the use of rodenticides were other important phases of plague prevention.

Dr. Niyazi Erzin in a private communication to the US Public Health Service reported a successful DDT campaign against plague in Turkey (Simmons and Hayes [519]).

Macchiavello and Mostajo [346] reported an epidemic of plague in Huacho City and Port in the Huaura River Valley of Peru, a country which according to Hoekenga [274] has had 21,727 cases since 1903 and 690 cases since 1940. Huacho has 13,762 inhabitants, and most of the houses are adobe or thatch and mud, and two fifths of them have earth floors.

Before DDT dusting was initiated, a survey in 1945 revealed that 63·7% of 785 *Rattus norvegicus* examined were plague-positive, and that 22 of 54 guinea pigs inoculated with fleas contracted the disease. *X. cheopis* was the predominant vector and averaged 6 per rat and 22 per nest, with a high degree of infection. The epizootic was extensive, but only five human cases of plague were diagnosed.

Control operations began in November 1945 and extended until April 1946, with partial control to July 1947. The first treatment was with 10 per cent DDT and involved the dusting of 90 blocks and 2,055 premises, using an average of 305 g of DDT per premises. A total of 2,858 rat burrows and 6,317 other sites frequented by rats were treated with an average of 48 g of DDT per burrow or other site dusted. The second treatment was in January and February 1946, and 640 of the original dusted premises did not need retreatment. The third treatment was applied to only those blocks where the flea index was still high. The campaign comprised 100 days of work, required 1,145 kg of 5 or 10 per cent DDT to make 6,443 applications to 3,800 burrows and 1,400 applications to 10,000 other sites.

Fleas were found on rats in all blocks before treatment but in only 29·4% of the blocks after treatment, flea-infested rats decreased from 100 to 17·6%, and the index of *X. cheopis* per rat dropped from 5·36 to 1·07. The average reduction of *X. cheopis* for the city was 80·5. A similar reduction of fleas from rat burrows also occurred, and the plague infection among rats decreased.

The campaign showed a residual effect which checked the epizootic and epidemic of plague and prevented reinfection of the city from neighboring foci for three years, at the end of which the flea index was at the pretreatment level and the first posttreatment case of plague was reported.

Macchiavello [345] described an epizootic of murine plague followed by a human epidemic beginning the latter part of 1945 in Tumbes, Peru. This city had 10,000 inhabitants. The majority of the houses were constructed of bamboo or of mud and sticks and there was no public water supply or sewerage system. A large population of *Rattus rattus alexandrinus* was present, with a flea index of 4·5 per rat (11·3 in the epizootic zone). Most of the fleas were *X. cheopis*, but *Ctenocephalides canis*, *Pulex irritans*, and *Tunga penetrans* were also common on the floors of houses and in the clothing of plague cases.

In the 21 epizootic foci, 27·3% of the rats were infected, as were 56% of the fleas found on rats or in their nests. There was a total of 40 human cases, all of the bubonic type, with a fatality of 35%. The only control method was the use of DDT dusting followed by 1080 (sodium fluoroacetate) poisoning of rats at the time of the third dust application. The DDT was diluted to 10% with pyrophyllite and to 5 and 2% with wheat flour.

The first application was from November 30 to December 10, 1945, and dwelling houses, churches, schools, theaters, warehouses, and other places of public business were treated. An average of 2·5 g of 10 per cent DDT was used per square meter treated. A 10 per cent DDT powder was applied with mechanical dusters to the floors, clothing, household goods, and furniture of 30 plague patients, as well as to the clothing of contacts. The epidemic was stopped four days after completion of the first application of DDT; the one case which occurred a month later was due to the incomplete treatment of a known focus. An 82 per cent reduction in the flea infestation of rats was obtained, together with an 88 per cent decrease in the number of fleas found in rat nests. The reduction of rat plague was 76% after this first treatment.

The second treatment was called a 'subsurface application' because it was applied beneath floors, in corridors and passageways used by rats, in spaces between double walls, and in other dead spaces accessible to rats. This treatment was carried out from December 11 to 19 with 5 per cent and 2 per cent DDT. No rat plague could be detected after the second treatment, and the accumulated percentage of control of fleas on rats reached 83·3.

The third application of DDT, in which 5 per cent DDT was used, was made December 20, 1945 to January 19, 1946. The treatment included governmental and public buildings, and dwellings located in the central part of the city. Special attention was given to rat burrows and harborages.

A rat control program using 1080 was begun simultaneously with the third application of DDT, which was after the epidemic of typhus had been broken. The 1080 was used both in water and in bait and good results were reported. The author concluded that 'the application of DDT, followed by poisoning with 1080, promises to be the procedure of choice in the control of epidemics of bubonic plague' (MACCHIAVELLO [345]).

During 1946, MACCHIAVELLO [344] conducted control studies on ten farms in rural Huacho and Huaura Valley in Peru, using 5 or 10 per cent DDT and 1080. On one farm, the El Carmen, there were 17 cases of bubonic plague, and the infection was found in both fleas and rats. One day's work in treating the farm with 10 per cent DDT and 1080 was sufficient to completely stop the transmission of the disease.

A large plague epizootic was detected on the Hacienda Laredo Farm in February 1947 followed by 18 human cases. This farm had 1,322 houses and 6,459 inhabitants. The place was infested with *Rattus frujivorus*, *R. alexandrinus*, and *R. rattus*. The houses also were infested with *Mus musculus* and the nearby fields with *Sigmodon peruanus* and *Oryzomys* species. *X. cheopis* and *L. segnis* were common on *R. rattus* and *Echidnophaga gallinacea* on

R. frujivorus. A dense rat population occurred along the banks of a stream dividing the farm houses from the industrial factory. Control measures consisted of the application of 5 per cent DDT to buildings and to rat burrows along the river. An average reduction in the rat flea index of 88·68% was obtained. The reduction was only 45% on *R. frujivorus* and in all instances was higher on domestic than on field rats. A marked rise in the flea index was observed simultaneously in untreated areas. Although 45·7% of the pools of rat viscera and 27 lots of fleas were positive for plague, the epidemic completely disappeared after the application of DDT. During the year following treatment, another epizootic started which was attributed to the difficulty in control of rats, and their parasites, breeding along the river banks. Good distribution of DDT in rat galleries in the river banks could not be obtained and plague-infected fleas were found after treatment. A new application of DDT and 1080, however, completely controlled the epizootic.

In the village of Hualmay near Huacho, plague occurred in several foci and attempts at control were futile until DDT was applied, after which plague disappeared.

MACCHIAVELLO [345] also reported the results of experiments to control sylvatic plague with 10 per cent DDT along the Peruvian-Ecuadorian border. The test was carried out on a mountain where tree squirrels and several species of wild rodents were found to be infected with plague. The results of the tests were inconclusive since there was no appreciable reduction in the flea population. This was attributed to the difficulty and almost impossibility of adequately treating the squirrel nests and rodent burrows in the area.

In 1946 and 1947, an outbreak of sylvatic plague occurred in the mountainous regions of Huancabamba. Several hundred cases of plague were reported from more than 30 villages in the area. Wild guinea pigs, rabbits, and wild rodents, particularly *Akodon mollis orophilus* and *Oryzomys longicaudatus stolzmanni*, were extensively infected and fleas of the genera *Polygenis* and *Pleochaetis* were involved as vectors. The human fatality rate in some of the villages was as high as 90%, and burning of huts and migration of the population from the plagued area resulted in general confusion. DDT dust was applied only to dwellings in the infected area and resulted in the complete control of the epidemic, although the field epizootic continued in full force. MACCHIAVELLO points out that residual insecticides and rodenticides cannot be applied for the control of sylvatic plague due to the extended area of hundreds of square kilometres over which the epizootic may occur.

DDT has been used in Ecuador since 1946 with excellent results in the control of rural plague (SAENZ-VERA [496]). DDT dusting is carried out every four months in areas considered to be foci of plague. The principal species of fleas involved are: *P. irritans, Nosopsyllus londiniensis, X. cheopis*, and various species of the genus *Polygenis*. The thorough treatment of breeding places of rodents is considered to be essential, and this consists of blowing DDT powder under pressure into all rodent burrows. This procedure has resulted in a noticeable decline in the human plague morbidity curve. Since 1950 it has been noted that a number of fleas survive treatment in those areas which have been systematically treated with DDT for several years. It has been concluded that DDT-resistant fleas have developed, and this has recently jeopardized the plague work in Ecuador. Furthermore, the phenomenon creates grave concern regarding the possibilities of plague control in other countries with residual insecticides, and this should be a stimulus to further research on the problem.

The WHO Expert Committee on plague, at its first session in Geneva, Switzerland, September 19–24, 1949, considered that the protection of port cities from the introduction of plague was an important international problem. They recommended that the following measures be carried out by the various national health-administrations, which it was thought should maintain permanent organizations for the purpose (WORLD HEALTH ORGANIZATION [664]):

'6.1 *Measures of Protection for Cities, Sea- and Airports*

(i) Application of 5 per cent DDT dusting powder every six months or at intervals compatible with the maintenance of a flea index under 2;

(ii) systematic deratting with sodium fluoracetate (compound 1080) or other methods of rodent destruction;

(iii) protection of merchandise with 5 per cent DDT dusting powder or other effective insecticides;

(iv) application of 5 per cent DDT dusting powder to vehicles, in case of epizootic recrudescence within an enzootic area;

(v) rat-proofing of buildings and outbuildings.

Additional measures appropriate for airports include:

(i) The maintenance of a clear zone within a radius of 200 m around the airport buildings and the ground used for the parking of aircraft;

(ii) application of DDT to aircraft;

(iii) application of DDT before loading to merchandise coming from enzootic zones and which, in the judgment of the local health-authorities, might contain infected fleas;

(iv) inspection of aircraft in order to avoid transportation of rodents;

(v) in case of an epidemic in the enzootic zone: application of 5 per cent DDT dusting powder to the garments and personal effects of the passengers coming from the infected zone, at port of departure.'

Murine or Endemic Typhus Fever

Murine typhus is a rickettsial disease of rats transmitted to man by the oriental rat flea, *Xenopsylla cheopis*. There has been much discussion on the relationship of the etiological agent, *Rickettsia typhi*, to that of epidemic typhus, but the two organisms apparently are serologically distinct.

The disease was first reported in America from Atlanta, Georgia, by PAULLIN [431], but it was not until 1931 that its transmission was elucidated.

On the basis of work in Montgomery, Alabama, and Savannah, Georgia, MAXCY [367] suggested rodents as a reservoir of the disease and stated that fleas, mites, or ticks might possibly be parasitic intermediaries.

DYER et al. [162] inoculated guinea pigs with fleas, X. cheopis and Nosopsyllus fasciatus, collected at a typhus focus in Baltimore, Maryland, and produced the disease. Monkeys and rabbits developed agglutinins for Bacillus proteus X$_{19}$

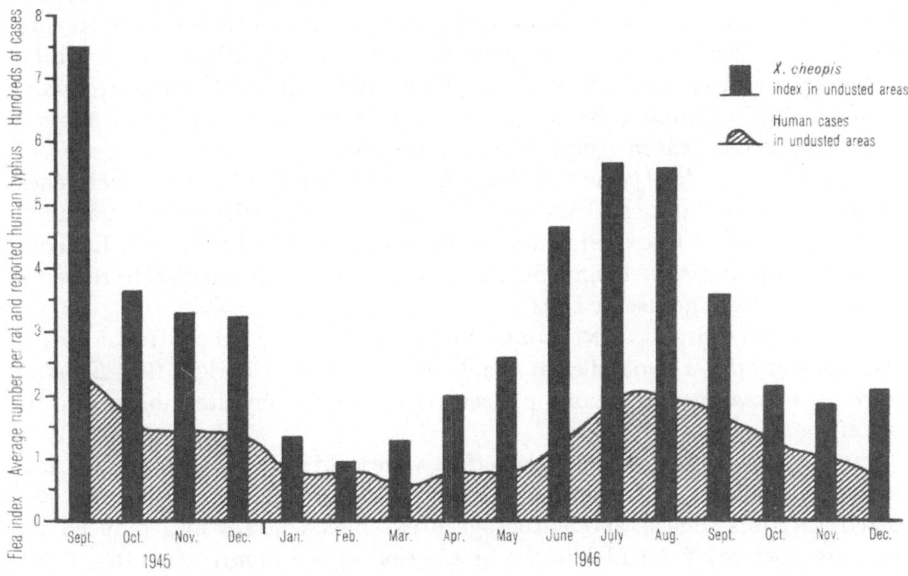

Figure 22

Correlation of murine typhus incidence in humans with *Xenopsylla cheopis* indices. Nine southeastern States.

(type 0) following inoculation with this type of virus. DYER and co-workers, as well as others, later confirmed the transmission of murine typhus by *X. cheopis*. Epidemiological and entomological studies by various workers of the US Public Health Service [595] further demonstrated a definite correlation of human murine typhus incidence and *X. cheopis* seasonal abundance (Figure 22). This correlation did not exist for seven additional rat parasite studies.

Other fleas, particularly *N. fasciatus* and *Leptopsylla segnis*, may rarely transmit the disease to man; and these fleas, as well as other ectoparasites, including various species of mites and the rat louse, are probably instrumental in transmitting the disease from rat to rat.

Since murine typhus is essentially an insect-borne disease, an attack on the vectors immediately becomes a first line of defense, and the advent of DDT permitted initiation of successful control activities. Typhus control work directed toward a reduction of the rat population had been practiced in endemic areas prior to the establishment of vector control procedures. Such campaigns did much to reduce the incidence of the disease but did not alone give satisfactory results.

Soon after the insecticidal properties of DDT were discovered, the material was tested against insects of importance in the transmission of human disease. DOMENJOZ [147] reported on tests conducted in 1941–1942 against the human flea, *Pulex irritans*, and the rat flea, *N. fasciatus*, at J. R. Geigy S.A., Basel, Switzerland. MOOSER [390] reported that mice dusted once with 5 per cent DDT (Neocid) remained free from fleas during observation periods of one to three weeks. On the basis of this work, he proposed the use of DDT in the control of both plague and typhus. 'Neocid', the first 5 per cent DDT powder prepared for the control of human parasites, was made available to the Swiss Army in 1942 and was used extensively, beginning in 1943.

MACCHIAVELLO [344] conducted laboratory tests in 1943 on the effectiveness of DDT (Gerasol 3 per cent) at the National Institute of Hygiene, Guayaquil, Ecuador. He refers to experiments conducted in 1943 in Riobamba, Ecuador, by the Ecuadorian Anti-Plague Service in which the rat flea index decreased to almost zero after the use of DDT.

In the United States, LINDQUIST *et al.* [320] reported good control of the dog flea, *Ctenocephalides canis*, the cat flea, *C. felis*, and the sticktight flea, *Echidnophaga gallinacea*, by the use of 5 per cent DDT in pyrophyllite rubbed into the hair of the animal.

DAVIS [124] demonstrated the effectiveness of DDT powder of various strengths in killing *X. cheopis* on treated rats in the laboratory. He also dusted rat runs, holes, and harborage areas in six stores and reduced the flea index per rat from 13·9 to 0·6 in a period of one month after treatment. Two of the stores had an index of 0·2 and 0·5 fleas per rat at four months after dusting.

After the effectiveness of DDT in controlling rat fleas had been established, control operations were gradually initiated throughout the major endemic typhus areas in the southern United States. This program was conducted by the Communicable Disease Center of the US Public Health Service in cooperation with the various State health departments concerned. After selection of the areas to be treated, rats were trapped and combed to determine their ectoparasite index, particularly the index of *X. cheopis*, the vector of the disease to man. Dusting was then initiated by treating all rat runways, harborage areas, and nesting sites with 10 per cent DDT in pyrophyllite. Particular attention was paid to rat nests, where rat fleas were found in abundance. Dust which is picked up by rats is carried back to harborage areas, and if holes and burrows are thoroughly dusted, rats may obtain DDT over their entire bodies. In runways, most of the powder is carried back on their feet and tails, but preening activities of rats transfer the DDT from their extremities to their fur where it can be contacted by the fleas. The amount of dust used per premises varies considerably. It may be as little as $^1/_4$ lb or as much as 15 lb or more, depending not only upon the size of the premises but upon the intensity of the infestation and the number of runways, holes, and harborage areas to be treated. In the southeastern United States an average of 5 lb of 10 per cent dust was used on rural premises and 2 to 3 lb on urban premises. The larger quantity used on rural premises was due to the treatment of outbuildings.

Figure 23

Foot pump duster for the treatment of rat harborages with DDT.

The selection of proper equipment for DDT dusting is important. A foot-pump duster (Figure 23) is particularly suitable for the blowing of DDT dust into burrows, between double walls and floors, and into other inaccessible places which might harbor rats. The thorough treatment of such suspected infested areas cannot be overemphasized. A hand-shaker duster was used by LUDWIG and NICHOLSON [340] to apply dust along runways. This proved to be very satisfactory for laying down an even layer of dust without billowing. These workers also devised a smaller hand duster fitted with an extension handle for dusting inaccessible places such as behind crates and overhead runways along beams and wall plates, which must be treated in areas where the roof rat,

                    S. W. Simmons

Rattus rattus, is present. These shakers are illustrated in Figure 24. Some workers have used only their hands for distributing the dust from receptacles which they carried.

Before considering results from the larger program, two smaller constituent programs deserve special mention, due to the unduly thorough study devoted to them and to the clear-cut results obtained.

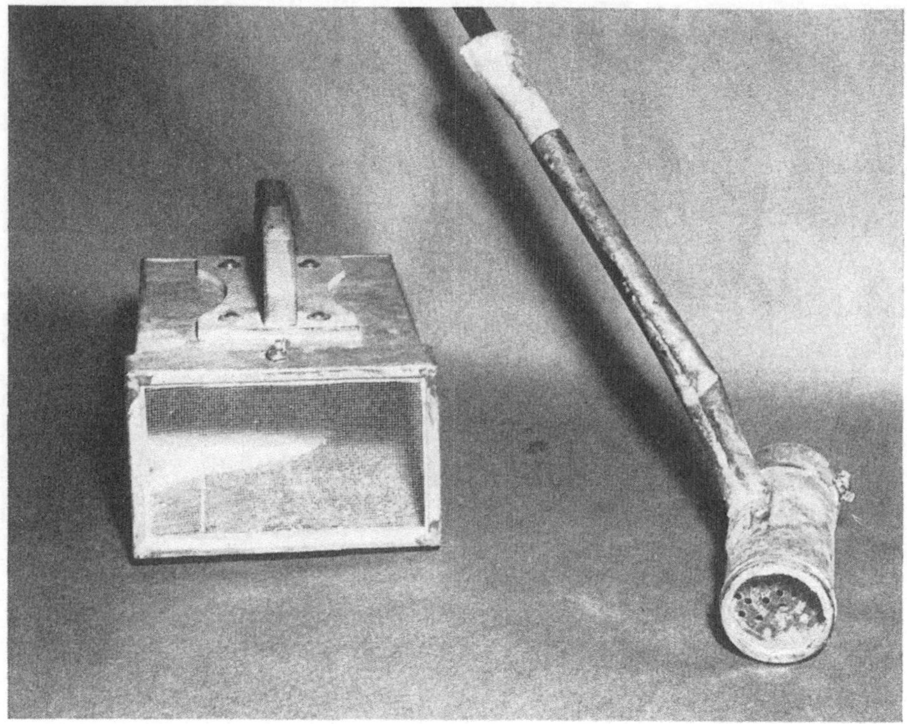

Figure 24

Hand dusters (shakers) for the treatment of overhead beams and rat runways with DDT.

During the five-year period ending in 1943, Chatham County, Georgia, had the highest human incidence of typhus fever of any county in the United States (SIMMONS and UPHOLT [520]). This trend continued in 1944 and 1945 with 129 and 132 cases, respectively. Most of the cases reported were from Savannah, a city of 150,000 people, the only large city in the county. On October 1, 1945, a DDT typhus control program was inaugurated which consisted of applying 10 per cent DDT in pyrophyllite to rat runways and harborage areas, as previously described. The rat population consisted almost entirely of *R. norvegicus*.

The ectoparasite population, especially *X. cheopis*, showed an almost instantaneous decrease, followed by a comparable reduction in human typhus cases. Table 32 shows the human typhus incidence in Savannah for the eight-year period 1944–1952. This appears to be a striking example of the efficacy of DDT dusting in the control of flea-borne typhus.

Although the control program did not begin until October 1945, a year in which there were 132 cases, it resulted in a reduction to 15 cases in 1946. This decline continued, with six cases in 1947, four in 1948, two in 1949, and none since that date. The control program was terminated in June 1950, but no human cases have been reported since that time.

Table 32

Human Typhus Cases in Savannah, Georgia, 1944–1952

Year	1944	1945	1946	1947	1948	1949	1950	1951	1952 (to July)
Typhus Cases No.	129	132	15	6	4	2	0	0	0

Table 33

Murine Typhus Fever Morbidity Rate, Prevalence of Antibodies in Domestic Rat Reservoir and Ectoparasite Abundance Prior to May, 1946

County	Human morbidity rate per 100,000 1945	Typhus in rats positive by complement fixation		Rats infested with *X. cheopis*	
		Rats examined No.	Positive %	Rats examined No.	Infested %
Grady [1])	274	241	46·5	260	45·4
Thomas [2])	202	454	53·7	584	38·2
Brooks [2])	218	434	48·4	657	31·5

[1]) Check County.
[2]) Selected for treatment with 10 per cent DDT in pyrophyllite.

Beginning in April 1946, an even more intensive study of the effects of DDT dusting was conducted in a three-county area of southeastern Georgia (HILL *et al.* [269]; HILL and MORLAN [268]). Grady County was used as an untreated check area, while Thomas and Brooks Counties were treated with 10 per cent DDT. Five rounds of dusting were completed in the period April 1, 1946 to September 30, 1947, with 60 to 96% of the premises receiving treatment.

Table 33 shows pretreatment data on typhus and ectoparasite prevalence in the study area.

Figure 25 demonstrates the reduction of rats with *X. cheopis* in the three counties, based on examination of 7,447 rats from May 1946 to November 1949.

S. W. Simmons

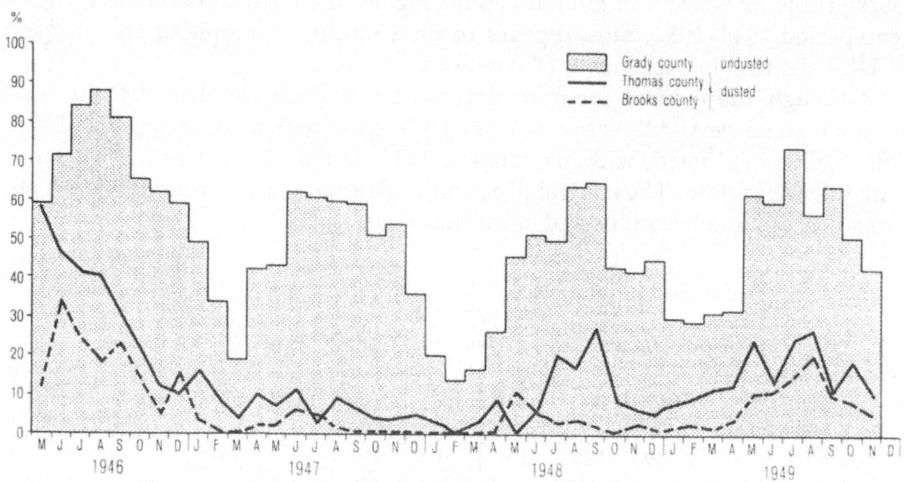

Figure 25

Percentage of commensal rats infested with *X. cheopis.*

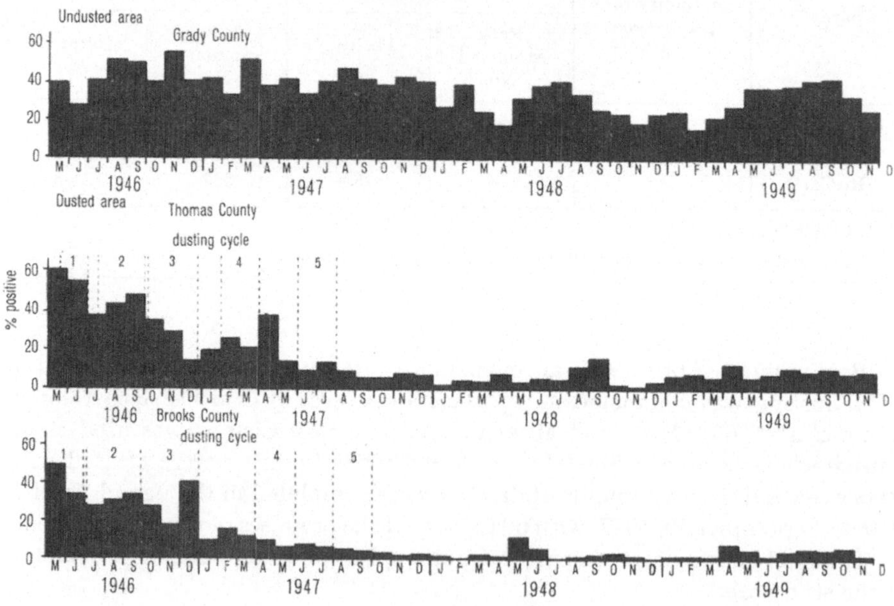

Figure 26

Percentage of rats positive to the murine typhus complement fixation test.

In the two treated counties, the percentage of rats with infestations of *X. cheopis* dropped from 28·4 to 5·2 and from 13·2 to 1·8, respectively, during the second year. There was also a drop in the check county from 60·7 to 41·4, but by 1949 the check county showed 59·0% of the rats examined to be infested, while the treated counties showed 19·1 and 11·6, respectively.

Coincident with the reduction of *X. cheopis*, there was a decrease in the percentage of rats showing positive complement fixation for typhus (Figure 26).

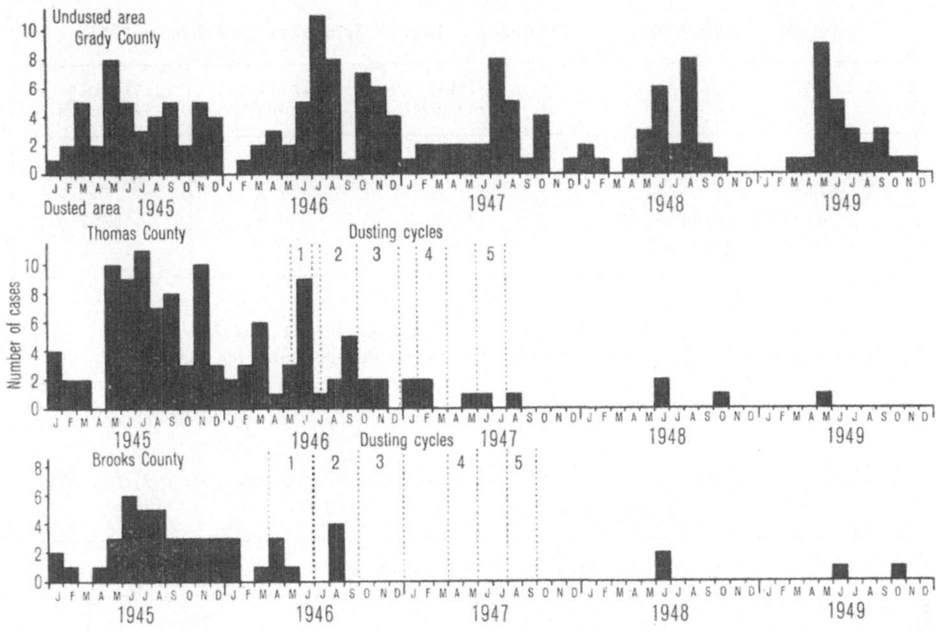

Figure 27

Incidence of human typhus fever by month of onset in the undusted area, Grady County, and in the dusted area, Thomas and Brooks Counties.

As would be expected, human typhus incidence showed a decided drop following the reduction of the vector and of the typhus incidence in rats (Figure 27).

Although the last dusting was completed in 1947, even in 1951 fewer rats from the treated counties were infested, and these showed a significantly lower *X. cheopis* index, as well as a lower infection rate, than rats from the check county (Tables 34 and 35).

In areas dusted prior to September 1947, *X. cheopis* abundance and the incidence of typhus antibodies in rats followed a downward trend until February 1948. The trend was gradually upward from March 1948 to August 1950 (MORLAN and HINES [393]).

This intensive study definitely showed that thorough DDT dusting would reduce significantly the prevalence of the vectors of endemic typhus, the prevalence of murine typhus complement fixing antibodies in rats, and the incidence of the disease in man. This reduction, moreover, has endured in large measure for at least five years following termination of the program.

The most extensive study of DDT dusting for the control of murine typhus was that conducted by the US Public Health Service in cooperation with eleven southeastern States. The program was organized in July 1945 and included those

Table 34

X. cheopis Abundance for All Rats for August, September, and October 1951

	Grady County[1]	Thomas County[2]	Brooks County[3]
Number rats examined	145	119	128
X. cheopis			
Percent of rats infested	55·2	25·2	29·7
Average number per rat	4·8	1·0	1·7

[1]) Untreated check county.
[2]) Last cycle of dusting with 10 per cent DDT completed July 30, 1947.
[3]) Last cycle of dusting with 10 per cent DDT completed September 30, 1947.

Table 35

Rat Typhus Incidence as Indicated by Positive Complement Fixation Test at 1:4 or Higher August, September, and October 1951

Species	Grady County			Thomas County			Brooks County		
	No. ex.	No. pos.	% pos.	No. ex.	No. pos.	% pos.	No. ex.	No. pos.	% pos.
R. rattus	40	7	17·5	22	0	0	76	0	0
R. norvegicus	105	27	25·7	90	1	1·1	43	0	0
Combined species	145	34	23·4	112	1	0·9	119	0	0

counties averaging 10 or more cases of human typhus per year for the five-year period, 1940–1944.

By March 1946 the program was essentially in full operation, involving 122 counties in the nine States of Alabama, Florida, Georgia, Louisiana, Mississippi, North Carolina, South Carolina, Tennessee, and Texas. Later expansion increased the number of counties to 188, including minor operations in Arkansas, California, and Virginia. Since 1932 about 97% of reported typhus cases in the United States occurred in the nine principal states, and in 1945, 81·7% occurred in the 188 counties (WILEY and FRITZ [627]).

Dusting was mainly with 10 per cent DDT in pyrophyllite, although in some test areas 5 per cent dust was employed, but was less effective in the control of *X. cheopis*. In 1945–1946, two, three, or even four dustings were made; but in succeeding years, one dusting per year was the general rule, although some counties received dust only on alternate years, and in a few instances even less frequently.

During the period, 1946–1949, a total of 1,815,266 premises were dusted utilizing 5,311,000 lb of 10 per cent DDT dust. During the same period, 20,199 premises were ratproofed and 918,912 premises were poisoned for rats.

Table 36

Correlation of Time of Dusting (10 per Cent DDT), with Duration and Degree of Control of the Oriental Rat Flea, X. cheopis, During the First 12 Months after Dusting in 11 Southern States, 1946–1949

February	March	April	May	June	July	August	September	October	November	December	January	February	March	April	May	June	July
D	e	e	e	e	s	s	s	s	e	e	e	e					
	D	e	e	e	s	s	s	s	e	e	e	e	e				
		D	e	e	e	e	e	e	e	e	e	e	e	e			
			D	e	e	e	e	e	e	e	e	e	e	e	e		
				D	e	e	e	e	e	e	e	e	e	e	s	u	
					D	e	e	e	e	e	e	e	e	e	s	u	u
						D	e	e	e	e	e	e	e	e	s	u	u
							D	e	e	e	e	e	e	e	s	u	u
								D	e	e	e	e	e	e	s	u	u
									D	e	e	e	e	e	s	u	u
										D	e	e	e	e	s	s	u
											D	e	e	e	e	s	u

D Month dust applied; *e* excellent control — 0 to 0·5 flea per rat; *s* satisfactory control — 0·6 to 1 flea per rat; *u* unsatisfactory results — over 1 flea per rat.

In the control area, *X. cheopis* was four times as abundant as *L. segnis* and *N. fasciatus* combined, and it was shown that dust should be applied in February through May, inclusive, in order to obtain satisfactory control of this species over a 12-month period (US PUBLIC HEALTH SERVICE [592]). When dust was applied earlier, it did not satisfactorily suppress the peak of *X. cheopis* during the following summer, and if applied late in the spring or summer, the carry-over to the next summer was not effective. Table 36 is a correlation of DDT dusting with duration and degree of control of *X. cheopis*, and Figure 28 illustrates actual results obtained over a 12-month period of dusting operations.

This information, although based on a specific vector in a restricted locality, may be a basis for arriving at the appropriate dusting cycles of other flea vectors in different areas. Although an *X. cheopis* index of over 1 per rat was considered

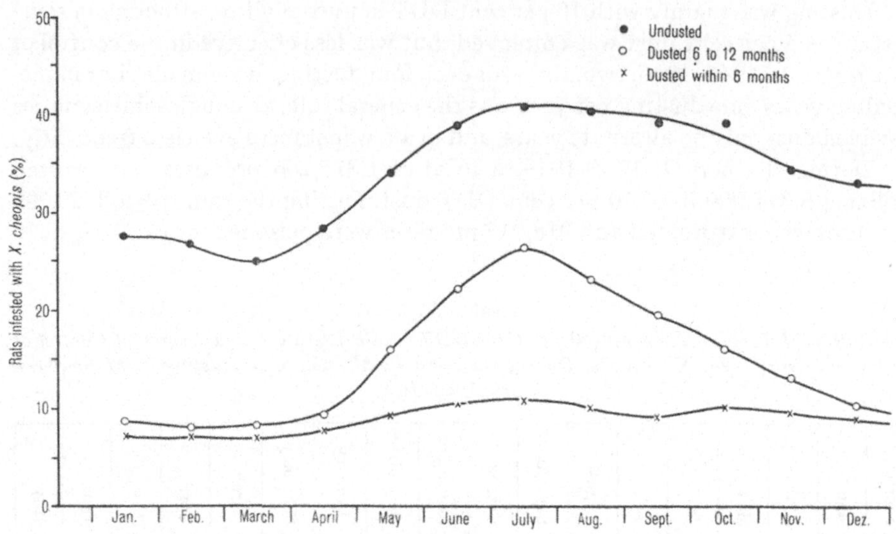

Figure 28

Duration of control of the rat flea *X. cheopis* by 10 per cent DDT dusting in eleven southern States.

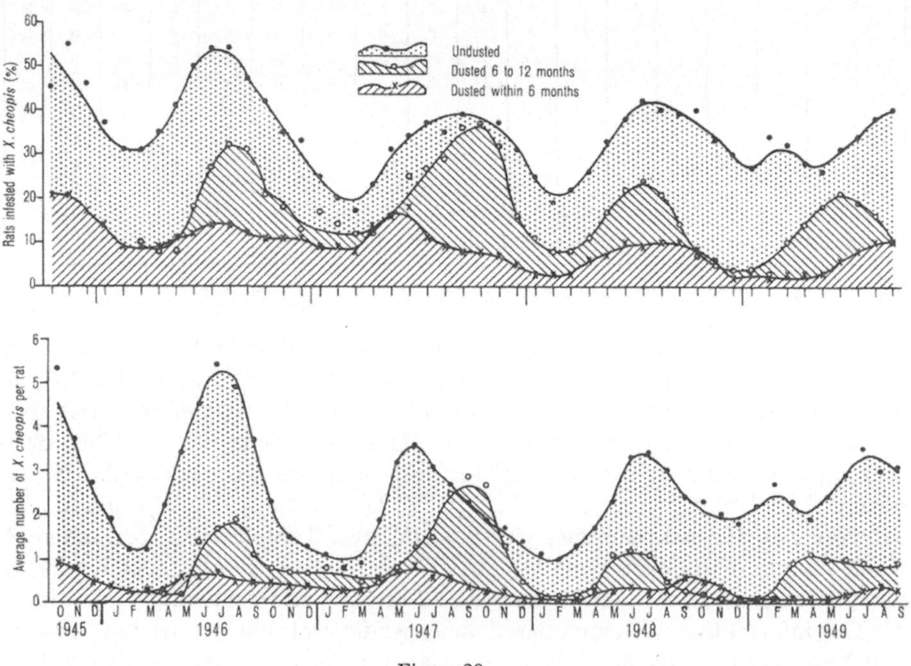

Figure 29

Control of the rat flea *X. cheopis* by 10 per cent DDT dusting in nine southern States.

unsatisfactory control in this program, this is not necessarily true under all conditions. A higher index might be tolerated in areas of very low typhus endemicity in both rats and humans, and in areas of low rat populations. Since considerable control is often present after an index of 1 flea per rat is reached, satisfactory disease control might continue for many months in such areas.

The control of *X. cheopis* from the beginning of the program in 1945 to September of 1949 was reported by GOOD [230] (Figure 29). Control of *X. cheopis* averaged 84, 79, 87, and 94%, respectively, for the first six months after

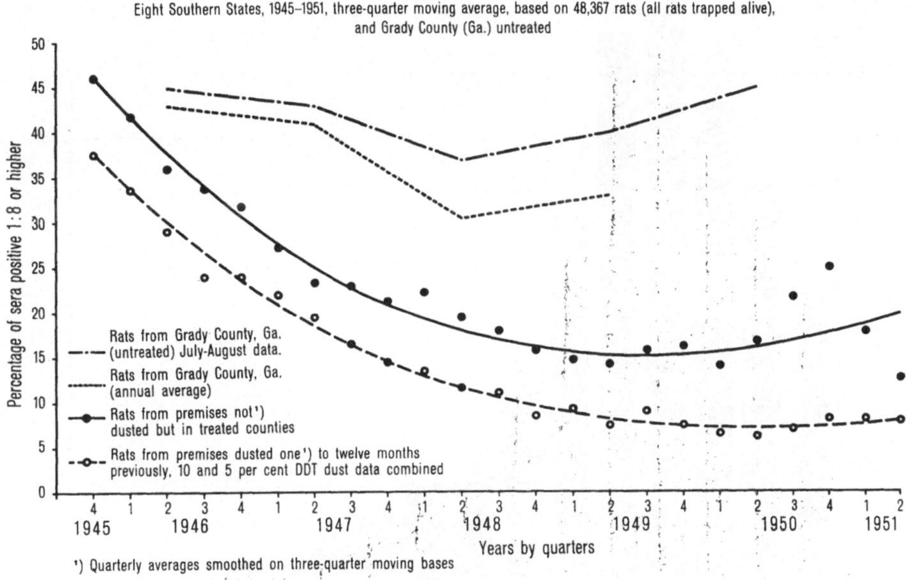

Figure 30

Percentage of rats with murine typhus fever antibodies.

dusting in 1946–1949. This excellent initial control is very significant, especially in a high endemic area. There was little change in the effectiveness of the 10 per cent DDT for the first four months after dusting, and control was entirely satisfactory for at least six months, and in many instances for a year or longer.

N. fasciatus is very susceptible to DDT, but *L. segnis* is more resistant. *E. gallinacea* is very difficult to control, due probably to the fact that they stick very closely to the skin where DDT may not reach them. Control of the mites *Liponyssus bacoti* and *Echinolaelaps echidninus* was very poor, but *Lealaps nuttali* proved very susceptible. Little or no control was achieved of the rat louse *Polyplax spinulosa*.

As would be expected, there was a reduction in rat and human typhus following control of the vector, *X. cheopis*; however, the reduction in rat in-

fections was less spectacular than in humans. Figure 30 illustrates the effect
of various DDT dusting procedures on typhus antibody prevalence in rats for
the period 1945–1951. There are at least three, and perhaps more, plausible reasons for the relatively slow reduction of typhus in rats: (1) The rat, once infected, maintains a positive serological reaction throughout life; therefore, an
appreciable decrease in positive rats would not occur soon after dusting. Young
rats born subsequent to dusting give, therefore, a truer picture of transmission
incidence; (2) rat parasites, such as the rat louse or mites, which are not readily

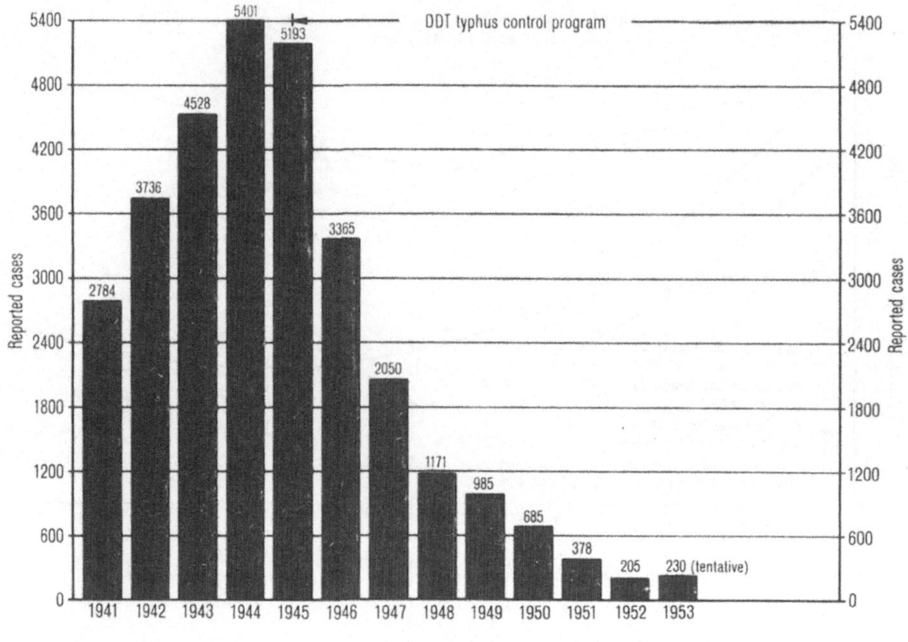

Figure 31

Annual totals of reported murine typhus fever cases in the United States, 1941–1953.

controlled by DDT, may continue to transmit the disease from rat to rat; (3) the
environment of the rat, such as nests and runways, may harbor viable rickettsia for a considerable period after dusting operations are begun.

Rats taken less than one month subsequent to DDT dusting almost assured-
ly acquire the infection prior to treatment and should not be considered in
evaluating the effects of a program on the incidence of rat typhus. In 1946, the
first full year of the program, 32% of rats examined from nondusted areas were
typhus positive to the murine typhus complement fixation test while from the
treated areas 23% examined during the first six months after dusting were
positive. There was little difference in the prevalence in the two areas in 1946,
but there was a 50 per cent reduction in both 1948 and 1949, when compared to

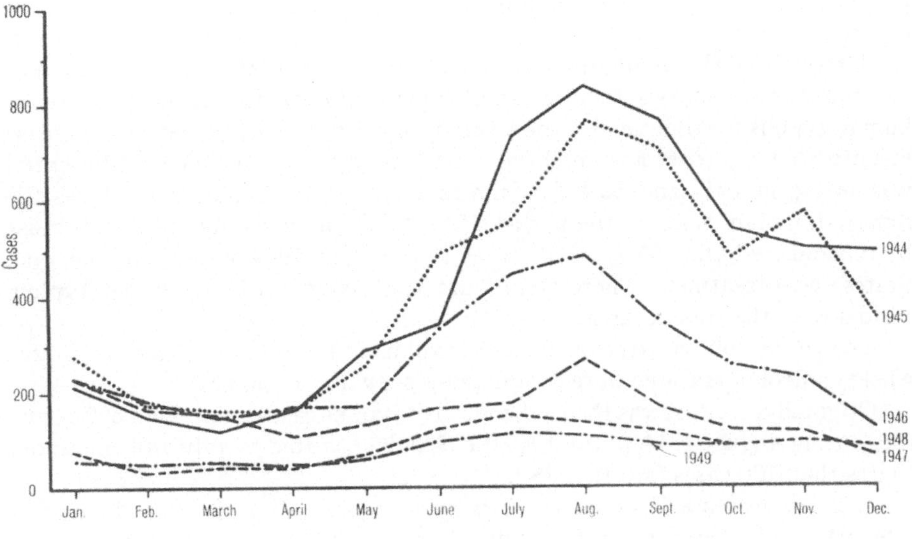

Figure 32

Yearly comparative incidence of typhus fever by months. Nine southeastern States.

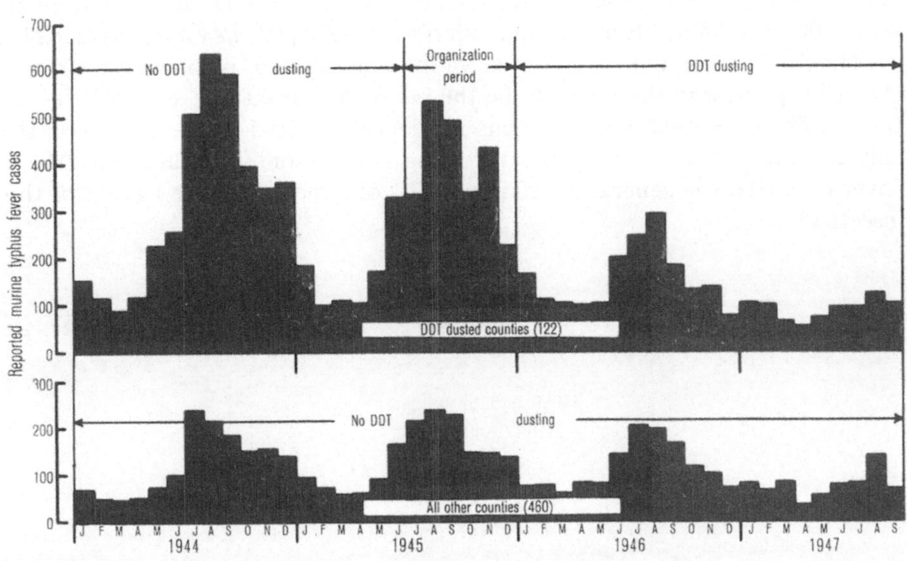

Figure 33

Reported murine typhus fever cases. Nine southeastern States, by months.

pretreatment prevalence. Little difference in prevalence between treated and nontreated areas was evident six months and longer subsequent to treatment.

Although DDT dusting did not begin until late in 1945 and involved a limited number of counties, there was a 10·7 per cent decrease in reported cases of human typhus for that year in the treated counties, and a 14·5 per cent increase in untreated counties; a differential of 25·2% (WILEY [626]). This differential was 44·1% in 1946 and 56·4% in the first half of 1947. Figure 31 shows the reported human cases for the period, 1941–1952, for the United States. It must be remembered that 97% of all cases reported in 1944 were from the nine States under treatment. Therefore, Figure 31 is essentially the picture of typhus incidence in the treated area.

The yearly comparative incidence of typhus in the nine southeastern States where control operations were conducted is shown in Figure 32.

Originally, August was the peak month of typhus cases, more than 800 occurring during August 1944, but in 1949 the peak had shifted to July and amounted to less than 100 cases (MOHR [385]).

A sharp decrease is noted in 1946 following initiation of DDT dusting in late 1945, and a further significant drop occurred in 1947. Since 1947 there has been a gradual but consistent decrease in incidence, to the point where the disease is only a fraction of its former importance.

That DDT dusting was the major factor in the rapid decline of typhus from 1945 to 1947 is well substantiated, even though there was a definite but less marked decrease in cases in undusted areas (Figure 33).

A striking example of the effectiveness of the program was the decrease in typhus incidence in the top 10 counties, all of which had DDT dusting projects, from 1,074 in 1944 to 395 in 1946, a reduction of 64·7% (WILEY and FRITZ [627]).

The ratproofing and poisoning programs contributed materially to the decline of typhus, and these should be the permanent measures adopted. Nevertheless, drastic reductions in typhus incidence occurred where DDT was the only control measure used, and ratproofing and poisoning in the pre-DDT era never produced the generally striking results obtained after the advent of this insecticide.

6.

HEMIPTERA-BORNE DISEASES

Chagas' Disease

Chagas' disease is a trypanosome infection prevalent in man in parts of Brazil, Chile, Venezuela, Argentina, and Central America. The etiologic agent is *Trypanosoma cruzi* and the organism is transmitted from various rodents and other small animals to man by several species of reduviid bugs. The organism has been recovered in several species of Triatomids and in various rodents and other animals of the southwestern United States (FAUST [179]).

LENT and DE OLIVEIRA [319] treated *Rhodnius prolixus, Panstrongylus megistus, Triatoma sordida,* and *T. infestans* with 'Neocid' (5 per cent DDT in talc) in Petri dishes and the adults were killed in 18 to 24 hours. Late larval stages of *T. sordida* were not killed and the eggs apparently were not affected. These workers fed *R. prolixus* on pigeons which several hours previously had been fed capsules containing 0·07–0·50 g of DDT and the bugs were killed in 12 to 20 hours.

An epidemiologic survey made by NEGHME *et al.* [410] revealed a human infection rate of 12 and 14·25%, respectively, in two areas of Chile. An infection rate of 44·1% was found for *T. infestans* and a 10·9 per cent infection rate for 4,006 domestic and wild animals.

Control of the disease was attempted by the spraying of houses in rural zones with a solution of 5 per cent DDT in kerosene. Based on the treatment of 1,132 houses, it was possible to control the breeding of *Triatoma* in 90% of the cases, with a low infestation of inhabitants up to 12 months after treatment. An epidemiological appraisal of the vector control campaign has not been published, but such a drastic reduction in the vector can hardly have but one course on the epidemiological picture.

The efficacy of employing combinations of chlorinated hydrocarbon insecticides against *T. infestans* was investigated by SILVA and NEGHME [511]. The period of exposure necessary to kill the various stages of *Triatoma* was shortened by using combinations of gammexane (gamma isomer of hexachlorocyclohexane) with chlordane, gammexane with DDT, and chlordane with DDT in the respective proportions of 1½ and 2½%. Results were based on a comparison of the killing time with each of the insecticides used alone.

Chagas' disease is an important problem in many parts of Brazil. PINOTTI [447] reported an infection rate in *Triatoma* of 22·9% following a survey in nine States of Brazil. This involved 45,310 examinations of six species of

Triatoma. The highest infection rate, 28·19%, was found in *T. infestans*, with the exception of a 50 per cent infection rate in *T. outros*. In the latter case, however, only six examinations were made. DDT was reported as inefficient against *Triatoma*, and best results were obtained with benzene hexachloride. A combination of BHC and parathion destroyed 90 to 95% of Triatomids following only one house spraying. A critical epidemiological evaluation is not available, but the incidence of acute Chagas' disease diminished remarkably following treatment of the cities of Uberaba and Bambui in the State of Minas Gerais.

FROES [192] reported the successful use of both DDT and gammexane against the Triatomid vectors of Chagas' disease in Brazil. It is probable that his experience was with bugs which had not acquired resistance. It is well known, however, that Triatomids do show a considerable degree of natural resistance to DDT and that this insecticide has not in general proved as effective as certain others, including, especially, benzene hexachloride.

DIAS [144] stated that the campaign against Chagas' disease must be directed at the vector since adequate methods of treatment and vaccination are not known, and since the control of the animal reservoirs would be limited to cats and dogs. DDT has not proved as effective in Brazil against *Triatoma* as against most other arthropod disease vectors. Gammexane has a residual action of about three months, but showed an appreciable decline during the first few weeks. Since the vector is a house dweller, residual treatment should be directed at all holes, cracks, thatched roofs, or other hiding places of *Triatoma*.

VILLELA [605] cited a report by DIAS and collaborators who sprayed 577 houses for the control of *Triatoma* in Brazil. Chlorinated hydrocarbons or thiophosphate compounds were used either alone or in combination with DDT. There was a drastic decrease in the *Triatoma* population immediately after the first application and a progressive decrease up to two months or longer after treatment. Nymphs were practically eliminated from houses which received two applications of spray. In the first eight days after treatment, 32,022 nymphs were recovered, and during the first eight weeks, a total of 55,000 nymphs were captured and died from their prior exposure to the treated surfaces. Similar results were obtained in Bambui in the city of Belo Horizonte. From these experiments, it was concluded that it was possible to obtain simultaneous control of Chagas' disease and malaria by the use of residual insecticides. It was said to be possible to destroy practically all of the insects by repeating the application two, four, or six months after the first one. It was presumed that because of the long life cycle of *Triatoma* that one treatment per year would give effective control of endemic Chagas' disease.

Following DIAS' work, VILLELA working in the States of Minas Gerais and Sao Paulo, Brazil, selected 25 to 30% or about 200,000 of the 710,000 houses in the region to be treated.

According to GALINDO [208], inhabitants of towns in Panama reported that *R. pallescens* virtually disappeared from areas treated with DDT for malaria control. This was stated to be at variance with reports from South America where *R. prolixus* was found to be relatively insusceptible to DDT.

TALICE [573] reported the use of DDT and gammexane in controlling the vectors *T. infestans* and *T. rubrovaria* in Uruguay, but epidemiological data are again lacking.

In Argentina, surveys of school children in specified areas revealed an infection rate of 48·5% among 1,263 serums tested. The infection rate was 40–50% in the vector *T. infestans*. Disinsectization was an important part of the prophylactic program, but DDT was found to be less effective than BHC (ROMAÑA [477]). The odor of BHC was a factor in its favor with the natives who believed it more potent than the comparatively nonodorous DDT.

MONTALVAN [387] made a survey in Ecuador to determine the extent of Chagas' disease and discovered it to be present in several provinces previously considered free from the infection. Only one *Triatoma* in Ecuador, *T. dimidiata*, has been found infected in nature, although six species have been recorded for the country. Contrary to the experience of several other workers, it is reported that DDT proved very effective in reducing the numbers of *Triatoma* in Guayaquil.

A part of the prophylactic program in Chagas' disease certainly should be the disinsectization of houses by the application of DDT, BHC (benzene hexachloride), or other suitable residual sprays. *Triatoma* unquestionably exhibits a comparatively high natural resistance to DDT, and although DDT has been used with some success, it is generally conceded that BHC is superior for the purpose.

7.

TICK-BORNE DISEASES

The importance of ticks as vectors of animal diseases of great economic significance has long been acknowledged. The role of ticks in veterinary medicine and the use of DDT in their control is discussed fully in Chapter VIII (KNIPLING). More recently, the role that ticks play in the transmission of human disease has been increasingly recognized. Among the diseases vectored by these arthropods are Rocky Mountain spotted fever and related rickettsioses, 'Q' fever, Bullis fever, tick-borne relapsing fever, Colorado tick fever, Russian spring-summer encephalitis, louping ill, tularemia, and perhaps, on infrequent occasions, western equine encephalomyelitis and plague (CRAIG and FAUST [119]). A list of the species of ticks which have been found naturally infected with the agents of these diseases is formidable. The virus of Rocky Mountain spotted fever is vectored by 5 species of *Dermacentor* ticks, 4 species of *Amblyomma*, 2 species of *Haemaphysalis*, and 1 species each of *Rhipicephalus*, *Ixodes*, and *Hyalomma*. About a dozen species of *Ornithodoros* have been found responsible for the transmission of tick-borne relapsing fever. 'Q' fever can result from the bite of species of *Dermacentor*, *Amblyomma*, *Ornithodoros*, *Rhipicephalus*, *Ixodes*, *Haemaphysalis*, *Hyalomma*, and *Otobius* ticks. *Dermacentor andersoni* and perhaps *D. variabilis* harbor the agents of Colorado tick fever and, along with certain other *Dermacentor* species, those of tularemia. The lone star tick, *Amblyomma americanum*, transmits Bullis fever. In Siberia, Russian spring-summer encephalitis is transmitted by *Ixodes persulcatus* and louping ill by *I. ricinus*. Louping ill has also been reported in man from other European countries. While there is no infective agent involved in tick paralysis, this disease resulting from the injection of toxin by the tick as it feeds, particularly when it feeds in sites near the base of the brain or the spinal cord, can be fatal. It occurs most commonly in young children, sheep, and dogs, but is becoming increasingly serious as a disease of cattle in British Columbia.

Preventive measures for these diseases should include, in addition to vaccines and chemotherapy, tick control. In 1944, SMITH and GOUCK [529] of the US Department of Agriculture undertook experiments to determine the effectiveness of DDT in tick-infested areas. Roadside plots were treated with DDT in various formulations, and the results were compared with those of adjacent plots which were treated with other compounds or left untreated. All the DDT sprays were equally, or more, effective than the comparison compounds; and by 21 to 22 days following application, the spray containing 0·1 per cent DDT was the most effective, even surpassing a 1 per cent DDT dust. Other studies that same year showed that emulsions containing 5 per cent DDT gave satisfactory control of the brown dog tick, *Rhipicephalus sanguineus*, and the lone star tick, *A. americanum*, when applied to dogs as washes (GOUCK and SMITH [235]). More recent investigations have shown that a 1 per cent water suspension of DDT and a 10 per cent dust prepared with pyrophyllite also give good results (BISHOPP *et al.* [52]). DDT sprays applied to woodland plots at the rate of 1 lb of DDT per acre gave excellent control of lone star tick nymphs and adults (SMITH and GOUCK [528]) as long as 125 days through-out the season of activity (GOUCK and SMITH [234]). DDT sprays and dusts at 9, 7, and 3 lb of DDT per acre controlled the American dog tick, *D. variabilis* (SMITH and GOUCK [528]), with a corresponding reduction in the local Rocky Mountain spotted fever hazard (GLASGOW and COLLINS [229]). At Camp Bullis, Texas, in 1946 and 1947, in order to prevent infections with Bullis fever, DDT was sprayed at the rate of 1 and 2 lb per acre to kill the vector, the lone star tick (McDUFFIE *et al.* [369]). This measure gave good control within 2 to 14 days; the insecticide remained effective for 4 to 6 weeks or more. Forty-three pounds of 10 per cent DDT dust per acre reduced the tick population 76% in one week, 97% in 6 weeks, and 99% in 9 months. Fourteen months after the application there were 30% fewer ticks in the treated areas than in the untreated ones. Airplane applications of DDT in oil solutions or water emulsions at the rate of 2·5 to 5 lb per acre did not give satisfactory results. In this area, chlordane and toxaphene applied as sprays at 1 and 2 lb/acre proved as effective as similar doses of DDT; BHC was less effective. At Kerrville, Texas, however, in 1948, BHC dust applied at the rate of 1 and 2 lb/acre gave about the same lasting control as 4 lb, and was as effective as DDT, chlordane, and toxaphene. Between 90 and 100 per cent control of *D. variabilis* was achieved in Jamestown, Rhode Island, by 50 per cent DDT wettable powder applied at the rate of 4 lb of DDT per acre to a 6 to 8 foot strip bordering several miles of roadside and paths (KNUTSON [315]).

In area control of ticks with DDT, a dosage that will provide several months of control if applied at the time ticks have reached their peak abundance, may provide only a few weeks' control if applied early in the season. As an example, the lone star tick, *A. americanum*, reaches peak abundance early in the season in the United States, so that an earlier treatment may last all season. The American dog tick, however, does not usually reach peak abundance until June, and treatments at that time last much longer than those applied in April. The variation in the habits of ticks between different localities may influence the effective-ness of area control. For instance, the lone star tick in Texas is more difficult to control than it is in Georgia. This is apparently because ticks in the semiarid Texas region tend to stay under leaves and other debris to conserve their body moisture until a host is available, whereas in Georgia where moisture is greater they spend most of their time on the surface of vegetation. In a wood-tick area control program, much time will be saved if an infestation survey is made with a flannel drag to delineate the areas that need treatment.

Five per cent DDT in a nondrying adhesive paste applied to the ears of cattle will kill the Gulf Coast ear tick and will prevent reinfestation for 3 to 6 weeks

(RUDE and SMITH [484]; RUDE [483]); however, full coverage of the animal with a DDT spray will control the ticks as effectively as ear smears and, in addition, will control other bloodsucking insects (BLAKESLEE *et al.* [55]).

Relapsing fever has been known in Africa for about 100 years (CURRAN and LUTZ [120]), but it was not until 1904 that the disease was shown to be tick-borne. Since that time some dozen species of *Ornithodoros* ticks have been incriminated as vectors in various parts of the world — Africa, Europe, the Middle East, Asia, and South, Central, and North America (CRAIG and FAUST [119]). In 1946, it was shown experimentally in the United States that DDT sprays of 10 and 20 per cent strengths were effective in killing *O. turicata* (RANDOLPH [465]). In East Africa, where the vector is *O. moubata*, which prior to DDT and other newer compounds had been resistant to the common insecticides, the lightest dusting with 0·5 per cent BHC of individual ticks in the laboratory

Table 37

Incidence of Relapsing Fever in a Military Garrison in Kashmir, India

	January	February	March	April	May	June	July	August	September	October	November	December
1948					1	5	17	53	39	21	16	1
1949	1	3	1	2	13	9	2	3	1	0	0	0

brought about an 80 to 100 per cent mortality within 8 to 10 days, and eggs from gravid female ticks failed to hatch (JEPSON [295]). Dusting with a 5 per cent DDT powder was slower and less effective, but mortalities of 50 to 80% were obtained after more than 20 days. In the field, 0·5 per cent Gammexane dust applied from a shaker to the floors and the first few inches of walls of infected houses at the rate of 3 to 4 lb of dust per 1,000 ft^2 of surface, over a period of nine months, reduced the tick infestation 99%.

In Kashmir, India, relapsing fever is transmitted by *O. crossi* and is found where there is close association with cattle, sheep, and goats (KALRA *et al.* [303]). In summer, these ticks are found in houses on the floors and behind the mud plaster on the walls. In winter, more of these ticks are found in animal sheds than in human dwellings, probably because of the greater warmth of the former. Consequently, the incidence of the disease among humans in this area is lower in the winter. Efforts to control the vectors with pyrethrum had been unsuccessful. However, both DDT (400 mg/ft^2) and BHC (200 to 250 mg/ft^2) dust mixtures controlled them on the floors in huts. The knockdown effect with BHC was greater than with DDT, and the dose required was smaller. Repeated laboratory trials showed that if mud surfaces are treated once a week for 6 weeks with BHC at 100 mg/ft^2, either as a powder or as a water suspension, there is a 100 per cent kill among ticks when exposed to the substance for 10 minutes (KALRA and JACOB [302]). This was compared with similar dosages of DDT water suspensions. Where BHC was applied, ticks disappeared 5 days after the first application, and the area remained free of them for 20 weeks after the last application. With DDT, ticks disappeared after the third treatment and reappeared 7 weeks after the last one. It was found that repeated sprayings with 100 mg/ft² of these insecticides was better than single sprayings of 200 and 400 mg. In practical use, DDT proved to be ineffective; its large-scale application in military areas did not reduce the incidence of relapsing fever. When it was replaced with BHC, the disease was reduced considerably as shown in Table 37. No ticks were found from September 1949 onwards.

8.

MITE-BORNE DISEASES

Scrub Typhus or Tsutsugamushi Fever

Scrub typhus or tsutsugamushi fever is a rickettsial disease transmitted from rats and other small animals to man by *Trombicula* mites.

Trombicula akamushi and *T. deliensis* are proved vectors of scrub typhus to man. BLAKE *et al.* [54] also reported that *T. fletcheri* was a vector in New Guinea, but according to KOHLS [313], it is not certain that *fletcheri* is a distinct species. It may be a variant of *akamushi*.

It was certainly not anticipated that the disease would be a major military problem during World War II, although it was conceded that epidemic typhus would be a disease of major consequence. Actually, the reverse was true. Through December 1945, there had been only 64 cases of louse-borne epidemic typhus in the United States Army, and 6,861 cases of scrub typhus, and an additional 613 cases in the US Navy.

An accurate appraisal of the number of cases of scrub typhus and deaths resulting therefrom during World War II is not available. PHILIP lists over 16,000 cases in the Asiatic-Pacific Region among Australians, British, Chinese, and US Forces, not including cases which occurred in India. The mortality varied considerably, but in one British battalion in North Burma 18% of a single battalion contracted the disease in a two-months' period and 5% of the total strength died of it (PHILIP [441]).

Consideration was given to dusting DDT by airplanes in areas of advance operations, but it was shown that only poor control could be achieved by such a procedure (PHILIP [441]).

BUSHLAND [82] tested 5 per cent DDT in calcium carbonate at the enormous rate of 320 lb/acre for the control of *Schongastia pusilla* and *S. blestowei* in a section of rain forest in the Cyclops Mountains near Hollandia, Dutch New Guinea. Treatment resulted in a considerable reduction, estimated at roughly 80% of the mite population up to 5 days after dusting. Mite activity was indicated, however, to have increased 8 to 12 days after treatment. The amount of DDT dust required is such that the treatment could be used only in cases of emergency.

Area treatment with 5 per cent DDT in diesel oil at the rate of 40 gal of the solution per acre gave a temporary cessation of mite activity, but the use of the same amount of diesel oil without DDT gave equally good results. It was concluded that the DDT added nothing to the effectiveness of the treatment.

JOHNSON and WHARTON [297] reported that soil could be disinfested by treatment with 5 to 10 lb of 10 per cent DDT powder per acre. This is much less than the dosage used by BUSHLAND, whose tests were conducted, however, in a rain forest under rigid conditions. It appears, nevertheless, that DDT is not effective for combatting scrub typhus; repellents such as benzyl benzoate, dimethylphthalate and dibutylphthalate, proved however to be a practical means of avoiding individual infestations. WELT [623] reported that the incidence of scrub typhus on a South Pacific island was highest in a group of troops with no protection, lower in those where the clothes were sprayed with dimethylphthalate, and lowest where the clothes had been impregnated with a 5 per cent emulsion of this repellent.

Scabies

A formula containing DDT was developed during World War II by the US Department of Agriculture [588] which has proved very effective in the control of the itch mite, *Sarcoptes scabiei*. The formula consists of benzyl benzoate 68 parts, DDT 6 parts, benzocaine 12 parts and a wetting agent 14 parts (all parts by weight). This formula should be diluted in 5 volumes of water before use. It is applied directly to the body, and about 60 to 75 ml is usually sufficient for one treatment. The formula has proved not only an effective treatment for scabies, but also for body lice (EDDY [164]).

454

9.

PEST INSECTS

The relation of insects to health is usually considered from the viewpoint of their role as carriers of disease agents. From this concept has come the phrase, 'insects of public health importance'. However, if it is considered that health is something more than the mere absence of disease, that it includes the general mental and physical well-being of individuals and communities, it is obvious that pestiferous insects, other than disease vectors, are actually 'insects of public health importance'. The effects of vector-borne disease on the health and economy of an area are known and recognized. The effects of pest insects on these elements have not been as carefully analyzed, although health authorities have been aware of them, in varying degrees, for some time. In many regions pest insects contribute substantially to lowered vitality among the inhabitants, and to inadequate development of natural resources and industry. For this reason it is appropriate to recognize, in a discussion of the use of DDT in human medicine, the part this insecticide has played and can assume in the control of pestiferous arthropods. The control of many species of insects and arachnid pests of man has already been discussed in relation to their role as disease vectors, or are considered in Chapter VIII (KNIPLING) in connection with their importance in veterinary medicine, and these will not be considered again here.

Bedbugs

The common bedbug, *Cimex lectularius*, is found throughout North America, South America, Europe, North Africa, South Africa, Egypt, the Sudan, Russia, China, and Australia, while its relative, *C. hemipterus*, lives in warmer areas of the globe (HARVEY and HILL [255]). The bedbug is most often found where housing is poor and overcrowding exists, but may be present under other conditions if proper cleanliness is not maintained. It may be found in bedding, crevices of walls, behind wallpaper, in furniture, and in books and papers. The bedbug is a nocturnal insect, obtaining its blood meal at night. It has often been suspected of transmitting disease, but this has not been proved. Experimentally, bedbugs have been shown capable of acting as carriers or intermediate hosts of sleeping sickness, infectious jaundice, malaria, relapsing fever, and yellow fever (HARVEY and HILL [255]). It is possible that they may act as mechanical carriers of several infectious diseases, but proof of the importance of this is lacking. The bedbug is recognized primarily as a pest, and its feeding habits contribute to the discomfort of its host through sleepless nights. Skin irritation resulting from the bite wound may sometimes be a source of secondary infection.

In the past, fumigation has been used to destroy bedbugs, but because the chemicals employed were for the most part very toxic to man, this procedure, while efficient, was not satisfactory for general use. Contact insecticides, such as DDT, usually give excellent control of bedbugs without the toxic hazards of the fumigants. The application of 5 per cent DDT oil solutions or emulsions to the baseboards, wall crevices, bedsteads, and mattresses of infested rooms will usually eliminate these insects (COMMUNICABLE DISEASE CENTER [103]). However, in certain parts of the United States and Japan, bedbugs have acquired resistance to DDT residues, which has limited the usefulness of this insecticide. Preliminary tests with a number of insecticides have shown lindane to be effective against these DDT-resistant strains. Since DDT resistance is at present limited geographically, this insecticide remains, in general, the one of choice for bedbug control.

Roaches

The distribution of cockroaches seems to be universal, and there are many pestiferous species. They are pests to householders, restaurant and hotel keepers, grocers, warehousemen, and others. They prefer darkness and usually seek shelter in warm, moist, unlit places such as cellars, sewage systems, behind furniture, baseboards, sinks, under rugs and similar places. Their diet is extremely varied; they prefer foodstuffs, but will eat paper, hair, and leather, and will act as scavengers and consume dead animal matter. Because of their association with filth, they have often been suspected of transmitting disease, but there is no direct evidence to date that they do, although pathogens have often been cultured from their bodies. Even so, their presence is undesirable as they contaminate foodstuffs, making it unpalatable, and are destructive of other items.

Insect powders, sprays, traps, poison baits, and fumigation have been used to destroy cockroaches. A spray of 5 per cent water emulsion or solution of DDT has been found useful in reducing infestations of these pests (COMMUNICABLE DISEASE CENTER [103]). However, a 2 to 2·5 per cent emulsion or oil solution of chlordane is the usual accepted chemical for spot spraying for roach control in homes. Dieldrin as a 1 per cent dust or a 0·5 per cent spray is also satisfactory for this purpose. Either substance should be applied only to harborage areas, to runways, and to openings through which roaches may gain entrance from the outdoors. Since 1952, there have been reports that under field conditions the German cockroach, *Blattella germanica*, has become resistant after repeated exposure to DDT as well as chlordane and lindane. While this problem is gradually increasing in extent, chlordane or dieldrin still continue to be the insecticides of choice for roach control. Limited experiments indicate that the organophosphorus compounds may offer promise in the temporary control of roaches that have become resistant to the chlorinated hydrocarbon insecticides.

Fleas

The chief concern of public health officials regarding fleas is the role that some of them play as vectors of diseases such as plague and typhus (SIMMONS and HAYES [519]), and the control of these vector species has already been discussed. However, the role that fleas play in inflicting discomfort to man should not be underestimated. The bite of the flea is extremely annoying to many people and may cause severe dermal reactions leading to secondary infections of the skin.

A 10 per cent DDT dust usually controls pestiferous fleas in their breeding places, but certain fleas, such as the cat flea *Ctenocephalides felis* and the dog flea *C. canis*, may be relatively resistant to DDT, and chlordane dusts are often preferred. A single application of 10 per cent chlordane dust has been found effective for reducing infestations of soil by cat and dog fleas. In areas where these fleas are very resistant to DDT and chlordane, experimental data indicate that malathion (0,0-dimethyl dithiophosphate of diethyl mercaptosuccinate) or lindane emulsions applied at the rate of 10 mg of toxicant per square foot are effective.

Flies

Flies are of greatest significance in their role as carriers of disease agents. However, they may have a direct effect on the general well-being of a community by annoying and irritating the population. Certain flies, in addition to other annoyance, inflict painful bites. During the larval stage, some flies are accidentally parasitic upon or within the human body where they cause anxiety, discomfort, and distress.

The control of certain disease through fly control has been described earlier and while these measures and those for the control of flies which are only pests are similar, they will bear brief enumerating here.

Fly control depends on good environmental sanitation combined with the use of insecticides. The prevention of fly breeding and of fly entry through sanitation and screening enhances the value of any chemical in reducing fly populations. The principal problem in fly control today is the resistance shown by *Musca domestica* to DDT and other chlorinated hydrocarbon compounds. In areas where they remain susceptible, excepting dairy barns or other situations where milk could become contaminated by the chemical, a DDT residual spray applied as a 5 per cent emulsion or suspension is the insecticide of choice. Where house flies are resistant to DDT, other chlorinated hydrocarbons or organophosphorus compounds may be used. Dieldrin applied as a 0·625 or 1·25 per cent emulsion to the exteriors of houses and in animal shelters, has controlled DDT-resistant house flies for varying lengths of time (COMMUNICABLE DISEASE CENTER [103]). Chlordane applied as a 5 per cent emulsion may also work for a time. Lindane applied at the rate of 25 mg/ft² or methoxychlor at 200 mg/ft² may be used in dairy barns where milk contamination is a risk, but at these dosages neither of these compounds is as effective as 5 per cent DDT against susceptible house flies. In certain instances a Dilan:DDT:cottonseed oil emulsion applied at the rate of 200:200:80 mg of Dilan:DDT:cottonseed oil per square foot have controlled flies for periods of about 8 weeks; but in other tests the results have been erratic. This mixture should not be used where food contamination could occur.

Organic phosphorus compounds offer some promise for the control of resistant flies. House flies are very susceptible to such compounds, but recently they have also developed resistance to certain of them in some areas.

Antilarval measures to control house fly populations can be useful in certain situations, such as in dirty garbage cans or in concentrated fly-breeding areas such as stockyards and exposed industrial refuse or waste (COMMUNICABLE DISEASE CENTER [103]). The use of chlorinated hydrocarbon insecticides for this purpose, however, can result in a more rapid development of resistance to these compounds when used as adulticides. Where flies are not resistant to them, chlordane, lindane, BHC, and dieldrin sprayed at the same dosages per square foot as residual sprays are effective larvicides. Greater penetration of the larval breeding medium can be obtained by diluting the spray to 3 to 5 times original concentration and using larger volumes.

Space sprays remain as an alternative method of chemical control of house flies which are no longer susceptible to residual adult sprays or larvicides. Nighttime spraying of known fly resting sites such as shrubs, high grasses, and the lower limbs of trees is effective at problem sites, such as insanitary city dumps [103]. DDT combined with DMC (Dimethyl Carbinol) at a ratio of 7·5:1 has given satisfactory field kills of flies resistant to several chlorinated hydrocarbons. Since space sprays are toxic only to the flies coming in contact with droplets, spraying must be done at frequent intervals.

Blow-fly infestation of animal carcasses can be prevented or controlled by the application of one of several chlorinated hydrocarbons or organic phosphorus compounds. One pint of 0·25 per cent aldrin, endrin, or Diazinon emulsion will protect a dog carcass from blow-fly infestation for 26 to 30 days [103, 104].

Smudges, repellents, and screens have been and are still used in order to obtain relief from sand flies or punkies, *Culicoides* sp. Since the advent of DDT, residual sprays and fogs have been employed with success. At the International Monetary Conference held in March 1946 at a resort hotel near Savannah, Georgia, DDT residual insecticide was sprayed or painted over the entire area with a resultant decrease in sand flies to almost zero (BRUCE and BLAKESLEE [69]). At a New York State resort area, a 5 per cent DDT solution in kerosene was dispersed by a truck-mounted thermal aerosol generator at the rate of 0·2 lb of DDT per acre, giving immediate relief which lasted 3 days. In a demonstration against *Culicoides* in Florida, a 200-acre swamp was treated by airplane with a 5 per cent DDT solution at the rate of 2 quarts/acre, which curbed sand flies for four days (MADDEN *et al.* [349]). In Alaska, however, attempts to control sand flies with heat-generated DDT aerosols gave only temporary relief, the flies reappearing immediately after the fog had disappeared (TRAVIS

458   S. W. Simmons

[584]). On Peleliu Island in the Southwest Pacific, treating large breeding areas
with a 10 per cent DDT dust at the rate of 12 to 15 lb/acre effectively controlled
C. peleiouensis (DORSEY [148]). In England, it was concluded that a solution of
5 per cent BHC (gamma isomer) in miscible oil at a dilution of 1 part to 500 parts
of water sprayed on breeding grounds at the rate of 100 mg/ft^2, prior to the expected emergence of a brood, would provide effective control of *C. obsoletus* and
C. impunctatus (HILL and ROBERTS [270]).

In 1948 the Broward County, Florida, Anti-mosquito District undertook a
program to control adult sand flies in an area of some 600 acres at Fort Lauder-
dale (BERTHOLF [45]). Two quarts of 2½ per cent DDT in oil applied by airplane
at an average interval of 5½ days between May and November gave excellent
relief to the middle section of the treated zone and relief for from 12 hours to
3 days on the edges which abutted the sand-fly producing marshes. Five per cent
DDT in oil was substituted in 1949 as the spray and was equally effective against
sand flies but had little effect on *Aedes sollicitans*. Sand-fly producing marshes
along a 20-mile stretch of the inland waterway were sprayed weekly in 1949
with a 2½ per cent DDT adulticide for 24 weeks from February to August. The
first four dispersals were made with a 2½ per cent DDT-xylene-Triton-water
emulsion, the others with 2½ per cent DDT in No. 2 fuel oil. While this operation destroyed a great many sand flies for periods ranging from 12 to 72 hours,
apparently it did not reduce sand-fly breeding. To attain effective relief from
sand-fly annoyance would require adulticiding every two days; therefore, it
was concluded that a more economic larviciding program should be developed.
Experiments showed that control of sand-fly breeding could be attained by soil
treatments in the producing marshes using any of several insecticides. Of these
insecticides, BHC, chlordane, dieldrin, and aldrin applied at 1 lb/acre or more
showed the greatest promise on plots 0·2 to 5 acres in size (GOULDING and CUR-
RAN [236]). Within one week after treatment larval populations were substantially reduced (GOULDING *et al.* [237]). Dieldrin provided the longest residual
toxicity, an application of 1 lb being effective for at least 44 weeks and 2 lb for
at least 65 weeks. Chlordane at 2 lb and BHC at 1 lb/acre varied considerably in
residual effectiveness, but higher dosages gave longer periods of control. Aldrin
at 1 lb/acre began to fail between 21 and 28 weeks. DDT at 2 and 4 lb/acre and
toxaphene at 2 lb/acre gave poor immediate control and were unreliable thereafter.

Mosquitoes

Mosquitoes, of all insects, are probably of greatest significance in their relation
to man and his environment. Since earliest times, they have been abominable
pests. Long before their connection with disease transmission was established,
they played a major role in limiting man's development of large areas of the
world. With the discovery of their disease vectoring potentialities, they assumed
additional importance. Of the more than 1,400 species of mosquitoes, only
about 130 have been incriminated as disease vectors of significance. Yet among

the nonvectoring species are some of the most vicious and persistent biters, which are the cause of much human misery and are frequently associated with lowered values in land and animals. However, it was for the purpose of disease reduction that organized mosquito control was first undertaken, and the many campaigns to eliminate a number of mosquito-borne diseases through mosquito control have already been described at some length. Awareness of the contribution to human welfare which might be made by decreasing pestiferous species also is becoming more general. With increasing frequency vector mosquito control programs are being expanded to include pest mosquitoes.

The classic techniques for mosquito control, drainage, larviciding, filling, clearing, screening, etc., have been largely displaced in many parts of the world since the advent of residual sprays.

In areas where resistance has not developed, DDT remains the insecticide of choice for space spray and residual treatments. Salt-marsh mosquitoes, *Ae. sollicitans* and *Ae. taeniorhynchus*, have been effectively controlled at open air theaters by nightly fog applications of 1 gal of 2·5 per cent DDT. Barrier strip spraying is often useful for premises control of pestiferous mosquitoes. A 1·25 per cent DDT emulsion applied at a rate of 5 lb of DDT per acre to the outside of houses and to shrubbery and other vegetation for about 100 ft around individual premises has given daytime relief from pest mosquitoes for periods of one to five weeks.

Effective larviciding of susceptible mosquitoes can be accomplished by either hand spray application of 0·05 lb of DDT in 1 gal of fuel oil per acre or by airplane application of 0·05 to 0·10 lb of DDT solution in methylated naphthalenes per acre [103].

In areas where resistance to chlorinated hydrocarbon insecticides has developed among pestiferous species of *Aedes*, the organic phosphorus compounds appear to be the most promising insecticide for their control.

460

10.

THE USE OF DDT IN AIRCRAFT AND SHIP DISINSECTION

The danger of importing exotic insects of public health and agricultural significance on board aircraft and ships engaged in international traffic has long been recognized. The International Sanitary Convention of 1933 contained provisions for the disinsection of aircraft for yellow fever control. The International Sanitary Convention for Aerial Navigation in 1944 modified the 1933 Convention decisions to include provisions relating to the disinsection of aircraft for diseases other than yellow fever. The provisions of this Convention continue to serve as the basis for measures to be applied for the control of insect vectors of malaria and other diseases on aircraft engaged in international traffic, although the World Health Organization is studying the desirability of adopting supplementary international regulations. The text of the relevant part of Article XVII of the International Sanitary Convention for Aerial Navigation, 1944, is as follows:

'In view of the special risk of conveying insect vectors of malaria and other diseases by aircraft on international flight, all such aircraft leaving affected areas will be disinsected. Notwithstanding the terms of Article 54 of the 1933 Convention as hereby amended, further disinsectization of the aircraft on or before arrival may be required if there is reason to suspect the importation of insect vectors.'

Disinsection of aircraft as a quarrantine measure has been in operation in some countries for more than 20 years. The tremendous increase in international air traffic during and since World War II and the discovery of DDT and other new synthetic insecticides have revolutionized the need for and practice of aircraft disinsection. In a study of the practices and methods in force for the disinsection of aircraft in 45 countries, Duguet [158] concluded that divergencies concerned principally details and not the methods used in aircraft disinsection.

Almost as soon as it became available for such uses, DDT was included as a component of most insecticidal formulations used in aircraft disinsection. The time of application, dosage rate, method of application, and formulation vary considerably from country to country. The great majority of countries, however, use pyrethrum-DDT aerosols which do not vary greatly from the basic formulation recommended by the World Health Organization Expert Committee on Insecticides at its Second and Fourth Sessions [659, 654, 645]. This formula is as follows:

DDT Technical. . . . . 3% w/w
Pyrethrum refined extract, total pyrethrins equivalent to . . . 0·4% w/w
Nonvolatile oil, suitable solvent and propellants . . . . 96·6% w/w

This formulation is generally used at a dosage rate of 10 g/28 m³ (1,000 ft³) of enclosed space for treatment of the interior of aircraft. For exterior parts of the aircraft which might constitute shelter for insects, treatments with the same materials and at the same rates as used for the interior of aircraft are recommended. Treatment of the aircraft before take-off, with all luggage and/or freight loaded, but without passengers, is recommended.

The US PUBLIC HEALTH SERVICE [594] has approved the following formulations for use on board aircraft entering the United States, and these formulations have also been endorsed by the Expert Committee on Insecticides [659]:

Formula G-382

DDT (aerosol grade)	3% w/w
Pyrethrum extract, purified, (20% pyrethrins, 1·5% maximum Freon-12 insoluble)	5% w/w
Cyclohexanone (water-free)	5% w/w
Lubricating oil (SAE-30, free from antioxidants and surface-active agents)	2% w/w
Freon-12 (refrigeration grade)	85% w/w

Formula G-651

DDT	2% w/w
Pyrethrum extract, purified (20% pyrethrins)	6% w/w
Solvent[1])	8% w/w
Freon-12	84% w/w

Both formulations may be used in the presence of passengers at a dosage rate of at least 5 g/28 m³ (1,000 ft³). Treatments may be applied prior to take-off or any time during flight but not later than 30 minutes prior to landing. In-flight treatments are not recommended by the World Health Organization Expert Committee on Insecticides.

This Committee [659] has also approved the following aerosol formulation for use in the absence of crew and passengers:

Formula S-43

DDT	3% w/w
Pyrethrum extract (20% pyrethrins)	2% w/w
Solvent[2])	1% w/w
Van Dyk 264	8% w/w
Piperonyl butoxide	1% w/w
Freon-12	85% w/w

For use in the absence of crew and passengers, this and other approved formulations may be used at rates of application greater than 10 g/28 m³ (1,000 ft³) for combined effect against public health and agricultural insects.

---

[1]) P.D. 544-G (Socony Vacuum Oil Co.), Velsicol AR-60 (Velsicol Corporation), or their equivalent may be used as the solvent.

[2]) Sovacide 544-G, or equivalent.

DDT residual treatments, applied as sprays or dusts, are employed by some countries as aircraft disinsection procedures, either separately or in combination with aerosols. The usual treatment involves the use of 5 per cent solutions or emulsions applied at a dosage rate of approximately 200 mg of DDT per square foot of treated surface area. Dusts containing as much as 15% of DDT have been used.

While the use of DDT residual deposits constitutes an acceptable aircraft disinsection technique in some countries, PIMENTEL and KLOCK [444] found that deposits of 200 mg of DDT per square foot in loaded baggage compartments of aircraft did not give satisfactory mortalities of free-flying house flies with 1½-hour exposures. Presumably the flies rested on the untreated simulated baggage, thereby avoiding contact with the insecticide. DDT applied in passenger compartments as a wax formula killed only a very low percentage of test house flies, although this treatment was effective when the flies were forced to rest on the treated surfaces in wall-cage tests [105].

The aerosol technique leaves much to be desired with regard to over-all effectiveness. Although the aerosol treatments in general use are effective against mosquitoes in passenger compartments, they kill only a low percentage of free-flying house flies [105].

Much developmental work is needed with regard to insecticidal fomulation and dispensing equipment to establish aircraft disinsection on a completely acceptable basis.

In addition to the disinsection of aircraft itself, most countries apply environmental controls in the vicinity of airports where insect vectors create hazards. In addition to the elimination of breeding areas by drainage and filling, DDT is frequently used as residual and space sprays and larvicides for international airport sanitation.

Residual spraying with DDT is recommended for the disinsectization of ships, particularly where they are plying frequently between mosquito-free ports and infested territory. DDT is normally used at the rate of 200 mg/ft², either as suspension solutions or emulsions, the choice depending upon the surface to be treated. The Expert Committee on Insecticides of the World Health Organization [659] recommended that treatment be carried out only in those ships found to be infested with malaria vectors and in those portions of the ships decided upon by health authorities. However, in the case of vessels of approximately 25 t or less, it was suggested that it would probably be easier to treat them routinely than to carry out an inspection service. The duration of effectiveness of a residual treatment will depend greatly upon the area treated. Surfaces exposed to weathering naturally have to be treated rather frequently as compared to more protected areas in the interior of the ship.

Since it is envisaged that treatment will normally be carried out only on infested vessels, and since any specific vessel might visit infested territory infrequently, space spraying with DDT formulations may at times prove appropriate. There may also be areas not well adapted to residual treatment where space spraying can be employed to advantage. The World Health Organization recommends a space spray containing a minimum concentration of 0·4% w/w pyrethrins and 3% w/w DDT, to be dispensed at the rate of 10 g/1,000 ft³. of enclosed space. Most DDT-pyrethrin-kerosene sprays dispensed through a fogging or mist nozzle are satisfactory for this purpose. In small boats, the use of DDT for the prevention of mosquito breeding in bilge water has been used with considerable success.

DDT and related insecticides have proved to be the most effective tools yet developed in preventing the introduction of exotic insect disease vectors into free areas.

11.

RESISTANCE TO DDT INSECTICIDES AMONG ARTHROPODS OF PUBLIC HEALTH IMPORTANCE

When DDT first came into prominence, it was believed by many public health workers to be the solution for all insect vector problems. While DDT was, and is, a highly effective insecticide, it is now known that insects can become resistant to its effects.

This fact was first reported from Italy in 1947 with regard to house flies, *Musca domestica* (MISSIROLI [380]; SACCA [492]). Resistance in flies had been developed experimentally in Sweden in 1946 and in the United States in 1947 (WIESMANN [625]; LINDQUIST and WILSON [323]). By 1948 the first resistant field strain in the United States was reported from New York State (BARBER and SCHMITT [32]). Since that time numerous reports of DDT resistance in the house fly have appeared from Canada (ROADHOUSE [474]) and Mexico, Central and South America (HESS [266]; UPHOLT [597]), Europe (MISSIROLI [380]; SACCA [492]; HESS [266]; KEIDING and VANDEURS [305]; HARRISON [253]; BROWN [67]; BUSVINE [89]) and Australia (HESS [266]). *Musca vicina* has shown resistance to DDT in the Near East (SILVERMAN and MER [512]) and to BHC (gammexane) in Egypt (GAHAN and WEIR [207]), but not in India, Ceylon, or Uganda (BUSVINE [90]; PAL et al. [423]; HESS [265]). Some resistance has been developed in the laboratory in *M. nebulo* (PAL et al. [423]).

Studies made with other halogenated hydrocarbon insecticides have revealed resistance in house flies to methoxychlor [2,2-bis(*p*-Methoxyphenyl)-1,1,1-trichloroethane], TDE [1,1-bis(*p*-Chlorophenyl)-2,2-dichloroethane), lindane (*Gamma*-1,2,3,4,5,6-hexachlorocyclohexane), chlordane (1,2,4,5,6,7,8,8-Octachloro-3a,4,7,7a-tetrahydro-4,7-methanoindane), heptachlor (1,4,5,6,7,8,8-Heptachloro-3a,4,7,7a-tetrahydro-4,7-methanoindene), dieldrin (1,2,3,4,10,10-Hexachloro-6,7-epoxy-1,4,4a,5,6,7,8,8a-octahydro-1,4,5,8-dimethano-naphthalene), toxaphene (chlorinated camphene), and Dilan [mixture of 2-Nitro-1,1-bis(*p*-chlorophenyl) propane (1 part) plus 2-Nitro-1,1-bis(*p*-chlorophenyl) butane (2 parts)], indicating that resistance to one insecticide leads to a more rapid development of resistance to a second related one. It is significant that where both larvae and adults of insects are subjected to treatment with DDT and related compounds, resistance in the population is acquired more rapidly.

At the time that house fly resistance to DDT was first evident in Italy, 1945–46, it was noted that certain strains of *Culex pipiens autogenicus* were also less affected by this residual spray (MOSNA [397]). By 1950 these strains had become resistant to combined residual deposits of DDT and chlordane (MOSNA [396]). The adults were resistant to DDT but the larvae remained quite susceptible (VEROLINI [603]). *C. molestus* (*C. p. autogenicus*) in Greece was reported as DDT-resistant (HESS [266]). Larvae of *C. pipiens* collected from heavily treated areas in the United States showed a significantly lower mortality when exposed to DDT than did larvae from untreated areas (HESS [265]). A laboratory strain of *C. fatigans* (*C. quinquefasciatus*) showed a greatly increased resistance to DDT and gammexane (BHC) (NEWMAN *et al.* [414]). In Venezuela, field strains of this species were also suspected of DDT resistance (HESS [266]). By 1951 *C. tarsalis* in the United States had developed resistance to DDT as well as to toxaphene, lindane, aldrin (1, 2, 3, 4, 10, 10-Hexachloro-1, 4, 4a, 5, 8, 8a-hexahydro-1, 4, 5, 8-dimethanonaphthalene), and heptachlor applied as larvicides (GJULLIN and PETERS [228]). Adults and larvae of the salt marsh species, *Aedes sollicitans* and *Ae. taeniorhynchus* developed not only DDT resistance but resistance to other halogenated hydrocarbons to a lesser degree (DEONIER *et al.* [141]; DEONIER and GILBERT [142]). During 1949, *Ae. nigromaculis* was proved to be resistant to DDT larvicides in California (BOHART and MURRAY [59]), and indications of DDT resistance were also reported for *Ae. dorsalis* (GEIB [219]).

Reports of resistance among anophelines have been scattered. Laboratory tests with a selected strain demonstrated that adults of *Anopheles quadrimaculatus* Say could develop some DDT resistance (FAY *et al.* [177]). *A. albimanus* have been observed in increasing numbers in DDT-treated houses in Panama, and it is suspected that this is due to the development of resistance (TRAPIDO [581]). Increases in *A. albimanus*, *A. albitarsis*, and *A. pseudopunctipennis* populations have been observed in DDT-treated areas of Venezuela (GABALDON [194]). *A. darlingi* presents a conflicting picture since in some regions it has been eradicated while in others it persists despite DDT-residual applications (HESS [265]; GABALDON [194]; PINOTTI [448]; BUSTAMANTE [86]). In treated areas of Greece, increasingly larger numbers of *A. sacharovi* (*elutus*) adults are able to survive and lay fertile eggs (LIVIDAS [327]). Reports indicate that *A. maculipennis* in Greece and *A. superpictus* in Turkey (HESS [265, 266]) are developing resistance to DDT. Anophelines in Africa (*A. gambiae* and *A. melas* in Nigeria and *A. gambiae* in *Uganda*) apparently enter DDT-treated houses only to feed and then rest on outside unsprayed surfaces (MUIRHEAD-THOMSON [400, 401]; HADAWAY [244]). It is still to be determined whether or not this behavior pattern is an inherent characteristic of the species not related to the use of insecticides.

Within the past two years resistance has been reported in the human body louse *Pediculus humanis corporis*, the German cockroach *Blatella germanica* L., the oriental roach *Blatta orientalis* L. (RICCI [473]), the common bedbug *Cimex lectularius* L., and various species of fleas. DDT resistance in the human body louse in Korea (HURLBUT *et al.* [284]) created a serious problem among prisoners of war, but it was found that this louse could be controlled for the time being by various halogenated hydrocarbons other than DDT and methoxychlor

466 S. W. Simmons

(EDDY [163]). Head lice have remained susceptible to DDT. Resistant body
lice have also been reported from Egypt (HURLBUT *et al.* [283]) and Japan
(KITSOKA [311]). In certain sections of the United States resistance in the German cockroach has been developed for chlordane and lindane (BEDINGFIELD
[39]; HEAL *et al.* [257]), and DDT- and BHC-resistant strains have been selected
in laboratory studies (GRAYSON [239]). DDT-resistant strains of the bedbug
have been reported from Greece (LIVIDAS [327]), Hawaii (JOHNSON and HILL
[299]), the Belgian Congo (VINCKE [606]), and the United States (HEAL [256]).
Fleas resistant to DDT dust applications have been mentioned in reports from
Greece (*Pulex irritans*), the Near East (*P. irritans*), South America (*P. irritans,
Xenopsylla cheopis, Ceratophyllus loninensis, Polygenis* spp.) (HESS [265, 266]),
and the United States (*Ctenocephalides* spp.) (FELLTON [180]). Laboratory
studies in the United States demonstrated that DDT resistance may be developed in the oriental rat flea, *X. cheopis*, through sublethal exposure to dust applications (KILPATRICK and FAY [307]). Experiments in Chile showed that the
effect of DDT on *Triatoma infestans* decreased considerably within 3 to 4
months, after which time its action was practically imperceptible (NEGHME
and SILVA [412]).

By the end of 1953, resistance to modern insecticides had been reported
from 33 countries or regions among 32 species of insects of vectoral or pestiferous significance, making it apparent that the occurrence of effective counter-
action to public health insecticides is widespread, both geographically and ento-
mologically.

Types and Mechanisms of Resistance

Natural resistance to certain insecticides and operational procedures may
exist in certain species of insects and may be physiologic, morphologic, or
ecologic in nature. Natural resistance does not arise as the result of population
selection pressures from insecticides but rather is already present at the time
the control measures are applied and leads to their failure. A majority of the
naturally resistant individual insects tested initially possessed a physiologic
mechanism for detoxifying DDT (STERNBURG and KEARNS [552]). Inherent
morphologic differences between strains have been credited with determining
the capabilities of certain insects to withstand exposure to insecticides (WIES-
MANN [625]; D'ALESSANDRO *et al.* [122]). The behavioristic and ecologic patterns of others may interfere with their control by residual spray.

In contrast to these types of natural resistance there are the various forms
of resistance induced in populations through intensive exposure to insecticides
which, as they progress, produce a change in the composition of the population.
Induced resistance may arise from the following types of mechanisms:

1. Physiologic mechanisms which permit detoxification of the insecticides within the insect.
2. Physiologic and morphologic mechanisms which permit the partial exclu-
   sion of the insecticides.

3. Behavioristic or physiologic mechanisms which act to separate the species ecologically from the insecticide applications before lethal dosages have been obtained.

The nature and causes of house-fly resistance to the newer organic insecticides are still highly uncertain. Numerous differences between susceptible and resistant flies have been suggested, among which the simplest is greater general vigor in those individuals which survive exposure and hence appears again in their progeny (Missiroli [377]). The term vigor, however, has never been clearly defined.

Only a few investigators have as yet accepted the challenge of fundamental research on the complex problem of resistance from a biochemical viewpoint; but these, in probing deeper into the physiology of the house fly, have obtained significant results. The ability of resistant flies to convert absorbed DDT to the relatively nontoxic derivative DDE [2,2-bis-(p-chlorophenyl)-1,1-dichloro-ethylene] has been demonstrated (Sternburg and Kearns [553]; Sternburg et al. [554]; Perry and Hoskins [437, 435]; March and Metcalf [361]; Lindquist et al. [322]; Fletcher [186]). A similar reaction occurred with the bromine analog of DDT (Winteringham et al. [634]).

Significant correlations found in recent studies between the degree of resistance or survival of house flies exposed to topical applications of DDT and the conversion of DDT to DDE led to the proposal of an index for measuring the degree of DDT resistance in various strains of flies. By means of this index it could be assumed that a strain having a low percentage of conversion of DDT to DDE was a susceptible strain and could be controlled with DDT (Perry et al. [438]).

The conversion of DDT to DDE *in vivo* appears to be controlled by an enzyme, for conversion is totally inhibited when flies are killed by heat (Perry and Hoskins [437]). This reaction is also interfered with by the application of certain chemicals such as piperonyl cyclonene (Winteringham et al. [634]; Perry and Hoskins [436]; Fullmer and Hoskins [193]), and by certain structural analogs of DDT, e.g., DMC [1,1-bis-(p-chlorophenyl$)$ methyl carbinol] (Perry et al. [438]) which reduces the ability of insects to convert DDT to DDE. Recently, an enzyme which can convert DDT to DDE *in vitro* has been isolated from resistant flies (Sternburg et al. [555]).

These studies present a more coherent and encouraging picture in explaining resistance from the physiologic viewpoint; yet this theory of detoxification of DDT, like many others, is not without its discrepancies. In some resistant flies the conversion rate to DDE seems too slow (Chadwick [94]). Amounts of unchanged DDT recovered from all tissues in the bodies of resistant flies are sufficient to kill several susceptible flies (Sternburg and Kearns [553]; Perry and Hoskins [437, 435]; Fletcher [186]; Winteringham et al. [634]; Hoskins [278]). It is not the presence of DDE which prevents this DDT from exerting its lethal effect, for the pretreatment of flies with DDE did not make them resistant (Chadwick [94]).

The rate of respiration in susceptible and DDT-resistant flies normally is equal; but when flies are exposed to DDT, this rate in susceptible strains increases markedly after delays of from a few minutes to four hours with high or low dosages, respectively (FULLMER and HOSKINS [193]). When piperonyl cyclonene is added to the DDT to inhibit the detoxification process, the respiration curves of the resistant strain resemble those of the susceptible flies but reach lower maxima. Hence, it is concluded that the conversion of DDT to DDE, although of major importance in survival, is not the only factor in resistance.

A somewhat different line of approach has shown (1) that adults of a DDT-resistant strain of flies had 50% more cytochrome oxidase activity than a susceptible strain, and (2) that pupae of the resistant strain had less cytochrome oxidase activity than those of the susceptible strain but performed a higher percentage of their respiration via a cyanide insensitive system, possibly flavin (SACKTOR [493, 494]). Since DDT inhibits the respiratory enzyme cytochrome oxidase (SACKTOR [495]), it would be evident that resistant flies which possess more of the enzyme would be in a better position to withstand a partial blocking of this vital metabolic pathway. Since the conversion of DDT to DDE *in vitro* suggests that oxidation-reduction reactions might play an important role in this change, the detoxification mechanism might well be interwoven with a biologic oxidation-reduction system of which cytochrome oxidase is a part (CHADWICK [94]).

While it has been claimed that structural differences, such as the thickness and pigmentation of the tarsi, pulvilli, and articular membranes of the tarsal joints, exist between susceptible and resistant flies (WIESMANN [625]; D'ALESSANDRO *et al.* [122]), other studies failed to show any noticeable morphologic differences among strains of varying susceptibility to DDT (MARCH and LEWALLEN [360]). While it has been determined by laboratory tests (PERRY and FAY [433]) that a slow rate of absorption contributes to apparent resistance in flies, it has not yet been established whether a thickened integument is involved. However, these factors do not seem to provide an answer as to the mechanism of resistance, especially since resistance is independent of the route of administration (BUSVINE [89]; MARCH and METCALF [362]; BETTINI [50]). Greater tolerance in resistant flies for high and low extremes in temperature (WIESMANN [625]), or an increase in the duration of the life cycle of DDT-resistant strains of flies (PIMENTEL *et al.* [443]), may contribute to some extent to the over-all mechanism of resistance.

Numerous observations have been made regarding the behavioristic responses of some insects to insecticide-treated surfaces, for example, the tendency of resistant flies to rest more on untreated horizontal surfaces than on treated walls (BRUCE [70]; KING and GAHAN [309]; BRUCE and DECKER [71]). Similarly, some flies have adopted the stable fly habit of sitting still on one spot for extended periods of time (DECKER and BRUCE [133]). Resistance due to changes in resting habits of *M. vicina* has been reported (SILVERMAN and MER [512]). Behavioristic resistance has also been shown in several species of malaria mosquitoes (TRAPIDO [581]; MUIRHEAD-THOMSON [400, 401]; HADAWAY [244]; LUDVIK *et al.* [339]; TRAPIDO [582]) and in certain agricultural insects (MORRISON [394]).

Evidence would seem to indicate the genetic inheritance of protective mechanisms, but the method of action is exceedingly varied and has been satisfactorily analyzed only in isolated cases. A single, incompletely dominant autosomal gene was found responsible for susceptibility and resistance to knockdown and DDT in house flies (HARRISON [252]). In studies with reciprocal crosses of resistant with susceptible strains of the house fly, autosomal multiple gene inheritance was thought responsible (BRUCE and DECKER [71]). The general complexity in the inheritance of resistance to DDT is illustrated in reciprocal crosses of various DDT-resistant strains of house flies which indicate considerable variation in the genetic composition of these strains (NORTON [417]). Studies on the transmission of DDT resistance in the German cockroach indicated that the inheritance mechanism was complex and involved both chromosomal and cytoplasmic factors; the chromosomal factors were considered as both autosomal and sex-linked (COCHRAN et al. [101]).

Field strains of DDT-resistant house flies, regardless of level of resistance, which have been reared in the laboratory for successive generations in the absence of DDT, have shown a gradual reversion to susceptibility (UPHOLT [597]; HARRISON [253]; MISSIROLI [377]; KING [308]; PIMENTEL et al. [446, 445]; MARCH and METCALF [361]). This reversion suggests the existence of an unknown factor which selects against resistant flies in an unexposed population. A similar trend has occurred with lindane-resistant flies when they were not exposed to any insecticide (PIMENTEL et al. [445]). Little evidence of any significant degree of reversion has been observed in the field (HESS [266]). This difference in field and laboratory experiences may be due to sublethal insecticide residues maintained through the widespread agricultural and household use of DDT and the persistence of DDT residuals over a number of years. The latter is further borne out by the fact that a strain of flies resistant to BHC, which is more volatile and shorter lasting than DDT, apparently reverted partially (HESS [266]).

Although a substantial amount of information is available on house-fly resistance, little is known about the mechanism of resistance to DDT in other insect vectors of disease, especially mosquitoes. Thorough study of the biochemical reaction in the insects is needed to support observations of acquired behavioristic patterns.

The majority of investigations on resistance have been with DDT, but some attention has been given to studies with related compounds. Investigations using susceptible house flies and *Aedes* mosquito larvae as test insects for bioassay (HOFFMAN and LINDQUIST [276]) showed that resistant flies could metabolize 5·9 μg of chlordane and 4·0 μg of toxaphene per fly within 24 hours following topical application of 10 μg per fly of either insecticide. Workers at the Technical Development Laboratories of the Communicable Disease Center in Savannah, Georgia, have demonstrated by chemical and colorimetric methods of analysis, as well as by microbioassay techniques, that certain strains of resistant flies are able to metabolize heptachlor, chlordane, lindane, and dieldrin.

Significance and Future Outlook of Resistance

The major problem in 1954 was the widespread resistance among house flies to a majority of the chlorinated hydrocarbon insecticides. Further reduction of enteric diseases and infant mortality through fly control, in areas where resistance occurs, cannot be assured by the use of present-day residual formulations. The success of residual sprays in fly control has been a major factor in the popularity of vector control programs and their endorsement by the public. Even though most of the new materials are still effective against malaria vectors, public support will not be as enthusiastic as before fly resistance occurred.

The potential importance of the resistance problem to the public health is a matter for speculation. From widely scattered areas there have been reports of resistance in mosquitoes, including several 'malaria vectors'. The significance of this will depend upon the degree of resistance which may subsequently be acquired by these insects and upon its geographical extent. The reporting of resistance, either suspected or confirmed, of malaria vectors from such widely distributed countries as Greece, Turkey, Panama, Venezuela, and Africa, could be an ominous warning of future trouble. The downward trend in the incidence of malaria could very well be reversed if vectors should become resistant in once highly paludic areas where only recently the disease has been virtually eliminated by the use of residual insecticides. Countries which have not yet undertaken the widespread use of insecticides may well wonder if they will obtain only temporary reductions in the occurrence of vector-borne diseases.

As a result of the early and phenomenal successes with DDT, many health authorities assumed the attitude that the panacea for all insect control problems had been found. However, resistance among insects to a variety of insecticides was an acknowledged fact and DDT has proven not to be unique in this respect. As a result the other extreme is heard, that resistance to insecticides by all insect vectors is inevitable, but it is too soon to view the problem so pessimistically. A good many important disease vectors remain quite susceptible to the effects of DDT. Of the 32 species of insects reported to be resistant in 1953, only 19 were proved disease vectors (Table 38). A 'middle-of-the-road' attitude toward the resistance problem would seem indicated. While some species are highly resistant and others will develop resistance in varying degrees, it appears that some species may never develop a significant degree of resistance.

Suggestions for the elucidation of this problem unfortunately cannot be specific. Currently, the utilization of various combinations of the available techniques affords the most satisfactory attack. Based on the experiences with resistance in house flies, the following recommendations may be made in regard to future vector control programs (SIMMONS [515]).

1. A widespread surveillance program using a simple field testing apparatus and procedure (FAY et al. [177]) should be initiated to determine the susceptibility status of vectors to DDT or other residual insecticides for the information

and consideration of program planners in initiating and revising vector control programs.

2. Fundamental research on new insecticides or techniques should be expanded. Studies of the bionomics of resistant and susceptible strains of insects may provide clues to methods for overcoming the resistance problem in some species. Basic research on insect physiology and morphology is needed to elucidate the mechanism of resistance and to devise approaches for nullifying the detoxification process, if this is the basis of resistance. Applied research on new chemicals, improved formulations, and new techniques may offer the most promise for immediate effectiveness in current control programs which must be kept in operation.

3. Past experience has shown that house flies and some mosquitoes develop a higher degree of resistance more quickly when both larval and adult stages are exposed to the insecticide (PIMENTEL *et al.* [443]; DECKER and BRUCE [132]). On the basis of this, it would seem desirable to limit malaria control operations to the use of residual sprays in homes and outbuildings, and to avoid the use of larvicides wherever possible. Where it is necessary to larvicide, chemicals should be selected which are as different as possible from those being used as residual sprays.

4. House flies resistant to DDT usually develop resistance rather rapidly to other chlorinated hydrocarbon insecticides. New materials, different chemically, such as the organic phosphorus compounds, show some promise as substitutes but must be explored further. Many of these formulations are too toxic to permit their use as over-all residual sprays. However, a few compounds have appeared which compare favorably with DDT insofar as toxicity to man is concerned. These do not have the long residual action of DDT, but it can be hoped that through research compounds of this type having a long residual action and a low order of toxicity for man may be developed. Two of these, malathion (0,0-dimethyl dithiophosphate of diethyl mercaptosuccinate) and NPD (tetra-*n*-propyldithionopyrophosphate) have proved to be effective larvicides against some resistant mosquito species in the United States at dosages of approximately 0·4 lb/acre (GJULLIN *et al.* [227]). Where resistance to chlorinated hydrocarbon residuals should make their continued use ineffective, larviciding with organic phosphorus compounds could be substituted for residual spraying for the control of mosquito vectors. Where larviciding and residual spraying are carried on simultaneously, it would seem desirable to substitute such organic phosphorus compounds for chlorinated hydrocarbon insecticides in the larviciding operation, with a view to delaying the development of resistance to the latter.

New techniques may also make possible the use of some of the more toxic organic phosphorus compounds. Parathion (0,0-Diethyl-0-*p*-nitrophenyl thiophosphate)-impregnated cotton twine festooned from the ceilings and rafters of barns has given excellent control of highly resistant house flies for several months without replacement of the treated cords (MAIER and MATHIS [350]).

Table 38
Reported Distribution of Insecticide Resistance

Mosquitoes	United States	Canada	Mexico	Puerto Rico	Trinidad	Panama	Costa Rica	British Guiana	Ecuador	Venezuela
Culex pipiens	P									
C. quinquefasciatus (fatigans)				p						P
C. tarsalis	P									
Aedes sollicitans	Pb									
A. taeniorhynchus	Pb									
A. dorsalis	P									
A. nigromaculis	P									
Anopheles elutus (sacharovi)										
A. maculipennis										
A. superpictus										
A. gambiae										
A. darlingi								b		b
A. quadrimaculatus	P or B[3])									
A. melas										
A. albimanus				O		P or B	O			b
A. albitarsis										b
A. pseudopunctipennis			O							b
Other flies										
Musca domestica	Pb	P	p	p	p	p				p
M. d. vicina										
M. nebulo										
Phlebotomus pappatasii										
Psychoda alternata	p									
Fleas										
Pulex irritans									P or B	
Ctenocephalides spp.	p									
Xenopsylla cheopis	P								P or B	
Ceratophyllus loninensis									P or B	
Polygenis spp.									P or B	
Other species										
Cimex lectularius	P									
Pediculus h. corporis										
Triatoma infestans										
Blattella germanica	P									
Blatta orientalis										

[1]) Syria, Lebanon, Trans Jordan, Palestine, Israel, and Cyprus.

[2]) This behavioristic pattern may be an inherent characteristic of the species and not related to insecticides.

[3]) Subsequent work by these investigators has revealed factors other than resistance which could account for the differences observed.

among Insects of Medical Importance, July 1953

Brazil	Peru	Argentina	Chile	Italy	Sardinia	Greece	Turkey	Sweden	Denmark	England	Egypt	Nigeria	Uganda	Belgian Congo	South Africa	Levantine States[1])	India	Ceylon	Australia	Korea	Japan	Hawaii
				P		P																
																	O					
						p																
						p																
							p or b															
												B[2])	B[2])									
B		B																				
												B[2])										
p	p	P		P	Pb	p		P	P	P							p		P	o	o	
											P				O		PB	O	O			
																	P	O				
						p																
	p					p											p					
						p								p								p
											P						p			P	P	
				P																		

Legend: *P* physiologic resistance; the ability through biochemical processes to withstand a toxicant after it has entered the body; *B* behavioristic resistance; the ability through protective habits of behavior to avoid injurious contact with a toxicant; *O* not resistant. Capital letters: Confirmed by experimentation; lower case: based on field observations only.

N.B. DDT-resistant lice have been confirmed in Japan, *P*.

The use of organic phosphorus compounds in fly baits exposed by several methods has shown promising results against resistant house flies in a number of localities in the southern United States (GAHAN *et al.* [205]; FARRAR and BRANNON [175]; THOMPSON *et al.* [578]).

5. The use of repellents in disease vector control is still largely an unexplored field. Some interesting results have been reported from the use of repellents for fly control around animal pens (HOWELL [279]). The possible application of repellents to dwelling surfaces, either indoors or outdoors or both, broadens the field of materials which could be used as compared to those with which skin contact is required.

6. One of the more important developments resulting from the resistance problem in house flies has been the realization of the importance of good sanitation practices. Aside from the value of such practices in the control of flies, good environmental sanitation has an over-all health value far beyond that inherent in the use of insecticides.

New chemicals are being developed continuously, new facts concerning the bionomics and physiology of arthropod disease vectors are being revealed, and it is believed that the capable army of scientists at work in this field will find solutions to the resistance problem.

REFERENCES

[1] ACKERKNECHT, E. H., *Malaria in the Upper Mississippi Valley, 1760–1900*, Suppl. Bull. Hist. Med. (Johns Hopkins Press, Baltimore 1945), 142 pp.

[2] AFENDOULIS, T. H., *On Malarious Fevers*, Proc. Congr. Greek Doctors (1887), Athens *1888* (cited by LIVADAS *et al.* [335]).

[3] AFRIDI, M. K., and BHATIA, M. L., *Malaria Control of Villages Around Quetta (Baluchistan) with DDT*, Indian J. Malar. *1*, 279–287 (1947).

[4] AFRIDI, M. K., and SINGH, D., *A Scheme for the Control of Malaria in Villages in Delhi Province*, Indian J. Malar. *1*, 423–440 (1947).

[5] AGRICULTURAL RESEARCH COUNCIL, LONDON. COLONIAL INSECTICIDE, FUNGICIDE & HERBICIDE COMMITTEE, *Insecticides-Research in Africa*, Inter-Dept. Insecticide Committees. Insect. Abst. and News Summary No. 13, 46–48 (1950).

[6] AHMAD, M. U., *Recent Outbreak of Plague in Calcutta*, Indian med. Gaz. *83* (3), 156 (1948).

[7] AITKEN, T. H. G., *A Study of Winter DDT House-Spraying and Its Concomitant Effect on Anophelines and Malaria in an Endemic Area*, J. nat. Malar. Soc. *5* (3). 169–187 (1946).

[8] ALVARADO, C. A., *Situación de la lucha antimalárica en el continente americano*, Org. sanit. pan-amer. Publ. No. 261, Anexo B., 1–35 (1951).

[9] ALVARADO, C. A., *Control de las enfermedades transmitidas por mosquitoes*, Bol. Ofic. sanit. pan-amer. *27* (12), 1105–1112 (1948).

[10] ALVARADO, C. A., and COLL, H. A., *Programa para la erradicación del paludismo en la república Argentina*, Bol. Ofic. sanit. pan-amer. *27* (7), 585–602 (1948).

[11] ALVARADO, C. A., COLL, H. A., and LAGUZZI, S. F., *El programa de erradicación del paludismo en la república Argentina. Las fallas de la campaña de dedetización y organización del servicio de vigilancia*, Bol. Ofic. sanit. pan-amer. *29* (1), 1–6 (1950).

[12] AMALFITANO, G., BENETAZZO, B., CIFARELLI, F., GAMBINI, G., SERVINO, V., and TARIZZO, M., *Lá leishmaniosi viscerale nella piana di fondi (Latina). (Relazione delle prime ricerche epidemiologiche e cliniche)*, Arch. ital. Sci. med. colon. Parassit. *29* (1–2), 1–14 (1948).

[13] AMERICAN MEDICAL ASSOCIATION, *Turkey: Control of Malaria*, J. Amer. med. Ass. *142*, 920–921 (1950).

[14] ANDERSON, A., *Iran and the Anglo-Iranian Oil Company*, in: *Industrial Council for Tropical Health, Industry and Tropical Health* (Harvard School of Public Health, Boston 1951).

[15] ANDREWS, J. M., *Nation-Wide Malaria Eradication Projects in the Americas. I. The Eradication Program in the U.S.A.*, J. nat. Malar. Soc. *10*, 99–123 (1951).

[16] ANDREWS, J. M., *Malaria Control in the Nearctic Region*, Chapter 63 in *Malariology*, ed. MARK F. BOYD (W. B. Saunders Co., Philadelphia 1949), pp. 1385–1399.

[17] ANDREWS, J. M., *What's Happening to Malaria in the U.S.A.?* Amer. J. publ. Hlth. *38*, 931–942 (1948).

[18] ANDREWS, J. M., and GILBERTSON, W. E., *Final Phases of Malaria Eradication in the United States*, J. nat. Malar. Soc. *9*, 5–9 (1950).

[19] ANDREWS, J. M., GRANT, J. S., and FRITZ, R. F., *Effects of Suspended Residual Spraying of Imported Malaria on Malaria Control in the United States*, Bull. World Hlth. Org. *11* (4–5), 839–848 (1954).

[20] ANDREWS, J. M., and GRANT, J. S., *Interruption of Residual Spraying After Several Years of Achieved Malaria Control in the United States*, World Hlth. Org./Mal./96, 10 August 1953.

[21] ANNECKE, S., *Malaria Control in Transvaal*, World Hlth. Org./Mal./48 — Afr./Mal./ Conf./4 — 3 October 1950.

[22] ANONYM, *Malaria Control in India*, Trop. Med. Hyg. News 2 (2), 14 (1953).

[23] ANONYM, *Afghanistan Triumphs Over Typhus*, World Hlth. Org. Newslett. 5 (6), 3 (1952).

[24] ANONYM, *Human Rickettsioses in Africa. Epidemic Typhus*, Epidem. vit. Stat. Rep. III (7–8), 163–177 (1950).

[25] AVERETT, W. L., Personal communication (1952).

[26] AYER, P. L., *Address of Plaza Leyó Ayer to Congress Aug. 11, 1952*, El Universo, Guayaquil.

[27] AZIZ, M., *The Island-Wide Anopheles Eradication Program in Cyprus 1946–48*, Proc. Fourth int. Congr. trop. Med. *1948*, 703–713.

[28] AZIZ, M., *Discussion of 'Anopheles Control in the Mediterranean Area' by Missiroli, A.*, Proc. Fourth Int. Congr. trop. Med. *1948*, 1575–1576.

[29] AZIZ, M., *Interim Report on Island-Wide Anopheles (Malaria) Eradication Programme for the Year 1947*, Cyprus Annu. med. sanit. Rep. *1947*.

[30] BAKER, W. C., Personal communication (1953).

[31] BANG, F. B., HAIRSTON, N. G., MAIER, J., and ROBERTS, F. H. S., *DDT Spraying Inside Houses as a Means of Malaria Control in New Guinea*, Trans. R. Soc. trop. Med. Hyg. *40*, 809–822 (1947).

[32] BARBER, G. H., and SCHMITT, J. B., *House Flies Resistant to DDT Residual Sprays*, Bull. N. J. agric. Expt. Sta., 742, 8 pp. (1948).

[33] BARBER, M. A., *The History of Malaria in the United States*, Publ. Hlth. Rep., Wash. *44*, 2575–2587 (1929).

[34] BARBER, M. A., and HAYNE, T. B., *Arsenic as a Larvicide for Anopheline Larvae*, Publ. Hlth. Rep., Wash. *36*, 3027 (1921).

[35] BARBER, M. A., and RICE, I. B., *Malaria Studies in Greece*, Ann. trop. Med. Parasit. *29* (3) (cited by LIVADAS et al. [335]).

[36] BARNARD, C. I., *The Rockefeller Foundation — A Review for 1949 (Malaria in Sardinia)* (New York 1949).

[37] BAYNE-JONES, S., *Commentary on Typhus Control in World War II*, Yale J. Biol. Med. *22*, 483–493 (1950).

[38] BAYNE-JONES, S., *Epidemic Typhus in the Mediterranean Area During World War II with Special Reference to the Control of the Epidemic in Naples in the Winter of 1943–1944*, in: *The Rickettsial Diseases of Man* (American Association for Advances in Science, Washington, D. C., 1948), pp. 1–15.

[39] BEDINGFIELD, W. D., *Insecticide Resistant Roaches?* Pest Control *1952* (April), 6.

[40] BERBERIAN, D. A., *The Use of DDT Residual Spray in Malaria Control and Its Effect on Sanitation in Rural Districts*, J. Palest. Arab med. Ass. *3* (3), 49–61 (1948).

[41] BERBERIAN, D. A., and DENNIS, E. W., *The Effect of Chloroquine Diphosphate on Malaria Splenomegaly*, Amer. J. trop. Med. *29*, 463–471 (1949).

[42] BERBERIAN, D. A., and DENNIS, E. W., *Field Experiments With Chloroquine Diphosphate*, Amer. J. trop. Med. *28*, 755–776 (1948).

[43] BERNARD, M. P., *La prophylaxie antipalustre au moyen des insecticides de contact à Madagascar*, Bull. Soc. Pat. exot. *43* (7/8), 505–512 (1950).

[44] BERNARD, M. P., *Malaria Control in Madagascar*, World Hlth. Org./Mal./52 — Afr./Mal./Conf./8 — 9 October 1952.

[45] BERTHOLF, J. H., *Report of Broward County Anti-Mosquito District*, Report of the 21st Annu. Mtg., Florida Anti-mosq. Ass., April 16–19, pp. 27–29 (1950).

[46] BERTI, A. L., Personal communication (1954).

[47] BERTI, A. L., *La ingenieria antimalarica en Venezuela*, Tercera Conf. Inter-amer. Agric.: Cuaderno Verde No. 24, 1–34 (1945).

[48] BERTRAM, D. M., *A Critical Evaluation of DDT and 'Gammexane' in Malaria Control in Upper Assam Over Five Years, With Particular Reference to Their Effect on Anopheles minimus*, Ann. trop. Med. Parasit. *44*, 242–254 (1950).

[49] BERTRAM, D. M., *Larvicidal Treatments With DDT and 'Gammexane' in Upper Assam, With Particular Reference to Their Effect on Anopheles minimus*, Ann. trop. Med. Parasit. *44*, 255–259 (1950).

[50] BETTINI, S., *Contributo allo studio della resistenza all'azione del DDT nelle mosche domestiche*, Riv. Parassit. *9*, 137–142 (1948).

[51] BEYE, H. K., EDGAR, S. A., MILLE, R., KESSEL, J. F., and BAMBRIDGE, B., *Preliminary Observations on the Prevalence, Clinical Manifestations and Control of Filariasis in the Society Islands*, Amer. J. trop. Med. Hyg. *1* (4), 637–661 (1952).

[52] BISHOPP, F. C., SMITH, C. N., and GOUCK, H. K., *The Brown Dog Tick, With Suggestions for Its Control*, Circ. U.S. Bur. Ent. E-292 (1946).

[53] BLAIR, D. M., *Report of Malaria Control in Southern Rhodesia*, Publ. Hlth., Johannesburg *15* (5), 100–103 (1951).

[54] BLAKE, F. G., MAXCY, K. F., SADUSK, J. F., jr., KOHLS, G. M., and BELL, E. J., *Trombicula fletcheri* Womersley and Heaslip 1943, *a Vector of Tsutsugamushi Disease (Scrub Typhus) in New Guinea*, Science *102* (2638), 61–64 (1945).

[55] BLAKESLEE, E. B., TISSOT, A. N., BRUCE, W. G., and SANDERS, D. A., *DDT to Control the Gulf-Coast Tick*, J. econ. Ent. *40*, 664–666 (1947).

[56] BLANTON, F. S., and TANI, T. G., *Typhus Control at Ports in Japan and Korea After World War II*, J. econ. Ent. *44* (5), 812–813 (1951).

[57] BLUM, R., and staff, *Report of Public Health Div., STEM-Indochina (Prepared for Southeast Asia Conference of Public Health Programs, Bangkok, Thailand, Aug. 6–11, 1951)*, Fed. Security Agency, Publ. Hlth. Serv., Div. of Int. Hlth., Wash. (1951).

[58] BODMAN, R. I., and STEWART, I. S., *Louse-Borne Relapsing Fever in Persia*, Brit. med. J. *116* (4545), 291–293 (1948).

[59] BOHART, R. M., and MURRAY, W. D., *DDT Resistance in Aedes nigromaculis Larvae*, Proc. Calif. Mosq. Contr. Ass. Annu. Conf. *18*, 20–21 (1950).

[60] BONNE-WEPSTER, J., and SWELLENGREBEL, N. H., *Anopheles sundaicus and DDT Spraying*, Docum. neerl. indones. Morb. trop. *2*, 155–160 (1950).

[61] BOYD, M. F., *A Historical Sketch of the Prevalence of Malaria in North America*, Amer. J. trop. Med. *21*, 223–244 (1941).

[62] BOYD, M. F., *Studies of the Epidemiology of Malaria in the Coastal Low Lands of Brazil, Made Before and After the Execution of Control Measures*, Amer. J. Hyg. Monogr. Ser. No. 5, 261 pp. (1926).

[63] BOYER, J., *Etude épidémiologique des cas de typhus exanthématique observés dans la région parisienne au retour des prisonniers et déportés*, Bull. Off. int. Hyg. publ. *38* (10–11–12), 855–870 (1947).

[64] BRADLEY, G. H., and LYMAN, F. E., *Discussion of Five Years' Use of DDT Residuals Against Anopheles quadrimaculatus*, J. nat. Malar. Soc. *9*, 113–118 (1950).

[65] BRADLEY, G. H., and LYMAN, F. E., *Recent Developments in the Antimosquito Work of the Communicable Disease Center, U.S. Public Health Service*, Proc. N. J. Mosq. Ext. Ass. *36*, 68–74 (1949).

[66] BRIERCLIFFE, R., *The Ceylon Malaria Epidemic, 1934–35*, Sessional papers XXII (Ceylon Government press 1935).

[67] BROWN, A. W. A., *The Development of Resistance of Insects to Insecticides*, 81st Annu. Rep. Ent. Soc. Ontario, 34–35 (1950).

[68] BROWN, H. W., and WILLIAMS, R. W., *Filariasis Control by DDT Residual House Spraying, Saint Croix, Virgin Islands. II. Results*, Publ. Hlth. Rep., Wash. *64* (27), 863–875 (1949).

[69] BRUCE, W. G., and BLAKESLEE, E. B., *Control of Salt-Marsh Sand Flies and Mosquitoes With DDT Insecticides*, Mosquito News *8*, 26–27 (1948).

[70] BRUCE, W. N., *Latest Report on Fly Control*, Pest Control *17*, 7, 28 (1949).

[71] BRUCE, W. N., and DECKER, G. C., *House Fly Tolerance for Insecticides*, Soap, N. Y. *26*, 122–125, 145–147 (1950).

[72] BRUCE-CHWATT, L. J., *Malaria in Nigeria*, Bull. World Hlth. Org. *4* (3), 301–327 (1951).

[73] BRUCE-CHWATT, L. J., Personal communication (1952).

[74] BRUCE-CHWATT, L. J., *Ilaro Experimental Eradication Scheme*, Nigeria-Malaria Service, Med. Dept., Third Annu. Rep., April 1951–December 1951.

[75] BRUCE-CHWATT, L. J., CAMBOURNAC, F. J. C., CHEVERTON, R. L., DAVIDSON, G., GARNHAM, P. C. C., MACDONALD, G., DEMEILLON, B., RUSSELL, P. F., VAUCEL, M. A., VINCKE, I., and WILSON, D. B., *Malaria Conference in Equatorial Africa.* Pt. II. *Report on the Debates*, Tech. Rep. World Hlth. Org., No. *38*, 9–44 (1951).

[76] BUONOMINI, G., SICCA, G. T., and MANFREDI, M., *La malaria all'isola d'Elba*, Riv. ital. Igiene *8* (1–2), 1–12 (1948).

[77] BUONOMINI, G., and MACCARRONE, A., *Istituto d'Igiene e microbiologia dell'Università di Pisa. Considerazioni sulla malaria dell'isola d'Elba. Sulla possibilità di eradicare l'A. (maculipennis) labranchiae*, Riv. ital. Igiene *8* (9–12), 352–362 (1948).

[78] BURMA, GOVERNMENT OF, *Report Submitted, Malaria Conference, Bangkok, Thailand, September 21–24, 1953*, World Hlth. Org., Sea./Mal./11, New Delhi, 3 July 1953.

[79] BURNET, F. M., *Murray Valley Encephalitis*, Amer. J. publ. Hlth. *42*, 1519–1521 (1952).

[80] BURROUGHS, A. L., *Sylvatic Plague Studies. The Vector Efficiency of Nine Species of Fleas Compared With Xenopsylla cheopis*, J. Hyg., Camb. *45* (3), 371–396 (1947).

[81] BURTON, G., Quoted in: 'News and Notes', Mosquito News *14*, 109 (1954).

[82] BUSHLAND, R. C., *Insecticides Applied to Forest Litter to Control New Guinea Chiggers*, J. econ. Ent. *39* (3), 344–347 (1946).

[83] BUSHLAND, R. C., MCALISTER, L. C., jr., EDDY, G. W., and JONES, H. A., *DDT for the Control of Human Lice*, J. econ. Ent. *37* (1), 126–127 (1944).

[84] BUSHLAND, R. C., MCALISTER, L. C., jr., EDDY, G. W., JONES, H. A., and KNIPLING, E. F., *Development of a Powder Treatment for the Control of Lice Attacking Man*, J. Parasit. *30* (6), 377–387 (1944).

[85] BUSHLAND, R. C., MCALISTER, L. C., jr., JONES, H. A., and CULPEPPER, G. H., *DDT Powder for the Control of Lice Attacking Man*, J. econ. Ent. *38* (2), 210–217 (1945).

[86] BUSTAMANTE, F. M., *Efeito das aplicações intradomiciliarias de DDT sôbre a densidade do Anopheles darlingi em várias regiões do Brasil*, Rev. bras. Malariol. *3*, 571–590 (1951).

[87] BUSTAMANTE, F. M., and FERREIRA, M. O., *Da aplicação intradomiciliar de DDT no combate à malária transmitida por Kerleszia*, Rev. bras. Malariol. *1*, 205–210 (1949).

[88] BUSVINE, J. R., *Forms of Insecticide Resistance in Houseflies and Body Lice*, Nature *171*, 118–119 (1953).

[89] BUSVINE, J. R., *Mechanism of Resistance to Insecticides in House Flies*, Nature, Lond. *168*, 193 (1951).

[90] BUSVINE, J. R., *A Test for DDT-Resistant Flies*, Ceylon J. Sci. (D) *6*, 181–183 (1949).

[91] CABRERA, M., ISIDRO, J., MCANALLY, W. J., jr., and MONTOYA, J. A., *Informe sobre las actividades en el control del tifo en la República de Guatemala: 1946–1951*, Bol. Ofic. sanit. pan-amer. *34* (3), 225–235 (1953).

[92] CAMERON, G. R., CHALKE, H. D., HILL, K. R., CASE, R. A. M., HELLIER, F. F., HACKETT, C. J., and CHESTERMAN, C. C., *Discussion on DDT*, Proc. R. Soc. Med. *39* (4), 165–168 (1946).

[93] CAVAILLON, M. M. A., ADAM, J. P., BERNARD, L., CAMAIN, R., JAUJOU, C., and TRINQUIER, E., *Deux années de lutte contre le paludisme en Corse: premiers résultats*, Commun. Acad. nat. Med. *1950*.

[94] CHADWICK, L. E., *The Current Status of Physiological Studies on DDT Resistance*, Amer. J. trop. Med. Hyg. *1*, 404–411 (1952).

[95] CHALKE, H. D., *Typhus: Experiences in the Central Mediterranean Force*, Brit. med. J. *1*, 977–980; *2*, 5–8 (1946).

[96] CHANDLER, A. C., and RICE, L., *Observations on the Etiology of Dengue Fever*, Amer. J. trop. Med. *3*, 233–262 (1923).

[97] CHAVARRIA, A. P., and ARGUEDAS, J. G., *La influencia del DDT en la incidencia del paludismo en Costa Rica*, Bol. Ofic. sanit. pan-amer. *35*, 487–493 (1953).

[98] CHINA, REPUBLIC OF, *Report Submitted, Malaria Conference, Bangkok, Thailand, September 21–24, 1953*, World Hlth. Org., Sea./Mal./10, New Delhi, 1 July 1953.

[99] CHRISTOPHERS, S. R., *Malaria in the Andamans*, Sci. Mem. med. sanit. Dep., Indian N. S., No. 56 (Govt. Press, Calcutta 1912).

[100] CLAVERO, G., and ROMEO VIAMONTE, J. M., *El paludismo en las huertas de Murcia y Orihuela, ensayes de aplicación de los insecticides modernos, D.D.T. y 666, en lucha antipalúdica*, Rev. Sanid. Hig. públ., Madr. *22* (3), 199–228 (1948).

[101] COCHRAN, D. G., GRAYSON, J. M., and LEVITAN, M., *Chromosomal and Cytoplasmic Factors in transmission of DDT Resistance in the German Cockroach*, J. econ. Ent. *45*, 997–1001 (1952).

[102] COCKBURN, T. A., PRICE, E. R., and ROWE, J. A., *Virus Encephalitis in the Missouri River Basin*, C.D.C. Bull. (Feb.) 13–24 (1950).

[103] COMMUNICABLE DISEASE CENTER, ATLANTA, GA., *Use of Pesticides for Controlling Pests of Public Health Importance, 1954 Report*, Pest Control *22* (3), 9–20 (1954).

[104] COMMUNICABLE DISEASE CENTER, ATLANTA, GA., Personal communication (1954).

[105] COMMUNICABLE DISEASE CENTER, ATLANTA, GA., *Techniques and Materials for the Disinsectization of Aircraft*, Bull. World Hlth. Org. *8*, 527 (1953).

[106] COMMUNICABLE DISEASE CENTER, ATLANTA, GA., *DDT Residual Spray Operations Handbook*, F.S.A. Publ. Hlth. Serv., Atlanta, Ga. (1947).

[107] CORBO, S., *La mosca domestica principale responsabile della mortalità infantile per malattie gastroenteriche*, Riv. Parassit. *12* (1), 37–45 (1951).

[108] CORBO, S., *La mortalità infantile per malattie intestinali in repporto con l'irrorazione di DDT e octaKlor*, Arch. ital. Pediat. *13*, 261–272 (1949).

[109] CORRADETTI, A., *The Epidemiology and Control of Oriental Sore in Abruzzo, Italy*, Amer. J. trop. Med. Hyg. *1* (4), 618–622 (1952).

[110] CORRADETTI, A., Personal communication (1952).

[111] CORRADETTI, A., *Esperimento di prevenzione della verruga peruviana e della malaria nella valle del Rio Santa Eulalia (Peru)*, Riv. Parassit. *10* (1), 53–58 (1949).

[112] CORRADETTI, A., *Studi sulla epidemiologia della leishmaniosi cutanea nella regione del medio Adriatico. II. Osservazioni sulla biologia del Phlebotomus perfiliewi*, Riv. Parassit. *11*, 111–116 (1949).

[113] CORRADETTI, A., *Studi sulla epidemiologia della leishmaniosi cutanea nella regione del medio Adriatico. I. Incidenza della leishmaniosi cutanea nella zona compresa tra il Tordino e il Vomano*, Riv. Parassit. *9*, 227–228 (1948).

[114] CORRADETTI, A., *Bases experimentales para la eliminación de la malaria en la Costa del Perú*, Publ. Dir. salud publ. 1–14 (1947).

[115] COVELL, G., *Malaria Incidence in the Far East*, Chapter 34, in: *Malariology*, ed. MARK F. BOYD (W. B. Saunders Co., Philadelphia 1949), pp. 810–819.

[116] COVELL, G., *Malaria Problems in the Oriental and Australian Regions*, Proc. 4th Int. Congr. trop. Med. and Malar. *1*, 932–939 (1948).

[117] COVELL, G., *Report of an Inquiry Into Malarial Conditions in the Andamans* (Government Press, Delhi 1927).

[118] COVELL, G., and SINGH, P., *Malaria in the Coastal Belt of Orissa*, J. Malar. Inst. India *4*, 457 (1942).

[119] CRAIG, C. F., and FAUST, E. C., *Clinical Parasitology* (Lea and Febiger, Philadelphia 1951), 1032 pp.

[120] CURRAN, C. H., and LUTZ, F. E., *Insects, Ticks, and Human Diseases*, Guide Leafl. Amer. Mus. nat. Hist., No. 113, New York, 38 pp. (1942).

[121] D'ALESSANDRO, G., BURGIO, G. R., and MARIANI, M., *Commento all'attuale recrudescenza della leishmaniosi viscerale a Palermo. Prospettive della lotta contro i flebotomi a mezzo del DDT*, Rif. med. *1947;* Nos. 21/22, 3–12.

480   S. W. Simmons

[122] D'ALESSANDRO, G., CATALANO, G., MARIANI, M., SCERRINO, E., SMIRAGLIA, C., and VALGUARNERA, G., *Sulla resistenza delle mosche al DDT*, Sicilia med. *6*, 5–16 (1949).

[123] DAVIDSON, G., *Results of Recent Experiments on the Use of DDT and BHC Against Adult Mosquitoes at Taveta, Kenya*, Bull. World Hlth. Org. *4* (3), 329–332 (1951).

[124] DAVIS, D. E., *The Control of Rat Fleas (Xenopsylla cheopis) by DDT*, Publ. Hlth. Rep., Wash. *60* (18), 485–489 (1945).

[125] DAVIS, D. H. S., *Current Methods of Controlling Rodents and Fleas in the Campaign Against Bubonic Plague and Murine Typhus*, J. R. sanit. Inst. *69* (3), 170–175 (1949).

[126] DAVIS, W. A., *Typhus at Belsen. I. Control of the Typhus Epidemic*, Amer. J. Hyg. *46* (1), 66–83 (1947).

[127] DAVIS, W. A., MALO-JUVERA, F., and HERNANDEZ-LIRA, P., *Studies on Louse Control in a Civilian Population*, Amer. J. Hyg. *39* (2); 177–188 (1944).

[128] DAVIS, W. A., *A Study of Birds and Mosquitoes as Hosts for the Virus of Eastern Equine Encephalomyelitis*, Amer. J. Hyg. *32* [Sect. C], 45–59 (1940).

[129] DECAIRES, P. F., *Filariasis Vector Control. A Study of the Control of DDT-Resistant Culex quinquefasciatus, Vector of Bancroftian Filariasis in British Guiana* · (in manuscript, 1952).

[130] DECAIRES, P. F., *The International Yellow Fever Problem in the Caribbean Islands*, W. Ind. med. J., Mon. Bull. *1952* (January).

[131] DECARVALHO, D. A., *Le typhus exanthématique au Portugal pendant les années de guerre*, Clin. Hig. e Hidrol., Lisbon *14* (7), 221–222 (1948).

[132] DECKER, G. C., and BRUCE, W. N., *House Fly Resistance to Chemicals*, Amer. J. trop. Med. Hyg. *1*, 395–403 (1952).

[133] DECKER, G. C., and BRUCE, W. N., *Where Are We Going With Fly Resistance?* Soap, N. Y. *27*, 139–143, 159 (1951).

[134] DE JONG, J. C. M., *Malaria-bestrijding — Friesland: Verslag over het jaar 1950*.

[135] DE JONG, J. C. M., *Malaria-bestrijding — Friesland: Verslag over het jaar 1949*.

[136] DE JONG, J. C. M., *Malaria-bestrijding — Friesland: Verslag over het jaar 1948*.

[137] DE JONG, J. C. M., *Malaria-bestrijding — Friesland: Verslag over het jaar 1947*.

[138] DE JONG, J. C. M., *Malaria-bestrijding — Friesland: Verslag over het jaar 1946*.

[139] DEKOCK, G., DUTOIT, R., and KLUGE, E., *Problems Confronting the Nagana Campaign in Zululand*, J. S. Afr. vet. med. Ass. *18*, 105–119 (1947).

[140] DEMEILLON, B., *Species and Varieties of Malaria Vectors in Africa and Their Bionomics*, Bull. World Hlth. Org. *4* (3), 419–441 (1951).

[141] DEONIER, C. C., CAIN, T. L., jr., and MCDUFFIE, W. C., *Aerial Spray Tests on Adult Salt-Marsh Mosquitoes Resistant to DDT*, J. econ. Ent. *43*, 506–510 (1950).

[142] DEONIER, C. C., and GILBERT, I. H., *Resistance of Salt-Marsh Mosquitoes to DDT and Other Insecticides*, Mosquito News *10*, 138–143 (1950).

[143] DHALIWAL, G. S., *Malaria in Arakan*, Indian J. Malar. *5*, 287 (1951).

[144] DIAS, EMMANUEL, *Contrôle das doenças transmitidas pelos Triatomas*, Bol. Ofic. sanit. pan-amer. *27* (12), 1160–1164 (1948).

[145] DICK, G. W. A., *The Relationship of Mengo Encephalomyelitis, Encephalomyocarditis, Columbia-SK and M.M. Viruses*, J. Immunol. *62*, 375–386 (1949).

[146] DICK, G. W. A., BEST, A. M., HADDOW, A. J., and SMITHBURN, K. C., *Mengo Encephalomyelitis, a Hitherto Unknown Virus Affecting Man*, Lancet *2*, 286–289 (1948).

[147] DOMENJOZ, R., *Experimentelle Erfahrungen mit einem neuen Insektizid (Neocid-Geigy), ein Beitrag zur Theorie der Kontaktgiftwirkung*, Schweiz. med. Wschr. *74* (36), 952–958 (1944).

[148] DORSEY, C. K., *Population and Control Studies of the Palau Gnat on Peleliu, Western Caroline Islands*, J. econ. Ent. *40*, 805–814 (1947).

[149] DOW, R. P., *Trip Report — Iran*, Fed. Security Agency, Publ. Hlth. Serv., Communicable Disease Center, Atlanta, Ga. (1952).

[150] DOWLING, M. A. C., *Malaria Eradication Scheme — Mauritius*, Annu. Rep. 1950 (J. E Felix, Government Printer, Port Louis, Mauritius, April 1951), pp. 1–25.

[151] DOWLING, M. A. C., *An Experiment in the Eradication of Malaria in Mauritius*, Bull. World Hlth. Org. *4* (3), 443–461 (1951).

[152] DOWLING, M. A. C., *Report on Progress of Malaria Eradication Scheme — Colony of Mauritius*, World Hlth. Org., Expert Committee on Malar., WHO/Mal./34, 1–9 (1950).

[153] DOWNS, W. G., and BORDAS, E., *Anopheles aztecus, Malaria, and Malaria Control in the Valley of Mexico*, J. nat. Malar. Soc. *10*, 350–358 (1951).

[154] DOWNS, W. G., and BORDAS, E., *Control of Anopheles pseudopunctipennis in Mexico With DDT Residual Sprays Applied in Buildings*. Part V. *Effectiveness of Residual Applications of DDT and Gammexane up to One Year After Application Under Controlled Conditions*, Amer. J. Hyg. *54*, 150–156 (1951).

[155] DOWNS, W. G., BORDAS, E., and COLORADO, M. C., *Malaria Control in the Southern Territory of Lower California*, Amer. J. trop. Med. *31*, 286–289 (1951).

[156] DOWNS, W. G., BORDAS, E., and CHAVEZ, A. E., *El control del paludismo en la región de Xochimilco, D. F.*, Rev. Inst. salubr. Enferm. trop., Mex. *11*, 3–10 (1950).

[157] DOWNS, W. G., CELIS, S., HELIODORO, and GAHAN, J. B., *Control of Anopheles pseudopunctipennis in Mexico With DDT Residual Sprays Applied in Buildings*. Part III. *Malariological Observations After Five Years of Annual Spraying*, Amer. J. Hyg. *52*, 348–352 (1950).

[158] DUGUET, J., *Disinsectization of Aircraft*, Bull. World Hlth. Org. *2*, 155 (1949).

[159] DUNN, C. L., *Malaria in Ceylon* (Baillière, Tyndall and Cox, London 1936).

[160] DuTOIT, R., *The Tsetse Fly Problem and Its Control in South Africa*, S. Afr. biol. Soc., Pamphlet No. 13, pp. 25–40 (1947).

[161] DuTOIT, R., *The Problems Presented by the Control of Tsetse Flies in the Union of South Africa*, S. Afr. J. Sci. *43*, 262–265 (1947).

[162] DYER, R. E., RUMREICH, A., and BADGER, L. F., *Typhus Fever — A Virus of the Typhus Type Derived From Fleas Collected From Wild Rats*, Publ. Hlth. Rep., Wash. *46* (7), 333–338 (1931).

[163] EDDY, G. W., *Effectiveness of Certain Insecticides Against DDT-Resistant Body Lice in Korea*, J. econ. Ent. *45*, 1043–1051 (1952).

[164] EDDY, G. W., *A Combination Treatment for Lice and Scabies*, J. invest. Derm. *7*, 85–91 (1946).

[165] EDDY, G. W., and SELHIME, A., Quart. Rep., Ent. Res., Bur. Ent. Plant Quarant., Agric. Res. Adm., U.S. Dept. Agric. Quarters ending March 31 and June 30, 1952.

[166] EGYPT, *General Report of the Functions of the Anti-Fly Campaign Committee Formed by the Ministry of Health of the Egyptian Government in Connection With the Problem of Fly Control During the Cholera Epidemic of 1947*, Minist. Publ. Hlth., Egypt (Government Press, Cairo 1948).

[167] EIDE, P. M., DEONIER, C. C., and BURRELL, R. W., *DDT as a Culicine Larvicide*, J. econ. Ent. *38* (5), 537–541 (1945).

[168] EJERCITO, A., HESS, A. D., and WILLARD, A., *The Six-Year Philippine-American Malaria Control Program*, Amer. J. trop. Med. Hyg. *3* (6), 971–980 (Nov. 1954).

[169] EJERCITO, A., HESS, A. D., and WILLARD, A., *Report on trip to Thailand, June 13–22, 1953*.

[170] ELMENDORF, J. E., jr., *Second and Supplementary Report on Field Experiments to Demonstrate Effectiveness of Various Methods of Malaria Control*, Amer. J. trop. Med. *28*, 425–436 (1948).

[171] ELMENDORF, J. E., jr., *Preliminary Report on Field Experiments to Demonstrate Effectiveness of Various Methods of Malaria Control*, Amer. J. trop. Med. *27*, 135–145 (1947).

[172] ELORDUY, C. C., *La campaña contra el paludismo en el puerte de Veracruz* (Talleres Gráficos de la Nación 1950).

[173] ERZIN, N., *Ikinci dünya savasi yillarinda yurdumuzun tifüs durumu*, Turk. Bull. Hyg. exp. Biol. *8* (3), 5–9 (1948).

[174] FAIRCHILD, G. B., and BARREDA, E. A., *DDT as a Larvicide Against Simulium*, J. econ. Ent. *38* (6), 694–699 (1945).

[175] FARRAR, M. S., and BRANNON, C. C., *TEPP in a Bait Against Resistant Flies*, Presented before Amer. Ass. econ. Ent., Dec. 1952.

[176] FAY, R. W., BAKER, W. C., and GRAINGER, M. M., *Laboratory Studies on the Resistance of Anopheles quadrimaculatus to DDT and Other Insecticides*, J. nat. Malar. Soc. *8*, 137–146 (1949).

[177] FAY, R. W., KILPATRICK, J. W., CROWELL, R. L., and QUARTERMAN, K. D., *A Method for Field Detection of Adult Mosquito Resistance to DDT Residues*, Bull. World Hlth. Org. *9*, 345–351 (1953).

[178] FAUST, E. C., *Malaria Incidence in North America*, Chapter 30 in: *Malariology*, ed. MARK F. BOYD (W. B. Saunders Co., Philadelphia 1949), pp. 749–763.

[179] FAUST, E. C., *The Etiologic Agent of Chagas' Disease in the United States*, Bol. Ofic. sanit. pan-amer. *28* (5), 455–461 (1949).

[180] FELLTON, H. L., Personal communication of 2 July 1953.

[181] FERGUSON, F. F., ARNOLD, E. H., and UPHOLT, W. M., *Control of Anopheline Mosquito Larvae by Use of DDT-Oil Mists*, Publ. Hlth. Rep., Wash. *62* (9), 296–302 (1947).

[182] FIELD, J. W., *Yaws*, The Institute for Med. Research 1900–1950, Kuala Lumpur, Malaya, pp. 292–295 (1951).

[183] FINLAY, C., *El mosquito hipotéticamente considerado como agente de transmissión de la fiebre amarilla*, An. Acad. Habana *18*, 147–169 (1881).

[184] FIRTOS, C., and DE JONG, J. C. M., *Gunstige uitkomsten van DDT-toepassing*, Ned. Tijdschr. Geneesk. *94*, 1105–1110 (1950).

[185] FLEGEL, A. F., and staff, *Report of the Public Health Division., STEM — Thailand (Prepared for Southeast Asia Conference on Public Health Programs, Bangkok, Thailand, Aug. 6–11, 1951)*, Fed. Security Agency, Publ. Hlth. Serv., Div. of Int. Hlth., Wash.

[186] FLETCHER, T. E., *DDT-Metabolism in two DDT-Resistant Strains and One Nonresistant Strain of Musca domestica L.*, Trans. R. Soc. trop. Med. Hyg. *46*, 6 (1952).

[187] FLOCH, H., *Lutte antiamarile et lutte antipaludique en Guyane française: quelques résultats enregistrés à ce jour*, Arch. Inst. Pasteur Guyane et du Territoire de L'Inini, Publ. No. 234 (1951).

[188] FLOCH, H., *Les résultats de la «dedetisation» en Guyane française: destruction de «Aedes aegypti», et réduction spectaculaire du paludisme*, Rev. Palud. Méd. trop. *9* (82), 49–57 (1951).

[189] FLOCH, H., *Lutte antiamarile et lutte antipaludique en Guyane française*, de L'Inst. Pasteur Guyane Francaise et du Territoire de L'Inini, Publ. No. 213, pp. 5–106 (1950).

[190] FOTHERGILL, L. D., DINGLE, J. H., FARBER, S., and CONNERLY, M. L., *Human Encephalitis Caused by the Virus of the Eastern Variety of Equine Encephalomyelitis*, New Engl. J. Med. *219*, 411 (1938).

[191] FOY, H., KONDI, A., DAMKAS, C., DEPANIAN, M., LEFCOPOULOU, T., BACH, L. G., DAX, R., PITCHFORD, J., SHIELE, P., and LANGTON, M., *Malaria and Blackwater Fever in Macedonia and Thrace in Relation to DDT*, Ann. trop. Med. Parasit. *42* (2), 153–172 (1948).

[192] FROES, H. P., *A campanha de dedetizacão no estado do Rio, Brazil*, Bol. Ofic. sanit., pan-amer. *26* (11–12), 954–956 (1947).

[193] FULLMER, O. H., and HOSKINS, W. M., *Effects of DDT Upon the Respiration of Susceptible and Resistant House Flies*, J. econ. Ent. *44*, 858–870 (1951).

[194] GABALDON, A., *The Effect of DDT on the Population of Anopheline Vectors in Venezuela*, Riv. Parassit. *13*, 29–41 (1952).

[195] GABALDON, A., *Nation-Wide Eradication Projects in the Americas, II. Progress of the Malaria Campaign in Venezuela*, J. nat. Malar. Soc. *10*, 124–141 (1951).

[196] GABALDON, A., *The Nation-Wide Campaign Against Malaria in Venezuela*, Trans. R. Soc. trop. Med. Hyg. *43*, 113–160 (1949).

[197] GABALDON, A., *Malaria Incidence in the West Indies and South America*, Chapter 31 in: *Malariology*, ed. MARK F. BOYD (W. B. Saunders Co., Philadelphia 1949), pp. 764–787.

[198] GABALDON, A., *Malaria Control in the Neotropic Region*, Chapter 64 in: *Malariology*, ed. MARK F. BOYD (W. B. Saunders Co., Philadelphia 1949), pp. 1400–1415.

[199] GABALDON, A., *Mortalidad por malaria en Venezuela: I. Bases estadísticas y distribución geográfica*, Tijeret. s. Malar. *10*, 191–237 (1946).

[200] GABALDON, A., and BERTI, A. L., *The First Large Area to Report Malaria Eradication in the Tropical Zone: North-Central Venezuela*, Amer. J. trop. Med. Hyg. *3* (5), 793–807 (1954).

[201] GAHAN, J. B., ANDERS, R. S., HIGHLAND, H., and WILSON, H. G., *Baits for the Control of Resistant Flies*, J. econ. Ent. *46*, 965–969 (1953).

[202] GAHAN, J. B., DOWNS, W. G., and CELIS S., HELIODORO, *Control of Anopheles pseudopunctipennis in Mexico With DDT Residual Sprays Applied in Buildings* (Part II), Amer. J. Hyg. *49*, 285–289 (1949).

[203] GAHAN, J. B., and LINDQUIST, A. W., *DDT Residual Sprays Applied in Buildings to Control Anopheles quadrimaculatus*, J. econ. Ent. *38* (2), 223–230 (1945).

[204] GAHAN, J. B., and PAYNE, G. C., *Control of Anopheles pseudopunctipennis in Mexico With DDT Residual Sprays Applied in Buildings*, Amer. J. Hyg. *45*, 123–132 (1947).

[205] GAHAN, J. B., TRAVIS, B. V., and LINDQUIST, A. W., *DDT as a Residual-Type Spray to Control Disease-Carrying Mosquitoes: Laboratory Tests*, J. econ. Ent. *38* (2), 236–240 (1945).

[206] GAHAN, J. B., TRAVIS, B. V., MORTON, F. A., and LINDQUIST, A. W., *DDT as a Residual-Type Treatment to Control Anopheles quadrimaculatus: Practical Tests*, J. econ. Ent. *38* (2), 231–235 (1945).

[207] GAHAN, J. B., and WEIR, J. M., *House Flies Resistant to Benzene Hexachloride*, Science *111*, 651–652 (1950).

[208] GALINDO, PEDRO, *Chagas' Disease Investigations*, Annu. Rep. Gorgas Memor. Lab., 1951, pp. 13–14.

[209] GARNHAM, P. C. C., *Modern Concepts in Malaria Control*, Copyright of the Royal Sanitary Institute, Exerpt from J. R. sanit. Inst. *69* (5), 617–623 (1949).

[210] GARNHAM, P. C. C., *The New Insecticides*, E. Afr. med. J. *25*, 5–10 (1948).

[211] GARNHAM, P. C. C., *The Control of Epidemic Malaria in the Kenya Highlands by DDT Impregnation of Huts—First Year's Results*, Rep. agric. Res. Coun., Inter-Dep. Insecticide Comm. pp. 1–22 (1947).

[212] GARNHAM, P. C. C., DAVIES, C. W., HEISCH, R. B., and TIMMS, G. L., *An Epidemic of Louse-Borne Relapsing Fever in Kenya*, Trans. R. Soc. trop. Med. Hyg. *41* (1), 141–170 (1947).

[213] GARNHAM, P. C. C., and McMAHON, J. P., *The Eradication of Simulium neavei, Roubaud, From an Onchocerciasis Area in Kenya Colony*, Bull. ent. Res. *37* (4), 619–628 (1947).

[214] GARNHAM, P. C. C., WILSON, D. B., and WILSON, M. E., *Malaria in Kigezi, Uganda*, J. trop. Med. Hyg. *51* (8), 156–159 (1948).

[215] GARRETT-JONES, C., *An Experiment in Trapping and Controlling Anopheles maculipennis in North Iran*, Bull. World Hlth. Org. *4*, 547–562 (1951).

[216] GAUD, J., and MÉCHALI, D., *Place de la lutte anti-imaginale dans l'action antipaludique au Maroc*, Congr. int. Hyg. Med. Méditerranéennes Alger. *3, 4, 5*, 247–251 (1950).

[217] GAUD, J., MÉCHALI, D., and CLIER, J. L., *Emploi et avenir des insecticides de contact au Maroc dans la prophylaxie des maladies épidémiques*, Bull. Inst. Hyg. Maroc *8*, 35–90 (1948).

[218] GEAR, J. H. S., and MURRAY, N. L., *Typhus Fever in the Eastern Transvaal, With Special Reference to an Epidemic Occurring in 1945*, S. Afr. med. J. *21* (7), 214–218 (1947).

[219] GEIB, A. F., *An Indication of Mosquito Larvae Resistance to DDT*, Proc. J. Meet. Amer. Mosq. Contr. Ass. and Virginia Mosq. Contr. Ass. Feb. 1950, pp. 68–71.

[220] GIGLIOLI, G., *Nation-Wide Eradication Projects in the Americas, III. Eradication of Anopheles darlingi From the Inhabited Areas of British Guiana by DDT Residual Spraying*, J. nat. Malar. Soc. *10*, 142–161 (1951).

[221] GIGLIOLI, G., *Malaria, Filariasis and Yellow Fever in British Guiana*, Mosq. Contr. Serv., Med. Dep., Brit. Guiana, 1948, pp. 153–182.

[222] GILBERTSON, W. E., *Sanitary Aspects of the Control of the 1943–1944 Epidemic of Dengue Fever in Honolulu*, Amer. J. publ. Hlth. *35* (3), 261–270 (1945).

[223] GILLETTE, H. P. S., *A Short Review of DDT Residual House Spraying for Malaria Control in Trinidad 1945–1948*, Caribbean med. J. *11*, 6–26 (1949).

[224] GILYARD, R. T., *Mosquito Transmission of Venezuelan Virus Equine Encephalomyelitis in Trinidad*, Bull. U.S. Army med. Dep. No. 75, pp. 96–107 (1944).

[225] GIMENO DE SANDE, A., *Campaña profiláctica contra el tifus exantemático en Motril, Salobreña y Almuñécar durante el año 1948*, Rev. Sanid. Hig. públ. Madr. *23* (5), 443–449 (1949).

[226] GIMENO DE SANDE, A., *Campaña profiláctica contra el tifus exantemático en Motril durante el año 1947*, Rev. Sanid. Hig. públ. Madr. *222* (4), 342–357 (1948).

[227] GJULLIN, C. M., ISAAK, L. W., and SMITH, G. F., *Effectiveness of EPN and Other Organic Phosphate Insecticides on Resistant Mosquitoes in California*, presented before Amer. Ass. econ. Ent., Dec. 1952.

[228] GJULLIN, C. M., and PETERS, R. F., *Recent Studies of Mosquito Resistance to Insecticides in California*, Mosquito News *12*, 1–6 (1952).

[229] GLASGOW, R. D., and COLLINS, D. L., *Ecological, Economic, and Mechanical Considerations Relating to the Control of Ticks and Rocky Mountain Spotted Fever on Long Island*, J. econ. Ent. *41*, 427–431 (1948).

[230] GOOD, N. E., *Effectiveness of DDT Dusting in Controlling Rat Ectoparasites and Typhus Infection in Rats*, CDC Bull., April 1950, pp. 5–11, Atlanta.

[231] GORDON, J. E., *Louse-Borne Typhus Fever in the European Theater of Operations, U.S. Army, 1945* in: The Rickettsial Diseases of Man (American Association for Advances in Science, Washington 1948), pp. 16–27.

[232] GORDON, J. E., and KNIES, P. T., *Flea Versus Rat Control in Human Plague*, Amer. J. med. Sci. *213*, 362–376 (March 1947).

[233] GORGAS MEMORIAL LABORATORY, *Annual Report, 1951, House Document No. 278, 82nd Congress, 2nd Session* (U.S. Government Printing Off., Washington 1952).

[234] GOUCK, H. K., and SMITH, C. N., *DDT to Control Wood Ticks*, J. econ. Ent. *40*, 303–308 (1947).

[235] GOUCK, H. K., and SMITH, C. N., *DDT in the Control of Ticks on Dogs*, J. econ. Ent. *37*, 130 (1944).

[236] GOULDING, R. L., jr., and CURRAN, R. F., *Activities of the Co-Operative Sand Fly Research Unit*, Report of the 22nd Annu. Meet., Fla. Anti-mosq. Ass., April 29–May 2, 1951 (pp. unnumbered).

[237] GOULDING, R. L., CURRAN, R. F., and LABRECQUE, G. C., *Field Tests of Insecticides for the Control of the Salt-Marsh Sand Fly, Culicoides furens Poey*, Rep. of the 23rd Annu. Meet., Fla. Anti-mosq. Ass., March 2–5, 1952, p. 67.

[238] GRAMICCIA, G., and SACCA, G., *Notes on Kala-Azar and Its Control in Iswarganj Thana, Mymensingh District, East Bengal*, Indian J. Malar. 7, 83–91 (1953).

[239] GRAYSON, J. McD., *Effects Upon the German Cockroach from Twelve Generations of Selection for Survival to Treatments With DDT and Benzene Hexachloride*, J. econ. Ent. *46*, 124–126 (1953).

[240] GREAVES, F. C., GEZON, H. M., and ALSTON, W. F., *Studies on Louse-Borne Relapsing Fever in Tunisia*, Nav. med. Bull., Wash. *45* (6), 1029–1048 (1945).

[241] HACKETT, L. W., *Conspectus of Malaria Incidence in Northern Europe, the Mediterranean Region, and the Near East*, Chapter 32 in: *Malariology*, ed. MARK F. BOYD (W. B. Saunders Co., Philadelphia 1949), pp. 788–799.

[242] HACKETT, L. W., *Malaria in Europe* (Oxford University Press, London 1937).

[243] HACKETT, L. W., and MOSHKOVSKI, S., *Malaria Control in the Palearctic Region*, Chapter 65 in: *Malariology*, ed. MARK F. BOYD (W. B. Saunders Co., Philadelphia 1949), pp. 1416–1431.

[244] HADAWAY, A. B., *Observations on Mosquito Behaviour in Native Huts*, Bull. ent. Res. *41*, 63–78 (1950).

[245] HAMMON, W. McD., *The Arthropod-Borne Virus Encephalitides*, Amer. J. trop. Med. *28*, 515–525 (1948).

[246] HAMMON, W. McD., *The Encephalitides of Virus Origin With Special Reference to Those of North America*, Clinics *4*, 485–503 (1945).

[247] HAMMON, W. McD., and REEVES, W. C., *Western Equine Encephalitis Control Studies in Kern County, California, 1945. II. An Evaluation of the Effectiveness of Certain Types of Mosquito Control Including Residual DDT on Virus Infection Rates in Culex Mosquitoes and in Chickens*, Amer. J. Hyg. *47*, 93–102 (1948).

[248] HAMMON, W. McD., and REEVES, W. C., *Recent Advances in the Epidemiology of the Arthropod-Borne Virus Encephalitides, Including Certain Exotic Types*, Amer. J. publ. Hlth. *35*, 994–1004 (1945).

[249] HAMMON, W. McD., and REEVES, W. C., *Laboratory Transmission of St. Louis Encephalitis Virus by Three Genera of Mosquitoes*, J. exp. Med. *78*, 241–253 (1943).

[250] HAMMON, W. McD., TIGERTT, W. D., SATHER, G., and SCHENKER, H., *Isolations of Japanese B Encephalitis Virus From Naturally Infected Culex tritaeniorhynchus Collected in Japan*, Amer. J. Hyg. *50*, 51–56 (1949).

[251] HARDY, A. V., WATT, J., and DeCAPITO, T., *Studies of the Acute Diarrheal Diseases. VI. New Procedures in Bacteriological Diagnosis*, Publ. Hlth. Rep., Wash. *57*, 521–523 (1942).

[252] HARRISON, C. M., *Inheritance of Resistance of DDT in the House Fly, Musca domestica L.*, Nature, Lond. *167*, 855–856 (1951).

[253] HARRISON, C. M., *DDT-Resistant House Flies*, Ann. appl. Biol. *37*, 306–309 (1950).

[254] HART, A., HART, J. H., and SARACHO-LOPEZ, E., *Malaria Control in Guayaramerin, Bolivia*, Mosquito News *8*, 21–25 (1948).

[255] HARVEY, W. C., and HILL, H., *Insect Pests* (Paul B. Hoeber, Inc., New York 1948), 347 pp.

[256] HEAL, R. E., Personal communication of 25 September 1952, to Dr. CARL O. MOHR.

[257] HEAL, R. E., NASH, K. B., and WILLIAMS, M., *An Insecticide-Resistant Strain of the German Cockroach From Corpus Christi, Texas*, J. econ. Ent. *46*, 385–386 (1953).

[258] HEDLEY, O., *Greece Wipes Out Malaria*, Sci. News Lett. *57* (25), 388 (1950).

[259] HEMPHILL, F. M., *Trends of Diarrheal Disease Mortality in the United States, 1941 to 1946, incl.*, Publ. Hlth. Rep., Wash. *63* (53), 1699–1711 (1948).

[260] HERTIG, M., *Observations on the Density of Phlebotomus Populations Following DDT Campaigns*, Bull. World Hlth. Org. *2* (4), 621–628 (1950).

[261] HERTIG, M., *Phlebotomus and Residual DDT in Greece and Italy*, Amer. J. trop. Med. *29* (5), 773–809 (1949).

[262] HERTIG, M., and FAIRCHILD, G. B., *Taxonomy and Biology of 'Phlebotomus' Sand Flies*, Annu. Rep. Gorgas Memor. Lab., 1951, pp. 15–17 (1952).

[263] HERTIG, M., and FAIRCHILD, G. B., *The Control of Phlebotomus in Peru With DDT*, Amer. J. trop. Med. *28* (2), 207–230 (1948).

[264] HERTIG, M., and FISHER, R. A., *Control of Sandflies With DDT*, Bull. U.S. Army med. Dep. No. 88, pp. 97–101 (1945).

[265] HESS, A. D., *Current Status of Insecticide Resistance in Insects of Public Health Importance*, Amer. J. trop. Med. Hyg. *2*, 311–317 (1953).

[266] HESS, A. D., *The Significance of Insecticide Resistance in Vector Control Programs*, Amer. J. trop. Med. Hyg. *1*, 371–388 (1952).

[267] HINMAN, E. H., *Criteria of Malaria Eradication, Committee Report*, J. nat. Malar. Soc. *10*, 195–196 (1951).

[268] HILL, E. L., and MORLAN, H. B., *Evaluation of County-Wide DDT Dusting Operations in Murine Typhus Control*, Publ. Hlth. Rep., Wash. *63* (51), 1635–1653 (1948).

[269] HILL, E. L., MORLAN, H. B., UTTERBACK, B. C., and SCHUBERT, J. H., *Evaluation of County-Wide DDT Dusting Operations in Murine Typhus Control (1946 Through 1949)*, Amer. J. publ. Hlth. *41* (4), 396–401 (1951).

[270] HILL, M. A., and ROBERTS, E. W., *An Investigation Into the Effects of 'Gammexane' on the Larvae, Pupae and Adults of Culicoides impunctatus* Goetghebuer *and on the Adults of Culicoides obsoletus* Meigen, Ann. trop. Med. Parasit. *41*, 143–163 (1947).

[271] HOCKING, K. S., Quoted in article: *Distribution and Fate of Anopheles gambiae and A. funestus in Two Different Types of Huts Treated With DDT and BHC in Uganda*, by P. R. WILKINSON, Bull. ent. Res. *42* (Pt. 1), 45–54 (1951).

[272] HOCKING, K. S., *The Residual Action of DDT Against Anopheles gambiae and funestus*, Trans. R. Soc. trop. Med. Hyg. *40* (5), 589–601 (1947).

[273] HODES, H. L., THOMAS, L., and PECK, J. L., *Complement-Fixing and Neutralizing Antibodies Against Japanese B Virus in the Sera of Okinawan Horses*, Science *103*, 357–359 (1946).

[274] HOEKENGA, M. T., *Plague in the Americas*, J. trop. Med. Hyg. *50* (10), 190–201 (1947).

[275] HOFFMAN, F. L., *A Plea for a National Committee on the Eradication of Malaria*, Sth. med. J., Nashville *9*, 413–420 (1916).

[276] HOFFMAN, R. A., and LINDQUIST, A. W., *Absorption and Metabolism of DDT, Toxaphene, and Chlordan by Resistant House Flies as Determined by Bioassay*, J. econ. Ent. *45*, 233–235 (1952).

[277] HOLDSWORTH, R. P., jr., *Local Control of Epidemic Dengue Fever*, Milit. Surg. *102* (5), 377–381 (1948).

[278] HOSKINS, W. M., *Recent Work on the Resistance Problem at Berkeley. Discussion. Conference on Insecticide Resistance and Insect Physiology, Dec. 8–9, 1951, University of Cincinnati*, Nat. Res. Coun. Publ. No. 219, p. 35 (1952).

[279] HOWELL, D. E., *Fly Control With Repellents*, presented before Amer. Ass. econ. Ent., Dec. 1952.

[280] HOWITT, B. F., DODGE, H. R., BISHOP, L. K., and GORRIE, R. H., *Recovery of the Virus of Eastern Encephalomyelitis From Mosquitoes (Mansonia perturbans) Collected in Georgia*, Science *110*, 141–142 (1949).

[281] HOWITT, B. F., DODGE, H. R., BISHOP, L. K., and GORRIE, R. H., *Virus of Eastern Equine Encephalomyelitis Isolated from Chicken Mites (Dermanyssus gallinae) and Chicken Lice (Eomenacanthus stramineus)*, Proc. Soc. exp. Biol., N. Y. *68*, 622–625 (1948).

[282] HOWITT, B. F., *Recovery of the Virus of Equine Encephalomyelitis From the Brain of a Child*, Science *88*, 455–456 (1938).

[283] HURLBUT, H. S., PEFFLY, R. L., and SALAH, A. Z., *DDT Resistance in Egyptian Body Lice*, Amer. J. trop. Med. Hyg. *3* (5), 922–929 (1954).

[284] HURLBUT, H. S., ALTMAN, R. M., and NIBLEY, C., jr., *DDT Resistance in Korean Body Lice*, Science *115* (2975), 11–12 (1952).

[285] ILPPÖ, A., HALLMAN, N., DONNER, M., LOUHIVUORI, K., and YLIRUOKANEN, A., *The Role of Insects in the Spreading of Infantile Diarrhea in Finland*, Ann. Med. intern. Fenn. *39* (2), 149–150 (1950).

[286] INDONESIA, GOVERNMENT OF, *Report, Malaria Conference, Bangkok, Thailand, Sept. 21–24, 1953*, World Hlth. Org., Sea./Mal./5, New Delhi, 25 June 1953.

[287] JACHOWSKI, L. A., jr., and OTTO, G. F., *Filariasis in American Samoa. II. Evidence of transmission Outside of Villages*, Amer. J. trop. Med. Hyg. *1* (4), 662–670 (1952).

[288] JACUSIEL, F., *Sandfly Control With DDT Residual Spray. Field Experiments in Palestine*, Bull. ent. Res. *38* (Part 3), 479–488 (1947).

[289] JADIN, J., FAIN, A., and RUPP, H., *Lutte antimalarienne étendue en zone rurale au moyen de D.D.T. à Astrida, Ruanda-Urundi*, Mém. Inst. colon. belge Sci. nat. *21* (1), 47 pp. (1952).

[290] Jaujou, C., Camain, R., and Adam, J. P., *Rapport sur L'Activité du Service Anti-paludique en 1949*, Rapp. Cons. Général Corse. Sess. extraordinaire de février 1950 (Malaria Control, Corsica, 1949).

[291] Jaujou, C., Michel, L., and Hamon, J., *Rapport sur L'Activité du Service Anti-paludique en 1950*, Rapp. Cons. Général Corse. Sess. ordinaire de mai 1951 (Malaria Control, Corsica, 1950).

[292] Jenkins, D. W., *Bionomics of Culex tarsalis in Relation to Western Equine Encephalomyelitis*, Amer. J. trop. Med. *30*, 909–916 (1950).

[293] Jenney, E. R., *Public Health in Indonesia*, Publ. Hlth. Rep., Wash. *68*, 409–415 (1953).

[294] Jenney, E. R., Ketterer, W. A., and Murray, L. J., *Report of Public Health Division, STEM-Indonesia (Prepared for Southeast Asia Conference on Public Health Programs, Bangkok, Thailand, Aug. 6–11, 1951)*, Fed. Security Agency, Publ. Hlth. Serv., Div. of Int. Hlth., Washington (1951).

[295] Jepson, W. F., *Economic Control of the Relapsing Fever Tick in African Houses*, Nature, Lond. *160*, 874 (1947).

[296] Jerace, F., *La malaria nell'Agro Romano nel quinquennio 1943–1947*, Riv. Malariol. *27* (3), 103–110 (1948).

[297] Johnson, D. H., and Wharton, G. W., *Tsutsugamushi Disease: Epidemiology and Methods of Survey and Control*, Nav. med. Bull., Wash. *46* (3), 459–472 (1946).

[298] Johnson, D. R., Personal communication dated April 26, 1954.

[299] Johnson, M. S., and Hill, A. J., *Partial Resistance of a Strain of Bedbug to DDT Residual*, Med. News Lett., Navy Bur. Med. Surg., Wash. *12*, 27–28 (1948).

[300] Jones, H. A., McAlister, L. C., jr., Bushland, R. C., and Knipling, E. F., *DDT Impregnation of Underwear for Control of Body Lice*, J. econ. Ent. *38* (2), 217–223 (1945).

[301] Jones, T. W. T., *Malaria Survey of Kyaukpyu, Ramree Island, Burma*, Indian J. Malar. *4*, 239–249 (1950).

[302] Kalra, S. L., and Jacob, V. P., *Effect of DDT and BHC on Ornithodorus Ticks*, Part II, Indian J. med. Res. *39*, 311–317 (1951).

[303] Kalra, S. L., Jacob, V. P., and Rao, K. N., *Effect of DDT and BHC on Ornithodorus Ticks*, Part I, Indian J. med. Res. *38*, 457–466 (1950).

[304] Kartman, L., and DaSilveira, M. M., *Effect of Short Contact With DDT Residues on Anopheles gambiae*, J. econ. Ent. *39* (3), 356–359 (1946).

[305] Keiding, J., and VanDeurs, H., *DDT Resistance in House Flies in Denmark*, Nature, Lond. *163*, 964–965 (1949).

[306] Ketterer, W. A., *Economic Benefits of Malaria Control in the Republic of Indonesia*, Publ. Hlth. Rep., Wash. *68*, 1056–1058 (1953).

[307] Kilpatrick, J. W., and Fay, R. W., *DDT-Resistance Studies With the Oriental Rat Flea*, J. econ. Ent. *45*, 284–288 (1952).

[308] King, W. V., *DDT-Resistant House Flies and Mosquitoes*, J. econ. Ent. *43*, 527–532 (1950).

[309] King, W. V., and Gahan, J. B., *Failure of DDT to Control House Flies*, J. econ. Ent. *42*, 405–409 (1949).

[310] Kitselman, C. H., and Grundmann, A. W., *Equine Encephalomyelitis Virus Isolated From Naturally Infected Triatoma sanguisuga* LeConte, Tech. Bull. Kans. agric. Exp. Sta. *50*, 1–5 (1940).

[311] Kitsoka, M., *DDT-Resistant Louse in Tokyo*, Japan. J. med. Sci. Biol. *5*, 75–88 (1952).

[312] Knipling, E. F., *The Greater Hazard — Insects or Insecticides*, J. econ. Ent. *46*, 1–7 (1953).

[313] Kohls, G. M., *Vectors of Rickettsial Diseases*, Ann. intern. Med. *26* (5), 713–719 (1947).

[314] Kosol, L. A., and Griffith, M. E., *Malaria Control in Thailand* (in manuscript, 1953).

[315] KNUTSON, H., *A Community Project for Control of the American Dog Tick*, J. econ. Ent. *45*, 889–890 (1952).

[316] KRATZ, F. W., BRIDGES, C. B., and HINTGEN, G. W., *Malaria Control in Turkey, 1951, 1952, 1953*, Total Project Rep. (TS 77–930) of Publ. Hlth. Group of Mutual Security Agency-Turkey to Chief of Mutual Security Agency, Special Mission to Turkey (1953).

[317] KUMM, H.W., *The Arthropod-Borne Virus Encephalitides*. From: *Rosenau Preventive Medicine and Hygiene*, ed. KENNETH F. MAXCY (Appleton-Century-Crofts, Inc., New York 1951), pp. 375–385.

[318] LAIRD, R. L., As quoted in personal communication from F. EDWARD COUGHLIN (1954).

[319] LENT, H., and DE OLIVEIRA, S. J., *Action of DDT (Dichlorodiphenyltrichloroethane) on Insects Which Transmit Chagas' Disease*, Rev. bras. Biol. *4*, 329–331 (1944).

[320] LINDQUIST, A. W., MADDEN, A. H., and KNIPLING, E. F., *DDT as a Treatment for Fleas on Dogs*, J. econ. Ent. *37* (1), 138 (1944).

[321] LINDQUIST, A. W., and McDUFFIE, W. C., *DDT Residual Spray Tests in Panama*, J. econ. Ent. *38*, 608 (1946).

[322] LINDQUIST, A. W., ROTH, A. R., YATES, W. W., HOFFMAN, R. A., and BUTTS, J. S., *Use of Radioactive Tracers in Studies of Penetration and Metabolism of DDT in House Flies*, J. econ. Ent. *44*, 167–172 (1951).

[323] LINDQUIST, A. W., and WILSON, H. S., *Development of a Strain of House Flies Resistant to DDT*, Science *107*, 276 (1948).

[324] LINDSAY, D. R., STEWART, W. H., and WATT, J., *Effect of Fly Control on Diarrheal Disease in an Area of Moderate Morbidity*, Publ. Hlth. Rep., Wash. *68* (4), 361–367 (1953).

[325] LIVADAS, G. A., *The Present Position of the Malaria Control Demonstration Project in Terai*, World Hlth. Org./Mal./76, 5 February 1952.

[326] LIVADAS, G. A., *Malaria Control in Greece During the Last Fifty-Year Period*, Riv. Malariol. *30* (1), 17–27 (1951).

[327] LIVADAS, G. A., *Do Anophelines Acquire Resistance to DDT?* World Hlth. Org. Expert Panel on Malar. 74, Dec. 18, 1951.

[328] LIVADAS, G. A., *Reduction and Disappearance of Anopheline Species Prevailing in Attica (Greece) as a Result of Control Programs Carried out During 1946–1949*, Riv. Malariol. *29* (2), 73–82 (1950).

[329] LIVADAS, G. A., *Les résultats de l'application du DDT en Grèce*, Pr. méd. *57* (33), 454–455 (1949).

[330] LIVADAS, G.A., *Malaria Control in Greece During the Four-Year Period 1946–1949*, Riv. Malariol. *28*, 247–278 (1949).

[331] LIVADAS, G. A., and BELIOS, G. B., *Postwar Malaria Control in Greece and Its Results on Basis of Epidemiological Data*, Proc. 4th Int. Congr. trop. Med., Wash., pp. 884–895 (1948).

[332] LIVADAS, G. A., and BELIOS, G. B., *La campagne antimalarique de 1946 en Grèce*, printed report presented to the Soc. Pat. exot. Feb. 12, 1947, p. 1–16.

[333] LIVADAS, G. A., BELIOS, G. B., and ISSARIS, P., *A Method of Spraying Shelters With the New Insecticide, DDT, for the Control of Anopheles Mosquitoes. Experimental Applications*, Malar. Sect., Athens School of Hygiene (1946).

[334] LIVADAS, G. A., and ISSARIS, P., *The New Insecticide DDT and Its Effect on Public Health From Its Applications*, Malar. Sect., Athens School of Hygiene (1945).

[335] LIVADAS, G. A., KOROYANNAKI, F. K., and ISSARIS, P. C., *Malaria Control in Greece During the Four-Year Period 1946–1949*, Riv. Malariol. *28* (5), 247–278 (1949).

[336] LOGAN, J. A., *The Sardinian Project — an Experiment in Malaria Control by Species Eradication*, Proc. Inst. Civ. Engrs. Part 3, April 1953, p. 1–14, London.

[337] LOGAN, J. A., *The Sardinian Project — an Experiment in the Eradication of an Indigenous Malarious Vector* (The Johns Hopkins Press, Baltimore 1953).

[338] Logan, J. A., *Anopheles labranchiae Eradication in Sardinia — an Interim Report*, Amer. J. trop. Med. *30* (2), 313–323 (1950).

[339] Ludvik, G. F., Snow, W. E., and Hawkins, W. B., *The Susceptibility of Anopheles quadrimaculatus to DDT After Five Years of Routine Treatment in the Tennessee River Valley*, J. nat. Malar. Soc. *10*, 35–43 (1951).

[340] Ludwig, R. G., and Nicholson, H. P., *The Control of Rat Ectoparasites With DDT*, Publ. Hlth. Rep., Wash. *62* (3), 77–84 (1947).

[341] Lyman, F. E., *Malaria Control in North Viet-Nam (Tonkin, Indo-China), 1950*, Communicable Disease Center Bull. *10* (8), 1–5 (1951).

[342] Macan, T. T., *Malaria Survey of the Arakan Region of Bengal and Burma*, Parasit. *40*, 290–297 (1950).

[343] Macan, T. T., *Mosquitoes and Malaria in the Kabaw and Kale Valleys, Burma*, Bull. ent. Res. *39*, 237–268 (1948).

[344] Macchiavello, A., Personal communication (1952).

[345] Macchiavello, A., *Plague Control With DDT and '1080'. Results Achieved in a Plague Epidemic at Tumbes, Peru, 1945*, Amer. J. publ. Hlth. *36* (8), 842–854 (1946).

[346] Macchiavello, A., and Mostajo, B., Personal communication (1946).

[347] MacDonald, G., and Davidson, G., *Dose and Cycle of Insecticide Applications in the Control of Malaria*, Bull. World Hlth. Org. *9*, 785–812 (1953).

[348] Mackie, T. T., Hunter, G. W., and Worth, C. B., *A Manual of Tropical Medicine* (W. B. Saunders Co., Philadelphia 1954).

[349] Madden, A. H., Lindquist, A. W., Longcoy, O. M., and Knipling, E. F., *Control of Adult Sand Flies by Airplane Spraying With DDT*, Florida Ent. *29*, 5–10 (1946).

[350] Maier, P. P., and Mathis, W., *Fly Control at Dairies With Parathion-Impregnated Twine*, presented before Amer. Ass. econ. Ent., Dec. 1952.

[351] Malaya, Government of, *Report submitted, Malaria Conference, Bangkok, Thailand, Sept. 21–24, 1953*, World Hlth. Org., Sea./Mal./9, New Delhi, 30 June 1953.

[352] Malaya, Federation of, *Institute for Medical Research, 1900–1950*, Jubilee vol. No. 25 (Government Press, Kuala Lumpur 1951), pp. 389.

[353] Malaya, Federation of, *Annual Report of the Institute for Medical Research for the Year 1950* (Government Press, Kuala Lumpur 1951), pp. 23–26.

[354] Manson-Bahr, P. H., *Manson's Tropical Diseases*, 13th ed. (Cassell and Co., Ltd., Lond. 1950), 1136 pp.

[355] Mandekos, A. G., *Can DDT Help Eradicate Anopheles and Malaria From Greece?* Practices of the Acad. Athens, 1946, vol. 21 (1950).

[356] Mandekos, A. G., *Dichlor-Diphenyl-Trichlora-Ethan (DDT, Neocid, Gesarol, Gix): A New Insecticide Which Will Revolutionize the Methods of Controlling Various Diseases Transmitted by Insects*, Malar. Sect., Athens School of Hygiene (1945).

[357] Mandekos, A. G., *Toxische Wirkung von Neocid auf Larven, Puppen und Imagines von Anopheles und Culex*, Dtsch. tropenmed. Z. *48*, 84–88 (1944).

[358] Mandekos, A. G., and Damkas, Chr., *Can DDT Help Eradicate Anopheles and Malaria From Greece?* Grèce méd., vol. 15, copy 12 (1946).

[359] Mandekos, A. G., Damkas, Chr., and Zaphiropoulos, M., *Old and New Methods of Controlling Malaria in Greek Macedonia*, Amer. J. trop. Med. *28* (1), 35–41 (1948).

[360] March, R. B., and Lewallen, L. L., *A Comparison of DDT-Resistant and Non-resistant House Flies*, J. econ. Ent. *43*, 721–722 (1950).

[361] March, R. B., and Metcalf, R. L., *Studies in California of Insecticide-Resistant Flies*, CSMA Proc., June 1950, Soap, N. Y., pp. 80–83.

[362] March, R. B., and Metcalf, R. L., *Laboratory and Field Studies of DDT-Resistant Flies in Southern California*, Bull. Calif. Dept. Agric. *38*, 93–101 (1949).

[363] Mark, R. S., and Brown, A. M., *Malaria Control in Greece*, Adm. Rep. U.S. Publ. Hlth. Serv. (1953).

[364] Mastbaum, O., *Field Experiments With DDT Emulsion and Wettable DDT With Special Reference to Malaria Incidence in Swaziland During the Transmission Season 1949–50*, E. Afr. med. J. *28* (2), 67–74 (1951).

490   S. W. Simmons

[365] MATHESON, R., *Medical Entomology*, 2nd ed. (Comstock Publ. Co., Inc., Ithaca, N. Y., 1950), 612 pp.

[366] MATHIS, W., and QUARTERMAN, K. D., *Field Investigations on the Use of Heavy Dosages of Several Chlorinated Hydrocarbons as Mosquito Larvicides*, Amer. J. trop. Med. Hyg. *2* (2), 318–324 (1953).

[367] MAXCY, K. F., *An Epidemiological Study of Endemic Typhus (Brill's Disease) in the Southeastern United States With Special Reference to Its Mode of Transmission*, Publ. Hlth. Rep., Wash. *41* (52), 2967–2995 (1926).

[368] McCAULEY, R. C., FAY, R. W., and SIMMONS, S. W., *The importance of Coverage in DDT Residual House Spraying for the Control of Anopheles quadrimaculatus Mosquitoes*, Publ. Hlth. Rep., Wash. *63* (13), 401–407 (1948).

[369] McDUFFIE, W. C., EDDY, G. W., CLARK, J. C., and HUSMAN, C. N., *Field Studies With Insecticides to Control the Lone-Star Tick in Texas*, J. econ. Ent. *43*, 520–527 (1950).

[370] MECHALI, M. D., *La lutte anti-paludique au Maroc*, Maroc. Méd. *29* (296), 112–117 (1950).

[371] MER, G. G., *Malaria Control in the State of Israel*, World Hlth. Org./Mal./33, 20 October 1949.

[372] MERCIER, S., *La prophylaxie du paludisme au moyen des insecticides, organiques de synthèse à Tananarive en 1950: Incidences épidémiologiques et démographiques*, Rev. Palud. Méd. trop. *10* (92), 21–31 (1952).

[373] MEXICO, D. F., *Secretaria de Salubridad y Asistencia. Memoria de la Campana Nacional contra el Paludismo*, 188 pp. (1949).

[374] MEXICO, D. F., *Morbilidad y mortalidad por paludismo en la republica mexicana. Mexico secretaria salubridad y asistencia*, Bol. epidem., Mex., No. 6, pp. 10–12 (1946).

[375] MEYER, K. F., HARING, C. M., and HOWITT, B. F., *The Etiology of Epizootic Encephalomyelitis of Horses in the San Joaquin Valley, 1930*, Science *74*, 227–228 (1931).

[376] MIRANDA-FRANCO, R., Personal communication (1952).

[377] MISSIROLI, A., *Resistenza agli insecticide di alcune razze di Musca domestica*, Riv. Parassit. *12*, 5–25 (1951).

[378] MISSIROLI, A., *Il controllo degli insetti della casa e dell'uomo*, Ann. Sanità pubbl. *10* (6), 3–30, Nov.–Dec. 1949.

[379] MISSIROLI, A., *Anopheles Control in the Mediterranean Area*, IVth Int. Congr. trop. Med., Wash., May 10–18, pp. 3–20 (1948).

[380] MISSIROLI, A., *Riduzione o eradicazione degli anofeli*, Riv. Parassit. *8*, 141–169 (1947),

[381] MISSIROLI, A., *La malaria nel 1945 e previsioni per il 1946*, Conf. Ist. sup. Sanit.. Febr. 1946.

[382] MISSIROLI, A., *La malaria nel 1944 e misure profilattiche previste per il 1945*, Estr. R.C. Ist. sup. Sanit., Rome *7*, 638 (1944).

[383] MISSIROLI, A., *Varieties of Anopheles maculipennis and the Malaria Problem in Italy*, Int. Congr. Ent., Berlin, Aug. 15–20, 1938.

[384] MOFFAT, A. L., and staff, *Report of Public Health Division, STEM-Burma (Prepared for Southeast Asia Conference of Public Health Programs, Bangkok, Thailand, Aug. 6–11, 1951)*, Fed. Security Agency, Publ. Hlth. Serv., Div. of Int. Hlth., Wash. (1951).

[385] MOHR, C. O., *Results of Cooperative State-Federal Typhus Control Programs*, Communicable Disease Center Bull., April 1950, pp. 1–5, Atlanta, Ga.

[386] MONTALVAN, J. A., *Campaña antipalúdica en el Ecuador. Metodos empleados y resultados obtenidos*, Rev. Ecuatoriana Hig. Méd. trop. *8/9*, 34–90 (1952).

[387] MONTALVAN, J. A., *Algunas consideraciones sobre el problema de la enfermedad de Chagas en el Ecuador*, Rev. Ecuatoriana Hig. Méd. trop. *7*, 1–6 (1950).

[388] MONTGOMERY, T. H. L., and BUDDEN, F. H., *Typhus in Northern Nigeria. I. Epidemiological Studies*, Trans. R. Soc. trop. Med. Hyg. *41* (3), 327–337 (1947).

[389] MONTOYA, J. A., and OSEJO, P. P., *Studies on the Use of DDT and Phenyl Cellosolve for Control of Pediculosis in Villages in Colombia*, Amer. J. Hyg. *47* (3), 247–258 (1948).

[390] Mooser, H., *Die Bedeutung des Neocid-Geigy für die Verhütung und Bekämpfung der durch Insekten übertragenen Krankheiten*, Schweiz. med. Wschr. *74* (36), 947–952 (1944).

[391] Morales, A. L., *Principios fundamentales de la desinsectación antipalúdica*, Rev. Sanid. Hig. públ., Madr. *25*, 1–16 (1951).

[392] Morales, A. L., and Juarez, E. J., *Consideraciones sobre una campaña de pulverizaciones residuales de la zona palúdica del rio Tiétar (Cáceres)*, Rev. Sanid. Hig. públ., Madr. *24*, 1–14 (1950).

[393] Morlan, H. B., and Hines, V. D., *Evaluation of County-Wide DDT Dusting Operations in Murine Typhus Control, 1950*, Publ. Hlth. Rep., Wash. *66* (33), 1052–1057 (1951).

[394] Morrison, F. O., *Morphological and Habit Changes Correlated With Resistance to Insecticidal Treatment*, 81st Annu. Rep. Ent. Soc., Ontario 1950, pp. 41–43.

[395] Mosher, W. E., jr., *Japanese 'B' Encephalitis; Epidemiological Report of the 1945 Outbreak on Okinawa*, Nav. med. Bull., Wash. *47*, 586–597 (1947).

[396] Mosna, E., *Il controllo con Octa-Klor delle mosche domestiche DDT-resistenti*, Riv. Parassit. *11*, 27–35 (1950).

[397] Mosna, E., *Su una carratteristica biologica del Culex pipiens autogenicus di Latina*, Riv. Parassit. *8*, 125–126 (1947).

[398] Mosna, E., and Alessandrini, M., *La lotta antianofelica con DDT nella provincia di Latina nel 1948 e 1949*, Riv. Parassit. *11* (1), 13–26 (1950).

[399] Muirhead-Thomson, R. C., *DDT and Gammexane as Residual Insecticides Against Anopheles gambiae in African Houses*, Trans. R. Soc. trop. Med. Hyg. *43* (4), 401–412 (1950).

[400] Muirhead-Thomson, R. C., *The Effects of House Spraying With Pyrethrum and with DDT on Anopheles gambiae and A. melas in West Africa*, Bull. ent. Res. *38*, 449–464 (1947).

[401] Muirhead-Thomson, R. C., *Recent Knowledge About Malaria Vectors in West Africa and Their Control*, Trans. R. Soc. trop. Med. Hyg. *40* (5), 511–527 (1947).

[402] Nair, C. P., *DDT as a Residual Insecticide Against A. letifer and A. maculatus in Malaya*, Nature, Lond. *167*, 74 (1951).

[403] Nair, C. P., *Effect of DDT Barrier Spray in A. sundaicus Area in Malaya*, Med. J. Malaya *3*, 198–203 (1949).

[404] Nair, C. P., *Investigations on DDT Barrier Spray in Multiple Vector Areas in Malaya*, Med. J. Malaya *3*, 250–258 (1949).

[405] Nair, C. P., *Investigations on Rural Malaria Control in Malaya by Residual Spraying With DDT*, Med. J. Malaya *2*, 93–124 (1947).

[406] Nasir-Ud-Din, M., *A Year's Work of the National Malaria Control Teams in East Pakistan (1950)*, Pakist. J. Hlth. *1*, 21–41 (1952).

[407] Near East Foundation, *Iranian Area Annual Report, July 1, 1948, to June 30, 1949*.

[408] Neghme, A., Personal communication (1952).

[409] Neghme, A., Albi, H., and Gutiérrez, J., *Campaña de erradicación del Aedes aegypti en Chile (resumen)*, Bol. parasit. chil. *7* (3), 41–42 (1952).

[410] Neghme, A., Gutiérrez, J., and Albi, H., *Control del anofelismo en las zonas maláricas Chilenas*, Bol. Ofic. sanit. pan-amer. *28* (2), 157–158 (1949).

[411] Neghme, A., Román, J., and Sotomayor, R., *Nuevos datos sobre la enfermedad de chagas en Chile*, Bol. Ofic. sanit. pan-amer. *28* (8), 808–817 (1949).

[412] Neghme, A., and Silva, R., *Ensayos sobre la acción de diversos insecticidas en contra del Triatoma infestans, Klug, 1834*, Congr. sudamer. de química, IV, March 5, 1948. Actas y trabajos; Seccion 5, Química en relación con medicamentos y toxicos, pp. 192–194.

[413] Neill, A. H., *Control de la malaria en Ponce*, Puerto Rico J. publ. Hlth. trop. Med. *22*, 300–307 (1946).

[414] NEWMAN, J. F., AZIZ, M. A., and KOSHI, T., *Studies in Contact Toxicity*. Part 1. *Resistance of Successive Generations of Culex fatigans* Wied. *to Contact Insecticides*, Proc. Indian Acad. Sci. *30*, 61–68 (1949).

[415] NOÉ, J., BERTIN, V., GUTIÉRREZ, J., and NEGHME, A., *Diez años de lucha antimalárica en Chile*, Riv. Parassit. *10* (1), 5–24 (1949).

[416] NORRIS, M., *Recovery of a Strain of Western Equine Encephalitis Virus From Culex restuans* Theobold (Diptera; *Culicidae*), Canad. J. Res. *24*, 63–70 (1946).

[417] NORTON, R. J., *Inheritance of DDT Tolerance in the House Fly*, Contr. Boyce Thompson Inst. *17*, 105–126 (1953).

[418] OLITSKY, P. K., and CASALS, J., *Viral Encephalitides*, Chapter 11 in: *Viral and Rickettsial Infections of Man*, ed. THOMAS M. RIVERS (J. B. Lippincott Co., Philadelphia 1952), pp. 214–266.

[419] OLIVIER, L. J. G., *Organized Malaria Control — Northern Transvaal*, Publ. Hlth., Johannesburg *14* (12), 375–379 (1950).

[420] ORTIZ-MARIOTTE, C., *Campaña nacional contra el tifo en México*, Bol. Ofic. sanit. pan-amer. *34* (3), 236–244 (1953).

[421] ORTIZ-MARIOTTE, C., MALO-JUVERA, F., and PAYNE, G. C., *Control of Typhus Fever in Mexican Villages and Rural Populations Through the Use of DDT*, Amer. J. publ. Hlth. *35* (11), 1191–1195 (1945).

[422] OUCHTERLONY, Ö., *Kolera-epidemien i Egypten 1947–1948*, Nord. med. *41* (4), 167–171 (1949).

[423] PAL, R., SHARMA, M. J. D., and KRISHNAMURTHY, B. S., *Studies on the Development of Resistant Strains of House Flies and Mosquitoes*, Indian J. Malar. *6*, 303–316 (1952).

[424] PAMPANA, E. J., *Lutte antipaludique par les insecticides à action rémanente: résultats des grandes campagnes*, Bull. World Hlth. Org. *3*, 557–619 (1951).

[425] PAMPANA, E. J., *Malaria in Europe 1938–1947*, Epidem. vit. Stat. Rep. *1* (18), 392–400 (1948).

[426] PAN AMERICAN SANITARY BUREAU, *Editorial. La erradicación del Aedes aegypti y el control de la fiebre amarilla: cuatro años de un programa continental*, Bol. Ofic. sanit. pan-amer. *31*, 386–392 (1951).

[427] PAN AMERICAN SANITARY BUREAU, *Demografia: México*, Bol. Ofic. sanit. pan-amer. *28*, 1329 (1949).

[428] PAN AMERICAN SANITARY CONFERENCE (13th), TRUJILLO CITY, DOMINICAN REPUBLIC, OCT. *1–10*, 1950, *Health Appraisal Forms*, Bol. Ofic. sanit. pan-amer. *30*, 191–219 (1951).

[429] PATEL, T. B., and RODDA, S. T., *Use of DDT as a Plague Control Measure in the Bombay State*, Indian med. Gaz. *87* (5), 217–221 (1952).

[430] PATRISSI, T., *L'impiego del DDT nelle lotta antimalarica*, Roma (cited by PAMPANA, [424]), Bull. World Hlth. Org. *3* (4), 557–619 (1949).

[431] PAULLIN, J. E., *Typhus Fever, With a Report of Cases*, Sth. med. J., Nashville, vol. 6, pp. 36–43 (1913).

[432] PELAEZ, D., *Mortalidad y morbilidad por paludismo en México*, Bol. epidem., Mex. *11*, 234–244 (1947).

[433] PERRY, A. S., and FAY, R. W., *Increased Resistance Arising From Reduced Absorption of DDT in a Strain of Musca domestica* L., In manuscript.

[434] PERRY, A. S., FAY, R. W., and BUCKNER, A. J., *Dehydrochlorination as a Measure of DDT-Resistance in House Flies*, J. econ. Ent. *46*, 972–976 (1953).

[435] PERRY, A. S., and HOSKINS, W. M., *Detoxification of DDT as a Factor in the Resistance of House Flies*, J. econ. Ent. *44*, 850–857 (1951).

[436] PERRY, A. S., and HOSKINS, W. M., *Synergistic Action With DDT Toward Resistant house Flies*, J. econ. Ent. *44*, 839–850 (1951).

[437] PERRY, A. S., and HOSKINS, W. M., *The Detoxification of DDT by Resistant House Flies and Inhibition of This Process by Piperonyl Cyclonene*, Science *111*, 600–601 (1950).

[438] PERRY, A. S., MATTSON, A. M., and BUCKNER, A. J., *Mechanism of Synergistic Action of DMC With DDT Against Resistant House Flies*, Biol. Bull., Wood's Hole *104*, 426–438 (1953).

[439] PETRIE, P. W. R., *Epidemic Typhus in Southwestern Arabia*, Amer. J. trop. Med. *29* (4), 501–526 (1949).

[440] PETRILLA, A., *O.K.I. járványügyi osztályáról. Az 1946. év járványügyi mérlege*, Orv. Lapja *3* (7), 217–218 (1947).

[441] PHILIP, C. B., *Tsutsugamushi Disease (scrub typhus) in World War II*, J. Parasit, *34* (3), 169–191 (1948).

[442] PHILIPPINES, GOVERNMENT OF, *Report submitted, Malaria Conference, Bangkok, Thailand, Sept. 21–24, 1953*, World Hlth. Org., Sea./Mal./12, Rev. 1, New Delhi, 8 July 1953.

[443] PIMENTEL, D., DEWEY, J. E., and SCHWARDT, H. H., *An Increase in the Duration of the Life Cycle of DDT-Resistant Strains of the House Fly*, J. econ. Ent. *44*, 477–481 (1951).

[444] PIMENTEL, D., and KLOCK, J. W., *Disinsectization of Aircraft by Residual Deposits of Insecticides*, Amer. J. trop. Med. Hyg. *3*, 191 (1954).

[445] PIMENTEL, D., SCHWARDT, H. H., and DEWEY, J. E., *Development and Loss of Insecticide Resistance in the House Fly*, J. econ. Ent. *46*, 295–298 (1953).

[446] PIMENTEL, D., SCHWARDT, H. H., DEWEY, J. E., and NORTON, L. B., *Studies on Insecticide Resistant House Flies in New York*, CSMA Proc., Dec. 1950, Soap, N. Y. pp. 94–96.

[447] PINOTTI, M., *Profilaxia da doença de Chagas' no Brazil*, Report read at Round Table on Chagas' disease at the 1st Int. Congr. Hyg., Havana, October 1952.

[448] PINOTTI, M., *Nation-Wide Malaria Eradication Projects in the Americas. IV. The Nation-Wide Malaria Eradication Program in Brazil*, J. nat. Malar. Soc. *10*, 162–182 (1951).

[449] PINOTTI, M., *The Biological Basis for the Campaign Against the Malaria Vectors of Brazil*, Trans. R. Soc. trop. Med. Hyg. *44*, 663–682 (1951).

[450] PIRES, F. A., *Relationship Between G. brevipalpis*, Newst., *and the Hippopotamus on the Maputo Valley—Portuguese East Africa*, Bur. perm. Interafr. Tsé-Tsé et Trypanosomiase, No. 143/0, Léopoldville (Congo Belge), 8 mimeographed pages (1950).

[451] POLLITZER, R., *Plague and Plague Control in Chinai*, Th. Chin. med. J. *66* (6), 328–333 (1948).

[452] POLLOCK, J. S. McK., *Plague Controlled in Haifa by the Use of DDT Alone*, Trans. R. Soc. trop. Med. Hyg. *41* (5), 647–656 (1948).

[453] PONS, J. A., Personal communication (1954).

[454] PRATT, H. D., *The ECA Malaria Control Program in Indochina: a Progress Report*, Trop. Med. Hyg. News *1* (2), 8–10 (1952).

[455] PRICE, D. E., *Relation of Pesticides to Health*, Agric. Chem. *9* (10), 34–37, 117 (1954).

[456] PURI, I. M., and BHATIA, M. L., *Studies on Some Insecticides Against Anopheline Adults and Larvae. Part III. Experiment With DDT as a Residual Spray in Some Villages in Baluchistan*, Indian J. Malar. *1*, 183–191 (1947).

[457] PURI, I. M., and KRISHNASWAMI, A. K., *Studies on Some Insecticides Against Anopheline Adults and Larvae. Part IV. Experiments With DDT and 666 as Residual Spray in Some Villages Around Delhi in 1946*, Indian J. Malar. *1*, 193–209 (1947).

[458] QUARAISHI, M. S., AHMED, M., and GRAMICCIA, G., *Pre-Monsoon Malaria Transmission in the District of Mymensingh, East Pakistan*, Bull. World Hlth. Org. *3*, 673–682 (1951).

[459] QUARTERMAN, K. D., *Field Investigations on the Effects of a DDT Residual Treatment on the Resting and Feeding Habits of Several Species of Mosquitoes* (in manuscript).

[460] QUARTERMAN, K. D., *Some Factors Influencing the Residual Effectiveness of DDT and Chlordane in Anopheline Mosquito Control*, J. nat. Malar. Soc. *7*, 300–306 (1948).

[461] Rachou, R. G., Bustamante, F. M., and Ferreira, M. O., *Da aplicação extra-domiciliar de DDT por helicóptero como meio de combate aos anofelinos do sub-genero Kerteszia em região com predominância de bromélias terrestres*, Rev. bras. Malariol. *1*, 264–281 (1949).

[462] Raffaele, G., and Coluzzi, E. A., *Relazione sulla malaria nella provincia di Frosinone negli anni 1945–1948 l'epidemia malarica di cassino*, Riv. Malariol. *28*, 1–48 (1949).

[463] Rajendram, S., and Jayewickreme, S. H., *Malaria in Ceylon. Part I. The Control and Prevention of Epidemic Malaria by the Residual Spray of Houses With DDT*, Indian J. Malar. *5*, 1–73 (1951).

[464] Ramakrishnan, S. P., Krishnan, K. S., and Ramakrishna, V., *Report on a Pilot Scheme for Malaria Control in the Betelnut-Growing Areas in Puthur Taluk, South Kanara District, Madras Province, 1947–48*, Indian J. Malar. *2*, 247–282 (1948).

[465] Randolph, N. M., *DDT to Control the Relapsing Fever Tick*, J. econ. Ent. *39*, 396 (1946).

[446] Rao, T. R., *Malaria Control Using Indoor Residual Sprays in the Eastern Province of Afghanistan*, Bull. World Hlth. Org. *3*, 639–661 (1951).

[467] Rao, V. V., *A Critical Review of Malaria Control Measures in India*, Indian J. Malar. *3*, 313–326 (1949).

[468] Rao, B. A., *A Short Account of the Malaria Problem in Mysore State*, Indian J. Malar. *2*, 285–305 (1948).

[469] Reeves, W. C., *The Encephalitis Problem in the United States*, Amer. J. publ. Hlth. *41*, 678–686 (1951).

[470] Reeves, W. C., Hammon, W. McD., Furman, D. P., McClure, H. E., and Brookman, B., *Recovery of Western Equine Encephalomyelitis Virus from Wild Bird Mites (Liponyssus sylviarum) in Kern County, California*, Science *105*, 411–412 (1947).

[471] Reeves, W. C., Washburn, G. E., and Hammon, W. McD., *Western Equine Encephalitis Control Studies in Kern County, California, 1945. I. The Effectiveness of Residual DDT Deposists on Adult Culex Mosquito Populations*, Amer. J. Hyg. *47*, 82–92 (1948)

[472] Renjifo, S., and deZulueta, J., *Five Years' Observations of Rural Malaria in Eastern Colombia*, Amer. J. trop. Med. Hyg. *1* (4), 598–611 (1952).

[473] Ricci, M., *Sull'azione del DDT su Blatta orientalis*, Riv. Parassit. *9*, 143–167 (1948).

[474] Roadhouse, L., *Laboratory Studies of DDT-Resistant Flies in Canada*, Canad. Ent. *85* (9), 340–346 (Sept. 1953).

[475] Rock, R. E., *Yaws on Guam: Treatment and Environmental Considerations*, Amer. J. trop. Med. *2* (1), 74–78 (1953).

[476] Rodriquez, J. A., *Ensayo de control del Aedes aegypti mediante la aplicación de DDT en los tanques de captación de aguas*, Bol. Ofic. sanit. pan-amer. *29* (11), 1150–1151 (1950).

[477] Romaña, C., *Informe sobre la enfermedad de Chagas en la Argentina: Situación actual del problema*, Bol. Ofic. sanit. pan-amer. *28* (10), 993–1004 (1949).

[478] Rose, G., *Fortschritte in der Bekämpfung des Läusefleckfiebers und der Malaria*, Acta trop., Basel *1* (2), 193–218 (1944).

[479] Rosen, G., *Public Health in Foreign Periodicals*, Amer. J. publ. Hlth. *42* (2), 213–217 (1952).

[480] Ross, R., *Proposals for Antimalaria Measures*, issued by Lpool Sch. trop. Med., cited by Savvas, *Malaria in Greece* (Athens 1907, cited by Livadas *et al.* [335]).

[481] Rouanet, P. L., *Methods Used by the National Yellow Fever Service in the Aedes aegypti Eradication Campaign in Brazil*, First Inter-Amer. Congr. Publ. Hlth., 26 Sept.–1 Oct., 1952, CIH/23 (Eng.).

[482] Rowe, J. A., *The Relation of Interested Agencies and Organizations in the Planning and Execution of Water Development Programs*, Proc. 18th Annu. Conf. Calif. Mosq. Control Ass., pp. 42–44 (1950).

[483] Rude, C. S., *DDT to Control the Gulf Coast Tick*, J. econ. Ent. *40*, 301–303 (1947).

[484] RUDE, C. S., and SMITH, C. L., *DDT for Control of Gulf Coast and Spinose Ear Ticks*, J. econ. Ent. *37*, 132 (1944).

[485] RUSSELL, P. F. *Italy, in the History of Malaria*, Riv. Parassit. *13* (1), 93–104 (1952).

[486] RUSSELL, P. F., *The Eradication of Malaria*, Sci. Amer., June 1952.

[487] RUSSELL, P. F., *The Present Status of Malaria in the World*, Amer. J. trop. Med. Hyg. *1* (1), 111–123 (1952).

[488] RUSSELL, P. F., *Epidemiology of Malaria in the Philippines*, Amer. J. publ. Hlth. *26*, 1–7 (1936).

[489] SABATINI, A., *La lotta antianofelica nell'agro romano negli anni 1947–1950*, Riv. Parassit. *12* (3), 185–192 (1951).

[490] SABIN, A. B., *Epidemic Encephalitis in Military Personnel. Isolation of Japanese B Virus on Okinawa in 1945, Serologic Diagnosis, Clinical Manifestations, Epidemiologic Aspects and Use of Mouse Brain Vaccine*, J. Amer. med. Ass. *133*, 281–293 (1947).

[491] SABIN, A. B., GINDER, D. R., and MATUMOTO, M., *Difference in Dissemination of the Virus of Japanese B Encephalitis Among Domestic Animals and Human Beings in Japan*, Amer. J. Hyg. *46*, 341–355 (1947).

[492] SACCA, G., *Sull'esistenza di mosche domestiche resistenti al DDT*, Riv. Parassit. *8* (2–3), 127–128 (1947).

[493] SACKTOR, B., *Some Aspects of Respiratory Metabolism During Metamorphosis of Normal and DDT-Resistant House Flies, Musca domestica L.*, Biol. Bull., Wood's Hole, *100*, 229–243 (1951).

[494] SACKTOR, B., *A Comparison of the Cytochrome Oxidase Activity of Two Strains of House Flies*, J. econ. Ent. *43*, 832–838 (1950).

[495] SACKTOR, B., *The Effect of DDT and Methoxychlor on Cytochrome Oxidase in Two Strains of House Flies, M.domestica L.* (unpublished thesis, Rutgers University, 1949).

[496] SAENZ VERA, C., *DDT in the Prevention of Plague*, World Hlth. Org. Expert Comm. on Plague, Second Session, Bombay, 5–10 Dec. 1952.

[497] SAMBASIVAN, C., ADAN, G. L., and SANTOS, M. N., *A Preliminary Report on Mindoro Pilot Project for the Period Febr. 16, 1952 to March 31, 1953* (unpublished report).

[498] SANI, A., *L'endemia malarica del basso ferrarese dal 1940 al 1948*, Ann. sanità pubb. *11* (4), 1149–61 (1950).

[499] SAPERO, J., *The Malaria Problem Today; Influence of Wartime Experience and Research*, J. Amer. med. Ass. *132*, 623–627 (1946).

[500] SASSE, B. E., *DDT in Alcohol as a Larvicide for Aedes aegypti Control*, Mosquito News *10* (3), Sept. 1950.

[501] SCHULZ, K. H., *Control of Plague in Taranto, Italy, 1945/1946. An Account of a Successful Programme of Rodent Extermination*, Bull. World Hlth. Org. *2* (4), 675–685 (1950).

[502] SCOVILLE, A. B., jr., *Epidemic Typhus Fever in Japan and Korea, The Rickettsial Diseases of Man* (American Association for Advances in Science, Washington 1948), pp. 28–35.

[503] SCUDDER, H. I., *A New Technique for Sampling the Density of Housefly Populations*, Publ. Hlth. Rep., Wash. *62*, 681–686 (1947).

[504] SEMPLE, A. B., *The Control of Phlebotomus Fever*, The Med. Offr. *79* (4), 35–37 (1948).

[505] SENIOR WHITE, R., and GHOSH, A. R., *House Spraying With DDT: Further Results*, J. Malar. Inst. India *6*, 489–508 (1946).

[506] SEVERO, O. P., *Eradication of the Aedes aegypti in the Americas*, First Inter-Amer. Congr. publ. Hlth., 26 Sept.–1 Oct. 1952, CIH/5 (Eng.), pp. 1–26, Annex I–III.

[507] SHELLEY, H., and AZIZ, M., *Anopheles Eradication in Cyprus*, Brit. med. J. *4606*, 767–768 (1949).

[508] SHOUSHA, A. T., *Species-Eradication. The Eradication of Anopheles gambiae from Upper Egypt 1942–1945*, Bull. World Hlth. Org. *1* (2), 309–352 (1948).

[509] SHOUSHA, A. T., *Cholera Epidemic in Egypt (1947). A Preliminary Report*, Bull. World Hlth. Org. *1* (2), 353–381 (1948).

[510] Sicault, G., *Considérations sur l'épidémie de la fièvre récurrente mondiale au Maroc.* Bull. Inst. Hyg. Maroc *4*, 5–25 (1944).

[511] Silva, R., and Neghme, A., *Estudio de la acción sinérgica de diversos insecticidas sobre el Triatoma infestans* Klug *1834*, congr. sudamericano de quimica, IV, March 5, 1948, Act., Sección 5, química en relación con medicamentos y tóxicos, pp. 195–196

[512] Silverman, P. H., and Mer, G. G., *Preliminary Report: Behavior of a DDT-Resistant Strain of Flies at an Agricultural Settlement in Israel,* Riv. Parassit. *13,* 123–128 (1952).

[513] Simmons, J. S., *Progress in the Development of Insecticides for Prevention of Insect-Borne Diseases,* Advanc. internal. Med. *2,* 228–261 (1947).

[514] Simmons, J. S., Whayne, T. F., Anderson, G. W., and Horack, H. M., *Global Epidemiology,* Vol. 1 (J. B. Lippincott Co., Philadelphia 1944), 504 pp.

[515] Simmons, S. W., *New Developments in Malaria Prevention and Control: Field Control by Residual Insecticides,* Proc. of Second Conf. of Indust. Council for trop. Hlth., Harvard School of Publ. Hlth., Boston *2,* 122–131 (1955).

[516] Simmons, S. W., *Public Health Significance of Insect Resistance to Insecticides and Future Orientation of Vector Control Programs,* 1st Internat. Symposium on Control of Insect Vectors of Disease, Supplemento ai R. C., sup. Sanit., Rome, pp. 97–125 (1954).

[517] Simmons, S. W., *Progress in the Development of Malaria Control Techniques,* J. nat. Malar. Soc. *5,* 157–163 (1946).

[518] Simmons, S. W., *Tests of the Effectiveness of DDT in Anopheline Control,* Publ. Hlth. Rep., Wash. *60* (32), 917–927 (1945).

[519] Simmons, S. W., and Hayes, W. J., jr., *Fleas and Disease,* Proc. Fourth Int. Congr. trop. Med. Malar., Wash. *2,* 1678–1689 (1948).

[520] Simmons, S. W., and Upholt, W. M., *Disease Control With Insecticides. A Review of the Literature,* Bull. World Hlth. Org. *3* (4), 535–556 (1951).

[521] Simmons, S. W., and staff, *Techniques and Apparatus Used in Experimental Studies on DDT as an Insecticide for Mosquitoes,* Suppl. No. 186, Publ. Hlth. Rep., Wash., pp. 3–20 (1945).

[522] Singh, J., Personal communication (1954).

[523] Singh, J., *Malaria and Its Control in India,* J. R. sanit. Inst. *72,* 515–525 (1952).

[524] Singh, J., and Kariapa, C. B., *Malaria Control in Coorg,* Indian J. Malar. *3,* 191–198 (1949).

[525] Singh, J., Pal, R., and Sharma, M. I. D., *Field Studies on the Comparative Effectiveness of DDT and BHC Against Mosquitoes When Applied Separately and in Combination,* Indian J. Malar. *5,* 235–248 (1951).

[526] Singh, J., and Singh, D., *Control of Rural Malaria With DDT Indoor Residual Spraying in Delhi Province During the Year 1948,* Indian J. Malar. *3,* 129–144 (1949).

[527] Sivalingam, V., and Rustomjee, K. J., *Spleen and Parasite Surveys in Ceylon,* J. Malar. Inst. India *4,* 155–173 (1941).

[528] Smith, C. N., and Gouck, H. K., *DDT to Control Ticks on Vegetation,* J. econ. Ent. *38,* 553–555 (1945).

[529] Smith, C. N., and Gouck, H. K., *Effectiveness of DDT in the Control of Ticks on Vegetation,* J. econ. Ent. *37,* 128–130 (1944).

[530] Smith, G. E., *Investigation and Evaluation of Water Development Programs Relative to Mosquito Production,* Proc. 18th Annu. Conf. Calif. Mosq. Control Ass., pp. 44–46 (1950).

[531] Smith, H. F., and Dy, F. J., *Can Malaria be Controlled in the Philippines by DDT Residual Spraying of Houses?* Acta med. philipp. *6,* 81–111 (1949).

[532] Smith, M. G., Blattner, R. J., Heys, F. M., and Miller, A., *Experiments on the Role of the Chicken Mite, Dermanyssus gallinae, and the Mosquito in the Epidemiology of St. Louis Encephalitis,* J. exp. Med. *87,* 119–138 (1948).

[533] Smith, M. G., Blattner, R. J., and Heys, F. M., *The Isolation of the St. Louis Encephalitis Virus from Chicken Mites (Dermanyssus galliane) in Nature,* Science *100,* 362–363 (1944).

[534] SNYDER, J. C., *Typhus Fever in the Second World War*, Calif. Med. *66* (1), 3–10 (1947).

[535] SOEPARMO, H. T., and STOKER, W. J., *Malaria Control in Indonesia*, J. Indones. med. Ass. *2* (7), 1–8 (1952).

[536] SOPER, F. L., Personal communication (1952).

[537] SOPER, F. L., *General Principles of the Eradication Programs in the Western Hemisphere*, J. nat. Malar. Soc. *10*, 183–194 (1951).

[538] SOPER, F. L., *Discussion on Infectious Disease Control*, Industry and Trop. Hlth., Proc. First Industr. Trop. Hlth. Conf., Harvard School Publ. Hlth., Boston, Dec. 8–10, 1950, pp. 254–257.

[539] SOPER, F. L., DAVIS, W. A., MARKHAM, F. S., and RIEHL, L. A., *Typhus Fever in Italy, 1943–1945, and Its Control with Louse Powder*, Amer. J. Hyg. *45* (3), 305–334 (1947).

[540] SOPER, F. L., DAVIS, W. A., MARKHAM, F. S., RIEHL, L. A., and BUCK, P., *Louse Powder Studies in North Africa (1943)*, Arch. Inst. Pasteur Algér. *23* (3), 183–223 (1945).

[541] SOPER, F. L., KNIPE, F. W., CASINI, G., RIEHL, L. A., and RUBINO, A., *Reduction of Anopheles Density Effected by the Pre-Season Spraying of Building Interiors With DDT in Kerosene, at Castel Volturno, Italy in 1944–1945 and in the Tiber Delta in 1945*, Amer. J. trop. Med. *27* (2), 177–200 (1947).

[542] SOPER, F. L., PENNA, H., CARDOSO, E., SERAFIM, J., jr., FROBISHER, M., jr., and PINHEIRO, J., *Yellow Fever Without Aedes aegypti. Study of a Rural Epidemic in the Valle do Chanaan, Espirito Santo, Brazil, 1932*, Amer. J. Hyg. *18*, 555–587 (1933).

[543] SOPER, F. L., and WILSON, D. B., *Anopheles gambiae in Brazil, 1930 to 1940* (The Rockefeller Foundation, New York 1943), 262 pp.

[544] SOPER, F. L., and WILSON, D. B., *Species Eradication; a Practical Goal of Species Reduction in the Control of Mosquito-Borne Disease*, J. nat. Malar. Soc. *1*, 5–24 (1942).

[545] SPANEDDA, A., *Disinfestazione massiva in Sardegna ed endemia tifoide*, Rass. med. Sarda *52* (4), 205–210 (1950).

[546] SPANEDDA, A., and MARCHINU, M., *Azione del DDT sul'andamento della morbilità per tifo, paratifi, dissenteria bacillare, gastroenteriti infantili*, Rass. med. Sarda *50*, 45–49 (1948).

[547] SRIVASTAVA, R. S., CHAND, D., and SINGH, M. V., *Malaria Control Measures in Ganga Khadar Colonization Scheme, Hastinapur, District Meerut, Uttar Pradesh (1948–1952)*, Nat. Soc. India Malar. and Other Mosquito-borne Dis. *1*, 111–119 (1953).

[548] SRIVASTAVA, R. S., and CHAND, D., *Control of Malaria in Sarda Hydel Power-House Construction*, Indian J. Malar. *5*, 579–594 (1951).

[549] SRIVASTAVA, R. S., *Malaria Control Measures in the Terai Area Under the Terai Colonization Scheme, Kichha, District Naini Tal: September 1947 to December 1948*, Indian J. Malar. *4*, 151–165 (1950).

[550] STAGE, H. H., and PRATT, H. D., *Observations on Mosquito and Malaria Control in the Caribbean Area. Part IV. Puerto Rico*, Mosquito News *10*, 54–57 (1950).

[551] STEPHENS, P. A., and PRATT, H. D., *Work With Residual DDT Spray in Puerto Rico*, Science *105* (2715), 32–33 (1947).

[552] STERNBURG, J., and KEARNS, C. W., *Metabolic Fate of DDT When Applied to Certain Naturally Tolerant Insects*, J. econ. Ent. *45*, 497–505 (1952).

[553] STERNBURG, J., and KEARNS, C. W., *Degradation of DDT by Resistant and Susceptible Strains of House Flies*, Ann. ent. Soc. Amer. *43*, 444–458 (1950).

[554] STERNBURG, J., KEARNS, C. W., and BRUCE, W. N., *Absorption and Metabolism of DDT by Resistant and Susceptible House Flies*, J. econ. Ent. *43*, 214–219 (1950).

[555] STERNBURG, J., VINSON, E. B., and KEARNS, C. W., *Enzymatic Dehydrochlorination of DDT by Resistant Flies*, J. econ. Ent. *46*, 513–515 (1953).

[556] STEWARD, J. S., *Action of Gammexane on Arthropods of Medical and Veterinary Importance*, Trans. R. Soc. trop. Med. Hyg. *40* (5), 559–565 (1947).

498 S. W. Simmons

[557] STEYN, J. J., *DDT Field Trials Against Tsetse Fly (G. palpalis) on Nkuzi Island,
Lake Victoria*, J. ent. Soc. S. Afr. *12*, 126–129 (1949).

[558] STIERLI, H., SIMMONS, S. W., and TARZWELL, C. M., *Operational Procedures and
Equipment Used in the Practical Application of DDT as a Residual House Spray*,
Publ. Hlth. Rep., Wash., Suppl. No. 186 (1945).

[559] STOLL, N. R., *This Wormy World*, J. Parasit. *33* (1), 1–18 (1945).

[560] STRODE, G. K., *Yellow Fever* (McGraw-Hill Book Co., Inc., New York 1951), 710 pp.

[561] STRONG, R. P., *Stitt's Diagnosis, Prevention and Treatment of Tropical Diseases*,
7th ed., vol. 1 (The Blakiston Co., Philadelphia 1944).

[562] SUBRAMANIAN, R., and VAID, B. K., *Some Observations on Rural Malaria Control
Using Residual Insecticides in Wettable Powder Form*, Indian J. Malar. *6*, 215–218
(1952).

[563] SUBRAMANIAN, R., VAID, B. K., and DIXIT, D. T., *Kaknar Nyay Panchayat Malaria
Control Co-Operative Scheme*, Indian J. Malar. *3*, 339–348 (1949).

[564] SUBRAMANIAN, R., and DIXIT, D. T., *Observations on the Vectors of Malaria in
Khandwa Tahsil*, Indian J. Malar. *2*, 313–322 (1948).

[565] SULKIN, S. E., *Recovery of Equine Encephalomyelitis Virus (Western Type) From
Chicken Mites*, Science *101*, 381–383 (1945).

[566] SULKIN, S. E., and IZUMI, E. M., *Isolation of Western Equine Encephalomyelitis
Virus From Tropical Fowl Mites, Liponyssus bursa* (Berlese), Proc. Soc. exp. Biol.,
N. Y. *66*, 249–250 (1947).

[567] SWELLENGREBEL, N. H., *The Malaria Epidemic of 1943–1946 in the Province of
North-Holland*, Trans. R. Soc. trop. Med. Hyg. *43*, 445–464 (1950).

[568] SWELLENGREBEL, N. H., and DEBUCK, A., *Malaria in the Netherlands* (Scheltema
and Holkema, Ltd., Amsterdam 1938), 267 pp.

[569] SWELLENGREBEL, N. H., and LODENS, J. G., *Anopheles aconitus and DDT Spraying*,
Docum. neerl. Indones. Morb. trop. *1*, 245–254 (1949).

[570] SYMES, C. B., *Synthetic Insecticides and Malaria*, Rev. agric. Maurice *27*, 113–121
(1948).

[571] SYMES, C. B., HADAWAY, A. B., BARLOW, F., and GALLEY, W., *Field Experiments
With DDT and Benzene Hexachloride Against Tsetse (Glossina palpalis)*, Bull. ent.
Res. *38* (part 4), 591–612 (1948).

[572] TAIWAN, *Proposed Malaria Eradication Program for Taiwan*, Trop. Med. Hyg.
News *1*, 12–13 (1952).

[573] TALICE, R. V., *Recherches sur la Maladie de Chagas en Uruguay*, Bull. Soc. Pat.
exot. *42* (7/8), 352–354 (1949).

[574] TARZWELL, C. M., and STIERLI, H., *The Evaluation of DDT Residual Sprays for the
Control of Anopheline Mosquitoes in Dwellings*, Suppl. No. 186, Publ. Hlth. Rep ,
pp. 35–48 (1945).

[575] TEN BROECK, C., *Transmission of Equine Encephalomyelitis*, Proc. 3rd Congr. Int.
Microbiol., New York, pp. 300–301 (1940).

[576] TEN BROECK, C., *Birds as Possible Carriers of the Virus of Equine Encephalomyelitis*,
Arch. Path. *25*, 759 (1938).

[577] THAILAND, GOVERNMENT OF, *Report, Malaria Conference, Bangkok, Thailand, Sept.
21–24, 1953*, World Hlth. Org., Sea./Mal./3, New Delhi, 16 June 1953.

[578] THOMPSON, R. K., WHIPP, A. A., DAVIS, D. L., and BATTE, E. G., *Fly Control With
a New Bait Application Method*, J. econ. Ent. *46*, 404–409 (1953).

[579] TONKING, H. D., and GÉBERT, S., *The Use of DDT Residual Sprays in the Control
of Malaria Over an Area of 16 Square Miles in Mauritius*, Med. and Hlth. Dept.,
Mauritius. Cent. Lab. Publ. No. 40, 23 pp. (1947).

[580] TONKING, H. D., LAVOIPIERRE, R., and COURTOIS, C. M., *A Small Scale Experiment
in the Use of DDT in Mauritius* (J. H. Bowkett, Government Printer, Port Louis
1946), 5 pp.

[581] TRAPIDO, H., *Modified Response of Anopheles albimanus to DDT Residual House
Spraying in Panama*, Amer. J. trop. Med. Hyg. *1*, 853–861 1952).

[582] Trapido, H., *The Toxicity of DDT to Anopheles claviger* (Meigen) *in Sardinia and on the Italian Mainland*, J. nat. Malar. Soc. *10*, 266–271 (1951).

[583] Trapido, H., *The Residual Spraying of Dwellings With DDT in the Control of Malaria Transmission in Panama With Special Reference to Anopheles albimanus*, Amer. J. trop. Med. *26*, 383–415 (1946).

[584] Travis, B. V., *Studies of Mosquito and Other Biting-Insect Problems in Alaska*, J. econ. Ent. *42*, 451–457 (1949).

[585] Trinquier, E., and Jaujou, C., *Rapport sur le fonctionnement du service anti-paludique en Corse en 1948*, Rapp. Cons. Général Corse. Sess. extraordinaire de janvier 1949 (Malaria Control, Corsica, 1949).

[586] Turkey, *Turkey: Fifth International Congress of Tropical Medicine and Malaria*, J. Amer. med. Ass. *154*, 80 (1954).

[587] United States Army, *The Prevention of Dengue Fever*, Bull. U. S. Army med. Dep. *4* (5), 535–539 (1945).

[588] United States Department of Agriculture, *DDT and Other Insecticides and Repellents Developed for the Armed Forces*, Misc. Publ. U. S. Dep. Agric. No. 606, pp. 55-58 (1946).

[589] United States Operations Mission to Liberia, *Malaria Control Activities, June 1954*, Mon. Rep., Monrovia, Liberia (1954).

[590] United States Operations Mission to Liberia, *Malaria Control Activities, December 1953*, Mon. Rep., Monrovia, Liberia (1953).

[591] United States Public Health Service, *Communicable Disease Center, Technology Branch, Summary of Investigations No. 4*, Atlanta, Ga. (1953).

[592] United States Public Health Service, *Operational Control Program*, Communicable Disease Center Activities 1949–50, pp. 2–21, Atlanta, Ga. (1951).

[593] United States Public Health Service, *Report of Philippine Public Health Rehabilitation Program, July 4, 1946–June 30, 1950* (Manila 1950), 598 pp.

[594] United States Public Health Service, *Foreign Quarantine Manual of Operations*, Fed. Security Agency, Wash., pp. 13/1–13/5 (1948).

[595] United States Public Health Service, *Evaluation Data for Murine Typhus Fever Control Activities, 1946*, Communicable Disease Center, Atlanta, Ga. (1947).

[596] United States War Department, *Technical Bulletin. Medical and Sanitary Data on French Indo-China*, Tech. Bull., Med. *86*, 18 August 1944, Wash.

[597] Upholt, W. M., *Significance to the Insecticide Industry of Fly Resistance*, CSMA Proc., Dec. 1950, Soap, N. Y., pp. 69–71.

[598] Upholt, W. M., Gaines, T. B., Simmons, S. W., and Arnold, E. H., *The Experimental Use of DDT in the Control of the Yellow Fever Mosquito Aedes aegypti* (L.), Publ. Hlth. Rep., Wash., Separate No. 8, Supplement No. 186 (1945).

[599] van Rooyen, C. E., and Rhodes, A. J., *Virus Diseases of Man* (Thomas Nelson and Sons, New York 1948), 1202 pp.

[600] van Thiel, P. H., and Winoto, R. M. P., *Control of Hyperendemic Malaria in Java, Caused by A. sundaicus, by DDT House-Spraying*, Docu. neerl. indones. Morb. trop. *3*, 295–319 (1951).

[601] Vedamanikkam, J. C., *Incidence of Anopheles fluviatilis James larvae in a DDT Sprayed Area in Wynaad, South India*, Indian J. Malar. *3*, 331–338 (1949).

[602] Verhaeghe, *Note sur l'action des bombes fumigènes D.D.T. sur les glossines*, Ann. Soc. belge Méd. trop. *28* (4), 445–448 (1948).

[603] Verolini, F., *Sul meccanismo di azione del DDT sulle larve di culicida*, R. C. Ist. sup. Sanit. *11*, 442–449 (1948).

[604] Viel, B., and Romero, H., *Dos epidemias de tifus exantemático*, Rev. méd. Chile *73* (10), 847–854 (1945).

[605] Villela, E. de A., *Moléstia de Chagas. Algumas aquisicões recentes, em especial relativas a profilaxia*, Rev. bras. Malariol. *3* (1), 101–121 (1951).

[606] Vincke. I. H., Personal communication of 21 November 1952.

[607] VINCKE, I. H., *Malaria Control by Means of DDT in Katanga* (*1947–1950*), World Hlth. Org./Mal./47 — Afr./Mal./Conf./3, 2 October 1950.

[608] VINE, M. J., *The Anti-Malaria Campaign in Greece 1946*, Bull. World Hlth. Org. *1* (1), 197–204 (1947).

[609] VISWANATHAN, D. K., Personal communication (1954).

[610] VISWANATHAN, D. K., *Malaria and Its Control in Bombay State* (D. K. Viswanathan Connaught House, Poona 1950), 263 pp.

[611] VISWANATHAN, D. K., *A Study of the Effects of Malaria and of Malaria Control Measures on Population and Vital Statistics in Kanara and Dharwar Districts as Compared With the Rest of the Province of Bombay*, Indian J. Malar. *3*, 69–107 (1949).

[612] VISWANATHAN, D. K., and RAO, T. R., *Control of Rural Malaria With DDT Indoor Residual Spraying in Kanara and Dharwar Districts, Bombay State: Third Year's Results*, Indian J. Malar. *3*, 269–312 (1949).

[613] VISWANATHAN, D. K., and RAO, T. R., *Control of Rural Malaria With DDT Indoor Residual Spraying in Kanara and Dharwar Districts, Bombay Province. First Year's Results*, Indian J. Malar. *1*, 503–542 (1947).

[614] VISWANATHAN, D. K., RAO, T. R., and JUNEJA, M. R., *Further Notes on the Use of Benzene Hexachloride as a Residual Insecticide Compared With Dichloro-diphenyl-Trichloroethane*, Indian J. Malar. *4*, 505–531 (1950).

[615] VON EMMEL, L., *Autotomie bei Anopheles maculipennis als Reaktion auf ein Kontaktgift (Gesarol)*, Z. hyg. Zool. *35*, 119–124 (1943).

[616] WALLACE, R. B., *Insecticides and A. maculatus*, Med. J. Malaya *3*, 5–33 (1948).

[617] WANSON, M., *Contribution à l'étude de l'onchocercose africaine humaine* (*Problèmes de prophylaxie à Léopoldville*), Ann. Soc. belge Méd. trop. *30* (4), 775–792 (1950).

[618] WARREN, J., *Infections of Minor Importance*, in: *Viral and Rickettsial Infections of Man*, ed. T. M. RIVERS (J. B. Lippincott Co., Philadelphia 1952), pp. 665–680.

[619] WATT, J., and LINDSAY, D. R., *Diarrheal Disease Control Studies. I. Effect of Fly Control in a High Morbidity Area*, Publ. Hlth. Rep., Wash. *63* (41), 1319–1334 (1948).

[620] WAYSON, N. E., *Plague—Field Surveys in Western United States During Ten Years* (*1936–1945*), Publ. Hlth. Rep., Wash. *62* (22), 780–791 (1947).

[621] WEBSTER, L. T., and FITE, G. L., *A Virus Encountered in the Study of Material From Cases of Encephalitis in the St. Louis and Kansas City Epidemic of 1933*, Science *78*, 463–465 (1933).

[622] WEBSTER, L. T., and WRIGHT, F. H., *Recovery of Eastern Equine Encephalomyelitis Virus From Brain Tissue of Human Cases of Encephalitis in Massachusetts*, Science *88*, 305–306 (1938).

[623] WELT, L. G., *Use of Dimethylphthalate Impregnated Clothing as Protection Against Scrub Typhus*, Amer. J. trop. Med. *27*, 221–224 (1947).

[624] WHEELER, C. M., *Control of Typhus in Italy 1943–1944 by Use of DDT*, Amer. J. publ. Hlth. *36* (2), 119–129 (1946).

[625] WIESMANN, R. VON, *Untersuchungen über das physiologische Verhalten von Musca domestica L. verschiedener Provenienzen*, Mitt. Schweiz. Ent. Ges. *20*, 484–504 (1947).

[626] WILEY, J. S., *A Preliminary Report Concerning DDT Dusting and Murine Typhus Fever in Nine Southeastern States*, Publ. Hlth. Rep., Wash. *63* (2), 41–43 (1948).

[627] WILEY, J. S., and FRITZ, R. F., *Tentative Report on Expanded Murine Typhus Fever Control Operations in Southern States*, Amer. J. trop. Med. *28* (4), 589–597 (1948).

[628] WILKINSON, P. R., *Distribution and Fate of Anopheles gambiae and A. funestus in Two Different Types of Huts Treated With DDT and BHC in Uganda*, Bull. ent. Res. *42* (Part 1), 45–54 (1951).

[629] WILLIAMS, L. L., jr., *Economic Importance of Malaria Control*, Proc. N. J. Mosq. Ext. Ass. *25*, 148–152 (1938).

[630] WILSON, D. B., *Control of Mosquitoes in Dar-es-Salaam With DDT Air Spray*, preliminary report E. Afr. Malar. Unit, P. O. Amani, Tanga, Tanganyika Territory, October 1951, pp. 1–8.

[631] WILSON, D. B., *Malaria incidence in Central and South Africa*, from: *Malariology— a Comprehensive Survey of all Aspects of This Group of Diseases from a Global Standpoint*, ed. MARK F. BOYD, vol. II, Section IV, *Intermediate Host*—Chapter 33, *Malaria Incidence in Central and South Africa*, pp. 800–809 (1949).

[632] WILSON, S. G., *The Feeding of 'Gammexane' and DDT to Bovines*, Bull. ent. Res. *39*, 423–434 (1949).

[633] WILSON, T., and REID, J. A., *Filariasis. The Institute for Medical Research, 1900–1950*, Kuala Lumpur, Malaya, pp. 209–227 (1951).

[634] WINTERINGHAM, F. P. W., LOVEDAY, P. M., and HARRISON, A., *Resistance of House Flies to DDT*, Nature *167*, 106–107 (1951).

[635] WISECUP, C. B., BURRELL, R. W., and DEONIER, C. C., *DDT Sprays Mechanically Dispersed for Control of Anopheline Mosquito Larvae*, J. econ. Ent. *38* (4), 434–436 (1945).

[636] WRIGHT, J. W., Personal communication of 7 May 1953.

[637] WORLD HEALTH ORGANIZATION, *Malaria Control in Afghanistan*, Chron. World Hlth. Org. *8*, 212–213 (1954).

[638] WORLD HEALTH ORGANIZATION, *Somewhere in India*, W.H.O. Newslett. *6* (5), Picture Sheet No. 2 (1953).

[639] WORLD HEALTH ORGANIZATION, *Increased Production of Antibiotics and Insecticides; a Venture in International Aid, Insecticides; Ceylon*, Chron. World Hlth. Org. *7*, 330–331 (1953).

[640] WORLD HEALTH ORGANIZATION, *Report of the First Asian Malaria Conference, Bangkok, September 21–24, 1953*, World Hlth. Org., Sea./Mal./15, Bangkok, 24 September 1953.

[641] WORLD HEALTH ORGANIZATION, *Malaria Control in Burma*, Chron. World Hlth. Org. *7*, 24 (1953).

[642] WORLD HEALTH ORGANIZATION, *Malaria*, Chron. World Hlth. Org. *7*, 157 (1953).

[643] WORLD HEALTH ORGANIZATION, *Report of the First Asian Malaria Conference, Bangkok, September 21–24, 1953*, World Hlth. Org., Sea./Mal./15, Bangkok, 24 September 1953.

[644] WORLD HEALTH ORGANIZATION, *Colombian Campaign Against Malaria and Yellow Fever*, Chron. World Hlth. Org. *7* (10), 280–281 (1953).

[645] WORLD HEALTH ORGANIZATION, *Control of Insect Vectors in International Air Traffic*, Comm. on Internat. Quarantine, World Hlth. Org. mimeographed bull. No. WHO/ IQ/1, June 30, 1953.

[646] WORLD HEALTH ORGANIZATION, *Iran Marches Against Malaria*, W.H.O. Newslett. *5* (1), 5 (1952).

[647] WORLD HEALTH ORGANIZATION, *Afghanistan . . . (2) Over Malaria*, W.H.O. Newslett. *5* (6), 3 (1952).

[648] WORLD HEALTH ORGANIZATION, *Final Comprehensive Report Pakistan, E. Bengal Malaria Control Demonstration Team, Mymemsingh District by United Nations W.H.O. Regional Office, Eastern Mediterranean*, Pakist. J. Hlth. *2*, 61–91 (1952).

[649] WORLD HEALTH ORGANIZATION, *Achievements of a Malaria-Control Team in India*, Chron. World Hlth. Org. *6*, 38–40 (1952).

[650] WORLD HEALTH ORGANIZATION, *Malaria-Control Project in Burma*, Chron. World Hlth. Org. *6*, 53–54 (1952).

[651] WORLD HEALTH ORGANIZATION, *Successful Malaria Control in Thailand*, Chron. World Hlth. Org. *6*, 140–141 (1952).

[652] WORLD HEALTH ORGANIZATION, *The Work of WHO, 1951: Annual Report of the Director-General*, Off. Rec. World Hlth. Org. No. 38, March 1952, p. 82.

[653] WORLD HEALTH ORGANIZATION, *The Work of WHO, 1951: Annual Report of the Director-General*, Off. Rec. World Hlth. Org. No. 38, March 1952, p. 83.

[654] WORLD HEALTH ORGANIZATION, *Fourth Report. Expert Committee on Insecticides*, World Hlth. Org. Tech. Rep. No. 54 (1952).

[655] WORLD HEALTH ORGANIZATION, *Malaria Control in Iran*, Chron. World Hlth. Org. *5*, 28 (1951).

[656] WORLD HEALTH ORGANIZATION, *Malaria-Control Team Completes Assignment*, Chron. World Hlth. Org. *5*, 307 (1951).

[657] WORLD HEALTH ORGANIZATION, *Malaria Conference in Equatorial Africa*, Tech. Rep. World Hlth. Org. No. 38, pp. 1–72 (1951).

[658] WORLD HEALTH ORGANIZATION, *WHO Nurse Tells of Two Years' Work in South India*, W.H.O. Newslett. No. 11–12, Nov.–Dec. 1951.

[659] WORLD HEALTH ORGANIZATION, *Report on the Second Session, Expert Committee on Insecticides*, Tech. Rep. World Hlth. Org. No. 34 (1951).

[660] WORLD HEALTH ORGANIZATION, *Antimalaria Campaign in Persia*, Chron. World Hlth. Org. *4*, 125 (1950).

[661] WORLD HEALTH ORGANIZATION, *WHO Antimalaria Activities in Pakistan*, Chron. World Hlth. Org. *4*, 153 (1950).

[662] WORLD HEALTH ORGANIZATION, *WHO Antimalaria Activities in India*, Chron. World Hlth. Org. *4*, 297 (1950).

[663] WORLD HEALTH ORGANIZATION, *Notes and News: Malaria*, Chron. World Hlth. Org. *4*, 192 (1950).

[664] WORLD HEALTH ORGANIZATION, *Report of the Expert Committee on Plague — First Session*, Tech. Rep. World Hlth. Org. No. 11, October 1950.

[665] WORLD HEALTH ORGANIZATION, *Notes and News: Malaria Control in Europe and Asia*, Chron. World Hlth. Org. *3*, 136 (1949).

[666] YUST, H. R., *DDT to Control Anopheles farauti on Espiritu Santo, New Hebrides Islands*, J. econ. Ent. *40*, 762–768 (1948).

[667] ZINSSER, H., *Rats, Lice and History* (The Cornwall Press, Inc., New York 1934).

[668] ZIONY, M., *Malaria Control in Iran: Resume of Reports Made by Dr. Justin M. Andrews and Lawrence B. Hall*, Publ. Hlth. Rep., Wash. *65*, 351–367 (1950).

VIII

THE USE OF DDT
IN VETERINARY MEDICINE

BY

E. F. KNIPLING

1.

INTRODUCTION

The value of DDT in veterinary medicine was first demonstrated when WIESMANN [200] reported that the application of DDT to fly resting surfaces in animal stables controlled the house fly, *Musca domestica* (L.) and the stable fly, *Stomoxys calcitrans* (L.), important pests of livestock in many parts of the world. The investigations followed in Europe, North America, Australia, Africa, and in other parts of the world during World War II on the use of DDT for controlling arthropods attacking man. This research made indirect but important contributions to the development of the insecticide for use in veterinary medicine, and clearly suggested the future role that DDT would play in the veterinary field.

Many species of arthropods attack animals. These arthropods vary greatly in their mode of living which complicates the development of methods for their control. They affect the health of animals through transmission of diseases, consumption of blood, destruction of tissues, and by their annoyance. Some of the most serious of the diseases of livestock are transmitted by arthropods.

Piroplasmosis and trypanosomiasis are among the most important of the diseases of animals transmitted by arthropods but there are many others. The economy of many nations is dependent in large measure on the production of livestock. The loss or gain by a few per cent of the income derived from animals can, therefore, influence rather markedly the welfare of nations.

Intensive research has been conducted since about 1944 by investigators in many parts of the world in efforts to determine for what purposes and how DDT should be used for maximum efficiency, safety, and economy for controlling external parasites of livestock, poultry, and pets.

In making this appraisal it has been recognized that the degree of susceptibility of a parasite to an insecticide is not necessarily the only or major criterion for determining the practicability of the agent in actual control operations. The susceptibility of the host to the insecticide, the effect on animal products of the insecticide composition used, habits of the parasite, animal management practices, cost of the treatment in relation to benefits, and other factors were considered and resolved before DDT was recommended for the control of specific animal pests.

Results of the investigations by hundreds of scientists have shown that many species of arthropods affecting animals which could not be controllep effectively and economically with older insecticides may now be controllep with DDT. The insecticide is especially useful for the control of certain flies, lice, ticks, fleas, sheep ticks (keds), and blow-fly strike. The degree of control achieved against the various external parasites named depends to a large extent on the conditions and method of employing the insecticide.

Certain important parasites of animals are not effectively controlled with DDT either because of inherent resistance of the species to the insecticide or because of their habits. The insecticide is of little or no value for the control of certain mites and bot fly larvae including *Hypoderma* spp., *Gasterophilus* spp., and *Oestrus ovis*. Other important species are controlled in part. Direct application of DDT to animals will not prevent attack by certain bloodsucking insects including the tabanids, mosquitoes, black flies, tsetse flies, and stable flies, even though some of the insects may succumb following contact with the DDT on the host. Such insects, if adequately controlled with DDT, require treatment of the parasite while off the host. However, except for stable flies, which may be concentrated around animal shelters, it is often not economically feasible to treat extensive areas to protect livestock from attack by these insects.

2.

MAMMALIAN TOXICITY OF DDT IN RELATION TO ITS USE IN VETERINARY MEDICINE

The toxicity of DDT to animals and man is discussed in Chapter VI. However, for proper reader orientation, the merits and limitations of DDT in veterinary medicine from the standpoint of mammalian toxicity will be briefly considered.

Generally speaking there is a wide margin of safety between the dosage of DDT, per se, required to destroy the animal parasite and the dosage that is injurious to the host. The common domestic cat is the only animal known which is so susceptible to DDT that it is not safe to apply the insecticide to the host for controlling external parasites. This danger to cats is, no doubt, due in part to the inherent susceptibility of the animal to DDT but danger of toxic effects is increased because of the cat's habit of licking the hair coat, leading to the ingestion of considerable amounts of DDT. Other animals, including dogs, cattle, horses, sheep, goats, and hogs may be safely treated for external parasite control without danger of acute or chronic toxic effects. Millions of these animals have been treated repeatedly with aqueous sprays containing up to 2% of DDT or dusts containing up to 10% of DDT without showing harmful effects either acutely or chronically. This wide safety margin as compared with arsenical dips and sprays which were employed extensively for controlling certain external parasites of livestock is an asset which further stimulated interest in the use of DDT in veterinary medicine. Poultry appear to be moderately susceptible to DDT and the use of the insecticide in excessive amounts may prove harmful, although when used as recommended the insecticide has not injured poultry in practical arthropod control operations.

There is, however, danger of harming animals through careless use of certain DDT insecticides, because of the toxic effects of solvents employed in formulations. In 1945 and 1946 when DDT first became available for use by the civilian public in the United States, some deaths of cattle were reported following the application of emulsion sprays prepared from improperly formulated DDT emulsifiable concentrates, or following the application of excessive amounts of DDT oil solutions to animals. Although the DDT may have contributed to the toxicity of the materials applied, the solvents commonly used in emulsions or solutions, such as xylene, petroleum oils, and methylated naphthalene, when used alone in excessive quantities will injure or cause the death of livestock. Because of improvements in commercial products and through proper education in the use of DDT insecticides, injury to animals has been avoided in the United States since about 1947.

In Australia, it was reported (NICHOLSON [129]) that DDT has shown no adverse effects to livestock, except in some early instances when combinations of DDT and arsenical solutions were employed in dipping vats. It is now generally recognized throughout the world that DDT used in accordance with prescribed practices is not injurious to livestock and other animals.

A problem of great concern to pharmacologists and health officials, however, is the potential hazards to man of DDT residues which may appear in animal products following use of the insecticide in the veterinary field.

The direct application of DDT to livestock for controlling external parasites results in the absorption of DDT through the skin and subsequent storage in the fat, and in the case of dairy animals, excretion of a small amount of DDT in the milk. When applied to dairy cows at intervals of about 1 month in the manner prescribed for horn-fly control, the milk may contain up to 2 ppm of DDT, with an average of about 0·5 ppm, during the period between treatments (CARTER *et al.* [38]). DDT is likewise excreted in milk when the insecticide is ingested by dairy cows. The application of DDT in a manner which may lead to ingestion of even small amounts of DDT by the animals may result in slight contamination of the milk.

There is no evidence that the consumption of milk from cows treated with DDT has had any adverse effects on the public during 3 years of extensive use of the chemical for insect control on dairy cows. Since milk constitutes the principal food of infants the Food and Drug Administration in the United States believes that there may be a calculated risk in the use of the chemical in a manner that would result in the contamination of milk (DUNBAR, [50]). Therefore, since 1949, federal and state agencies in the United States have recommended that DDT not be used on dairy cows, or in dairy barns or milk rooms, for controlling animal parasites. DDT, however, is still recommended for application to other livestock for controlling external parasites and for use in and outside of other farm installations for controlling flies and other insects. Its use on dairy animals is also still recommended in other countries.

3.

TYPES OF FORMULATIONS

Many kinds of DDT formulations have been employed for controlling arthropods of veterinary importance. However, there are four major types of preparations in extensive use in the veterinary field. These are solutions, emulsions, suspensions, and dusts.

Solutions

Kerosene, both crude and partially refined, is the most commonly used solvent for DDT solutions employed in veterinary medicine. Unrefined kerosene will dissolve 5% or more of DDT. If refined kerosene is used the maximum that can be dissolved ranges between 2 and 3%. If higher concentrations of DDT are required, an auxiliary solvent such as the methylated naphthalenes or xylene must be added. The amount of such auxiliary solvent necessary to make a 5 per cent DDT solution may range from 5 to 10%.

DDT solutions in general are not intended for direct application to animals because of the danger of injuring the animals with the solvent. However, if oil base sprays are applied as a light mist uniformly in quantities of about 1 or 2 oz per animal the oil does not wet the skin and no injury follows.

DDT solutions are often employed, however, for the control of parasites while off the host, particularly as residual or surface sprays. DDT solutions are also employed extensively as insect sprays for space treatments, particularly for fly control. Such sprays are atomized by mechanical means or by utilizing liquefied gases under pressure, such as in the Freon aerosol container. DDT space sprays may contain added insecticides for knockdown effect, such as pyrethrum, allethrin, or certain organic thiocyanate compounds.

Emulsions

DDT emulsifiable concentrates employed in the veterinary field usually contain about 25% of DDT, 65% of a suitable solvent such as xylene, toluene, or benzene, and 10% of an oil miscible emulsifier. Other solvents such as pine oil may also be used, although the maximum amount of DDT that such a solvent will hold in solution is about 15%. There is a wide choice of emulsifiers suitable for such preparations.

The emulsifiable concentrates are diluted with water to prepare the desired concentration of insecticide in the finished spray or dip. Properly formulated and mixed emulsion sprays may be applied safely to the animals, provided they will mix uniformly with water. Emulsifiable concentrates have been known to deteriorate in containers to the extent that emulsification, particularly in hard waters, does not take place readily. Emulsions are also used extensively for application as a contact or residual treatment to control external parasites while off the host.

Suspensions

DDT suspension sprays and dips consists of finely ground or precipitated particles of the insecticide in water. The wettable powder concentrates which contain 50 or 75% of DDT are most widely used to make suspensions. The DDT is finely ground with an inert diluent such as talc. Suitable conditioners and wetting and anti-caking agents are added so that the product will disperse readily in water and will remain fairly well suspended. For proper performance of wettable powder suspensions they should be ground to a fineness of 5 to 10 μ. The maximum amount of DDT that can be employed in wettable powder suspension sprays is about 2·5% if the 50 per cent concentrate is employed and about 3·5% if the 75 per cent concentrate is used. Concentrations in excess of these amounts may lead to clogging of spray nozzles, especially when spray equipment is not provided with good agitators.

Another product consisting of 50% of DDT in concentrated form in a suitable 'solubilizer' is used extensively in Australia to prepare dips and sprays. The product is melted and then added to soft water to make an aqueous concentrate containing from 4 to 12% of DDT. This is further diluted before using. The DDT in the finished spray or dip is dispersed in extremely small particles, perhaps colloidal. This type of preparation is reported to be more effective than other DDT suspensions (JENKINS and FORTE [90]).

Suspensions are used extensively both as sprays and dips for direct treatment of animals. Suspension sprays are also widely used as a contact or residual spray where the parasites exist while off the host. Because of their uniform efficacy, safety and economy the suspensions are often the preparations of choice for controlling arthropods of veterinary importance.

Dusts

DDT in dust form is employed to a considerable extent in the veterinary field although probably less so than are the other formulations discussed. Up to 10% of DDT is used for application to animals or as a contact and residual treatment to control certain parasites while off the host. The dusts are finely ground and the inert part consists primarily of talc or similar diluents.

4.

METHODS OF APPLICATION. GENERAL

As previously stated, the use of DDT or other insecticides to control arthropods attacking animals involves two methods, namely, direct application of the material to the host and application of the material to the environment where the parasites exist while off the host, or a combination of both treatments.

For parasites such as lice and sheep ticks, which spend their entire life on the animal, the application of the insecticide directly to the animal is obviously the only way to achieve control. However, most parasites spend much of their life off the host. Others visit animals occasionally and may contact the host for only a few minutes. The problem then is to determine the most effective and practical way to achieve control. The method of choice may depend on susceptibility of the different stages of the parasite to the insecticide and degree of concentration of the parasite while off the host. However, both host treatment and environmental control efforts may be necessary to achieve maximum results.

A wide range of equipment and various methods of using such equipment are employed in the application of DDT. In general, however, the application of DDT poses no difficult problem. For some purposes the insecticide may be applied effectively with a sponge, although this method is obviously time consuming. Specific directions and details for using DDT insecticides will be given when necessary in connection with the discussion of the various parasites against which DDT is used. It seems expedient, however, to discuss in a general way the methods of application, equipment most generally employed and some of the problems encountered in connection with the use of DDT insecticides.

Host Treatment

Sprays

Application of DDT to the host by means of sprays is the most widely employed method of external parasite control. A wide range of equipment is available for this purpose. The most practical types of equipment to employ will vary, depending on such factors as kind of parasite, availability, and cost of specific types of insecticides, number and type of animals involved, cost of labor, livestock management followed, and type of insecticide formulation used.

Large volume wet sprays: DDT insecticides are most commonly used in quantities and in a manner which will permit complete treatment of the host. For satisfactory control of parasites such as lice and ticks, it is important to apply sufficient spray to wet the animal, thereby assuring that the insecticide contacts the parasite. The quantity of spray material required will vary, depending on the size of the animal and the amount of hair. Mature cattle, depending on size, breed, and season, may require less than 2 quarts or as much as 2 gal or more to wet the animal.

Spray equipment suitable for this method of application varies from the hand-operated compressed air pressure sprayers having a capacity of a few gallons (see Figure 12) to large power equipment (see Figure 11) developing several hundred pounds pressure which delivers several gallons of spray per minute. In some parts of the world fog machines are employed to apply DDT to animals but their value for this purpose has not been firmly established.

The portable compressed-air sprayers are satisfactory for treating a few animals, or if spray equipment is not available the insecticide may be applied as a wash. For small herds compressed-air sprayers, wheelbarrow-type sprayers, bucket pumps, or small power units may be most practical. For large herds of cattle, power units delivering the spray at a rapid rate are generally most practical because they permit the treatment of large numbers of animals in a short period of time.

Either DDT suspensions or emulsions may be applied using power sprayers with good agitators. Smaller sprayers not equipped with agitators perform more satisfactorily with emulsions than with wettable powder suspensions.

Light mist sprays: When the application of small quantities of a light mist spray to the animals will achieve practical control of parasites such as the horn fly or buffalo fly, and if only a few animals are involved, it may be most economical to employ hand-operated sprayers commonly used in homes for space treatment. Such sprayers atomize the liquid by means of air blast over a small siphon tube leading from the spray tank. DDT oil solutions or emulsions may be employed in such equipment but DDT suspensions cannot be used because the suspended material will cause clogging of the small siphon tube. Mist sprays of this type are usually applied in quantities of about 1 oz (about 28 cm³) per animal per treatment. The concentration of DDT used in the spray material may range from less than 1 to as much as 5%. If oil solutions are employed the quantity of spray should be limited so as not to wet the skin.

Parasites such as lice and ticks usually cannot be controlled satisfactorily unless sufficient spray is applied to wet the animal thoroughly, thereby assuring contact with the parasite. Oil solutions of DDT, or any other insecticide for that matter, are not generally recommended for controlling parasites such as lice and ticks, although because of their tolerance for oils and the thin hair coat, swine are sometimes treated with DDT oil solutions for the control of lice.

Dips

DDT dips are employed less extensively than sprays. Dipping, however, is the preferred method of treating animals for external parasite control in many parts of the world. Dipping accomplishes complete and uniform wetting of all parts of the host, thus assuring better parasite control than is possible with sprays. However, DDT and other chlorinated insecticides at present have a serious limitation when employed as dips. The insecticide is not in solution as is the case with arsenical dips which have been employed so extensively in the past. DDT emulsions or suspensions may settle to the bottom of the vats on standing or, in the case of certain types of emulsions having a density less than water, may rise to the surface.

Thus, the emulsions or suspensions may not disperse uniformly throughout the dipping solution, particula·ly when used repeatedly during the course of several months. This may prevent uniform treatment of the host and erratic control of the parasite. In some instances, primarily due to toxic action of the solvent used in emulsions, the dips may prove hazardous to the animals.

Another problem, which is perhaps of greater significance, is brought about by the selective pick-up of the suspended DDT particles by the animal, resulting in gradual lowering of the concentration of insecticide as more and more animals are dipped. The lowering of DDT concentration in dips following the dipping of animals was reported by CRAGG [44] in Great Britain as early as 1945. Other investigators have since made similar observations, including SHAW [163], NICOL [130], GRAHAM and SCOTT [74], BEKKER et al. [17], and WHITNALL et al. [198].

To compensate for the depletion of insecticide from the dip, it is necessary to start with a higher initial concentration than would normally be required and to recharge with a concentration higher than the initial concentration. No simple and rapid field method for determining the concentration of DDT in dips has yet been developed. The addition of extra DDT to compensate for the loss is not a satisfactory procedure, for the rate of exhaustion of the DDT will likely vary, depending on the type, age, and stability of the dip used and the type of animals dipped. RADELEFF [142] working with toxaphene emulsions, showed that the amount of insecticide deposited on animal hair varies depending on the degree of stability of the emulsion. SPARR et al. [174] subsequently correlated the rate of selective absorption of toxaphene in emulsion form with the particle size of the oil phase, which varies depending on the stability of the emulsion. Emulsion having colloidal or extremely small particles are not exhausted from dips as rapidly as emulsions having larger particles, and their behavior is more uniform.

DDT itself appears not to deteriorate to a significant degree when used as a dip. In the United States, dips used for horn-fly control showed no loss in effectiveness after 1 year (LAAKE [99]). Continued insecticidal efficacy of a DDT suspension dip after several years' use has been reported by NICHOLSON [129].

Müller II/33

 E. F. Knipling

Dusts

Little need be said about the method of applying dusts to the host for controlling external parasites. Any suitable dusting equipment may be used. Ordinary containers with a number of small holes to permit application of dusts are satisfactory. It is difficult to obtain satisfactory penetration of dust in the hair. Rubbing of the hair coat with the hand while the dust is applied is usually necessary for satisfactory penetration and dispersion of the insecticide.

5.

CONTROL OF PARASITES WHILE OFF THE HOST

Parasites such as fowl ticks, poultry mites, and bed-bugs normally do not stay on the host for long periods. They are, however, usually concentrated in the animal houses or places where the host animals roost. Since they remain on the host for only a few minutes while feeding, the treatment of the host usually does not result in satisfactory control. The application of DDT sprays or dusts to the restricted places where the parasites live will, therefore, achieve most effective control through direct contact with the parasite, and by leaving a residual deposit which the parasite may contact while hiding or moving about in search of the host.

Parasites such as fleas and dog ticks may spend considerable time on the host and also may be concentrated in restricted areas while off the host. Thus either, or preferably both, host treatment and treatment of the premises may achieve effective control.

Mosquitoes, black flies, horse and deer flies, and tsetse flies spend only a few minutes on the host while feeding. It is, therefore, difficult to obtain long lasting control by treating the host. Moreover, such pests, unlike the fleas and dog ticks mentioned, are widely dispersed while off the host. Therefore, it is often not economically feasible to concentrate efforts on environmental control. However, if such procedures are feasible, the area-control methods applicable for protecting man from their attack can be used. Treatment of restricted areas where animals are concentrated may be feasible in some instances. The number of stable flies and mosquitoes may be reduced in animal stables or corrals by using residual treatments or space treatments in the stables or applying residual treatment to vegetation and other surfaces where the animals concentrate.

6.

CONTROL OF SPECIFIC ARTHROPODS

Flies

House Flies (Musca domestica and Related Species)

The house fly is of greater significance as a disease carrier and pest of man than of animals; therefore, the use of DDT for its control will be discussed only briefly. When first employed as a residual or surface treatment in and around stables, DDT provided a higher degree of fly control than was possible to achieve with older insecticides or any other method of control.

Resistance to DDT which has been acquired by the house fly in many areas has limited the usefulness of DDT for controlling this insect. Because of the resistance problem other chlorinated hydrocarbon insecticides have been employed as substitutes. However, where chlorinated hydrocarbon insecticides, methoxychlor (1,1,1-trichloro-2,2-bis-[p-methoxy-phenyl]-ethane), lindane (gamma isomer of benzene hexachloride of not less than 99 per cent purity), and chlordane (1,2,4,5,6,7,8,8-octachloro-2,3,3a,4,7,7a-hexahydro-4,7-methano-indene), have been substituted for DDT, the flies have also, within a season or two, developed resistance to them.

Considerable investigation is now under way in various countries to find materials which, when used in combination with DDT, will restore the efficacy of the compound against resistant strains. If this approach proves successful, DDT may again come into extensive use where flies have become resistant.

Although a number of the new insecticides developed during the last decade are highly efficient against nonresistant flies, none of the materials now in general use possess the long lasting residual effect shown by DDT. Dieldrin (1,2,3,4,10,10-hexachloro-6,7-epoxy-1,4,4a,-5,6,7,8,8a-octahydro-1,4,5,8-dimethano-naphthalene) is comparable with DDT from the standpoint of lasting effect but it is a greater toxic hazard to man and animals.

In employing DDT for fly control to protect animals, it must be emphasized that complete reliance should not be placed on the action of the insecticide. It is generally agreed that the disposal of manure and other animal or vegetable matter in which the larvae develop, as well as materials which attract the adult flies, is necessary to maintain good fly control. Screens to exclude flies should also be a standard practice, wherever practical.

Stable Flies [Stomoxys calcitrans (L.)]

The stable fly is primarily a pest of livestock, although it is capable of transmitting animal diseases including anthrax and equine infectious anemia. The insect is widely distributed throughout the world. It is a painful bloodsucking pest and the annoyance and loss of blood caused by the insect, especially during serious outbreaks, adversely affect the productivity of dairy cows and meat animals. The insects are also serious pests of work animals.

As previously mentioned, the value of DDT as a residual spray in and around animal sheds for the control of the stable fly was first determined by WIESMANN [200]. Other investigators, including BLAKESLEE [24], OLSEN [135], SCHOEN [152], BLAXTER [28], and SILVA LEITAO [164], also reported successful control of the insect when DDT was employed in a similar manner. Most practical experience in the use and value of DDT for controlling the stable fly has been obtained in connection with the use of DDT for house-fly control around animal barns and shelters.

Treatment of resting places: In general, the habits of the stable fly around animal houses are similar to those of the house fly. There are some differences, however, which if kept in mind in applying treatments may result in greater efficacy. The stable fly is not normally attracted to waste foods for feeding purposes and is not, therefore, highly concentrated in a few specific locations, as is often the case with the house fly. Scattered breeding places such as in fields and pastures, and the attraction of flies to animal hosts, are other factors favoring a dispersed population. During the summer stable flies may, however, concentrate on walls and other surfaces on shaded sides of buildings or on trees and other vegetation.

The identification of these resting places and their thorough treatment will usually result in satisfactory control.

In applying the DDT for stable-fly control the same type of formulations, concentrations, dosages, and equipment employed for house-fly control may be used. See Chapter VII.

Host treatment: It is still questionable whether host treatment with DDT is sufficiently effective to aid in the control of the stable fly. The insect is an intermittent feeder on animals. It may light on the animal and take a full blood meal within about 2 to 8 minutes (SIMMONS [165]), thus remaining in contact with a residual insecticide for only a short time. The application of DDT wettable powder or emulsion sprays to animals does not repel the insect. The flies feed on treated animals within a few minutes after application of the spray. However, observations by BLAKESLEE [24], SUTER [179], and EDDY and McGREGOR [59] have shown that most of the flies feeding on animals during the first 2 to 3 days after treatment are killed by contact with the DDT. The destruction of the feeding insects, thus preventing subsequent attack, may significantly reduce the annoyance, especially when most of the animals subjected to attack in an area are treated. Unfortunately, it has not been determined in carefully controlled experiments how much control is obtained over a period of time by regular treatments of animals with DDT.

Treatment of breeding places: The stable fly may create a serious nuisance problem to both animals and man near seashores where marine grasses accumulate and provide extensive breeding places for the insect. BLAKESLEE [25] found that 0·5 per cent DDT-emulsion sprays were effective in preventing emergence of flies from the marine deposits. QUARTERMAN *et al.* [141] reported on the value of DDT sprays applied to marine grass in large scale operations along the Gulf coastal areas in Florida of the United States. Spraying the surface of the marine grass deposits with 1·5 per cent DDT-emulsion sprays prepared by diluting emulsion concentrates with sea water, resulted in excellent control of the insect at one third the cost of creosote sprays used prior to the availability of DDT. The DDT treatment may destroy some of the younger larvae; however, control was achieved primarily because the adults were killed when they contacted the DDT upon emergence. SIMMONS and WRIGHT [166] also found that the application of emulsion sprays containing 0·25 and 0·5% of DDT to peanut litter at the rate of 1 gal per cubic foot of litter reduced emergence of flies about 98 to 99%. This type of treatment may also control the stable fly in waste celery or other waste agricultural products which often serve as breeding media.

Tsetse Flies (Glossina spp.)

The tsetse flies are among the most important of the insects which attack animals. Several species are involved in the transmission of the causative organisms for trypanosomiasis, or sleeping sickness. The disease is so serious that cattle, horses, sheep, and swine cannot be produced profitably in vast areas in Africa. Because of the economic importance of the tsetse flies, extensive research has been undertaken by investigators in Africa in an effort to develop ways to control these vectors with DDT.

The possibility of controlling the tsetse flies by applying DDT to the host was suggested by SUTER [179] on the basis of his studies on the stable fly. The tsetse flies are highly susceptible to DDT, as shown by VANDERPLANK [190], who reported on results of tests in the laboratory against 13 species. Tests were also conducted on animals using a spray consisting of equal parts of 5 per cent kerosene solution of DDT and ox serum. The ox serum was used as a sticker for the DDT. Flies feeding on treated oxen were killed for 2 days and then the treatment became less effective although considerable numbers were killed for a period of 5 days. WILKINSON [201] found that *Glossina fuscipes palpalis* (Newst.) feeding on cattle sprayed with 1 pint of 5 per cent DDT emulsion were killed for 4 days or longer. Emulsions containing coumarone resins were more persistent than those without the sticker. WHITESIDE [196] reported that spraying oxen once each week with a light application of peanut oil containing 9% of DDT resulted in about 70 per cent kill of flies. Two sprayings each week effected a 95 per cent mortality. In an experiment in which there were about six times as many oxen as large wild game in an area, the treatment of all the oxen at intervals during a period of 2 months caused an 80 per cent reduction in the tsetse-fly population in the area.

The feasibility of employing DDT to control tsetse flies in Africa by area-control measures has also been explored by several investigators. SYMES *et al.* [182] reported on the value of oil solutions of DDT and BHC sprays applied to vegetation on an island at the rate of about 100 mg of the insecticides per square foot of leaf surface. One treatment with DDT reduced the fly population by 50 to 80% for 2 weeks. Four treatments at 10- to 14-day intervals resulted in a reduction of about 98%.

Reports of HEILBRON [81] and HALL [79], [80] summarize extensive investigations with DDT and BHC applied in various ways to control tsetse flies in Africa. Aerosols (smokes and fogs) using ground generators and aircraft, sprays applied with airplanes, and residual treatments applied to vegetation are the methods of application which have been investigated most intensively.

It was shown (HEILBRON [81]), that both DDT and BHC smokes applied with ground generators are lethal to tsetse flies but smokes present difficulties in application. Smokes and sprays applied with aircraft have given excellent control in some experiments. HALL [79] reported over 99 per cent control of *Glossina morsitans* (Westw.) and *G. swynnertoni* (Austen) when DDT smokes were applied to an area eight times at intervals of 2 weeks. The so-called smokes actually contained spray particles ranging in size from 5 to 250 μ, so that the results achieved were no doubt similar to those one might expect when sprays are applied. Airplane sprays applied at intervals of 2 weeks in light brush areas at the rate of 0·25 lb of DDT per acre resulted in over 99 per cent control of *G. morsitans*, *G. swynnertoni*, and *G. pallidipes* (Austen) after the 4th or 5th treatment.

The investigations have demonstrated that tsetse flies are highly susceptible to DDT applied in one of several ways. The utilization of the insecticide in practical control will be largely a problem of developing suitable and economical methods of application because of the extensive infested areas involved. In critical areas the use of DDT will no doubt prove valuable as a control method for the insects and the diseases transmitted by them. HALL [80] states that an experiment is under way in efforts to control sleeping sickness by applying DDT suspension to vegetation along tributaries of the Nyado river in Kenya. Results on disease reduction have not been reported, but a marked reduction in fly populations occurred immediately after the first treatment.

Horse Flies and Deer Flies (Tabanidae)

The Tabanidae, comprising many species, are widely distributed throughout the world. They are vicious bloodsucking parasites which attack all kinds of livestock, especially horses and cattle. The Tabanidae are large flies, thus they consume considerable quantities of blood. When present in large numbers the annoyance and loss of blood can seriously affect the health and productivity of the animals. BRUCE and DECKER [34] have shown that horse flies may reduce the production of dairy cows by 10 to 20%, and reduce weight gains in beef cattle by as much as 1 lb per day.

In addition to the annoyance and loss of blood caused by horse and deer flies, the insects are important as transmitters of animal diseases. Anaplasmosis, anthrax, surra, African sleeping sickness, and tularemia are among the diseases that these insects are capable of transmitting.

The horse and deer flies are intermittent feeders, remaining on the animals for only a short time while taking their blood meals. The insects normally do not concentrate around animal buildings and shelters, and they appear to be well dispersed in woodlands and other vegetated areas.

Although the Tabanidae are susceptible to DDT when contacted with it, there is no known economical way to apply the insecticide to obtain effective control of the insects.

The tabanids, like stable flies and tsetse flies, are not repelled by DDT treatment of the host (GJULLIN and MOTE [68]). EDDY [55] found that tabanid species feeding on treated animals for the first few days are killed. Little is known about the number of times the insects feed on animals and the time interval between blood meals. There is also very little information regarding rate and extent of dispersion. Some livestock owners in the midwestern and southeastern sections of the United States have reported marked reduction in the number of tabanids attacking cattle following the treatment of the animals with sprays for controlling other external parasites. However, carefully designed and controlled experiments to evaluate the degree of control obtained have not been conducted.

The possibility of applying chlorinated hydrocarbon insecticides by means of aircraft to wooded areas for the control of *Tabanus abactor* (Philip) and *Tabanus sulcifrons* (Macq.) was explored in the United States (HOWELL *et al.* [89]). DDT, methoxychlor, toxaphene (chlorinated camphene having a chlorine content of 67–69%), and chlordane were used in the experiments. Wooded plots ranging in size from 10 to 60 acres were sprayed with solutions consisting of 10% of the insecticides named, dissolved in fuel oil using cyclohexonone as an auxiliary solvent. The materials were applied at the rate of 2 lb of insecticide per acre which is near the maximum considered safe and practical for general area control. During the first week after treatment horses were led into the treated plots and the numbers of tabanids attracted to the animals were compared with the numbers on the animals in untreated control plots. The authors concluded that none of the treatments showed appreciable reduction in numbers of tabanids in the area.

GJULLIN and MOTE [68] have found that first-instar larvae of *Chrysops discalis* (Will.) in Oregon were destroyed by DDT at concentrations of 1 ppm. The possibility of employing DDT as a larvicide for tabanids seems to offer little promise, however, because of cost and the danger of destroying fish and other beneficial aquatic organisms.

Black Flies or Buffalo Gnats (Simuliidae)

The use of DDT for the control of black flies, important pests and disease vectors of man is also discussed in Chapter VII. These insects are also important pests of livestock and poultry and transmit certain animal diseases.

Horses, mules, cattle, and poultry are severely attacked by the bloodsucking parasites. When they feed, they inject a toxin into the animal which causes severe reactions and during outbreaks they may even cause death of the hosts. As disease carriers, the black flies are noted primarily for the transmission of a protozoan disease caused by the *Leucocytozoon* parasite. This organism causes a severe disease sometimes called blackhead, in turkeys, chickens, ducks, and geese. Black flies also transmit the filarial worm, *Onchocerca*, which causes a fistula-like growth in horses.

The black flies usually breed in swift flowing streams and are thus often localized in abundance. In some of the northern areas where streams are numerous they seem to disperse widely and are generally distributed in vast areas in the arctic and subarctic regions.

The Simuliidae are intermittent feeders on animals and thus contact DDT or other insecticide deposits on animals for only a short time. Observations (DEONIER [47]), have shown that treatment of cattle with 1 per cent DDT-suspension sprays will not prevent attack by the gnats after the spray dries. It is possible, however, that the insects feeding for several days after treatment may be killed, thereby reducing the black-fly population.

An experiment was conducted by DEONIER [47] in efforts to control adults of *Simulium pecuorum* (Riley), which were attacking livestock severely along the Navasota River in Texas. DDT applied as an oil solution by means of an airplane at the rate of 0·4 lb per acre failed to protect animals in a treated area of 400 acres. The spray destroyed the adults present at the time of treatment, but the insects from untreated areas migrated into the treated portion within 24 hours.

DDT applied by means of aircraft and ground aerosol machines failed to provide satisfactory control of adults in arctic areas (GOLDSMITH *et al.* [70] and WILSON *et al.* [202]). The insecticide apparently killed the insects present in the area at the time of treatment, but black flies from surrounding untreated areas quickly migrated into the treated area. The investigations by HOCKING and RICHARDS [85] have shown that *Simulium venustum* (Say) will migrate 4 to 6 miles, but that *Prosimulium hirtipes* (Fries) does not migrate in large numbers more than 2 miles. The degree of control that might be achieved by concentrating efforts on the adults may vary, therefore, depending on the species and other factors.

The larvae of black flies are highly susceptible to DDT. This has been reported by a number of investigators, including FAIRCHILD and BARREDA [61], GARNHAM and MCMAHON [64], GJULLIN *et al.* [67], ARNASON *et al.* [3], TRAVIS *et al.* [186], COLLINS *et al.* [41], and others. On the basis of the excellent results reported by the many investigators, it seems entirely practical to

522   E. F. Knipling

consider destruction of black-fly larvae as a practical way to protect livestock
from attack by these pests.

In treating streams for black-fly control, the following procedure is suggested (BUREAU OF ENTOMOLOGY AND PLANT QUARANTINE [37]): Fuel oil,
Diesel oil, or kerosene containing 5% of DDT may be applied by means of
airplanes or with any type of spray equipment or sprinkling can if streams
are accessible to vehicles or on foot. In applying the oil solution enough is
applied to give an average dosage of 0·025 lb of DDT per acre of stream
surface. It is not necessary to treat the entire surface. Application of oil at
intervals of several hundred feet is satisfactory. The moving water will adequately disperse the insecticide. It is recommended that all infested streams
be treated within a radius of 5 miles from the area to be protected.

Horn Fly [*Siphona irritans* (L.)] *and the Buffalo Fly* [*Siphona exigua* (De Meijere)]

These two related bloodsucking insects, about half as large as the stable
fly and similar in appearance, are widely distributed. They are considered
among the most important pests of the bovines although they have not been
incriminated as disease vectors. Infestations averaging 1,000 or more flies
per animal are not uncommon when no effort is made to control them. BRUCE
and DECKER [33] have shown that milk production may be reduced by 10 to
20% when dairy cows are subjected to annoyance by the flies. LAAKE [98]
showed that beef cattle protected from horn-fly attack by DDT treatments
gained more weight than untreated cattle.

The horn fly is primarily a pest of cattle and only occassionally attacks
sheep and horses. Most of the adult life of the insect is spent on the host where
the insects congregate in groups (Figure 1). The preferred location on the animal
depends on temperature and time of day. Normally, the flies prefer to rest and
feed on the backs and shoulders of the host but during midday, intense sun-
shine may force them to the underside of the animal. On occassion the horn
flies will concentrate in a band near the base of the horns, hence the common
name. However, this name is largely a misnomer for such habits are not
frequently observed.

The habit of the flies of remaining on the host, their high degree of susceptibility to DDT, together with the long lasting residual action of this
insecticide makes treatment of the host an efficient and economical method
of controlling both the horn fly and the buffalo fly. WELLS [195] was among the
first to publish a study on the value of DDT against the horn fly in the United
States. During the same year the Australian workers (AUSTRALIAN COUNCIL
FOR SCIENTIFIC AND INDUSTRIAL RESEARCH [8]) reported that DDT controlled
the buffalo fly. The efficacy of DDT for controlling the flies was soon confirmed
by other investigators, including VORHIES and WEHRLE [192], LAAKE [98],
MATTHYSSE [115], PEAIRS [140], BRUCE and BLAKESLEE [31], and NORRIS

[132]. Since 1946, various workers have published results of additional investigations on DDT and other chlorinated hydrocarbon insecticides.

Because of the early widespread and successful use of DDT for controlling the horn and buffalo flies, much of the research on formulations and methods of application of DDT to livestock was done in connection with the development of control measures for these insects.

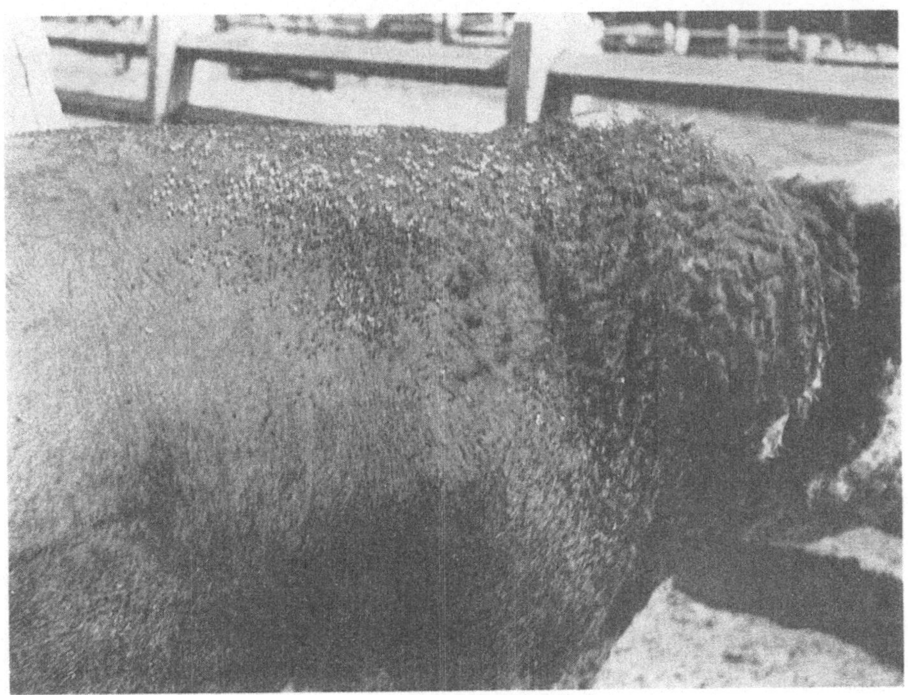

Figure 1

A bull heavily infested with the horn fly in the United States. Such heavy infestations on range cattle were common before DDT came into use for controlling the pest. Photo courtesy of the National Livestock Loss Prevention Board, Kansas City, Missouri.

Methods of applying DDT: It was soon recognized that DDT could be applied to cattle in many ways and yet obtain effective control of the flies. It is not necessary to completely saturate the animal to achieve the desired result. In fact, DDT applied to such parts of the animal as the head, hindquarters, and legs probably contributes little towards control of the insects. Moreover, treatment of all animals in a herd is not essential for good control, as the flies migrate freely from one animal to another and within a short time all or most of the flies from untreated animals may contact DDT on treated animals. It should be emphasized, however, that it is often desirable to treat all animals

                E. F. Knipling

thoroughly in order to control lice, ticks, and other pests when the cattle are treated for horn-fly control.

In the United States, DDT is generally used for horn-fly control in the manner described by the Bureau of Entomology and Plant Quarantine [35]. The cattle are sprayed with hand operated compressed air sprayers or with various types of power sprayers. Special spray equipment is not essential. Excellent results can be obtained by applying the DDT with a sponge. A

Figure 2
Spraying cattle with DDT for control of the horn flies. USDA photograph.

concentration of 0·5 per cent DDT is most commonly applied, using either wettable powder suspensions or emulsions. The animals are treated on the backs and sides at the rate of about 2 quarts per mature animal. This leaves a theoretical deposit of about 10 g of actual DDT per animal. However, in some parts of the United States, particularly in the Southeast, where the cattle breeds are predominately short haired, most livestock growers use concentrations up to 1·5% and apply an average of about 1½ pt per animal. Thus, the amount of DDT applied is about equal to the amount using the lower concentration and larger volume of spray. The application of about 10 g of DDT per animal will usually provide control for 3 to 4 weeks, although the period will vary, depending on many factors subsequently discussed.

In treating herds of cattle with power sprayers the animals are commonly crowded into pens and sprayed as a group as shown in Figure 2.

In Canada, 0·25 per cent DDT-emulsion or -suspension sprays are recommended for horn-fly control (TWINN [187]). This concentration is also advocated by some of the Experiment Stations in the United States. In areas where horn-fly populations do not normally become too high, the lower concentration appears to be adequate for satisfactory results.

Figure 3

A wire cable wrapped with cloth and treated with DDT solutions is being used successfully as a self treatment device for horn fly control. USDA photograph.

It has been shown (McGREGOR (121)], that the application of 10 g of DDT in a narrow strip about 6 in. wide along each side of the back will protect cattle from horn flies equally as well as the same amount of DDT applied as a spray over the entire body. McGREGOR [120] also found that the application of 2 quarts of a 0·5 per cent spray all over the body did not provide any better control than 2 quarts applied to the back.

The method of applying DDT to animals is likewise not a critical matter in obtaining control of the buffalo fly in Australia. It is reported (AUSTRALIAN COUNCIL FOR SCIENTIFIC AND INDUSTRIAL RESEARCH [9]), that application of DDT to an area of 12 × 18 in. on each side of the shoulders gave adequate control of the buffalo fly. In earlier tests (AUSTRALIAN COUNCIL FOR SCIENTIFIC

AND INDUSTRIAL RESEARCH [8]), a light application of kerosene containing 4% of DDT provided excellent control for 2 to 3 weeks and the pretreatment index was not regained until 3 to 4 weeks after treatment. NORRIS [133] reported that the application of about ¼ pt of 0·5 per cent DDT spray, either in the form of an emulsion or a suspension, to backs of cattle following routine arsenical dipping, controlled the buffalo fly for 3 weeks.

There has been considerable interest in the United States in the use of a self-treatment device recently investigated by ROGOFF [146]. A cable made of burlap wrapped around a core consisting of several strands of barbed wire is suspended between two posts as shown in Figure 3. This device is installed near watering places, salt licks or other places where cattle congregate. The burlap is treated at about 2-week intervals with fuel oil containing 5% of DDT. The animals rub themselves on the treated burlap and enough DDT is deposited on the animals to keep horn flies under control. The cost of this method of treatment is very low and no adverse effects to animals have been reported. HOWELL [88], LINDQUIST [107], and MCGREGOR [122] have reported that such devices are effective on ranges where brush and trees are scarce. In brushy areas the cattle apparently do not always rub on the treated burlap frequently enough to obtain control. The method, where effective, provides an extremely economical way to control an important livestock pest.

Some factors influencing horn or buffalo-fly control with DDT. The amount of DDT applied to cattle is obviously an important factor in determining the length of time that satisfactory horn-fly control can be maintained. This was demonstrated by MATTHYSSE [115], BRUCE and BLAKESLEE [32], SMITH [167], LAAKE [100], and others. Generally speaking, dosages of 5 g per animal will give significantly less control than 10 g. The lower dosage may control flies for a period of 2 to 3 weeks, whereas the higher dosage usually remains effective for 3 to 4 weeks. However, increasing the dosage to 20 g does not add greatly to the duration of control (BRUCE and BLAKESLEE [32]).

The type of formulation may influence the control; however, in practical control operations other factors may be more important. MATTHYSSE [115] found wettable powder-suspension sprays to be more effective in Florida than emulsion sprays. EDDY and GRAHAM [58] also found that DDT wettable powder sprays were more effective than an emulsion spray. The tests by the latter workers were conducted by exposing flies on animals in screened stalls, thus ruling out effects of population density of the flies and other factors. LAAKE [100] found DDT wettable powder sprays under Kansas (USA) conditions to be only slightly more effective than an emulsion spray, the former giving an average of 30 days' protection as compared with 28 days for the latter.

The amount of rainfall following treatment of animals with DDT is obviously a factor in the duration of control. There is a rapid loss, however, of DDT from the animal in the absence of rainfall. It was observed that DDT applied to living cattle in Australia became ineffective after 2 weeks, whereas when applied to hair on a removed hide and exposed to weathering the DDT remained effective for 5 months (AUSTRALIAN COUNCIL FOR SCIENTIFIC AND INDUSTRIAL RESEARCH [8]). LINDQUIST [106], working in Oregon, has obtained similar results. HACKMAN [78] also noted the rapid loss of effectiveness of DDT when applied to cattle. EDDY and GRAHAM [58] conducted tests in screened animal stalls and found that DDT applied as commonly used in practical control failed to kill within 48 hours all of the horn flies introduced on animals in stalls after 2 weeks.

The evidence seems to be fairly well founded that the DDT applied actually destroys the flies for only about 2 to 3 weeks. In practical use, however, treatments often give satisfactory control for 4 weeks or longer. The longer period of control is undoubtedly due to the delay in population increase after the treatment becomes ineffective.

It has not been determined why DDT loses its effectiveness so much more quickly when applied to hair on living animals than when applied to hair on hides. Some of the DDT is known to be absorbed because of its appearance in the fat of treated animals. No doubt there is also considerable mechanical loss and loss due to weathering. It is believed that licking accounts for a minor loss. This opinion is based on data which showed that DDT treated dairy cows excreted as much DDT in the milk when they were muzzled as did animals which were not muzzled.

The breed of cattle apparently influences the duration of DDT sprays. EDDY and GRAHAM [58] found that DDT was somewhat longer lasting on Herefords than on Jerseys or Brahman cattle. Apparently breeds with longer and denser hair retain the DDT longer.

The foregoing factors are, no doubt, important in determining the duration of control obtained with DDT. However, almost every investigator working on horn flies or buffalo flies has noted that the density of the fly population in the vicinity of the treated animals is one of the most important factors in determining the duration of control beyond a period of 2 to 3 weeks after treatment of the animals.

Community effort in which all animals in a large area are treated is encouraged wherever possible. SMITH and GATES [169] reported on experiments in Kansas in which an attempt was made to treat all cattle in a county of 750 sq. miles. Unfortunately, it was not possible to get 100 per cent cooperation. About 1% of the cattle were not treated. However, the cattle in the center area about 10 to 12 miles from the edge of the treated zone were protected for an average of 11 weeks with one treatment. In the outer portion of the county the average protection obtained per treatment ranged from 6 to 7 weeks. During the previous year a single treatment on individual herds in the county protected animals for about 4 weeks.

DDT compared with other insecticides for horn-fly control. Since DDT came into use for controlling the horn fly, a number of other insecticides have been developed and have been tested extensively in the United States. These include methoxychlor, TDE (1,1-dichloro-2,2-bis-[P-Chlorophenyl]-ethane), chlordane, lindane, and the pyrethrum-piperonyl butoxide combination, known commercially as Pyrenone. Results of investigations with various insecticides have been published by SMITH [167], LAAKE [100], McGREGOR [120], LAAKE *et al.* [102], and DENNING and PFADT [46].

In all cases sprays were applied at the rate of 2 quarts per animal. When the horn-fly population reached an average of 25 per animal, re-treatments were made. The average protection in days for each treatment with DDT, toxaphene, methoxychlor, TDE, and chlordane as reported by SMITH [167] and LAAKE [100] is tabulated below:

*Average Number of Days Protection from Horn-Fly
Attack for Each Treatment During the Season*

Insecticide	Spray concentration	SMITH [167] Texas	LAAKE [100] Kansas
DDT	0·5 per cent wettable powder	29	30
DDT	0·25 per cent wettable powder	17	19
DDT	0·5 per cent emulsion	—	28
Methoxychlor	0·5 per cent wettable powder	20	24
Toxaphene	0·5 per cent wettable powder	31	27
TDE	0·5 per cent wettable powder	24	25
Chlordane	0·5 per cent wettable powder	32	—
Chlordane	0·25 per cent wettable powder	—	17

It appears that there is little difference in the degree of control obtained with DDT, toxaphene, and chlordane wettable powder sprays. Methoxychlor and TDE are indicated to be somewhat less effective than the three named, although good control can also be obtained with these materials. In the report of LAAKE *et al.* [102] lindane emulsion at a concentration of 0·03% gave an average season's control of about 14 days as compared with about 18 days for the methoxychlor wettable powder spray. An emulsion spray consisting of 0·05% pyrethrins and 0·5% piperonyl butoxide controlled the horn fly for a period of about 1 week. Subsequent data obtained in Texas showed that lindane at a concentration of 0·03% provided control for a period of about 1 week, as compared with about 18 days for methoxychlor wettable powder used at a concentration of 0·5% (EDDY [57]).

Mosquitoes

Mosquitoes are of such great significance in human medicine, their importance in veterinary medicine is often ignored. Mosquito annoyance may, however, seriously affect the health and productivity of livestock. In addition, certain diseases such as equine encephalomyelitis and heartworm (filarial worm) disease in dogs are transmitted by mosquitoes.

The value of DDT for controlling mosquitoes attacking man is discussed thoroughly in chapter VII. The methods of control described for the protection of man will in general apply to circumstances where it is economically feasible to control the insects for the protection of livestock. It is unnecessary, therefore, to discuss equipment, formulations, and rates of application. Some comments on the use of DDT to protect livestock from mosquito attack are warranted, however.

It is recognized that mosquito control programs on a community-wide basis to protect livestock are not feasible under many circumstances. The protection of livestock may, however, be regarded an added benefit in many areas where programs to protect man are under consideration, and should, therefore, be taken into account in determining the value of such programs.

The individual livestock owner can do more to protect livestock from mosquito attack than is generally realized. It is known that many mosquitoes migrate very little during daylight hours. Suspension or emulsion residual sprays applied to tree trunks, lower branches of trees and other resting surfaces to a height of 5 to 10 ft at the rate of 2 to 3 lb of DDT per acre will provide areas relatively free from mosquito attack during daylight hours, for several days to several weeks. This barrier strip treatment method might be practical under special conditions where mosquitoes create serious annoyance to livestock.

The application of DDT residual sprays to mosquito resting places in and around animal shelters will reduce annoyance caused by the pests.

The value of DDT applied to animals as a means of protecting livestock has not been adequately studied. Field observations clearly show that DDT applied to the host does not repel mosquitoes but the contact with the DDT on the animal will cause the death of the insects for some time after treatment. No published data have been noted which show how long the treatment will prove fatal to feeding mosquitoes. This might be expected to vary depending on the species of mosquito involved. TRAVIS [185] found that most *Anopheles quadrimaculatus* (Say), feeding on a calf treated with 0·5 per cent DDT spray, were killed for 3 days following treatment.

In general, we have the same problem with mosquitoes that has already been mentioned in connection with the discussion on stable flies, tsetse flies, and the horse flies, and deer flies. It has not been determined in carefully controlled experiments whether substantial reduction in mosquito annoyance can be achieved by treating individual herds with DDT. BRUCE and BLAKESLEE [31] report, however, that annoyance to animals due to freshwater mosquitoes was eliminated after about 2 weeks by treating all livestock in a county in Florida with DDT in connection with horn-fly-control experiments.

Sheep Tick [Melophagus ovinus (L.)]

The sheep tick, also known as the sheep ked, is a common pest on sheep throughout the world. This wingless fly is regarded as one of the important external parasites of sheep. In addition to its adverse effect on the health of the animals, the annoyance of the sheep tick causes the host animals to rub and bite themselves, thus lowering the yield and quality of the fibers.

Many investigators have found that DDT is an excellent insecticide for controlling this insect. The parasite normally spends its entire life on the host. Host treatment is, therefore, the only method involved in its control.

The sheep tick is highly susceptible to a number of insecticides. Rotenone dips at very low concentrations have been used successfully for many years to control this insect. The chief advantage of DDT over rotenone is its longer lasting protective action. In addition, DDT is superior to rotenone for the control of lice attacking sheep. Since infestations of lice and sheep ticks may occur among the flocks simultaneously, there is an obvious advantage in treating the animals with a material which will effectively control both kinds of parasites.

Much of the investigations on the use of DDT for sheep-tick control have been conducted in Australia, Great Britain and the United States, although workers in other countries have contributed to the available information on the subject.

In Australia it was found that infestations were destroyed when sheep were dipped in aqueous suspensions containing 0·02% of DDT. The animals were protected from artificial reinfestations for 4 weeks (AUSTRALIAN COUNCIL FOR SCIENTIFIC AND INDUSTRIAL RESEARCH [9]). COBBETT and SMITH [40] reported that DDT suspension dips ranging in concentration from 0·1 to 1·0% controlled the insect. A concentration of 0·1% was also reported to control the insects (EINARSSON [60]). RUDE and PARISH [147] reported complete control of sheep ticks with a single dipping in a 0·2 per cent DDT emulsion. KEMPER *et al.* [93] found that emulsion or suspension dips containing 0·2% of DDT eradicated the sheep ticks but concentrations of 0·1 and 0·15% did not. NICOL [130] found sheep free of sheep ticks 2 months after dipping in a suspension dip containing an initial concentration of 0·18% of DDT. The concentration dropped to 0·11% after 250 sheep were dipped. COOP [42] reported a concentration of 0·125% effective, but some sheep ticks survived a concentration of 0·025%. FAIRCHILD *et al.* [62] dipped small flocks in 0·05 per cent DDT suspensions and obtained an estimated control of 97% after 42 days. Dips containing 0·2% of DDT effected complete control for at least 110 days. GRAHAM and SCOTT [74] obtained complete control with a suspension dip when the concentration was maintained between 0·056 and 0·09% of DDT. COOP and McLEOD [43] obtained complete control with suspension dips containing 0·03 to 0·045% of para-para DDT.

SHANAHAN [160] of Australia reported on results of sprays containing 0·025, 0·05, and 0·075% of DDT. The spray equipment was designed so that the liquid draining from the sheep returned to the spray tank and thus was reused. Although the concentrations of DDT did not remain constant, all three concentrations gave excellent control of sheep ticks on shorn sheep. No living adults were found on sheep sprayed with 0·075 per cent DDT, but a few living adults and developing pupae were present on sheep sprayed with the lower concentrations. Some dead adults were present on the sheep after 28 days, however, indicating that the DDT was still toxic at the time of examination even on the sheep sprayed with 0·025 per cent DDT. In a later publication of SHANAHAN [161] it is recommended that the initial concentration of DDT be 0·1% to compensate for loss of the insecticide coincident with the progressive treatment of a flock.

FAIRCHILD *et al.* [62] also investigated the efficacy of DDT sprays and obtained good but incomplete control with sprays containing 0·5% of DDT. SEGHETTI and FIREHAMMER [156] obtained excellent control with 0·5 per cent DDT using high-pressure sprays. Unsatisfactory results were obtained, however, when sheep passed through a rectangular spray frame set in a chute. HOFFMAN and LINDQUIST [87] obtained excellent control on sheep with short wool using low pressure but high volume sprays containing 0·5% of DDT. Sprays applied to recently shorn ewes and short wooled lambs resulted in almost complete control for 6 months. It is of particular interest that these investigators found that the treatments were more effective when wetting agents were added to the spray material.

Use of DDT in Practical Control

Dips: The results of investigations reviewed show that DDT is highly efficient as an insecticide for controlling the sheep tick. It is noted, however, that there is considerable variation in the efficacy of concentrations in the range of 0·1 per cent DDT or less.

The type of preparation, type of sheep, and other factors no doubt influence the results. It is apparent, however, that one of the most important factors to consider in using DDT dips is the matter of allowing for the lowering of the concentration of DDT after large numbers of sheep have gone through the vat. This problem, as previously discussed, is also of great importance in connection with the dipping of cattle.

Since there is no rapid field method for determining the concentration of DDT in the dipping fluid, it is important with currently available formulations to charge the vat with an excess of DDT to compensate for the progressive loss during the treatment of large numbers of animals. For practical control operations the UNITED STATES DEPARTMENT OF AGRICULTURE [188] recommends dips, either emulsions or suspensions, containing an initial concentration of 0·25% of DDT. TWINN [187], also recommends this concentration for sheep-tick control in Canada. A concentration of 0·25% provides reasonable assurance that the strength of the dip will remain high enough to eliminate infestations even after large numbers of animals have gone through the dipping fluid. The results reported by various investigators clearly indicate, however, that complete control of the insect can be expected if the DDT is maintained at 0·1 per cent DDT or even lower in some instances. In Australia an initial concentration of 0·1% is recommended (SEDDON [154] and MELDRUM [125]).

Many types of dipping vats are being used to dip sheep. Figure 4 shows the dipping of sheep in a small plunge dip used on farms with flocks up to 600 sheep.

Dipping sheep soon after shearing while the wool is short conserves material and animals are less likely to be injured while they are dipped.

Figure 4

Dipping sheep in a small plunge dip in Australia for controlling sheep ticks and lice. (Photograph courtesy Southern Australia Department of Agriculture).

Sprays: In many areas in the world, sheep are grown on marginal lands where the sheep range over large acreages. Many stock farmers also maintain only small flocks. An insecticide which can be used effectively as a spray is therefore highly desirable.

Proper spraying with DDT is now recognized as an effective procedure and is now practiced widely in many parts of the world. The use of 0·5 per cent DDT, or two times the concentration recommended in dips, is advocated by the UNITED STATES DEPARTMENT OF AGRICULTURE [188]. The addition of 1 lb of detergent of the polyphosphate or surface-active emulsifier type to each 100 gal of spray will increase the efficacy of the treatment. Animals should preferably be sprayed before the wool grows too long in order to conserve material and to permit more thorough application. Low pressure spray equipment as shown in Figure 5 will permit satisfactory control on sheep with short fleece. High-pressure sprays as shown in Figure 6 are essential for satisfactory results when the wool exceeds 3 to 4 in. in length.

Fog-producing machines have been tested by SHANAHAN [162]. When freshly shorn sheep were treated, good control was achieved. SEDDON [154] indicated that additional experience with fogs would be required, however, before this method of treatment can be recommended.

Figure 5

Low-pressure sprays can be used effectively for the control of sheep ticks if treatments are made before wool is too long. USDA photograph.

Dusts: DDT dusts apparently have not proved too effective for sheep-tick control. MATTHYSSE [114] failed to obtain satisfactory control using 5 per cent DDT dusts applied with a power duster. Commercial operators in the United States have attempted to develop this method of treatment without success.

Relative Efficacy of DDT and Other Insecticides

Other chlorinated hydrocarbon insecticides including BHC and lindane, methoxychlor, TDE, toxaphene, and chlordane have been tested in comparison with DDT for controlling the sheep tick. All of these materials, as well as

534   E. F. Knipling

rotenone, are effective for controlling the insect. All of them produce kills
through action on the adult stage. No effect on the pupae was noted by
HOFFMAN [86].

SHANAHAN [160], [162], GRAHAM and SCOTT [74], FAIRCHILD *et al.* [62],
COOP [42], COOP and McLEOD [43], and HOFFMAN and LINDQUIST [87] have
investigated BHC or lindane. The gamma isomer of BHC is highly effective
and initial control can be achieved at lower concentration than with DDT.

Figure 6

High-pressure sprays containing DDT properly applied are effective for controlling sheep ticks even
when wool is long and dense. Photo courtesy of the Livestock Loss Prevention Board, Kansas City,
Missouri.

It does not possess the persistent action of DDT, although all of the chlorinated
insecticides appear to persist on the wool of sheep much longer than on the
hair of cattle. This is thought to be due, in part at least, to the wool grease
which prevents deterioration and loss of the insecticide.

Investigations by FAIRCHILD *et al.* [62] using dips and sprays indicate that
DDT, methoxychlor, and TDE are similar in effectiveness. However, BHC,
chlordane, and toxaphene are more effective when compared directly with
DDT. In sprays particularly, BHC and chlordane proved superior. It is be-

lieved that these two compounds cause death of sheep ticks through effects of the vapors even if the insects do not make contact with the insecticide deposits on the animals.

Other Hippoboscidae

SILVA-LEITAO [164] reported that *Hippobosca equina* (L.) was effectively controlled with an unstated concentration of DDT powder. A dust containing 5% of the insecticide was found to be effective by PAVLOV and GUEORGUIEV [139]. EADS [54] reported that *Lipoptena mazamae* (Rondani) was controlled on deer with 1 to 2 per cent DDT suspension sprays applied at the rate of 1 pt per animal.

Ticks

A great need in the veterinary field, prior to the development of the new chlorinated hydrocarbon insecticides, was an effective agent for the control of ticks. The ticks are the most important of the external parasites of domestic animals. The annoyance and loss of blood caused by these parasites are of sufficient concern to warrant control measures. The parasites are of most significance, however, because of the many diseases they transmit.

Piroplasmosis, a disease caused by protozoa of the genus *Babesia*, affect livestock in many parts of the world. When the tick vectors and disease are prevalent it is difficult to produce cattle profitably unless control measures are instituted. Other important tick-borne diseases are the theilerial diseases, anaplasmosis, spirochetosis and tick paralysis. In view of the economic significance of the ticks, the possibilities of utilizing DDT for their control have been investigated by many workers in various parts of the world.

Arsenical solutions employed as dips were the standard treatment but they are highly toxic to animals and the concentration necessary for tick control allows only a narrow margin of safety. Under certain conditions the arsenicals injure animals even at concentrations approved by regulatory agencies. Moreover, the arsenicals provide little protection against reinfestations, especially against nymphs and adult ticks. The continuous use of arsenicals has also resulted in the development of arsenic-resistant strains of ticks.

Investigations soon established that DDT does not meet all of the requirements of an effective tick control agent. It has distinct advantages over the arsenicals, however, because of its safety to animals, longer protective properties, and usefulness against a wider range of species. Against some species of ticks, DDT alone provides good control but for most species it is relatively ineffective against the engorged forms. When DDT is combined with the gamma isomer of benzene hexachloride, an effective tick control agent is formed for use against a wide range of species. Benzene hexachloride possesses a high degree of toxicity to all stages of ticks, including engorged forms. DDT, on the other hand, protects animals from reinfestation for a longer period than does BHC.

There is a strong likelihood that ticks will develop resistance to the chlorinated hydrocarbon insecticides. WHITNALL *et al.* [199] have recently reported that strains of the blue tick *Boophilus decoloratus* (Koch) have been found resistant to BHC in South Africa. Strains of the cattle tick *Boophilus annulatus microplus* (Can.) resistant to BHC have also been reported from Australia (MACKERRAS [111]) and from Jamaica (ARNOLD [5]). Thus far no DDT-resistant strains have been reported.

The first reports on the value of DDT for tick control in the United States were published by GOUCK and SMITH [72], SMITH and GOUCK [171], and RUDE and SMITH [148]. GOUCK and SMITH demonstrated that a 5 per cent DDT emulsion used as a wash gave good control of the brown dog tick *Rhipicephalus sanguineus* (Latr.) and the lone-star tick *Amblyomma americanum* (L.). SMITH and GOUCK demonstrated by laboratory and field tests that DDT applied to vegetation along roadsides controlled the black-legged tick *Ixodes ricinus scapularis* (Say). RUDE and SMITH found that DDT at a concentration of 5%, incorporated in a nondrying adhesive and applied to the ears of cattle, controlled the Gulf-Coast tick *Amblyomma maculatum* (Koch) and the spinose ear tick, *Otobius megnini* (Duges). In Australia, good results were reported with DDT in 1944 against *Boophilus annulatus microplus* (AUSTRALIAN COUNCIL FOR SCIENTIFIC AND INDUSTRIAL RESEARCH [8]). VARGAS and COLORADO [191] found DDT to be effective against *Ornithodoros nicollei* (Moosen) and *Ornithodoros turicata* (Duges).

These early reports are cited to show that DDT was indicated to be useful against a wide range of species in various parts of the world. Extensive studies have since been conducted in Australia, Africa, North America, South and Central America, Europe, and other areas to determine the value and limitations of DDT for tick control.

Much of the research has dealt with the use of DDT in dips, the conventional method of tick control. Difficulty in developing dips uniform in performance has, however, stimulated research on the use of DDT-insecticide sprays for controlling ticks as well as other external parasites.

Cattle-Fever Tick [Boophilus annulatus microplus (Can.)]

The cattle-fever tick, *Boophilus annulatus microplus*, is the most important of the ticks affecting animals (Figures 7 and 8). These ticks and some closely related species are the vectors of bovine piroplasmosis. This disease causes untold losses in tropical and subtropical regions.

In view of the importance and widespread distribution of these ticks, many workers have investigated DDT for their control. The first report on the efficacy of DDT for controlling the fever ticks is that of the AUSTRALIAN COUNCIL FOR SCIENTIFIC AND INDUSTRIAL RESEARCH [8]. Emulsion sprays ranging in concentration from 0·5 to 2·0 per cent DDT resulted in effective control. Even the higher concentration did not destroy the engorged females but due to the greater susceptibility of the larval stage, animals were protected from reinfestation for periods up to about 2 weeks. Other investigators soon confirmed the value and limitations of DDT against the fever tick in other parts of the world. These included SQUIBB [175], SONI [172], COBBETT [39], HITCHCOCK and MACKERRAS [82], STAGE [176], BARONI [13], MARTINEX MORENO [113], BLAKESLEE and BRUCE [26], BARONI and EGLI [15], GARCIA

Figure 7

A cow infested with the fever tick, *Boophilus annulatus microplus*, vector of anaplasmosis and other important diseases of cattle. DDT alone or in combination with other insecticides is used to control this important and widespread tick (Photograph courtesy of Department of Agriculture and Stock, Busbane, Australia).

[63], LAAKE [99], MUSKUS and PACHECO TORRES [128], MAUNDER [116], NORRIS *et al.* [134], and SEDDON [155].

The various investigators report similar findings in regard to the efficacy of DDT. Concentrations as high as 3% of DDT apparently will not kill all of the engorged ticks. Therefore, it is impractical to employ concentrations which will destroy all stages of the tick. Satisfactory control is obtained by directing efforts toward the protection of animals from reinfestation. A concentration of 0·5% appears to be the most satisfactory concentration to use. DDT may be used either as a spray or dip. Figure 9 shows cattle being dipped.

However, spraying, if thoroughly done, will give about the same degree of control that is achieved by dipping.

As previously indicated, an improved tick treatment can be made by combining benzene hexachloride with DDT. Concentrations of 0·025% of gamma benzene hexachloride will destroy all stages of the tick, including engorged forms. Sprays containing 0·025% of gamma benzene hexachloride

Figure 8

A close-up of Figure 7 showing fever ticks in various stages of engorgement on the host.

plus 0·5% of DDT were found to be effective in South America (LAAKE [99]). Commercial preparations of this combination are used in North and South America for controlling various species of ticks. ARNOLD [4] reported good control with a combination containing 0·25% of DDT and 0·0065% of gamma BHC. The author expressed the opinion that the combination exhibits some synergistic action.

The application of 1 to 2 lb of DDT per acre to infested grounds will greatly reduce the number of ticks. This was demonstrated by MARTINEX MORENO

[113] in Cuba, by BLAKESLEE and BRUCE [26] in the United States, and by workers in Australia (AUSTRALIAN COMMONWEALTH SCIENTIFIC AND INDUSTRIAL RESEARCH ORGANISATION [8]). The treatment of infested areas, however, is not considered a practical way to control this tick, because of the high cost and possible toxic effects of the treatment to animals foraging on the treated range.

Figure 9

Dipping cattle in Transvaal, Africa. DDT alone or in combination with other insecticides is employed for controlling ticks, lice, and other external parasites. Photo courtesy of Cooper Technical Bureau, England.

Blue Tick [*Boophilus decoloratus* (Koch)]

This species of tick, common in Africa, is another vector of bovine piro-plasmosis. It is also a vector of spirochetosis in cattle, sheep, and horses. BEKKER and GRAF [16] reported that dips containing 0·5% of DDT protected animals in South Africa from ticks for 12 days following the second treatment. DDT is not highly effective against the engorged nymphs or adults as shown by WHITNALL and BRADFORD [197]. Good protection against reinfestation is obtained, however, against the larval stage. The work of WHITNALL *et al.* [199], previously referred to, gives a detailed account of the work on BHC-resistant strains of this tick. After about 2 years' use of BHC on certain tick-infested areas, strains of the tick about 20 times as resistant as normal susceptible ticks appeared. The BHC-resistant strains also appear to be some-what resistant to toxaphene and chlordane. No resistance to DDT was indi-cated, however, since a dip containing 0·15% of DDT was found to be effective in protecting animals from BHC-resistant strains.

Bont Tick [*Amblyomma hebraeum* (Koch)]

BEKKER *et al.* [17] state that 0·5 per cent DDT dips will control the bont tick, the carrier of heartwater disease of cattle and sheep. WHITNALL *et al.* [198] report excellent control of this species in South Africa by applying weekly treatments of a wettable powder spray containing 0·005% of gamma benzene hexachloride and 0·1% of DDT. Sprays were reported to be more satisfactory than dips because of the rapid depletion of the insecticide when employed in dips.

Lone-Star Tick [*Amblyomma americanum* (L.)]

Investigations on the control of this important pest of cattle and other livestock in the United States have been summarized by KNIPLING [96]. Engorged adults of this tick are not destroyed with either emulsions or wettable powder sprays containing up to 2% of DDT. However, the unengorged nymphs and adult stages which attack livestock are susceptible to DDT concentrations as low as 0·5%. Sprays containing 0·025% of gamma benzene hexachloride and 0·5% of DDT are used successfully for controlling the lone-star tick in the United States. Treatment of animals at intervals of 3 weeks will usually provide satisfactory control of this species, although if the ticks are numerous, a 2-weeks' treatment schedule is recommended (BUREAU OF ENTOMOLOGY AND PLANT QUARANTINE [36a]).

The application of DDT at rates of 0·5 and 1 lb per acre to infested areas has been shown to control the lone-star tick (GOUCK and SMITH [73]). McDUFFIE *et al.* [119] have also reported on the efficacy of DDT and several other insecticides when applied as dusts and sprays to infested areas. 1 to 4 lb of DDT per acre in dust form provided 85 to 97 per cent control for 6 weeks after application. Similar results were obtained with sprays.

Gulf-Coast Tick [*Amblyomma maculatum* (Koch)]

This three-host tick is prevalent along the Gulf Coast in the United States and in Mexico. It attaches itself in and around the ears of cattle and other livestock. The tick is an important pest because of the irritation caused by the bites. However, stockmen are more concerned about the tick because it produces wounds that are subject to infestation with screw-worms [*Callitroga hominivorax* (Coquerel)]. Control efforts against this tick are therefore primarily for the purpose of protecting animals from attack by the screw-worm.

RUDE and SMITH [148] reported that a nondrying adhesive smear containing 5% of DDT applied to the ears protected cattle in South Texas from tick attack for 3 weeks. MATTHYSSE [115] reported sufficient control of the Gulf-Coast tick in Florida with a smear containing 5% of DDT to prevent screw-worm infestations for 6 weeks.

Because of the difficulty of applying ointments to ears of cattle, the application of sprays to the entire animal, with particularly heavy treatments of the ears, head, and neck, is now the most common practice in the United States. For this purpose the combination of benzene hexachloride-DDT sprays used for controlling the lone-star tick is also used with success. As with other species, DDT alone will not kill the engorged forms of this tick at concentrations practical to use as sprays or dips, whereas the tick is highly susceptible to benzene hexachloride. A treatment schedule at intervals of 2 to 3 weeks with the DDT-BHC treatment (0·5% of DDT plus 0·025% of gamma BHC) will adequately protect animals from serious attack by the Gulf-Coast tick. BLAKESLEE *et al.* [27] found that sprays containing 2·3% of DDT applied at the rate of about 1 pt per animal protected cattle in Florida for 3 to 4 weeks. Cattle in this area generally have a thinner coat of hair during the summer months than in other parts of the United States. It is therefore the general practice to apply higher concentrations of spray but to reduce the volume applied per animal. Sprays containing 1·5% of DDT are most commonly used for tick control by stockmen in Florida.

The Winter Tick [*Dermacentor albipictus* (Pack.)]

This tick attacks horses and cattle in North America. It is most abundant during the fall and winter months. DDT at high concentrations will not control engorged forms. However, the DDT deposits on animals provide excellent protection from attack by the tick larvae, the stage of the tick which starts infestation on the host. PARISH and RUDE [138] found that a wash containing 0·8% of DDT made from a soluble pine oil concentrate containing 15% of DDT gave good protection to horses for an average of about 6 weeks. TWINN [187] recommends sponging horses with 1 per cent DDT emulsions or suspensions. McDuffie and Clark [118] found that sprays containing from 0·25 to 1·5% of DDT did not kill engorged ticks but protected animals from reinfestation for periods ranging from 3 to 8 weeks.

The combination of 0·025% of gamma BHC and 0·5% of DDT is considered a satisfactory preparation for controlling this tick in the United States.

Since the winter tick is an important pest during the colder months, the application of sprays or dips may be objectionable. The observations by McDuffie and Clark [118] that DDT dusts may also be used successfully are therefore significant. Concentration up to 5% did not kill engorged ticks but a 5 per cent dust protected horses from reinfestation for several weeks.

Castor-Bean Tick [Ixodes ricinus (L.)]

Shaw [163] reported that dipping sheep in emulsions containing 0·5% of DDT reduced infestation of *Ixodes ricinus* for 7 weeks, whereas arsenical dips remained effective for 2 weeks. Arthur [6], however, reported that a wash containing 0·5% of DDT in the form of a suspension protected cattle for only 3 days. Derris at the rate of 1 lb/gal of water employing soft soap as a wetting agent protected the animals for 8 days.

Fowl Ticks [Argas persicus (Oken) and Argas reflexus (F.)]

The fowl tick (*Argas persicus*) inhabits poultry houses and other places where poultry roost. It is a serious pest of poultry and it also transmits spirochetosis of fowls in many parts of the world. The ticks, in all stages, remain hidden in cracks, crevices, and other hiding places. The nymphs and adult stages remain on the host for only a short period of time while feeding. Treatment of the premises where the poultry roost is, therefore, the most practical and effective way to control this pest.

Legg [104] of Australia stated that a 2 per cent DDT-oil solution applied as a residual treatment in poultry houses prevented reinfestation by larvae of *Argas persicus* for at least 3 months. Some nymphs and adults could be found during this period, although greatly reduced in numbers.

Stage [176] found that the application of 2·5 per cent wettable powder controlled the fowl tick in chicken coops in Surinam, South America. Likerman and Viegas Aurelio [105] reported that 5 per cent DDT dust applied to poultry protected them from attack by the larvae. They advocate treating poultry with 5 per cent DDT dust and spraying the premises with 5 per cent spray.

A 5 per cent kerosene or emulsion spray applied in poultry houses in Southwest Texas will control the fowl tick (McDuffie [117]). Seddon [155], in Australia, also reported successful control of the tick with DDT, although BHC is considered more effective. The species, *Argas reflexus*, which commonly infests pigeon houses is also susceptible to DDT, according to investigations by Geurden [65]. The treatment of pigeon houses with a 10 per cent DDT dust is suggested by the author for its control.

It is apparent from these reports that excellent control of fowl ticks can be obtained with sprays containing from 2 to 5% of DDT. In view of the habits of the tick, it is important to make careful inspection of infested premises and apply sprays thoroughly. The ticks are known to live for months or years without feeding. This factor must also be considered in attempts to eliminate infestations.

Brown-Dog Tick [*Rhipicephalus sanguineus* (Latr.)]

This species, a primary pest on dogs, is widely distributed in temperate and tropical parts of the world. This species carries canine piroplasmosis. GOUCK and SMITH [72] found that DDT was effective for the control of this species. SERGENT *et al.* [157] also demonstrated that this tick could be controlled by treating dogs with DDT dissolved in petroleum oil. The amount of oil used in the experiments would likely cause injury to the host, which suggested the desirability of using other types of DDT formulations.

The brown-dog tick concentrates in homes, kennels, small animal hospitals, and other places where dogs spend much of their time. Host treatment as well as environmental control are therefore indicated. Although the tick does not attack man, its presence in homes is of great concern to most occupants. Washes containing 1% of DDT applied on dogs and the thorough treatment of homes, kennels, etc. are advocated for satisfactory control (BISHOPP *et al.* [22] and SEDDON [155]). Because of the many places where the ticks may hide in homes, and because of the long life of the tick, the control of the pest requires thorough application of DDT to corners, baseboards, cracks, crevices, in furniture, back of any objects against walls, and similar places. A 5 per cent DDT spray may be used. If the ticks are hidden in baseboards or other spaces not readily treated with sprays, both the surface spray and a 5 or 10 per cent DDT dust for cracks may be indicated.

Brown Tick [*Rhipicephalus appendiculatus* (Neum.)]

This tick, the chief vector of East-Coast fever of Africa, attaches mainly to the ears of cattle, sheep, horses, and goats. WILSON [203] found that a light treatment with 5 per cent DDT-emulsion spray killed the brown tick within 24 hours, but after 4 days additional ones attached. Spraying at weekly intervals, however, resulted in destruction of ticks before they completed development. Dips containing 0·5% of DDT will also control the tick (BEKKER *et al.* [17]).

The American Dog Tick [*Dermacentor variabilis* (Say)]

This is one of the most important ticks in the United States. It is primarily of concern because it readily attacks man and is the principal vector of Rocky Mountain spotted fever in the East. It is of veterinary interest because it commonly attacks dogs and livestock.

DDT has been found useful for the control of the American dog tick around homes, resort areas and other places where the tick is abundant (BISHOPP *et al.* [21], SMITH *et al.* [170], GLASGOW and COLLINS [69], GOUCK and SMITH [73], GOUCK and FLUNO [71]). DDT fogs and airplane sprays have also been found to reduce the tick populations greatly although results are more variable than those obtained when sprays or dusts are applied. Where large infested areas are involved, it has been found that application of DDT sprays to vegetation along roadways, paths, borders of lawns, and fields will provide satisfactory results. The ticks concentrate in such places, where man and animals are also more likely to be exposed. The application of 1 to 2 lb of DDT per acre, using wettable powder or emulsion sprays on vegetation and ground litter will control the ticks. Dusts may also be used, although from 2 to 3 lb per acre are required for satisfactory control.

Other Ticks

DDT has been reported useful for the control of certain other species of ticks. SEDDON [155] of Australia has stated that *Ixodes holocyclus* Neumann, the paralysis tick in Australia, can be controlled with DDT sprays and dips. This author also reports that routine dipping of cattle for *Boophilus annulatus microplus* will control the bush tick, *Haemaphysalis bispinosa* (Neum.). SERGENT *et al.* [157] found that good control of *Hyalomma mauritanicum* (Senevet) on calves was obtained with 5 per cent DDT-oil solutions. BEKKER *et al.* [17] found that 0·5 per cent DDT dips controlled *Rhipicephalus evertsi* (Neum.) and *Hyalomma aegyptium* (L.) in South Africa.

NIKOL'SKII [131] reporting on laboratory studies conducted in Russia, found that larvae of *Boophilus annulatus calcaratus* (Bir.), *Dermacentor marginatus* (Sulz.), *Haemaphysalis otophila* (Schulze), and *Hyalomma scupense* (Schulze) were killed when dusted with 5 per cent DDT or when immersed in suspensions containing from 0·01 to 0·5% of DDT. In tests on cattle DDT showed some promise for the control of *Hyalomma scupense*.

PAVLOV and GUEORGUIEV [139] found DDT applied as a 5 per cent DDT powder to be effective against *Rhipicephalus bursa* (Can. & Fanz.) and *Hyalomma marginatum* (C. L. Koch). The authors advocate the use of DDT for preventing piroplasmosis in domestic animals by controlling the tick vectors.

Relative Efficacy of DDT and Other Tick-Control Agents

Gamma benzene hexachloride is much more toxic to engorged ticks than is DDT, but it lacks the residual action. Therefore, the two insecticides make an excellent combination for control of ticks on animals.

TDE and methoxychlor are in general somewhat less effective than DDT for tick control. Toxaphene and chlordane are reported more toxic to engorged ticks of certain species and equal in residual effect (KNIPLING [96]). LAAKE [101] and MUSKUS and PACHECO TORRES [128] found toxaphene to be more effective than DDT against the cattle-fever tick in South America.

Lice on Livestock

Almost all mammals and birds are parasitized by one or more species of lice. The lice are either of the bloodsucking type (Anopleura) or the biting type (Mallophaga). Some domestic animals become infested with several species. Poultry are attacked by a number of species of biting lice. Cattle are known to become infested with 5 species of sucking lice and 2 species of biting lice. Lice are not regarded as disease vectors. Under certain conditions, especially in the winter, when animals are closely confined in stables or on open ranges during periods of food scarcity, infestations may become extremely severe (Figure 10). The annoyance and loss of blood caused by lice interferes with the health and productivity of domestic animals. The success of DDT for controlling lice attacking man suggested its potential value for controlling lice on livestock. Results by various investigators in Europe, North America, and elsewhere soon established its value, however, for controlling lice on horses, cattle, hogs, sheep, goats, and poultry.

DDT was readily adopted by livestock owners for controlling lice. Insecticides in use before DDT became available required two treatments to achieve satisfactory control. The use of DDT does not necessarily assure eradication of lice with a single treatment when it is employed as a spray or wash or even when used as a dip in practical control operations. A single treatment, however, with the proper concentration usually provides excellent practical control. The chief advantage of DDT over other materials is its long residual qualities.

Cattle Lice

Among the early reports on the use of DDT for controlling cattle lice is that of ANNAND [2]. Sprays containing 0·5% of DDT were stated to destroy all motile forms of the sucking louse *Haematopinus eurysternus* (Nitz.) and the insecticide was reported to persist on the animals long enough to destroy the young lice hatching from the eggs. DOMENJOZ [48] also reported that DDT was an effective insecticide for controlling this species, although the concentration was not stated.

A dust containing 5% of DDT was found effective against *Linognathus vituli* (L.) (DYRENDAHL [52]). Two treatments at 14-day intervals were necessary for satisfactory control. MUNRO and KNAPP [127] found that a 10 per cent DDT dust controlled *Haematopinus eurysternus* and the biting louse *Bovicola bovis* (L.). LYLE and STRONG [109] also found that a 10 per cent DDT dust controlled *Bovicola bovis* and the sucking louse *Linognathus vituli*.

EINARSSON [60] reported effective control of cattle lice with two treatments at 8-to 10-day intervals with 5 or 10 per cent DDT dust. The author also found spraying short haired cattle with 2 per cent DDT in acetone to be effective. KNAPP and AANESTAD [95] found that cattle sprayed for horn-fly control

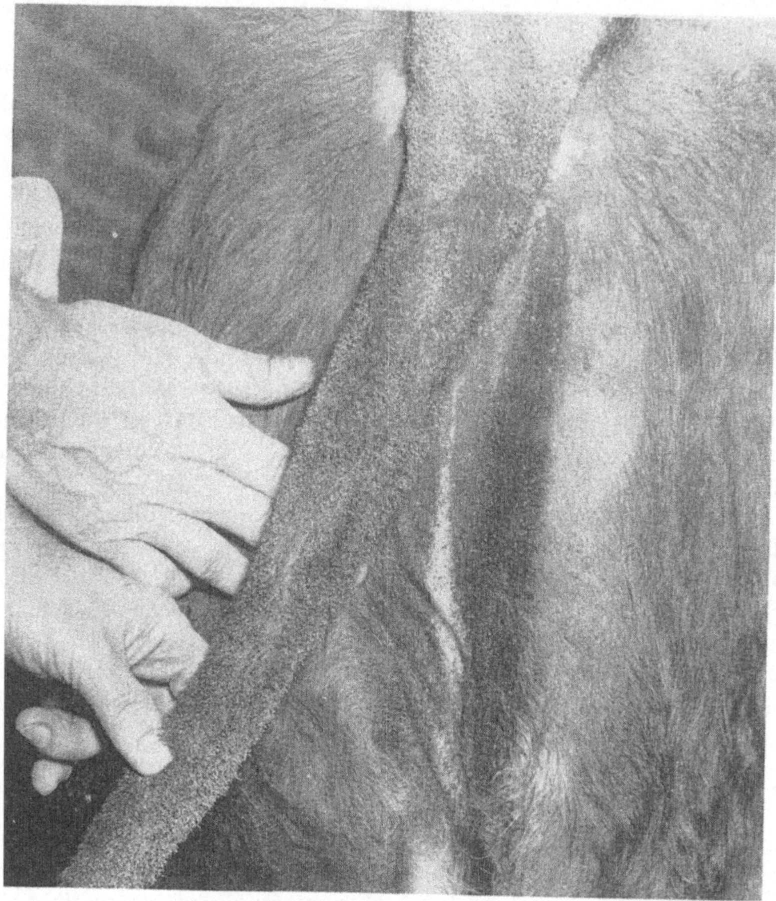

Figure 10

Cattle lice are effectively controlled with DDT sprays or dips. This is an unusually heavy infestation of the ox louse *Haematopinus eurysternus*. Photo courtesy of the National Livestock Loss Prevention Bureau, Kansas City, Missouri.

during the summer caused a marked decrease in infestations of *Bovicola bovis* and *Linognathus vituli*. BRUCE [30] reported that DDT sprays containing 1·5% of DDT controlled the tail louse *Haematopinus quadripertusus* (Fahrenh.). NIKOL'SKII [131] stated that DDT was effective for controlling lice on cattle in Russia. LAAKE [101] found that sprays containing 0·025% of gamma BHC and 0·5% of DDT eradicated the louse *Haematopinus tuberculatus* (Burm.) (probably *Haematopinus quadripertusus*) on cattle in Brazil. BERTRAM and ROBERTS [18] reported almost complete control of *Bovicola bovis* when cattle were sponged with a 0·1 per cent DDT suspension. The authors employed a

Figure 11

When large numbers of animals are to be treated high-pressure and large-volume DDT sprays are desirable for controlling lice and ticks. Photo courtesy of the Livestock Loss Prevention Board, Kansas City, Missouri.

unique and accurate method of determining the degree of control which involved the use of heated pads applied to definite areas on the treated animals.

DDT has no ovicidal effect on any of the lice attacking cattle but persists long enough to destroy most of the lice that hatch from eggs present at the time of treatment.

DDT, to a large extent, has replaced rotenone for louse control on cattle in the United States. Rotenone insecticides are equally or more effective than DDT against the motile forms, but unlike DDT the toxic action does not persist. Therefore, two treatments with rotenone are necessary.

It is apparent from the findings of the investigations cited that DDT may be used effectively for cattle-lice control when used as a dip, spray, wash, or dust. The method of choice will depend on circumstances. Dipping (Figure 9) is the most effective method. A concentration of 0·5 per cent DDT using emulsions or suspensions appears to be a satisfactory concentration for use in dips. Facilities for dipping cattle are not available to most cattle owners. Sprays and washes may be used, however, with excellent results. The degree of control that will be obtained depends to a great extent on the thoroughness of the treatment. Lice or their eggs are found on any part of the animals and often infestations are most severe in the winter months when the hair coat is long and thick. Thorough treatment of the animal is essential for good results. In practical control operations, involving large numbers of animals, a single treatment with sprays cannot be expected to eliminate infestations. If eradication of lice from the animals is to be attempted, at least two thorough treatments are necessary. Emulsion or suspension sprays containing 0·5% of DDT are generally advocated in the United States (UNITED STATES DEPARTMENT OF AGRICULTURE [189]). A suspension spray containing from 0·25 to 0·5% of DDT is recommended in Australia (SEDDON [154)]. Suspension or emulsion sprays containing 0·25% of DDT are recommended in Canada (TWINN [187]). For controlling the tail louse *Haematopinus quadripertusus* in Florida (USA), a concentration of 1·5% is recommended (BRUCE [30]).

Sprays are employed extensively for louse control in many parts of the world. The use of power sprayers, as shown in Figure 11, is desirable when large numbers of animals are to be treated. Many types of sprayers may be used, including compressed-air sprayers having a capacity of 3 to 5 gal (see Figure 12).

Dust treatments are preferred during winter months in cold climates. Apparently either a 5 or 10 per cent dust can be used with satisfactory results. The lower concentration will require more thorough and uniform treatment in order to achieve the desired degree of control. In making the treatment, the dust should be worked in among the hairs by hand and distributed uniformly. A second treatment should be made 2 weeks after the initial treatment to assure a high degree of louse control.

The relative efficacy of DDT and other chlorinated hydrocarbon insecticides has been investigated. (KNIPLING [96]) and in general, DDT, methoxychlor, TDE, toxaphene, BHC (10 to 12 per cent gamma), and chlordane are of the same general order of effectiveness. LANCASTER [103] has reported that chlordane proved more effective than DDT.

Hog Louse

The species *Haematopinus adventicius* (Neum.), is the only louse attacking hogs. It is a large bloodsucker which is prevalent on swine throughout the world.

OLSEN [135] reported control of the hog louse with a spray containing 1% of DDT. DYRENDAHL [52] reported control of the louse in Sweden with 5 per cent DDT dust, but found that several treatments were necessary to achieve eradication from herds. SILVA LEITAO [164] also reported satisfactory control with an unstated concentration of DDT dust. Complete control of lice was achieved in Norway by thoroughly treating hogs with 1 per cent DDT sprays (BERGE [17a]). KEMPER and ROBERTS [92] found spraying or dipping in a 0·75 per cent concentration achieved complete control. They also found that concentrations of 0·1 and 0·5% gave excellent control but failed to eliminate the lice. LEGG [104] reported that a 2 per cent suspension controlled the lice and offered good protection against reinfestation. SWEETMAN [181] found that lice were eliminated with two treatments of a 0·2 per cent emulsion

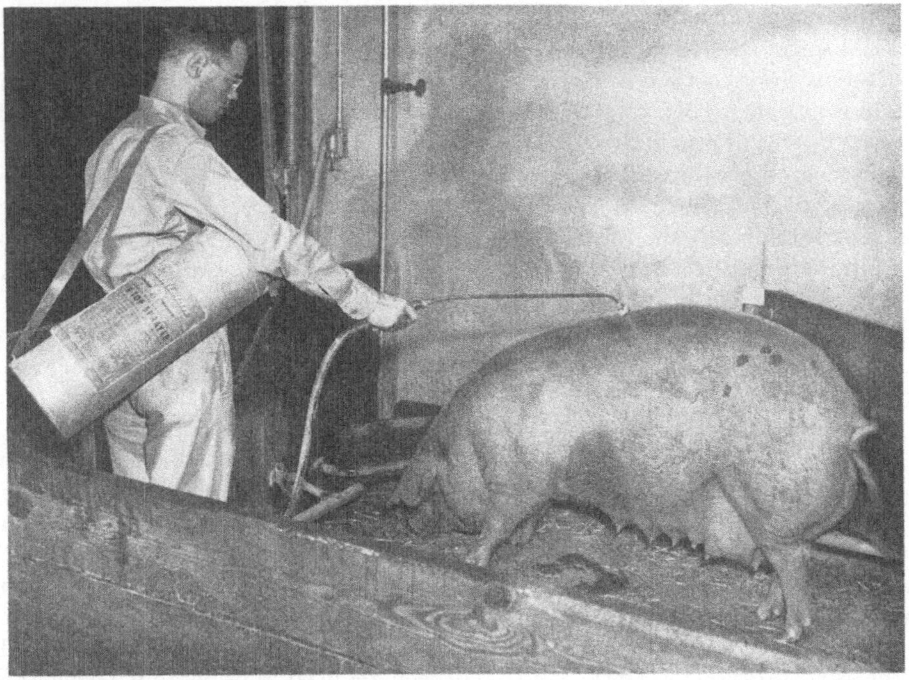

Figure 12

Hog lice are readily controlled with DDT sprays employing various kinds of employment. The compressed-air sprayer of the type shown is satisfactory for this purpose. USDA photograph

spray. KNIPLING reported that a single treatment with a 0·2 per cent emulsion spray gave excellent control but did not eliminate the infestation.

It is apparent from the investigations that control of the hog louse is readily accomplished with DDT. Two thorough sprayings at an interval of about 2 weeks, using a concentration of 0·2% or perhaps even less DDT, will in all probability eliminate infestations. To obtain eradication with one treatment, 0·5 to 0·75 per cent concentration should be used. The spray may be applied with any suitable spray equipment. A compressed-air sprayer of the type shown in Figure 12 is commonly employed.

If dusts are preferred, two treatments should be applied at an interval of about 2 weeks, using a minimum of 5% of DDT.

Several of the chlorinated hydrocarbon insecticides, including chlordane, BHC, toxaphene, methoxychlor, and TDE appear to be of the same general order of effectiveness as DDT for controlling hog lice (KNIPLING [96]).

Sheep and Goat Lice

A number of investigators have shown that the several species of biting and bloodsucking lice on goats and sheep can be completely controlled with DDT dips. BABCOCK [11] obtained complete control of *Bovicola crassipes* (Rudow), *Bovicola caprae* Gurlt, and *Bovicola limbatus* (Gerv.) with an emulsion dip containing 0·3% of DDT. *Linognathus stenopsis* (Burm.) were eliminated with dips containing 0·07 or higher percentages of DDT, but 0·04% failed to accomplish complete control. PARISH and RUDE [136] also reported eradication of *Bovicola crassipes*, *Bovicola caprae*, and *Linognathus stenopsis* with dips containing 0·2% of DDT made from a soluble pine oil emulsion concentrate. Complete control was achieved whether goats were dipped shortly after shearing or in full fleece.

It was reported in Australia that an emulsion dip containing 0·04% of DDT eliminated *Bovicola ovis* (L.) on sheep but 0·02% failed to do so (AUSTRALIAN COUNCIL FOR SCIENTIFIC AND INDUSTRIAL RESEARCH [9]). GRAHAM and SCOTT [74] reported complete control of the body louse of sheep (*Bovicola ovis*) with an emulsion dip containing 0·04% of DDT and a suspension dip containing 0·01% of DDT. PARISH and RUDE [137] in the United States found that a dip containing 0·2% of DDT made from a soluble pine oil emulsion concentrate eradicated the body louse on sheep.

BROWN [29] reported failure to eliminate lice on goats with an emulsion dip containing 0·25% of DDT. However, a large number of animals were treated and the author concluded that by the time the last animals were dipped the concentration of DDT may have been reduced considerably below the initial 0·25 per cent strength.

COOP and McLEOD [43] reported emulsion dips containing 0·1% of DDT to be effective for 3 months against lice on sheep but 0·01% was unsatisfactory.

Results of the investigations cited show the high degree of effectiveness of DDT for controlling lice on sheep and goats. In practical dipping operations it is advisable to allow for the reduction in concentration of insecticide when considerable numbers of animals are to be dipped. Either emulsions or wettable powder suspensions dips containing 0·25% of DDT are recommended in the United States (UNITED STATES DEPARTMENT OF AGRICULTURE [188]), and in Canada (TWINN [187]). Figure 13 shows goats being dipped for louse control. In Australia, a concentration of 0·1% (para-para DDT) is recommended (SEDDON [154] and McKENNA and FEARN [123]). This concentration of para-para DDT, which is equivalent to about 0·125% of technical DDT, is adequate for complete elimination of lice from herds. However, dips that lose the DDT rapidly when animals are put through the vats would soon be reduced in concentration below that required to assure elimination of lice with a single dipping.

Excellent control of lice on sheep and goats can also be obtained by thoroughly spraying the animals with DDT sprays. Either emulsion or suspension sprays containing 0·5% of DDT are recommended (UNITED STATES DEPARTMENT OF AGRICULTURE [188]). More thorough treatment can be made with less spray if the treatments are applied soon after the animals are sheared.

Figure 13

Angora goats are being dipped in a DDT bath for louse control on a southwest range in the USA. This vat is designed for cattle as well as sheep and goats. USDA photograph.

The use of DDT fogs for louse control on sheep has been investigated in Australia by SHANAHAN [162]. The treatment failed to eliminate the lice, although the infestations were greatly reduced. Similar results were obtained in tests conducted on goats in the United States by SMITH [168].

Horse Lice

The effectiveness of DDT for controlling the horse louse, *Haematopinus asini* (L.), was reported by SCHMIDT as early as 1943. The author reported excellent control with 5 per cent DDT powder. SCHNEIDER [151] also reported on the effectiveness of DDT powder and DDT-emulsion treatments. Investigations by DOMENJOZ [48] and DYRENDAHL [52] confirmed the efficacy of 5 per cent DDT dusts. The latter author recommended two treatments at an interval of 14 days.

BATTE and GAINES [15] reported DDT to be effective against the biting louse *Bovicola equi* (L.). A concentration of 0·2% killed all of the lice. The authors do not state if reinfestations occurred after treatment. MELDRUM [125] reports that a 0·5 per cent DDT spray is employed for controlling lice on horses in Australia. Sprays or washes containing 0·5% of DDT are also advocated for controlling lice on horses in the United States. In Canada, TWINN [187] recommends 0·25 per cent DDT emulsion or suspension sprays or 3 to 5 per cent DDT dusts for controlling *Haematopinus asini* and *Bovicola equi*.

Poultry Lice

Several workers have investigated DDT for controlling lice on poultry. TELFORD [183] reported that 4 per cent DDT dust was equal to 33 per cent sodium fluoride when applied to poultry infested with *Eomenacanthus stramineus* (Nitz.) and *Menopon gallinae* (L.). The same author [184] found that an emulsion dip containing 0·03% of DDT destroyed the motile forms and prevented reinfestation for several weeks. An emulsion containing 2·5% of DDT and applied as a spray at the rate of 32 to 39 cm³ per fowl also controlled the two species mentioned, and *Goniocotes gallinae* (Deg.).

SEDDON [154] recommends 5 to 10 per cent DDT powder applied to the poultry or a 1 per cent powder applied to nests or dust bath.

A 5 per cent DDT dust applied to the host appears to be about the desired concentration for effective and practical control of the lice attacking poultry. In applying the dust a hand shaker container is satisfactory. One person holds the fowl while the other applies the dust as the feathers are ruffed. The pinch method used extensively for applying sodium fluoride may also be used. This method consists of the application of a pinch of the powder to various places on the fowl, including the head, neck, back, breast, wings, thighs and vent.

Fleas

Dogs, cats, and poultry are commonly infested with fleas. Occasionally fleas may also infest hogs and other livestock. Fleas are intermediate hosts for the dog and cat tapeworm *Dipylidium caninum*, and the heartworm *Dirofilaria immitis*.

LINDQUIST *et al.* [108] reported control of *Ctenocephalides canis* (Curt.), *Ctenocephalides felis* (Bouche), and *Echidnophaga gallinacea* (Westw.) on dogs with a 5 per cent DDT powder. The treatment prevented reinfestations for 4 to 7 days. Derris powder containing 4·8 per cent rotenone under the same conditions prevented reinfestations for 2 days. DOMENJOZ [48] and DYREN-

Figure 14

A 5 to 10 per cent DDT powder is an effective treatment for controlling fleas on dogs.
USDA photograph.

DAHL [52] reported that DDT powder was effective against *Ctenocephalides canis*. DDT is reported to be highly effective in destroying both larvae and adults of the cat flea, *Ctenocephalides felis* (KERR [94]). SWEETMAN [180] reported excellent control of fleas on dogs and cats with a 10 per cent DDT powder applied to the animals, and a light application of the dust to the floors. EADS [53] obtained control of *Echidnophaga gallinacea* by applying 10 per cent DDT in poultry houses. The poultry were free of fleas within 1 week and no reinfestations occurred within 1 month. ROBERTS *et al.* [145] obtained excellent control of *Echidnophaga gallinacea* in Australia by dipping fowls in DDT suspensions containing 0·5 to 2% of DDT. The poultry houses and infested grounds were treated with the same material. Dogs were protected from reinfestations when dipped in 0·5, 1, and 2 per cent DDT suspensions. The duration of protection was respectively 2, 7, and 3 to 16 days. Cats dipped in 0·5 per cent suspensions became ill in some instances.

DDT dusts and sprays are employed extensively in practical control of fleas affecting dogs and poultry. Dusts containing 5 to 10% of DDT are most commonly employed for application to dogs. A shaker can is used to apply the dust to the dog (Figure 14). The dust is worked in among the hair with the hand. It is especially important to apply the dust thoroughly on the back and neck, since fleas are often concentrated on these parts of the host. Treated animals probably will become reinfested within a few days to a week after treatment unless the infested premises are also treated. In the United States a 10 per cent DDT dust applied to dogs and a 2·5 to 5 per cent spray applied to infested areas are advocated for controlling the dog flea (BISHOPP [20]). The application of dusts or sprays to floors of poultry houses and other places where the fleas might be developing while off the host will provide excellent control for the poultry flea. In Australia, SEDDON [154] advocates either a 5 per cent dust or a 1 per cent DDT wash applied to dogs and an application of 50 mg of DDT per square foot of surface on the breeding areas. For controlling the sticktight flea, *Echidnophaga gallinacea* and to prevent its spread, the dipping of poultry, dogs, and goats in 2 per cent DDT suspension was practiced in Australia. This high concentration, however, resulted in a fairly high mortality of chickens and ducks. Consequently the recommended concentration was reduced to 1% for routine control and eradication programs.

The human flea [*Pulex irritans* (L.)] and the dog and cat fleas may also infest hogs, goats, and other livestock. Treatment of the host with 0·5 per cent DDT spray or 5 to 10 per cent dust, and thorough spraying of the infested premises with DDT spray or dust as suggested for controlling dog fleas, will undoubtedly control infestations of the species named as well as other species which attack domestic animals and poultry.

Bedbugs

The value of DDT for controlling bedbugs [*Cimex lectularius* (L.) and other species] to protect man from attack is well-known. MADDEN *et al.* [112] published a study on the value of the insecticide for this purpose. Bedbugs, as pointed out by BACK [12], are of some importance in poultry houses. KULASH [97] reported complete control and protection against reinfestation in poultry houses a year after applying 5 per cent DDT in kerosene as a residual treatment. The insect also attacks rats, guinea pigs, and rabbits and may be serious as a pest in laboratories and other establishments where these animals are raised for commercial or experimental purposes. DDT is commonly employed as a residual treatment to control the pest in such animal quarters.

Mites

Although DDT is not a satisfactory agent for the control of most of the mites attacking animals, some species are effectively controlled.

EINARSSON [60] states that 10 per cent DDT powder or 5 per cent DDT in a bland petroleum oil will control mites in ears of rabbits, foxes, and cats. He also obtained control of foot scab of horses with a 10 per cent acetone solution of DDT. The sheep-scab mite was not controlled, however. DOWNING [49] also failed to control the sheep-scab mite, even with two dippings at 2-week intervals, employing 0·5 per cent emulsions or suspensions. BITYUKOV [23] reported control of mange mites on sheep and horses with one treatment of an unstated concentration of DDT powder. It was further reported that mange mites on cattle were controlled with 2 to 3 treatments. HIXSON and MUMA [83] failed to control *Sarcoptes scabiei suis* Gerl. on hogs with 0·5 per cent DDT emulsion.

The poultry mite, *Dermanyssus gallinae* (Deg.), one of the most important parasites of poultry in many parts of the world, is effectively controlled with DDT. HIXSON and MUMA [83] obtained good control in the United States by treating poultry houses with DDT-residual spray. MCQUEEN [124] obtained good control in Australia with DDT-suspension sprays containing 0·5, 1·0, and 1·5% of DDT. MELDRUM [125] reports that residual treatment with 5% of DDT in kerosene is recommended in Tasmania, Australia. MCDUFFIE [117] obtained effective control of the poultry mite in Texas (USA) with residual treatments in poultry houses using either a 5 per cent DDT emulsion or the same concentration in kerosene.

The mites are often well hidden in cracks, behind loose boards, or in any other suitable hiding place. The degree of control obtained with the insecticide will therefore depend on the thoroughness of the treatment. Infested premises should be carefully inspected and all infested areas treated thoroughly with DDT residual sprays, employing solutions, emulsions or suspensions. A dosage of 200 mg of DDT per square foot appears to be a satisfactory rate of application. The removal of debris, loose boards or other objects which furnish harborages will permit maximum efficacy of the insecticide applied and will further aid in preventing reinfestations.

The northern fowl mite, *Bdellonyssus sylviarum* (C. & F.), apparently is not susceptible to control with DDT. RITCHER and INSKO [143] reported some initial control but very little protection against reinfestation when poultry were treated with a 5 per cent DDT powder.

Infestations of the scaly-leg mite, *Knemidokoptes mutans* (R. & L.), on poultry were reported controlled (GHIDINI [66]) when the infested legs were treated with 5% of DDT in kerosene. The author reported treating the feet with olive oil following the DDT-kerosene treatment in order to reduce irritation.

DDT for the Prevention of Myiasis

Several species of fly larvae infest animals, destroying tissues or otherwise affecting the health of the host. Some of the most important of these insects are the ox warbles, *Hypoderma* spp.; horse bots, *Gasterophilus* spp.; the tropical warble fly, *Dermatobia hominis* (L. Jr.); the screw worm, *Callitroga hominivorax* (Coquerel); the gadfly of sheep, *Oestrus ovis* (L.), and wool maggot flies, *Phaenicia* spp. There are a number of other species of Diptera throughout the world which cause myiasis. However, no references to investigations on their control with DDT have been noted.

Hypoderma spp.

It has been stated earlier that DDT is ineffective against the larvae of the ox warble or cattle grubs. This has been reported by STEWART [177], SONI [173], MILLS *et al.* [126], SERGENT and SERGENT [158], and BEVAN and EDWARDS [19]. In addition to these reports, it is known that the widespread use of DDT for many years on cattle for controlling lice, ticks, flies, and other external parasites has not caused any detectable reduction in warble infestations in cattle.

Even though DDT is not effective against the grub larvae in the backs of cattle, the possibility of interrupting the life cycle of the parasite by directing control efforts against the ovipositing adults has been investigated. SONI [173] applied DDT sprays ranging in concentration from 0·5 to 5% to cattle during the egg-laying season for the warble fly but failed to obtain control.

Similar experiments were conducted by GRAHAM [75] in efforts to control cattle grubs [*Hypoderma lineatum* (De Vill.)] by applying DDT and certain other chlorinated hydrocarbon insecticides to cattle at 2-week intervals during the 4-months' season when the adult flies (heel flies) were laying eggs on the host. Sprays containing 2% of DDT and the same concentration of certain other insecticides were applied thoroughly to animals for 8 treatments. The following fall and winter there was no significant reduction in grub infestations in the animals treated with DDT, benzene hexachloride, toxaphene, or chlordane. Failure to obtain control shows that the insecticides did not discourage oviposition by the adults, were not ovicidal, and had no significant effect on the newly hatched larvae.

Gasterophilus spp.

No experiments with DDT for controlling *Gasterophilus* larvae while in the stomachs of equines have been noted. However, laboratory data reported by GRAHAM and ALFORD [76] showed that the larvae of *Gasterophilus intestinalis* (Deg.) were not killed within 48 hours when immersed for 1 hour in an acetone solution containing 5% of DDT. Several other chlorinated hydrocarbon insecticides including lindane, chlordane, and dieldrin also failed to kill the larvae when tested in the same way. Moreover, these materials failed to kill larvae of *Gasterophilus intestinalis* and *Gasterophilus nasalis* (L.) in the stomachs of horses when administered as emulsions by stomach tube at the rate of 25 to 50 mg of body weight of the host.

Oestrus ovis (L.)

The sheep bot or gadfly is world-wide in distribution. The only available reference to tests with DDT for its control is that of SALCES and CALVO [149]. These authors reported that DDT was not effective against the larvae when tested *in vitro* and *in vivo*.

Dermatobia hominis (L. Jr.)

DDT has little or no direct effect on larvae of the warble fly, *Dermatobia hominis*, infesting animals when applied in concentrations used for tick control. Very little information has been published on the subject. SQUIBB [175] stated that the application of about 400 cm³ of a kerosene emulsion spray containing about 0·25% of DDT and the extract from 40 g of chopped derris root caused a good reduction in infestation of the warble fly. It is not known if DDT contributed to this reduction.

It is possible that DDT has not been adequately investigated to determine its value for *Dermatobia*-warble-fly control. A recent report by ADAMS *et al.* [1] shows that toxaphene used repeatedly as a 0·5 per cent spray or dip during the course of several months will control the warble fly. The authors conclude that reduction in warble-fly infestations is due to the control of the arthropod carriers of the warble-fly eggs (principally flies and mosquitoes). Since DDT is, in general, at least equal to toxaphene in effectiveness against horn flies, stable flies, house flies and mosquitoes, the use of proper concentrations of DDT sprays or dips at suitable intervals during a period of several months might prove effective in reducing the number of warble-fly larvae in animals. It is possible however, that warble-fly control is accomplished in part or largely because of action on larvae before or soon after they penetrate the host. If this is true, DDT might be less effective because of its lower toxicity to dipterous larvae.

Callitroga hominivorax (Coquerel)

The new world screw worm is one of the most destructive of the pests attacking livestock in the southern United States and in parts of Mexico and Central and South America. The fly deposits eggs on wounds and the larvae which hatch feed on the flesh. The indirect value of DDT in controlling this pest has already been mentioned in discussing ticks, particularly the Gulf-Coast tick, *Amblyomma maculatum*. This, and other ticks, provide wounds which become infested by the screw worm. Control of the ticks with sprays containing DDT, therefore, greatly reduces the incidence of screw worm infestations in livestock, principally cattle, along the Gulf-Coast region of the United States and Mexico where the Gulf-Coast tick is prevalent.

DDT is not particularly effective against the larvae of the screw worm. EDDY [56] reported that DDT at a concentration of 10% in acetone failed to kill larvae near maturity. Therefore, it would be of little value in destroying larvae in established infestations. However, when added to media in which the larvae are grown in the laboratory, the minimum lethal concentration ranged between 0·005 and 0·01%, which indicates that DDT is at least as toxic as diphenylamine. Diphenylamine has been used extensively in wound dressing to prevent screw-worm infestations. Other chlorinated hydrocarbon insecticides are more effective than DDT against the larvae. Lindane, the gamma isomer of BHC, is highly effective against both mature and immature larvae.

DDT applied to wounds leaves a residual deposit which will destroy adults that frequent wounds for ovipositing or feeding but the flies succeed in ovipositing before the insecticide takes effect. A commercial preparation is being sold in the United States as a screw-worm remedy which contains 20% of DDT, 5% of lindane, an emulsifying agent, and Xylene. In view of the high degree of efficacy of lindane, it is not known whether the DDT is contributing to the value of the treatment in the destruction of larvae in the wounds and in protecting wounds from reinfestation.

A related primary wound infesting fly, *Chrysomyia bezziana* Vill., occurs in South Africa, India, and possibly in other parts of the world but no references to investigations with DDT for controlling the insect have been noted.

Blow Flies

The larvae of certain calliphorid flies infest sheep, producing a cutaneous type myiasis. Losses caused by these parasites are very great throughout the world. Animals which are infested become more attractive to flies and the infestation rapidly becomes more severe. The host may be killed in a few days unless measures are taken to destroy the infestation and prevent further attack. *Phaenicia cuprina* (Wied.) and *Phaenicia sericata* (Meig.) are of the most economic importance. *Phaenicia cuprina* is of most importance in Australia and South Africa. *Phaenicia sericata* commonly attacks sheep in Great Britain and New Zealand, but has been recorded as infesting sheep in other parts of the world. The black blow fly, *Phormia regina* (Meig.) and the secondary screw-worm fly, *Callitroga macellaria* (F.) also cause maggot infestations in sheep in North America. The factors predisposing attack by these flies, commonly referred to as blow flies, are decomposition products in wool which are formed during warm, humid or rainy periods. Soiled wool in the crotch region and wounds may also predispose attack by blow flies.

The incidence of wool maggot infestations in sheep is subject to extreme variations depending on the air temperature and humidity. A wave of strikes may occur during the course of several days or weeks and then subside within a few days. The evaluation of the efficacy of insecticides in preventing strikes is, therefore, sometimes difficult. However, DDT has been shown by a number of investigators in Australia, Great Britain, South Africa, and other areas to be useful in reducing the incidence of strike among sheep.

Full-grown larvae of *Phaenicia cuprina*, the most important sheep maggot fly in Australia, are not affected by DDT (WATERHOUSE [194]). The adult flies, however, were found to be killed by contact with 50 mg per square foot deposits of DDT. In North Wales, CRAGG [44] found that sheep dipped in an emulsion dip containing 0·37% of DDT remained free of strikes due to *Phaenicia sericata* for 42 days. In another experiment 215 sheep treated in an emulsion dip containing an initial concentration of 0·45 per cent DDT developed 3 strikes in 44 days. Another comparable untreated group of sheep developed 21 strikes. CRAGG [45] also reported on experiments in which 1061 sheep were dipped in a 0·5 per cent DDT emulsion. A total of 14 infestations occurred during a period of 4 to 6 weeks. A like number of untreated controls developed 55 infestations. LYLE-STEWART [110] also obtained excellent control of strikes due to *Phaenicia sericata* by dipping sheep in a 0·5 per cent DDT emulsion.

SHANAHAN [159] found that DDT mixed with ground liver at a concentration of 0·1% prevented maturity of a high percentage of larvae of *Phaenicia cuprina*. WATERHOUSE [193] reported that larvae of this species in contact with DDT were not affected; however, when the insecticide is consumed by the larvae feeding on a DDT-contaminated medium, growth is inhibited at a 0·03 per cent concentration. The larvae become increasingly difficult to kill by DDT as they become older. WATERHOUSE and SCOTT [194] also reported on tests conducted in an insectary, using artificial attractants. The authors concluded that DDT applied to the sheep prevented or reduced the number of strikes due to *Phaenicia cuprina* for several months. STONES [178] also reported on tests in an insectary against *Phaenicia sericata* in England. He found that a 0·5 per cent DDT suspension was superior to an emulsion using benzene as a solvent. ROBERTS and MOULE [144] and JOHNSTONE and SCOTT (91) obtained excellent results in field trials in Australia. They used 1 per cent DDT sprays to prevent body strike due to *Phaenicia cuprina*. DuTOIT and GOOSEN [51] found DDT effective in preventing blow-fly strike among sheep in South Africa. SCHWARDT [153] found that infestations due to *Lucilia caesar* L. (probably *Lucilia illustris*) were controlled by dipping sheep in 0·5 per cent DDT suspension. GRAHAM and EDDY [77] found that 2 per cent DDT applied to artificially induced infestations of *Phormia regina* and *Callitroga macellaria* in sheep provided excellent protection against reinfestations.

Practical Control of Sheep Maggots

The control of maggot infestations involves first of all a program of prevention, if possible. This is accomplished in large measure by removing wool from the crotch region, the area on sheep which is most susceptible to attack by wool maggots. The removal of wool by clipping the crotch region is widely

Figure 15

Spraying sheep for external parasite control. To prevent blow-fly strike, a fine mist is applied to
the backs by means of a centrally placed nozzle. For sheep ticks and lice nozzles are placed on the
ends of the rotating arms and in fixed positions on the floor. Photo courtesy of Australia Commonwealth Scientific and Industrial Research Organization.

practiced wherever sheep are grown. Operations involving the removal of
skin folds on lambs, called the Mules operation, are also practiced.

Even though these preventive measures are highly effective, the animals
are often subject to infestations by the wool maggots. Various insecticide
treatments have been employed in the past. Arsenical solutions applied as
dips or sprays were used most extensively.

DDT dips or sprays are far superior to the arsenicals for the prevention of
strike among sheep. DDT protects the host primarily through action on the
adult flies. The DDT deposits on the wool prevent or discourage oviposition.

It has been shown by several investigators, however, that DDT possesses some toxicity to young larvae as a stomach poison. The presence of DDT in wool, therefore, tends to inhibit growth of newly hatched larvae even if some adults do succeed in ovipositing. The combined action against adults and young larvae and the long residual action of the DDT on the wool are, therefore, the factors responsible for the efficacy of DDT treatments.

A Joint-Blow-Fly-Control Committee, a committee appointed by several of the government agencies in Australia, issued recommendations, in September 1950, on the use of DDT and BHC for the prevention of body strike due to *Phaenicia cuprina*.

The committee report advocates spraying the sheep from poll to rump with a 1 per cent DDT spray, using from 1 to 2 quarts per animal. This treatment is stated to prevent body strike for 4 to 10 weeks. Dipping is not advocated except under severe fly-wave conditions. Figure 15 shows equipment that is used in treating the animals.

In Great Britain, dipping sheep in DDT preparations is considered the most practical method of control for *Phaenicia sericata*. A concentration of 0·25% has been found to be almost as effective as 0·5%. However, it is considered important to start with an excess of DDT in order to allow for the lowering of DDT concentration by the straining action of the wool. The use of sprays at the rate of about 2 quarts per animal is also advocated. This method of treatment provides only about 4 weeks protection as compared with 8 weeks for the dips.

When blow-fly infestations are present in animals it is necessary to destroy the larvae and also to protect the animal from reinfestation. Animals which have been infested are very susceptible to reinfestation because of the presence of attractive substances in and around the infested area. Because of the low toxicity of DDT to full-grown larvae, the insecticide is not an effective material to kill the larvae infesting animals although its presence in dressing for struck sheep may assure better protection against reinfestation. BHC appears to be much superior in destroying all sizes of larvae and this material is recommended for this purpose in Australia, Great Britain, the United States, and elsewhere.

Relative Efficacy of DDT and Other Insecticides

BHC is also an excellent insecticide for preventing and controlling blow-fly strikes. This is reported by several investigators, including SHANAHAN [159], GRAHAM and EDDY [77], WATERHOUSE and SCOTT [194], COOP and McLEOD [43], and others. At practical field dosages, BHC does not persist in the wool as long as DDT. Therefore, it does not provide equal protection against reinfestation. It is, however, superior to DDT as a larvicide and is used in dressings for blow-fly strikes and at the same time offers good protection against reinfestation. Toxaphene, as reported by GRAHAM and EDDY [77], is also an excellent insecticide for the protection of sheep from strike by *Phormia regina* and *Callitroga macellaria*.

562

REFERENCES

[1] ADAMS, P. G., CASTILLO, C. H., and SALMERON, R., *Application of Toxaphene for Torsalo Fly Control*, Agric. Chemic. *7* (12), 33–35, 119, 121, 123 (1952).

[2] ANNAND, P. N., *Investigations of Insects Affecting Man and Animals*, U. S. Bur. Ent. Plant Quar. Ann. Rpt. *1943*, 23 (1944).

[3] ARNASON, A. P., BROWN, A. W. A., FREDEEN, F. J. H., HOPEWELL, W. W., and REMPEL, J. G., *Experiments in the Control of Simulium arcticum Malloch by Means of DDT in the Saskatchewan River*, Sci. Agric. *29* (11), 527–537 (1949).

[4] ARNOLD, R. M., *Tick Control Measures. Assessment of the Value of Chemical Ticki-cides for Boophilus (Margoropus) annulatus var. micro-plus in Jamaica*, Jamaica Dept. agric. Bull. *38*, 33 pp. (1951).

[5] ARNOLD, R. M., *Laboratory Assessment of Tick-Killing Agents*, Vet. Rec. *64* (29), 426–429 (1952).

[6] ARTHUR, D. R., *Acaricidal control of the Tick, Ixodes ricinus (L.) on cattle*, Bull. ent. Res. *41* (3), 555–562 (1951).

[7] AUSTRALIAN COMMONWEALTH SCIENTIFIC AND INDUSTRIAL RESEARCH ORGANIZA-TION, *First Annual Report for the Year Ended 30th June, 1949*. VII. *Sheep*, VIII. *Cattle*, pp. 37–49.

[8] AUSTRALIAN COUNCIL FOR SCIENTIFIC AND INDUSTRIAL RESEARCH, *Eighteenth Annual Report for the Year Ended 30th June, 1944*. III. *Entomological Investigations*, pp. 14–19.

[9] AUSTRALIAN COUNCIL FOR SCIENTIFIC AND INDUSTRIAL RESEARCH, *Nineteenth Annual Report for the Year Ended 30th June, 1945*. III. *Entomological investigations*, pp. 19–30.

[10] [AUSTRALIA] JOINT BLOWFLY CONTROL COMMITTEE (Appointed by Commonwealth Sci. Industr. Res. Organ., N. S. Wales Dept. Agric., Queensland Dept. Agric. Stock) . *Control of Body-Strike in Sheep*, Interim recommendations, 4 pp. (1950).

[11] BABCOCK, O. G., *DDT for the Control of Goat Lice*, J. econ. Ent. *37* (1), 138 (1944).

[12] BACK, E. A., *Bedbugs*. U. S. Dept. Agric. Leaf. 146, 8 pp. (1937).

[13] BARONI, O., *O tratamento do carrapato do gado bovino com DDT* [The Treatment of Cattle with DDT Against the Tick], Biológico *13* (9), 157–161 (1947).

[14] BARONI, O., and EGLI, R., *Combate ao carrapato do gado bovino com DDT*, Biológico *14* (5), 109–112 (1948).

[15] BATTE, E. G., and GAINES, J. C., *Comparison of DDT and Chlordan for the Control of the Horse Louse*, J. econ. Ent. *41* (5), 830–831 (1948).

[16] BEKKER, P. M., and GRAF, H., *Gammexane and DDT for Control of Arsenic-Resistant Blue Tick*, Fumig. South Africa *21* (243), 361–366, 420 (1946).

[17] BEKKER, P. M., GRAF, H., MALAN, J. R., and VAN DER MERWE, S. W., *Die vergely-kende waarde van arseen, BHC. en DDT indien gebruik in dipbakke teen bosluise op beeste* [The Comparative Value of Arsenicals, BHC. and DDT, when Used in Dips for Ticks on Cattle in Afrikaans], J. S. Afr. vet. med. Ass. *20* (3), 110–124 (1949).

[17a] BERGE, S., *Nye og effektive midler i kampen mot snyltere og skadeinsekter hos hus-dyrene*, Norsk Landbruk *12* (3/4), 26–30 (1946).

[18] BERTRAM, D. S., and ROBERTS, E. W., *The Control of Mallophaga on Cattle by Contact Insecticides*, Int. Congr. Ent. [Stockholm] Proc. *8*, 848–849 (1950).

[19] BEVAN, W. J., and EDWARDS, E. E., *Studies on Ox Warble Flies, Hypoderma lineatum and Hypoderma bovis*, Bull. ent. Res. *41* (4), 639–662 (1951).

[20] BISHOPP, F. C., *How to Control Fleas*, U. S. Dept. Agric. Leaf. 152, 4 pp. (1946).

[21] BISHOPP, F. C., SMITH, C. N., and GOUCK, H. K., *Combating the American Dog Tick, Carrier of Rocky Mountain Spotted Fever in the Central and Eastern States*, U. S. Bur. Ent. Plant Quar. E–454, Rev., 5 pp. (1946).

[22] BISHOPP, F. C., SMITH, C. N., and GOUCK, H. K., *The Brown Dog Tick, with Suggestions for its Control*, U. S. Bur. Ent. Plant Quar. E–292, Sec. Rev., 4 pp. (1946).

[23] BITYUKOV, P. A., *The Application of the Dust DDT to Cure Mange on Agricultural Animals* [In Russian]. Veterinariya *25* (9), 34–35 (1948).

[24] BLAKESLEE, E. B, *DDT as a Barn Spray in Stablefly Control*, J. econ. Ent. *37* (1), 134–135 (1944).

[25] BLAKESLEE, E. B., *DDT Surface Sprays for Control of Stablefly Breeding in Shore Deposits of Marine Grass*, J. econ. Ent. *38* (5), 548–552 (1945).

[26] BLAKESLEE, E. B., and BRUCE, W. G., *DDT to Control the Cattle Tick — Preliminary Tests*. J. econ. Ent. *41* (1), 104–105 (1948).

[27] BLAKESLEE, E. B., TISSOT, A. N., BRUCE, W. G., and SANDERS, D. A., *DDT to Control the Gulf Coast Tick*, J. econ. Ent. *40* (5), 664–666 (1947).

[28] BLAXTER, K. L., *Control of Flies in Cowsheds. Experiments with DDT*, Agriculture, London, *52* (3), 112–115 (1945).

[29] BROWN, A. W. A., *A DDT Emulsion Dip to Control Goat Lice*, J. econ. Ent. *40* (4), 605 (1947).

[30] BRUCE, W. G., *The Tail Louse, a New Pest of Cattle in Florida*. J. econ. Ent. *40* (4), 590–591 (1947).

[31] BRUCE, W. G., and BLAKESLEE, E. B., *DDT to Control Insect Pests Affecting Livestock*, J. econ. Ent. *39* (3), 367–374 (1946).

[32] BRUCE, W. G., and BLAKESLEE, E. B., *Factors Affecting Tests with DDT Sprays to Control Horn Flies*, J. econ. Ent. *41* (1), 39–42 (1948).

[33] BRUCE, W. N., and DECKER, G. C., *Fly Control and Milk Flow*, J. econ. Ent. *40* (4), 530–536 (1947).

[34] BRUCE, W. N., and DECKER, G. C., *Control of Horse Flies on Cattle*. Ill. Nat. Hist. Surv. Div., Biol. Notes 24, 8 pp. (1951).

[35] BUREAU OF ENTOMOLOGY AND PLANT QUARANTINE, *Horn Fly Control on Beef Cattle*, U. S. Dept. Agric. Leaf. 291, 6 pp. (1950).

[36] BUREAU OF ENTOMOLOGY AND PLANT QUARANTINE, *New Sprays for Ticks on Livestock*, U. S. Bur. Ent. Plant Quar. EC–10, 7 pp. (1950).

[37] BUREAU OF ENTOMOLOGY AND PLANT QUARANTINE, *Control of Black Flies*, U. S. Bur. Ent. Plant Quar. EC–20, 6 pp. (1952).

[38] CARTER, R. H., WELLS, R. W., RADELEFF, R. D., SMITH, C. L., HUBANKS, P. E., and MANN, H. D., *The Chlorinated Hydrocarbon Content of Milk from Cattle Sprayed for Control of Horn Flies*, J. econ. Ent. *42* (1), 116–118 (1949).

[39] COBBETT, N. G., *Preliminary Tests in Mexico with DDT, Cube, Hexachlorocyclohexane (Benzene hexachloride), and Combinations Thereof for the Control of the Cattle Fever Tick, Boophilus annulatus*, Amer. J. vet. Res. *8* (28), 28–283 (1947).

[40] COBBETT, N. G., and SMITH, C. E., *DDT has Possibilities of Becoming a Remedy for Sheep Ticks*, J. Amer. Vet. Med. Ass. *107* (822), 147–148 (1945).

[41] COLLINS, D. L., Travis, B. V., and JAMNBACK, H., *The Application of Larvicides by Airplane for Control of Blackflies*, Mosquito News *12* (2), 75–77 (1952).

[42] COOP, I. E., *Sheep Dipping Trials with Derris, DDT, and Gammexane*, Canterbury Chamber Commerce Bull. 222, 4 pp. (1948).

[43] COOP, I. E., and McLEOD, G. B., *Sheep Dipping Trials wih Derris, Bentonite Sulphur, DDT, and Benzene Hexachloride*, N. Z. J. Sci. Tech. [A] *30* (5), 292–304 (1950).

[44] CRAGG, J. B., *'DDT' as a Sheep Blowfly Dip*, Nature, Lond. *155* (3935), 394 (1945).

[45] CRAGG, J. B., *The Control of Sheep Blowflies by DDT Dips*, Ann. appl. Biol. *33* (1), 127–129 (1946).

[46] DENNING, D. G., and PFADT, R. E., *Evaluation of Certain Insecticides in Horn Fly Control*, J. econ. Ent. *43* (4), 557–558 (1950).

[47] DEONIER, C. C., U. S. Bur. Ent. Plant Quar., Orlando, Fla. Personal communication (1946).

[48] DOMENJOZ, VON R., *Experimentelle Erfahrungen mit einem neuen Insektizid (Neocid-Geigy), ein Beitrag zur Theorie der Kontaktgiftwirkung*, Schweiz. med. Wschr. *74* (36), 952–958 (1944).

[49] DOWNING, W., *The Control of Psoroptic Scab on Sheep by Benzene Hexachloride and DDT* (Cooper. Tech. Bureau, Berkhamstead), Vet. Rec. *59* (42), 581–582 (1947).

[50] DUNBAR, P. B., *The Food and Drug Administration Looks at Insecticides*, Food Drug Cosmet. Law Quart. *1949* (June), 233–239.

[51] DUTOIT, R.; and GOOSEN, P. J. *DDT for the Protection of Sheep Against Blow Fly Strike*, Onderstepoort J. vet. Sci. *22*, 285–290 (1949).

[52] DYRENDAHL, S., *DDT (Alltox) som medel mot chyra hos d'jur*, Skand. Vet. Tidskr. *35* (4), 202–210 (1945).

[53] EADS, R. B., *Control of the Sticktight Flea on Chickens*, J. econ. Ent. *39* (5), 659–660 (1946).

[54] EADS, R. B., *A Second Record of Lipoptena mazamae from Cattle*, J. econ. Ent. *42* (1), 158 (1949).

[55] EDDY, G. W., U. S. Bur. Ent. Plant Quar., Kerrville, Tex., Personal communication (1950).

[56] EDDY, G. W., *The Toxicity of Some Organic Insecticides to Screw-Worm Larvae*, J. econ. Ent. *44* (2), 254 (1951).

[57] EDDY, G. W., U. S. Bur. Ent. Plant Quar., Kerrville, Tex., Personal communication (1951).

[58] EDDY, G. W., and GRAHAM, C. H., *Tests to Control Horn Flies with New Insecticides*, J. econ. Ent. *42* (2), 265–268 (1949).

[59] EDDY, G. W., and MCGREGOR, W. S., *Use of White Mice for Testing Materials Used as Repellents and Toxicants for Stable Flies*, J. econ. Ent. *42* (3), 461–463 (1949).

[60] EINARSSON, A., *DDT lusa og flum aeitur*, Freyr *41* (15/16), 240 (1946).

[61] FAIRCHILD, G. B., and BARREDA, E. A., *DDT as a larvicide Against Simulium*, J. econ. Ent. *38* (6), 694–699 (1945).

[62] FAIRCHILD, H. E., HOFFMAN, R. A., LINDQUIST, A. W., and MOTE, D. C., *A Comparison of the Chlorinated Hydrocarbon Insecticides for Control of the Sheep Tick*, J. econ. Ent. *42* (3), 410–414 (1949).

[63] GARCIA MATA, E., *Acción del DDT sobre la salud de los bovinos*, Bol. Cám. agríc., Guayaquil. *3* (14), 6–10 (1949).

[64] GARNHAM, P. C. C., and MCMAHON, J. P., *The Eradication of Simulium neavei Roubaud, From an Onchocerciasis Area in Kenya Colony*, Bull. ent. Res. *37* (4), 619–628 (1947).

[65] GEURDEN, L. M. G., *Comparative Effect of DDT and Similar Products on Argas reflexus*, Int. Congr. Plant Protect, 1st. Heverlee, Belgium, Gen. Rpt., pp. 316–318 (1946).

[66] GHIDINI, G. M., *Disinfestazione con DDT da Sarcoptes mutans C*, Rob. Soc. Ent. Ital. Mem. *26* (Suppl.), 70–71 (1947).

[67] GJULLIN, C. M., COPE, O. B., QUISENBERRY, B. F., and DECHANOIS, F. R., *The Effect of Some Insecticides on Black Fly Larvae in Alaskan Streams*, J. econ. Ent. *42* (1), 100–105 (1949).

[68] GJULLIN, C. M., and MOTE, D. C., *Notes on the Biology and Control of Chrysops discalis Williston (Diptera, Tabanidae)*, Ent. Soc. Wash. Proc. *47* (8), 236–244 (1945).

[69] GLASGOW, R. D., and COLLINS, D. L., *Control of the American Dog Tick, a Vector of Rocky Mountain Spotted Fever: Preliminary Tests*, J. econ. Ent. *39* (2), 235–240 (1946).

[70] GOLDSMITH, J. B., HUSMAN, C. N., BROWN, A. W. A., McDUFFIE, W. C., and SHARP, J. F., *Exploratory Studies on the Control of Adult Mosquitoes and Blackflies with DDT Under Arctic Conditions*, Mosquito News *9* (3), 93–97, 3 refs. (1949).

[71] GOUCK, H. K., and FLUNO, J. A., *Fields Tests on Control of American Dog Tick in Massachusetts*, J. econ. Ent. *43* (5), 698–701 (1950).

[72] GOUCK, H. K., and SMITH, C. N., *DDT in the Control of Ticks on Dogs*, J. econ. Ent. *37* (1), 130 (1944)

[73] GOUCK, H. K., and SMITH, C. N., *DDT to Control Wood Ticks*, J. econ. Ent. *40* (3), 303–308 (1947).

[74] GRAHAM, N. P. H., and SCOTT, M. T., *Observations on the Control of Some Ectoparasites of Sheep. 2. Progress Report on the Use of DDT and Benzene Hexachloride for the Control of the Sheep Ked (Melophagus ovinus) and the Sheep Body Louse (Damalinia ovis)*, Austral. Council Sci. Industr. Res. J. *21* (4), 266–274 (1948).

[75] GRAHAM, O. H., *An Attempt to Protect Cattle from Grub Infestation by Use of Insecticides*, J. econ. Ent. *42* (5), 837 (1949).

[76] GRAHAM, O. H., and ALFORD, H. I., jr., *The Occurence of Gasterophilus spp. in Texas and Some Tests with insecticides for Their Control*, J. econ. Ent. *44* (4), 577–579 (1951).

[77] GRAHAM, O. H., and EDDY, G. W., *Persistence of Chlorinated Camphene as a Fleece Worm Larvicide*, J. econ. Ent. *41* (3), 521 (1948).

[78] HACKMAN, R. H., *The Persistence of DDT on Cattle*, Austral. Council. Sci. Industr. Res. J. *20* (1), 56–65 (1947).

[79] HALL, W. J., *Colonial Insecticides, Fungicides and Herbicides Committee Third Annual Report (1949–1950)*, [Commonwealth Inst. Ent., Brit. Mus.], pp. 126–149.

[80] HALL, W. J., *Colonial Insecticides, Fungicides and Herbicides Committee Fourth Annual Report (1950–1951)* [Commonwealth Inst. Ent., Brit. Mus.], pp. 172–199.

[81] HEILBRON, I., *Colonial Insecticides Committee Second Annual Report (1948–1949)* [Imperial Coll. Sci. Tech.], pp. 101–129.

[82] HITCHCOCK, L. F., and MACKERRAS, I. M., *The Use of DDT in Dips to Control Cattle Tick*, Austral. Council Sci. Industr. Res. J. *20* (1), 43–55 (1947).

[83] HIXSON, E., and MUMA, M. H., *Hog Mange Control Tests*, J. econ. Ent. *40* (3), 451 (1947).

[84] HIXSON, E., and MUMA, M. H., *Toxicity of Certain Insecticides to the Chicken Mite*, J. econ. Ent. *40* (4), 596–598 (1947).

[85] HOCKING, B., and RICHARDS, W. R., *Biology and Control of Labrador Black Flies*, Bull. ent. Res. *43* (2), 237–257 (1952).

[86] HOFFMAN, R. A., *Effect of Several Insecticides on Sheep Tick Pupae*, J. econ. Ent. *42* (6), 984 (1949).

[87] HOFFMAN, R. A., and LINDQUIST, A. W., *Control of the Sheep Tick with Low-Pressure Sprays*, J. econ. Ent. *44* (6), 928–930 (1951).

[88] HOWELL, D. E., *Oklahoma Agric. mech. Coll., Stillwater, Okla. Personal communication (1952).

[89] HOWELL, D. E., EDDY, G. W., and CUFF, R. L., *Effect on Horse Fly Populations of Aerial Spray Applications to Wooded Areas*, J. econ. Ent. *42* (4), 644–646 (1949).

[90] JENKINS, C. F. H., and FORTE, P. N., *1945–1946 Experiments with DDT and 666 as Agricultural insecticides*, Dept. Agric. West. Austral. J. [2], *23* (4), 307–317 (1946).

[91] JOHNSTONE, I. L., and SCOTT, M. T., *Field Trials on the Prevention of Body-Strike in Sheep by the Use of DDT and BHC*, Austral. Vet. J. *27* (4), 79–82 (1951).

[92] KEMPER, H. E., and ROBERTS, I. H., *Preliminary Tests with DDT for Single Treatment Eradication of the Swine Louse, Haematopinus suis.*, J. Amer. Vet. Med. Ass. *108* (829), 252–254 (1946).

[93] KEMPER, H. E., ROBERTS, I. H., SMITH, C. E., and COBBETT, N. G., *DDT Dips for the Control of Sheep Ticks, Melophagus ovinus*, J. Amer. Vet. Med. Ass. *111* (846), 196–199 (1947).

[94] KERR, R. W., *Control of Fleas: Laboratory Experiments with DDT and Certain Other Insecticides*, Austral. Council. Sci. Industr. Res. J. *19* (3), 233–240 (1946).

[95] KNAPP, R. B., and AANESTAD, A., *Summer Fly Spraying with DDT Controls Cattle Lice*, N. Dak. Agric. Expt. St. Bimonthly Bull. *9* (4), 119 (1947).

[96] KNIPLING, E. F., *The New Insecticides for Controlling External Parasites of Livestock*, U. S. Bur. Ent. Plant Quar. E–762, Sec. Rev., 28 pp. (1951).

[97] KULASH, W. M., *DDT for Control of Bedbugs in Poultry Houses*, Poult. Sci. *26* (1), 44–47 (1947).

[98] LAAKE, E. W., *DDT for the Control of the Horn Fly in Kansas*, J. econ. Ent. *39* (1), 65–68 (1946).

[99] LAAKE, E. W., U. S. Bur. Ent. Plant Quar., Kerrville, Tex. Personal communication (1946).

[100] LAAKE, E. W., *Chlorinated Hydrocarbon Insecticides for the Control of the Horn Fly on Beef Cattle in Kansas and Missouri*, J. econ. Ent. *42* (1), 143–144 (1949).

[101] LAAKE, E. W., *Livestock Parasite Control Investigations and Demonstrations in Brazil*, J. econ. Ent. *42* (2), 276–280 (1949).

[102] LAAKE, E. W., HOWELL, D. E., DAHM, P. A., STONE, P. C., and CUFF, R. L., *Relative Effectiveness of Various Insecticides for Control of House Flies in Dairy Barns and Horn Flies on Cattle*, J. econ. Ent. *43* (6), 858–861 (1950).

[103] LANCASTER, J. L., jr., *One Application Control for Cattle Lice*, J. econ. Ent. *44* (5), 718–724 (1951).

[104] LEGG, J., *Report of the Acting Director, Division of Animal Industry*, Queensland Dept. Agric. Stock, Ann. Rpt. *1945–46*, 31–43 (1946).

[105] LIKERMAN, J. E., and VIEGAS AURELIO, J. I., *Acción del DDT sobre el Argas persicus*, Rev. Med. vet., B. Aires *29* (Jan./Mar.), 550–556 (1947).

[106] LINDQUIST, A. W., U. S. Bur. Ent. Plant Quar., Corvallis, Ore. Personal communication (1950).

[107] LINDQUIST, A. W., U. S. Bur. Ent. Plant Quar., Corvallis, Ore. Personal communication (1952).

[108] LINDQUIST, A. W., MADDEN, A. H., and KNIPLING, E. F., *DDT as a Treatment for Fleas on Dogs*, J. econ. Ent. *37* (1), 138 (1944).

[109] LYLE, C., and STRONG, R. G., *Dusting for Cattle Lice*, J. econ. Ent. *38* (5), 611–612 (1945).

[110] LYLE-STEWART, W., *Sheep Maggots and DDT*, Agriculture *53* (4), 178–180 (1946).

[111] MACKERRAS, I. M., Queensland Inst. med. Res., Brisbane, Queensland, Australia. Personal communication (1952).

[112] MADDEN, A. H., LINDQUIST, A. W., and KNIPLING, E. F., *DDT as a Residual Spray for the Control of Bedbugs*, J. econ. Ent. *37* (1), 127–128 (1944).

[113] MARTINEX MORENO, J. I., *El uso de DDTOL en el control de la garrapata*, Control Plagas, Habana *10* (7), 101 (1948).

[114] MATTHYSSE, J. G., *Large Scale Power Dusting of Feeder Lambs for Winter Control of the Sheep Tick*, J. econ. Ent. *38* (3), 285–290 (1945).

[115] MATTHYSSE, J. G., *DDT to Control Hornflies and Gulf Coast Ticks on Range Cattle in Florida*, J. econ. Ent. *39* (1), 62–65 (1946).

[116] MAUNDER, J. C. J., *Cattle Tick Control: Results Achieved in the Field with DDT and BHC*, Queensland Agric. J. *69* (3), 160–167 (1949).

[117] McDUFFIE, W. C., U. S. Bur. Ent. Plant Quar., Kerrville, Tex. Personal communication (1949).

[118] McDUFFIE, W. C., and CLARK, J. C., U. S. Bur. Ent. Plant Quar., Kerrville, Tex. Personal communication (1950).

[119] McDUFFIE, W. C., EDDY, G. W., CLARK, J. C., and HUSMAN, C. N., *Field Studies with Insecticides to Control the Lone Star Tick in Texas*, J. econ. Ent. *43* (4), 520–527 (1950).

[120] McGREGOR, W. S., *Field Tests of Insecticides and Spraying Methods to Control Horn Flies in Dairy Herds*, J. econ. Ent. *42* (4), 641–643 (1949).

[121] McGREGOR, W. S., U. S. Bur. Ent. Plant Quar., Kerrville, Tex. Personal communication (1950).

[122] McGregor, W. S., U. S. Bur. Ent. Plant Quar., Kerrville, Tex. Personal communications (1952).

[123] McKenna, C. T., and Fearn, J. T., *Sheep Lice and Ked. Their Habits and Control*, S. Austral. Dept. Agric. Bull. *425*, 24 pp. (1952).

[124] McQueen, D., *Control of Red Mite on Poultry. Trials with DDT*, Victoria Dept. Agric. J. *46* (7), 292 (1948).

[125] Meldrum, G. K., Dept. Agric., Tasmania, Australia. Personal communication (1952).

[126] Mills, H. B., Marsh, H., and Willson, F. S., *Cattle Grubs in Montana and Their Control*, Mont. Agric. Expt. Sta. Bull. *437*, 16 pp. (1946).

[127] Munro, J. A., and Knapp, R., *Effect of DDT on Cattle Lice*, N. Dak. Agric. Expt. Sta. Bimo. Bull. *7* (4), 21 (1945).

[128] Muskus, G. S., and Pacheco Torres, M. F., *El uso de diversos agentes insecticidas. Estudio comparativo del DDT, BHC (gammexano) y toxafeno, empleados como agentes garrapaticidas*, Rev. grancolomb. Zootec. *3* (7/9), 560–566 (1949).

[129] Nicholsen, A. J., Dept. Agric. Stock, Brisbane, B. 7 Queensland, Australia. Personal communication (1952).

[130] Nicol, G., *Control of the Sheep Ked. Field Trial with DDT*, Victoria Dept. Agric. J. *45* (5), 211–212 (1947).

[131] Nikol'skii, S. N., *The Action of DDT and Hexachlorane on Ixodid Ticks*, Veterinariya *25* (9), 29–33 (1948).

[132] Norris, K. R., *The Testing of the American Horn Fly Trap, and a New Type of Trap Involving the Use of DDT Against the Buffalo Fly, Siphona exigua (de Meijere)*, Austral. Council Sci. Industr. Res. J. *19* (1), 65–76 (1946).

[133] Norris, K. R., *The Use of DDT for the Control of the Buffalo Fly [Siphona exigua (de Meijere)]*, Austral. Council Sci. Industr. Res. J. *20* (1), 25–42 (1947).

[134] Norris, K. R.; Roulston, W. J., and Snowball, G. J., *Observations on the Control of the Cattle Tick with Preparations of DDT and Benzene Hexachloride (BHC) in Dips*, Austral. J. Agric. Res. *1* (2) 165–167 (1950).

[135] Olsen, N. J. H., *Bekaempelse af fluer* [Control of Flies], Ugeskr. for Landmaend *89* (43), 646–647 (1944).

[136] Parish, H. E., and Rude C. S., *DDT for the Control of Goat Lice*, J. econ. Ent. *38* (5), 612–613 (1945).

[137] Parish, H. E., and Rude, C. S., *DDT for the Control of the Biting Sheep Louse*, J. econ. Ent. *39* (4), 546 (1946).

[138] Parish, H. E., and Rude, C. S., *DDT to Control the Winter Horse Tick*, J. econ· Ent. *39* (1), 92–93 (1946).

[139] Pavlow, P., and Gueorguiev, B., *Recherches sur ixodicides HCH, DDT, gesarol et dedetex puder* [In Bulgarian], Bulgaria Min. na Zemedel i Gorite. Nauch-Izsled. Inst. Spis. *18* (1), 107–118 (1950).

[140] Peairs, L. M., *DDT and Hornfly Populations*, J. econ. Ent. *39* (1), 91–92 (1946).

[141] Quarterman, K. D., Parkhurst, J. D., and Dunn, W. J., *DDT for Control of Stable Fly, or Dog Fly, in Northwestern Florida*, J. econ. Ent. *44* (1), 61–65 (1951).

[142] Radeleff, R. D., U. S. Bur. Ent. Plant Quar., Kerrville, Tex. Personal communications (1952).

[143] Ritcher, P. O., and Insko, W. M., jr., *Control of the Northern Fowl Mite*, J. econ. Ent. *41* (1), 123–124 (1948).

[144] Roberts, F. H. S., and Moule, G. R., *A Preliminary Report on the Value of DDT and BHC for the Control of Body-Strike in Sheep*, Austral. vet. J. *27* (2), 35–39 (1951).

[145] Roberts, F. H. S., O'Sullivan, P. J., Rumball, P., and McLaughlan, A. W., *Observations on the Value of DDT for the Control of the Poultry Stickfast Flea, Echinophaga gallinacea Westwood*, Austral. vet. J. *23* (6), 148–152 (1947).

[146] Rogoff, W. M., *Cable-Type Backrubbers for Horn Fly Control on Cattle*, S. Dak. Agric. Expt. Sta. Bull. *418*, 12 pp. (1952).

[147] RUDE, C. S., and PARISH, H. E., *Control of the Sheep Tick by a Single Dipping in DDT Emulsion*, U. S. Bur. Ent. Plant Quar. E-679, 2 pp. (1946).

[148] RUDE, C. S., and SMITH, C. L., *DDT for Control of Gulf Coast and Spinose Ear Ticks*, J. econ. Ent. *37* (1), 132 (1944).

[149] SALCES, F., and CALVO, S. P., *Ensayo experimental con DDT sobre Oestrus ovis*, Rev. Med. vet., B. Aires, *27* (3/4), 137-142 (1945).

[150] SCHMIDT, G., *Versuche über Läusebekämpfung beim Pferd*, Schweiz. Arch. Tierheilk. *85* (5), 206-209 (1943).

[151] SCHNEIDER, R., *Neocidol als Läusebekämpfungsmittel*, Schweiz. Arch. Tierheilk. *85* (11), 453-455 (1943).

[152] SCHOEN, C., *Gesarol tegen vliegen in veestallen*, Tijdschr. Plantenziekten *50* (3), 69-71 (1944).

[153] SCHWARDT, H. H., *Control of Northeastern Livestock Parasites by Use of Efficient Insecticides*, Agric. Chem. *5* (3), 33-35, 87-88 (1950).

[154] SEDDON, H. R., *Diseases of Domestic Animals in Australia. 2. Fly, Louse and Flea Infestations*, Austral. Div. Vet. Hyg. Serv. Publ. *6*, 173 pp. (1951).

[155] SEDDON, H. R., *Diseases of Domestic Animals in Australia. 3. Tick and Mite Infestations*, Austral. Div. Vet. Hyg. Serv. Publ. 7, 200 pp. (1951).

[156] SEGHETTI, L., and FIREHAMMER, B. D., *High Pressure Sprays in Control of the Sheep Tick (Melophagus ovinus)*, J. Amer. Vet. Med. Ass. *117* (885), 447-449 (1950).

[157] SERGENT, E., DONATIEN, A., and PARROT, L., *Expériences de protection des animaux domestiques contre la piqûre des tiques par la poudre insecticide DDT*, Inst. Pasteur d'Algérie, Arch. *23* (4), 249-259 (1945).

[158] SERGENT, ED., and SERGENT, ET., *Expériences de destruction des varrons (Hypoderma bovis de Geer) par la poudre insecticide DDT)* Inst. Pasteur d'Algérie, Arch. *24* (2), 110-111 (1946).

[159] SHANAHAN, G. J., *DDT and '666' as Larvicides. Experiment with the Primary Sheep Blowfly (Lucilia cuprina)*, Agric. Gaz. N. S. Wales *57* (7), 382 (1946).

[160] SHANAHAN, G. J., *The Control of Sheep Ked (Melophagus ovinus). Experiments with DDT and '666' Used in a Power Spray Unit*, Agric. Gaz. N. S. Wales *57* (12), 632-635 (1946).

[161] SHANAHAN, G. J., *DDT as an Anti-Ked Dipping Agent*, Agric. Gaz. N. S. Wales, Misc. Publ. 3,321, 2 pp. (1947).

[162] SHANAHAN, G. J., *Fogging with DDT and BHC for the Control of Sheep Lice and Ked*, Agric. Gaz. N. S. Wales Misc. Publ. 3,406, 8 pp. (1951).

[163] SHAW, H., *Some Uses of DDT in Agriculture*, Nature, Lond., *157* (3984), 285-287 (1946).

[164] SILVA LEITAO, J. L., *Parasitologia pratica. Alguns resultados com o emprego do DDT*, Rev. Med. vet., Lisboa, *40* (314), 321-323 (1945).

[165] SIMMONS, S. W., *Observations on the Biology of the Stablefly in Florida*, J. econ. Ent. *37* (5), 680-686 (1944).

[166] SIMMONS, S. W., and WRIGHT, M., *The Use of DDT in the Treatment of Manure for Fly Control*, J. econ. Ent. *37* (1), 135 (1944).

[167] SMITH, C. L., *Chlorinated Hydrocarbon Insecticides for Control of the Horn Fly on Dairy Cattle in Texas*, J. econ. Ent. *41* (6), 981-982 (1948).

[168] SMITH, C. L., U. S. Bur. Ent. Plant Quar., Kerrville, Tex. Personal communication (1952).

[169] SMITH, C. L., and GATES, D. E., *County-Wide Control of the Horn Fly with DDT*, J. econ. Ent. *42* (5), 847-848 (1949).

[170] SMITH, C. N., COLE, M. M., and GOUCK, H. K., *Biology and Control of the American Dog Tick*, U. S. Dept. Agric. Tech. Bull. 905, 74 pp. (1946).

[171] SMITH, C. N., and GOUCK, H. K., *Effectiveness of DDT in the Control of Ticks on Vegetation*, J. econ. Ent. *37* (1), 128-130 (1944).

[172] SONI, B. N., *DDT and Cattle Ticks*, Curr. Sci. *14* (12), 334 (1945).

[173] SONI, B. N., *DDT and Ox-Warble Control*. Current Sci. *15* (7), 197 (1946).

[174] Sparr, B. I., Clark, J. C., Vallier, E. E., and Baumhover, A. H., *Studies of the Stability of Toxaphene Emulsions in Dipping Vats*, U. S. Bur. Ent. Plant Quar. E–849, 11 pp. (1952).

[175] Squibb, R. L., *Studies on the Control of the Nuche Fly and Cattle Tick*, J. Anim. Sci. *4* (3), 291–296 (1945).

[176] Stage, H. H., *DDT to Control Insects Affecting Man and Animals in a Tropical Village*, J. econ. Ent. *40* (6), 759–762 (1947).

[177] Stewart, M. A., *Rotenone and Ox Warble Control*, J. econ. Ent. *37* (6) 756–760 (1944).

[178] Stones, L. C., *A Comparative Method for Testing Agents Used in Sheep Maggot Fly Control*, Bull. ent. Res. *41* (3), 503–521 (1951).

[179] Suter, H. E., *Untersuchung über die Wirkung von DDT auf Stomoxys calcitrans in Hinsicht auf die Bekämpfung der Tsetsefliege* (Zürich 1946), 88 pp.

[180] Sweetman, H. L., *DDT to Control Cat and Dog Fleas and Dog Lice*, J. econ. Ent. *39* (3), 417–418 (1946).

[181] Sweetman, H. L., *New Organic Insecticides to Control Hog Lice*, J. econ. Ent. *40* (3), 454–455 (1947).

[182] Symes, C. B., Hadaway, A. B., Barlow, F., and Galley, W., *Field Experiments with DDT and Benzene Hexachloride Against Tsetse (Glossina palpalis)*, Bull. ent. Res. *38* (4), 591–612 (1948).

[183] Telford, H. S., *Chicken Louse Control*, Soap, N. Y. *20* (8), 113, 139 (1944).

[184] Telford, H. S., *DDT as a Chicken Louse Control.* J. econ. Ent. *38* (6), 700–703 (1945).

[185] Travis, B. V., U. S. Bur. Ent. Plant Quar., Orlando, Fla. Personal communication (1947).

[186] Travis, B. V., Collins, D. L., de Foliart, G., and Jamnback, H., *Strip Spraying by Helicopter to Control Black Fly Larvae*, Mosquito News *11* (2), 95–98 (1951).

[187] Twinn, C. R., *Livestock Insect Control with DDT*, Sci. Service, Dept. Agric. Div. Ent. Publ. 65, 12 pp. (1947).

[188] United States Department of Agriculture, *Control of Lice and Sheep Ticks on Sheep and Goats*, U. S. Dept. Agric. Leaf. 308, 11 pp. (1951).

[189] United States Department of Agriculture, *Control of Lice on Cattle*, U. S. Dept Agric. Leaf. 319, 11 pp. (1951).

[190] Vanderplank, F. L., *Experiments with DDT on Various Species of Tsetse Flies in the Field and Laboratory*, Roy. Soc. trop. Med. Hyg. Trans. *40* (5), 603–620 (1947).

[191] Vargas, L., and Colorado Iris, R., *Acción del DDT sobre algunos artropodos domésticos*, Soc. Mex. Hist. Nat. Rev. *5* (3/4), 229–235 (1944).

[192] Vorhies, C. T., and Wehrle, L. P., *Preliminary Tests of DDT Applications to Crop Plants and Livestock with Navy's Fog Generator*, Ariz. Agric. Expt. Sta. Rpt. No. 75, 13 pp. (Mimeographed 1945).

[193] Waterhouse, D. F., *Studies of the Physiology and Toxicology of Blowflies. 12. The Toxicity of DDT as a Contact and Stomach Poison for Larvae of Lucilia cuprina.* Austral. Council Sci. Industr. Res. Bull. *218*, 7–18 (1947).

[194] Waterhouse, D. F., and Scott, M. T., *Insectary Tests with Insecticides to Protect Sheep Against Body Strike*, Austral. J. Agric. Res. *1* (4), 440–445 (1950).

[195] Wells, R. W., *DDT as a Flyspray on Range Cattle*, J. econ. Ent. 37 (1), 136–137 (1944).

[196] Whiteside, E. F., *An Experiment in Control of Tsetse with DDT-Treated Oxen*, Bull. ent. Res. *40* (1), 123–124 (1949).

[197] Whitnall, A. B. M., and Bradford, B., *An Arsenic-Resistant Tick and its Control with Gammexane Dips*, Bull. ent. Res. *38* (2), 353–372 (1947).

[198] Whitnall, A. B. M., McHardy, W. M., Whitehead, G. B., and Meerholz, F., *Some Observations on the Control of the Bont Tick, Amblyomma hebraeum Koch*, Bull. ent. Res. *41* (3), 577–591 (1951).

[199] Whitnall, A. B. M., Thorburn, J. A., McHardy, W. M., Whitehead, G. B., and Meerholz, F., *A BHC-Resistant Tick*, Bull. ent. Res. *43* (1), 51–65 (1952).

[200] WIESMANN, R., *Eine neue Methode der Bekämpfung der Fliegenplage in Ställen*, Anz. Schädlingsk. *19* (1), 5–8 (1943).

[201] WILKINSON, P. R., *Tests of Insecticides on Cattle*, E. Afr. agric. J. *14* (1), 12–17 (1948).

[202] WILSON, C. S., APPLEWHITE, K. H., and REDLINGER, L. M., *Heavy Ground Aerosol Generators for the Control of Adult Biting Insects in Alaska*, Mosquito News *9* (3), 97–101 (1949).

[203] WILSON, S. G., *A Method of Assessing the Acaricidal Properties of DDT and 'Gammexane' Preparations in Field Trials*, Bull. ent. Res. *39* (2), 269–276 (1948).